Element	Symbol	Atomic Number	Atomic Mass	Element	Symbol	Atomic Number	Atomic Mass
Mercury	Hg	80	200.59	Scandium	Sc	21	44.9559
Molybdenum	Mo	42	95.94	Selenium	Se	34	78.96
Neodymium	Nd	60	144.24	Silicon	Si	14	28.086
Neon	Ne	10	20.179	Silver	Ag	47	107.868
Neptunium	Np	93	237.0482[c]	Sodium	Na	11	22.98977
Nickel	Ni	28	58.71	Strontium	Sr	38	87.62
Niobium	Nb	41	92.9064	Sulfur	S	16	32.06
Nitrogen	N	7	14.0067	Tantalum	Ta	73	180.9479
Nobelium	No	102	(254)[a]	Technetium	Tc	43	98.9062[c]
Osmium	Os	76	190.2	Tellurium	Te	52	127.60
Oxygen	O	8	15.9994	Terbium	Tb	65	158.9254
Palladium	Pd	46	106.4	Thallium	Tl	81	204.37
Phosphorus	P	15	30.97376	Thorium	Th	90	232.0381[c]
Platinum	Pt	78	195.09	Thulium	Tm	69	168.9342
Plutonium	Pu	94	(242)[a]	Tin	Sn	50	118.69
Polonium	Po	84	(210)[a]	Titanium	Ti	22	47.90
Potassium	K	19	39.102	Tungsten	W	74	183.85
Praseodymium	Pr	59	140.9077	Uranium	U	92	238.029
Promethium	Pm	61	(147)[a]	Vanadium	V	23	50.9414
Protactinium	Pa	91	231.0359[c]	Xenon	Xe	54	131.30
Radium	Ra	88	226.0254[c]	Ytterbium	Yb	70	173.04
Radon	Rn	86	(222)[a]	Yttrium	Y	39	88.9059
Rhenium	Re	75	186.2	Zinc	Zn	30	65.38
Rhodium	Rh	45	102.9055	Zirconium	Zr	40	91.22
Rubidium	Rb	37	85.4678				
Ruthenium	Ru	44	101.07				
Samarium	Sm	62	150.4				

[a] Mass number of most-stable or best-known isotope.
[b] Tentative name.
[c] Mass of most commonly available, long-lived isotope.

PHYSICAL CHEMISTRY
with Applications to
Biological Systems

2nd edition

Raymond Chang Williams College

PHYSICAL CHEMISTRY
with Applications to
Biological Systems

MACMILLAN PUBLISHING CO., INC.
New York

Collier Macmillan Publishers
London

Earlier edition copyright © 1977 by Macmillan Publishing Co., Inc.

Macmillan Publishing Co., Inc.
866 Third Avenue, New York, New York 10022

Collier Macmillan Canada, Ltd.

Library of Congress Cataloging in Publication Data

Chang, Raymond.
 Physical chemistry with applications to biological systems.

 Includes bibliographies and index.
 1. Chemistry, Physical and theoretical.
2. Biological chemistry. I. Title. [DNLM:
1. Chemistry, Physical. 2. Biochemistry.
QD453.2 C456p]
QD452.2.C48 1981   541.3   80-12025
ISBN 0-02-321040-0 (Hardbound Edition)
ISBN 0-02-979050-X (International Edition)

PRINTING 10 11 12 13 14 15 YEAR   0 1 2 3 4 5

ISBN 0-02-321040-0

For E and M

Preface

This text is designed for use in a one-semester or one-quarter physical chemistry course at the junior level. In writing the second edition, I have retained the original aim of placing the emphasis on understanding the physical concepts rather than on precise mathematical development or on actual experimental detail. The principles of physical chemistry are presented from the viewpoint of their applications to chemical and biochemical problems. The text is also suitable for a full-year physical chemistry course where a more rigorous text would be inappropriate.

Many topics from the first edition have been extensively rewritten and a number of changes, corrections, and additions have been made. For example, the chapter on thermodynamics has been expanded into three separate chapters. In response to the request of many instructors, I have altered the sequence of the chapters so that thermodynamics and related material now precede chemical bonding and spectroscopy. The number of end-of-chapter problems has been more than doubled. These problems are divided into two categories: those marked with a star are more challenging, while the unmarked ones are relatively straightforward. In most chapters the references have been expanded to include more up-to-date articles and texts. A new feature in this edition is the appendices at the end of some chapters, which provide mathematical derivations of equations and/or extensions of material discussed in the chapters.

The case of SI units versus CGS units has not changed noticeably since the first edition. For this reason, I have decided to continue with the practice of presenting most physical quantities in both units. An important change is the replacement of angstrom (Å) with nanometer (nm) for wavelength.

I have greatly benefited from the suggestions and criticisms of many instructors who have used the text, as well as a number of reviewers and colleagues. In particular I would like to thank the following individuals: Jesse S. Binford, Jr. (University of South Florida), Robert E. Blankenship (Amherst College), Donald Boerth (Southeastern Massachusetts University), Luther K.

Brice, Jr. (Virginia Polytechnic Institute and State University), B. J. Chapman (University of Southampton, England), John N. Cooper (Bucknell University), Allen A. Denio (University of Delaware), George E. Ewing (Indiana University), James Franzen (University of Pittsburgh), Chien Ho (Carnegie-Mellon University), Gary W. Hunt (Shorter College), Richard S. Myers (Delta State University), Gerald Nagahashi (Williams College), Richard F. Olivo (Smith College), Reeves B. Perry (Southwest Texas State University), Douglas D. Radtke (University of Wisconsin), John S. Ricci, Jr. (Williams College), and Edmund C. Shearer (Fort Hays State University). Thanks are also due Bolesh J. Skutnik (Fairfield University) and William J. Zaks (Williams College) for help in checking the accuracy of many problems.

Finally, I would like to express my appreciation to Gregory W. Payne, the chemistry editor at Macmillan, for his assistance in general; to Elaine W. Wetterau for editorial supervision; and to Eileen Sprague and Holly Andrews for typing the manuscript.

I welcome comments and suggestions from readers.

R. C.

Contents

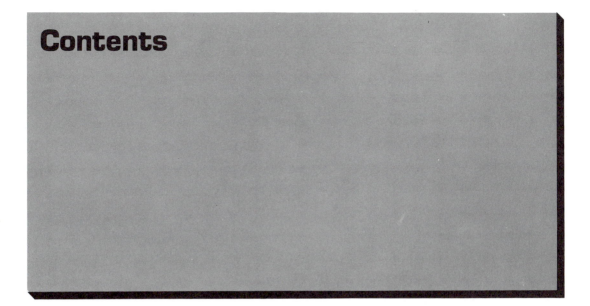

10 Chemical Equilibrium

11 Electrochemistry

Intermolecular Forces 17

487

Spectroscopy 18

505

PHYSICAL CHEMISTRY
with Applications to Biological Systems

Introduction

1

And it's hard, and it's hard, ain't it hard, good Lord.
Woody Guthrie*

Nature of Physical Chemistry 1.1

Physical chemistry can be described as a set of characteristically quantitative approaches to the study of various chemical problems. A physical chemist is a person who seeks to predict chemical events using certain models and postulates. Because the problems encountered are often both diversified and complex, a number of different approaches must be employed. For example, in the study of thermodynamics and rates of chemical reactions, we employ the phenomenological, macroscopic approach. On the other hand, a microscopic, molecular approach is required for an understanding of the kinetic behavior of molecules and reaction mechanism. Ideally, it would be best to study all the phenomena at the molecular level, because it is here that one learns what really occurs. Now, however, this is not possible—our knowledge of atoms and molecules is neither extensive enough nor thorough enough. Fortunately, there are some areas in which we are beginning to have good, semiquantitative understanding. As each topic is developed, it is well to keep in mind the scope and limitation of the approaches involved.

To see how physiochemical principles can be applied to study a biochemical system, let us consider the binding of oxygen with hemoglobin. This is one of the most important biochemical reactions and may be the most extensively studied. Hemoglobin is a protein molecule that has a molecular mass of about 65,000. The molecule contains four subunits, made up of two α chains (141 amino acids each) and two β chains (146 amino acids each). Each of the chains contains a heme group to which an oxygen molecule can be bound. The main function of hemoglobin is to carry oxygen in the blood from the lungs to the tissues, where it unloads the oxygen molecules to myoglobin.

* "Hard, Ain't It Hard." Words and Music by Woody Guthrie. TRO—© Copyright 1952 Ludlow Music, Inc., New York, N.Y. Used by permission.

Myoglobin, which possesses only one polypeptide chain (153 amino acids) and one heme group, stores oxygen for metabolic processes.

The hemoglobin system has many interesting aspects. To begin with, we are interested in the size, shape, and detailed structure of the molecule. A number of techniques that have been developed in recent years, such as viscosity measurement, electrophoresis, ultracentrifugation, and light scattering, enable us to measure the molecular mass and/or to estimate the shape of macromolecules. The best technique for determining structure is X-ray diffraction, but this can be quite difficult. Nevertheless, the complete or almost complete structure of a number of protein molecules has been determined by the X-ray method. Two such molecules are myoglobin and hemoglobin. A detailed understanding of the three-dimensional structure of a protein molecule is perhaps the single most important factor in revealing the secrets of its various functions.

The next questions concern the binding of oxygen. To understand how oxygen and other molecules, such as carbon monoxide and nitric oxide, bind to the heme group, we need to investigate the coordination chemistry of transition-metal ions in general and complexes of iron in particular. For example, it is important to know which orbitals are involved in the complex, as well as the reasons why the binding constant for CO is some 200 to 300 times stronger than that for oxygen. Knowledge of the molecular orbitals involved will also help to explain the spectral properties.

A very important phenomenon is the cooperative nature of binding. It was noticed many years ago that oxygen molecules did not bind to the four heme groups independently; rather, the presence of the first molecule facilitates the binding of the second, and so on. The biological function of cooperative binding is that it results in the more efficient transport and release of oxygen. The kinetic and thermodynamic details of this phenomenon have been successfully accounted for, at least in broad outline, by current theories based on the allosteric model, which is also applicable to many other regulatory enzymes.

Another function of hemoglobin is the transportation of carbon dioxide from the tissues to the lungs. The pH dependence of the oxygen equilibrium, the Bohr effect, is coupled to this role. The maintenance of proper physiological pH is of paramount importance, because the function and efficiency of most proteins and enzymes depend critically on the hydrogen-ion concentration. The CO_2–O_2 transport process in blood is buffered by the bicarbonate–carbonic acid system. Being amphoteric,* hemoglobin itself can also act as a buffer. Here the problem is dealt with in terms of acid–base equilibria.

Finally, we may raise the following point: Of the very large number of possible structures that a molecule this size can assume, why is it that only one predominant structure is observed for hemoglobin? We must realize that in addition to the normal chemical bonds, many other types of molecular interaction, such as electrostatic forces, hydrogen bonding, and van der Waals forces, are also present. In principle, a macromolecule can fold up in many different ways; the native conformation represents the minimum-energy structure or is in the neighborhood of the minimum-energy structure. The specificity in binding depends precisely on the environment at and near the active site which is maintained by the rest of the three-dimensional molecule.

---

* An amphoteric substance can react either as an acid or as a base, depending on reaction conditions.

To appreciate how delicate the balance of these forces must be in some cases, consider the replacement of a glutamic acid by valine in the β chains:

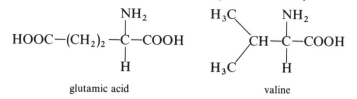

glutamic acid   valine

This seemingly small alternation is sufficient to result in a significant conformational change—an increase in the attraction between protein molecules, resulting in polymerization. Insoluble polymers so formed will then distort red blood cells into a sickle shape, causing the disease that we call sickle-cell anemia.

All these problems can be understood, at least in theory, by application of the principles of physical chemistry. Obviously, very different approaches are needed if a thorough investigation of the chemistry of hemoglobin is desired. We could easily have chosen another example, such as photosynthesis, to demonstrate our point. It is not the purpose of this text to present a detailed explanation for each of the phenomena described above; however, the hemoglobin example serves to illustrate that a student must first understand the basic principles of physical chemistry before embarking on the study of many exciting biochemical phenomena.

Units 1.2

Students are frequently confused by the variety of units used in physical chemistry. Since people in vastly different disciplines have contributed to the development of this science, it is not surprising that different units are often used to express the same quantity. A case in point relates to the quantity of heat, which can be expressed in terms of calories, joules, or British thermal units (Btu). To add to the confusion, the calorie unit used to measure the fuel value of food is 1000 times greater than that for measuring bond energies and heats of chemical reactions.

The CGS (centimeter–gram–second) system, which was developed in France after the French Revolution, has been widely adopted throughout the world, with the notable exception of the United States. The advantage of metric units lies in their convenience and simplicity, in contrast to such native units as the foot and the pound. Physicists, on the other hand, have long favored the MKS (meter–kilogram–second) system.

In 1960, the General Conference of Weights and Measures, an international authority on units, agreed to adopt the International System of Units (SI). This system has now been endorsed by scientists in many countries. The advantage of the SI system is that many of the units employed are derivable from natural constants. For example, the CGS unit of length is the meter (strictly the centimeter; 1 meter is exactly equal to 100 cm), which is defined to be the distance between two marked lines on a bar kept in Sèvres, near Paris. On the other hand, the SI system defines meter as the length equal to 1,650,763.73 wavelengths of radiation corresponding to a particular electronic transition from the $6d$ to the $5p$ orbital in krypton. The unit of time, second, is defined as 9,192,631,770 cycles of the radiation associated with a certain elec-

Table 1.1

Physical Quantity	CGS Unit	Symbol	SI Unit	Symbol
Length	Centimeter	cm	Meter	m
Mass	Gram	g	Kilogram	kg
Time	Second	s	Second	s
Thermodynamic temperature	Kelvin	°K	Kelvin	K

tronic transition of the cesium atom. The fundamental unit of mass, the kilogram, is defined as the mass of a platinum–iridium alloy cylinder kept by the International Bureau of Weights and Measures in Sèvres. Thus, in contrast to both length and time, the unit of mass is defined in terms of an artifact, not in terms of a naturally occurring phenomenon.

At the time when the first edition of this text was being written, the situation in the United States was rather unsettled. Chemistry textbooks ranged from complete adoption of SI units, to their partial adoption, to no mention of them at all. Now, four years later, there has not been any significant change in one direction or another. Although general acceptance of SI units in the United States seems inevitable, it will probably be quite a few years before everyone agrees to the change. For this reason, both CGS and SI units will be used in this text. Whenever a physical quantity is first introduced, both units will be presented and a conversion factor given. We shall discuss the basic and some of the derived SI units briefly here, and we present a more extensive tabulation in Appendix 1 at the end of the book.

Table 1.1 gives some of the fundamental physical quantities. Note that in SI units, temperature is given as K without the superscript °. A number of physical quantities can be derived from those listed in Table 1.1. We shall discuss only a few of them here.

FORCE

In the CGS system, *dyne* (abbreviated dyn) is defined as the force required to give a mass of 1 g an acceleration of 1 cm s^{-2}; that is,

$$1 \text{ dyn} = 1 \text{ g cm s}^{-2}$$

The name for force in the SI system is *newton* (N), defined as the force required to give a mass of 1 kg an acceleration of 1 m s^{-2}, that is,

$$1 \text{ N} = 1 \text{ kg m s}^{-2}$$

The conversion factor is

$$1 \text{ N} = 10^5 \text{ dyn}$$

PRESSURE

Pressure is defined as force/area, that is,

$$P = \frac{F}{A}$$

In the CGS system, pressure has the units dyn cm^{-2}. By definition, the standard atmosphere is the pressure exerted by a 76-cm-high column of mercury of density 13.595 g cm^{-3} in a place where the acceleration due to gravity is 980.67 cm s^{-2}.
Since

$$\text{pressure} = \frac{\text{force}}{\text{area}}$$

$$= \frac{(\text{volume})(\text{density})(\text{acceleration due to gravity})}{\text{area}}$$

$$= (\text{length})(\text{density})(\text{acceleration due to gravity})$$

we write

$$1 \text{ atm} = (76 \text{ cm})(13.595 \text{ g cm}^{-3})(980.67 \text{ cm s}^{-2})$$

$$= 1.01325 \times 10^6 \text{ g cm}^{-1} \text{ s}^{-2}$$

$$= 1.01325 \times 10^6 \text{ dyn cm}^{-2}$$

In the SI system, pressure has the units N m^{-2}. To express the standard atmospheric pressure, we need the following conversion factors:

$$\text{density of mercury} = 1.3595 \times 10^4 \text{ kg m}^{-3}$$

$$\text{acceleration due to gravity} = 9.8067 \text{ m s}^{-2}$$

Thus

$$1 \text{ atm} = (0.76 \text{ m})(1.3595 \times 10^4 \text{ kg m}^{-3})(9.8067 \text{ m s}^{-2})$$

$$= 1.01325 \times 10^5 \text{ kg m}^{-1} \text{ s}^{-2}$$

$$= 1.01325 \times 10^5 \text{ N m}^{-2}$$

The SI unit of pressure is the *pascal* (**Pa**), where

$$1 \text{ Pa} = 1 \text{ N m}^{-2}$$

so that 1 atm is equal to 1.01325×10^5 Pa. As we shall see, the non-SI unit of pressure, atm, is very useful in defining thermodynamic systems. For this reason, its use will be retained in this book.

ENERGY

Since energy is the ability to do work and work is force × distance, we have

CGS units: 1 dyn cm = 1 erg

SI units: 1 N m = 1 J (joule)

Conversion factor: 1 J = 1 × 10^7 ergs

Although the calorie has been a long-cherished unit for chemists and biologists, its use will soon be eliminated in most scientific literature. For this reason, energy is expressed in terms of joules rather than calories or ergs in this book. The conversion factor is

$$1 \text{ cal} = 4.184 \text{ J}$$

1.3 The Chemical Mole

The ratio of the mass of an atom of an element to the mass of some chosen standard is called the atomic weight of that element. The standard chosen is $\frac{1}{12}$ times the mass of the carbon-12 isotope ($_6^{12}C$), which, by definition, has an atomic weight exactly equal to 12. Similarly, the ratio of the mass of a molecule of a substance to $\frac{1}{12}$ times the mass of a $_6^{12}C$ atom is called the *molecular weight* of that substance. Thus when we say that the molecular weight of carbon dioxide is 44.01, we mean that a carbon dioxide molecule has a mass that is 44.01/12 times the mass of a $_6^{12}C$ atom. Because molecular weights are really molecular masses and the fact that they are defined as ratios of two quantities, the proper term for molecular weight is *relative molecular mass*. For simplicity, we use the term *molecular mass* instead of molecular weight in this book. The same argument applies to the term *atomic weight*, which we call *atomic mass*.

A *mole* (abbreviated mol) of any substance is the mass of that substance which contains as many atoms or molecules as there are atoms in exactly 12 g of $_6^{12}C$. It has been determined experimentally that the number of atoms in this quantity of $_6^{12}C$ is 6.02217×10^{23}. This number is called *Avogadro's number*. Avogadro's number has no units. On the other hand, when this number is divided by mol, we obtain the Avogadro constant (N_0), where

$$N_0 = 6.02217 \times 10^{23} \text{ mol}^{-1}$$

For most purposes, N_0 can be taken as $6.022 \times 10^{23} \text{ mol}^{-1}$. The following examples indicate the number and kind of particles in 1 mol of any substance.

1. One mole of helium atoms contains 6.022×10^{23} He atoms.
2. One mole of water molecules contains 6.022×10^{23} H_2O molecules or $2 \times (6.022 \times 10^{23})$ H atoms and 6.022×10^{23} O atoms.
3. One mole of NaCl contains 6.022×10^{23} NaCl units or 6.022×10^{23} Na^+ ions and 6.022×10^{23} Cl^- ions.

In fact, the mole concept applies to any elementary entities, whether they are atoms, molecules, ions, electrons, radicals, photons, or whatever. The *molar mass* of a substance is the mass in g (CGS) or kg (SI) of 1 mol of the substance. Thus the molar mass of atomic hydrogen is 1.0079 g mol^{-1}, of molecular hydrogen 2.0158 g mol^{-1}, of hemoglobin 65,000 g mol^{-1}.

Suggestions for Further Reading

The following standard texts are useful references. They contain mathematical derivations of equations and provide experimental details for a number of topics covered in this book.

PHYSICAL CHEMISTRY

ADAMSON, A. W. *A Textbook of Physical Chemistry*. Academic Press, Inc., New York, 1973.

ATKINS, P. W. *Physical Chemistry*. W. H. Freeman and Company, San Francisco, 1978.

BARROW, G. M. *Physical Chemistry*, 4th ed. McGraw-Hill Book Company, New York, 1979.

BROMBERG, J. P. *Physical Chemistry*. Allyn and Bacon, Inc., Boston, 1980.

CASTELLAN, G. W. *Physical Chemistry*, 2nd ed. Addison-Wesley Publishing Company, Inc., Reading, Mass., 1971.

DANIELS, F., and R. A. ALBERTY. *Physical Chemistry*, 5th ed. John Wiley & Sons, Inc., New York, 1979.

LEVINE, I. N. *Physical Chemistry*, McGraw-Hill Book Company, New York, 1978.

MOORE, W. J. *Physical Chemistry*, 4th ed. Prentice-Hall, Inc., Englewood Cliffs, N.J., 1972.

BIOPHYSICAL CHEMISTRY

BULL, H. B. *An Introduction to Physical Biochemistry*, 2nd ed. F. A. Davis Company, Philadelphia, 1971.

EDSALL, J. T., and J. WYMAN. *Biophysical Chemistry*, Vol. 1. Academic Press, Inc., New York, 1958.

FREIFELDER, D. *Physical Biochemistry*. W. H. Freeman and Company, San Francisco, 1976.

MARSHALL, A. G. *Biophysical Chemistry*. John Wiley & Sons, Inc., New York, 1978.

TANFORD, C. *Physical Chemistry of Macromolecules*. John Wiley & Sons, Inc., New York, 1961.

VAN HOLDE, K. E. *Physical Biochemistry*. Prentice-Hall, Inc., Englewood Cliffs, N.J., 1971.

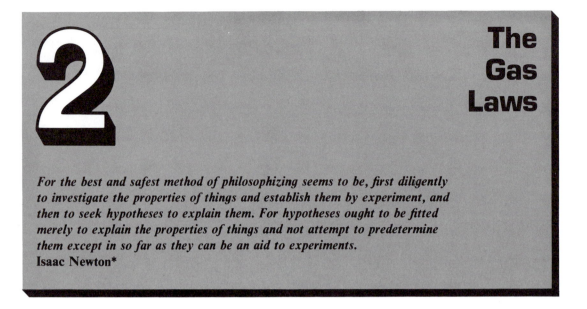

For the best and safest method of philosophizing seems to be, first diligently to investigate the properties of things and establish them by experiment, and then to seek hypotheses to explain them. For hypotheses ought to be fitted merely to explain the properties of things and not attempt to predetermine them except in so far as they can be an aid to experiments.
Isaac Newton*

Studying the behavior of gases has been very valuable in the development of a number of chemical and physical theories. In many ways, the gaseous state provides the simplest system for investigation. In this chapter we examine a number of gas laws based on experimental observations and introduce the concept of temperature.

2.1 Some Basic Definitions

Before we discuss the gas laws, it is useful to define a few basic terms that will be used throughout the book. We often speak of the *system* in reference to a particular part of the universe in which we are interested. Thus a system could be a collection of gas molecules in a container, a saturated NaCl solution, a tennis ball, or a Siamese cat. Having defined a system, we call the rest of the universe *surroundings*. A *macroscopic* system is one whose physical properties, such as mass, volume, and pressure, can be suitably studied by using laboratory instruments. There are three types of macroscopic systems, described below. An *open system* is one that can exchange both mass and energy with its surroundings. A *closed system* is one that does not exchange mass with its surroundings but can exchange energy. An *isolated system* is one that can exchange neither mass nor energy with its surroundings (Figure 2.1). To completely define a macroscopic system, we need to understand a minimum number of experimental variables, such as pressure, volume, temperature, and composition, which are referred to as the *state* of the system.

Most of the properties that we measure quantitatively may be divided into

* *Principia*, Motte's translation reviewed by Cajori, Vol. 2, University of California Press, Berkeley, Calif., 1934, p. 673. Copyright © 1962 by The Regents of the University of California, reprinted by permission of the University of California Press.

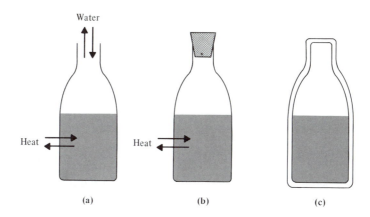

(a) An open system allows exchange of both mass and energy; (b) a closed system allows exchange of energy but not of mass; and (c) an isolated system allows exchange of neither mass nor energy.

Figure 2.1

two classes. Consider, for example, two beakers containing the same amounts of water at the same temperature. If we combine these two systems by pouring the water from one beaker to the other, we find that the volume of the water is doubled and so is its mass. On the other hand, we find that both the temperature and the density of water remain unchanged. Properties whose values depend on the amount of the material present in the system are called *extensive properties*; those that do not are called *intensive properties*. Examples of the former include mass, area, volume, energy, and electrical charge. As already mentioned, temperature and density are both intensive properties, and so are pressure and electrical potential. Note that intensive properties are normally defined in terms of the ratio of two extensive properties: for example,

$$\text{pressure} = \frac{\text{force}}{\text{area}}$$

$$\text{density} = \frac{\text{mass}}{\text{volume}}$$

An Operational Definition of Temperature 2.2

Temperature is a very important quantity in many branches of science, and not surprisingly, it can be defined in a number of different ways. Daily experience tells us that temperature is a measure of coldness and hotness. Here we shall give an operational definition of temperature. Consider the following system of gas A in a container. The walls of the container are flexible so that its volume can expand and contract. Being a closed system, heat but not mass can flow into and out of the container. Initially, the pressure and volume are P_A and V_A. Now we bring the container in contact with a similar container of gas B at P_B and V_B. Heat exchange will take place until thermal equilibrium is reached. At equilibrium the pressure and volume of A and B will be altered to P'_A, V'_A and P'_B, V'_B. It is possible to remove container A temporarily, readjust its properties to P''_A and V''_A, and still have A in thermal equilibrium with B at

P'_B and V'_B. In fact, an infinite set of such values (P'_A, V'_A), (P''_A, V''_A), (P'''_A, V'''_A), ... can be obtained that will satisfy the equilibrium conditions. Figure 2.2 shows a plot of these points.

For all these states of A to be in thermal equilibrium with B, they must possess the same value of a certain variable which we call temperature. It follows from the discussion above that if two system are in thermal equilibrium with a third system, they must also be in thermal equilibrium with each other. This statement is generally known as the *zeroth law of thermodynamics*. The curve in Figure 2.2 is the locus of all the points that represent the states that can be in thermal equilibrium with system B. Such a curve is called an *isotherm*, or "same temperature." At another temperature a different isotherm is obtained.

Figure 2.2

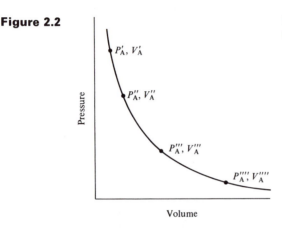

Plot of pressure versus volume at constant temperature for a given amount of gas.

2.3 Boyle's Law

In his study of gas behavior in 1662, Boyle found that the volume of a given amount of gas at constant temperature is inversely proportional to the pressure:

$$V \propto \frac{1}{P}$$

or

$$PV = \text{constant} \tag{2.1}$$

Equation (2.1) is known as *Boyle's law*.* A plot of P versus V at a given temperature gives a hyperbola, which is the isotherm discussed in Figure 2.2.

Boyle's law is used to predict the pressure of a gas when its volume is changed, or vice versa. Letting the initial values of pressure and volume be P_1 and V_1 and the final values be P_2 and V_2, we have

$$P_1 V_1 = P_2 V_2 \quad \text{(constant temperature)} \tag{2.2}$$

---

* For an unusual demonstration of Boyle's law, see T. C. Loose, *J. Chem. Educ.* **48**, 390 (1971).

At constant pressure the volume of a given amount of gas is directly proportional to temperature:

$$V \propto T$$

or

$$\frac{V}{T} = \text{constant} \tag{2.3}$$

The quantitative study of the thermal expansion of gases was first made by Charles in 1787 and later in greater detail by Gay-Lussac in 1802. Equation (2.3) is known as *Charles' law* or the *law of Charles and Gay-Lussac*. An alternative form of Charles' law is that at constant volume, the pressure of a given amount of gas is directly proportional to temperature:

$$P \propto T$$

or

$$\frac{P}{T} = \text{constant} \tag{2.4}$$

Equations (2.3) and (2.4) now permit us to relate the volume–temperature and pressure–temperature values of a gas in states 1 and 2 as follows:

$$\frac{V_1}{T_1} = \frac{V_2}{T_2} \qquad \text{(constant pressure)} \tag{2.5}$$

$$\frac{P_1}{T_1} = \frac{P_2}{T_2} \qquad \text{(constant volume)} \tag{2.6}$$

Figure 2.3 shows plots of the volume of a gas versus temperature at several pressures. By extrapolating the straight lines to zero volume, it is found that they all converge to the same point on the temperature axis, which is found to be $-273.15°C$. This interesting behavior suggests that $-273.15°C$ is absolutely the lowest attainable temperature, which is appropriately called the *absolute zero*. Thus it is more convenient to measure temperature from this point rather than from the zero degree on the Celsius scale. A new temperature scale, called the *absolute-zero temperature scale*, was therefore devised by

Plots of volume of a given amount of gas versus temperature at constant pressures. By extrapolation it is found that the lines all converge on the temperature axis at $-273.15°C$.

Figure 2.3

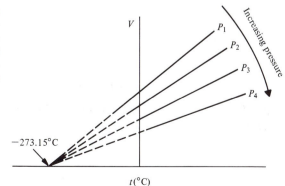

Lord Kelvin. The relation between the two scales is

$$T(K) = t(°C) + 273.15 \qquad (2.7)$$

where K denotes Kelvin. It is important to note that the only difference between the two scales is that the zero is shifted; the size of the degree remains the same. As we shall see later, the absolute zero of temperature does have a major theoretical significance; absolute temperatures generally must be employed in gas laws and thermodynamic calculations and absolute temperature also appears in the Boltzmann distribution law.

2.5 Avogadro's Law

Another important gas law was formulated by Avogadro in 1811. He proposed that, under the same conditions of temperature and pressure, equal volumes of gases contain the same number of molecules. This means that

$$V \propto n$$

or

$$\frac{V}{n} = \text{constant} \qquad (2.8)$$

where n is the number of moles. Equation (2.8) is known as *Avogadro's law*.

2.6 Ideal Gas Equation

According to Eqs. (2.1), (2.3), and (2.8), the volume of a gas depends on the pressure, temperature, and number of moles as follows:

$$V \propto \frac{1}{P} \text{ (at constant } T \text{ and } n) \qquad \text{(Boyle's law)}$$

$$V \propto T \text{ (at constant } P \text{ and } n) \qquad \text{(Charles' law)}$$

$$V \propto n \text{ (at constant } T \text{ and } P) \qquad \text{(Avogadro's law)}$$

Therefore, V must be proportional to the product of these three terms, that is,

$$V \propto \frac{nT}{P}$$

$$= R\frac{nT}{P}$$

or

$$PV = nRT \qquad (2.9)$$

where R, a proportionality constant, is the *gas constant*. Equation (2.9) is called the *ideal gas equation*.

From Eq. (2.9), we write

$$R = \frac{PV}{nT}$$

The value of R can be obtained as follows. In the CGS system we deal with
13

Sec. 2.7

**Dalton's Law
of Partial
Pressures**

the knowledge that 1 mol of an ideal gas occupies 22.414 liters at 1 atm and 273.15 K (the standard temperature and pressure). Thus

$$R = \frac{(1 \text{ atm})(22.414 \text{ liters})}{(1 \text{ mol})(273.15 \text{ K})} = 0.08206 \text{ liter atm K}^{-1} \text{ mol}^{-1}$$

From Section 1.2, we find that

$$1 \text{ atm} = 1.01325 \times 10^6 \text{ dyn cm}^{-2}$$

To express R in terms of ergs and calories, we proceed as follows:

$$R = 0.08206 \text{ liter atm K}^{-1} \text{ mol}^{-1}$$
$$= (0.08206 \times 10^3 \text{ cm}^3)(1.01325 \times 10^6 \text{ dyn cm}^{-2})\text{K}^{-1} \text{ mol}^{-1}$$
$$= 8.314 \times 10^7 \text{ dyn cm K}^{-1} \text{ mol}^{-1}$$
$$= 8.314 \times 10^7 \text{ ergs K}^{-1} \text{ mol}^{-1}$$
$$= 8.314 \text{ J K}^{-1} \text{ mol}^{-1}$$
$$= 1.987 \text{ cal K}^{-1} \text{ mol}^{-1}$$

In SI units we start with (see Section 1.2)

$$1 \text{ atm} = 1.01325 \times 10^5 \text{ N m}^{-2}$$

so that

$$R = \frac{PV}{nT}$$
$$= \frac{(1.01325 \times 10^5 \text{ N m}^{-2})(22.414 \times 10^{-3} \text{ m}^3)}{(1 \text{ mol})(273.15 \text{ K})}$$
$$= 8.314 \text{ N m K}^{-1} \text{ mol}^{-1}$$
$$= 8.314 \text{ J K}^{-1} \text{ mol}^{-1}$$

The conversion factor for volume is 1 liter = 1×10^{-3} m^3.

Dalton's Law of Partial Pressures 2.7

In the discussion of pressure–volume–temperature (PVT) behavior so far we have considered only pure gases. For a system containing two or more different gases, the total pressure (P_T) is the sum of the individual pressures that each gas would exert if it were alone and occupied the same volume. Thus

$$P_T = P_1 + P_2 + \cdots$$
$$= \sum_i P_i \qquad (2.10)$$

where P_1, P_2, ... are the individual or *partial* pressures of components 1, 2, ... Equation (2.10) is known as *Dalton's law of partial pressures*.

Consider a system containing two gases (1 and 2) at temperature T and volume V. The partial pressures of the gases are P_1 and P_2, respectively. From Eq. (2.9),

$$P_1 V = n_1 RT \quad \text{or} \quad P_1 = \frac{n_1 RT}{V}$$

$$P_2 V = n_2 RT \quad \text{or} \quad P_2 = \frac{n_2 RT}{V}$$

where n_1 and n_2 are the numbers of moles of the two gases present. According to Dalton's law,

$$P_T = P_1 + P_2$$

$$= n_1 \frac{RT}{V} + n_2 \frac{RT}{V}$$

$$= (n_1 + n_2) \frac{RT}{V}$$

Dividing the partial pressures by the total pressure, we get

$$P_1 = \frac{n_1}{n_1 + n_2} P_T = X_1 P_T$$

and

$$P_2 = \frac{n_2}{n_1 + n_2} P_T = X_2 P_T$$

where X_1 and X_2 are the mole fractions of gases 1 and 2. Because mole fractions are defined as the ratio of number of moles, they are dimensionless quantities. Furthermore, by definition, the sum of all the mole fractions in a mixture must be unity, that is,

$$\sum_i X_i = 1$$

E X A M P L E 2.1   The oxygen gas generated in an *in vitro* photosynthesis experiment (by shining visible light on extracted chloroplast) is collected over water. The volume of the gas collected at 22.0°C and an atmospheric pressure of 758.0 mm Hg is 186 ml. Calculate the mass of oxygen obtained. The vapor pressure of water at 22.0°C is 19.8 mm Hg.

Answer: Our first step is to calculate the partial pressure of oxygen. Since

$$P_T = P_{O_2} + P_{H_2O}$$

or

$$P_{O_2} = P_T - P_{H_2O}$$

$$= (758.0 - 19.8) \text{ mm Hg}$$

$$= 738.2 \text{ mm Hg}$$

$$= 0.971 \text{ atm}$$

$$PV = nRT$$

$$= \frac{m}{M} RT$$

where m and M are the mass of O_2 collected and molar mass of O_2, respectively. Thus

$$m = \frac{PVM}{RT}$$

$$= \frac{(0.971 \text{ atm})(0.186 \text{ liter})(32.0 \text{ g mol}^{-1})}{(0.08206 \text{ liter atm K}^{-1} \text{ mol}^{-1})(273.2 + 22.0) \text{ K}}$$

$$= 0.239 \text{ g}$$

As mentioned earlier, the gas laws played an important role in the development of atomic theory. There are many practical illustrations of the gas laws in our everyday life experiences. Here we shall briefly mention two examples that are particularly important to scuba divers. Seawater has a slightly higher density than fresh water—about 1.03 g cm^{-3} compared to 1.00 g cm^{-3}. Therefore, the pressure exerted by a column of 33 ft (10 m) of seawater is equivalent to 1 atm pressure.

What would happen if the diver were to rise to the surface rather quickly, holding his breath? If the ascent starts at 40 ft under water, the decrease in pressure from this depth to the surface is (40 ft/33 ft) × 1 atm, or 1.2 atm. When the diver reaches the surface, the volume of air trapped in his lungs has increased by a factor of (1 + 1.2) atm/1 atm, or 2.2 times! This sudden expansion of air can rupture the membranes of his lungs, seriously injuring or even killing the diver.

Dalton's law has a simple application to scuba diving. As Problem 2.6 shows, the partial pressure of oxygen in air is about 0.2 atm. Because oxygen is so essential for our survival, it is sometimes hard to believe that it could be harmful if we breathe more than our normal share. In fact, the toxicity of oxygen is well established.* Physiologically, our bodies function best when oxygen has a partial pressure of 0.2 atm. For this reason, the composition of air needs to be changed when the diver is submerged. For example, at a depth where the total pressure is 4 atm, the oxygen content in the air supply should be reduced to 5% by volume to maintain the same partial pressure (0.05 × 4 atm = 0.2 atm). At a greater depth, the oxygen content must be made even smaller. Although nitrogen seems to be the obvious choice for mixing with oxygen, there is a serious problem. When its partial pressure exceeds 1 atm, a sufficient amount of nitrogen will be dissolved in the blood, causing a condition known as *nitrogen narcosis*. Its symptoms include light-headedness and impairment of judgment, similar to alcoholic intoxication.† Divers suffering

* At partial pressure above 2 atm, oxygen becomes toxic enough to produce convulsions and coma. Also, newborn infants placed in oxygen tents often develop *retrolental fibroplasia*, damage of the retinal tissues by excess oxygen. This damage usually results in partial or total blindness.

† For this reason, nitrogen narcosis is frequently called *rapture of the deep*.

from nitrogen narcosis have been known to do strange things, such as dancing on the sea floor and chasing sharks bare-handed. For this reason, helium is usually employed to dilute oxygen gas. Helium is an inert gas, much less soluble in blood than nitrogen, and it does not produce narcotic effects.

2.8 Real Gases

The ideal gas equation holds only for gases that have the following properties: (1) the gas molecules possess no finite volume, and (2) there is no interaction, attractive or repulsive, among the molecules. Such gases are obviously nonexistent. Nevertheless, Eq. (2.9) is quite useful for many gases at high temperatures or moderately low pressures (≤ 10 atm).

When a gas is being compressed, the molecules are brought closer and closer to one another, and appreciable deviation will result. One way to measure the deviation from ideality is to plot the compressibility factor (Z) of a gas versus pressure. Starting with the ideal gas equation, we write

$$PV = nRT$$

or

$$Z = \frac{PV}{nRT} = \frac{P\overline{V}}{RT} \tag{2.11}$$

where $\overline{V}$ is the molar volume of the gas or the volume of 1 mol of the gas at the specified temperature and pressure. For an ideal gas, we have $Z = 1$ for any value of P. However, as Figure 2.4 shows, the compressibility factors of real gases show fairly divergent dependence on pressure. At low pressures (≤ 10 atm), the compressibility factors of most gases are close to unity. In fact, in the limit of P approaching to zero, we have $Z = 1$ for all gases. This is what we would expect, since all real gases behave ideally at low pressures. As pressure increases some gases have $Z < 1$ (that is, $P\overline{V} < RT$), which means that they are easier to compress than an ideal gas. As pressure increases

Figure 2.4

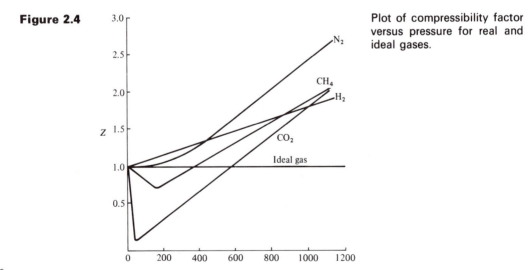

Plot of compressibility factor versus pressure for real and ideal gases.

16

further, all gases have $Z > 1$ (that is, $P\bar{V} > RT$). Over this region, the gases are harder to compress than an ideal gas. These behaviors are consistent with our understanding of intermolecular forces. When molecules are far apart (for example, at low pressures), the predominant intermolecular interaction is that of attraction. As the average distance of separation between molecules decreases, the repulsive interaction among molecules also becomes important. In general, attractive forces are long-range forces, whereas repulsive forces operate only within a short range (more on this topic in Chapter 17).

There has been considerable effort over the years to modify the ideal gas equation for real gases. Of the numerous such equations proposed, two will be considered.

The first equation is called the *van der Waals equation of state*:

$$\left(P + \frac{an^2}{V^2}\right)(V - nb) = nRT \tag{2.12}$$

This equation attempts to account for the finite volume of individual molecules and the attractive forces between them. The pressure exerted by the individual molecules on the walls of the container depends on both the frequency of molecular collisions with the walls and the momentum imparted by the molecules to the walls. Both contributions are diminished by the attractive intermolecular forces. In each case, the reduction in pressure depends on the number of molecules present or the density of the gas, n/V so that:

$$\begin{array}{c}\text{reduction in pressure} \\ \text{due to attractive forces}\end{array} \propto \left(\frac{n}{V}\right)\left(\frac{n}{V}\right)$$

$$= a\frac{n^2}{V^2}$$

where a is a proportionality constant.

Note that in Eq. (2.12), P is the pressure of the gas we measure experimentally and $(P + an^2/V^2)$ is the pressure of the gas if there were no intermolecular forces present. Since an^2/V^2 must have the unit of pressure, a is expressed as atm liter^2 mol^{-2}. To allow for the finite volume of molecules, we replace V in the ideal gas equation with $(V - nb)$, where nb represents the total effective volume of n moles of the gas. Thus nb must have the unit of volume or b has the units liters mol^{-1}. Both a and b are constants characteristic of the gas under study. Table 2.1 lists the values of a and b for a number

Van der Waals Constants and Boiling Points of Some Substances

Table 2.1

Substance	a (atm liter^2 mol^{-2})	b (liters mol^{-1})	Boiling Point (K)
He	0.0353	0.0241	4.2
Ne	0.208	0.0169	27.2
H_2	0.246	0.0267	20.3
N_2	1.35	0.0386	77.4
O_2	1.36	0.0319	90.2
CO	1.46	0.0393	83.2
CO_2	3.60	0.0427	195.2
CH_4	2.25	0.0428	109.2
H_2O	5.43	0.0303	373.15
NH_3	4.19	0.0373	239.8

Table 2.2

**Comparison of the Pressure (atm) Calculated Using the Ideal
Gas Equation and the van der Waals Equation with the
Observed Pressure (atm) at 298 K**[a]

Gas	Observed	Calculated Using the Ideal Gas Equation	Calculated Using the van der Waals Equation
H_2	50	48.7	50.2
	75	72.3	75.7
	100	95.0	100.8
CO_2	50	57.0	49.5
	75	92.3	73.3
	100	133.5	95.8

[a] Data taken from S. H. Maron and J. B. Lando, *Fundamentals of Physical Chemistry*, Macmillan Publishing Co., Inc., New York, 1974. Used by permission.

of gases. The value of a is somehow related to the magnitude of attractive forces. Using the boiling point as a measure of the strength of intermolecular forces (the higher the boiling point, the stronger are the intermolecular forces), we see that there is a rough correlation between the values of a and the boiling point of these gases. The quantity b is more difficult to interpret. Although b is proportional to the size of the molecule, the correlation is not always what we expect it to be. For example, the value of b for helium is 0.0241 liter mol^{-1}; that for neon is 0.0169 liter mol^{-1}. Based on these values, we might expect that helium is larger than neon, a fact we know to be untrue. The values of a and b of a gas can be determined by several methods. The common practice is to apply the van der Waals equation to the gas in the critical state. We shall return to this point in Section 2.9.

The van der Waals equation is valid over a much wider range of pressure and temperature than is the ideal gas equation. Furthermore, it provides a molecular interpretation for the equation of state. Table 2.2 compares the observed pressure of real gases with those calculated using the ideal gas equation and the van der Waals equation. From the values of n, T, and V as well as the van der Waals constants, we can calculate the pressure of the gas using Eqs. (2.9) and (2.12). At very high pressures and low temperatures, the van der Waals equation also becomes unreliable.

Another useful equation of state has the following expression:

$$\frac{PV}{nRT} = \frac{P\overline{V}}{RT} = 1 + \frac{Bn}{V} + \frac{Cn^2}{V^2} + \frac{Dn^3}{V^3} + \cdots \qquad (2.13)$$

Equation (2.13) is known as the *virial equation of state* and $B, C, D, \ldots$ are the second, third, fourth, ... virial coefficients. The values of the coefficients are such that $B \gg C \gg D$ so that at ordinary pressures the series converges rapidly.* (For an ideal gas, the virial coefficients are all equal to zero.) For most purposes, the terms beyond the one containing B can be neglected. The values of virial coefficients are determined experimentally by a curve-fitting procedure. Knowing how the compressibility factor of a gas varies with pressure at a given temperature, we can construct theoretical curves by guessing the values of $B, C, D, \ldots$ using Eq. (2.13) and compare them with the experi-

* The symbol $\gg$ means "much greater." Thus B is at least an order of magnitude (or a factor of 10) greater than C, which is an order of magnitude greater than D.

mental curve shown in Figure 2.4. The best matching then gives us the most reliable values of the virial coefficients. Note that these coefficients are all temperature-dependent, so that a set of different values has to be determined for every desired temperature.

Equations (2.12) and (2.13) exemplify two rather different approaches used in physical chemistry. The van der Waals equation deals with the nonideality of gases by correcting for the finite molecular volume and intermolecular forces. Although these corrections do result in a definite improvement over the ideal gas equation, Eq. (2.12) is still an approximate equation. The reason is that our present knowledge of intermolecular forces is insufficient to quantitatively explain molecular behavior. We could, of course, further improve this equation by adding more corrective terms; indeed, numerous other equations of state have been proposed since van der Waals first presented his analysis. On the other hand, Eq. (2.13) is an accurate equation for real gases although it does not provide us with any molecular interpretation. The nonideality of the gas is accounted for mathematically by a series expansion in which the coefficients B, C, ... can be determined experimentally. These coefficients do not have any physical meanings although they can be related to intermolecular forces in an indirect way. Thus our choice in this case is between an approximate equation that gives us some physical insight or an equation that describes the gas behavior accurately (if the coefficients are known) but tells us nothing about molecular behavior.

Condensation of Gases and the Critical State **2.9**

The condensation of gas to liquid is a familiar phenomenon. The first quantitative study of the pressure–volume relationship of this process was made by Andrews in 1869 on carbon dioxide. He measured the volume of a given amount of the gas as a function of pressure at various temperatures and obtained a series of isotherms like those shown in Figure 2.5. At high temperatures the curves are roughly hyperbolic, indicating that the gas obeys Boyle's

Isotherms of carbon dioxide at various temperatures. The critical temperature is T_5. At and above the critical temperature, carbon dioxide cannot be liquified no matter how great the pressure.

Figure 2.5

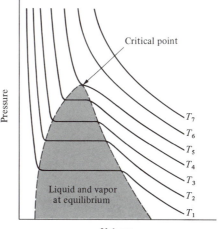

Volume

law. As the temperature is lowered, deviations become evident and a drastically different behavior is observed at T_4. Moving along the isotherm from right to left, we see that although the volume of the gas decreases with pressure, the product PV is no longer a constant (because the curve is no longer a hyperbola). Increasing the pressure further, we reach a point that is the intersection between the isotherm and the dashed curve on the right. If we could observe this process, we would note the formation of liquid carbon dioxide at this pressure. With the pressure held constant, the volume will now continue to decrease (as more and more vapor is converted to liquid) until all the vapor has been condensed. Beyond this point (the intersection between the horizontal line and the dashed curve on the left) the system is entirely liquid, and any further increase in pressure will only result in a very small decrease in volume, since liquids are much less compressible than gases. The pressure corresponding to the horizontal line (region in which vapor and liquid coexist) is called the *vapor pressure* of the liquid at the temperature of the experiment. The length of the horizontal line decreases with temperature. At a particular temperature (T_5 in our case) the isotherm is tangential to the dashed curve and only one phase is present. Thus the previous horizontal line has now become a point that is called the *critical point*. The corresponding temperature, pressure, and volume at this point are called the critical temperature (T_c), critical pressure (P_c), and critical volume (V_c). Critical temperature is the temperature above which no condensation of vapor to liquid can occur no matter how great the pressure. The critical constants of several gases are listed in Table 2.3. Note that the critical volume value is usually expressed as a molar quantity, called the molar critical volume ($\bar{V}_c$), which is the volume of 1 mol of the substance at the critical point.

The phenomenon of condensation and the existence of a critical temperature are direct consequences of the nonideal behavior of gases. After all, if molecules do not attract one another, no condensation would occur and if molecules had no volumes, then we would not be able to see liquids. As mentioned earlier, the nature of molecular interaction is such that the force among molecules is attractive when they are relatively far apart, but as they approach closer to one another (for example, liquid molecules under pressure) this force becomes repulsive, because of electrostatic repulsions between nuclei and between electrons. In general, the attractive force goes through a

Critical Constants of Some Substances[a]

Table 2.3

Substance	P_c (atm)	$\bar{V}_c$ (liters mol^{-1})	T_c (K)
He	2.26	0.0576	5.3
Ne	26.9	0.0417	44.4
H_2	12.8	0.0650	33.3
N_2	33.5	0.0900	126.1
O_2	49.7	0.0744	154.4
CO	35.0	0.0900	134.2
CO_2	73.0	0.0957	304.2
CH_4	45.6	0.0988	190.2
H_2O	219.5	0.0450	647.6
NH_3	111.5	0.0724	405.6

[a] Data taken mostly from S. H. Maron and J. B. Lando, *Fundamentals of Physical Chemistry*, Macmillan Publishing Co., Inc., New York, 1974. Used by permission.

maximum at a certain finite intermolecular distance. At temperatures below T_c it is possible to compress the gas and bring the molecules to within this attractive range where condensation will occur. Above T_c the kinetic energy of the gas molecules is such that they will always be able to break away from this attraction and no condensation can take place.

An interesting relation exists between the van der Waals constants a and b and the critical constants. Consider 1 mol of a gas $(n = 1)$. Rearranging Eq. (2.12), we obtain

$$\overline{V}^3 - \left(b + \frac{RT}{P}\right)\overline{V}^2 + \frac{a\overline{V}}{P} - \frac{ab}{P} = 0 \tag{2.14}$$

where $\overline{V}$ is the molar volume. This is a cubic equation and the solution yields three value of $\overline{V}$. At temperatures below T_c, $\overline{V}$ has three real roots; two of them correspond to the intersections of the horizontal line with the dashed curve in Figure 2.5, but the third root has no physical significance. Above T_c, $\overline{V}$ has one real root and two imaginary roots. At T_c, however, all three roots of $\overline{V}$ are real and identical. That is,

$$(\overline{V} - \overline{V}_c)^3 = 0 \tag{2.15}$$

or

$$\overline{V}^3 - 3\overline{V}_c\overline{V}^2 + 3\overline{V}_c^2\overline{V} - \overline{V}_c^3 = 0$$

Comparison of the coefficients for $\overline{V}^3$, $\overline{V}^2$, and $\overline{V}$ in Eqs. (2.14) and (2.15) gives the following:

For $\overline{V}^2$:

$$3\overline{V}_c = b + \frac{RT_c}{P_c} \tag{2.16}$$

For $\overline{V}$:

$$3\overline{V}_c^2 = \frac{a}{P_c} \tag{2.17}$$

Also,

$$\overline{V}_c^3 = \frac{ab}{P_c} \tag{2.18}$$

From Eqs. (2.16), (2.17), and (2.18), we obtain

$$a = 3P_c\overline{V}_c^2 \tag{2.19}$$

$$b = \frac{\overline{V}_c}{3} \tag{2.20}$$

$$R = \frac{8a}{27T_cb} \tag{2.21}$$

Therefore, if the critical constants are measured for a substance, we can calculate both a and b. Actually, any two of the foregoing three equations can be used to obtain a and b. Now if the van der Waals equation were accurately obeyed in the critical region, the choice would be unimportant. However, this is not the case. The values of a and b depend significantly on whether we use the $P_c - T_c$, $T_c - \overline{V}_c$, or $P_c - \overline{V}_c$ combinations. It is customary to choose P_c and T_c. From Eq. (2.20), we have

$$\overline{V}_c = 3b$$

which, when substituted into Eq. (2.19), yields

$$a = 3P_c(3b)^2 = 27P_c b^2 \tag{2.22}$$

Equation (2.22) into Eq. (2.21) gives

$$b = \frac{RT_c}{8P_c} \tag{2.23}$$

and

$$a = \frac{27R^2 T_c^2}{64P_c} \tag{2.24}$$

Although Eqs. (2.23) and (2.24) should be viewed as approximate relations because of the unreliability of the van der Waals equation in the critical region, they are frequently used to obtain the van der Waals constants. Thus the values of a and b listed in Table 2.1 are calculated from the critical constants (see Table 2.3) and Eqs. (2.23) and (2.24).

Suggestions for Further Reading

HILDEBRAND, J. H. *An Introduction to Molecular Kinetic Theory*, Chapman & Hall Ltd., London, 1963 (Van Nostrand Reinhold Company, New York).

The topics discussed in this chapter can be found in the standard physical chemistry texts listed in Chapter 1.

Reading Assignments

"Concept of Empirical Temperature for Introductory Chemistry," P. Ander, *J. Chem. Educ.* **48**, 325 (1971).

"The Critical Temperature: A Necessary Consequence of Gas Nonideality," F. L. Pilar, *J. Chem. Educ.* **44**, 284 (1967).

"The van der Waals Gas Equation," E. S. Swinbourne, *J. Chem. Educ.* **32**, 366 (1955).

"A Simple Model for van der Waals," S. S. Winter, *J. Chem. Educ.* **33**, 459 (1959).

"The Discovery of Boyle's Law," R. G. Neville, *J. Chem. Educ.* **39**, 356 (1962).

"Effective Insect Fogging—The Origin of Sea Breezes—Hot Air Balloons," R. C. Plumb, *J. Chem. Educ.* **52**, 104 (1975).

"The Lung," J. H. Comroe, *Sci. Am.*, Feb. 1966.

"Scuba Diving and the Gas Laws," E. D. Cooke, *J. Chem. Educ.* **50**, 425 (1973).

"The Cabin Atmosphere in Manned Space Vehicles," W. H. Bowman and R. M. Lawrence, *J. Chem. Educ.* **48**, 152 (1971).

"Comparisons of Equations of State in Effectively Describing PVT Relations," J. B. Ott, J. R. Goates, and H. T. Hall, *J. Chem. Educ.* **48**, 515 (1971).

Problems

2.1 Some gases, such as NO_2 and NF_2, do not obey Boyle's law at any pressure. Explain.

2.2 The saturated vapor pressure of mercury is 0.002 mm Hg at 300 K and the density of air at 300 K is 1.18 g liter^{-1}. (a) Calculate the concentration of mercury

vapor in air in mol liter^{-1}. (b) What is the number of parts per million (ppm) by
weight of mercury in air?

23

Chap. 2

Problems

2.3 A very flexible balloon initially of volume 1.2 liters at 1.0 atm and 300 K is
allowed to rise up to the stratosphere, where the temperature and pressure are 250 K
and 3.0×10^{-3} atm, respectively. What is the final volume of the balloon? Assume
ideal gas behavior.

2.4 The relative humidity in a closed room of volume 645.2 m^3 is 87.6% at 300 K
and the vapor pressure of water at 300 K is 0.0313 atm. Calculate the mass of water in
the air. [*Hint*: The relative humidity is defined as $(P/P_s) \times 100\%$, where P and P_s are
the partial pressure and saturated partial pressure of water, respectively.]

2.5 Death by suffocation in a sealed container is normally due to CO_2 poisoning
(which occurs at about 7% CO_2 by volume), not to oxygen deficiency. For what length
of time would it be safe to be in a sealed room $10 \times 10 \times 20$ ft? [*Source*: "Eco-Chem,"
J. A. Campbell, *J. Chem. Educ.* **49**, 538 (1972).]

2.6 A sample of dry air near sea level has the following composition by volume:
nitrogen: 78.08%; oxygen: 20.94%; argon: 0.95%; carbon dioxide: 0.03%. Assuming
ideal gas behavior and that the atmospheric pressure is 1.00 atm, calculate: (a) the
partial pressure of each gas in atm, and (b) the concentration of each gas in mol liter^{-1}
at 273 K.

2.7 A mixture containing nitrogen and hydrogen weighs 3.50 g and occupies a
volume of 7.46 liters at 300 K and 1.0 atm. Calculate the weight percent of these two
gases. Assume ideal gas behavior.

2.8 A sample of air occupies 2.5 liters when the pressure is 1.2 atm. (a) What volume
does it occupy at 6.5 atm? (b) What pressure is required to compress it to 2.5 cm^3?

2.9 A vessel contains a mixture of two ideal gases A and B. Show graphically how the
total pressure of the system depends on the amount of A present; that is, plot the total
pressure versus the mole fraction of A. Do the same for B on the same graph.

2.10 A diver quickly ascends to the surface of the water from a depth of 4.08 m
without exhaling gas in her lungs. By what factor would the volume of her lungs
increase by the time she reaches the surface? Assume constant temperature and ideal
gas behavior. The density of seawater is 1.03 g cm^{-3} and acceleration due to gravity is
980.67 cm s^{-2}.

2.11 An air bubble with a radius of 1.5 cm at the bottom of a lake where the
temperature is 8.4°C and the pressure is 2.8 atm rises to the surface, where the temperature is 25.0°C and the pressure is 1.0 atm. Calculate the radius of the bubble when it
reaches the surface. Assume ideal gas behavior. (*Hint*: The volume of a sphere is given
by $\frac{4}{3}\pi r^3$, where r is the radius.)

*2.12 The density of dry air at 1.00 atm and 34.4°C is 1.15 g liter^{-1}. Calculate the
composition of air (percent by weight) assuming only nitrogen and oxygen to be
present and ideal gas behavior. (*Hint*: First calculate the "molar mass" of air, then the
mass fractions of O_2 and N_2.)

2.13 A gas evolved during the fermentation of glucose has a volume of 0.78 liter when
measured at 20.1°C and 1.0 atm. What was the volume of this gas at the fermentation
temperature of 36.5°C? Assume ideal gas behavior.

⋆**2.14** Two bulbs of volumes V_A and V_B are connected by a stopcock. The number of moles of gases in the bulbs are n_A and n_B and initially the gases are at the same pressure P and temperature T. Show that the final pressure of the system, after opening the stopcock, is equal to P. Assume ideal gas behavior.

2.15 Two bulbs, 1.00 liter and 1.50 liters in volume, connected by a stopcock, are filled, respectively, with argon at 0.75 atm and helium at 1.20 atm at the same temperature. Calculate the total pressure and partial pressure of each gas after the stopcock is opened and also the mole fraction of each gas. Assume ideal gas behavior.

2.16 Deep-sea divers usually breathe a mixture of oxygen and helium gases underwater. Estimate the percent composition by volume of oxygen in the mixture at a depth of 20.5 m below sea level such that the partial pressure of oxygen is 0.20 atm. Assume ideal gas behavior. The density of seawater is 1.03 g cm^{-3}. (*Hint*: First calculate the hydrostatic pressure at the depth of 20.5 m. The acceleration due to gravity is 980.67 cm s^{-2}.)

2.17 Calculate the column of water that can be supported by 1 atm pressure. What conclusion can you draw regarding pumping water from a well using a hand pump?

2.18 The equation for metabolic decomposition of glucose is the same as that for combustion of glucose in air:

$$C_6H_{12}O_6(s) + 6O_2(g) \longrightarrow 6CO_2(g) + 6H_2O(l)$$

Calculate the volume of carbon dioxide produced at 37.0°C and 1 atm when 5.60 g of glucose is consumed in the combustion.

2.19 A common unit for pressure is pounds per square inch (psi). Show that 1 atm = 14.7 psi. An automobile tire is inflated to 28.0 psi gauge pressure when cold, at 18°C. (a) What will be the pressure if the tire is heated to 32°C by running? (b) What percent of the air in the tire would then have to be let out in order to reduce the pressure to the original 28.0 psi? Assume that the volume of the tire remains constant with temperature. (A tire gauge measures not the pressure of the air inside but its excess over the external pressure, which is 14.7 psi.)

2.20 An ultra-high-vacuum pump can reduce the pressure of air from 1.0 atm to 1.0×10^{-10} mm Hg. Calculate the number of air molecules in a liter at this pressure and 298 K. Compare your results with the number of molecules in 1.0 liter at 1.0 atm and 298 K. Assume ideal gas behavior.

⋆**2.21** A stockroom supervisor measured the contents of a partially filled 25.0-gallon acetone drum on a day when the temperature was 18.0°C and atmospheric pressure was 780 mm Hg, and found that 15.4 gallons of the solvent remained. After tightly sealing the drum, a student assistant dropped the drum while carrying it upstairs to the organic laboratory. The drum was dented and its internal volume was decreased to 20.4 gallons. What is the total pressure inside the drum after the accident? The vapor pressure of acetone at 18.0°C is 400 mm Hg. (*Hint*: At the time the drum was sealed, the pressure inside the drum, which is equal to the sum of the pressures of air and acetone, is equal to the atmospheric pressure.)

2.22 A mixture of helium and neon weighing 5.50 g occupies a volume of 6.8 liters at 300 K and 1 atm. Calculate the composition of the mixture in weight percent.

2.23 Starting with the ideal gas equation, show how you can calculate the molar mass of a gas from a knowledge of its density.

2.24 The critical temperature and critical pressure of naphthalene are 474.8 K and 40.6 atm, respectively. Calculate the van der Waals constants a and b for naphthalene.

2.25 The van der Waals constants a and b for benzene are 18.00 atm liters^2 mol^{-2} and 0.115 liter mol^{-1}, respectively. Calculate the critical constants for benzene.

2.26 Using the data shown in Table 2.1, calculate the pressure exerted by 2.500 mol of carbon dioxide confined in a volume of 1.000 liters at 450 K. Compare the pressure with that calculated, assuming ideal gas behavior.

★2.27 In terms of the hard-sphere gas model, molecules are assumed to possess finite volume, but there is no interaction among the molecules. (a) Compare the P–V isotherm for an ideal gas and that for a hard-sphere gas. (b) Let b be the effective volume of the gas. Write an equation of state for this gas. (c) From this equation, derive an expression for $Z = PV/RT$ for the hard-sphere gas and make a plot of Z versus P for two values of T (T_1 and T_2, $T_2 > T_1$). Be sure to indicate the value of the intercepts on the Z axis. (d) Make a plot of Z versus T for fixed P for an ideal gas and for the hard-sphere gas. (Assume that $n = 1$ mol for all parts of this problem.)

2.28 A sample of zinc is allowed to completely react with an excess of hydrochloric acid. The hydrogen gas generated is collected over water at 25.0°C. The volume of the gas is 7.80 liters and the pressure of the gas is 0.98 atm. Calculate the quantity of zinc metal consumed. The vapor pressure of water at 25.0°C is 23.8 mm Hg. Assume ideal gas behavior.

★2.29 Derive the van der Waals constants a and b in terms of critical constants by recognizing the fact that for the van der Waals equation at the critical point, $(\partial P/\partial \overline{V})_T = 0$ and $(\partial^2 P/\partial \overline{V}^2)_T = 0$, where $\overline{V}$ is the molar volume.

2.30 From the relationships among the van der Waals constants and the critical constants, show that $Z_c = P_c\overline{V}_c/RT_c = 0.375$, where Z_c is the compressibility factor at the critical point.

2.31 Assume that the distance of closest approach between two similar spherical molecules is the sum of their radii. Calculate the "volume of influence" per molecule.

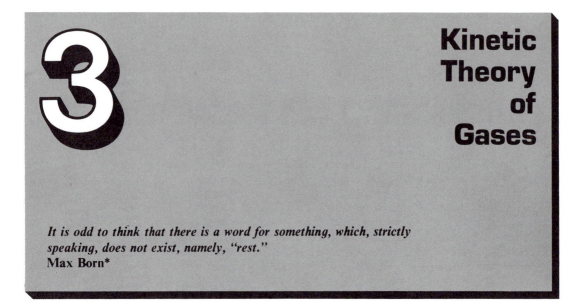

It is odd to think that there is a word for something, which, strictly speaking, does not exist, namely, "rest."
Max Born*

The study of gas laws exemplifies the phenomenological, macroscopic approach to physical chemistry. Equations describing gas laws are relatively simple; experimental data are readily accessible. Yet the study of gas laws offers no real physical insight into processes that occur at the molecular level. Although the van der Waals equation attempts to account for the nonideal behavior in terms of intermolecular interactions, it does so in a rather vague manner. How is the pressure of a gas related to the motion of individual molecules? Why do real gases obey Boyle's law at low pressures? A number of such questions need to be answered in more detail. The next logical step, then, is to attempt an understanding of the behavior of gases in terms of the dynamics of molecular motion. The kinetic theory of gases is one such approach which allows us to interpret various properties of gas molecules in a more quantitative manner.

3.1 The Model

To develop a theory to account for experimental observations, we first need to define our system. If we do not understand all the properties of a system, as is usually the case, we must make a number of assumptions. Our *model for the kinetic theory of gases* is based on the following assumptions:

1. A gas is made up of a great number of atoms or molecules, separated by distances that are large compared to their size.
2. The molecules have mass, but their volume is negligibly small.
3. The molecules are constantly in random motion.
4. Collisions among molecules and between molecules and the walls of the container are elastic; that is, kinetic energy may be transferred from one

* *The Restless Universe*, 2nd ed., Dover Publications, Inc., New York, 1951. Used by permission.

molecule to another, but it is not converted to other forms of energy, such as heat.

5. There is no interaction, attractive or repulsive, between the molecules.

Note that assumptions 2 and 5 are the same as those for gases discussed in Chapter 2. The difference between the ideal gas laws and the kinetic theory of gases is that here we are using the foregoing assumptions in an explicit manner to derive expressions for macroscopic properties such as pressure and temperature in terms of individual molecular motions.

Pressure of a Gas **3.2**

We may now derive an expression for the *pressure of a gas* in terms of its molecular properties, using the model given in Section 3.1. Consider a certain gas made up of N molecules each of mass m confined in a cubic box of length l. At any instant, the molecular motion inside the container is completely random. Let us analyze the motion of a particular molecule with velocity v. Since velocity is a *vector* quantity, that is, it has both magnitude and direction, v can be resolved into three mutually perpendicular components v_x, v_y, and v_z. These three components give the rates at which the molecule is moving along the x, y, and z directions, respectively, and v is simply the resultant velocity (Figure 3.1). The projection of the velocity vector on the xy plane is OA, which, according to Pythagoras' theorem, is given by

$$\overline{OA}^2 = v_x^2 + v_y^2$$

Similarly,

$$v^2 = \overline{OA}^2 + v_z^2$$
$$= v_x^2 + v_y^2 + v_z^2 \tag{3.1}$$

Let us for the moment consider the motion of a molecule only along the x direction. Figure 3.2 shows the changes that take place when the molecule collides with the wall of the container (the yz plane) with velocity component v_x. Since the collision is elastic, the velocity after collision is the same as before but opposite in direction. The momentum of the molecule is mv_x, where m is its mass, so that the *change* in momentum is given by

$$mv_x - m(-v_x) = 2mv_x$$

Velocity vector v and its components.

Figure 3.1

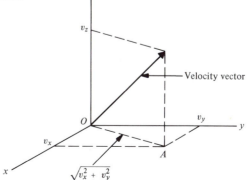

Figure 3.2

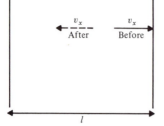

Change in velocity upon collision of a particle moving with v_x.

The sign of v_x is such that it is positive when the molecule moves from left to right and negative when it moves in the opposite direction. Immediately after the collision the molecule will take time l/v_x to collide with the other wall, and in time $2l/v_x$ the molecule will strike the same wall again.* Thus the frequency of collision between the molecule and a given wall (that is, the number of collisions per unit time) is $v_x/2l$, and the change in momentum per unit time is $(2mv_x)(v_x/2l)$, or mv_x^2/l. According to Newton's second law of motion,

$$\text{force} = \text{mass} \times \text{acceleration}$$

$$= \text{mass} \times \text{distance} \times (\text{per unit time})^2$$

$$= \text{momentum per unit time}$$

Therefore, the force exerted by one molecule on one wall as a result of the collision is mv_x^2/l and the total force due to N molecules is Nmv_x^2/l. Since pressure is force/area and area is l^2, we can now express the total pressure exerted on one wall as

$$P = \frac{F}{A}$$

$$= \frac{Nmv_x^2}{l(l^2)} = \frac{Nmv_x^2}{V}$$

or

$$PV = Nmv_x^2 \tag{3.2}$$

where V is the volume of the cube. When we are dealing with a large collection of molecules (for example, when N is of the order of 6×10^{23}), there is a tremendous spread of molecular velocities. It is more appropriate, therefore, to replace v_x^2 in Eq. (3.2) with the *average* quantity $\overline{v_x^2}$. The relation between the average of the square of the velocity components and the average of the square of the velocity, $\overline{v^2}$, is still

$$\overline{v^2} = \overline{v_x^2} + \overline{v_y^2} + \overline{v_z^2}$$

The quantity $\overline{v^2}$ is called the *mean-square velocity*, defined as

$$\overline{v^2} = \frac{v_1^2 + v_2^2 + \cdots + v_N^2}{N} \tag{3.3}$$

* We assume that the molecule does not collide with other molecules along the way. A more rigorous treatment including molecular collision gives exactly the same result.

When N is a large number, it is correct to assume that molecular motions along the x, y, and z directions are equally probable. This means that

$$\overline{v_x^2} = \overline{v_y^2} = \overline{v_z^2} = \frac{\overline{v^2}}{3}$$

and Eq. (3.2) can now be written as

$$P = \frac{Nm\overline{v^2}}{3V}$$

Multiplying top and bottom by 2 and recalling that the average kinetic energy of the molecule $\overline{E}_{\text{trans}}$ is given by $\frac{1}{2}m\overline{v^2}$ (where the subscript trans denotes translational motion), we obtain

$$P = \frac{2N}{3V}\left(\tfrac{1}{2}m\overline{v^2}\right) = \frac{2N}{3V}\overline{E}_{\text{trans}} \tag{3.4}$$

This is the pressure exerted by N molecules on one wall. The same result can be obtained regardless of the direction (x, y, or z) we describe for the molecular motion. We see that the pressure is directly proportional to the average kinetic energy or more explicitly, to the mean-square velocity of the molecule. The physical meaning of this dependence is that the larger the velocity, the more frequent is the collision and the larger the velocity, the greater is the change in momentum. Thus these two independent terms give us the quantity $\overline{v^2}$.

Kinetic Energy and Temperature **3.3**

An interesting and important result is obtained when we compare Eq. (3.4) with the ideal gas equation [Eq. (2.9)]:

$$PV = nRT$$

$$= \frac{N}{N_0}RT$$

or

$$P = \frac{NRT}{N_0 V} \tag{3.5}$$

where N_0 is the Avogadro constant. Identifying the pressures in Eqs. (3.4) and (3.5), we get

$$\frac{2}{3}\frac{N}{V}\overline{E}_{\text{trans}} = \frac{N}{N_0}\frac{RT}{V}$$

or

$$\overline{E}_{\text{trans}} = \frac{3}{2}\frac{RT}{N_0} = \frac{3}{2}kT \tag{3.6}$$

where $R = kN_0$ and k is *Boltzmann's constant*, equal to 1.38054×10^{-16} erg K^{-1} (CGS units) or 1.38054×10^{-23} J K^{-1} (SI units). Now we can see that mean kinetic energy is proportional to absolute temperature. This is a conclusion drawn by combining the kinetic theory and the ideal gas equation. The significance of Eq. (3.6) is that it provides an interpretation in terms of molecular motion of the temperature concept. For this reason, random molecular motion is sometimes referred to as *thermal motion*. It is important to keep in mind that kinetic theory is a *statistical* treatment of our model; hence it is meaningless to associate temperature with the kinetic energy of a few molecules. Further, this simple result tells us that whenever two ideal gases are at the same temperature T, they must have the same average kinetic energy. The reason is that the term $\frac{3}{2}kT$ in Eq. (3.6) is independent of molecular properties or amount of the gas present, as long as N is a large number.

It is easy to see that $\overline{v^2}$ will be a very difficult quantity to measure, if indeed it can be measured at all. To do so, we would need to measure each individual velocity, square it, and then take the average [see Eq. (3.3)]. Fortunately, $\overline{v^2}$ can be obtained quite directly from other quantities. From Eq. (3.6), we write

$$\overline{v^2} = \frac{3RT}{mN_0} = \frac{3kT}{m}$$

or

$$\sqrt{\overline{v^2}} = v_{rms} = \sqrt{\frac{3kT}{m}} = \sqrt{\frac{3RT}{M}} \tag{3.7}$$

where v_{rms} is the *root-mean-square velocity** and M is the molar mass of the gas. Note that v_{rms} is directly proportional to the square root of temperature and inversely proportional to the square root of molar mass of the molecule.

3.4 Maxwell Distribution

So far we have considered the velocity in an average way; that is, we have used the root-mean-square quantity. When we are studying 1 mol of a gas, say, it is impossible to know the velocity of each individual molecule for two reasons. First, the number of molecules is so huge that there is no way we can follow all their motions. Second, although molecular motion is a well-defined quantity, we cannot measure its value exactly. For example, suppose that we were able to conduct a careful experiment and determine the velocity of a molecule to be 3025.4 cm s^{-1}. But this is at best an estimate, since the actual velocity is not 3025.4 but 3025.3298462 ... cm s^{-1}. Therefore, it is not appropriate to concern ourselves with individual molecular velocities in this manner. Instead, we ask the question: For a given system at some known temperature, how many molecules are moving between velocities v and $v + \Delta v$ at any moment? Referring to our example above, we might ask this question: How many molecules are moving between velocities 3025 cm s^{-1} and 3026 cm s^{-1} at any moment? Because the total number of molecules is very large, there is a continuous spread or *distribution* of velocities (from zero to very large values) as a result of collision. We can therefore make the

---

* Since velocity is a vector quantity (it has both magnitude and direction), the average molecular velocity $\bar{v}$ must be zero—there are just as many molecules moving in the positive direction as there are in the negative direction. On the other hand, v^2 is a scalar quantity; that is, it has magnitude but no direction.

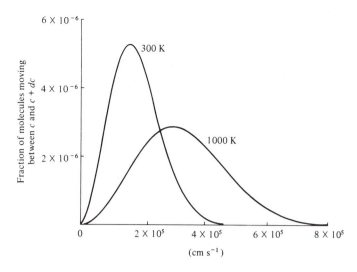

Maxwell speed distribution curves for hydrogen gas at 300 and 1000 K. **Figure 3.3**

velocity range smaller and smaller, and in the limit it becomes dv. This fact has great significance, since it allows us to replace the summation sign with the integral sign in calculating the number of molecules whose velocities fall between v and $v + dv$. Mathematically speaking, it is considerably easier to integrate than to sum a large series. This distribution-of-velocities approach was first employed by Maxwell in 1860 and later refined by Boltzmann. They showed that for a system containing N ideal gas molecules at thermal equilibrium with its surroundings, the fraction of molecules dN/N moving with velocities v and $v + dv$ is given by

$$\frac{dN}{N} = 4\pi \left(\frac{m}{2\pi kT}\right)^{3/2} e^{-mv^2/2kT} \, dv \qquad (3.8)$$

where m is the mass of the molecule, k the Boltzmann constant, and T the temperature. Equation (3.8) is known as the *Maxwell distribution* or *Maxwell–Boltzmann distribution of velocities*.* As we mentioned earlier, velocity is a vector quantity. In many cases we prefer to deal with the speed of molecules (c), which is a scalar quantity. The corresponding Maxwell equation for distribution of speeds is given by

$$\frac{dN}{N} = 4\pi c^2 \left(\frac{m}{2\pi kT}\right)^{3/2} e^{-mc^2/2kT} \, dc \qquad (3.9)$$

From now on in this chapter we shall be interested primarily in the quantity speed and will not discuss Eq. (3.8) further.

Plots of the speed distribution of H_2 at 300 K and 1000 K are shown in Figure 3.3. At low temperatures the distribution has a rather narrow range; as the temperature increases, the curve becomes flatter, meaning that there are now more fast-moving molecules (sometimes called "hot" molecules). This temperature dependence of the distribution curve has important implications in chemical reaction rates. As we shall see in Chapter 13, in order to react, a molecule must possess a minimum amount of energy, called *activation energy*.

* The mathematical derivations of Eqs. (3.8) and (3.9) are quite lengthy. The interested reader is referred to the standard physical chemistry texts listed in Chapter 1.

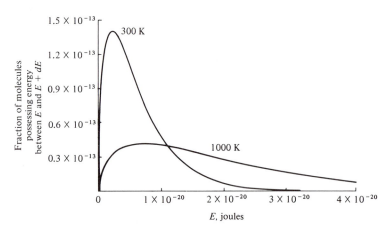

Figure 3.4 Maxwell (kinetic) energy distribution curves for an ideal gas at 300 and 1000 K.

At low temperatures the number of fast-moving molecules is small; hence most reactions proceed at a slow rate. Raising the temperature enhances the number of hot molecules, resulting in an increase in the reaction rate.

Since energy is related to speed by $E = \frac{1}{2}mc^2$, it is not surprising that there is also an energy distribution function. A corresponding relation for the (kinetic) energy distribution is

$$\frac{dN}{N} = 2\pi E^{1/2} \left(\frac{1}{\pi kT}\right)^{3/2} e^{-E/kT} \, dE \qquad (3.10)$$

where dN/N is the fraction of molecules having kinetic energies between E and $E + dE$. Figure 3.4 shows the energy distribution curves at 300 K and 1000 K. Keep in mind that these curves apply to *any* ideal gas that fits the kinetic theory description. As mentioned earlier, two different gases at the same temperature possess the same average kinetic energy and hence the same energy distribution curve.

The v_{rms} discussed earlier is the same as the root-mean-square speed, c_{rms}.*
In addition, there are also *average speed* and the *most probable speed*. The average speed $\bar{c}$ is defined as

$$\bar{c} = \frac{c_1 + c_2 + \cdots + c_N}{N} \qquad (3.11)$$

Using the distribution function [Eq. (3.9)], we can obtain an explicit expression for $\bar{c}$:

$$\bar{c} = \sqrt{\frac{8kT}{\pi m}} = \sqrt{\frac{8RT}{\pi M}} \qquad (3.12)$$

The most probable speed c_{mp} is simply the speed possessed by the largest fraction of molecules. It is obtained by differentiating the distribution expression in Eq. (3.9) with respect to speed, setting the result to zero, and determining the value of c that corresponds to this relation. This yields

$$c_{mp} = \sqrt{\frac{2kT}{m}} = \sqrt{\frac{2RT}{M}} \qquad (3.13)$$

---

* Since the square of the average velocity is a scalar quantity, it follows that $\overline{v^2} = \overline{c^2}$; hence $v_{rms} = c_{rms}$.

EXAMPLE 3.1 Calculate c_{mp}, $\bar{c}$, and c_{rms} for O_2 at 300 K.

Answer

CGS units

$$R = 8.314 \times 10^7 \text{ ergs K}^{-1} \text{ mol}^{-1}$$

$$M = 32 \text{ g mol}^{-1}$$

$$T = 300 \text{ K}$$

$$c_{mp} = \sqrt{\frac{2 \times 8.314 \times 10^7 \text{ ergs K}^{-1} \text{ mol}^{-1} \times 300 \text{ K}}{32 \text{ g mol}^{-1}}}$$

$$= \sqrt{1.56 \times 10^9 \text{ ergs g}^{-1}}$$

$$= \sqrt{1.56 \times 10^9 \text{ cm}^2 \text{ s}^{-2}} = 3.95 \times 10^4 \text{ cm s}^{-1}$$

SI units

$$R = 8.314 \text{ J K}^{-1} \text{ mol}^{-1}$$

$$M = 0.032 \text{ kg mol}^{-1}$$

$$T = 300 \text{ K}$$

$$c_{mp} = \sqrt{\frac{2 \times 8.314 \text{ J K}^{-1} \text{ mol}^{-1} \times 300 \text{ K}}{0.032 \text{ kg mol}^{-1}}}$$

$$= \sqrt{1.56 \times 10^5 \text{ J kg}^{-1}}$$

$$= \sqrt{1.56 \times 10^5 \text{ m}^2 \text{ s}^{-2}}$$

$$= 395 \text{ m s}^{-1}$$

Similarly, we can show that

$$\bar{c} = \sqrt{\frac{8RT}{\pi M}} = 4.46 \times 10^4 \text{ cm s}^{-1} = 446 \text{ m s}^{-1}$$

and

$$c_{rms} = \sqrt{\frac{3RT}{M}} = 4.84 \times 10^4 \text{ cm s}^{-1} = 484 \text{ m s}^{-1}$$

Thus

$$c_{rms} > \bar{c} > c_{mp}$$

Molecular Collision and the Mean Free Path **3.5**

Now that we have an explicit expression for the average speed, we can proceed to study some dynamic processes involving gases in terms of this quantity. We know that the speed of a molecule is not a constant but changes frequently as a result of collision. Therefore, the question we ask is: How often do molecules collide with one another? As it turns out, this collision rate is found to depend on the density of the gas and the molecular speed and therefore on

Figure 3.5

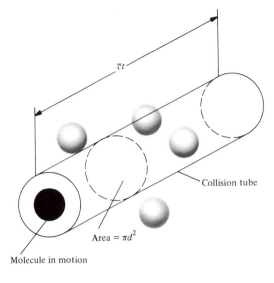

The collision cross section and the collision tube. Any molecule whose center lies within or touches the tube will collide with the moving molecule (sphere at the left).

Collision tube

Area $= \pi d^2$

Molecule in motion

the temperature of the system. In the kinetic theory model we assume each molecule to be a hard sphere of diameter d. A molecular collision is one in which the separation between the two spheres (measured from each center) is d or less. Let us consider the motion of a particular molecule. A simple approach is to assume that at a given instant, all except this molecule are standing still. In time t this molecule moves a distance $\bar{c}t$ (where $\bar{c}$ is the average speed) and sweeps out a collision tube of area πd^2 (Figure 3.5). The volume of the cylinder is $(\pi d^2)(\bar{c}t)$. Any molecule whose center lies within this cylinder will collide with the molecule. If there are altogether N molecules in volume V, then the density of the gas is N/V, the number of collisions in time t is $\pi d^2 \bar{c}t(N/V)$, and the number of collisions per unit time or the *collision frequency* (Z_1) is $\pi d^2 \bar{c}(N/V)$. The expression for the collision frequency needs a correction. If we assume that the rest of the molecules are frozen in position, the quantity $\bar{c}$ should be replaced by the average relative speed. Figure 3.6 shows three different approaches for two colliding molecules. The relative speed for the case shown in (c) is $\sqrt{2}\,\bar{c}$ so that

$$Z_1 = \sqrt{2}\,\pi d^2 \bar{c}\,\frac{N}{V} \qquad \text{collisions s}^{-1} \qquad (3.14)$$

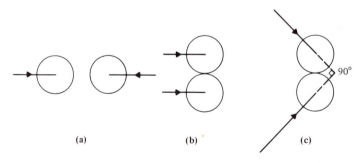

(a)   (b)   90°   (c)

Figure 3.6 Three different approaches for two colliding molecules. The situations shown in (a) and (b) represent the two extreme cases, while that shown in (c) may be taken as the "average" case for molecular encounter.

This is the number of collisions a *single* molecule makes in 1 s. If there are N molecules in volume V and each makes Z_1 collisions s^{-1}, the total number of binary collisions per unit volume per unit time Z_{11} is given by

$$Z_{11} = \frac{1}{2} Z_1 \left(\frac{N}{V} \right)$$

$$= \frac{\sqrt{2}}{2} \pi d^2 \bar{c} \left(\frac{N}{V} \right)^2 \qquad \text{collisions cm}^{-3} \text{ s}^{-1} \qquad (3.15)$$

The factor $\frac{1}{2}$ is introduced in Eq. (3.15) to ensure that we are only counting each collision between two molecules once. The probability of three or more molecules colliding together is very small except at high pressures. Since the rate of a chemical reaction generally depends on how often reacting molecules come into contact with one another, Eq. (3.15) is of considerable importance to chemical kinetics. We shall return to this equation in Chapter 13.

A quantity closely related to the collision number is the average distance traveled by a molecule between successive collisions. This distance, called the *mean free path* (λ), is defined as

$$\lambda = (\text{average speed}) \times (\text{average time between collisions})$$

Since average time between collisions is the reciprocal of the collision frequency, we have

$$\lambda = \frac{\bar{c}}{Z_1} = \frac{\bar{c}}{\pi d^2 \sqrt{2} \, \bar{c}(N/V)} = \frac{1}{\sqrt{2} \, \pi d^2 (N/V)} \qquad (3.16)$$

We see that the mean free path is inversely proportional to the density of the gas (N/V). This behavior is reasonable because in a dense gas, a molecule makes more collisions per unit time and hence travels a shorter distance between successive collisions.

There are many everyday examples that can be understood in terms of the kinetic theory of gases. Here we shall consider molecular motion inside a light bulb. Ordinary incandescent light bulbs are filled with argon, an inert gas, to prevent the evaporation of the tungsten filament as it heats to a temperature of about 3000°C. Experts have known for many years that the light-bulb efficiency and longevity can be increased by replacing argon with krypton, another inert gas. The molar masses of Ar and Kr are 39.95 g mol^{-1} and 83.80 g mol^{-1}, respectively. According to Eq. (3.12), the ratio of their average speeds is given by

$$\frac{\bar{c}_{Ar}}{\bar{c}_{Kr}} = \sqrt{\frac{M_{Kr}}{M_{Ar}}} = \sqrt{\frac{83.80}{39.95}} = 1.45$$

Since the lighter atoms move more swiftly, they collide with the bulb's glass wall more frequently. When colliding, the molecules transfer heat generated by the filament to the surrounding air. Because krypton is a less efficient heat conductor than argon, the filament's temperature can be maintained higher for the same amount of electricity flowing through it. A higher temperature means a brighter light. Replacing Ar with Kr also prolongs the life of a light bulb from about 1000 to 3000 hours. When heated to 3000°C, some of the tungsten atoms on the surface of the filament will sublime, eventually condensing on the inner glass wall.* This is why an old light bulb shows dark spots. When tungsten atoms collide with the inert gas atoms, part of their kinetic energy is transferred to the latter molecules. By reducing the speed of the tungsten atoms, this process gives the filament a chance to take back its atoms with another collision. Being a heavier gas, krypton retards filament evaporation more efficiently than argon. Filaments heated in a krypton atmosphere do not burn out as quickly and the glass wall will show fewer tungsten deposits.

The reason why krypton-filled light bulbs are not in common use today is purely economic—krypton costs about $10 per cubic foot while argon costs only 5 cents. However, considering the steadily increasing cost of electricity, it seems likely that at least some household light bulbs will contain krypton instead of argon during the next decade.

EXAMPLE 3.2 The density of dry air at 1 atm and 298 K is about 2.5×10^{19} molecules cm^{-3}. Assuming that air contains only nitrogen molecules, calculate the collision frequency, the binary collision number, and the mean free path of nitrogen molecules under these conditions. The collision diameter of nitrogen is 3.75 Å.

Answer: Our first step is to calculate the average speed of nitrogen. From Eq. (3.12) we get $\bar{c} = 4.8 \times 10^4$ cm s^{-1}. The collision frequency is given by

$$Z_1 = \sqrt{2}\,\pi(3.75 \times 10^{-8} \text{ cm})^2(4.8 \times 10^4 \text{ cm s}^{-1})$$
$$\times (2.5 \times 10^{19} \text{ molecules cm}^{-3})$$
$$= 7.5 \times 10^9 \text{ collisions s}^{-1}$$

Note that we have used the conversion factor 1 Å = 10^{-8} cm and replaced the units molecules with collisions. This is valid since, in the derivation of Z_1, every molecule in the collision volume represents a collision. The binary collision number is

$$Z_{11} = \frac{Z_1}{2}\left(\frac{N}{V}\right)$$
$$= \frac{(7.5 \times 10^9 \text{ collisions s}^{-1})}{2}(2.5 \times 10^{19} \text{ molecules cm}^{-3})$$
$$= 9.4 \times 10^{28} \text{ collisions cm}^{-3} \text{ s}^{-1}$$

Again, we converted molecules to collisions in calculating the total number of binary collisions. Finally, the mean free path is given by

$$\lambda = \frac{\bar{c}}{Z_1} = \frac{4.8 \times 10^4 \text{ cm s}^{-1}}{7.5 \times 10^9 \text{ collisions s}^{-1}}$$
$$= 6.4 \times 10^{-6} \text{ cm collision}^{-1}$$
$$= 640 \text{ Å collision}^{-1}$$

Note that it is usually sufficient to express mean free path only in terms of distance rather than distance per collision. Thus, in our case, the mean free path of nitrogen becomes 6.4×10^{-6} cm.

---

* Sublimation is the direct passage of a substance from solid to vapor, skipping the liquid phase.

So far we have concentrated on the properties of gas molecules in the equilibrium state. Although microscopically gas molecules may fluctuate tremendously because of random motion, macroscopically a system remains unchanged with time. Now let us consider a different situation in which the rate of flow of gases along a tube, for example, is measured. At a given temperature and density the rate will vary from gas to gas because each gas has a different viscosity.

A simple equation for gas viscosity can be derived from the kinetic theory as follows. The flow motion of molecules can be analyzed in terms of a set of laminars shown in Figure 3.7. Laminars are layers with negligible thickness.

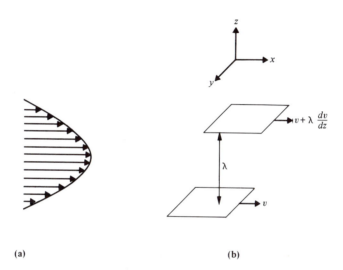

(a)   (b)

(a) Profile of the frontal motion of a gas along a tube. (b) Motion of two   **Figure 3.7**
laminars separated by distance λ, the mean free path.

The layer immediately adjacent to the surface of the inner wall is stationary, but the velocity of the layers increases as we move away from the surface, with the result that there is a velocity gradient set up along the z axis.* Consider two layers separated by a distance λ, where λ is the mean free path. If v is the velocity of the slower-moving layer, then the velocity for the faster-moving layer is $v + \lambda(dv/dz)$, where λ is the mean free path. In addition to the flow along the x axis, there is random motion along the z axis. When a molecule moves from a faster layer to a slower one, it transports some additional momentum to the latter and tends to speed it up. The reverse is true when a molecule moves to a quickly flowing layer from one traveling more slowly. Consequently, there is a drag or frictional force between these two layers, causing a viscosity effect. However, to maintain the velocity gradient, an external force F must be applied along x; this force is directly proportional

* A gradient is a measure of how a certain parameter changes with distance. Other examples are temperature gradient, concentration gradient, and electric field gradient.

to both the area of the layer, A, and the velocity gradient. Thus

$$F \propto A \frac{dv}{dz}$$

$$= \eta A \frac{dv}{dz} \tag{3.17}$$

where η, the constant of proportionality, is known as the *coefficient of viscosity* or simply the *viscosity*. This quantity is measured in the unit *poise* (P).

CGS units: $\quad\quad 1 \text{ P} = 1 \text{ g cm}^{-1} \text{ s}^{-1} = 1 \text{ dyn s cm}^{-2}$

SI units: $\quad\quad\quad 1 \text{ P} = 0.1 \text{ kg m}^{-1} \text{ s}^{-1} = 0.1 \text{ N s m}^{-2}$

Conversion factor: $\quad 1 \text{ dyn s cm}^{-2} = 0.1 \text{ N s m}^{-2}$

Next we shall obtain an explicit expression for η. Consider a particular laminar which is at height h cm above the stationary plane (that is, the wall of the tubing). All molecules coming from a distance λ below height h will make their first collisions, and hence their first transfer of momentum, when they reach the latter level. If the velocity of the molecules moving in the plane is $h(dv/dz)$, then the velocity of molecules moving in a plane λ cm below this plane is $(h - \lambda)(dv/dz)$. The momentum transferred by each arriving molecule from the slower-moving plane to the faster one is $m(h - \lambda)(dv/dz)$. Similarly, for a molecule arriving from plane $(h + \lambda)$ cm above the wall, the momentum transferred is $m(h + \lambda)(dv/dz)$. Although in a viscosity experiment we are measuring the rate of flow of gas, within the gas sample molecular motions are nevertheless random. Therefore, we may assume, as an approximation, that one-third of the molecules are moving along the x, y, and z directions. Consequently, one sixth of the molecules will be moving upward and one-sixth downward, at any instant along one particular axis. The number of molecules going from one layer to the next per unit area per second is $(\frac{1}{6})(N/V)\bar{c}$, so that the rate of transfer of momentum by molecules moving upward per unit area is $(\frac{1}{6})(N/V)\bar{c}m(h - \lambda)(dv/dz)$. Similarly, for the downward movement, the rate of transfer of momentum per unit area is $(\frac{1}{6})(N/V)\bar{c}m(h + \lambda)(dv/dz)$. The difference of these two quantities, $(\frac{1}{3})(N/V)\bar{c}m\lambda(dv/dz)$, gives the *net* transfer of momentum per unit area per unit time, or F/A. Thus

$$\frac{F}{A} = \frac{1}{3}\left(\frac{N}{V}\right)\bar{c}m\lambda\left(\frac{dv}{dz}\right)$$

From Eq. (3.17),

$$\frac{F}{A} = \eta\left(\frac{dv}{dz}\right)$$

Combining these two equations, we obtain

$$\eta = \frac{1}{3}\left(\frac{N}{V}\right)m\lambda\bar{c} \tag{3.18}$$

Substituting Eq. (3.16) into Eq. (3.18), we obtain

$$\eta = \frac{m\bar{c}}{3\sqrt{2}\,\pi d^2} \tag{3.19}$$

This is an interesting and rather unexpected result, for it tells us that viscosity is *independent* of density and hence of pressure. As we can see from Eq. (3.18),

Viscosity and Collision Diameter of Some Gases at 288 K[a]

Table 3.1

Gas	Viscosity[b]		Collision Diameter (Å)
	$\eta \times 10^4$ (dyn s cm^{-2}) CGS Units	$\eta \times 10^4$ (N s m^{-2}) SI Units	
Air	1.796	0.1796	3.72
Ar	2.196	0.2196	3.64
CH_4	1.077	0.1077	4.14
CO_2	1.448	0.1448	4.59
H_2	0.871	0.0871	2.74
He	1.943	0.1943	2.18
Hg	4.700 (at 492.6 K)	0.4700	4.26
H_2O	0.926	0.0926	4.60
Kr	2.431	0.2431	4.16
N_2	1.734	0.1734	3.75
NH_3	0.970	0.0970	4.43
O_2	2.003	0.2003	3.61

[a] From *Kinetic Theory of Gases* by E. H. Kennard. Copyright 1938 by McGraw-Hill Book Company. Used with permission of McGraw-Hill Book Company.
[b] The reader should keep in mind that when units are expressed in this manner, each value should be multiplied by the factor 10^{-4}. For example, the viscosity of air is 1.796×10^{-4} dyn s cm^{-2} or 0.1796×10^{-4} N s m^{-2}.

η increases with density of the gas (N/V). However, we must realize that in a more dense gas, the number of molecular collisions is correspondingly higher and hence there is a smaller mean free path. These two opposing contributions to viscosity exactly cancel each other. From Eqs. (3.18) and (3.19), we see that the viscosity of a gas *increases* with temperature because the average speed $\bar{c}$ is proportional to $\sqrt{T}$ [see Eq. (3.12)]. This result is certainly contrary to our experience of the viscosity of liquids.* We can understand this behavior by realizing that the viscosity of gases arises as a result of transfer of momentum. At higher temperatures, the rate of transfer is greater and hence a larger force is required to maintain the motion of the layers of gas. Experiments have generally confirmed the validity of Eqs. (3.18) and (3.19). Table 3.1 summarizes the viscosity and molecular collision diameter of a number of gases. Note that the collision diameter of a gas molecule can be calculated by employing Eq. (3.19).

EXAMPLE 3.3 Calculate the viscosity of oxygen gas at 288 K.

Answer

CGS units: From Eq. (3.19),

$$\eta = \frac{m\bar{c}}{3\sqrt{2}\,\pi d^2}$$

---

* The viscosity of liquids decreases with temperature (see Chapter 5). Most of us know, for example, that hot syrup pours more easily than cold syrup.

Since

$$\bar{c} = 4.37 \times 10^4 \text{ cm s}^{-1}$$

$$d = 3.61 \text{ Å} = 3.61 \times 10^{-8} \text{ cm} \qquad \text{(from Table 3.1)}$$

$$m = \frac{32.0 \text{ g}}{6.022 \times 10^{23}} = 5.31 \times 10^{-23} \text{ g}$$

we have

$$\eta = \frac{(5.31 \times 10^{-23} \text{ g})(4.37 \times 10^4 \text{ cm s}^{-1})}{3\sqrt{2}\,\pi(3.61 \times 10^{-8} \text{ cm})^2}$$

$$= 1.34 \times 10^{-4} \text{ g cm}^{-1} \text{ s}^{-1} = 1.34 \times 10^{-4} \text{ P}$$

SI units

$$\bar{c} = 437 \text{ m s}^{-1}$$

$$d = 3.61 \text{ Å} = 3.61 \times 10^{-10} \text{ m}$$

$$m = 5.31 \times 10^{-26} \text{ kg}$$

$$\eta = \frac{(5.31 \times 10^{-26} \text{ kg})(437 \text{ m s}^{-1})}{3\sqrt{2}\,\pi(3.61 \times 10^{-10} \text{ m})^2}$$

$$= 1.34 \times 10^{-5} \text{ kg m}^{-1} \text{ s}^{-1} = 1.34 \times 10^{-5} \text{ N s m}^{-2}$$

Comment: The calculated viscosity differs from that given in Table 3.1. The reason is that Eq. (3.19) is only an approximate equation. A more rigorous derivation gives $\eta = m\bar{c}/2\sqrt{2}\,\pi d^2$. If we had used this equation, we would find that $\eta = 2.00 \times 10^{-4}$ P.

3.7 Graham's Laws of Diffusion and Effusion

In this section we consider two other types of molecular motion in terms of kinetic theory of gases—diffusion and effusion.

The phenomenon of gas diffusion offers a direct demonstration of molecular motion. Without it, the perfume industry would not exist and skunks would be much less feared. Removing a partition separating two different gases in a container or replacing the partition with a porous wall quickly leads to a complete mixing of molecules. This is a spontaneous process, the thermodynamic basis of which will be discussed in Chapter 7. Effusion, on the other hand, describes the streaming of a gas from a high-pressure region to a low-pressure one through a pinhole or orifice. The condition for effusion is that the mean free path of the molecules must be large compared to the diameter of the orifice. Although the basic molecular mechanisms for diffusion and effusion are quite different (the former involves *bulk* flow while the latter involves *molecular* flow), these two phenomena obey laws of the same form. Both laws were discovered by Thomas Graham, his law of diffusion in 1831 and his law of effusion in 1864. These laws state that under the same conditions of temperature and pressure, the rates of diffusion (or effusion) of gases are inversely proportional to the square roots of their molar mass. Thus, for two gases 1 and 2, we have

$$\frac{r_1}{r_2} = \sqrt{\frac{M_2}{M_1}} \qquad (3.20)$$

where r_1 and r_2 are the rates of diffusion (or effusion) of the two gases.

There are many practical examples of effusion and diffusion. For example, we often find that a helium-filled balloon deflates much faster than an air-filled balloon. This is essentially an effusion process, the rate of which is given by Eq. (3.20). The pressure of the gas inside the balloon is greater than atmospheric pressure and the stretched surface of the rubber has many tiny holes that allow the gas molecules to escape. Perhaps the best-known application of effusion is the separation of uranium isotopes ^{235}U and ^{238}U, whose natural abundances are approximately 0.72% and 99.28%, respectively. Only uranium-235 is a fissionable material. Although uranium is a solid, it can be readily converted to uranium hexafluoride (UF_6), which is easily vaporized above room temperature. Thus these two isotopes can be separated from each other by effusion, since $^{238}UF_6$ is heavier than $^{235}UF_6$ and therefore effuses more slowly.* We define a quantity s, called the separation factor, such that

$$s = \frac{\text{rate of effusion of } ^{235}UF_6}{\text{rate of effusion of } ^{238}UF_6}$$

From Eq. (3.20), we have

$$s = \sqrt{\frac{238 + 6 \times 19}{235 + 6 \times 19}} = \sqrt{\frac{352}{349}} = 1.0043$$

Thus, in a single-stage effusion, the separation factor is very close to unity. The situation is not hopeless, however. After a second effusion process, the overall separation factor becomes $(1.0043)^2$, or 1.0086, a slight improvement. In general, then, the value of s for a n-stage process is $(1.0043)^n$. If n is a large number, say 2000, then it is indeed possible to obtain uranium with about 99% enrichment of the ^{235}U isotope.

Equipartition of Energy **3.8**

In this section we consider a theorem, based on ideas of classical mechanics, for the distribution of energy within a molecule. We saw in Section 3.3 that the average kinetic energy is $\frac{3}{2}kT$ for one molecule or $\frac{3}{2}RT$ for 1 mol of a gas. According to the *principle of equipartition of energy*, the energy of a molecule is equally divided among all types of motions or *degrees of freedom*. For monatomic gases each atom needs three coordinates (x, y, and z) to completely define its position. It follows that an atom has three translational degrees of freedom, each of which has energy $\frac{1}{2}kT$.

Other motions, such as rotation and vibration, are also present in molecules, so we must consider various type of degrees of freedom. For a molecule containing N atoms, we would need a total of $3N$ coordinates to describe its motion. Three coordinates are required to describe the translational motion (for example, the motion of the center of mass of the molecule),

---

* Note that fluorine has only one stable isotope so that the separation involves only two species, $^{235}UF_6$ and $^{238}UF_6$.

Table 3.2

Species	Translation	Rotation	Vibration
Atom	$\frac{3}{2}RT$	—	—
Linear molecule	$\frac{3}{2}RT$	RT	$(3N - 5)RT$
Nonlinear molecule	$\frac{3}{2}RT$	$\frac{3}{2}RT$	$(3N - 6)RT$

leaving $3N - 3$ coordinates for rotation and vibration. Three angles are needed to define the rotation of the molecule about the three mutually perpendicular axes through its center mass, leaving $3N - 6$ degrees of freedom for vibration. If the molecule is linear, only two angles will suffice. Since rotation of the molecule about the internuclear axis does not change the positions of the nuclei; therefore, this motion does not constitute a rotation.* Thus a linear molecule such as CO_2 or C_2H_2 has $3N - 5$ degrees of freedom for vibration. Table 3.2 shows the equipartition of energy for different types of molecules.

For 1 mol of a gas, each translational and rotational degree of freedom possesses energy $\frac{1}{2}RT$, whereas each vibrational degree of freedom possesses energy RT. This is true because the vibrational energy contains two terms, one kinetic ($\frac{1}{2}RT$) and one potential ($\frac{1}{2}RT$).

The concept of equipartition of energy is based on classical mechanics. Its correctness can be tested by *heat-capacity* measurements. In physics we have learned that the specific heat of a substance is the energy required to raise the temperature of 1 g of the substance by 1 degree. For chemists a more convenient unit of mass is the mole and the corresponding specific heat is called molar heat capacity ($\overline{C}$), or the energy required to raise 1 mol of the substance by 1 degree, that is,

$$\overline{C} = \frac{\Delta U}{\Delta T}$$

where ΔU is the amount of energy required to raise its temperature by ΔT.† It turns out that the value of $\overline{C}$ not only depends on the nature of the substance but also on the manner in which the temperature is raised. For example, $\overline{C}$ can have quite different values if the heating is carried out by keeping the volume of the gas constant or by keeping the pressure of the gas constant. In the latter case, the gas will expand. Although both conditions are important in heat capacity measurements (see Chapter 6), we need only be concerned with the heat capacity at constant volume ($\overline{C}_V$) here. It is defined by the equation

$$\overline{C}_V = \left(\frac{\partial U}{\partial T}\right)_V \tag{3.21}$$

where the ∂ sign denotes the partial derivative (see Appendix 2 at the end of the book). For 1 mol of a monatomic gas, we have

$$\overline{C}_V = \left[\frac{\partial(\frac{3}{2}RT)}{\partial T}\right]_V = \frac{3}{2}R = 12.47 \text{ J K}^{-1} \text{ mol}^{-1}$$

---

* This is also true for the spinning motion of an atom about an axis through its center.
† The symbol Δ (delta) means "change of."

Table 3.3

**Calculated and Measured Heat Capacities of
Gases at 298 K**

Gas	$\overline{C}_V$ (J K^{-1} mol^{-1}) Calculated	$\overline{C}_V$ (J K^{-1} mol^{-1}) Measured
He	12.47	12.47
Ne	12.47	12.47
Ar	12.47	12.47
H_2	29.10	20.50
N_2	29.10	20.50
O_2	29.10	21.05
CO_2	54.06	28.82
H_2O	49.87	25.23
SO_2	49.87	31.51

For a diatomic gas, we have

$$U = \underset{\text{(translation)}}{\tfrac{3}{2}RT} + \underset{\text{(rotation)}}{RT} + \underset{\text{(vibration)}}{RT} = \tfrac{7}{2}RT$$

so that

$$\overline{C}_V = \left[\frac{\partial(\tfrac{7}{2}RT)}{\partial T}\right]_V = \frac{7}{2}R = 29.10 \text{ J K}^{-1} \text{ mol}^{-1}$$

Similar quantities can be obtained for polyatomic molecules. Table 3.3 compares the predicted values with the experimentally measured heat capacities for several gases. The agreement is excellent for monatomic gases, but considerable discrepancies are found for molecules.

The discrepancies can be explained as follows. According to quantum mechanics, the electronic, vibrational, and rotational energies of a molecule are quantized.* That is to say that in molecules there are energy levels associated with each type of motion, shown in Figure 3.8. We see that the spacing between successive electronic energy level is much larger than that between the vibrational energy levels, which in turn is much larger than that between the rotational energy levels. The spacing between successive translational energy levels is so small that the levels practically merge into a continuum of energy. In fact, for most practical purposes, they can be treated as a continuum. Translational motion, then, is treated as a classical rather than a quantum-mechanical phenomenon. What have these energy levels to do with heat capacities? When a system (for example, a sample of gas molecules) absorbs heat from the surroundings, the energy is used to promote various kinds of motion. In this sense the term *heat capacity* really means energy capacity for its value tells us the capacity of the system to store energy. Energy may be partly stored in vibrational motion; that is, the molecules may be promoted to a higher vibrational energy level, or they may be partly stored in electronic or rotational motion. In each case, the molecules are promoted to a higher energy level. From Figure 3.8 we expect that it is much easier to excite a molecule to a higher rotational energy level than to a higher vibrational or electronic energy level. This is indeed the case. Quantitatively, the ratio of the

---

* We shall discuss the quantum-mechanical treatment of atoms and molecules in more detail in Chapters 15 and 16.

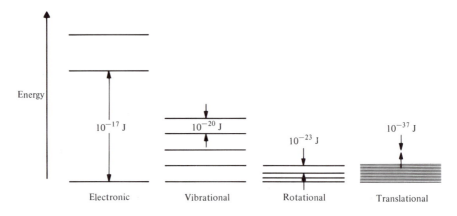

Figure 3.8 Energy levels associated with translational, rotational, vibrational, and electronic motions.

populations (that is, the number of molecules) in any two energy levels E_2 and E_1, N_2/N_1, is given by the Bolzmann distribution law:

$$\frac{N_2}{N_1} = e^{-\Delta E/kT} \tag{3.22}$$

where $\Delta E = E_2 - E_1$ (we assume here that $E_2 > E_1$), k is the Boltzmann constant, and T is the absolute temperature. Equation (3.22) tells us that at a finite temperature $N_2/N_1 < 1$ or that the number of molecules in the upper level is *always* less than that in the lower level. We can make some simple estimates using Eq. (3.22). For translational motion ΔE is about 10^{-37} J, so that $\Delta E/kT$ at 298 K is

$$\frac{10^{-37} \text{ J}}{(1.38 \times 10^{-23} \text{ J K}^{-1})(298 \text{ K})} = 2.4 \times 10^{-17}$$

This is such a small number that the exponential term on the right-hand side of Eq. (3.22) is essentially unity. Thus the number of molecules in a higher level is the same as the one below it.

Similarly, with rotational motion we find that ΔE is small compared to the kT term; therefore, the ratio N_2/N_1 is close to (although smaller than) unity. This means that the molecules are distributed fairly evenly among the energy levels.

The situation is quite different when we consider vibrational motion. Here the spacing between levels becomes quite large (that is, $\Delta E \gg kT$), so that the ratio N_2/N_1 is much smaller than 1. This means that at 298 K, we find most of the molecules in the lowest energy level with only a few in the higher energy levels. Finally, because the spacing between electronic energy levels is so very large that almost all the molecules are in the lowest energy level.

From the discussion above, we can draw several conclusions about heat capacities. We would expect that for molecules at room temperature, only the translational and rotational motions can contribute to the heat capacity. Take the O_2 molecules as an example. If we neglect both the vibrational and electronic motions, the energy of the system becomes

$$U = \tfrac{3}{2}RT + RT = \tfrac{5}{2}RT$$

and

$$\overline{C}_V = \tfrac{5}{2}R = 20.79 \text{ J K}^{-1} \text{ mol}^{-1}$$

This calculated value is indeed quite close to the measured value of $21.05 \text{ J K}^{-1} \text{ mol}^{-1}$ (see Table 3.3). Good agreement is also obtained for other diatomic as well as polyatomic molecules by including only translational and rotational motions. Furthermore, we predict that as temperature increases, more molecules will be promoted to higher vibrational levels. Therefore, at elevated temperatures, vibrational motions should begin to make an appreciable contribution to the heat capacity. This prediction is also correct. The heat capacities at constant volume of O_2 at several temperatures are as follows:

T(K)	$\overline{C}_V$(J K^{-1} mol^{-1})
298	21.05
600	23.78
800	25.43
1000	26.56
1500	28.25
2000	29.47

We see that as temperature increases, the measured $\overline{C}_V$ for O_2 gets closer and closer to the calculated value based on the equipartition of energy. The value at 1500 K is quite close to $29.10 \text{ J K}^{-1} \text{ mol}^{-1}$. In fact, at 2000 K the measured value is greater than the calculated value! The only way we can explain this behavior is to assume that at 2000 K, electronic motion begins to contribute to the heat capacity.

In summary, we should keep in mind that at room temperature, only translational and rotational motions contribute to heat capacity. At elevated temperatures, vibrational motions must also be taken into account. Only at very high temperatures $(T \gtrsim 1500 \text{ K})$ does electronic motion play a role in determining the $\overline{C}_V$ values.

Suggestions for Further Readings

The topics covered in this chapter can be found in any of the standard physical chemistry texts listed in Chapter 1.

INTRODUCTORY–INTERMEDIATE

HILDEBRAND, J. H. *An Introduction to Molecular Kinetic Theory.* Chapman & Hall Ltd., London, 1963 (Van Nostrand Reinhold Company, New York).
An excellent introductory text.

KAUZMANN, W. *Thermal Properties of Matter*: Vol. 1, *Kinetic Theory of Gases.* W. A. Benjamin, Inc., Menlo Park, Calif., 1966.
A suitable text for those interested in more (but not too many more) mathematical details.

Reading Assignments

SECTION A

"Kinetic Energies of Gas Molecules," J. C. Aherne, *J. Chem. Educ.* **42**, 655 (1965).

"Space Vehicle Reentry and Thermal Effects of High Winds," R. C. Plumb, *J. Chem. Educ.* **48**, 119 (1971).

"Sea-Lab Experiment," R. C. Plumb, *J. Chem. Educ.* **47**, 175 (1970).

"Faster Dinner via Molecular Potential Energy," S. A. Modestino, *J. Chem. Educ.* **49**, 706 (1972).

"Scuba Diving and the Gas Laws," E. D. Cooke, *J. Chem. Educ.* **50**, 425, (1973).

"Effective Insect Fogging—The Origin of Sea Breezes—Hot Air Balloons," R. C. Plumb, *J. Chem. Educ.* **52**, 104 (1975).

"Light Bulbs Filled with Krypton Gas," R. C. Plumb, *J. Chem. Educ.* **52**, 388 (1975).

SECTION B

"Kinetic Theory, Temperature, and Equilibrium," D. K. Carpenter, *J. Chem. Educ.* **43**, 332 (1966).

"Are We Teaching the Most Useful Ideas About Transport?" P. Goldberg, *J. Chem. Educ.* **49**, 112 (1972).

"The Assumption of Elastic Collisions in Elementary Gas Kinetic Theory," B. Rice and C. J. G. Raw, *J. Chem. Educ.* **51**, 139 (1974).

"Graham's Laws of Diffusion and Effusion," E. A. Mason and B. Kronstadt, *J. Chem. Educ.* **44**, 740 (1967).

"Heat Capacity and the Equipartition Theorem," J. B. Dence, *J. Chem. Educ.* **49**, 798 (1972).

Problems **3.1** Apply the kinetic theory of gases to explain Boyle's law, Charles' law, and Dalton's law.

3.2 Is temperature a microscopic or macroscopic concept?

3.3 Calculate c_{rms}, c_{mp}, and $\bar{c}$ of argon at 298 K.

3.4 Calculate the temperature at which the c_{rms} of ozone molecules is 1.50×10^3 m s^{-1}.

3.5 The escape velocity v from the earth's gravitational field is given by $(2GM/r)^{1/2}$, where G is the universal gravitational constant (6.67×10^{-11} m^3 kg^{-1} s^{-2}), M is the mass of the earth (6.0×10^{24} kg), and r is the distance from the center of the earth to the object, in meters. Compare the average speeds of He and O_2 molecules in the thermosphere (altitude about 100 km, $T = 250$ K). Which of the two molecules will have a greater tendency to escape? The radius of the earth is 6.4×10^6 m.

3.6 Calculate the mean free path and the binary number of collisions per liter per second between HI molecules at 300 K and 1 atm. The collision diameter of the HI molecules may be taken to be 5.10 Å. Assume ideal gas behavior.

3.7 Calculate the mean free path of helium molecules at 25.0°C and 1 mm Hg. The collision diameter of helium is 2.18 Å.

*3.8 Calculate the number of collisions per second for one molecule for (a) air (molar mass = 29.0 g), and (b) water at 298 K and 1.00 atm. From your results and from the knowledge that water is a nonlinear triatomic molecule while both N_2 and O_2 are diatomic, explain why hot, humid air feels more uncomfortable than hot, dry air. (*Hint*: First calculate the collision frequency for air and water molecules and then compare their $\overline{C}_V$ values listed in Table 3.3.)

3.9 Plot the speed distribution function curves for He and O_2 at 300 K on the same graph.

3.10 Suppose that in a sealed container all the helium atoms start with the same speed, 2.74×10^4 cm s^{-1}. The atoms are then allowed to collide with one another until Maxwell distribution is established. What is the temperature of the gas at equilibrium? Assume that there is no heat exchange between the gas and its surroundings.

3.11 Calculate the average speed and collision diameter for ethylene at 288 K. The viscosity of ethylene is 998×10^{-7} dyn s cm^{-2} or 99.8×10^{-7} N s m^{-2} at the same temperature.

3.12 Compare the collision number and the mean free path for air molecules at (a) sea level ($T = 300$ K and density $= 1.2$ g liter^{-1}), and (b) in the stratosphere ($T = 250$ K and density $= 5.0 \times 10^{-3}$ g liter^{-1}.) The molar mass of air may be taken as 29 g.

3.13 The viscosity of sulfur dioxide at 21.0°C and atmospheric pressure is 1.25×10^{-5} N s m^{-2}. Calculate (a) the collision diameter of a SO_2 molecule, and (b) the collision frequency and mean free path for SO_2 at the given temperature and pressure.

3.14 The speeds of 12 particles in cm s^{-1} are 0, 1.5, 2.0, 3.0, 3.5, 3.5, 3.5, 3.5, 4.0 4.5, 5.0, and 5.5. Find (a) the average speed, (b) the root-mean-square speed, and (c) the most probable speed of these particles. Explain your results.

3.15 A bag containing 20 marbles is being shaken vigorously. Calculate the mean free path of the marbles if the volume of the bag is 850 cm^3. The diameter of each marble is 1.0 cm.

3.16 In the kinetic theory of gases we have assumed that the walls of the container are elastic for molecular collisions. Actually, it makes no difference whether these collisions are elastic or inelastic as long as the walls are at the same temperature as the gas. Explain.

3.17 If 2.0×10^{23} nitrogen molecules strike 4.0 cm^2 of wall per second at an angle 90° with the wall when moving with a speed of 45,000 cm s^{-1}, what pressure (in atm) do they exert on the wall?

3.18 Derive Eq. (3.20) from Eq. (3.12).

3.19 An inflammable gas is generated in marsh lands and sewage by a certain anaerobic bacterium culture. A pure sample of this gas was found to effuse through an orifice in 12.6 min. Under identical conditions of temperature and pressure, it takes oxygen 17.8 min to effuse through the same orifice. Calculate the molecular mass of the gas and suggest what this gas might be.

3.20 Uranium-235 can be separated from uranium-238 by the effusion process using UF_6. Assuming a 50–50 mixture at the start, what is the percentage of enrichment after a single stage of separation.

*3.21 Suppose that you are traveling in a space vehicle on a journey to the moon. The atmosphere consists of 20% oxygen and 80% helium by volume. It was noticed before takeoff that there is a small leakage which, if unchecked, would lead to a continual loss of gas at a rate of 0.05 atm day^{-1} through effusion. If the temperature in the space vehicle is maintained at 22°C and the volume of the vehicle is 15,000 liters, calculate the amounts of helium and oxygen in grams that must be stored on a 5-day journey to allow for the leakage. (*Hint*: First calculate the quantity of gas lost each day, using $PV = nRT$. Note that the rate of effusion is proportional to the pressure of the gas. Assume that effusion does not affect the pressure or the mean free path of the gas.)

3.22 A balloon contains an equal mixture of nitrogen and helium gas at a total pressure of 1 atm. How many times faster will helium leak through the small holes in the membranes of the balloon? Compare your result for a mixture in which the mole fraction of helium is 0.95.

3.23 Calculate the mean kinetic energy ($\frac{1}{2}m\overline{c^2}$) in joules of the following molecules at 350 K: (a) He, (b) CO_2, and (c) UF_6. Explain your results.

★3.24 Consider the following two situations: (a) a person (body temperature 310 K) standing on top of Mt. Rainier, where the mean square speed of air molecules is 1.6×10^9 cm^2 s^{-2}, and (b) a space vehicle entering the earth's atmosphere at 18,000 mph (8.2×10^5 cm s^{-1}). The mean square speed of air molecules in this region is the same as that in (a). Using the appropriate equations and arguments based on the kinetic theory of gases, show that the person will lose heat to the surroundings while the space vehicle will heat up to a temperature of about 90,000 K.

3.25 Calculate the various degrees of freedom for the following molecules: (a) Xe, (b) HCl, (c) CS_2, (d) C_2H_2, (e) C_6H_6, and (f) a protein molecule containing 43,670 atoms.

3.26 Calculate $\overline{C}_V$ for H_2, N_2, CO_2, H_2O, and SO_2, assuming that only translational and rotational motions contribute to the heat capacities. Compare your results with those listed in Table 3.3.

3.27 The typical energy differences between successive rotational, vibrational, and electronic energy levels are 5.0×10^{-22} J, 0.50×10^{-19} J, and 1.0×10^{-17} J, respectively. Calculate the ratios of the number of molecules in the two adjacent energy levels (higher to lower) in each case at 298 K.

★3.28 The first excited electronic energy level of helium atom is 3.13×10^{-18} J above the ground level. Estimate the temperature at which the electronic motion will begin to make a significant contribution to the heat capacity. (*Hint*: Use Boltzmann's equation and set $\Delta E \simeq kT$.)

The Solid State

Since every solid substance contains parts that are crystalline, and since in many of them the whole is an aggregation of crystals, it will be readily understood that a knowledge of crystal structure often affords an explanation of the properties of the substance.
Sir William Bragg*

The gaseous state is characterized by complete randomness. To the opposite extreme, a crystalline solid has a highly ordered structure. There are four different types of crystals: ionic crystals, metallic crystals, covalent crystals, and molecular crystals. Although a biochemist does not ordinarily deal with solid-state chemistry, he or she should understand the factors governing the arrangement of atoms and molecules in these crystals. For one thing, three-dimensional structural determination by the X-ray diffraction technique has played a central role in advancing our knowledge of the stability and function of proteins and nucleic acids.

Classification of Crystal Systems 4.1

What is a crystal? It is a substance in which the atoms or molecules are packed closely together in such a way that the total potential energy is at a minimum. These atoms and molecules make up a highly ordered structure, called the *crystal lattice*, in which the atoms are arranged periodically in three dimensions. Let us start by considering a one-dimensional lattice, shown in Figure 4.1. The only geometric parameter is the repeating distance between the atoms, or lattice points,† which can be imagined to extend infinitely both to the left and right. In this case, each lattice point represents a *unit cell*, the basic repeating unit.

By definition, a two-dimensional lattice is a planar system; five different arrangements, or unit cells, can be constructed from the lattice points (Figure 4.1).

---

* *The Universe of Light*, Dover Publications, Inc., New York, 1940. Used by permission of G. Bell & Sons, Ltd., London, England.
† Generally, a lattice point can be an atom, an ion, or a molecule.

Figure 4.1

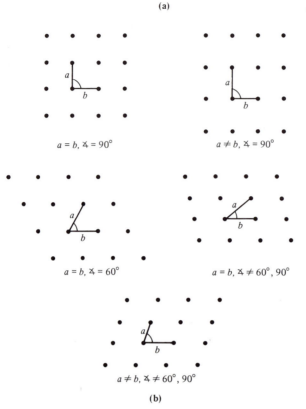

(a) One-dimensional lattice. (b) Five different arrangements of a two-dimensional lattice.

(a)

$a = b, \angle = 90°$

$a \neq b, \angle = 90°$

$a = b, \angle = 60°$

$a = b, \angle \neq 60°, 90°$

$a \neq b, \angle \neq 60°, 90°$

(b)

Figure 4.2 shows a three-dimensional lattice. The unit cell is characterized by lengths a, b, and c and angles α, β, and γ. A total of seven possible lattice types can be constructed from these parameters. The results are summarized in Table 4.1. Each of the unit cells defined in the table is called a *primitive cell*, because all lattice points are at the corners of the unit cell. In 1850, Bravais showed that there should be altogether 14 different unit cells to account for lattice points at the center of the cell as well as at the centers of some of the faces. These 14 unit cells are known as the *Bravais lattices* (Figure 4.3).

The first task that confronts a crystallographer is measuring the size and shape of the unit cells. If a crystal is well formed, it is possible to determine its

Figure 4.2

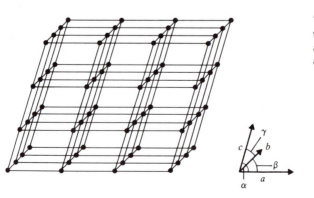

Three-dimensional lattice with lengths a, b, and c and angles α, β, and γ.

Crystal Lattices

System	Axes	Angles	Table 4.1
Cubic	$a = b = c$	$\alpha = \beta = \gamma = 90°$	
Rhombohedral (or trigonal)	$a = b = c$	$\alpha = \beta = \gamma \neq 90°$	
Tetragonal	$a = b \neq c$	$\alpha = \beta = \gamma = 90°$	
Hexagonal	$a = b \neq c$	$\alpha = \beta = 90°, \ \gamma = 120°$	
Orthorhombic	$a \neq b \neq c$	$\alpha = \beta = \gamma = 90°$	
Monoclinic	$a \neq b \neq c$	$\alpha = \gamma = 90°, \ \beta \neq 90°$	
Triclinic	$a \neq b \neq c$	$\alpha \neq \beta \neq \gamma \neq 90°$	

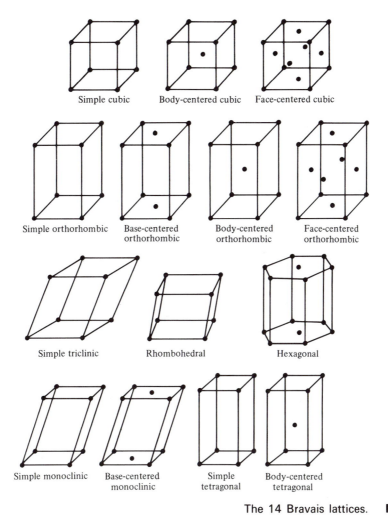

The 14 Bravais lattices.   **Figure 4.3**

external symmetry by measuring the interfacial angles and hence classify the crystal according to one of the seven types listed in Table 4.1. The cell constants, or length and angle of the unit cell, must be determined by X-ray diffraction technique, to be discussed shortly.

In X-ray crystallography, it is convenient to characterize a given crystal in terms of a set of planes. Consider a two-dimensional crystal lattice shown in Figure 4.4. A number of planes with different orientations (*AA'*, *BB'*, *CC'*, etc.)

51

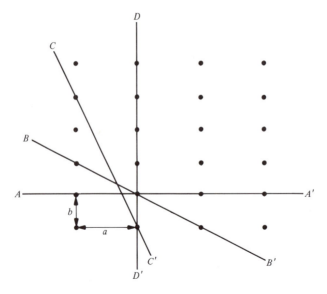

Figure 4.4 Characterization of a two-dimensional lattice in terms of a set of planes. The origin is at the extreme lower left-hand side.

may be drawn, with each plane containing some of the lattice points. Parallel to any plane there is a whole set of planes that can be generated from it by application of the unit lattice translation. The AA' BB', ... planes are labeled according to their intercepts on the a, b, and c axes, measured from an arbitrarily chosen origin. Thus the CC' plane intercepts the axes at a, $4b$, and ∞ (since we are dealing with a two-dimensional crystal, this plane is parallel to the c axis and therefore the intercept is at infinity). Next we take the reciprocals of the intercepts, which are $1/a$, $1/4b$, and $1/\infty$. The fractions are then cleared by multiplying each quantity by the least common denominator, which is 4 in this case (we exclude infinity). This gives us three numbers (410), which are known as the *Miller indices* of the plane. Generally, the Miller indices (*hkl*) of any plane gives the orientation of the plane in the crystal with reference to its three internal axes. Table 4.2 summarizes the steps for obtaining the Miller indices for the planes shown in Figure 4.4.

Miller Indices for a Two-Dimensional Lattice

Table 4.2

Crystal Face or Plane	Intercepts	Reciprocals of Multiples	Miller Indices (*hkl*)
AA'	$\infty a, b, \infty c$	$\dfrac{1}{\infty}, \dfrac{1}{1}, \dfrac{1}{\infty}$	010
BB'	$2a, 2b, \infty c$	$\dfrac{1}{2}, \dfrac{1}{2}, \dfrac{1}{\infty}$	110
CC'	$a, 4b, \infty c$	$\dfrac{1}{1}, \dfrac{1}{4}, \dfrac{1}{\infty}$	410
DD'	$a, \infty b, \infty c$	$\dfrac{1}{1}, \dfrac{1}{\infty}, \dfrac{1}{\infty}$	100

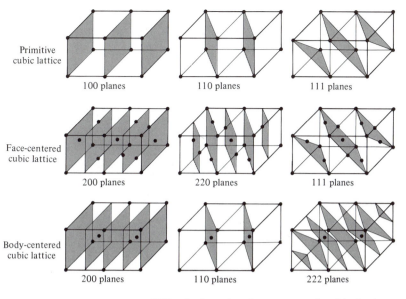

Primitive cubic lattice		
100 planes	110 planes	111 planes
Face-centered cubic lattice		
200 planes	220 planes	111 planes
Body-centered cubic lattice		
200 planes	110 planes	222 planes

Miller indices for three types of cubic lattices. **Figure 4.5**

The same procedure can be extended to a three-dimensional crystal. Consider the three types of cubic crystals shown in Figure 4.5. Each set of the Miller indices describes a series of parallel, equally spaced planes.

Bragg's Equation **4.2**

Let us now consider what happens when a monochromatic X-ray beam of wavelength λ strikes the face of a crystal (Figure 4.6). Because of the penetrating power of the X ray, the incident radiation can interact with atoms in many layers. On the first layer, the beam is reflected by the atoms in much the same manner as a mirror reflects ordinary light.

How does the reflected beam from the first plane interact with a reflected beam from a different plane? The condition for the scattered X-ray beams to interfere with one another constructively, that is, for the waves to be in phase with each other, is*

$$2d \sin \theta = n\lambda \qquad n = 1, 2, \ldots \tag{4.1}$$

where $2d \sin \theta$ is the difference in path length between two waves. A *diffraction pattern* of alternating intensities of reflected X-ray beams is obtained by placing a photographic film at a certain angle. The number n represents the order of diffraction; if $n = 1$, we have first-order diffraction; $n = 2$, second-order diffraction; and so forth. However, from Eq. (4.1) we see that the angle θ for a certain value of n and d is the same as the first-order diffraction for a set of planes with spacing d/n. For example, second-order diffraction from a 111 plane can be viewed as if it were a first-order diffraction from the 222

* The wave properties of light are discussed in Chapter 15.

planes, even though this plane may not be present in a particular crystal. The Miller indices readily show that $d_{222} = d_{111}/2$ (see Figure 4.5) so that

$$\sin \theta = \frac{2\lambda}{d_{111}} = \frac{1\lambda}{d_{222}}$$

For this reason, we can always set $n = 1$ in Eq. (4.1) and treat higher-order diffractions as if they were first-order diffractions from planes with reduced spacings. Thus we have

$$2d \sin \theta = \lambda \tag{4.2}$$

which is known as the *Bragg equation*.

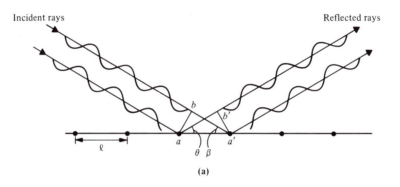

(a)

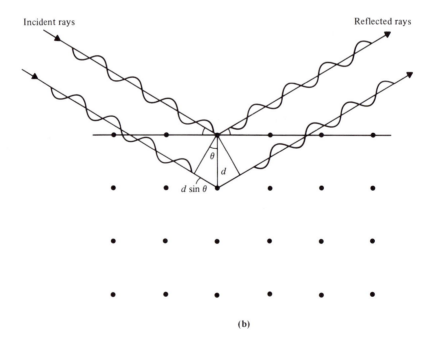

(b)

Figure 4.6 (a) Reflection of X rays from the first layers of atoms. (b) Reflection of X rays from different layers of atoms.

The Bragg equation immediately provides us with a way for measuring cell dimensions. For a cubic lattice, the perpendicular distance d_{hkl} between adjacent members of the set of parallel planes represented by the Miller indices, (hkl) can be obtained as follows. From Table 4.1, $\alpha = \beta = \gamma = 90°$ and $a = b = c$. The three-dimensional form of the Pythagorean theorem gives*

$$\frac{1}{d_{hkl}^2} = \frac{h^2}{a^2} + \frac{k^2}{b^2} + \frac{l^2}{c^2}$$
$$= \frac{h^2 + k^2 + l^2}{a^2}$$

or

$$d_{hkl} = \frac{a}{\sqrt{h^2 + k^2 + l^2}} \tag{4.3}$$

The reader should apply this equation to the cubic lattices shown in Figure 4.5. From Eq. (4.2), we have

$$\sin \theta_{hkl} = \frac{\lambda}{2d_{hkl}} = \frac{\lambda}{2a} \sqrt{h^2 + k^2 + l^2}$$

or

$$\sin^2 \theta_{hkl} = \frac{\lambda^2}{4a^2} (h^2 + k^2 + l^2) \tag{4.4}$$

The quantity $(h^2 + k^2 + l^2)$ is determined by the various planes:

hkl	100	110	111	200	210	211	220	221	$\cdots$
$(h^2 + k^2 + l^2)$	1	2	3	4	5	6	8	9	$\cdots$

If we know the angle θ_{hkl} for each plane, then a set of lines can be constructed on the $\sin \theta_{hkl}$ scale (Figure 4.7). Note that the seventh line is missing, since we cannot have $(h^2 + k^2 + l^2) = 7$. Similarly, the fifteenth line is also missing, as are others in the sequence.

Similar plots for body-centered and face-centered lattices are also shown in Figure 4.7. Here, fewer lines are observed in comparison to the primitive cubic lattice. To see why some of the lines are missing, let us consider the body-centered cube. As Figure 4.5 shows, the 110 planes pass through all lattice points and a strong diffraction pattern is observed. The situation is different for the 100 planes because they are interleaved by another layer, the 200 plane, of atoms. The X rays diffracted by the 100 planes are in phase with one another but out of phase by half a wavelength with those diffracted by the 200 planes. Since a crystal contains many, many unit cells, there are as many atoms in the 200 planes as there are in the 100 planes. Consequently, there is a total destructive interference and the diffraction line from the 100 planes will be absent. On the other hand, diffraction from the 200 planes will be present

---

* For a derivation of Eq. (4.3), see G. M. Barrow, *Physical Chemistry*, 4th ed., McGraw Hill Book Company, New York, 1979, p. 507.

Figure 4.7

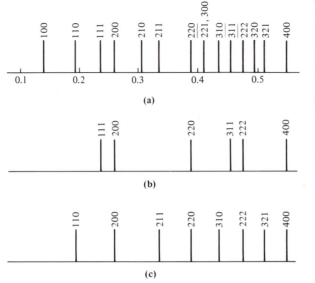

(a)

(b)

(c)

The theoretical plots of $(\lambda/2a)\sqrt{h^2 + k^2 + l^2}$ on the sin θ scale for (a) primitive, (b) face-centered, and (c) body-centered lattices. The $(\lambda/2a)$ term is assumed to be 0.137. Each line represents a specific set of h, k, and l values. In reality, these lines have quite different intensities.

because *all* atoms lies in these planes. The absence of other lines can be explained in a similar manner.

The X-ray diffraction pattern of a crystal can be recorded in one of two ways. In the first method, a small single crystal is mounted with a particular axis perpendicular to the direction of the X-ray beam. The crystal is situated at the center of a table that is calibrated with an angular scale (Figure 4.8). The detector is usually a Geiger counter. The procedure is to measure the intensity of the reflected rays as a function of the angle θ—the intensity reaching a maximum whenever the angle satisfies the Bragg equation.* This method yields a number of lines, each one representing a particular plane, and measures the corresponding angles.

The crystal method was originally employed by Bragg in the early development of X-ray diffraction studies. Since then, a number of improvements have been introduced, greatly aiding data recording. The single-crystal method is

Figure 4.8

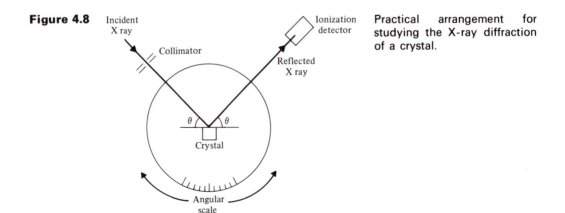

Practical arrangement for studying the X-ray diffraction of a crystal.

* The intensity depends on the number of atoms in the plane and the type of atoms present. X rays are almost entirely scattered by the electrons of an atom.

essential for analyzing complex structures, but it requires careful growing and mounting of the crystal. An alternative approach, introduced by Debye and Scherrer and independently by Hull, enables structural determination on powder samples rather than single crystals. In this arrangement, a beam of X rays is directed at a mass of finely ground powder of the substance under study. The powder sample is actually a large number of small crystals or *crystallites*. Since these crystallites are randomly oriented, the X-ray beam will meet the crystal planes at every possible value of θ for which the Bragg equation is satisfied. The reflected beams from each plane actually form a cone, as shown in Figure 4.9. The most convenient way to record the diffraction pattern is to use a cylindrical photographic film so arranged that its axis is perpendicular to the incident X-ray beam. From the distances between the lines and the dimensions of the film, the angle θ can be calculated for each line.

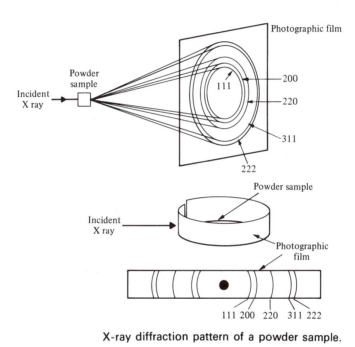

X-ray diffraction pattern of a powder sample. **Figure 4.9**

Let us now briefly consider a specific example, the structural determination of NaCl. Table 4.3 gives the Bragg angles for some of the lines observed.

X-ray Diffraction Data for NaCl

Table 4.3

θ_{hkl}	$\sin^2 \theta_{hkl}$	$\dfrac{\sin^2 \theta_{hkl}}{0.0188}$	hkl
13°41′	0.0560	3	111
15°50′	0.0743	4	200
22°42′	0.1489	8	220
26°50′	0.2051	11	311
28°20′	0.2253	12	222
33°08′	0.2990	16	400

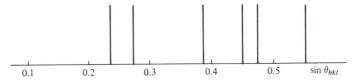

Figure 4.10 Bragg diffraction lines plotted on the $\sin \theta_{hkl}$ scale for the NaCl powder sample. In reality, these lines have different intensities.

These lines are positioned on the $\sin \theta_{hkl}$ scales as in Figure 4.10. Comparing this pattern with Figure 4.7, we see that NaCl has a face-centered lattice. To determine the length of the cube, we need to find a common factor that divides all values of $\sin^2 \theta_{hkl}$. This factor turns out to be 0.0188, which is equal to $\lambda^2/4a^2$ in Eq. (4.4). The experimental data in Table 4.3 are obtained using X rays generated when copper is employed as a target. The characteristic wavelength of the X ray is 1.542 Å (0.1542 nm) so that

$$0.0188 = \frac{\lambda^2}{4a^2} = \frac{(1.542 \text{ Å})^2}{4a^2}$$

or

$$a = 5.623 \text{ Å}$$

where a is the length of the cube.

Also, from Eq. (4.4) we can index each line as follows:

$$\frac{\sin^2 \theta_{hkl}}{0.0188} = h^2 + k^2 + l^2$$

Since the ratio on the left-hand side is known, each line can be indexed in a straightforward way.

The NaCl structure is still not completely described because we do not know how many atoms are in each unit cell. The density of NaCl is 2.16 g cm^{-3} and the formula mass is 58.44 g; thus the molecular volume is 58.44/2.16, or 27.06 cm^3, and the volume of one NaCl unit is

$$\frac{27.06 \text{ cm}^3}{6.022 \times 10^{23}} = \frac{27.06 \times (10^8)^3}{6.022 \times 10^{23}} \text{ Å}^3 = 44.9 \text{ Å}^3$$

The volume of the unit cell is a^3 or 177.8 Å^3, so that there must be 177.8/44.9, or 4 NaCl per unit cell. Figure 4.11 shows the crystal structure of sodium chloride. At first, it might appear that there are more than four units of NaCl present in a unit cell. Except for the center sodium ion, however, every other ion is shared among adjacent cells. For example, each of the eight chloride ions at the corners is shared among eight unit cells, and each of the six chloride ions in the center of the outer faces is shared between two unit cells. Thus the total number of chloride ions per unit cell is given by $(8 \times \frac{1}{8} + 6 \times \frac{1}{2}) = 4$. Similarly, we can show that the total number of sodium ions per unit cell is given by $(12 \times \frac{1}{4} + 1) = 4$.

The powder method is most useful for crystals with only one or two parameters to be determined—cubic, tetragonal, and rhombohedral crystals, for example. For other crystal systems, the task of indexing the lines becomes very difficult if not impossible.

Sodium chloride crystal lattice. The large spheres represent chloride ions and the small spheres represent sodium ions.

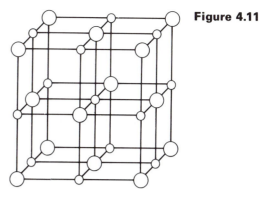

Figure 4.11

For simple crystals, for example, the alkali halide crystals, knowledge of the symmetry and unit cell dimensions enables us to determine the exact structure. However, for most crystals we must determine the array of atoms or ions within each unit cell. To do this, we need to measure and relate each observed intensity of the X rays reflected from a known set of planes (*hkl*) to the distribution of atoms within this set of planes. Theory shows that the measured intensity I_{hkl} is proportional to the square of the *structure factor* F_{hkl}. The significance of F_{hkl} is that it can also be calculated once the positions and scattering power of the atoms within the unit cell are known. An obvious approach, therefore, would be to guess the positions of all the atoms and then calculate the F_{hkl} values. Good agreement between calculation and experimental data would confirm the correct structure. Unfortunately, this trial-and-error method is not practical because there are many possible positions for each atom in the unit cell, even though some intelligent guesses can be made. In practice, the reverse procedure is adopted; that is, the atomic positions are determined from the measured intensities. This is aided by a well-known mathematical technique, called *Fourier synthesis*, which maps out the electron density distribution within a unit cell. The positions of atoms can be determined by noting where the electron density rises to peak values.

A major obstacle in structural determination by Fourier synthesis is the *phase problem*. In order to construct the electron density map, we must know both the signs and magnitudes of F_{hkl}. Experimentally, only the magnitudes of F_{hkl} are measured. The *isomorphous replacement technique*, introduced by Robertson in 1936 in his study of phthalocyanine, helps to circumvent this difficulty in some instances. The intensities I_{hkl} from various planes are first measured photographically. By substituting a heavy atom (for example, mercury or gold) at the center of the phthalocyanine molecule, the intensities are redetermined. The signs of F_{hkl} can now be deduced as follows. If a spot on the photographic film becomes stronger (that is, if I_{hkl} increases) when the heavy atom is introduced, its original F_{hkl} value with respect to the center of the molecule must have been positive; if the spot becomes weaker, the F_{hkl} value must have been negative. In this manner all the signs of F_{hkl} can be determined.* Figure 4.12 shows the Fourier electron density map of phthalocyanine.

* This technique is based on the assumption that the crystal structure is not altered by the presence of the heavy atom.

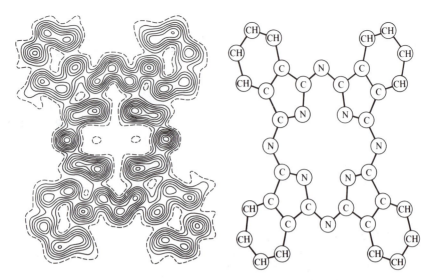

Figure 4.12   Electron density contour map (left) and molecular structure of phthalocyanine (right). [From J. M. Robertson, *J. Chem. Soc.* 1195 (1936). Used by permission.]

Phthalocyanine is a relatively small molecule, containing only 56 atoms. When Dorothy Hodgkin in 1955 carried out the complete X-ray analysis of vitamin B_{12} (181 atoms), it seemed that we might have reached the limit of complexity in structural determination by X-ray. However, as early as 1953 Perutz realized that the isomorphous replacement technique was equally applicable to protein molecules which contain thousands or even tens of thousands of atoms. Figure 4.13 shows the photograph of a diffraction pattern for the enzyme lysozyme. We usually speak of the *resolution* of the X-ray diffraction data. At 4.6 Å resolution, the electron density map provides an overall shape of the protein molecule. At 3.5 Å resolution, it is often possible to follow the backbone, that is, the polypeptide chain. At 3.0 Å resolution, we begin to be able to identify the amino acid side chains and can therefore determine, in favorable cases, the primary sequence* of the protein. At 2.5 Å resolution, the positions of atoms may be located with an accuracy of ±0.4 Å. Finally, at 1.5 Å resolution, which is the present limit of protein crystallography, the positions of atoms can be located to about ±0.1 Å.

Generally, the process of determining a protein structure consists of the following steps: (1) crystallization of the native protein (that is, protein in its functional state) and collection of diffraction data; (2) obtaining and collecting diffraction data from heavy atom derivatives; (3) obtaining phases for the native data; (4) computing an electron density map from the data of steps 1, 2, and 3; and (5) building a structural model and checking it against the observed data. To determine the structure of such an enormously complex system, about 500,000 intensities must be accurately measured and perhaps 1 million calculations performed! Largely as a result of the recent rapid development in high-speed digital computers and the improvement in in-

---

* The primary sequence of a protein molecule tells us how the amino acids are linked to one another to form the polypeptide chain.

X-ray diffraction pattern obtained from the lysozyme crystal. (Courtesy of J. R. Knox.)

Figure 4.13

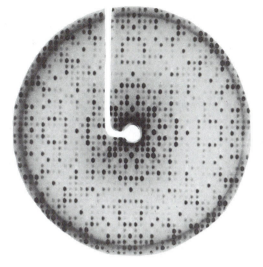

strumentation for recording and measuring intensities, the task is not as hopeless as it may seem at first. At present, the complete or almost complete structure of a number of proteins (myoglobin, hemoglobin, cytochrome c, and so forth) and enzymes (lysozyme, carboxypeptidase, ribonuclease, and so forth) have been elucidated. In fact, the present status of the art in X-ray crystallography is such that as long as we can grow suitable crystals of proteins for diffraction measurements, not an easy task by any means, it is possible in most cases to unravel their three-dimensional structure. Knowledge of the three-dimensional structure has probably made the greatest contribution to our understanding of the stability and function of proteins and nucleic acids.

Types of Crystals 4.4

Having discussed the classification of crystal systems and the determination of cell constants in the last two sections, we shall now consider systematically the factors affecting crystal structure.

The structure of any crystal is primarily determined by the size of its atoms or molecules and by the forces that hold them together. Before we discuss different types of crystals, let us first examine the general geometric requirements for crystal formation.

In the simplest case, the internal structure of the crystalline form may be represented by using a number of spheres having the same radius r. By placing the spheres together in contact with one another and then stacking one layer over the other, a number of geometric structures can be generated. One arrangement is that shown in Figure 4.14. When the second layer is placed over the first in such a way that each sphere in the second layer is directly over a sphere in the first layer, a primitive cell is formed. The primitive-cell arrangement is not the most efficient way of packing spheres

Figure 4.14

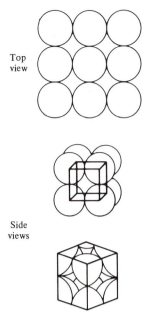

Top view

Side views

Three representations of a primitive cube. The total volume of spheres within a unit cell is equal to that of one complete sphere.

within a given volume. We can calculate the fraction of the cell space occupied by the spheres as follows. Let the primitive-cell dimension be a so that

$$\text{volume of one sphere} = \frac{4\pi}{3}r^3 = \frac{4\pi}{3}\left(\frac{a}{2}\right)^3$$

$$\text{volume of the cube} = a^3$$

The ratio of these two volumes gives

$$\text{fraction of space occupied by one sphere} = \frac{(4\pi/3)(a/2)^3}{a^3} = 0.523$$

or the packing efficiency is 52.3%. Since each sphere has six immediate neighbors (four spheres in the same layer, one in the upper layer, and one in the lower layer), it has a coordination number of 6.

A different arrangement is that shown in Figure 4.15. The second layer fits in the depressions or notches of the first layer and the third layer into the depressions of the second layer, and so on. This arrangement forms a *body-*

Figure 4.15

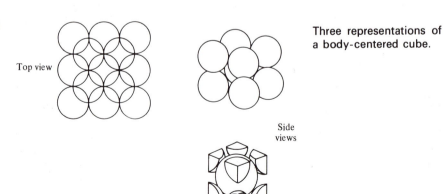

Top view

Side views

Three representations of a body-centered cube.

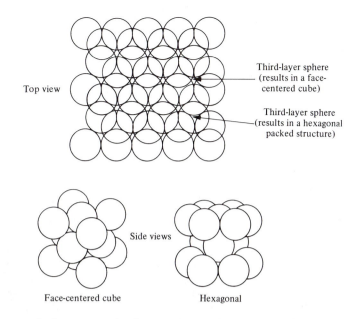

Top view

Third-layer sphere (results in a face-centered cube)

Third-layer sphere (results in a hexagonal packed structure)

Side views

Face-centered cube

Hexagonal

Two different close-packed structures leading to either a face-centered cube or a hexagonal structure. The third-layer spheres are not shown in the top diagram. **Figure 4.16**

centered cube (bcc). The coordination number of each sphere is 8 and the packing efficiency is increased to 68%.

We could further increase the efficiency of the packing by starting with the close-packed structure shown in Figure 4.16. Again, the second layer is stacked into the first layer in such a way that each sphere fits in a depression formed by the contact of three spheres on the first layer. There are two ways the third layer may cover the second layer. The spheres can fit into the depressions so that each third-layer is directly over a first-layer sphere. We call this arrangement the *hexagonal close-packed* (hcp) structure. Alternatively, the spheres can fit into the depressions that lie directly over the depressions in the first layer. In this case we obtain the *face-centered cube* (fcc). In both the hcp and fcc structures, each sphere has a coordination number of 12. The packing efficiency is the same, 74%. We are now ready to consider the specific types of crystals that follow.

METALLIC CRYSTALS

Metals are characterized by their tensile strength and the ability to conduct electricity. Both properties are the result of the special nature of the metallic bond. The understanding of bonding in metals requires a thorough knowledge of quantum mechanics and will not be discussed here. Nevertheless, it is worth noting that the bonding electrons in metals are highly delocalized over the entire crystal. Metal atoms can be imagined as an array of positive ions immersed in a sea of delocalized valence electrons. The great cohesive force resulting from the delocalization is responsible for the strength noted in metals. The mobility of the delocalized electrons accounts for electrical conductivity.

Metal crystals all have a high density. This fact means that they usually have the hcp or fcc structure. Metals such as magnesium, scandium, titanium, cobalt, zinc, and cadmium have the hcp structure; aluminum, calcium, nickel, copper, palladium, silver, platinum, gold, and lead have the fcc structure; and the alkali metals and iron, chromium, barium, and tungsten possess the bcc structure.

IONIC CRYSTALS

Ionic crystals are hard and brittle solids. They possess high melting points. These solids are poor conductors of electricity, but their ability to conduct electricity increases drastically in melt. The packing of spheres in *ionic crystals* is complicated by two factors: (1) charged species are present, and (2) anions and cations are generally quite different in size. Consider the alkali halides. In such an ionic crystal, the number of anions surrounding a cation must be equal to the number of cations surrounding an anion. A quantity of great interest is the ionic or crystal radius of the ion. As we saw in Section 4.2, the length of the unit cell for NaCl is 5.623 Å. The packing of Na^+ and Cl^- is shown in Figure 4.17. It is clear, therefore, that the length of the cell is twice the sum of the ionic radii of Na^+ and Cl^-. Usually, there is no way to measure the radius of individual ions, but it is always possible to estimate the radius of a few ions in some favorable cases. For example, the radius of I^- in LiI is estimated to be 2.16 Å. Using this value, we can proceed to determine the radius of K^+ in KI, the radius of Cl^- in KCl, and so on. Table 4.4 lists the ionic radii of a number of ions. Note that these are the *average* values obtained from a large body of data. Thus the sum of the cationic and anionic radii is not usually equal to a particular cell dimension. For example, from Table 4.4 we have $2(r_{Na^+} + r_{Cl^-}) = 2(0.98 + 1.81) = 5.58$ Å, which is different from 5.623 Å, the length of the NaCl unit cell.

Some general conclusions can be drawn from these ionic radii. Within the same period the anions always have larger radii than the cations, and the radius of the trivalent cation is smaller than that of the divalent cation, which is smaller than that of the monovalent cation. It should be realized that the value of any ionic radius only serves as a useful but approximate size of the ion. The fact that the ionic radius of Na^+ is 0.98 Å does not mean that the

Table 4.4

Crystal Radii of Some Ions (Å)[a,b]

						H⁻ 2.08
Li⁺ 0.60	Be²⁺ 0.31	B³⁺ 0.20	C⁴⁺ 0.15	N⁵⁺ 0.11	O²⁻ 1.40	F⁻ 1.36
Na⁺ 0.98	Mg²⁺ 0.65	Al³⁺ 0.50	Si⁴⁺ 0.41	P⁵⁺ 0.34	S²⁻ 1.84	Cl⁻ 1.81
K⁺ 1.33	Ca²⁺ 0.99	Ga³⁺ 0.62	Ge⁴⁺ 0.53	As⁵⁺ 0.47	Se²⁻ 1.98	Br⁻ 1.95
Rb⁺ 1.48	Sr³⁺ 1.13	In³⁺ 0.81	Sn⁴⁺ 0.71	Sb⁵⁺ 0.62	Te²⁻ 2.21	I⁻ 2.16
Cs⁺ 1.69	Ba²⁺ 1.35	Tl³⁺ 0.95	Pb⁴⁺ 0.84	Bi⁵⁺ 0.74		

[a] Reprinted from Linus Pauling, *The Nature of the Chemical Bond*, 3rd ed. Copyright 1939 and 1940, third edition © 1960, by Cornell University. Used by permission of Cornell University Press.
[b] The radii of first-row transition-metal ions are given in Table 16.10.

Sum of ionic radii in a sodium chloride crystal.

Figure 4.17

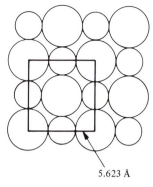

5.623 Å

electron cloud of the ion never extends beyond this value. This value is significant because when it is added together with the radius of an anion, say Cl^-, the sum is approximately equal to the observed interionic distance.

COVALENT CRYSTALS

Covalent crystals are hard solids that possess very high melting points. They are poor conductors of electricity. In covalent crystals, atoms are held together by covalent bonds. Well-known examples are two allotropic forms of carbon, graphite and diamond (Figure 4.18). The structure of a diamond is based on an fcc lattice. There are eight carbon atoms at the center of the cube, six carbon atoms in the face center, and four more carbon atoms within the unit cell. Each atom is tetrahedrally bonded to four other atoms. This tightly held lattice contributes to diamond's unusual hardness. The C—C distance is 1.54 Å, which is similar to the C—C distance in ethane. In graphite, each carbon atom is bonded to three carbon atoms. The C—C bond length is 1.42 Å, which is close to the C—C bond length in benzene. The layers are held together by rather weak forces. Consequently, graphite is easily deformed in directions parallel to the layers. Within a plane, graphite, like diamond, is a covalently bonded crystal. Another important example of a covalent crystal is quartz or SiO_2, also shown in Figure 4.18.

Structures of diamond, graphite, and quartz. In quartz, the filled circles represent silicon and the open circles represent oxygen.

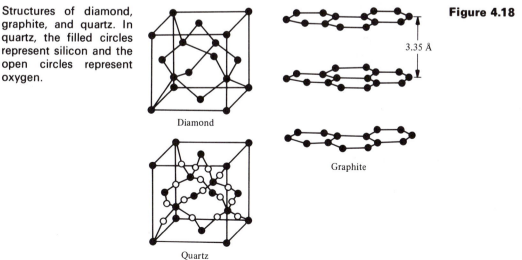

Figure 4.18

3.35 Å

Diamond

Graphite

Quartz

Molecular crystals are very soft solids that possess low melting points. They are poor conductors of electricity. Molecular crystals consist of such substances as N_2, CCl_4, I_2, and benzene. Generally, the molecules are packed together as closely as their size and shape allow. The attractive forces are mainly van der Waals interactions (see Chapter 17).

Finally, we should keep in mind that not all solids exist in crystalline forms. Solids that do not possess a regular structure are called *amorphous solids*, glass being a well-known example. The structure of such a substance is much more difficult to study, because regular patterns are lacking.

Suggestions for Further Reading

The solid state is discussed in most standard physical chemistry texts. For more details the reader should consult any of the texts listed in Chapter 1 as well as the following references.

INTRODUCTORY–INTERMEDIATE

GALWEY, A. K. *Chemistry of Solids.* Chapman & Hall Ltd. (Barnes & Noble, Inc., New York), 1967.
   Discusses the chemistry of solids as well as crystallography.

HOLDEN, A. *The Nature of Solids.* Columbia University Press, New York, 1965.
   A readable, descriptive text.

WHEATLY, P. J. *The Determination of Molecular Structure.* Oxford University Press, Inc., New York, 1959, Chapters 6–8.
   A clear and detailed discussion of the X-ray diffraction technique.

WORMALD, J. *Diffraction Methods.* Oxford University Press, Inc., New York, 1973.
   A useful introductory text with many examples.

INTERMEDIATE–ADVANCED

BLUNDELL, T. L., and L. N. JOHNSON. *Protein Crystallography.* Academic Press, Inc., New York, 1976.
   A valuable text for advanced students.

TANFORD, C. *Physical Chemistry of Macromolecules.* John Wiley & Sons, Inc., New York, 1961, Chapter 2.
   Presents a detailed discussion of the principles and applications of X-ray diffraction.

WALTON, A. G., and J. BLACKWELL. *Biopolymers.* Academic Press, Inc., New York, 1973, Chapter 3.
   Presents a detailed discussion of the same topic as that of Tanford's book.

Reading Assignments

SECTION A

"X-ray Crystallography," Sir L. Bragg, *Sci. Am.,* July 1968.

"An Introduction to X-ray Structure Determination," J. A. Kapecki, *J. Chem. Educ.* **49**, 231 (1972).

"X-ray Crystallography Experiment," F. P. Baer and T. H. Jordan, *J. Chem. Educ.* **42**, 76 (1965).

"X-ray Crystallography as a Tool for Structural Chemists," W. M. MacIntyre, *J. Chem. Educ.* **41**, 526 (1964).

"X-ray Analysis of Crystal Structures," M. H. Harding, *Chem. Brit.* **4**, 548 (1968).

"X-Ray Diffraction and Nucleic Acid," D. R. Davies, *Chemistry* **40**(2), 8, (1967).

"The Solid State," Sir N. Mott, *Sci. Am.*, Sept. 1967.

"Ionic Bonding in Solids," J. E. House, Jr., *Chemistry* **43**(2), 18 (1970).

"X-ray Crystallography and Enzyme Structure," D. Eisenberg, in P. D. Boyer, ed., *The Enzymes*, Vol. 1, Academic Press, Inc., New York, 1970, Chapter 1.

SECTION B

"Protein Molecular Weight by X-ray Diffraction," J. R. Knox, *J. Chem. Educ.* **49**, 476 (1972).

"X-ray Studies of Protein Mechanisms," R. E. Dickerson, *Ann. Rev. Biochem.* **41**, 807 (1972).

"The Revolution in Crystallography," W. C. Hamilton, *Science* **169**, 133 (1970).

"Macromolecules, the X-ray Contribution," C. Bunn, *Chem. Brit.* **11**, 171 (1975).

"Protein Crystallography in a Molecular Biophysics Course," P. Argos, *Am. J. Phys.* **45**, 31 (1977).

Problems

4.1 Aluminum has a face-centered cubic lattice. The cell dimension is 4.05 Å. Calculate the closest interatomic distance and the density of the metal.

4.2 Account for the fact that the nonmetallic network solids are usually quite hard.

4.3 Calculate the number of spheres in the simple cubic, body-centered cubic and face-centered cubic cells. Also calculate the packing efficiency for each type of cell.

4.4 Construct a table that lists the h, k, l, and $h^2 + k^2 + l^2$ values for the simple cubic, fcc, and bcc lattices. How would you use this table to deduce the nature of a crystal lattice from a series of experimentally determined θ_{hkl} values?

4.5 Explain why diamond is a harder substance than graphite.

4.6 Which of the following has a greater density: crystalline SiO_2 or amorphous SiO_2?

4.7 Silver crystallizes in a face-centered cubic lattice; the edge length of the unit cell is 4.08 Å and the density of the metal is 10.5 g cm^{-3}. From these data, calculate Avogadro's number.

4.8 When X rays with a wavelength of 0.85 Å are diffracted by a metallic crystal, the angle of first-order diffraction ($n = 1$) is measured to be 14.8°. What is the distance between the layers of atoms responsible for the diffraction?

4.9 Barium crystallizes in the body-centered arrangement. Assuming a hard-sphere model, calculate the "radius" of a barium atom if the unit cell edge length is 5.015 Å.

4.10 Metallic iron can exist in the β form (bcc, cell dimension = 2.90 Å) and the γ form (fcc, cell dimension = 3.68 Å). The β form can be converted into the γ form by applying high pressures. Calculate the ratio of the densities of the β form to the γ form.

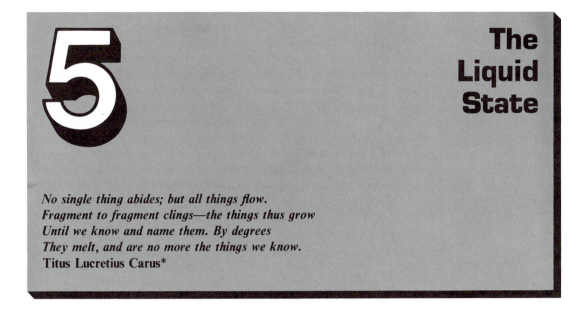

No single thing abides; but all things flow.
Fragment to fragment clings—the things thus grow
Until we know and name them. By degrees
They melt, and are no more the things we know.
Titus Lucretius Carus*

The structure of liquids lies somewhere between the completely disordered gaseous state and the highly ordered crystalline state. This "in-between" quality makes accurate accounting for the intermolecular interactions difficult. In this chapter we discuss the structure of liquids and consider three important topics: viscosity, surface tension, and diffusion.

Structure of Liquids 5.1

The word *structure* applied to liquids seems strange at first. A given amount of liquid has a fixed volume but assumes the shape of its container. At the molecular level, however, liquids do possess some degree of structure or order, as evidenced by a number of physical measurements.

Figure 5.1a shows the X-ray diffraction pattern of liquid argon. The plots of intensity of scattered X rays versus $\sin \theta / \lambda$, where θ is the angle of reflection and λ the wavelength of the X rays, exhibit a number of maxima and minima, implying that there must be some order present in liquid argon. A more informative approach is to plot the *radial distribution function* versus distance r (see Figure 5.1b). Imagine two concentric spheres of radii r and $(r + dr)$. The volume of the spherical shell between the two spheres is $4\pi r^2 \, dr$.† The radial distribution functions, $4\pi r^2 \rho(r)$, gives the probability of finding argon atoms within this shell at a distance r from the center of an argon atom. The term $\rho(r)$ is the density function of argon measured from the center. Crystalline argon is face-centered; each argon atom has a coordination number of 12. In the crystalline state, the plot shows a series of sharp lines at different values of

---

* *No Single Thing Abides*, translated by W. H. Mallock, A & C Black Ltd., Publishers, London, England, 1900. Used by permission.
† The volume of the spherical shell is given by $(4\pi/3)(r + dr)^3 - (4\pi/3)r^3 = 4\pi r^2 \, dr$ if we ignore the $(dr)^2$ and $(dr)^3$ terms.

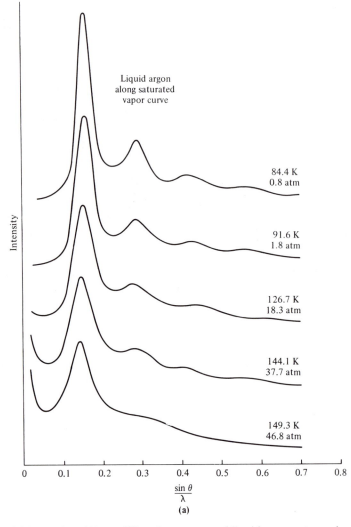

Liquid argon
along saturated
vapor curve

84.4 K
0.8 atm

91.6 K
1.8 atm

126.7 K
18.3 atm

144.1 K
37.7 atm

149.3 K
46.8 atm

Intensity

0   0.1   0.2   0.3   0.4   0.5   0.6   0.7   0.8

$\dfrac{\sin \theta}{\lambda}$

(a)

Figure 5.1   (a) Intensity of X-ray diffraction pattern of liquid argon at a series of temperatures.

r, which represent the distance of separation between nearest neighbors, next nearest neighbors, and so on.

These lines are broadened at higher temperatures, owing to increased vibrations of the atoms about their lattice positions. When argon melts, its volume increases by about 10% and its coordination number decreases to between 11 and 10. However, the maxima still have roughly the same r values as in solid argon, although there is a rapid damping out of the amplitudes as distance increases from the center. As the temperature is raised further, these peaks become even broader and shift toward larger values of r. These observations are consistent with the fact that liquid argon, like all other liquids, possesses short-range order but lacks long-range order. Even its short-range order is disrupted by an increase in the kinetic energy of the atoms caused by the increase in temperature. The word *order*, when applied to liquids, has a different meaning than when it is used to describe the solid state. In liquids, molecules are constantly moving from one point to another, and the X-ray diffraction pattern corresponds to the *time-averaged* positions of the atoms.

When the substance goes from the solid to the liquid state, the volume increases, but the average distance between the atoms remains nearly the same. How do we account for this? A simple model assumes that holes or vacancies are created in the liquid state. This would explain the increase in volume while interatomic distances remain constant. The holes would negate long-range order, because they disrupt the pattern of repeating units. Unlike the vacant sites in solids, these holes are highly mobile, and their behavior is conveniently described as reversing the behavior of molecules in a gas. In a gas molecules move among holes; in a liquid, the holes move among molecules.

Let V_L and V_S be the molar volumes of the liquid and solid of the same

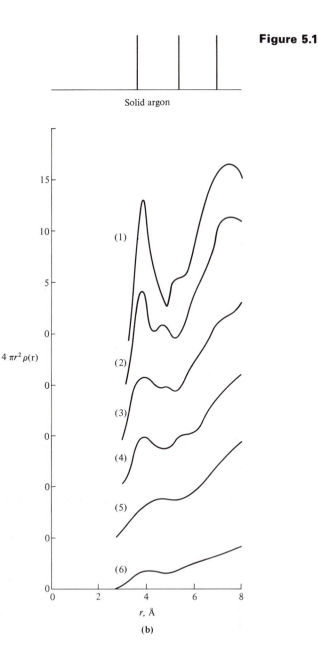

(b) Radial distribution function of argon at a series of temperatures. Curves (1), (2), and (5) refer to liquid argon at 84 K and 0.8 atm, 92 K and 1.8 atm, and 149 K and 46.8 atm, respectively. Curve (6) is the radial distribution curve for gaseous argon at 149 K and 43.8 atm. The vertical lines on top represent the diffraction lines for solid argon. [Modified from A. Eisenstein and N. S. Gingrich, *Phys. Rev.* **62**, 261 (1940).]

Figure 5.1

substance. The excess volume, that is, the increase in volume upon melting,* is given by $V_L - V_S$. In 1 mol of the liquid there are $(V_L - V_S)/V_S$ or $(V_L/V_S) -$ 1 mol of holes present. Since the total number of moles of liquid and holes present is $1 + (V_L/V_S) - 1$ or V_L/V_S, the fraction of holes present is given by

$$\frac{(V_L/V_S) - 1}{V_L/V_S} = \frac{V_L - V_S}{V_L}$$

As mentioned earlier, these holes possess gaslike properties. The remaining fraction, given by

$$1 - \frac{V_L - V_S}{V_L} = \frac{V_S}{V_L}$$

then reveals the fraction of molecules having solidlike properties. This treatment enables us to deal with the various properties of a liquid substance in terms of the known properties of its gaseous and solid states.

The paragraphs above briefly outline the *significant structure theory* formulated by Eyring. Explaining the details of this theory involves statistical mechanics and will not be undertaken here. Agreement between the theory and experimental data ranges from good to excellent for a number of liquids tested.

5.2 Viscosity of Liquids

The *viscosity* of a fluid, that is, a gas, a pure liquid, or a solution, is an index of its resistance to flow. In Chapter 3 we derived an expression for the viscosity of gases using the simple kinetic theory [Eq. (3.19)]. This section will consider the viscosity of liquids.

Most viscometers monitor the ease with which fluids flow through capillary tubing. Let us now derive an expression relating the viscosity of a liquid η to the experimental parameters. Consider a certain liquid flowing through a capillary tube of radius R and length L under constant pressure P (Figure 5.2). The velocity of the liquid is zero at the walls and increases toward the center of the tube, reaching a maximum at the center. Imagine two concentric cylinders of radii r and $(r + dr)$. According to Eq. (3.17), the frictional drag F between these two cylindrical layers is

$$F = -\eta(2\pi rL)\frac{dv}{dr} \tag{5.1}$$

where $2\pi rL$ is the surface area of the inner cylinder and dv/dr the velocity gradient. Because the velocity decreases as r increases, dv/dr is a negative quantity, and a negative sign has been included in Eq. (5.1) to make F a positive number. For steady-state flow, the frictional drag must be exactly balanced by the downward force, which is given by the product of pressure P and area πr^2. Thus

$$P(\pi r^2) = -\eta(2\pi rL)\frac{dv}{dr}$$

$$dv = -\frac{P}{2\eta L}r\,dr$$

* Water is one of very few substances whose molar solid volumes are *greater* than molar liquid volumes.

Flow of a liquid through a capillary tube of radius R.

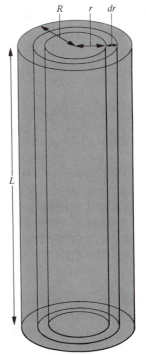

R r dr **Figure 5.2**

Integration between $v = 0$ (at $r = R$) and $v = v$ (at $r = r$) yields

$$\int_0^v dv = -\frac{P}{2\eta L}\int_R^r r\, dr$$

Hence

$$v = \frac{P}{4\eta L}\left(R^2 - r^2\right) \qquad (5.2)$$

We see that the velocity of flow anywhere in the tube is a parabolic function of r. It is important to keep in mind that Eq. (5.2) holds only for *laminar flow*, which depends on small diameters and low flow rates. Without these conditions, Eq. (5.2) is no longer valid and *turbulent flow* may result.* A useful distinction between turbulent and laminar flow is provided by the *Reynolds number*, defined as

$$\text{Reynolds number} = \frac{2Rv\rho}{\eta} \qquad (5.3)$$

where ρ is the density of the liquid. If the number is less than about 2000, we have laminar flow; if the number is greater than 3500, the flow is turbulent. Intermediate values (between 2000 and 3500) may accompany either type of flow and must be determined experimentally.

Our next step is to calculate the total flow rate of the liquid through the capillary as a function of viscosity. The volume of liquid that flows through a cross-sectional element $2\pi r\, dr$ per second is simply $(2\pi r\, dr)v$, and the total

---

* For laminar flow, all particles of the liquid move parallel to the tube, and the velocity increases regularly from zero at the wall to a maximum at the center. These conditions are not satisfied if the flow is turbulent.

73

volume of the liquid flowing per second, Q, is given by

$$Q = \frac{V}{t} = \int_0^R v(2\pi r \, dr) = \frac{2\pi P}{4\eta L} \int_0^R (R^2 - r^2) r \, dr$$

$$= \frac{\pi P R^4}{8\eta L} \tag{5.4}$$

where V is the total volume and t the flow time. Equation (5.4), known as *Poiseuille's law*, applies to liquids as well as to gases.

A relatively simple apparatus for measuring viscosity is the Ostwald viscometer, shown in Figure 5.3. In a typical experiment, the time of flow for a given volume V (between markings a and b) through a vertical capillary tube under the influence of gravity is recorded. The quantity P is really the difference in pressure between the two ends of the U tube and is assumed to be directly proportional to the density of the liquid.

Rearrangement of Eq. (5.4) gives

$$\eta = \frac{\pi P R^4 t}{8VL} \tag{5.5}$$

The accurate measurement of η using this equation is difficult because of the uncertainties in determining the radius of the capillary tubing, R (note that the radius appears as R^4). In practice, the viscosity of a liquid is most conveniently determined by comparison with a reference liquid of accurately known viscosity as follows. The ratio of the viscosities of a sample and a reference liquid is given by

$$\frac{\eta_{\text{sample}}}{\eta_{\text{reference}}} = \frac{\pi R^4 (Pt)_{\text{sample}}}{8VL} \times \frac{8VL}{\pi R^4 (Pt)_{\text{reference}}}$$

Since V, L, and R are the same if we use the same viscometer and the pressure acting on the liquid is directly proportional to its density, that is, $P = \text{constant} \times \rho$, the preceding equation reduces to

$$\frac{\eta_{\text{sample}}}{\eta_{\text{reference}}} = \frac{(\rho t)_{\text{sample}}}{(\rho t)_{\text{reference}}}$$

Figure 5.3

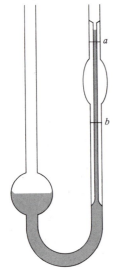

Ostwald viscometer.

Viscosity of Some Common Liquids at 293 K

Table 5.1

Liquid	Viscosity (P)[a] CGS Units	Viscosity (N s m^{-2}) SI Units
Acetone	0.00316 (298 K)	0.000316
Benzene	0.00652	0.000652
Carbon tetrachloride	0.00969	0.000969
Ethanol	0.01200	0.001200
Ethyl ether	0.00233	0.000233
Glycerine	14.9	1.49
Mercury	0.01554	0.001554
Water	0.0101	0.00101
Blood plasma	0.015 (310 K)	0.0015
Whole blood	0.04 (310 K)	0.004

[a] 1 poise (P) = 1 dyn s cm^{-2} = 0.1 N s m^{-2}.

Thus the viscosity of the sample can be readily obtained from the densities of the liquids and the times of flow if $\eta_{reference}$ is known. Table 5.1 lists the viscosity values of a number of common liquids.*

Equation (5.5) can be applied to the study of blood flow in our bodies. Figure 5.4 shows a schematic diagram of the various routes for blood circulation. The heart is, in effect, a single pump powering a double circuit. It contains four chambers—two atria and two ventricles—and four sets of valves. Freshly oxygenated blood is conducted away from the heart through

Schematic diagram of the blood circulation system. Arterial blood, rich in oxyhemoglobin (hemoglobin combined with oxygen), is pumped through the left ventricle of the heart to the tissues where oxygen is released and carbon dioxide taken up. The venous blood, rich in dissolved carbon dioxide, is pumped by the right ventricle of the heart to the lungs, in which the carbon dioxide is released and oxygen is taken up.

Figure 5.4

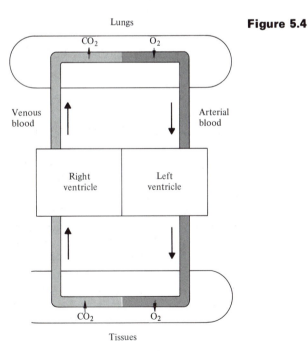

* The units of viscosities were discussed in Section 3.6.

the large artery, called the *aorta*, from the left chamber of the heart, and branches into smaller arteries, which lead to the various parts of the body. These, in turn, branch into still smaller arteries, the smallest of which are called *arterioles*, which break up into a complex network of capillaries. These are the minute structures that thread their way to every part of the body to carry out the vital function of the blood, trading oxygen and other materials for carbon dioxide and waste with the cells. The capillaries join into tiny veins called *venules*, which in turn merge into larger and larger of veins that lead the deoxygenated blood back to the right atrium of the heart.

In the space of a single beat, first the atria contract, forcing blood into the ventricles; then the ventricles contract, forcing the blood out of the heart. Because of the pumping action of the heart, blood enters the arteries in spurts or pulses. The maximum pressure at the peak of the pulse is called the *systolic pressure*; the lowest blood pressure between pulses is called *diastolic pressure*. In a healthy young adult the systolic pressure is about 120 mm Hg (120 torr) and the diastolic pressure is about 80 mm Hg (80 torr). These are the excess pressures over atmospheric pressure. Thus the absolute systolic and diastolic pressures are 880 and 840 mm Hg, respectively. Thus the average value of the blood pressure is about 100 mm Hg.

Because the radius of the aorta is sufficiently large (about 1 cm), a small pressure difference is required to sustain normal blood flow through it. At rest, the rate of blood flow is roughly 0.08 liter s^{-1}. Equation (5.5) can be rewritten as

$$Q = \frac{\pi \, \Delta P R^4}{8\eta L} \tag{5.6}$$

where ΔP is the difference in pressure at two points along the aorta and L is the distance between the points. Setting $L = 1$ cm and converting Q to 80 cm^3 s^{-1}, we find that

$$\Delta P = \frac{8(0.04 \text{ dyn s cm}^{-2})(1 \text{ cm})(80 \text{ cm}^3 \text{ s}^{-1})}{\pi(1 \text{ cm})^4}$$

$$= 8 \text{ dyn cm}^{-2} = 6 \times 10^{-3} \text{ mm Hg}$$

(The conversion factor is 1 dyn cm^{-2} = 7.5 × 10^{-4} mm Hg.) A drop of pressure equal to 6 × 10^{-3} mm Hg per cm is negligibly small compared to the total blood pressure. The situation is different as the blood enters the major arteries. Because these vessels have much smaller radii than the aorta, a pressure drop of about 20 mm Hg is required to maintain the flow through them. Therefore, the pressure is only 80 mm Hg when the blood enters the arterioles. Since these vessels have still smaller radii, there is another drop in pressure of about 50 mm Hg. A further drop of 20 mm Hg results when the blood flows through the capillaries. Note that although the capillaries have much smaller radii than arterioles, their number is very large, so that the amount of blood passing through each one is very small. By the time the blood reaches the veins, its pressure is reduced to about 10 mm Hg. Fortunately, veins are fitted with cup-shaped valves to prevent backflow at this low pressure. The movement of blood in veins is promoted by the massaging effect of surrounding skeletal muscle or by the adjacent arteries. Finally, the blood returns to the right atrium, ready to be circulated again.

An interesting comparison can be made between Eq. (5.6)

$$Q = \frac{\Delta P}{8L\eta/\pi R^4}$$

and Ohm's law:

$$\text{current} = \frac{\text{voltage}}{\text{resistance}}$$

By analogy, the resistance to flow is given by $8L\eta/\pi R^4$. Equation (5.6) can be applied to study the flow of blood in arteries, arterioles, and capillaries. Since resistance is inversely proportional to the fourth power of the radius, a decrease from 2×10^{-4} cm, a typical capillary radius, to 1.5×10^{-4} cm would result in an increase in the resistance by a factor of 3. Normal blood flow is then sustained by higher blood pressure, causing a condition known as *hypertension*. On the other hand, if the resistance is lowered while the blood pressure is unchanged, the blood flow Q is increased. During vigorous exercise, there is both an increase in blood pressure and an increase in the radii of blood vessels, known as *vasodilation*. These two changes facilitate a greater blood flow to meet the enhanced metabolic rate of the body.

Generally, the viscosity of a solution is greater than that of the pure solvent.* The presence of solute molecules disrupts the smooth flow pattern or velocity gradient of the fluid, resulting in an increase in viscosity. This is particularly true for solutions that contain macromolecules. As we would expect, the viscosities of such solutions must also depend on the conformation of the macromolecules. The viscosity of a DNA solution, for example, can vary greatly depending on whether the solute molecules possess the native double-helical conformation or are shaped in random coils. Often the kinetics of denaturation from helix to random coil can be conveniently followed by measuring changes in the solution's viscosity over a period of time.

The viscosities of most liquids decrease with increasing temperature. A molecular interpretation is as follows. As mentioned in Section 5.1, liquids possess a number of holes or vacancies. Molecules are continually moving into these vacancies. This process permits flow of a liquid but requires energy. To be able to move into a vacancy, a molecule must possess an activation energy to overcome the repulsion by the molecules that surround the vacancy. At higher temperatures more molecules possess the necessary activation energy, so that the liquid flows more easily. In contrast to liquids, the viscosity of a gas *increases* with temperature.† In the kinetic molecular treatment of gases, the origin of the viscous drag between two adjacent layers is the momentum transfer of the molecules from one layer to the other. The rate of transfer increases with temperature; hence so does the viscosity of the gas.

In Chapter 21 we shall discuss the determination of molar masses of macromolecules from viscosity measurements.

---

* There are a number of cases in which the reverse is true. For example, the viscosities of many aqueous solutions containing alkali metal and ammonium ions and certain anions are *lower* than that of water (more on this in Section 9.2).

† From Eq. (3.19), we write

$$\eta = \frac{m\bar{c}}{2\sqrt{2}\,\pi d^2} = \frac{1}{d^2}\sqrt{\frac{mkT}{\pi^3}}$$

5.3 Surface Tension

When the surface of a liquid is expanded, molecules that were originally in the interior region are brought out to the exterior. Work must be done to counteract the attractive forces among these molecules and their neighbors. This process is somewhat similar to vaporization of a liquid. However, in vaporization the molecules are completely removed from the liquid, whereas molecules in a surface layer are still under the influence of strong intermolecular forces, except along the direction toward the vapor phase (Figure 5.5). This unbalanced interaction experienced by surface-layer molecules results in a tendency for the liquid to minimize its surface area. For this reason, a small drop of liquid would assume a spherical shape. We can define the *surface tension* (γ) of a liquid as the amount of energy required to expand the surface by unit area. In CGS units γ is in ergs cm^{-2} or dyn cm^{-1} (1 erg = 1 dyn cm); in SI units, γ is in N m^{-1}. These two units are related by the equation 1 dyn cm^{-1} = 1 $\times$ 10^{-3} N m^{-1}.

The *capillary-rise method* provides a simple means for measuring the surface tension of liquids. In this arrangement, a capillary tube of radius r is dipped into the liquid under study (Figure 5.6). The force acting downward is the gravitational pull on the liquid, given by $\pi r^2 h \rho g$, where $\pi r^2 h$ is the volume,* ρ the density of the liquid, and g the acceleration due to gravity. Now this weight must be balanced by an upward force caused by the liquid's surface tension. This force, which acts along the periphery of the cylindrical bore, between the liquid and the glass wall, is given by $2\pi r \gamma \cos \theta$, where $2\pi r$ is the circumference of the bore, θ the angle of contact between the liquid and the capillary tube in the meniscus, and $\cos \theta$ gives the vertical component of the force (acting upward). Thus we have

$$\pi r^2 h \rho g = 2\pi r \gamma \cos \theta$$

or

$$\gamma = \frac{rhg\rho}{2 \cos \theta} \tag{5.7}$$

The surface tension of several common liquids are listed in Table 5.2.

Although the rise of liquids up a capillary tube is commonly observed, it is by no means a universal phenomenon. For example, when a capillary tube is dipped into liquid mercury, the upper level of the liquid in the tube is actually

Figure 5.5

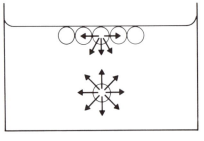

Intermolecular forces acting on a molecule in the surface layer and on a molecule in the interior region of a liquid.

---

* We ignore here the small amount of liquid above the top of the meniscus. For accurate work, a correction term of $r/3$ can be added to h in the calculation.

Table 5.2

Liquid	Surface Tension	
	(CGS Units) dyn cm^{-1}	(SI Units) N m^{-1}
Acetic acid	27.6	0.0276
Acetone	23.7	0.0237
Benzene	28.9	0.0289
Carbon tetrachloride	26.6	0.0266
Chloroform	27.1	0.0271
Ethanol	22.3	0.0223
Ethyl ether	17.0	0.0170
n-Hexane	18.4	0.0184
Mercury	476 (298 K)	0.476
Water	72.75	0.07275

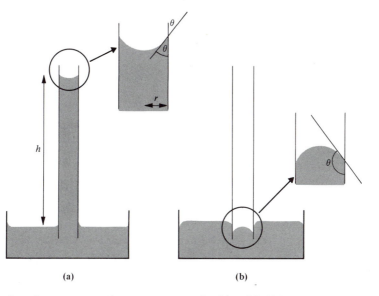

(a) Capillary-rise phenomenon of most common liquids. (b) When cohesion is greater than adhesion, a depression of the liquid in the capillary tube results.   **Figure 5.6**

lower than the surface of free liquid (Figure 5.6b). These two divergent behaviors can be understood by considering the intermolecular attraction between like molecules in the liquid, called *cohesion*, and the attraction between the liquid and the glass wall, called *adhesion*. If adhesion is stronger than cohesion, the walls become wettable and the liquid will rise along the walls. Since the vapor–liquid interface resists being stretched, the liquid in the lumen also rises. Conversely, if cohesion is greater than adhesion, a depression of the liquid in the capillary would result.

EXAMPLE 5.1   The typical radius of a xylem vessel is about 0.02 cm. How high will water rise in such a vessel at 293 K?

Answer

CGS units: From Eq. (5.7),

$$h = \frac{2\gamma \cos \theta}{rg\rho}$$

Since the contact angle is usually quite small, we assume that $\theta = 0$. Now

$$\gamma = 72.75 \text{ dyn cm}^{-1}$$

$$\cos \theta = 1$$

$$r = 0.02 \text{ cm}$$

$$g = 980.7 \text{ cm s}^{-2}$$

$$\rho = 1 \text{ g cm}^{-3}$$

We have

$$h = \frac{2(72.75 \text{ dyn cm}^{-1})}{(0.02 \text{ cm})(980.7 \text{ cm s}^{-2})(1 \text{ g cm}^{-3})}$$

$$= 7.4 \text{ dyn s}^2 \text{ g}^{-1}$$

$$= 7.4 \text{ cm}$$

$$1 \text{ dyn} = 1 \text{ g cm s}^{-2}$$

SI units

$$\gamma = 0.07275 \text{ N m}^{-1}$$

$$r = 0.0002 \text{ m}$$

$$g = 9.807 \text{ m s}^{-2}$$

$$\rho = 1 \times 10^3 \text{ kg m}^{-3}$$

Hence

$$h = \frac{2(0.07275 \text{ N m}^{-1})}{(0.0002 \text{ m})(9.807 \text{ m s}^{-2})(1 \times 10^3 \text{ kg m}^{-3})}$$

$$= 0.074 \text{ N s}^2 \text{ kg}^{-1}$$

$$= 0.074 \text{ m}$$

$$1 \text{ N} = 1 \text{ kg m s}^{-2}$$

Example 5.1 shows that the capillary-rise phenomenon is partially responsible for the rise of water in plants and soils but cannot wholly account for it. The water is helped up by another major mechanism, osmosis, to be discussed in Chapter 8.

The surface tension of aqueous solutions is generally close to that of pure water if the solutes are salts, such as NaCl, or sucrose and other substances that do not preferentially collect at the air–water interface. On the other hand, a dramatic decrease in surface tension can result if the dissolved substance is a fatty acid or a lipid. These molecules consist of two regions: at one end, a polar group such as —COOH, which is hydrophilic (water-liking); at the

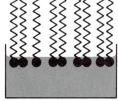

Figure 5.7

other end, a long hydrocarbon chain that is nonpolar and is therefore hydrophobic (water-fearing). The nonpolar groups tend to line up together along the surface of water with the polar groups pointing toward the interior of the solution (Figure 5.7). Molecules of this type are often said to exhibit "schizophrenic" behavior. Consequently, surface tension decreases. This effect depends on the nature of solute molecule. Thus although a 0.01 M solution of caproic acid [$CH_3-(CH_2)_4-COOH$] lowers the surface tension by about 15 dyn cm^{-1} (0.015 N m^{-1}), a decrease of about 25 dyn cm^{-1} (0.025 N m^{-1}) in surface tension is observed in a 0.0005 M solution of capric acid [$CH_3-(CH_2)_8-COOH$]. Any substance that causes a reduction in surface tension in this manner is called a *surfactant*. Among the most effective surfactants are soaps (salts of long-chain fatty acids) and denatured proteins.

The action of surfactants also plays an important role in the breathing process. By far the most extensive surface of the human body in contact with the surroundings is the moist interior surface of the lungs. To carry on the active exchange of carbon dioxide and oxygen between the circulating blood and the atmosphere in an average adult requires a lung surface area roughly that of a tennis court. Such an area is encompassed in the relatively small volume of the chest by compartmentation of the lungs into hundreds of millions of tiny air spaces or sacs called *alveoli*. The alveoli have an average radius of about 0.005 cm; they are connected by confluent passages through the bronchial tree and the trachea to the atmosphere (Figure 5.8).

During normal inhalation the pressure in the alveoli is about 3 mm Hg below atmospheric pressure, or we say that they have a gauge pressure* of −3 mm Hg, which enables air to flow into them through the bronchial tubes. The alveoli are lined with mucous tissue fluid that normally has a surface tension of 50 dyn cm^{-1}. During inhalation, the radius of the alveoli expands by about a factor of 2; the pressure difference required to inflate an alveolus is given by the equation (see Appendix 5.1 for derivation):

$$P_i - P_o = \frac{2\gamma}{r} \qquad (5.8)$$

where P_i and P_o are the gauge pressures inside and outside of alveoli, γ the surface tension of mucous fluid, and r the radius of alveoli. To carry out such an expansion, the pressure difference must be at least

$$P_i - P_o = \frac{2(50 \text{ dyn cm}^{-1})}{0.005 \text{ cm}} = 2 \times 10^4 \text{ dyn cm}^{-2}$$

$$= 15 \text{ mm Hg}$$

* Gauge pressure is the difference between the absolute pressure of a fluid (gas or liquid) and atmospheric pressure. For example, when we measure the pressure of a tire, the value corresponds not to the pressure of air inside the tire but its excess over the atmospheric pressure. The same applies to blood pressure, discussed earlier. (Also see Problem 2.19.)

The quantity P_o is the pressure in the space between the lungs and the pleural cavity that holds the lungs. However, since this pressure is only -4 mm Hg (or 756 mm Hg in absolute value), we have

$$P_i - P_o = (-3 \text{ mm Hg}) - (-4 \text{ mm Hg})$$

$$= 1 \text{ mm Hg}$$

which is only $\frac{1}{15}$ of the pressure required to expand an alveolus. To overcome this difficulty, the aveolar cells secrete a special type of surfactant (a phospholipoprotein complex), which effectively reduces the surface tension, so that the alveolus can be expanded without difficulty in the course of the 15,000 or so breaths that are drawn into the lungs of an adult each day. A striking example of what occurs when insufficient surfactant is present is demonstrated by the disease known as "respiratory-distress syndrome of the newborn," which frequently afflicts premature infants in whom the surfactant synthesizing cells do not yet function adequately. Even in a normal healthy baby the alveoli in his or her lungs are so collapsed at birth that a pressure difference as large as 25 to 30 mm Hg is required to expand them the first time. Therefore, the first breath of life requires extraordinary effort to overcome the surface tension in the alveoli.

The surface activity discussed above bears also on water conservation. Spreading a thin film of a certain material over the surface of water can cut down the evaporation rate of water in reservoirs. Cetyl alcohol $[CH_3(CH_2)_{14}CH_2OH]$ has been found to be particularly suitable for this purpose. A solid, it is insoluble in water. However, it has a surface solubility

Figure 5.8

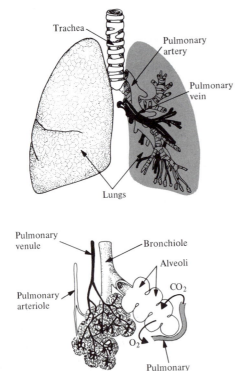

Relationships between respiratory airways and blood vessels. Alveoli are the air spaces in the lungs through which oxygen enters the blood and carbon dioxide leaves. An average alveolus expands and contracts about 15,000 times a day during breathing. (From A. J. Vander, J. H. Sherman, and D. S. Luciano, *Human Physiology,* 2nd ed., p. 286. Copyright 1975 by McGraw-Hill Book Company. Used with permission of McGraw-Hill Book Company.)

in the sense that its molecules float on water, forming a thin film that spreads to cover the surface. If the film is disrupted by weather or other disturbances, it can readily be reformed. Only 30 g of the material is sufficient to cover about 10,000 m² (about 3 acres) of water surface.

Diffusion **5.4**

Diffusion is the process by which concentration gradients in a solution spontaneously decrease until a uniform, homogenous distribution is obtained. The diffusion process is important to many chemical and biological systems. It is the major mechanism by which carbon dioxide reaches the sites of photosynthesis in chloroplasts. Understanding the transportation of solute molecules across cell membranes requires a detailed knowledge of diffusion. The diffusion phenomenon plays an important role in determining the molar mass of macromolecules (see Chapter 21). In this section we describe some basics of diffusion in solution.

Imagine a container with a solution in the lower region and the pure solvent on top, as shown in Figure 5.9a. Initially, there is a sharp boundary between the solution and the solvent. As time progresses, solute molecules gradually move upward by diffusion. This process continues until the entire system becomes homogeneous. In 1855, Fick studied the diffusion phenomenon and found that the *flux* (*J*), that is, the net amount of solute that

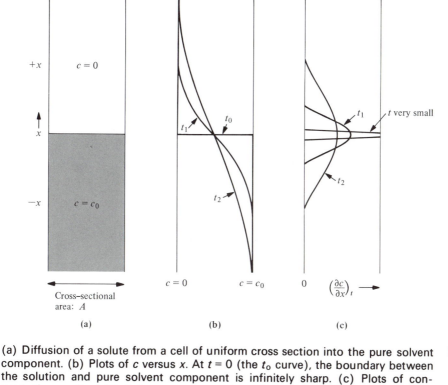

(a) Diffusion of a solute from a cell of uniform cross section into the pure solvent component. (b) Plots of *c* versus *x*. At *t* = 0 (the t_0 curve), the boundary between the solution and pure solvent component is infinitely sharp. (c) Plots of concentration gradient $(\partial c/\partial x)_t$ versus *x* at various times *t* after diffusion has begun. At *t* = 0, the gradient is a vertical line of infinite height and no width centered at *x* = 0.

Figure 5.9

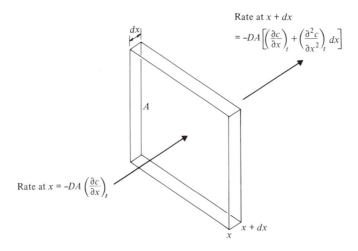

Rate at $x + dx$

$$= -DA\left[\left(\frac{\partial c}{\partial x}\right)_t + \left(\frac{\partial^2 c}{\partial x^2}\right)_t dx\right]$$

A

Rate at $x = -DA\left(\frac{\partial c}{\partial x}\right)_t$

$x + dx$

x

Figure 5.10   Rate of accumulation of solute in a volume element $A\, dx$ during a diffusion process.

diffuses through unit area per unit time, is proportional to the concentration gradient. Expressing this mathematically in one dimension along the x axis, we write

$$J \propto -\left(\frac{\partial c}{\partial x}\right)_t$$

$$= -D\left(\frac{\partial c}{\partial x}\right)_t \qquad (5.9)$$

Equation (5.9) is known as *Fick's first law of diffusion* in one dimension. The quantity $(\partial c/\partial x)_t$ is the concentration gradient of the diffusing substance (c in mol liter^{-1}) after time t of diffusion, and D is the diffusion coefficient of the diffusing substance in the medium concerned. The negative sign indicates that the diffusion proceeds from higher to lower concentration since the concentration gradient is negative in the direction of diffusion. Thus the flux is a positive quantity. The units of D are cm^2 s^{-1} (CGS units) or m^2 s^{-1} (SI units).

Let us investigate the diffusion process in a little more detail. A question of importance is: What is the change of concentration with time at a given point along the x axis? Consider a volume element $A\, dx$ (where A is the area of cross section) shown in Figure 5.10. At distance x measured from the original boundary, the rate of solute molecules entering the volume element is $-DA(\partial c/\partial x)_t$. Since the rate at which the concentration gradient changes with x is given by

$$\frac{\partial}{\partial x}\left(\frac{\partial c}{\partial x}\right)_t = \left(\frac{\partial^2 c}{\partial x^2}\right)_t$$

the rate of solute molecules leaving the volume element, after having traveled distance dx, is

$$-DA\left(\frac{\partial c}{\partial x}\right)_t - DA\left(\frac{\partial^2 c}{\partial x^2}\right)_t dx = -DA\left[\left(\frac{\partial c}{\partial x}\right)_t + \left(\frac{\partial^2 c}{\partial x^2}\right)_t dx\right]$$

Thus the rate of accumulation of solute in the volume element is the difference of the foregoing two quantities:

rate of accumulation of   rate of solute   rate of solute
solute in the volume   = entering the   − leaving the
element   volume element   volume element

$$= -DA\left(\frac{\partial c}{\partial x}\right)_t + DA\left[\left(\frac{\partial c}{\partial x}\right)_t + \left(\frac{\partial^2 c}{\partial x^2}\right)_t dx\right]$$

$$= DA\left(\frac{\partial^2 c}{\partial x^2}\right)_t dx$$

Now, there is another way of arriving at an expression for the rate of accumulation. As time goes on, the concentration of solute in the volume element is steadily increasing as a result of diffusion. The rate of this increase is given by the product of the volume element and the change of concentration with time, that is, $(\partial c/\partial t)_x(A\,dx)$. Equating these two rates of solute accumulation, we obtain

$$\left(\frac{\partial c}{\partial t}\right)_x = D\left(\frac{\partial^2 c}{\partial x^2}\right)_t \tag{5.10}$$

Equation (5.10) is known as *Fick's second law of diffusion*. It says that the change of concentration with time at a certain distance x from the origin is equal to the product of diffusion coefficient and the change of concentration gradient in the direction of x at time t.

Equation (5.10) is a fundamental equation of diffusion; however, it must be integrated before we can apply it to practical systems. To obtain the value of D from suitable experimental measurements, the appropriate *boundary condtions** must be applied. If the columns of liquids shown in Figure 5.9 are effectively of infinite length so that the concentrations of solute at top and bottom remain zero and c_0, respectively, throughout the experimental run, the following boundary conditions will hold:
at $t = 0$,

$$c = 0 \qquad \text{for } x > 0$$

$$c = c_0 \qquad \text{for } x < 0$$

at $t > 0$,

$$c \to c_0 \qquad \text{as } x \to -\infty$$

$$c \to 0 \qquad \text{as } x \to +\infty$$

The solution of Eq. (5.10) with the boundary conditions given above is†

$$c = \frac{c_0}{2}\left[1 - \frac{2}{\sqrt{\pi}}\int_0^\beta e^{-\beta^2}\,d\beta\right] \tag{5.11}$$

---

* In physics we encounter many equations that, mathematically speaking, have an infinite number of solutions. However, to be physically meaningful, these equations must satisfy a set of specific conditions, called boundary conditions, which often reduce the solutions to a very small number, usually 1.
† For details, see C. Tanford, *Physical Chemistry of Macromolecules*, John Wiley & Sons, Inc., New York, 1961, p. 354.

where

$$\beta = \sqrt{\frac{x^2}{4Dt}}$$

Equation (5.11) now enables us to calculate the concentration of solute at a distance x from the origin after it has diffused for time t. Figure 5.9b shows a graphical representation of Eq. (5.11) corresponding to various values of time t. We can also express Eq. (5.11) in the differential form as

$$\left(\frac{\partial c}{\partial x}\right)_t = -\frac{c_0}{\sqrt{4\pi Dt}} e^{-x^2/4Dt} \tag{5.12}$$

Equation (5.12) allows us to plot the concentration gradient $(\partial c/\partial x)_t$ versus x at different times t (Figure 5.9c).

The diffusion coefficient is rather difficult to determine accurately. Optical methods such as refractive index measurements are normally employed to monitor the concentration gradients at various distances from the origin after diffusion has started. In recent years a powerful technique for measuring diffusion coefficients of biomacromolecules from laser light scattering has been developed.* Here we shall describe a simple, although less accurate method for determining D. From Eq. (5.12), we see that the concentration gradient at the origin $(x = 0)$ is given by

$$\left(\frac{\partial c}{\partial x}\right)_t = -\frac{c_0}{\sqrt{4\pi Dt}} \tag{5.13}$$

Further, we can rewrite Fick's first law [Eq. (5.9)] as

$$\frac{n}{At} = -D\left(\frac{\partial c}{\partial x}\right)_t$$

where n is the number of moles of solute that has diffused across the boundary (area A) in a time t. Thus

$$\left(\frac{\partial c}{\partial x}\right)_t = -\frac{n}{ADt} \tag{5.14}$$

From Eqs. (5.13) and (5.14), we arrive at the result

$$\frac{n}{ADt} = \frac{c_0}{\sqrt{4\pi Dt}}$$

or

$$D = \frac{4n^2}{c_0^2 A^2}\frac{\pi}{t}$$

By using a specially constructed cell, the solvent column is totally removed after a diffusion experiment and stirred to produce a homogeneous solution of concentration c. If the height of the solvent column is h, it follows that

$$n = cAh$$

Substituting this expression for n in the equation above, we get

$$D = \left(\frac{2ch}{c_0}\right)^2\frac{\pi}{t}$$

* See S. B. Dubin, J. H. Lunacek, and G. Benedek, *Proc. Nat. Acad. Sci. U.S.* **57**, 1164 (1967).

Thus by determining the concentration c after time t and knowing the original concentration c_0, we can calculate the value of D. Two points are worth noting. First, our boundary conditions assume that the liquids columns are infinitely long, whereas in practice relatively short columns are employed. However, if we keep t short, the concentrations of the solute at the extreme ends are still close to c_0 and zero at the end of the experiment. Second, strictly, the diffusion coefficient is concentration-dependent so that it is preferable to work with dilute solutions. Table 5.3 lists the diffusion coefficients of a number of molecules. We would expect that the larger the molecule, the slower is its motion. The data in Table 5.3 qualitatively confirm this expectation.

An important quantity in diffusion studies is the distance traveled by solute molecules from their place of origin in a given time t. Although diffusion occurs in one definite direction, the movement of each individual molecule is completely random and unpredictable. It follows that the average or net distance, $\bar{x}$, traveled by the molecules is zero. For this reason, we need to consider the mean-square distance $\overline{x^2}$, defined as

$$\overline{x^2} = \frac{\int_{-\infty}^{+\infty} x^2 \left(\frac{dc}{dx}\right) dx}{\int_{-\infty}^{+\infty} \left(\frac{dc}{dx}\right) dx} \tag{5.15}$$

This standard integral is tabulated in most handbooks of mathematics. The result is

$$\overline{x^2} = 2Dt$$

Hence the root-mean-square distance, $\sqrt{\overline{x^2}}$, is given by

$$\sqrt{\overline{x^2}} = \sqrt{2Dt} \tag{5.16}$$

In Eq. (5.16) we have a simple, yet useful relation for estimating mean diffusion distances.

We would expect the frictional force exerted by a solvent medium to affect

Diffusion Coefficients of Some Molecules in Water at 298 K[a]

Table 5.3

	$D \times 10^5$ (cm^2 s^{-1}) CGS Units	$D \times 10^9$ (m^2 s^{-1}) SI Units
Ethanol	1.10	1.10
Urea	1.18	1.18
Glucose	0.57	0.57
Sucrose	0.46	0.46
Myoglobin	0.113	0.113
Hemoglobin	0.069	0.069
DNA (calf thymus)	0.0013	0.0013

[a] The reader should keep in mind, when units are expressed in this manner, that each value should be multiplied by the factor 10^{-5} or 10^{-9}. For example, the diffusion constant of ethanol is 1.10×10^{-5} cm^2 s^{-1} or 1.10×10^{-9} m^2 s^{-1}.

the diffusion of a solute molecule. Einstein proposed the following quantitative relationship:

$$D = \frac{kT}{f} \tag{5.17}$$

where k is Boltzmann's constant and f is the *frictional coefficient* of the solute molecule. In CGS units f is expressed as dyn s cm^{-1}; in SI units f is N s m^{-1}. Thus the product of f and the velocity of the solute molecule gives the frictional force of resistance (in dyn or N) exerted on the particle by the solvent. Stokes showed that for a spherical particle,

$$f = 6\pi\eta r \tag{5.18}$$

where η is the viscosity of the solvent and r the radius of the molecule. Equation (5.17) now becomes

$$D = \frac{kT}{6\pi\eta r} \tag{5.19}$$

Either Eq. (5.17) or (5.19) provides a physical interpretation of the diffusion coefficient. The quantity kT is a measure of the thermal or kinetic energy of the molecule, while f or η is a measure of the viscous resistance to diffusion. The ratio of these two opposing quantities determines how easily a solute molecule diffuses in solution.

EXAMPLE 5.2 Estimate the diffusion coefficient of a spherical molecule of radius 1.5 Å in water at 300 K.

Answer

CGS units: From Eq. (5.19),

$$D = \frac{kT}{6\pi\eta r}$$

$$k = 1.38 \times 10^{-16} \text{ erg K}^{-1}$$

$$T = 300 \text{ K}$$

$$\eta = 0.0101 \text{ dyn s cm}^{-2}$$

$$r = 1.5 \times 10^{-8} \text{ cm}$$

Hence

$$D = \frac{(1.38 \times 10^{-16} \text{ erg K}^{-1})(300 \text{ K})}{6\pi(0.0101 \text{ dyn s cm}^{-2})(1.5 \times 10^{-8} \text{ cm})}$$

$$= 1.46 \times 10^{-5} \text{ erg dyn}^{-1} \text{ s}^{-1} \text{ cm}$$

Since

$$1 \text{ erg} = 1 \text{ dyn cm}$$

$$D = 1.46 \times 10^{-5} \text{ cm}^2 \text{ s}^{-1}$$

$$k = 1.38 \times 10^{-23} \text{ J K}^{-1}$$

$$T = 300 \text{ K}$$

$$\eta = 0.00101 \text{ N s m}^{-2}$$

$$r = 1.5 \times 10^{-10} \text{ m}$$

Hence

$$D = \frac{(1.38 \times 10^{-23} \text{ J K}^{-1})(300 \text{ K})}{6\pi(0.00101 \text{ N s m}^{-2})(1.5 \times 10^{-10} \text{ m})}$$

$$= 1.46 \times 10^{-9} \text{ J N}^{-1} \text{ s}^{-1} \text{ m}$$

Since

$$1 \text{ J} = 1 \text{ N m}$$

$$D = 1.46 \times 10^{-9} \text{ m}^2 \text{ s}^{-1}$$

Equation (5.19) suggests a way to measure the radius of the molecule if both D and η are known. We must realize, however, that Stokes law is an idealized expression. Further, even if a molecule is sufficiently symmetrical to be treated like a sphere, the radius measured may not necessarily correspond to the true radius because most solute molecules are solvated to a certain extent in solution. A measured radius, then, might often be greater than the true radius.

Liquid Crystals **5.5**

Ordinarily, a sharp distinction can be made between the highly ordered crystalline state of a solid and the more random molecular arrangement of liquids. Crystalline ice and liquid water, for example, differ from each other in this respect. One class of substances, however, tends so greatly toward an ordered arrangement that a melting crystal first forms into a milky liquid, called the *mesomorphic* or *paracrystalline state*, exhibiting certain characteristic crystalline properties. At higher temperatures, this milky fluid changes sharply into a clear liquid that behaves like an ordinary liquid. Such substances are known as *liquid crystals*.

Molecules that exhibit liquid crystallinity are usually long, straight, and rodlike. An example is 4,4′-dimethoxyazoxybenzene,

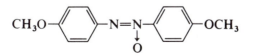

which has the following "melting" or transition points:

$$\text{solid} \xrightarrow{\text{391 K}} \text{mesomorphic state} \xrightarrow{\text{409 K}} \text{liquid}$$

In the mesomorphic state, the molecules are constrained to lie parallel to one

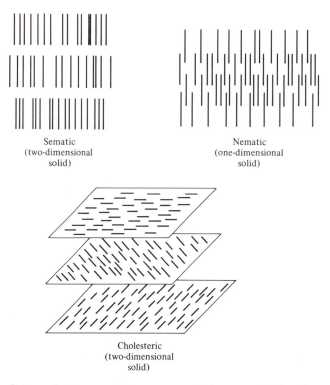

Sematic
(two-dimensional
solid)

Nematic
(one-dimensional
solid)

Cholesteric
(two-dimensional
solid)

Figure 5.11   Schematic diagram of three types of thermotropic liquid crystals.

another, with rotation being permitted only about the long axes. Materials of this type are said to be *anisotropic* because they show different properties along different directions. In an ordinary liquid, there is no preferred orientation and the substance is *isotropic*.

There are two known types of liquid crystals, called thermotropic and lyotropic. *Thermotropic liquid crystals* are formed by heating the compounds and they can be subdivided into three classes, called sematic, nematic, and cholesteric. Figure 5.11 shows a schematic diagram of these three structures. In *sematic liquid crystals*, the long axes of the molecules are perpendicular to the plane of the layers. The layers are free to slide over one another so that the substance has the mechanical properties of a two-dimensional solid. Optically, a sematic liquid crystal also behaves like a three-dimensional crystal such as quartz. It has been demonstrated by using polarized light (see Chapter 19) that the velocity of light traveling perpendicular to the layers is slower than that traveling parallel to the layers. *Nematic liquid crystals* are less ordered. Although the molecules are still aligned with their long axes parallel to one another, they are not separated into layers. *Cholesteric liquid crystals* resemble sematic liquid crystals in that molecules are arranged in layers, except that the long axes of the molecules are parallel to the layers.

Lyotropic liquid crystals are formed by mixing two or more compounds together, one of which is usually a polar molecule such as water or some other polar solvent. Relatively little is known about lyotropic liquid crystals although it is believed that they may occur in living systems. A number of synthetic polypeptides, such as poly-γ-benzyl-L-glutamate and poly-β-benzyl-L-aspartate, when dissolved in water, dimethylformamide, or pyridine, form structures that resemble cholesteric liquid crystals.

90

Thermotropic liquid crystals have many applications in science, technology, and medicine. As a result of the changes in color of cholesteric liquid crystals over very small temperatures ranges, a number of techniques have been developed whereby these liquid crystals are used as sensitive thermometers. In metallurgy, for example, they are used to detect metal stress, heat sources, and conduction paths. Medically, the temperature of the body at specific sites can be determined by using liquid crystals. This has become an important diagnostic tool in treating infection and tumor growth (for example, breast tumors). Since localized infections and tumors increase the metabolic rate at that area and hence the temperature, a thin film of liquid crystal could tell a physician visually if an infection or tumor were present by responding to a temperature difference with a change in color.

Under the influence of an electric field, nematic liquid crystals exhibit an interesting optical phenomenon called *dynamic scattering* (the transparent liquid crystal becomes opaque).*

Appendix 5.1
Derivation of Equation (5.8)

Consider a soap bubble of radius r. For the bubble to maintain its spherical shape, the internal force must be balanced by the external force. Figure 5.12 shows the bubble divided into two hemispheres. Consider the upper hemisphere. In addition to the outside force, there is a downward force F as a result of the surface tension. This surface force is given by the surface tension multiplied by the circumference. (Remember that surface tension has the units dyn cm^{-1}.) The total downward force is then

$$F = P_o(\pi r^2) + 2(2\pi r\gamma)$$

where P_o is the outside pressure. The extra factor 2 arises because the bubble has both an inner and an outer layer. The total upward force is $P_i(\pi r^2)$, where P_i is the inside pressure so that at equilibrium we have

$$P_i(\pi r^2) = P_o(\pi r^2) + 4\pi r\gamma$$

or

$$P_i - P_o = \frac{4\gamma}{r}$$

The forces acting on a soap bubble. P_0 **Figure 5.12**

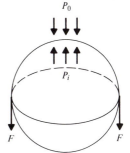

P_i

F F

---

* "Liquid Crystal Display Devices," G. H. Heilmeier, *Sci. Am.*, Apr. 1970.

For bubbles having only one surface, the equation above becomes

$$P_i - P_o = \frac{2\gamma}{r}$$

which is Eq. (5.8). Note that the difference $(P_i - P_o)$ is large for small bubbles (when r is small) but decreases as r increases.

Suggestions for Further Reading

The standard physical chemistry texts listed in Chapter 1 give a fairly detailed discussion of viscosity, surface tension, and diffusion.

SECTION A

"The Structure of Liquids," J. D. Bernal, *Sci. Am.*, Aug. 1960.

"The Significant Structure Theory of Water," H. Eyring and M. S. Jhon, *Chemistry* **39**(9), 8 (1966).

"Liquid Crystals," J. L. Ferguson, *Sci. Am.*, Aug. 1964.

"Liquid Crystals," G. H. Brown, *Chemistry* **40**(9), 10 (1967).

"Liquid Crystals and Their Roles in Inanimate and Animate Systems," G. H. Brown, *Am. Sci.* **60**(1), 64 (1972).

"Liquid Crystals for Electro-optical Displays," G. Elliot, *Chem. Brit.* **9**, 213 (1973).

"Surface Tension in the Lungs," J. A. Clements, *Sci. Am.*, Dec. 1962.

"Life at Low Reynolds Number," E. M. Purcell, *Am. J. Phys.* **45**, 3 (1977).

"Molecular Volumes and the Stokes–Einstein Equation," J. T. Edward, *J. Chem. Educ.* **47**, 261 (1970).

SECTION B

"Significant Structure Theory of Liquids," H. Eyring and R. P. Marchi, *J. Chem. Educ.* **40**, 562 (1963).

"The Diffusion Coefficient of Sucrose in Water," P. W. Linder, L. R. Nassimbeni, A. Polson, and A. L. Rodgers, *J. Chem. Educ.* **53**, 330 (1976).

"Measurement and Interpretation of Diffusion Coefficients of Proteins," L. J. Gosting, *Advan. Protein Chem.* **11**, 429 (1956).

"A Quantitative Diffusion Experiment for Students," M. De Paz, *J. Chem. Educ.* **46**, 784 (1969). [Also see *J. Chem. Educ.* **47**, A204 (1970).]

"Three Liquid-Crystal Teaching Experiments," J. R. Lalanne and F. Hare, *J. Chem. Educ.* **53**, 793 (1976).

Problems

5.1 The viscosity of a gas increases with increasing temperature [see Eq. (3.18)], yet the viscosity of a liquid decreases with increasing temperature. Explain.

5.2 The viscosity of a liquid usually decreases with increasing temperature. An empirical equation is $\log \eta = A/T + B$. Determine the constants A and B for water from the following data:

T(K)	273.15	293.15	310.15	373.15
η (P)	0.01787	0.0101	0.00719	0.00283

5.3 At 293 K, the time of flow for water through an Ostwald viscometer is 342.5 s; for the same volume of an organic solvent, the time of flow is 271.4 s. Calculate the viscosity of the organic liquid relative to that of water. The density of the organic solvent is 0.984 g cm^{-3}.

5.4 For blood flowing in a capillary of radius 2.0×10^{-4} cm, estimate the maximum velocity for laminar flow at 37°C. (The density of whole blood is about 1.2 g cm^{-3}.)

5.5 An arteriole has a diameter of 2.4×10^{-5} m with blood flowing at 2.6×10^{-3} m s^{-1}. Calculate the pressure drop, ΔP, from one end to the other if the length of the arteriole is 5.0×10^{-3} m.

5.6 Water has an unusually large surface tension. Explain.

5.7 Give a molecular interpretation for the decrease in surface tension of a liquid with temperature.

5.8 A glass capillary of diameter 0.10 cm is dipped into (a) water (contact angle 10°) at 293 K and (b) mercury (contact angle 170°) at 298 K. Calculate the level of the liquid in the capillary in each case.

5.9 Both ethanol and mercury are used in thermometers. Explain the difference between the meniscus of the liquids in these two types of thermometers.

5.10 The surface tension of liquid naphthalene at 127°C is 28.8 dyn cm^{-1} and its density at this temperature is 0.96 g cm^{-3}. What is the radius of the largest capillary that will permit the liquid to rise 3.0 cm? Assume the angle of contact to be zero.

5.11 Two soap bubbles of radii r_1 and r_2 $(r_2 > r_1)$ are connected by a piece of tubing. Predict how the bubbles will change in size.

5.12 The surface tension of quinoline is twice that of acetone at 20°C. If the capillary rise is 2.5 cm for quinoline, what is the rise for acetone in the same capillary? Assume the angles of contact to be zero. The densities of quinoline and acetone at 20°C are 1.09 g cm^{-3} and 0.79 g cm^{-3}, respectively.

5.13 Swimming coaches sometimes suggest that a drop of alcohol placed in an ear plugged with water "draws out the water." Comment from a molecular point of view. [*Source:* "Eco-Chem," J. A. Campbell, *J. Chem. Educ.* **52**, 655 (1975).]

5.14 The diffusion coefficient of glucose is 5.7×10^{-10} m^2 s^{-1}. Calculate the time required for a glucose molecule to diffuse through (a) 10,000 Å, and (b) 0.1 m.

∗ 5.15 The diffusion coefficient of oxygen in air is 0.20 cm^2 s^{-1}; the diffusion coefficient of the same gas in water is about 10^4 times smaller. (a) Explain the huge difference in magnitude in these two cases. (b) Most cells in animals are bathed in fluids, so that a hemoglobin-type molecule and a circulatory system are necessary in order to transport O_2 to their cells and carry CO_2 away. (The diffusion coefficients of CO_2 in air and in water are of comparable magnitude to those of oxygen.) Since plants do not have a circulating system, explain how the O_2 and CO_2 gases are transported efficiently in these systems. (c) Insects do possess a circulating system but lack a hemoglobin-type molecule. In view of the diffusion coefficients of CO_2 and O_2 in water, do you think it likely that ants, bees, and cockroaches can grow to the size of human beings, as shown in horror movies?

5.16 The diffusion coefficient of sucrose in water at 298 K is 0.46×10^{-5} cm^2 s^{-1}, and the viscosity of water at the same temperature is 0.010 dyn s cm^{-2}. From these data estimate the effective radius of a sucrose molecule.

5.17 From the diffusion coefficients listed in Table 5.3, estimate the radius and molecular volume for myoglobin and hemoglobin. What conclusion can you draw from the results?

5.18 Diffusion coefficients in solids have been measured for a number of systems. If the diffusion coefficient of bismuth in lead is 1.1×10^{-16} cm^2 s^{-1} at 20°C, calculate how long it will take (in years) for a bismuth atom to travel by 1.0 cm.

5.19 The carbon monoxide–hemoglobin complex has a diffusion coefficient of 0.062×10^{-9} m^2 s^{-1} water at 298 K. In more viscous cytoplasm, the diffusion coefficient is only 0.013×10^{-9} m^2 s^{-1}. How long would it take for such a complex to travel the 3-μm length of a bacterial cell?

5.20 Show that the Reynolds number [see Eq. (5.3)] is dimensionless.

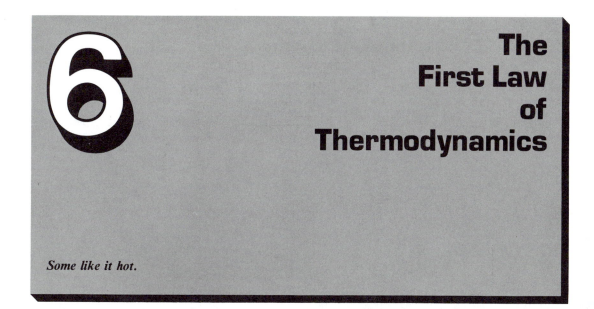

The First Law of Thermodynamics

6

Some like it hot.

Thermodynamics is the science of heat and temperature and, in particular, of the laws governing the conversion of heat into mechanical, electrical, or other forms of energy. It is a central branch of science with important applications in chemistry, physics, biology, and engineering. What makes thermodynamics such a powerful tool? It is a completely logical discipline and can be applied without any sophisticated mathematical techniques. The immense practical value of thermodynamics lies in the fact that it systemizes the information obtained from experiments performed on systems and enables us to draw conclusions about other aspects of the same systems, and similar aspects of other systems, without further experimentation. It allows us to predict whether a certain reaction may proceed and what maximum yield may be obtained.

Thermodynamics is a macroscopic science, concerning quantities such as pressure, temperature, and volume. Unlike quantum mechanics, thermodynamics is not based on any specific molecular models, and therefore it is unaffected by our changing concepts of atoms and molecules. Indeed, the major foundations of thermodynamics were laid long before detailed atomic theories became available. This fact is one of its major strengths. On the negative side, equations derived from laws of thermodynamics do not provide us with a molecular interpretation of complex phenomena. Furthermore, while thermodynamics helps us predict the direction and extent of chemical reactions, it says nothing about the *rate* of a process. The topic of chemical kinetics is dealt with in Chapter 13.

In this chapter we introduce the first law of thermodynamics and discuss some examples of thermochemistry.

6.1 Work and Heat

In classical mechanics, work is defined as force × distance. In thermodynamics, work becomes a more subtle concept, encompassing a broader range,

Table 6.1

Type of Work	Expression[a]	Meaning of Symbols
Mechanical work	$f\,dx$	f: force; dx: distance traveled
Surface work	$\gamma\,dA$	γ: surface tension; dA: change in area
Electrical work	$\mathscr{E}\,dQ$	$\mathscr{E}$: potential difference; dQ: electric charge
Gravitational work	$Mg\,dh$	M: mass; g: acceleration due to gravity; dh: change in height
Expansion work	$P\,dV$	P: pressure; dV: change in volume

[a] The work done in each case corresponds to an infinitesimal process, as indicated by the d symbol.

which includes surface work, electrical work, work of magnetization, and so on (Table 6.1). Let us consider a particularly important example of a system doing work—the expansion of a gas. A sample of a gas is placed in a cylinder fitted with a weightless and frictionless piston; then the entire apparatus is immersed in a thermostat at temperature T. The gas is allowed to expand from its initial state—P_1, V_1, T—to its final state—P_2, V_2, T—as shown in Figure 6.1. We assume that there is no atmospheric pressure present, so that the gas is only expanding against the weight of an object of mass M placed on the piston. The work done (W) in lifting the mass from the initial height h_1 to the final height h_2 is given by

$$W = -\text{force} \times \text{distance}$$
$$= -\text{mass} \times \text{acceleration} \times \text{distance}$$
$$= -Mg(h_2 - h_1)$$
$$= -Mg\,\Delta h \tag{6.1}$$

where g is the acceleration due to gravity and $\Delta h = h_2 - h_1$. From the units of M (kg), g (m s^{-2}), and Δh (m), we can readily show that W has the units kg m^2 s^{-2} or J. The minus sign in Eq. (6.1) has the following meaning. In an expansion, $h_2 > h_1$ and W is negative. This notation follows the now accepted convention that when a system does work on its surroundings, the work performed is a negative quantity. In a compression, Δh is negative, work is done on the system, and W becomes a positive quantity.

Figure 6.1

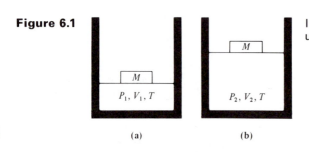

(a) (b)

Initial (a) and final (b) states of a gas undergoing an isothermal expansion.

so that

$$P_{ex} = \frac{Mg}{A}$$

or

$$W = -P_{ex} A \, \Delta h = -P_{ex}(V_2 - V_1)$$
$$= -P_{ex} \, \Delta V \tag{6.2}$$

where A is the area of the piston and the product $A \, \Delta h$ gives the change in volume. Equation (6.2) makes clear the fact that the amount of work done in expansion depends on the value of P_{ex}. Depending on experimental conditions, the amount of work performed by a gas in expanding from V_1 to V_2 at T can vary a great deal from one case to another. In one extreme we can have gas expanding against a vacuum (for example, if the mass M is removed from the piston). Since $P_{ex} = 0$, the work done, $-P_{ex} \, \Delta V$, is also zero. A more common arrangement is to have some mass resting on the piston so that the gas is expanding against a *constant* external pressure. As we saw earlier, the amount of work performed by the gas in this case is $-P_{ex} \, \Delta V$, where $P_{ex} \neq 0$. Note that as the gas expands, the pressure of the gas, P_{in}, decreases constantly. However, for the gas to expand, we must have $P_{in} > P_{ex}$ at every stage of expansion. For example, if initially $P_{in} = 5$ atm and the gas is expanding against a constant external pressure of 1 atm ($P_{ex} = 1$ atm), then the piston will finally come to a halt when P_{in} decreases to exactly 1 atm.

We now ask the question: Is it possible to have the gas perform a greater amount of work for the same increase in volume? Suppose that we now have an infinite number of identical weights which exert a total pressure on the piston of 5 atm. Since $P_{in} = P_{ex}$, the system is at equilibrium. Removing one such weight will decrease the external pressure by an infinitesimal amount so that $P_{in} > P_{ex}$ and there will be a very slight expansion until P_{in} is again equal to P_{ex}. When the second weight is removed, the gas expands a bit further and so on until enough weights have been lifted from the piston such that the external pressure becomes 1 atm. At this point we have completed the expansion process. How do we calculate the amount of work done in this case? At every stage of expansion (that is, when one such weight is lifted) the infinitesimal amount of work done is given by $-P_{ex} \, dV$, where dV is the infinitesimal increase in volume. The total work done in expanding from V_1 to V_2 is therefore

$$W = -\int_{V_1}^{V_2} P_{ex} \, dV \tag{6.3}$$

Since P_{ex} is no longer a constant value, the integral cannot be evaluated in this form.* However, we note that at every instant, P_{in} is only infinitesimally greater than P_{ex}, that is,

$$P_{in} - P_{ex} = dP$$

so that we can rewrite Eq. (6.3) as

$$W = -\int_{V_1}^{V_2} (P_{in} - dP) \, dV$$

---

* If P_{ex} were constant, this integral becomes $-P_{ex}(V_2 - V_1)$ or $-P_{ex} \, \Delta V$.

Realizing that $dP \, dV$ is a product of two infinitesimal quantities, we have $dP \, dV \simeq 0$ and write

$$W = -\int_{V_1}^{V_2} P_{in} \, dV \tag{6.4}$$

Equation (6.4) is a more manageable form, since P_{in} is the pressure of the system (that is, the gas) and we can express it in terms of a particular equation of state. For an ideal gas

$$P_{in} = \frac{nRT}{V}$$

so that

$$W = -\int_{V_1}^{V_2} \frac{nRT}{V} \, dV$$

$$= -nRT \ln \frac{V_2}{V_1} = -nRT \ln \frac{P_1}{P_2} \tag{6.5}$$

since $P_1 V_1 = P_2 V_2$.

Equation (6.5) looks quite different from our earlier expression for work done $(-P_{ex} \, \Delta V)$, and in fact it represents the *maximum* amount of work of expansion from V_1 to V_2. The reason for this is not difficult to see. Since in expansion the work performed is that against external pressure, we can maximize the work done by adjusting the external pressure so that it is only infinitesimally smaller than the internal pressure at every stage, as described above. An expansion carried out under these conditions is called a *reversible* process. By reversible we mean that if we increase the external pressure by an infinitesimal amount dP, we can bring the system to an equilibrium. A further increase in P_{ex} by dP would actually result in a compression. Thus the movement of the piston can go either way by a slight change in pressure.

A truly reversible process would take an infinite amount of time to complete,* and therefore it can never be realized in practice. We could, of course, arrange matters so that the gas does expand very slowly and try to approach reversibility, but we shall never be able to attain it. In the laboratory we must deal with *real* processes, which are always irreversible. The reason we are interested in a reversible process is that it enables us to calculate the *maximum* amount of work that could possibly be extracted. This quantity is important in estimating the efficiency of chemical and biological processes, as we shall see in Chapter 10.

EXAMPLE 6.1 A quantity of 0.850 mol of an ideal gas initially at a pressure of 15.0 atm and 300 K is allowed to expand isothermally until its final pressure is 1.00 atm. Calculate the work done if the expansion is carried out (a) against a vacuum, (b) against a constant external pressure of 1.00 atm, and (c) reversibly.

Answer

(a) Since $P_{ex} = 0$, $-P_{ex} \, \Delta V = 0$, so that no work is performed in this case.

(b) $W = -P_{ex}(V_2 - V_1)$.

---

* It would take an infinite amount of time to remove an infinite number of weights from the piston at the rate of one weight at a time.

Since $PV = nRT$, we have

$$V_1 = \frac{nRT}{P_1}, \qquad V_2 = \frac{nRT}{P_2}, \quad \text{and} \quad P_{ex} = P_2$$

CGS units: The CGS units of pressure and volume are dyn cm^{-2} and cm^3, respectively. In general, it is more convenient to express pressure in atmospheres and volume in liters. Although atmospheres and liters are not strictly CGS units (they are more appropriately called classical or historical units), we shall group them under the heading of CGS units. Thus we write R as 0.08206 liter atm K^{-1} mol^{-1}, P in atmospheres, and V in liters so that

$$W = -P_{ex}(V_2 - V_1) = -P_2(V_2 - V_1)$$

$$= -nRTP_2\left(\frac{1}{P_2} - \frac{1}{P_1}\right)$$

$$= -(0.850 \text{ mol})(0.08206 \text{ liter atm K}^{-1} \text{ mol}^{-1})$$

$$\times (300 \text{ K}) \times (1.00 \text{ atm}) \times \left(\frac{1}{1.00 \text{ atm}} - \frac{1}{15.0 \text{ atm}}\right)$$

$$= -19.5 \text{ liters atm}$$

$$= -1980 \text{ J}$$

Conversion factor: 1 liter atm = 101.34 J

SI units: Here we must use

$$R = 8.314 \text{ J K}^{-1} \text{ mol}^{-1}$$

$$1 \text{ atm} = 101{,}325 \text{ N m}^{-2}$$

and proceed as above.

(c)

$$W = -nRT \ln \frac{P_1}{P_2}$$

$$W = -(0.85 \text{ mol})(8.314 \text{ J K}^{-1} \text{ mol}^{-1})(300 \text{ K}) \ln \frac{15}{1}$$

$$= -5740 \text{ J}$$

As we can see, the largest amount of work performed is one carried out reversibly. Figure 6.2 shows graphically the work done for cases (b) and (c). In an irreversible process (Figure 6.2a), the amount of work done is given by $P_2(V_2 - V_1)$, which is the area under the curve. For a reversible process, the amount of work is also given by the area under the curve (Figure 6.2b); however, because the external pressure is no longer held constant, the area is considerably greater.

From the foregoing discussion we may draw several conclusions about work. We see that work should be thought of as a mode of energy transfer. Gas expands because there is a pressure difference. When the internal and external pressure are equalized, the word *work* is no longer applicable. Further, the amount of work done depends on how the process is carried out or the *path* (for example, reversible versus irreversible), even though the initial

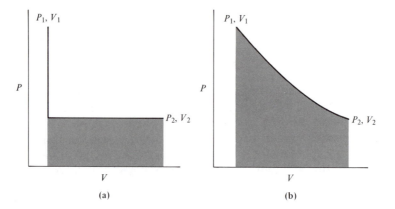

Figure 6.2 Isothermal gas expansion from P_1V_1 to P_2V_2. (a) An irreversible process. (b) A reversible process. In each case the shaded area represents the work done in expansion.

and final states are the same. Thus work is not a state function and we cannot say that a system has, within itself, so much work or work content.

An important property of state functions is that when the state* of a system is altered, a change in any state function depends only on the initial and final states of the system, not on how the change is accomplished. Let us assume that the change involves the expansion of a gas from an initial volume V_1 (2 liters) to a final volume V_2 (4 liters). The change or the increase in volume is given by

$$\Delta V = V_2 - V_1 = 2 \text{ liters}$$

Now the change can be brought about in many, in fact in an infinite number of, ways. We can let the gas expand directly from 2 liters to 4 liters as described above, or first allow it to expand to 6 liters and then compress the volume down to 4 liters, and so on. No matter how we carry out the process, the change in volume is always 2 liters.

Another mode of energy transfer involves the quantity heat. Like work, heat appears only at the boundary of the system and is defined by a process. Heat flows from a hotter object to a colder one because there is a temperature difference. When the temperatures of the two objects are equal, the word *heat* is no longer applicable. Heat is also not a state function and change in the quantity of heat from state 1 to 2 depends on the path taken. Suppose that we raise the temperature of 100 g of water initially at 20.0°C and 1 atm to 30.0°C and 1 atm. What is the heat change for this process? We do not know the answer because the process is not specified. One way to raise the temperature is to heat the water with a Bunsen burner or electrically using an immersion heater. The heat change Q (transferred from the surroundings to the system) is given by

$$Q = mC \, \Delta T$$
$$= (100 \text{ g})(4.184 \text{ J K}^{-1} \text{ g}^{-1})(10.0 \text{ K})$$
$$= 4184 \text{ J}$$

* See Section 2.1 for some of the terms used in thermodynamics.

where C is the specific heat of water. Alternatively, we can bring about the temperature change by doing mechanical work on the system; for example, by rubbing two small stones against each other in water until the desired temperature is reached. The heat change in this case is zero. Or we could first raise the temperature of water from 20.0°C to 25.0°C by direct heating and then do enough rubbing of stones to bring it up to 30°C. In this case Q is somewhere between zero and 4184 J. It is clear that there are, in fact, an infinite number of ways to increase the temperature of the system by the same amount, but the heat change in each case depends on the path of the process. Heat and work are measures of energy transfer, and both have the same units as energy. The conversion factor between the *thermochemical calorie* and the joule, which is the unit for work, is

$$1 \text{ cal} \equiv 4.184 \text{ J} \quad \text{exactly}$$

The First Law of Thermodynamics **6.2**

Since heat and work are not state functions, it is meaningless to ask how much heat or work a system possesses. On the other hand, the internal energy of a system is a state function, since it depends only on the thermodynamic coordinates of the state, such as temperature, pressure, and composition. Note that the adjective "internal" implies that other kinds of energy may be associated with the system. For example, the whole system may be in motion and therefore possess kinetic energy (K.E.). Further, the system may also possess potential energy (P.E.). Thus the total energy of the system E_{total} is given by

$$E_{total} = \text{K.E.} + \text{P.E.} + U$$

where U denotes internal energy. This internal energy consists of molecular translational, rotational, and vibrational energies; electronic energy; and nuclear energy. In most cases that we shall consider, the system will be at rest and external fields (for example, electric or magnetic fields) will not be present. Thus both K.E. and P.E. are zero and $E_{total} = U$. As mentioned earlier, thermodynamics is not based on any particular model, and therefore there is no need for us to know the exact nature of U. In fact, we have no way to calculate this quantity accurately. All we are interested in, as we shall see below, are methods for measuring the *change* in U for a process. For simplicity, we shall now refer to internal energy as energy and write its change as ΔU.

Energy differs from both heat and work in that it always changes by the same amount in going from one state to another, irrespective of the nature of the path. According to the first law of thermodynamics, we write

$$\Delta U = Q + W \tag{6.6}$$

or, for an infinitesimal change,

$$dU = dQ + dW \tag{6.7}$$

In words, Eqs. (6.6) and (6.7) tell us that an increase in the energy of a system is equal to the heat absorbed by the system from its surroundings plus the work done on the system by its surroundings. The sign convention for Q and W are summarized in Table 6.2. Note that we have deliberately omitted the Δ

Table 6.2

Process	Sign
Work done by the system on the surroundings	−
Work done on the system by the surroundings	+
Heat absorbed by the system from the surroundings (endothermic process)	+
Heat absorbed by the surroundings from the system (exothermic process)	−

(delta) sign for Q and W, since this notation represents the difference between the final and initial states and is therefore not applicable for heat and work, which are not state functions. Similarly, while dU is an *exact differential* (see Appendix 2 at the end of the book), that is, an integral of the type $\int_1^2 dU$ is independent of the path, we have employed the $đ$ notation to remind us that $đQ$ and $đW$ are *inexact differentials*, and, therefore, path-dependent.

The first law of thermodynamics is a law of conservation of energy; its formulation is based on our vast experience in the study of relationships between different forms of energies. Conceptually, the first law is quite easy to comprehend, and it can be readily applied to any practical system. Consider, for example, the thermochemical changes in a constant-volume adiabatic bomb calorimeter (Figure 6.3). This device allows us to measure the com-

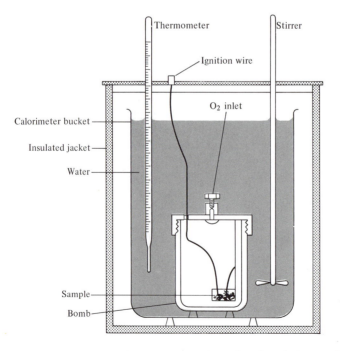

Figure 6.3 Schematic diagram of a constant-volume bomb calorimeter. The entire apparatus is immersed in a water bath whose temperature is adjusted to be the same as that of the water surrounding the bomb calorimeter throughout the experiment.

bustion heat of substances. It is a tightly sealed, heavy-walled, stainless steel container, thermally isolated from its surroundings. (The temperature of water surrounding the air jacket is adjusted to be the same as that of the water surrounding the bomb calorimeter throughout the experiment. In this manner, no heat will flow into or out of the inner jacket and hence the description *adiabatic*, or no heat exchange.) The substance under investigation is placed inside the container, which is filled with oxygen at about 30 atm. The combustion is started by an electrical discharge through a pair of wires making contact with the substance. The heat released can then be measured by registering the rise in the temperature of water filling the inner jacket of the calorimeter. From a knowledge of the specific heat of the calorimeter, we can calculate the change in energy as a result of combustion ΔU, given by

$$\Delta U = Q_V + W = Q_V - P\,\Delta V = Q_V \qquad (6.8)$$

The negative sign for $P\,\Delta V$ has the following meaning. If the work is expansion, then $\Delta V > 0$ and W is a negative quantity; for a compression process, $\Delta V < 0$ and W is positive, in accordance with our convention. However, in our arrangement, the volume is kept constant so that $\Delta V = 0$, $P\,\Delta V = 0$, and $\Delta U = Q_V$. We have used the subscript V on Q to remind us of this condition. Equation (6.8) may seem strange at first; ΔU is equated to the heat released, which, as we said earlier, is not a function of state. However, we have restricted ourselves to a particular process or path, that is, the one that takes place at constant volume; hence Q_V can have only one value for a given amount of the particular compound.

Enthalpy **6.3**

In the laboratory, most chemical reactions are carried out at constant pressure rather than constant volume. In such cases we write*

$$\Delta U = Q_P + W = Q_P - P\,\Delta V$$

or

$$U_2 - U_1 = Q_P - P(V_2 - V_1)$$

where the subscript P denotes the constant-pressure process. Rearrangement of the equation above gives

$$Q_P = (U_2 + PV_2) - (U_1 + PV_1) \qquad (6.9)$$

We *define* a function called enthalpy (H) as follows:

$$H = U + PV \qquad (6.10)$$

where U, P, and V are the energy, pressure, and volume of the system. Like U, P, and V, enthalpy is a function of state, because it is defined in terms of quantities that are all functions of state; it also has the units of energy. From Eq. (6.10), we can now write

$$\Delta H = H_2 - H_1 = (U_2 + P_2 V_2) - (U_1 + P_1 V_1)$$

* Unless otherwise stated, we shall assume that all work done is of the P–V type.

Setting $P_2 = P_1 = P$ for a constant-pressure process, we obtain

$$\Delta H = (U_2 + PV_2) - (U_1 + PV_1) = Q_P$$

Again we have restricted the change to take place along a specific path—this time at constant pressure—so that the heat change Q_P can be directly related to the change in the state function ΔH.

In general, when a system undergoes a change from state 1 to 2, the change in enthalpy is given by

$$\Delta H = \Delta U + \Delta(PV)$$
$$= \Delta U + P\,\Delta V + V\,\Delta P + \Delta P\,\Delta V \qquad (6.11)$$

This equation applies if neither pressure nor volume is kept constant. The last term, $\Delta P\,\Delta V$, is not negligible.* It is important to keep in mind that both P and V in Eq. (6.11) refer to the system. If the change is carried out, say, at constant atmospheric pressure and if the pressure exerted by the system on the surroundings (P_{in}) is equal to the pressure exerted by the surroundings on the system, that is,

$$P_{in} = P_{ext} = P$$

then we have $\Delta P = 0$ and Eq. (6.11) now becomes

$$\Delta H = \Delta U + P\,\Delta V$$

Similarly, for an infinitesimal change that occurs under constant-pressure conditions with the external pressure equal to the internal pressure,

$$dH = dU + P\,dV$$

The difference between ΔH and ΔU for a constant-pressure process involving chemical reaction of ideal gases is given by $\Delta(PV) = P\,\Delta V$ or $\Delta(PV) = \Delta(nRT) = RT\,\Delta n$ (if T is kept constant). As an estimate, let $\Delta n = 1$ and $T = 300$ K; this gives a value of 2.5 kJ mol^{-1} for the difference between ΔH and ΔU. This is a small but not negligible quantity in accurate work. On the other hand, for chemical reactions occurring in the condensed phases, ΔV is usually a small number (about 0.1 liter per mole of reactant converted to product or less) so that $P\,\Delta V = 0.1$ liter atm mol^{-1} or 10 J mol^{-1} (we assume that $P = 1$ atm), which can be neglected in comparison with ΔU and ΔH, which are of the order of 100 kJ mol^{-1} or greater. Thus changes in enthalpy and energy in solution for chemical reactions are one and the same for all practical purposes.

EXAMPLE 6.2 A quantity of 0.2590 g of benzoic acid was burned in a constant-volume bomb calorimeter. Consequently, the temperature of the water in the inner jacket (see Figure 6.3) rose from 20.17°C to 22.22°C. If the effective heat capacity of the bomb calorimeter plus water is 5267.8 J K^{-1}, calculate ΔU and ΔH for the combustion of benzoic acid in kJ mol^{-1}.

Answer: The reaction is

$$2C_6H_5COOH(s) + 15O_2(g) \longrightarrow 14CO_2(g) + 6H_2O(l)$$

* Note that for an infinitesimal change, we write $dH = dU + d(PV) = dU + P\,dV + V\,dp + dP\,dV$. Since $dP\,dV$ is negligibly small, we have $dH = dU + P\,dV + V\,dP$.

Since this is a constant-volume process,

$$\Delta U = Q_V + W$$
$$= Q_V - P \, \Delta V$$
$$= Q_V$$

The amount of heat evolved is given by

$$(2.05 \text{ K})(5267.8 \text{ J K}^{-1}) = 10.80 \text{ kJ}$$

From the molar mass of benzoic acid (122.12 g), we write

$$Q_V = \Delta U = - \frac{(10.80 \text{ kJ})(122.12 \text{ g mol}^{-1})}{0.2590 \text{ g}}$$

$$= -5092 \text{ kJ mol}^{-1}$$

The negative sign incorporated here denotes the exothermic nature of the reaction.

To calculate ΔH, we start by writing $\Delta H = \Delta U + \Delta(PV)$. When all reactants and products are condensed phases, the $\Delta(PV)$ term is negligible in comparison with ΔH and ΔU. When gases are involved, the $\Delta(PV)$ term cannot be neglected. Assuming ideal gas behavior, we have $\Delta(PV) = \Delta(nRT) = RT \, \Delta n$, where Δn is the change in the number of moles of gas in the reaction. Note that the temperature T here refers to the *initial* temperature, since we are comparing reactants and products under the same conditions. For our reaction here $\Delta n = -1$ so that

$$\Delta H = \Delta U + RT \, \Delta n$$

$$= -5092 \text{ kJ mol}^{-1} + \frac{(8.314 \text{ kJ K}^{-1} \text{ mol}^{-1})}{1000}$$

$$\times (293.32 \text{ K})(-1)$$

$$= -5094 \text{ kJ mol}^{-1}$$

Comments: (1) We see that the difference between ΔU and ΔH is quite small for this reaction. The reason is that the $\Delta(PV)$ term (which in this case is equal to ΔnRT) is small compared to either ΔU or ΔH. Since the reaction involves ideal gases only (we neglect the volume change of condensed phases), ΔU has the same value $(-5092 \text{ kJ mol}^{-1})$ whether the process is carried out at constant V or constant P. Similarly, $\Delta H = -5094 \text{ kJ mol}^{-1}$ whether the process is carried out at constant V or constant P. The heat change, however, is $-5092 \text{ kJ mol}^{-1}$ at constant V and $-5094 \text{ kJ mol}^{-1}$ at constant P, because it is path-dependent. (2) In our calculation we have ignored the heat capacities of water and carbon dioxide. Because the amounts of these substances formed are small compared to the bomb calorimeter itself, this omission does not introduce any serious errors.

EXAMPLE 6.3 Compare the difference between ΔH and ΔU for the following physical changes: (a) 1 mol ice $\rightarrow$ 1 mol water at 273 K and 1 atm and (b) 1 mol water $\rightarrow$ 1 mol steam at 373 K and 1 atm. The molar volumes of ice and water at 273 K are 0.0196 liter mol^{-1} and 0.0180 liter

mol^{-1}, respectively, and the molar volumes of water and steam at 373 K
are 0.0188 liter mol^{-1} and 30.61 liters mol^{-1}, respectively.

Answer: Since in both cases we have a constant-pressure process

$$\Delta H = \Delta U + \Delta(PV)$$

$$= \Delta U + P\,\Delta V$$

or

$$\Delta H - \Delta U = P\,\Delta V$$

(a)

$$\Delta V = \overline{V}(\text{l}) - \overline{V}(\text{s})$$

$$= (0.0180 - 0.0196) \text{ liter mol}^{-1}$$

$$= -0.0016 \text{ liter mol}^{-1}$$

Hence

$$P\,\Delta V = (1 \text{ atm})(-0.0016 \text{ liter mol}^{-1})$$

$$= -0.0016 \text{ liter atm mol}^{-1}$$

$$= -0.16 \text{ J mol}^{-1}$$

(b) Since

$$\Delta V = \overline{V}(\text{g}) - \overline{V}(\text{l})$$

$$= (30.61 - 0.0188) \text{ liter mol}^{-1}$$

$$= 30.59 \text{ liters mol}^{-1}$$

Hence

$$P\,\Delta V = (1 \text{ atm})(30.59 \text{ liters mol}^{-1})$$

$$= 30.59 \text{ liters atm mol}^{-1}$$

$$= 3100 \text{ J mol}^{-1}$$

Comment: This example clearly shows that $(\Delta H - \Delta U)$ is negligibly small
for condensed phases but can be quite appreciable if the process involves
gases. Further, in (a) we have $\Delta U > \Delta H$ because when ice melts, there is a
decrease in volume. Consequently, work is done on the system by the
surroundings. The opposite situation holds for (b) because in this case
steam is doing work on the surroundings.

6.4 Heat Capacities

The importance of ΔH is that, for a constant-pressure process, it is equal to
the heat change, Q_P. As we can well imagine, most chemical reactions are
accompanied by either evolution or absorption of heat. The ΔH for any such
process can be calculated from the temperature change and the *heat capacities*
of the reactants and products. This section introduces two important heat
capacities (at constant volume and at constant pressure) and their
applications.

When heat is added to a substance, its temperature will rise. This fact we know too well. But just how much the temperature will rise depends on (1) the amount of heat delivered, (2) the amount of the substance present, (3) the chemical nature and physical state of the substance, and (4) the conditions under which heat is added to the substance. In general, the temperature rise ΔT for a given amount of substance is directly proportional to the heat added:

$$\Delta T \propto Q$$
$$= C'Q$$

or

$$C = \frac{Q}{\Delta T} \quad \text{J K}^{-1} \tag{6.12}$$

where $C(=1/C')$, a proportionality constant, is called the heat capacity. Because the increase in temperature depends on the amount of substance present, it is often convenient to speak of the heat capacity of 1 mol of a substance or molar heat capacity $\overline{C}$, where

$$\overline{C} = \frac{C}{n} = \frac{Q}{n \, \Delta T} \quad \text{J K}^{-1} \text{ mol}^{-1} \tag{6.13}$$

where n is the number of moles of the substance present in a particular measurement.

Heat capacities vary greatly from substance to substance. Liquid water, for example, has a large heat capacity (75.3 J K^{-1} mol^{-1}), so that it requires more heat to raise the temperature of 1 mol of water by 1 degree than it would for 1 mol of copper (molar heat capacity: 24.47 J K^{-1} mol^{-1}) under similar conditions. Heat capacity is a directly measurable quantity. Knowing the amount of the substance present, the heat added, and the temperature rise, we can readily calculate the value $\overline{C}$ using Eq. (6.13). However, it turns out that the value we calculate also depends on how this amount of heat is added to the substance. Although many different conditions can be realized in practice, we shall consider only two important cases here: constant volume and constant pressure. We have already seen that for a constant-volume process, the heat absorbed by the system is equal to the increase in internal energy, that is, $\Delta U = Q_V$. Hence the heat capacity at constant volume, C_V, is given by*

$$C_V = \frac{Q_V}{\Delta T} = \frac{\Delta U}{\Delta T}$$

or, expressed in partial derivatives,

$$C_V = \left(\frac{\partial U}{\partial T}\right)_V \tag{6.14}$$

Similarly, for a constant-pressure process we have $\Delta H = Q_P$ so that the heat capacity at constant pressure is

$$C_P = \frac{Q_P}{\Delta T} = \frac{\Delta H}{\Delta T}$$

---

* This quantity was introduced in Section 3.8.

or, expressed in partial derivatives,

$$C_P = \left(\frac{\partial H}{\partial T}\right)_P \tag{6.15}$$

Because many processes we encounter in thermodynamics involve heat input to or withdrawn from a system, accompanied by either a rise or fall in temperature, a knowledge of heat capacities of the substances involved is essential for calculating changes in thermodynamic quantities. Integrating Eqs. (6.14) and (6.15) between temperatures T_1 and T_2, we obtain

$$\Delta U = \int_{T_1}^{T_2} C_V \, dT = C_V(T_2 - T_1) = C_V \, \Delta T \tag{6.16}$$

$$\Delta H = \int_{T_1}^{T_2} C_P \, dT = C_P(T_2 - T_1) = C_P \, \Delta T \tag{6.17}$$

We have assumed here that both C_V and C_P are independent of temperature. This is not always true. In Section 3.8 we saw that there are several different types of contributions to heat capacity (translational motion as well as molecular rotation and vibration). At low temperatures only the translational and rotational motions make a major contribution. At high temperatures when transitions among vibrational energy levels become appreciable, the heat capacity will increase accordingly. In general, the heat capacity of a substance is a function of temperature; it increases with increasing temperature. If the temperature change in a process is small, say 50 K or less, we can often treat C_V and C_P as if they were independent of temperature.

The heat capacities at constant volume and constant pressure for a given substance are generally not equal to each other. Since work has to be done by the system on the surroundings in a constant-pressure process, *more* heat is required to raise the temperature by a definite amount in a constant-pressure process than that in a constant-volume process. It follows, therefore, that $C_P > C_V$. The difference between C_P and C_V is obviously greater for gases than for either liquids or solids, because the volume of a condensed phase does not change appreciably with temperature; consequently, the work done on expansion is quite small. Let us now evaluate the quantity $(C_P - C_V)$ for n moles of an ideal gas. We start by writing

$$H = U + PV = U + nRT$$

When the temperature of the system is raised from T to $T + dT$, the corresponding enthalpy change, dH, is given by

$$dH = dU + d(nRT)$$
$$= dU + nR \, dT$$

Since $dH = C_P \, dT$ and $dU = C_V \, dT$, we write

$$C_P \, dT = C_V \, dT + nR \, dT$$
$$C_P = C_V + nR$$
$$C_P - C_V = nR \tag{6.18a}$$

or

$$\overline{C}_P - \overline{C}_V = R \tag{6.18b}$$

Thus, for an ideal gas, the molar constant-pressure heat capacity is greater than the molar constant-volume heat capacity by R, the gas constant.

EXAMPLE 6.4 Calculate ΔU and ΔH in heating 55.40 g of Xe from 300 K to 400 K. Assume the heat capacities at constant volume and constant pressure to be independent of temperature and ideal gas behavior.

Answer: Xenon is a monatomic gas. From Section 3.8 we observe that $\overline{C}_V = \frac{3}{2}R = 12.47$ J K^{-1} mol^{-1}. Thus from Eq. (6.18b), we have $\overline{C}_P = \frac{3}{2}R + R = \frac{5}{2}R = 20.79$ J K^{-1} mol^{-1}. The quantity 55.40 g of Xe corresponds to 0.4219 mol. From Eqs. (6.16) and (6.17),

$$\Delta U = C_V \, \Delta T = n\overline{C}_V \, \Delta T$$

$$= (0.4219 \text{ mol})(12.47 \text{ J K}^{-1} \text{ mol}^{-1})(100 \text{ K})$$

$$= 526 \text{ J}$$

$$\Delta H = C_P \, \Delta T = n\overline{C}_P \, \Delta T$$

$$= (0.4219 \text{ mol})(20.79 \text{ J K}^{-1} \text{ mol}^{-1})(100 \text{ K})$$

$$= 877 \text{ J}$$

EXAMPLE 6.5 The heat capacity at constant pressure of copper is given by $\overline{C}_P = (22.65 + 6.30 \times 10^{-3} \, T)$ J K^{-1} mol^{-1}. Calculate the enthalpy change when 1 mol of copper is heated from 300 K to 400 K.

Answer: From Eq. (6.17),

$$\Delta H = \int_{T_1}^{T_2} n\overline{C}_P \, dT = \int_{300 \text{ K}}^{400 \text{ K}} (22.65 + 6.30 \times 10^{-3} T) \, dT$$

$$= \left[22.65T + \frac{6.30 \times 10^{-3}T^2}{2} \right]_{300}^{400}$$

$$= 2485 \text{ J}$$

Ideal Gas Expansions 6.5

Up to this point we have introduced the quantities work, heat, energy, and enthalpy. To see how these quantities are applied to a simple process, let us consider the expansion of an ideal gas. Although ideal gas expansions do not have much biological significance, it is instructive to see how some of the equations derived in the previous sections are employed to calculate changes in thermodynamic quantities. We consider the following two special cases.

ISOTHERMAL EXPANSION

The work done in isothermal reversible and irreversible expansions has been discussed in some detail in Section 6.1 and will not be repeated here. Since the temperature is kept constant in an isothermal process, the change in energy is zero, that is, $\Delta U = 0$. This follows from the fact that in an ideal gas molecules do not attract or repel one another. Consequently, their total energy is independent of the distance of separation and therefore the volume. Mathematically, this is expressed as

$$\left(\frac{\partial U}{\partial V}\right)_T = 0$$

At constant temperature, the energy of the ideal gas is independent of its volume. From Eq. (6.6),

$$\Delta U = Q + W = 0$$

or

$$Q = -W$$

Thus, in an isothermal expansion, the heat absorbed by the gas is equal to the work done by the ideal gas on its surroundings; that is, $Q = -W$. Referring to Example 6.1, we see that the heat absorbed by the ideal gas in expanding from a pressure of 15 atm to 1 atm at 300 K is zero in (a), 1980 J in (b), and 5740 J in (c).

Finally, we also need to calculate the enthalpy change for such an isothermal process. Starting from

$$\Delta H = \Delta U + \Delta(PV)$$

we see that $\Delta U = 0$ as mentioned above, and since PV is a constant at constant T and n (Boyle's law), $\Delta(PV) = 0$, so that $\Delta H = 0$. Alternatively, we could write $\Delta(PV) = \Delta(nRT)$. Since the temperature is unchanged and no chemical reaction occurs, both n and T are constant and $\Delta(nRT) = 0$; hence $\Delta H = 0$.

ADIABATIC EXPANSION

Suppose that we now isolate the cylinder (see Figure 6.1) thermally from its surroundings so that there is no heat exchange during the expansion. For such an adiabatic process we have $Q = 0$. Consequently, there will be a temperature drop and T will no longer be a constant, as in the isothermal expansion case. Let us first suppose that the expansion is reversible. The two questions we ask are: What are the P–V relations between the initial and final states, and how much work is done in the expansion?

For an infinitesimal adiabatic expansion, the first law takes the form

$$dU = dQ + dW$$

$$= dW = -P\,dV = \frac{-nRT}{V}\,dV$$

or

$$\frac{dU}{nT} = \frac{-R\,dV}{V}$$

Note that $dQ = 0$ and we have replaced the external, opposing pressure with the internal pressure of the gas, since we are dealing with a reversible process. Since

$$C_V = \left(\frac{\partial U}{\partial T}\right)_V$$

so that

$$dU = C_V \, dT$$

Substituting this relation in the equation above, we obtain

$$\frac{C_V \, dT}{nT} = \overline{C}_V \frac{dT}{T} = -R \frac{dV}{V} \tag{6.19}$$

Integrating Eq. (6.19) between the initial and final states, we obtain

$$\int_{T_1}^{T_2} \overline{C}_V \frac{dT}{T} = -R \int_{V_1}^{V_2} \frac{dV}{V}$$

$$\overline{C}_V \ln \frac{T_2}{T_1} = R \ln \frac{V_1}{V_2}$$

Since $\overline{C}_P - \overline{C}_V = R$ for an ideal gas, we have

$$\overline{C}_V \ln \frac{T_2}{T_1} = (\overline{C}_P - \overline{C}_V) \ln \frac{V_1}{V_2}$$

Dividing by $\overline{C}_V$ on both sides, we obtain

$$\ln \frac{T_2}{T_1} = \left(\frac{\overline{C}_P}{\overline{C}_V} - 1 \right) \ln \frac{V_1}{V_2}$$

$$= (\gamma - 1) \ln \frac{V_1}{V_2}$$

where

$$\gamma = \frac{\overline{C}_P}{\overline{C}_V} \tag{6.20}$$

For a monatomic gas, $\overline{C}_V = \frac{3}{2}R$ (see Section 3.8) and $\overline{C}_P = \frac{5}{2}R$, so that $\gamma = \frac{5}{3}$, or 1.67. For diatomic molecules, we have $\overline{C}_V = \frac{5}{2}R$ and $\overline{C}_P = \frac{7}{2}R$, so that $\gamma = \frac{7}{5}$, or 1.4.* Finally, we arrive at the following useful results:

$$\left(\frac{V_1}{V_2} \right)^{\gamma-1} = \frac{T_2}{T_1} = \frac{P_2 V_2}{P_1 V_1}$$

or

$$\left(\frac{V_1}{V_2} \right)^{\gamma} = \frac{P_2}{P_1}$$

Thus, for an adiabatic process, the P–V relation becomes

$$P_1 V_1^{\gamma} = P_2 V_2^{\gamma} \tag{6.21}$$

It is useful to keep in mind the conditions under which this equation was derived: (1) It applies to an ideal gas, and (2) it applies to a reversible adiabatic change. Equation (6.21) differs from Boyle's law ($P_1 V_1 = P_2 V_2$) in the exponent γ, since temperature is *not* kept constant in an adiabatic expansion.

---

* We assume that the principle of equipartition of energy to hold in this case. This is usually not strictly correct (see Table 3.3), but the value can be used in approximate cases.

The work done in this case is given by

$$W = \int_1^2 dU = \Delta U = \int_{T_1}^{T_2} C_V \, dT$$

$$= C_V(T_2 - T_1) \tag{6.22}$$

where $T_2 < T_1$, since the gas expands during the process.*

EXAMPLE 6.6 Referring to Example 6.1(c), how much work is done in an adiabatic expansion if the process is carried out reversibly?

Answer: Our first task is to calculate the final temperature T_2. This is done in three steps. First, we need to evaluate V_1, given by $V_1 = nRT_1/P_1$.

CGS units:

$$P_1 = 15 \text{ atm}$$

$$R = 0.08206 \text{ liter atm K}^{-1} \text{ mol}^{-1}$$

Thus

$$V_1 = \frac{(0.850 \text{ mol})(0.08206 \text{ liter atm K}^{-1} \text{ mol}^{-1})(300 \text{ K})}{15.0 \text{ atm}}$$

$$= 1.40 \text{ liters}$$

Next we calculate V_2 using the following relation:

$$P_1 V_1^\gamma = P_2 V_2^\gamma$$

$$V_2 = \left(\frac{P_1}{P_2}\right)^{1/\gamma} V_1 = \left(\frac{15}{1}\right)^{3/5} (1.40 \text{ liters}) = 7.1 \text{ liters}$$

We assume the gas to be monatomic, so that $\gamma = \frac{5}{3}$.
Finally, we have

$$T_2 = \frac{P_2 V_2}{nR} = \frac{(1 \text{ atm})(7.1 \text{ liters})}{(0.850 \text{ mol})(0.08206 \text{ liter atm K}^{-1} \text{ mol}^{-1})}$$

$$= 102 \text{ K}$$

Hence

$$\Delta U = W = n\overline{C}_V(T_2 - T_1)$$

$$= (0.850 \text{ mol})(12.47 \text{ J K}^{-1} \text{ mol}^{-1})(102 - 300) \text{ K}$$

$$= -2100 \text{ J}$$

SI Units: Here we must express

$$P_1 = 15 \times 101325 \text{ N m}^{-2}$$

$$R = 8.314 \text{ J K}^{-1} \text{ mol}^{-1}$$

and proceed as above.

---

* It may seem strange that the quantity C_V appears in Eq. (6.22), since the volume is not held constant. However, the adiabatic expansion (from $P_1V_1T_1$ to $P_2V_2T_2$) can be imagined to take place in two steps. First, the gas is isothermally expanded from P_1V_1 to P_2V_2 at T_1. Since temperature is constant, $\Delta U = 0$. Next the gas is cooled at constant volume from T_1 to T_2. In this case we have $\Delta U = C_V(T_2 - T_1)$ which is Eq. (6.22).

Figure 6.4

P–V plots of an adiabatic, reversible, and an isothermal, reversible expansion of an ideal gas. In each case, the work done in expansion is represented by the area under the curve.

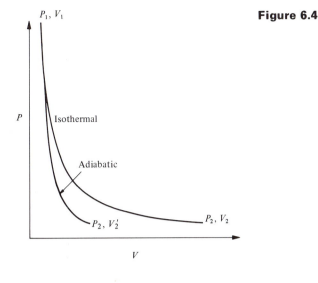

Examples 6.1 and 6.6 show that less work is performed in an adiabatic expansion. During an isothermal expansion, heat is absorbed from the surroundings to make up for the work done by the gas (we can think of this as foreign aid), but this does not occur for the adiabatic process so that the temperature drops. Plots of isothermal and adiabatic reversible expansions are shown in Figure 6.4.

Adiabatic irreversible expansion is more complicated and will not be dealt with here (see Problem 6.19). The decrease in temperature or the cooling effect as a result of an adiabatic expansion has some interesting practical consequences. No doubt we are all familiar with the formation of fog when the caps of soft drinks or corks of champagne bottles are removed. Initially, the bottles are pressurized with carbon dioxide and air, and the space above the liquid is saturated with water vapor. When the cap is removed, the gases inside undergo an expansion which takes place so rapidly that it can be approximated as adiabatic. As a result, there is a temperature drop and water vapor condenses to form the observed fog.*

Liquefaction of gases is also based on the same principle. Normally, the steps are (1) compress a gas isothermally, (2) let the compressed gas expand adiabatically, (3) recompress the cooled gas isothermally, and so on, until condensation from gas to liquid occurs.

Thermochemistry **6.6**

STANDARD ENTHALPY OF FORMATION

Chemical reactions almost always involve some changes in heat. The *heat of reaction* can be defined as the heat withdrawn from the surroundings in the transformation of reactants at some temperature and pressure to products at the same temperature and pressure. For a constant-pressure process, the heat of reaction Q_P is equal to the enthalpy change ΔH. An *exothermic* reaction is

---

* For other interesting illustrations of adiabatic cooling, see R. C. Plumb, *J. Chem. Educ.* **47**, 176 (1970), and P. E. Stevenson, *J. Chem. Educ.* **47**, 272 (1970).

a process that gives off heat to its surroundings and ΔH is negative; an *endothermic* reaction absorbs heat and ΔH is positive. Consider the following reaction:

$$C(\text{graphite}) + O_2(g) \longrightarrow CO_2(g)$$

Experimentally, we find that when the reaction is carried out at 1 atm and 298 K, the formation of 1 mol of gaseous CO_2 from 1 mol of graphite and 1 mol of gaseous O_2 gives off 393.51 kJ of heat, or $\Delta H = -393.51$ kJ. The quantity ΔH can be expressed as

$$\Delta H = (1 \text{ mol})\overline{H}_{CO_2} - (1 \text{ mol})\overline{H}_{\text{graphite}} - (1 \text{ mol})\overline{H}_{O_2}$$

where $\overline{H}_{CO_2}$, $\overline{H}_{\text{graphite}}$, and $\overline{H}_{O_2}$ are the molar enthalpies of the individual components at the same temperature and pressure.

If it were possible to have available the molar enthalpy of all the elements and their compounds, we would be able to calculate the heat of reactions without having actually to perform an experiment. This, unfortunately, we cannot do, since there is no known way of measuring or calculating the *absolute* values of molar enthalpies. Except for the quantity entropy (to be discussed in Chapter 7), in thermodynamics we are interested only in the changes of U and H rather than in their absolute values. A convenient way of dealing with this problem is to arbitrarily assign a zero value for every element in its most stable state of aggregation at 1 atm and 298 K (25°C).* The choice of 1 atm defines what we call the *standard-state* condition. On the other hand, the choice of 298 K has nothing to do with the standard state. We could, for example, easily have chosen the combination of 1 atm and 310 K (37°C) and assign zero molar enthalpies for stable elements under these conditions. Because 298 K corresponds to room temperature, chemists have decided to tabulate the molar enthalpies of elements and compounds at this temperature. From our definition, then, we can write that at 298 K,

$$\overline{H}^{\circ}_{O_2} = 0$$

$$\overline{H}^{\circ}_{\text{graphite}} = 0$$

Since neither ozone nor diamond is the more stable allotropic form at 1 atm and 298 K, we have

$$\overline{H}^{\circ}_{O_3} \neq 0$$

$$\overline{H}^{\circ}_{\text{diamond}} \neq 0$$

where the superscript ° denotes the standard state. Returning to our earlier example of combustion of graphite in oxygen, the standard enthalpy of reaction $\Delta H°$ is given by

$$\Delta H° = (1 \text{ mol})\overline{H}^{\circ}_{CO_2} - (1 \text{ mol})\overline{H}^{\circ}_{\text{graphite}} - (1 \text{ mol})\overline{H}^{\circ}_{O_2}$$

$$= (1 \text{ mol})\overline{H}^{\circ}_{CO_2} = -393.51 \text{ kJ}$$

What this relation says is that the standard enthalpy of reaction is equal to the molar enthalpy of CO_2. Since this reaction also represents the formation of CO_2 from its elements [O_2 and C(graphite)], it follows that

$$\overline{H}^{\circ}_{CO_2} = \Delta \overline{H}^{\circ}_f(CO_2)$$

* Strictly it should be 298.15 K. However, the difference, 0.15 K, is quite small and is therefore omitted for simplicity.

where $\Delta \overline{H}_f^\circ$ denotes the standard molar enthalpy of formation of a compound. A note on the units for ΔH° and $\Delta \overline{H}_f^\circ$ is helpful here. In words, the following equation

$$C(\text{graphite}) + O_2(g) \longrightarrow CO_2(g) \qquad \Delta H^\circ = -393.51 \text{ kJ}$$

says that when 1 mol of graphite and 1 mol of gaseous oxygen in their standard states at 298 K are converted to 1 mol of gaseous carbon dioxide in its standard state at 298 K, 393.51 kJ of heat is released. Therefore, the standard enthalpy of reaction is given by $\Delta H^\circ = -393.51$ kJ. On the other hand, $\Delta \overline{H}_f^\circ$ is expressed in kJ mol^{-1}. Thus we have $\Delta \overline{H}_f^\circ(CO_2) = -393.51$ kJ mol^{-1}. The enthalpies of formation of a number of compounds are tabulated in Appendix 3 at the end of the book. Generally, compounds having large negative $\Delta \overline{H}_f^\circ$ values (called exothermic compounds) tend to be more stable than those having large positive $\Delta \overline{H}_f^\circ$ values (called endothermic compounds). The reason is that energy has to be supplied to the former compounds in order to decompose them into elements while the latter compounds decompose with the evolution of heat (Figure 6.5).

There is really no mystery regarding the conventions about zero molar enthalpy for elements in the standard state (and 298 K.) Our choice of zero here is just as arbitrary (and as useful) as choosing sea level as the zero point for terrestrial altitudes. Two questions might arise from this procedure. First, why use zero? The answer to this is that we constantly add and subtract enthalpies—zero makes this task easy. The second question is that since we know very well that every element has a nonzero molar enthalpy, are we introducing serious errors by this assumption? No, because in an ordinary chemical reaction, the same number of elements appear on both sides of a chemical equation, so that whatever their molar enthalpies may be, they are canceled out when we calculate the enthalpy of a reaction. In general, the standard enthalpy ΔH° of any reaction is given by the difference between the $\Delta \overline{H}_f^\circ$ values of all the products and the $\Delta \overline{H}_f^\circ$ values of all the reactants. Consider the reaction

$$a\text{A} + b\text{B} \longrightarrow c\text{C} + d\text{D}$$

where a, b, c, and d are stoichiometric coefficients, we have

$$\Delta H^\circ = \sum \Delta \overline{H}_f^\circ(\text{products}) - \sum \Delta \overline{H}_f^\circ(\text{reactants})$$
$$= c\,\Delta \overline{H}_f^\circ(\text{C}) + d\,\Delta \overline{H}_f^\circ(\text{D}) - a\,\Delta \overline{H}_f^\circ(\text{A}) - b\,\Delta \overline{H}_f^\circ(\text{B}) \qquad (6.23)$$

Of course, if any of the reactants (A and B) and the products (C and D) are

(a) Exothermic compounds, that is, compounds that have negative $\Delta \overline{H}_f^\circ$ values. (b) Endothermic compounds, that is, compounds that have positive $\Delta \overline{H}_f^\circ$ values. See Appendix 3 for examples.

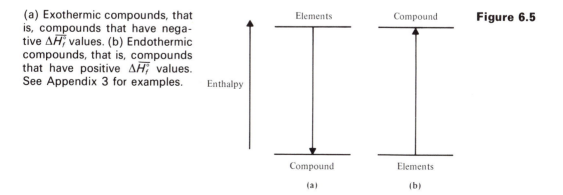

Figure 6.5

elements in their most stable form, we can immediately set their $\Delta \overline{H}_f^{\circ}$ values to be zero. Equation (6.23) will be much used later.

EXAMPLE 6.7 The combusion of α-D-glucose is an important biological process:

$$C_6H_{12}O_6(s) + 6O_2(g) \longrightarrow 6CO_2(g) + 6H_2O(l)$$

From the standard enthalpies of formation listed in Appendix 3 at the end of book, calculate the enthalpy of reaction.

Answer: From Appendix 3, we find that

$$\Delta \overline{H}_f^{\circ}(H_2O) = -285.84 \text{ kJ mol}^{-1} \quad \Delta \overline{H}_f^{\circ}(CO_2) = -393.51 \text{ kJ mol}^{-1}$$

$$\Delta \overline{H}_f^{\circ}(C_6H_{12}O_6) = -1274.45 \text{ kJ mol}^{-1} \quad \Delta \overline{H}_f^{\circ}(O_2) = 0$$

From Eq. (6.23), we have

$$\Delta H^{\circ} = (6 \text{ mol})(-393.51 \text{ kJ mol}^{-1}) + (6 \text{ mol})(-285.84 \text{ kJ mol}^{-1})$$

$$- (1 \text{ mol})(-1274.45 \text{ kJ mol}^{-1}) - (6 \text{ mol})(0 \text{ kJ mol}^{-1})$$

$$= -2801.65 \text{ kJ}$$

Comments: (1) The enthalpy of reaction in this case is also called the enthalpy of combustion, and (2) in looking up the $\Delta \overline{H}_f^{\circ}$ values, it is not only important to find the correct compound, but also the proper physical state. For example, the standard enthalpy of formation of liquid water at 298 K is $-285.84 \text{ kJ mol}^{-1}$, while that of water vapor is $-241.83 \text{ kJ mol}^{-1}$. The difference, $44.01 \text{ kJ mol}^{-1}$, is the enthalpy of vaporization

$$H_2O(l) \longrightarrow H_2O(g)$$

measured at 1 atm and 298 K.*

HESS'S LAW

In favorable cases the enthalpies of formation of compounds can be directly measured from reactions involving their elements as in the case of carbon dioxide. However, in some instances such an approach does not work because of experimental difficulties, and we must evaluate standard enthalpies of formation by an indirect method. Two examples will now be discussed.

Suppose that we need to know the standard enthalpy of formation for carbon monoxide. A seemingly obvious approach is to measure the heat of combustion of the following reaction in the standard state:

$$C(\text{graphite}) + \tfrac{1}{2}O_2(g) \longrightarrow CO(g)$$

However, it would be very difficult, if indeed possible, to burn carbon in oxygen without forming some carbon dioxide as well. A way to circumvent this difficulty is to carry out the following two separate reactions:

(1) $C(\text{graphite}) + O_2(g) \longrightarrow CO_2(g) \qquad \Delta H^{\circ} = -393.51 \text{ kJ}$

(2) $CO(g) + \tfrac{1}{2}O_2(g) \longrightarrow CO_2(g) \qquad \Delta H^{\circ} = -282.99 \text{ kJ}$

* The enthalpy of vaporization depends on temperature so that its value at 373 K (100°C) is 40.79 kJ mol^{-1}.

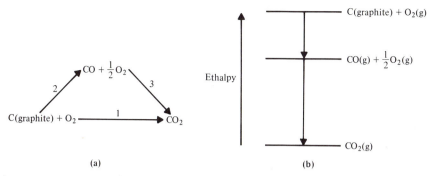

(a)   (b)

(a) Diagram illustrating Hess's law. Enthalpy change for step 1 is equal to the **Figure 6.6** sum of the enthalpy changes for steps 2 and 3. (b) Formation of carbon dioxide from graphite and oxygen as shown on the enthalpy scale.

Since chemical equations can be added and subtracted just like algebraic equations; therefore, $(1) - (2)$ gives us

$$C(\text{graphite}) + \tfrac{1}{2}O_2(g) \longrightarrow CO(g) \qquad \Delta H° = -110.52 \text{ kJ}$$

This $\Delta H°$ value corresponds to the heat of combustion of graphite to form carbon monoxide. By definition, the standard molar enthalpies of graphite and oxygen are zero. Thus

$$\Delta \overline{H}°_f(CO) = -110.52 \text{ kJ mol}^{-1}$$

Figure 6.6 shows the overall scheme of our procedure.

As another example, let us consider the standard enthalpy of formation of acetylene. The following reaction will obviously not work in the laboratory:*

$$2C(\text{graphite}) + H_2(g) \longrightarrow C_2H_2(g)$$

However, it is possible to determine the enthalpies of combusion of the following:

(1) $C(\text{graphite}) + O_2(g) \longrightarrow CO_2(g)$ $\qquad\qquad \Delta H° = -393.51 \text{ kJ}$

(2) $H_2(g) + \tfrac{1}{2}O_2(g) \longrightarrow H_2O(l)$ $\qquad\qquad\quad \Delta H° = -285.84 \text{ kJ}$

(3) $2C_2H_2(g) + 5O_2(g) \longrightarrow 4CO_2(g) + 2H_2O(l) \quad \Delta H° = -2599.18 \text{ kJ}$

Carrying out the operation of $4 \times (1) + 2 \times (2) - (3)$, we obtain

$$4C(\text{graphite}) + 2H_2(g) \longrightarrow 2C_2H_2(g) \quad \Delta H° = 453.46 \text{ kJ}$$

or

$$2C(\text{graphite}) + H_2(g) \longrightarrow C_2H_2(g) \quad \Delta H° = 226.73 \text{ kJ}$$

Thus we have $\Delta \overline{H}°_f(C_2H_2) = 226.73 \text{ kJ mol}^{-1}$.

DEPENDENCE OF ENTHALPY OF REACTION ON TEMPERATURE

How does the enthalpy of reaction vary with temperature? A simple relation can be derived as follows. For any reaction the change in enthalpy is

$$\Delta H = H_{\text{products}} - H_{\text{reactants}}$$

---

* Heating graphite in hydrogen may form some C_2H_2, but the reaction will also produce a number of other hydrocarbons.

117

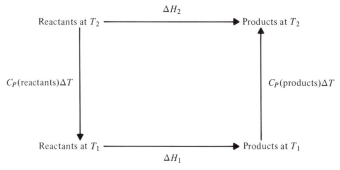

Figure 6.7 Schematic diagram showing Kirchhoff's equation [Eq. (6.25)]. The enthalpy change is

$$\Delta H_2 = \Delta H_1 + C_P(\text{products})\ \Delta T + C_P(\text{reactants})\ \Delta T$$

or $\Delta H_2 = \Delta H_1 + \Delta C_P(T_2 - T_1)$, where ΔC_P is the difference in heat capacities between the products and reactants.

Differentiating this equation with respect to temperature at constant pressure, we obtain

$$\left(\frac{\partial\ \Delta H}{\partial T}\right)_P = \left(\frac{\partial H_{\text{products}}}{\partial T}\right)_P - \left(\frac{\partial H_{\text{reactants}}}{\partial T}\right)_P$$

$$= C_P(\text{products}) - C_P(\text{reactants})$$

$$= \Delta C_P \tag{6.24}$$

Upon integration, the result is

$$\int_1^2 d\ \Delta H = \Delta H_2 - \Delta H_1 = \int_{T_1}^{T_2} \Delta C_P\ dT = \Delta C_P(T_2 - T_1) \tag{6.25}$$

where ΔH_1 and ΔH_2 are the enthalpies of reaction at T_1 and T_2. We have assumed here that C_P's are all independent of temperature. Equation (6.25) is known as Kirchoff's equation (Figure 6.7).

EXAMPLE 6.8 The enthalpy change for the reaction

$$\tfrac{1}{2}N_2(g) + \tfrac{1}{2}O_2(g) \longrightarrow NO(g)$$

is given by $\Delta H^\circ = 90.37$ kJ at 298 K and 1 atm. What is the enthalpy of reaction at 558 K? Assume that C_P's are all independent of temperature.

Answer: From Appendix 3 the heat capacities at constant pressure for N_2, O_2, and NO are 29.12, 29.36, and 29.86 J K^{-1} mol^{-1}, respectively. From Eq. (6.25),

$$\Delta H^\circ_{558} - \Delta H^\circ_{298}$$

$$= \frac{[(1\ \text{mol})29.86 - (\tfrac{1}{2}\ \text{mol})29.12 - (\tfrac{1}{2}\ \text{mol})29.36]\ \text{J K}^{-1}\ \text{mol}^{-1}}{1000}$$

$$\times\ (558 - 298)\ \text{K}$$

$$= 0.16\ \text{kJ}$$

118

$$\Delta H^\circ_{558} = (90.37 + 0.16)\ \text{kJ}$$
$$= 90.53\ \text{kJ}$$

BOND ENERGIES

Because chemical reactions involve the breaking and making of chemical bonds in the reactant and product molecules, it is clear that a proper understanding of the thermochemical nature of reactions requires a detailed knowledge of the strength of chemical bonds. It is important to first distinguish two terms that often cause confusion: *bond dissociation energy* (*D*) and *bond energy* (BE). Bond dissociation energy is the energy required to break a specific bond in a diatomic molecule to produce two neutral species; the molecule may be heteronuclear or homonuclear:

$$AB(g) \longrightarrow A(g) + B(g)$$
$$A_2(g) \longrightarrow A(g) + A(g)$$

The enthalpy change for this process is called the bond enthalpy. Consider the reaction

$$H_2(g) \longrightarrow H(g) + H(g)$$

The enthalpy change for this reaction at 298 K is 436.0 kJ, which is the energy that has to be supplied to break the H—H bond in 1 mol of H_2 molecules. Figure 6.8 shows the *potential-energy curve* of the H_2 molecule. Let us start by asking how the molecule is formed. At first the two hydrogen atoms are far apart and exert no influence on each other. As the distance of separation is shortened, both Coulombic attraction (between electron and nucleus) and Coulombic repulsion (between electron and electron and nucleus and nucleus) begin to affect each atom. Since attraction outweighs

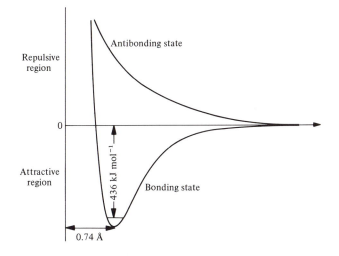

Figure 6.8

Potential-energy curve for a diatomic molecule. The horizontal line of the bonding-state potential energy curve represents the lowest vibrational energy level of the molecule (for example, H_2); the intercepts of this line with the curve represents the minimum and maximum bond lengths during a vibration.

repulsion, the potential energy of the system decreases with decreasing distance of separation. This process continues until the net attraction force reaches a maximum leading to the formation of a hydrogen molecule. Further shortening of the distance increases the repulsion and the potential rises steeply. The reference state (zero potential energy) corresponds to the case of two infinitely separated H atoms; potential energy is a negative quantity in the attractive region and a positive quantity in the repulsive region. Thus energy in the form of heat is given off as a result of the bond formation.

The most important feature in Figure 6.8 is that the potential-energy curve for the bonding state has a minimum that represents the most stable state for the molecule, and the corresponding distance of separation is called the *equilibrium distance*. However, molecules are constantly executing vibrational motions that persist even at the absolute zero. Furthermore, the energies associated with vibration, like the energies of an electron in an atom, are quantized. The lowest vibrational energy is not zero but equal to $\frac{1}{2}hv$, where v is the frequency of vibration. Consequently, the two hydrogen atoms cannot be rigidly held in the molecule, as is the case when a molecule is situated at the minimum point. Instead, the lowest state for H_2 is represented by the horizontal line. The intercepts between this line and the potential-energy curve represent the two extreme bond lengths during a vibration. It is still meaningful to speak of equilibrium distance in this case, although it has now become an averaged quantity (of the two extreme bond lengths). The bond dissociation energy of H_2 is the vertical distance from lowest vibrational energy level to the reference state of zero potential energy. The same description also applies to a heteronuclear diatomic molecule.

The situation is somewhat different when we examine polyatomic molecules. Consider the C—C bond in ethane. We might at first be tempted to define the bond dissociation energy as the energy required to dissociate ethane into two methyl radicals:

$$C_2H_6(g) \longrightarrow CH_3(g) + CH_3(g)$$

However, this definition involves the assumption that the dissociation does not alter the energies attributable to the C—H bonds, that is, the energies of the C—H bonds are the same in CH_3 as they are in ethane. If these energies do not remain the same, then some of the energy change in the homolytic fission of C_2H_6 to $2CH_3$ will be caused by changes in the C—H bonds, and the term "bond dissociation energy" is no longer applicable. An indication of the constancy of the C—H bond energies in different environments is given by the stepwise breakup of C—H bonds in methane, shown in Table 6.3. The variation indicated in Table 6.3 can be understood by realizing that the bonding scheme between C and H is different in each of the four species (the

Table 6.3

**Energies of the C—H Bond in Different
Environments**

Reaction	ΔH (kJ mol^{-1})
$CH_4(g) \rightarrow CH_3(g) + H(g)$	435
$CH_3(g) \rightarrow CH_2(g) + H(g)$	370
$CH_2(g) \rightarrow CH(g) + H(g)$	385
$CH(g) \rightarrow C(g) + H(g)$	339

Table 6.4

Average Bond Energies (kJ mol^{-1})^a

Bond	Bond Energy	Bond	Bond Energy
H—H	436.0	C—S	255
H—N	393	C=S	477
H—O	460	N—N	393
H—S	368	N=N	418
H—P	326	N≡N	941.4
H—F	568.2	N—O	176
H—Cl	430.9	N—P	209
H—Br	366.1	O—O	142
H—I	298.3	O=O	498.8
C—H	414	O—P	502
C—C	347	O=S	469
C=C	619	P—P	197
C≡C	812	P=P	490
C—N	276	S—S	268
C=N	615	S=S	351
C≡N	891	F—F	150.6
C—O	351	CL—Cl	243
C=O	724	Br—Br	192.5
C—P	264	I—I	151.0

^{a} Bond energies for diatomic molecules have more significant figures than those for polyatomic molecules because for diatomic molecules these are directly measurable quantities and not averaged over many compounds.

hybridization of C atom changes from CH_4 to CH). Although there is a spread in the values of the energy that can be associated with the C—H bond, it is often useful in approximate calculations to assume that all C—H bonds have the same energy, regardless of their particular environments. This energy, called bond energy, represents an average quantity obtained by studying a large number of molecules. If we only use the average value listed in Table 6.3, then the bond energy for the C—H bond is 382 kJ mol^{-1}. Table 6.4 lists the bond energy values for a number of common bonds. Of course, the bond energies of diatomic molecules are equal to their bond dissociation energies.

EXAMPLE 6.9 Estimate the enthalpy of combustion for methane,

$$CH_4(g) + 2O_2(g) \longrightarrow CO_2(g) + 2H_2O(g)$$

at 298 K and 1 atm using the average bond energies in Table 6.4. Compare your result with that calculated from the enthalpies of formation of products and reactants.

Answer: We start by counting the number of bonds broken for reactants and the number of bonds made for products.

Reactants: 4 C—H bonds   C—H = 414 kJ mol^{-1}

2 O=O bonds   O=O = 498.8 kJ mol^{-1}

total energy input = (4 mol)(414 kJ mol^{-1})

+ (2 mol)(498.8 kJ mol^{-1})

= 2653.6 kJ

121

Products: 2 C=O bonds $C=O = 724 \text{ kJ mol}^{-1}$

4 O—H bonds $O-H = 460 \text{ kJ mol}^{-1}$

total energy released $= (2 \text{ mol})(724 \text{ kJ mol}^{-1})$

$$+ (4 \text{ mol})(460 \text{ kJ mol}^{-1})$$

$$= 3288 \text{ kJ}$$

The standard enthalpy of reaction is given by

$$\Delta H^\circ = \sum BE(\text{reactants}) - \sum BE(\text{products})$$

$$= 2653.6 \text{ kJ} - 3288 \text{ kJ}$$

$$= -634.4 \text{ kJ}$$

From Appendix 3, we find that

$$\Delta \overline{H}_f^\circ(CO_2) = -393.51 \text{ kJ mol}^{-1} \qquad \Delta \overline{H}_f^\circ(H_2O) = -241.83 \text{ kJ mol}^{-1}$$

$$\Delta \overline{H}_f^\circ(CH_4) = -74.85 \text{ kJ mol}^{-1}$$

Thus the standard enthalpy of reaction is

$$\Delta H^\circ = (2 \text{ mol})(-241.83 \text{ kJ mol}^{-1}) + (1 \text{ mol})(-393.51 \text{ kJ mol}^{-1})$$

$$- (1 \text{ mol})(-74.85 \text{ kJ mol}^{-1})$$

$$= -802.3 \text{ kJ}$$

Comment: We see that the agreement between estimated ΔH° using bond energies and the actual ΔH° value is only qualitative. In general, the more exothermic (or more endothermic) the reaction, the better the agreement. If the actual ΔH° is only a small positive or negative quantity, then the value obtained from bond energies becomes unreliable. It may even give the wrong sign for the reaction.

Suggestions for Further Reading

INTRODUCTORY–INTERMEDIATE

BENT, H. A. *The Second Law.* Oxford University Press, Inc., New York, 1965.
    Written in a refreshingly different style from many books on thermodynamics.

KLOTZ, I. M. *Energy Changes in Biochemical Reactions.* Academic Press, Inc., New York, 1967.

LEHNINGER, A. L. *Bioenergetics*, 2nd ed. W. A. Benjamin, Inc., Menlo Park, Calif., 1971.
    These two texts present excellent elementary treatment of bioenergetics.

MAHAN, B. H. *Elementary Chemical Thermodynamics.* W. A. Benjamin, Inc., New York, 1964.
    A readable introductory text.

NASH, L. K. *Chemthermo: A Statistical Approach to Classical Chemical Thermodynamics.* Addison-Wesley Publishing Company, Inc., Reading, Mass., 1971.
    A very useful introductory text.

WASER, J. *Basic Chemical Thermodynamics.* W. A. Benjamin, Inc., New York, 1966.
    Another well-written text.

ATKINSON, D. E. *Cellular Energy Metabolism and Its Regulation.* Academic Press, Inc., New York, 1977.

A useful text.

BLUM, H. F. *Time's Arrow and Evolution,* 3rd ed. Princeton University Press, Princeton, N.J., 1958.

Although out of date, this book still presents an interesting and philosophical approach to the second law of the thermodynamics and its application in biology.

DENBIGH, K. *The Principles of Chemical Equilibrium.* Cambridge University Press, Cambridge, England, 1961.

A standard text.

DICKERSON, R. E. *Molecular Thermodynamics.* W. A. Benjamin, Inc., Menlo Park, Calif., 1969.

Provides a molecular interpretation of thermodynamic properties and discusses bioenergetics.

LEWIS, G. N., M. RANDALL, K. S. PITZER, and L. BREWER, *Thermodynamics,* 2nd ed. McGraw-Hill Book Company, New York, 1961.

Not particularly easy to read, but it has everything you ever wanted to know about thermodynamics.

MOROWITZ, H. J. *Energy Flow in Biology.* Academic Press, Inc., New York, 1968.

An interesting book that relates thermodynamics to the biosphere.

WALL, F. T. *Chemical Thermodynamics,* 2nd ed. W. H. Freeman and Company, San Francisco, 1965.

Another standard text.

SECTION A   **Reading Assignments**

"What Is Heat?" F. J. Dyson, *Sci. Am.,* Sept. 1954.

"The Definition of Heat," T. B. Tripp, *J. Chem. Educ.* **53,** 782 (1976).

"What the Standard State Doesn't Say About Temperature and Phase," H. Carmichael, *J. Chem. Educ.* **53,** 695 (1976).

"Thermodynamics, Folk Culture, and Poetry," W. L. Smith, *J. Chem. Educ.* **52,** 97 (1975).

"Tire Inflation Thermodynamics," J. J. Connors, *J. Chem. Educ.* **48,** 837 (1971).

"The Convertible Effect," R. C. Plumb, *J. Chem. Educ.* **49,** 285 (1972).

"Auto Air Conditioning Without Refrigerant," A. Nergaraian, *J. Chem. Educ.* **49,** 285 (1972).

"Hydrogen Bonding and Heat of Solution," N. Friedman, *J. Chem. Educ.* **54,** 248 (1977).

"Bond Energies," S. Benson, *J. Chem. Educ.* **42,** 423 (1965).

SECTION B

"Perpetual Motion Machines," S. W. Augrist, *Sci. Am.,* Jan. 1968.

"The Scope and Limitations of Thermodynamics," K. G. Denbigh, *Chem. Brit.* **4,** 339 (1968).

"The Use and Misuse of the Laws of Thermodynamics," M. L. McGlashan, *J. Chem. Educ.* **43**, 226 (1966).

"The Use and Misuse of Bond Energies," B. E. Knox and H. B. Palmer, *J. Chem. Educ.* **38**, 292 (1961).

"The Human Thermostat," T. H. Benzinger, *Sci. Am.*, Jan. 1961.

Problems

6.1 A thermobottle contains milk. The bottle is vigorously shaken. Consider the milk as the system. (a) Will the temperature rise as a result of the shaking? (b) Has heat been added to the system? (c) Has work been done on the system? (d) Has the system's internal energy changed?

6.2 When a bicycle tire is inflated with a hand pump, the temperature of the gas inside rises. Explain.

6.3 State the requirements a system must meet in order for it to be at thermal equilibrium.

★6.4 It is stated in some driver's test manuals that the stopping distance quadruples as the velocity doubles. Justify this statement by using mechanics and thermodynamic arguments.

6.5 Predict whether Q, W, ΔU, and ΔH are positive, zero, or negative for each of the following processes: (a) melting of ice at 1 atm and 273.15 K, (b) melting of solid cyclohexane at 1 atm and the normal melting point, (c) reversible isothermal expansion of an ideal gas, and (d) reversible adiabatic expansion of an ideal gas.

6.6 Show that for the adiabatic reversible expansion of an ideal gas

$$T_1^{C_V/R}V_1 = T_2^{C_V/R}V_2$$

6.7 Calculate the work done by the reaction

$$\text{Zn(s)} + \text{H}_2\text{SO}_4(\text{aq}) \longrightarrow \text{ZnSO}_4(\text{aq}) + \text{H}_2(\text{g})$$

when 1 mol of hydrogen gas is collected at 273 K and 1 atm. (Neglect volume changes other than gas.)

6.8 At 373.15 K and 1 atm, the molar volume of liquid water and steam are 1.88×10^{-5} m^3 and 3.06×10^{-2} m^3, respectively. Given that the heat of vaporization of water is 40.79 kJ mol^{-1}, calculate $\Delta H°$ and $\Delta U°$ for the following process:

$$\text{H}_2\text{O}(\text{l, 373.15 K, 1 atm}) \longrightarrow \text{H}_2\text{O}(\text{g, 373.15 K, 1 atm})$$

6.9 When 1 mol of naphthalene is completely burned in a constant-volume bomb calorimeter at 298 K, 5150 kJ of heat is evolved. Calculate ΔU and ΔH for the reaction.

6.10 The constant-pressure heat capacity of nitrogen is given by the expression

$$C_P = (27.0 + 5.90 \times 10^{-3}T - 0.34 \times 10^{-6}T^2) \text{ J K}^{-1} \text{ mol}^{-1}$$

Calculate ΔH for heating 1 mol of nitrogen from 25.0°C to 125°C.

6.11   A quantity of 4.50 g of CaC_2 is reacted with an excess of water at 298 K and atmospheric pressure:

$$CaC_2(s) + 2H_2O(l) \longrightarrow Ca(OH)_2(aq) + C_2H_2(g)$$

Calculate the work done by the acetylene gas against the atmospheric pressure.

★**6.12**   Einstein's special relativity equation is $E = mc^2$, where E is energy, m is mass, and c is the velocity of light. Does this equation invalidate the law of conservation of energy, and hence the first law of thermodynamics?

6.13   The convention of arbitrarily assigning a zero enthalpy value for all the (most stable) elements in the standard state and (usually) 298 K is a convenient way of dealing with the enthalpy changes of chemical processes. However, there is one kind of process for which this convention is not applicable. What is it? Why?

6.14   The standard molar enthalpy of formation of molecular oxygen at 298 K is zero. What is its value at 315 K? (*Hint*: Look up its $\bar{C}_P$ value in Appendix 3.)

★**6.15**   The equation of state for a certain gas is given by $P[(V/n) - b] = RT$. Obtain an expression for the maximum work done by the gas in an isothermal reversible expansion from V_1 to V_2.

6.16   Two moles of an ideal gas is compressed isothermally and reversibly at 298 K from 1 atm to 200 atm. Calculate Q, W, ΔU, and ΔH for the process.

6.17   Referring to Problem 6.16, calculate Q, W, ΔU, and ΔH if the compression is carried out by applying an external pressure of 300 atm.

6.18   Calculate Q, W, ΔU, and ΔH for the adiabatic and reversible expansion of 1 mol of a monatomic ideal gas from 5.00 m^3 to 25.0 m^3. The temperature of the gas is initially at 298 K.

★**6.19**   One mole of an ideal monatomic gas initially at 300 K and a pressure 15.0 atm is expanded to a final pressure of 1 atm. The expansion can occur via any one of the four different paths: (a) isothermal and reversible, (b) isothermal and irreversible. (c) adiabatic and reversible, and (d) adiabatic and irreversible. In irreversible processes, the expansion is against an external pressure of 1 atm. In each case calculate Q, W, ΔU, and ΔH.

★**6.20**   A molecule has the molecule formula X_2Y. The heat capacity ratio $(\bar{C}_P/\bar{C}_V)$ for the gaseous molecule is found to be 1.17. Assume ideal behavior. What conclusion can you draw about the structure of the molecule? (*Hint*: See Table 3.2 for $\bar{C}_V$ and use $\bar{C}_P = \bar{C}_V + R$.)

6.21   Indicate which of the following properties are extensive and which are intensive: energy, enthalpy, molar heat capacity, volume, force, and pressure.

6.22   The enthalpy of combustion of benzoic acid is commonly used as the standard for calibrating constant-volume bomb calorimeters; its value has been accurately determined to be -3226.7 kJ mol^{-1}. (a) When 0.9862 g of benzoic acid was oxidized, the temperature rose from 21.84°C to 25.67°C. What is the heat capacity of the calorimeter? (b) In a separate experiment, a quantity of 0.4654 g of α-D-glucose was oxidized in the same calorimeter and the temperature rose from 21.22°C to 22.28°C. Calculate the enthalpy of combusion of glucose, ΔU for the combustion, and the molar enthalpy of formation of glucose.

*6.23 The fuel value of hamburger is about 3.6 kcal g^{-1}. If a person eats 1 pound of hamburger for lunch and if none of the energy is stored in his body, estimate the amount of water that would have to be lost in perspiration in order to keep his body temperature constant. (1 lb = 454 g.)

6.24 The enthalpies of formation of trinitrotoluene (TNT) and nitroglycerine are -35.4 kJ mol^{-1} and -363.8 kJ mol^{-1}, respectively. The equations for the decompositions are

$$2C_7H_5(NO_2)_3(s) \longrightarrow 7C(s) + 7CO(g) + 3N_2(g) + 5H_2O(g)$$

$$2C_3H_5(NO_3)_3(l) \longrightarrow 6CO_2(g) + O_2(g) + 3N_2(g) + 5H_2O(g)$$

Compare the enthalpies of decomposition of these two explosives. Despite the smaller value of TNT, explain why it is often preferred over nitroglycerine.

6.25 From the reaction

$$C_{10}H_8(s) + 12O_2(g) \longrightarrow 10CO_2(g) + 4H_2O(l) \qquad \Delta H° = -5153.0 \text{ kJ}$$

and the enthalpies of formation of CO_2 and H_2O (see Appendix 3), calculate the enthalpy of formation of naphthalene.

*6.26 The oxyacetylene flame is often used in the welding of metals. Estimate the flame temperature produced by the reaction

$$2C_2H_2(g) + 5O_2(g) \longrightarrow 4CO_2(g) + 2H_2O(g)$$

Assume that the heat generated from this reaction is all used to heat the products. (*Hint*: First calculate $\Delta H°$ for the reaction. Next, look up the heat capacities of the products. Assume the heat capacities to be temperature-independent.)

6.27 The hydrogenation of ethylene is

$$C_2H_4(g) + H_2(g) \longrightarrow C_2H_6(g)$$

Calculate the change in the heat of hydrogenation of ethylene from 298 K to 398 K. Assume the heat capacities to be temperature-independent.

6.28 You are given that
(a) $C(\text{graphite}) + O_2(g) \longrightarrow CO_2(g)$ $\qquad \Delta H° = -393.5 \text{ kJ}$
(b) $H_2(g) + \frac{1}{2}O_2(g) \longrightarrow H_2O(l)$ $\qquad \Delta H° = -285.8 \text{ kJ}$
(c) $2C_2H_6(g) + 7O_2(g) \longrightarrow 4CO_2(g) + 6H_2O(l)$ $\quad \Delta H° = -3119.6 \text{ kJ}$
Calculate the heat of formation of ethane from its elements:

$$2C(\text{graphite}) + 3H_2(g) \longrightarrow C_2H_6(g).$$

6.29 From the following enthalpy of combustion of ethylene,

$$C_2H_4(g) + 3O_2(g) \longrightarrow 2CO_2(g) + 2H_2O(l) \qquad \Delta H° = -1411 \text{ kJ}$$

and the enthalpies of formation of CO_2 and H_2O (see Appendix 3), calculate the enthalpy of formation of ethylene from its elements at (a) a constant pressure of 1 atm and 298 K, and (b) at a constant volume and 298 K.

6.30 The $\Delta \overline{H_f°}$ values listed in Appendix 3 all refer to 1 atm and 298 K. Suppose that a student wishes to set up a new table listing the $\Delta H°$ values at 1 atm and 273 K. Show how he or she should proceed on the conversion, using acetone as an example.

6.31 The enthalpies of hydrogenation of ethylene and benzene have been determined at 298 K:

$$C_2H_4(g) + H_2(g) \longrightarrow C_2H_6(g) \qquad \Delta H° = -132 \text{ kJ}$$

$$C_6H_6(g) + 3H_2(g) \longrightarrow C_6H_{12}(g) \qquad \Delta H° = -246 \text{ kJ}$$

What would be the enthalpy of hydrogenation for benzene if it contained three isolated, unconjugated double bonds? How would you account for the difference between the calculated value based on this assumption and the measured value?

6.32 The molar enthalpies of fusion and vaporization of water are 6.01 kJ mol^{-1} and 44.01 kJ mol^{-1} (at 298 K), respectively. From these values estimate the molar enthalpy of sublimation of ice.

6.33 From the following data, calculate the enthalpies of allotropic transformations from the more stable form to the less stable form:

(a) C(graphite) + O_2(g) $\longrightarrow$ CO_2(g) $\Delta H° = -393.51$ kJ
  C(diamond) + O_2(g) $\longrightarrow$ CO_2(g) $\Delta H° = -395.41$ kJ
(b) S(rhombic) + O_2(g) $\longrightarrow$ SO_2(g) $\Delta H° = -296.06$ kJ
  S(monoclinic) + O_2(g) $\longrightarrow$ SO_2(g) $\Delta H° = -296.36$ kJ

6.34 The enthalpy of formation at 298 K of HF(aq) is -320.1 kJ mol^{-1}, of OH$^-$(aq) is -229.94 kJ mol^{-1}, of F$^-$(aq) is -329.11 kJ mol^{-1}, and of H_2O(l) is -285.84 kJ mol^{-1}. (a) Calculate the enthalpy of neutralization of HF(aq),

$$HF(aq) + OH^-(aq) \longrightarrow F^-(aq) + H_2O(l)$$

(b) Using the value of -55.83 kJ mol^{-1} as the enthalpy change from the reaction

$$H^+(aq) + OH^-(aq) \longrightarrow H_2O(l)$$

calculate the enthalpy change for the dissociation

$$HF(aq) \longrightarrow H^+(aq) + F^-(aq)$$

6.35 It was stated in the text of the chapter that for reactions in condensed phases, the difference between ΔH and ΔU is usually negligibly small. This statement holds for processes carried out under atmospheric conditions. For certain geochemical processes the external pressures may be so great that ΔH and ΔU can differ by a significant amount. A well-known example is the slow conversion of graphite to diamond under the earth's surface. Calculate the quantity $(\Delta H - \Delta U)$ for the conversion of 1 mol of graphite to 1 mol of diamond at a pressure of 20,000 atm. The densities of graphite and diamond are 2.25 g cm^{-3} and 3.52 g cm^{-3}, respectively.

6.36 Metabolic activity in the human body releases about 1.0×10^4 kJ of heat per day. Assuming that the body is 50 kg of water, how fast would the body temperature rise if it were an isolated system? How much water must the body eliminate as perspiration to maintain the normal body temperature (98.6°F)? Comment on your results. The heat of vaporization of water may be taken as 2.41 kJ g^{-1}.

6.37 The standard enthalpies of combustion of fumaric acid and maleic acids (to form carbon dioxide and water) are -1336.0 kJ mol^{-1} and -1359.2 kJ mol^{-1}, respectively. Calculate the enthalpy of the following isomerization process:

maleic acid   fumaric acid

6.38 Calculate the difference between ΔU and ΔH for the oxidation of α-D-glucose at 298 K:

$$C_6H_{12}O_6(s) + 6O_2(g) \longrightarrow 6CO_2(g) + 6H_2O(l)$$

6.39 Glycolysis is the process of breaking down complex carbohydrates into simpler molecules. The energy released in the process is used for biosynthesis and doing useful work. In the absence of oxygen, glycolysis produces lactic acid in our bodies:

$$C_6H_{12}O_6(s) \longrightarrow 2CH_3CH(OH)COOH(l)$$

From the enthalpies of combustion of glucose (-2801.65 kJ mol^{-1}) and lactic acid (-1364.43 kJ mol^{-1}), calculate the enthalpy change for the process.

6.40 Alcoholic fermentation is the process in which carbohydrates are broken down into ethanol and carbon dioxide. The reaction is very complex and involves a number of enzyme-catalyzed steps. The overall change is

$$C_6H_{12}O_6(s) \longrightarrow 2C_2H_5OH(l) + 2CO_2(g)$$

Calculate the standard enthalpy change for this reaction, assuming the carbohydrate to be α-D-glucose.

6.41 Calculate ΔU and ΔH for the following processes: (a) One mole of water freezes at 0°C and 1 atm, and (b) 1 mol of water boils at 100°C and 1 atm. Given that the molar enthalpies of fusion and vaporization of water are 6.01 kJ and 40.79 kJ and molar volumes of liquid water and ice are 0.0180 liter and 0.0195 liter, respectively. Assume ideal gas behavior.

6.42 From the molar enthalpy of vaporization of water at 373 K and the bond energies of H_2 and O_2 (see Table 6.4), calculate the average O—H bond energy in water given that

$$H_2(g) + \tfrac{1}{2}O_2(g) \longrightarrow H_2O(l) \qquad \Delta H° = -285.84 \text{ kJ}$$

6.43 Using the average bond energy values in Table 6.4 to calculate the enthalpy of combustion for ethane,

$$2C_2H_6(g) + 7O_2(g) \longrightarrow 4CO_2(g) + 6H_2O(g)$$

Compare your result with that calculated from the enthalpy of formation values of the products and reactants listed in Appendix 3.

6.44 The molar enthalpy of vaporization of water at 373.15 K and 1 atm is 40.79 kJ mol^{-1}. Calculate the work done ($P \Delta V$) for this process and ΔU for vaporization. Assume ideal gas behavior.

6.45 (a) Explain why the bond dissociation energy of a molecule is always defined in terms of a gas-phase reaction. (b) The bond dissociation energy of F_2 is 150.6 kJ mol^{-1}. Calculate $\Delta \overline{H}_f°$ for F(g).

6.46 The molar heat of vaporization of water is 44.01 kJ mol^{-1} at 298 K and 40.79 kJ mol^{-1} at 373 K. Give a qualitative explanation of the difference in these two values.

6.47 The *thermodynamic equation of state* is given by $(\partial U/\partial V)_T = T(\partial P/\partial T)_V - P$ (for derivation of this equation, see the standard texts listed at the end of Chapter 1). Apply this equation to (a) an ideal gas, and (b) a van der Waals gas. Comment on your results.

7

The Second Law of Thermodynamics

Humpty Dumpty sat on a wall
Humpty Dumpty had a great fall
All the King's horses and all the King's men
Couldn't put Humpty together again

In Chapter 6 we discussed the first law of thermodynamics and how it can be applied to study the energetics of chemical reactions. The first law is based on the law of conservation of energy. It says that energy is neither created nor destroyed in processes, but flows from one part of the universe to another, or is converted from one form into another. The total amount of energy in the universe remains constant. Despite its immense value, the first law does have a major limitation; that is, it cannot predict the direction of change. Once we specify a particular process or change, the first law helps us to do the bookkeeping of energy balance such as the energy input, heat released, work done, and so forth. But it says nothing about whether such a process can indeed occur in reality. The answer is provided by the second law of thermodynamics. We have seen how the changes in internal energy (ΔU) and enthalpy (ΔH) help us understand work done on and by the system and endothermic and exothermic processes. In this chapter we introduce two new thermodynamic functions, called entropy (S) and Gibbs free energy (G); we shall see that the changes in these quantities—ΔS and ΔG—provide the necessary criterion for predicting the direction of any process. However, before defining these functions, we must first take a closer look at the nature of processes that occur on their own accord without the influence of any outside force, or spontaneous processes.

7.1 Spontaneous Processes

A lump of sugar dissolves in a cup of coffee, an ice cube melts in your hand, and an autumn leaf gravitates toward the ground—we witness so many of these *spontaneous* processes in everyday life that it is almost impossible to list them individually. The interesting aspect about any spontaneous process is that the reverse process never happens on its own accord. Ice melts at 20°C,

Figure 7.1

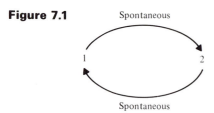

Spontaneous

Spontaneous

An impossible situation. If the change from state 1 to state 2 occurs spontaneously, then the reverse step, that is, 2 to 1, cannot also be a spontaneous process.

but water at the same temperature will not spontaneously turn into ice. A leaf lying on the ground will not spontaneously rise up in the air and return to the branch from which it came. A movie of a baseball smashing a window to pieces run backward seems hilariously funny to the audience because everyone knows that it is an impossible event and therefore can never happen on its own. But why not? Surely for any of the changes just described (and for countless more) we can demonstrate that the process can occur in *either* direction in accord with the first law of thermodynamics; yet, in fact, each process occurs in only one direction. After many observations we can conclude that processes occurring spontaneously in one direction cannot also take place spontaneously in the opposite direction; otherwise, nothing would ever happen in the world (Figure 7.1).

The question we ask is: What changes accompany a spontaneous process? Before we answer this question, however, it is helpful to see why the reverse of any spontaneous process cannot also occur by itself. Consider a rubber ball held at some distance above the floor. When the ball is released, it falls. The impact between the ball and the floor causes the ball to bounce upward, and when it has reached a certain height, it repeats its downward motion. In the process of falling, the potential energy of the ball is converted to kinetic energy. Experience tells us that after every bounce the ball does not rise quite as high as before. The reason is that the collision between the ball and the floor is inelastic, so that upon each impact some of the ball's kinetic energy is dissipated among the molecules constituting the floor. After each bounce, the floor becomes a little bit hotter.* This intake of energy results in an enhancement of the rotational and vibrational motions of the molecules in the floor (Figure 7.2a). Eventually, the ball comes to a complete rest as its kinetic energy is totally lost to the floor. To describe this process in another way, we say that the original potential energy of the ball, through its conversion to kinetic energy, is degraded into heat. All this is fine, but let us now consider what would be necessary for the reverse process to occur spontaneously; that is, a ball originally sitting on the floor spontaneously rises to a certain height in the air by absorbing some heat from the floor. That such a process will not violate the first law is clear. Letting the mass of the ball be m and the height above the floor to which it rises be h, we have

$$\text{energy extracted from the floor} = mgh$$

where g is acceleration due to gravity. The thermal energy of the floor is random molecular motion. To impart an amount of energy large enough to raise the ball from the floor, we would need to line up most of the molecules under the ball and have their vibrations in phase with one another, as shown

---

* Actually, the temperature of the ball also rises slightly after each impact. But we are only concentrating on what happens to the floor here.

in Figure 7.2b. At the instant the ball leaves the floor, the atoms in these molecules must all be moving upward for proper energy transfer. Now, it is conceivable that 200 or even 200,000 molecules may be executing this kind of motion at a given instant. But because of the magnitude of energy transfer, the number of molecules involved would need to be on the order of Avogadro's number, or 6×10^{23}. Considering the random nature of molecular motion, this is such an improbable event that it is virtually impossible. Indeed, no one has ever witnessed the spontaneous rising of a ball from the floor, and we can safely conclude that no one ever will either.

The discussion above helps us understand the nature of many spontaneous processes. Consider our familiar example of a gas in cylinder fitted with a movable piston. If the pressure of the gas is greater than the external pressure, then the gas will expand until the internal and external pressures are equal. This is another example of a spontaneous process. What would it take for the gas to contract spontaneously in this case? It would require most of the molecules to move away from the piston and into the other part of the cylinder at the same time. Now at any moment many molecules are indeed doing this, but we will never find 6×10^{23} molecules executing this unidirectional motion, because translation motions of molecules is totally random. By the same token, a metal bar originally at a uniform temperature will not suddenly become hotter at one end and colder at the other. To establish this temperature gradient, the collisions between randomly vibrating molecules would have to quench the thermal motion at one end and raise it at the other—a highly improbable event.

We now return to the question posed earlier; that is, what changes accompany a spontaneous process? It is logical to assume that all spontaneous processes occur in such a way so as to decrease the energy of the system. This assumption helps us explain why things fall downward, why springs unwind, and so forth. But energy consideration alone is not enough for predicting whether a process would be spontaneous. For example, in Chapter 6 we saw that the isothermal expansion of an ideal gas against vacuum does not result in a change in its internal energy. Yet the process is spontaneous. When ice melts spontaneously at 20°C to form water, the internal energy of the system actually increases. In fact, many endothermic physical and chemical processes

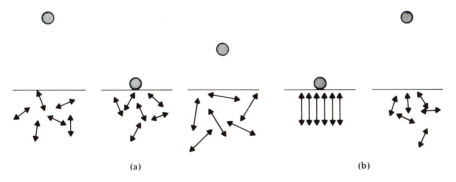

(a)   (b)

(a) A spontaneous process. A falling ball strikes the floor and loses some of its kinetic energy to the molecules in the floor. As a result, the ball does not rise up quite as high and the floor heats up a little. The length of arrow indicates the amplitude of molecular vibration. (b) An impossible event. The ball originally resting on the floor cannot spontaneously rise in the air by absorbing thermal energy from the floor.

Figure 7.2

are spontaneous. If energy change cannot be used to indicate the direction of a spontaneous process, then we must look for another thermodynamic function to help us. This function turns out to be entropy (S).

7.2 Entropy

Our discussion of spontaneous processes is based on events occurring in macroscopic systems. Referring to our earlier case of gas contraction, we see that it is not inconceivable to have several thousands or perhaps even several millions of molecules moving in a particular direction simultaneously, but certainly not an Avogadro's number of molecules. When we are dealing with an enormously large number such as 6×10^{23}, motions along every direction are represented by virtually the same number of molecules. There is no reason why all of the molecules would, in the absence of any external influence, choose a specific direction at the same time. In trying to understand spontaneous processes, our attention should therefore be focused on the *statistical* behavior of a very large number of molecules, not the motion of just a few of them. In this section we first derive a statistical definition of entropy and then define entropy in terms of thermodynamic quantities.

THE STATISTICAL DEFINITION OF ENTROPY

Consider a cylinder containing helium atoms shown in Figure 7.3. The probability of finding any one He atom in V_2, the entire volume of the cylinder, is 1, since all He atoms are known to be inside the cylinder. On the other hand, the probability of finding a helium atom in half of the volume of the cylinder, V_1, is only $\frac{1}{2}$. Now if the number of He atoms is increased to two, the probability of finding both of them in V_2 is still 1, but that of finding both of them in V_1 becomes $(\frac{1}{2})(\frac{1}{2})$, or $\frac{1}{4}$.* Since $\frac{1}{4}$ is an appreciable quantity, it would not be surprising to find them in the same region at a given time. However, it is not difficult to see that as the number of He atoms increases, the probability W of

Figure 7.3

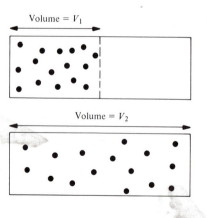

Volume = V_1

Volume = V_2

Schematic diagram showing N molecules occupying volumes V_1 and V_2 of a container.

---

* This follows from the fact that the probability of both events occurring is a *product* of probabilities of two independent events. We assume that the He gas behaves ideally so that the presence of one He atom in V_1 does not affect the presence of another He atom in the same volume in any way.

finding *all* the atoms in V_1 becomes progressively smaller:

$$W = \left(\frac{1}{2}\right)\left(\frac{1}{2}\right)\left(\frac{1}{2}\right)\cdots$$

$$= \left(\frac{1}{2}\right)^N$$

where N is the total number of atoms present. If $N = 100$, we have

$$W = \left(\frac{1}{2}\right)^{100} = 8 \times 10^{-31}$$

a very small number. If N is of the order of 6×10^{23}, the probability becomes $(\frac{1}{2})^{6 \times 10^{23}}$, a quantity so small that for all practical purposes it can be regarded as zero.* The results of these simple calculations have a most important message. If initially we had compressed all the He atoms in V_1 and allowed the gas to expand on its own, we would find that eventually the atoms would be evenly distributed over the entire volume V_2 because this situation corresponds to the most probable state. Thus the direction of spontaneous change is from a situation where the gas is in V_1 to one in which it is in V_2, or from a state with low probability of occurring to one of maximum probability.

Now that we know how to predict the direction of a spontaneous change in terms of probabilities of the initial and final states, it may seem appropriate to equate entropy to probability simply as $S = kW$, where k is a proportionality constant. But this relation is invalid for the following reason. As we shall see later, entropy is an extensive property (like U and H). Doubling the number of He atoms from N to $2N$ leads to a twofold increase in entropy. But probability is proportional to the volume raised to the power of N (that is, $W \propto V^N$)†; therefore, changing N to $2N$ gives us W^2. Thus the increases in entropy (from S to $2S$) and probability (from W to W^2) are not related to each other as predicted by the simple equation given above. A way out of this dilemma is to express entropy as a natural logarithmic function of probability as follows:

$$S = k \ln W \tag{7.1}$$

This equation tells us that as W increases to W^2, S increases to $2S$ because $\ln W^2 = 2 \ln W$. Equation (7.1) is known as the *Boltzmann equation* and k is the *Boltzmann constant*, given by 1.3805×10^{-23} J K^{-1}. Because the quantity $\ln W$ is dimensionless, the units of entropy are therefore J K^{-1}.

Equation (7.1) can be employed to calculate changes in entropy when a system changes from an initial state 1 to a final state 2. The entropies of the system in these two states are given by

$$S_1 = k \ln W_1$$

$$S_2 = k \ln W_2$$

---

* To borrow the analogy used by the noted thermodynamicist Henry A. Bent, this probability is less than that for the production of Shakespeare's complete works 15 quadrillion times in succession without error by a tribe of wild monkeys punching randomly on a set of typewriters.

† Since $W \propto V^N$, we have $W = C_N V^N$, where C is a proportionality constant. Referring to the situation shown in Figure 7.3, this constant is given by $1/V_2$. Thus the probability of finding a He atom in volume V_2 is given by $W = (1/V_2)(V_2) = 1$ and that of finding a He atom in volume V_1 is $W = (1/V_2)(V_1) = (V_1/V_2) = \frac{1}{2}$, since $V_1 = V_2/2$.

Because entropy is a state function (it depends only on the probability of a state occurring but not on the manner in which the state is created), the change in entropy ΔS for the $1 \rightarrow 2$ process is

$$\Delta S = S_2 - S_1 = k \ln \frac{W_2}{W_1} \tag{7.2}$$

Equation (7.2) can be applied to calculate the entropy change when an ideal gas expands isothermally from V_1 to V_2. As we saw earlier, the probabilities W_1 and W_2 are related to the volumes V_1 and V_2 as follows:

$$W_1 = (CV_1)^N$$

$$W_2 = (CV_2)^N$$

Substituting these relations in Eq. (7.2), we obtain

$$\Delta S = k \ln \frac{(CV_2)^N}{(CV_1)^N}$$

$$= Nk \ln \frac{V_2}{V_1}$$

Since $kN_0 = R$ where R is the gas constant and N_0 is Avogadro's constant, we write

$$\Delta S = \frac{N}{N_0} R \ln \frac{V_2}{V_1} = nR \ln \frac{V_2}{V_1} \tag{7.3}$$

where n is the number of moles of the gas present. It is important to remember that Eq. (7.3) holds only for an isothermal expansion because entropy of a system is also affected by changes in temperature. Note also that we do not have to specify the manner in which the expansion was brought about, since S is a state function.

EXAMPLE 7.1 A quantity of 2.0 mol of an ideal gas is allowed to expand isothermally from an initial volume of 1.5 liters to 2.4 liters. Calculate the entropy change for this process. Estimate the probability for the gas to contract spontaneously from the final volume to the initial one.

Answer: From Eq. (7.3), we write

$$\Delta S = (2.0 \text{ mol})(8.314 \text{ J K}^{-1} \text{ mol}^{-1}) \ln \frac{2.4}{1.5}$$

$$= 7.8 \text{ J K}^{-1}$$

To estimate the probability for spontaneous contraction, we note that this process must be accompanied by a *decrease* in entropy equal to -7.8 J K^{-1}. Since the process is now defined as $2 \rightarrow 1$, we have, from Eq. (7.2),

$$\Delta S = k \ln \frac{W_1}{W_2}$$

$$-7.8 \text{ J K}^{-1} = (1.3805 \times 10^{-23} \text{ J K}^{-1})(2.303) \log \frac{W_1}{W_2}$$

$$\log \frac{W_1}{W_2} = -2.5 \times 10^{23}$$

or

$$\frac{W_1}{W_2} = 10^{-2.5 \times 10^{23}}$$

This exceedingly small ratio means that the probability of state 1 occurring is so much smaller than that of state 2 that there is virtually no possibility for the process to occur on its own accord. This does not mean, of course, that the gas cannot be compressed from 2.4 liters to 1.5 liters, but it must be done with the aid of an external force.

THE THERMODYNAMIC DEFINITION OF ENTROPY

Equation (7.1) is a statistical formulation of entropy; defining entropy in terms of probability provides us with a molecular interpretation. However, this equation is not used for calculating changes in entropy in general. Calculating W for complex systems, where chemical reactions occur, for example, is too difficult, and our knowledge is insufficient to estimate intermolecular forces accurately. On the other hand, changes in entropy can be most conveniently measured from other thermodynamic quantities. In Section 6.5 we saw that the heat absorbed by an ideal gas in an isothermal reversible expansion is given by

$$Q = nRT \ln \frac{V_2}{V_1}$$

or

$$\frac{Q}{T} = nR \ln \frac{V_2}{V_1}$$

Since the right-hand side of the equation above is equal to ΔS [see Eq. (7.3)], we have

$$\Delta S = \frac{Q_{rev}}{T} \tag{7.4}$$

In words, Eq. (7.4) says that the entropy change of a system in a reversible process is given by the heat absorbed divided by the temperature at which the process occurs. If the temperature changes during the process, then the calculations of ΔS must be carried out to each small increment of the process, during which the temperature of the system can be considered practically constant. Thus, for an infinitesimal process, we can write

$$dS = \frac{dQ_{rev}}{T} \tag{7.5}$$

Both Eqs. (7.4) and (7.5) are the thermodynamic definition of entropy. Although these equations were derived by considering gas expansions, they are applicable to any type of process. It is important to note that the definition holds only for a reversible process, and hence the inclusion of the subscript "rev." While S is a state function, Q obviously is not and we must specify the reversible path in defining entropy. If the expansion were irreversible, then the work done by the gas on the surroundings would be smaller and so would the heat absorbed by the gas from the surroundings. Although the entropy change is still the same, that is, $\Delta S_{rev} = \Delta S_{irrev} = \Delta S$, we now have $\Delta S > Q_{irrev}/T$. We shall return to this point in the next section.

7.3 The Second Law of Thermodynamics

Thus far our discussion of entropy changes refers only to the system. For a proper understanding of entropy, we must also examine what happens to the surroundings. Because of its size and the amount of material contained in it, the surroundings can be considered as an infinitely large thermal reservoir. Therefore, we can assume that any amount of heat exchange between the system and the surroundings will not affect the latter's temperature or volume. For an infinitesimal process, the change in the surroundings' energy is given by

$$dU_{surr} = dQ_{surr} + dW_{surr}$$

Since there is no volume change, the P–V-type work done is zero; therefore, $dW_{surr} = 0$ and we have

$$dU_{surr} = dQ_{surr}$$

Since dU_{surr} is an exact differential (U is a state function), it follows that the quantity dQ_{surr} is independent of whether the heat exchange occurs reversibly or irreversibly. They are both equal to dU_{surr} and therefore also equal to each other:

$$(dQ_{surr})_{rev} = (dQ_{surr})_{irrev} = dQ_{surr}$$

For this reason, we shall not bother to specify the path for dQ_{surr}. Thus the change of entropy of the surroundings is

$$dS_{surr} = \frac{dQ_{surr}}{T_{surr}}$$

and for a finite isothermal process

$$\Delta S_{surr} = \frac{Q_{surr}}{T_{surr}}$$

Returning to the ideal gas isothermal expansion, we saw earlier that for a reversible process the heat absorbed by the gas from the surroundings is $nRT_{sys} \ln(V_2/V_1)$, where T_{sys} is the temperature of the system. Since the system is at equilibrium with its surroundings throughout the expansion, $T_{sys} = T_{surr} = T$. The heat lost by the surroundings to the system is therefore $nRT \ln(V_2/V_1)$ and the corresponding change in entropy is

$$\Delta S_{surr} = \frac{Q_{surr}}{T}$$

The total change in entropy of the universe, ΔS_{univ}, is given by

$$\Delta S_{univ} = \Delta S_{sys} + \Delta S_{surr}$$

$$= \frac{Q_{sys}}{T} + \frac{Q_{surr}}{T}$$

$$= \frac{nRT \ln(V_2/V_1)}{T} + \frac{[-nRT \ln(V_2/V_1)]}{T} = 0$$

Thus for a reversible process, the total change in entropy of the universe is equal to zero.

Now let us consider what happens if the expansion is irreversible. In the extreme case we can assume that the gas is expanding against a vacuum. Again, the change in entropy of the system is given by $\Delta S_{sys} = nR \ln(V_2/V_1)$ since S is a state function. However, since no work is done in this process, there is no heat exchange between the system and the surroundings. Therefore, we have $Q_{surr} = 0$ and $\Delta S_{surr} = 0$. The change in entropy of the universe is now given by

$$\Delta S_{univ} = \Delta S_{sys} + \Delta S_{surr}$$
$$= nR \ln(V_2/V_1) > 0$$

Combining these two expressions for ΔS_{univ}, we obtain

$$\Delta S_{univ} = \Delta S_{sys} + \Delta S_{surr} \geq 0 \qquad (7.6)$$

where the equality sign applies to a reversible process and the larger-than ($>$) sign applies to an irreversible (that is, spontaneous) process. Equation (7.6) is a mathematical statement of the second law of thermodynamics. In words, the second law is stated as follows: *The entropy of an isolated system increases in an irreversible process and remains unchanged in a reversible process. It can never decrease.* Thus either ΔS_{sys} or ΔS_{surr} can be a negative quantity for a particular process, but their sum can never be less than zero.

Entropy Changes **7.4**

Having defined entropy statistically and thermodynamically and presented the second law of thermodynamics, we are now ready to study entropy changes in various processes. We have already seen that the entropy change for the isothermal reversible expansion of an ideal gas is given by $nR \ln(V_2/V_1)$. This section will consider two other examples of entropy change: phase transitions and heating.

ENTROPY CHANGE AS A RESULT OF PHASE TRANSITIONS

A familiar phase change is the melting of ice. At 0°C and 1 atm ice and water are in equilibrium. Therefore, under these conditions heat is absorbed reversibly by the ice during the melting process. Furthermore, since this is a constant-pressure process, the heat absorbed is equal to the enthalpy change of the system, so that $Q_{rev} = \Delta H_{fus}$, where ΔH_{fus} is called the *heat* or *enthalpy of fusion*. Note that because H is a state function, it is no longer necessary to specify the path. The entropy of fusion, ΔS_{fus}, is given by

$$\Delta S_{fus} = \frac{\Delta H_{fus}}{T_f} \qquad (7.7)$$

where T_f is the fusion or melting point (273.15 K for ice). Similarly, we can write the entropy of vaporization, ΔS_{vap}, as

$$\Delta S_{vap} = \frac{\Delta H_{vap}}{T_b} \qquad (7.8)$$

where ΔH_{vap} and T_b are the enthalpy of vaporization and boiling point of the liquid, respectively.

EXAMPLE 7.2 The molar enthalpies of fusion and vaporization of water are 6.01 kJ mol⁻¹ and 40.79 kJ mol⁻¹, respectively. Calculate the entropy changes for the fusion and vaporization of 1 mol of water at its normal melting point and boiling point.

Answer: From Eq. (7.7), we write

$$\Delta \bar{S}_{\text{fus}} = \frac{6.01 \times 1000 \text{ J mol}^{-1}}{273.15 \text{ K}} = 22.0 \text{ J K}^{-1} \text{ mol}^{-1}$$

and, from Eq. (7.8),

$$\Delta \bar{S}_{\text{vap}} = \frac{40.79 \times 1000 \text{ J mol}^{-1}}{373.15 \text{ K}} = 109.3 \text{ J K}^{-1} \text{ mol}^{-1}$$

Comments: (1) Because entropy values are generally much smaller than enthalpy values, we express them as J K⁻¹ rather than kJ K⁻¹. (2) In both cases we find an increase in entropy. This may seem strange at first, since at 273.15 K ice is in equilibrium with water and at 373.15 K water is in equilibrium with its vapor, so that we would expect the change in entropy in each case to be zero. However, the entropy changes we have calculated refer only to the system. Because heat is absorbed reversibly in an equilibrium process, the changes in entropy in the surroundings for the fusion and vaporization are given by $-\Delta \bar{H}_{\text{fus}}/T_f$ and $-\Delta \bar{H}_{\text{vap}}/T_b$, respectively. Therefore, the total change in entropy of the universe is zero in each case. (3) As our calculation shows, ΔS_{vap} is invariably greater than ΔS_{fus} for the same substance. Since both solids and liquids are condensed phases, they possess a great deal of structure or order.* Consequently, the solid → liquid transition results in a relatively small disorder of molecular arrangement. On the other hand, the gaseous state is completely random, so that the liquid → vapor process is accompanied by a large increase in disorder.

Entropy can be interpreted as a measure of the randomness that exists in a system; therefore, the change in entropy would be much greater from an ordered state (liquid) to a highly disordered state (gas) than the case from one ordered state (solid) to another (liquid).

ENTROPY CHANGE AS A RESULT OF HEATING

When the temperature of a system is raised from T_1 to T_2, its entropy also increases. We can calculate this entropy increase as follows. Let S_1 and S_2 be the entropies of the system in states 1 and 2 (characterized by T_1 and T_2). If heat is transferred reversibly to the system, then the increase in entropy for an infinitesimal amount of heat transfer is given by Eq. (7.5).

$$dS = \frac{dQ_{\text{rev}}}{T}$$

The entropy at T_2 is given by

$$S_2 = S_1 + \int_{T_1}^{T_2} \frac{dQ_{\text{rev}}}{T}$$

* This is particularly true for water because of the intermolecular hydrogen bonds in ice and in liquid water.

If we define this to be a *constant-pressure* process, as is usually the case in practice, then $dQ_{rev} = dH$ so that

$$S_2 = S_1 + \int_{T_1}^{T_2} \frac{dH}{T}$$

But from Eq. (6.15), we have $dH = C_P\, dT$. Therefore, we write

$$S_2 = S_1 + \int_{T_1}^{T_2} \frac{C_P}{T}\, dT \tag{7.9}$$

If we assume C_P to be independent of temperature, then Eq. (7.9) becomes

$$S_2 = S_1 + C_P \ln \frac{T_2}{T_1} \tag{7.10}$$

and the increase in entropy ΔS as a result of heating is

$$S_2 - S_1 = \Delta S = C_P \ln \frac{T_2}{T_1} \tag{7.11}$$

EXAMPLE 7.3   A quantity 200 g of water is heated from 10°C to 20°C at constant pressure. Calculate the increase in entropy for this process. The heat capacity at constant pressure for water is 75.3 J K^{-1} mol^{-1}.

Answer: The number of moles of water present is 200/18.015, or 11.1 mol. The increase in entropy is given by

$$\Delta S = (11.1 \text{ mol})(75.3 \text{ J K}^{-1} \text{ mol}^{-1}) \ln \frac{293.15}{283.15}$$

$$= 29.0 \text{ J K}^{-1}$$

In this calculation we have assumed that C_P is independent of temperature and that water does not expand on heating.

Suppose that the heating of water in Example 7.3 had been carried out irreversibly, say with a Bunsen burner. What would be the increase in entropy? We note that regardless of the path, the initial and final states are the same, that is, 200 g of water heated from 10°C to 20°C. Therefore, the integral on the right-hand side of Eq. (7.9) gives ΔS for the irreversible heating. This follows from the fact that ΔS depends only on T_1 and T_2 and not on the path. Thus ΔS for this process is 29.0 J K^{-1}, whether the heating is done reversibly or irreversibly. For an example of an irreversible phase transition, the reader is referred to Problem 7.12.

The Third Law of Thermodynamics **7.5**

In thermodynamics we are normally interested only in the changes of properties, such as ΔU and ΔH. As mentioned earlier, there is no known way of measuring the absolute values of internal energy and enthalpy. The only exception is entropy, and Eq. (7.9) provides us a clue as to how this can be achieved. As written, this equation enables us to measure the change in entropy over a suitable temperature range between T_1 and T_2. Suppose that we now set the lower temperature to absolute zero, that is, $T_1 = 0$, and call the upper temperature (T_2) T; Eq. (7.9) then becomes

$$\Delta S = S_T - S_0 = \int_0^T \frac{C_P}{T} \, dT \tag{7.12}$$

Since entropy is an extensive property, its value at any temperature T is given by the sum of the contributions from 0 K to the specified temperature. We can carry out heat-capacity measurements as a function of temperature to evaluate the integral in Eq. (7.12). But we encounter two obstacles in this approach. First, what is the entropy of a substance at absolute zero; that is, what is S_0? Second, how do we account for that part of the contribution to the total entropy which lies between absolute zero and the lowest attainable temperature?

According to Boltzmann's equation [Eq. (7.1)], entropy is related to the probability of a certain state occurring. The quantity W can also be used to denote the number of *microstates* of a macroscopic system. The meaning of microstate can be understood by referring to a hypothetical perfect crystalline substance. In such a crystal, there can be only one particular arrangement of atoms or molecules, so that there is only one microstate. This means that $W = 1$ and

$$S = k \ln W = k \ln 1 = 0$$

This is the principle known as the *third law of thermodynamics*, which states that: *Every substance has a finite positive entropy, but at the absolute zero of temperature the entropy may become zero, and does so become in the case of a perfect crystalline substance.* Mathematically, we have*

$$\lim_{T \to 0 \, K} S = 0 \qquad \text{(perfect crystalline substance)}$$

The converse is always true; that is, if there is more than one way to arrange the atoms or molecules in a crystal (that is, if there is more than one microstate so that $W > 1$), then there will be a *residual* entropy even at the absolute zero. An example of this phenomenon is the carbon monoxide crystal. The two ends of CO molecule are similar enough that upon packing the molecules into a crystal, it does not greatly matter which of the two possible ways each molecule is oriented. Figure 7.4 shows the arrangements of CO molecules in a perfect crystalline solid and also the completely random arrangements of CO molecules. If there is no preferred arrangement, then for one molecule we

Figure 7.4

CO	CO	CO	CO		CO	OC	CO	OC
CO	CO	CO	CO		OC	CO	OC	CO
CO	CO	CO	CO		OC	CO	OC	CO
	(a)					(b)		

(a) Perfect arrangement of carbon monoxide molecules in the crystal; $S_0 = 0$ at 0 K. (b) Imperfect arrangement of carbon monoxide molecules; $S_0 > 0$ at 0 K.

* At temperatures above the absolute zero, various thermal motions contribute to the entropy of the system so that its entropy is no longer zero even if it remains perfectly crystalline.

have two, or 2^1, choices. For two molecules there are four, or 2^2, choices, and for 1 mol of carbon monoxide the choice becomes 2^{N_0}. Thus we have

$$S_0 = k \ln W$$

$$= (1.38 \times 10^{-23} \text{ J K}^{-1}) \ln 2^{6 \times 10^{23}}$$

$$= 5.7 \text{ J K}^{-1}$$

From spectroscopic and thermodynamic measurements, the residual entropy of 1 mol of the CO crystal is found to be 4.2 J K^{-1}. The difference between these two values is taken to mean that the packing of the CO crystal is random, but the randomness is not complete.* On the other hand, we expect that molecular nitrogen crystal should have a zero residual entropy. This fact is indeed confirmed by experiment.

THIRD-LAW OR ABSOLUTE ENTROPIES

The third law of thermodynamics now enables us to measure the entropy of a substance at temperature T. For a perfect crystalline substance, $S_0 = 0$. Equation (7.12) becomes

$$S_T = \int_0^T \frac{C_P}{T} dT = \int_0^T C_P \, d \ln T \qquad (7.13)$$

We can then proceed to measure the heat capacity over the desired temperature range. At very low temperatures (below 10 K), where such measurements are difficult to carry out, we can make use of Debye's theory of heat capacity:

$$C_P = aT^3 \qquad (7.14)$$

where a is a constant for a given substance. Equation (7.14) is applicable near absolute zero. Table 7.1 lists the various steps taken in measuring the entropy of the HCl gas at 298.15 K. In applying Eq. (7.13), we must always keep in

Entropy of HCl at 298.15 K from Its Heat-Capacity Measurements[a]

Table 7.1

Contribution	S_T(J K^{-1} mol^{-1})
1. Extrapolation from 0 to 16 K [Eq. (7.14)]	1.3
2. $\int C_P \, d \ln T$ for solid I from 16 K to 98.36 K	29.5
3. Phase transition at 98.36 K, solid I $\rightarrow$ solid II, $\Delta H/T = 1190$ J mol^{-1}/98.36 K	12.1
4. $\int C_P \, d \ln T$ for solid II from 98.36 K to 158.91 K	21.1
5. Fusion, 1992 J mol^{-1}/158.91 K	12.6
6. $\int C_P \, d \ln T$ for liquid from 158.91 K to 188.07 K	9.9
7. Vaporization, $\dfrac{16{,}150 \text{ J mol}^{-1}}{188.07 \text{ K}}$	85.9
8. $\int C_P \, d \ln T$ for gas from 188.07 K to 298.15 K	13.5
$S^\circ_{298.15} =$	185.9

[a] Walter J. Moore, *Physical Chemistry*, 4th ed., © 1972 Prentice-Hall, Inc., Englewood Cliffs, N.J., p. 111.

* The residual entropy can be viewed as the price one pays for fooling nature.

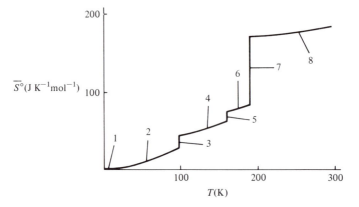

Figure 7.5 Graphical computation of the standard entropy of hydrogen chloride at 298.15 K. The numbers refer to the steps described in Table 7.1.

mind that it only holds for a perfectly ordered substance. Figure 7.5 shows the graphical computation of the entropy of HCl.

Entropies calculated by using Eq. (7.13) are called the third-law entropies or absolute entropies. Appendix 3 lists the standard absolute entropies of various elements and compounds at 298 K. Note that because these are absolute values, we omit the Δ sign and the subscript f for $S°$.

EXAMPLE 7.4 Calculate the standard molar entropy changes for the following reactions at 298 K:

(a) $CaCO_3(s) \longrightarrow CaO(s) + CO_2(g)$.

(b) $2H_2(g) + O_2(g) \longrightarrow 2H_2O(l)$.

(c) $N_2(g) + O_2(g) \longrightarrow 2NO(g)$.

Answer: The entropy change of a reaction is given by

$$\Delta S = \sum S_{\text{product}} - \sum S_{\text{reactant}}$$

From the $S°$ values listed in Appendix 3, we write

(a) $\Delta S = (\overline{S}°_{CaO} + \overline{S}°_{CO_2}) - (\overline{S}°_{CaCO_3})$
$= (1 \text{ mol})(39.75 + 213.64) \text{ J K}^{-1} \text{ mol}^{-1}$
$- (1 \text{ mol})(92.89 \text{ J K}^{-1} \text{ mol}^{-1})$
$= 160.50 \text{ J K}^{-1}$

(b) $\Delta S = (2\overline{S}°_{H_2O}) - (2\overline{S}°_{H_2} + \overline{S}°_{O_2})$
$= (2 \text{ mol})(69.94 \text{ J K}^{-1} \text{ mol}^{-1})$
$- [(2 \text{ mol})(130.59) + (1 \text{ mol})(205.03)] \text{ J K}^{-1} \text{ mol}^{-1}$
$= -326.33 \text{ J K}^{-1}$

(c) $\Delta S = (2\overline{S}°_{NO}) - (\overline{S}°_{N_2} + \overline{S}°_{O_2})$
$= (2 \text{ mol})(210.62 \text{ J K}^{-1} \text{ mol}^{-1})$
$- (1 \text{ mol})(191.49 + 205.03) \text{ J K}^{-1} \text{ mol}^{-1}$
$= 24.72 \text{ J K}^{-1}$

Comment: The results are consistent with our expectation that reactions producing a net increase in the number of gas molecules are accompanied by an increase in entropy as in reaction (a) while the reverse holds true for reaction (b). In (c), there is no change in the total number of gas molecules; therefore, the change in entropy is relatively small. Note that all the entropy changes apply to the system.

The Gibbs Free Energy 7.6

With the first law of thermodynamics to take care of energy balance and the second law to help us decide which processes can occur spontaneously, we might reasonably expect that we now have enough thermodynamic quantities to deal with any situation. This is true in principle, but the equations we have derived so far are not the most convenient to apply in practice. For example, in order to apply Eq. (7.6), we need to calculate the entropy change in both the system and the surroundings. Since we are generally interested only in what happens in the system and are not concerned with events occurring in the surroundings, it would be more convenient if we could obtain criteria for equilibrium and spontaneity only in terms of the change of a certain thermodynamic function of the system and not of the entire universe, as is the case of ΔS_{univ}.

Consider a system in thermal equilibrium with its surroundings at temperature T. Let there be a process occurring in the system such that an infinitesimal amount of heat, dQ, is transferred from the system to the surroundings. Thus we have $-dQ_{sys} = dQ_{surr}$. The total change in entropy, according to Eq. (7.6), is

$$dS_{univ} = dS_{sys} + dS_{surr} \geq 0$$

$$= dS_{sys} + \frac{dQ_{surr}}{T} \geq 0$$

$$= dS_{sys} - \frac{dQ_{sys}}{T} \geq 0$$

Note that every quantity on the right-hand side of the equation above now refers to the system. If the process takes place at constant pressure, then $dQ_{sys} = dH_{sys}$ or

$$dS_{sys} - \frac{dH_{sys}}{T} \geq 0$$

Multiplying the equation above by $-T$, we obtain*

$$dH_{sys} - T \, dS_{sys} \leq 0$$

We now *define* a function, called the *Gibbs free energy*, G, as

$$G = H - TS \tag{7.15}$$

At constant temperature, the change in the Gibbs free energy of the system is given by

$$dG_{sys} = dH_{sys} - T \, dS_{sys}$$

* The reversal of the inequality sign follows from the fact that if $2 > 0$, then $-2 < 0$.

We can now apply dG_{sys} as criteria for equilibrium and spontaneity as follows:

$$dG_{sys} \leq 0$$

where the $<$ sign denotes a spontaneous process while the equality sign denotes the equilibrium situation under conditions of constant temperature and pressure. Unless otherwise indicated, from now on we shall only consider the Gibbs free energy changes of a system. For this reason we shall omit the subscript "sys" for simplicity. From Eq. (7.15), we see that because H, T, and S are all state functions, G is also a state function. Further, like enthalpy, G has the units of energy. For a finite isothermal process $1 \to 2$, the change of Gibbs free energy is given by

$$\Delta G = \Delta H - T \, \Delta S \qquad (7.16)$$

where, if the pressure is also held constant,

$$\Delta G = G_2 - G_1 = 0 \qquad \text{system at equilibrium}$$

$$\Delta G = G_2 - G_1 < 0 \qquad \text{spontaneous process from 1 to 2}$$

If ΔG is negative, the process is said to be *exergonic*; if positive, *endergonic*.

A similar thermodynamic function, called the *Helmholtz free energy*, A, is defined as

$$A = U - TS \qquad (7.17)$$

Following the same procedure described above, we can show that at constant temperature and volume, the criteria for equilibrium and spontaneity are given by

$$dA_{sys} \leq 0$$

However, because many of the processes of interest occur at constant temperature and pressure, in this text we shall be mainly concerned with the Gibbs free energy.

The Gibbs free energy is a useful quantity because it enables us to consider both the enthalpy and entropy factors involved in a process. In some reactions the contributions of enthalpy and entropy reinforce each other. For example, if ΔH is negative (an exothermic reaction) and ΔS is positive, then $(\Delta H - T \, \Delta S)$ or ΔG is a negative quantity and the process is favored from left to right. In other reactions, enthalpy and entropy may work against each other; that is, ΔH and $(-T \, \Delta S)$ have different signs. Here the sign of ΔG is determined by the *magnitudes* of ΔH and $T \, \Delta S$ terms. If $|\Delta H| \gg |T \, \Delta S|$, then the reaction is said to be enthalpy driven because the sign of ΔG is predominantly determined by ΔH.* Conversely, if $|T \, \Delta S| \gg |\Delta H|$, then the process is an entropy-driven one. Table 7.2 shows the various situations encountered in practice.

As in the case of enthalpy, there is no way for us to measure the absolute values of Gibbs free energies. The convention we adopt here is analogous to that used for enthalpy discussed in Chapter 6; that is, we assign the standard Gibbs free energy of formation of an element in its most stable state of aggregation at 1 atm and 298 K to be zero. Again using the combustion of graphite as an example (see Section 6.6):

$$C(\text{graphite}) + O_2(g) \longrightarrow CO_2(g)$$

---

* The symbols $|x|$ denote the absolute magnitude of x. For example, if $x = -3$, then $|x| = 3$.

ΔH	ΔS	ΔG
$+$	$+$	Positive at low temperatures; negative at high temperatures. Reaction spontaneous in the forward direction at high temperatures and spontaneous in the reverse direction at low temperatures.
$+$	$-$	Positive at all temperatures. Reaction spontaneous in the reverse direction at all temperatures.
$-$	$+$	Negative at all temperatures. Reaction spontaneous in the forward direction at all temperatures.
$-$	$-$	Negative at low temperatures; positive at high temperatures. Reaction spontaneous at low temperatures; tends to reverse at high temperatures.

[a] We assume that both ΔH and ΔS are independent of temperature.

we write the standard free energy change $\Delta G°$ for the reaction as

$$\Delta G° = (1 \text{ mol}) \, \overline{G}°_{CO_2} - (1 \text{ mol}) \, \overline{G}°_{\text{graphite}} - (1 \text{ mol}) \, \overline{G}°_{O_2}$$

$$= (1 \text{ mol}) \, \overline{G}°_{CO_2}$$

where $\overline{G}°$ denotes the standard molar Gibbs free energy.

Since this reaction also represents the formation of CO_2 from its elements (C and O_2), it follows that

$$\overline{G}°_{CO_2} = \Delta \overline{G}°_f(CO_2)$$

where $\Delta \overline{G}°_f$ denotes the standard molar Gibbs free energy of formation. How do we determine this value? Note that both $\Delta \overline{H}°_f$ and $\Delta \overline{S}°_f$ for the reaction above can be calculated by using the values listed in Appendix 3.

$$\Delta \overline{H}°_f(CO_2) = -393.51 \text{ kJ mol}^{-1}$$

$$\Delta \overline{S}°_f(CO_2) = \overline{S}°(CO_2) - \overline{S}°(\text{graphite}) - \overline{S}°(O_2)$$

$$= (213.64 - 5.69 - 205.03) \text{ J K}^{-1} \text{ mol}^{-1}$$

$$= 2.92 \text{ J K}^{-1} \text{ mol}^{-1}$$

From Eq. (7.16), we have

$$\Delta \overline{G}°_f = \Delta \overline{H}°_f - T \, \Delta \overline{S}°_f$$

$$= -393.51 \text{ kJ mol}^{-1} - (298.15 \text{ K})\left(\frac{2.92}{1000} \text{ kJ K}^{-1} \text{ mol}^{-1}\right)$$

$$= -394.38 \text{ kJ mol}^{-1}$$

In this manner, we can evaluate the $\Delta \overline{G}°_f$ values of most substances, some of which are listed in Appendix 3.

EXAMPLE 7.5 Calculate the standard free energy changes for the following reactions at 298 K, using the thermodynamic data in Appendix 3:

(a) $C_6H_{12}O_6(s) + 6O_2(g) \longrightarrow 6CO_2(g) + 6H_2O(l)$

(b) $CH_4(g) + 2O_2(g) \longrightarrow CO_2(g) + 2H_2O(l)$

145

Answer: (a) This is the combustion of α-D-glucose. From Appendix 3, we write

$$\Delta H° = 6\ \Delta \overline{H}_f°(CO_2) + 6\ \Delta \overline{H}_f°(H_2O) - [\Delta \overline{H}_f°(C_6H_{12}O_6) + 6\ \Delta \overline{H}_f°(O_2)]$$

$$= [(6\ mol)(-393.51) + (6\ mol)(-285.84)]\ kJ\ mol^{-1}$$

$$- [(1\ mol)(-1274.45) - (6\ mol)(0.0)]\ kJ\ mol^{-1}$$

$$= -2801.65\ kJ$$

Similarly,

$$\Delta S° = 6\overline{S}_{CO_2}° + 6\overline{S}_{H_2O}° - (\overline{S}_{C_6H_{12}O_6}° + 6\overline{S}_{O_2}°)$$

$$= (6\ mol)(213.64 + 69.94)\ J\ K^{-1}\ mol^{-1}$$

$$- [(1\ mol)(212.13) + (6\ mol)(205.03)]\ J\ K^{-1}\ mol^{-1}$$

$$= 259.17\ J\ K^{-1}$$

Hence

$$\Delta G° = \Delta H° - T\ \Delta S°$$

$$= -2801.65\ kJ - (298.15\ K)\left(\frac{259.17}{1000}\right)\ kJ\ K^{-1}$$

$$= -2878.92\ kJ$$

An alternative way to calculate $\Delta G°$ is to write

$$\Delta G° = 6\ \Delta \overline{G}_f°(CO_2) + 6\ \Delta \overline{G}_f°(H_2O) - \Delta \overline{G}_f°(C_6H_{12}O_6) - \Delta \overline{G}_f°(O_2)$$

$$= (6\ mol)(-394.38\ kJ\ mol^{-1}) + (6\ mol)(-237.19\ kJ\ mol^{-1})$$

$$- (1\ mol)(-910.56\ kJ\ mol^{-1})$$

$$= -2878.88\ kJ$$

The small difference between these two values comes from rounding off. The large negative value of $\Delta G°$ indicates that the combustion of glucose is a highly spontaneous process when reactants and products are in their standard states. Because $\Delta H°$ is negative and $\Delta S°$ positive, the reaction is spontaneous at any temperature (if $\Delta H°$ and $\Delta S°$ are independent of temperature). Figure 7.6a shows the change of $\Delta H°$, $-T\ \Delta S°$, and $\Delta G°$ represented on a "vector diagram."

(b)   From Appendix 3, we can show that the standard enthalpy and entropy for the combustion of methane are

$$\Delta H° = -890.36\ kJ$$

$$\Delta S° = -242.67\ J\ K^{-1}$$

so that

$$\Delta G° = \Delta H° - T\ \Delta S°$$

$$= -890.36\ kJ - (298.15\ K)\left(\frac{-242.67}{1000}\right)\ kJ\ K^{-1}$$

$$= -818.01\ kJ$$

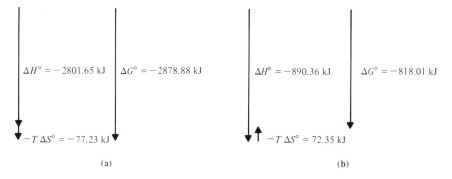

$\Delta H^\circ = -2801.65 \text{ kJ}$ $\Delta G^\circ = -2878.88 \text{ kJ}$ $\Delta H^\circ = -890.36 \text{ kJ}$ $\Delta G^\circ = -818.01 \text{ kJ}$

$-T \Delta S^\circ = -77.23 \text{ kJ}$ $-T \Delta S^\circ = 72.35 \text{ kJ}$

(a)   (b)

Vector diagrams showing the changes of ΔH°, $-T \Delta S^\circ$, and ΔG° at 298 K for (a) the combustion of α-D-glucose in air, and (b) the combustion of methane in air.   **Figure 7.6**

The reader should calculate ΔG° using the alternative method outlined in (a). The changes of ΔH°, $-T \Delta S^\circ$, and ΔG° are shown in the vector diagram in Figure 7.6b. Note that because ΔS° is negative, $-T \Delta S^\circ$ is a positive quantity, so that ΔG° is less negative than ΔH°. Nevertheless, ΔG° is a large negative quantity, indicating that the reaction above is spontaneous when reactants and products are in their standard states.

A question may be raised here: If the combustions of glucose and methane are such spontaneous processes, how is it that these substances can be kept in air for very long periods of time without any apparent change? Here lies the limitation of thermodynamics, for it tells us only which direction a reaction will go; it says nothing about the *rate* at which it occurs. For any reaction to start, the reactants must first acquire sufficient energy to get over the activation energy barrier. Glucose (or methane) molecules in a jar at room temperature lack this energy and are therefore perfectly stable. More will be said of this in Chapter 13.

Properties of the Gibbs Free Energy 7.7

Because the Gibbs free energy is such a useful thermodynamic function, we would like to know how it is affected by experimental conditions. In particular, we are interested in the dependence of G on temperature and pressure. However, before we derive the necessary equations relating G to temperature and pressure, it is useful to first examine some general properties of the Gibbs free energy.

We start with the definition of G:

$$G = H - TS$$

For an infinitesimal process, we write

$$dG = dH - T \, dS - S \, dT$$

Now since

$$H = U + PV$$

$$dH = dU + P \, dV + V \, dP$$

147

According to the first law of thermodynamics,

$$dU = dQ + dW$$

and

$$dU = dQ - P \, dV$$

If we assume the process to be reversible, then

$$dQ = T \, dS$$

so that

$$dU = T \, dS - P \, dV \tag{7.18}$$

and

$$dH = (T \, dS - P \, dV) + P \, dV + V \, dP$$
$$= T \, dS + V \, dP$$

Finally, we have

$$dG = (T \, dS + V \, dP) - T \, dS - S \, dT$$
$$= V \, dP - S \, dT \tag{7.19}$$

Equation (7.18) incorporates both the first and second laws, while Eq. (7.19) gives the dependence of G on pressure and temperature. Both are important fundamental equations of thermodynamics.

DEPENDENCE OF *G* ON TEMPERATURE

Suppose that we have measured the ΔG for a biochemical reaction at 298 K and would like to know whether the reaction would become more favored at the physiological temperature of 310 K (37°C). One way to find out is to run the same reaction at the higher temperature. Fortunately, the way in which ΔG changes with temperature at constant pressure for a reaction is predictable if the enthalpy of the reaction is known. We start with

$$G = H - TS$$

or

$$S = \frac{H - G}{T}$$

From Eq. (7.19), we find that varying G with respect to temperature at constant pressure gives

$$\left(\frac{\partial G}{\partial T}\right)_P = -S \tag{7.20a}$$

Because the entropy of a system is always positive, what Eq. (7.20a) says is that $(\partial G/\partial T)_P$ is always negative or that G always decreases with increasing temperature at constant pressure. Equating the two expressions for S above, we obtain

$$\left(\frac{\partial G}{\partial T}\right)_P = -\frac{H - G}{T}$$

Now, it is frequently important to know not only how G, but also how G/T depends on temperature. By the ordinary rule of differentiation, we obtain

$$\left[\frac{\partial(G/T)}{\partial T}\right]_P = \frac{1}{T}\left(\frac{\partial G}{\partial T}\right)_P - \frac{G}{T^2}$$

$$= -\frac{1}{T}\left(\frac{H-G}{T}\right) - \frac{G}{T^2}$$

$$= -\frac{H}{T^2} \tag{7.20b}$$

Equation (7.20b) is known as the Gibbs–Helmholtz equation. As it stands, it is not particularly informative. This equation takes a much more useful form when it is applied to a chemical reaction:

$$\text{reactants} \longrightarrow \text{products}$$

The change in Gibbs free energy for the reaction is given by

$$\Delta G = G_{\text{products}} - G_{\text{reactants}}$$

Expressed in terms of *changes* in G and H, Eq. (7.20b) now takes the form

$$\left[\frac{\partial(\Delta G/T)}{\partial T}\right]_P = \frac{-\Delta H}{T^2}$$

or

$$d\left(\frac{\Delta G}{T}\right) = -\frac{\Delta H}{T^2}\,dT$$

Integrating between temperatures T_1 and T_2, we obtain

$$\int_1^2 d\left(\frac{\Delta G}{T}\right) = -\int_{T_1}^{T_2}\frac{\Delta H}{T^2}\,dT$$

$$\frac{\Delta G_2}{T_2} - \frac{\Delta G_1}{T_1} = \Delta H\left(\frac{1}{T_2} - \frac{1}{T_1}\right) = -\Delta H\left(\frac{T_2 - T_1}{T_1 T_2}\right) \tag{7.21}$$

Equation (7.21) is the integrated form of the Gibbs–Helmholtz equation. The quantities ΔG_1 and ΔG_2 are the changes in the Gibbs free energy at T_1 and T_2. In deriving Eq. (7.21), we have assumed that the enthalpy change for the reaction, ΔH, is independent of temperature. This assumption usually holds true if the temperature range is small, say 50 K or less.

EXAMPLE 7.6 The overall reaction for the metabolic breakdown of glucose in our bodies is the same as the combustion of glucose in air. Using the $\Delta G°$ and $\Delta H°$ values calculated in Example 7.5, determine whether the reaction will be more favored at 37°C.

Answer: From Example 7.5, we have

$$\Delta G° = -2878.92 \text{ kJ}$$

and

$$\Delta H° = -2801.65 \text{ kJ}$$

Assuming $\Delta H°$ to be independent of temperature, from Eq. (7.21) we write

$$\frac{\Delta G°_{310}}{310 \text{ K}} - \frac{-2878.92 \text{ kJ}}{298 \text{ K}} = -(-2801.65 \text{ kJ})\left(\frac{310 - 298}{298 \times 310 \text{ K}}\right)$$

or

$$\Delta G°_{310} = -2882.03 \text{ kJ}$$

Since $\Delta G°_{310}$ is more negative than $\Delta G°_{298}$, the reaction is more favored at 310 K than at 298 K.

DEPENDENCE OF *G* ON PRESSURE

To see how the Gibbs free energy depends on pressure, we again employ Eq. (7.19). At constant temperature,

$$\left(\frac{\partial G}{\partial P}\right)_T = V \tag{7.22}$$

Since volume must be a positive quantity, Eq. (7.22) says that the Gibbs free energy of a system always increases with pressure at constant temperature. What we are interested in is how G increases when the pressure of the system increases from P_1 to P_2. Since at constant temperature

$$dG = V \, dP$$

we can write the change in G as

$$\Delta G = G_2 - G_1 = \int_{G_1}^{G_2} dG = \int_{P_1}^{P_2} V \, dP$$

For an ideal gas $V = nRT/P$ so that

$$G_2 - G_1 = \int_{P_1}^{P_2} \frac{nRT}{P} \, dP$$

$$= nRT \ln \frac{P_2}{P_1} \tag{7.23}$$

If we set $P_1 = 1$ atm so that G_1 is replaced by the symbol for the standard state $G°$, and replace G_2 by G and P_2 by P, we obtain the useful equation

$$G = G° + nRT \ln P$$

Expressed as molar quantities,

$$\bar{G} = \bar{G}° + RT \ln P \tag{7.24}$$

where $\bar{G}°$ is a function of temperature only. Equation (7.24) relates the molar Gibbs free energy of an ideal gas to its pressure. It is important to keep in mind that the quantity $\ln P$ should strictly be expressed as $\ln (P \text{ atm}/1 \text{ atm})$, so that P becomes a dimensionless quantity in Eq. (7.24). Later we shall see a similar equation relating the Gibbs free energy of a substance to its concentration in the condensed phase.

EXAMPLE 7.7 A 0.590-mol sample of an ideal gas initially at 300 K and 1 atm is compressed isothermally until its final pressure becomes 6.90 atm. Calculate the change in the Gibbs free energy for this process.

$$P_1 = 1.00 \text{ atm} \qquad P_2 = 6.90 \text{ atm}$$

so that

$$\Delta G = (0.590 \text{ mol})(8.314 \text{ J K}^{-1} \text{ mol}^{-1})(300 \text{ K}) \ln \frac{6.90}{1.00}$$

$$= 2840 \text{ J}$$

Thus far we have discussed the dependence of G on pressure for gases. Because the volume of a liquid or a solid is practically independent of applied pressure, we write

$$G_2 - G_1 = \int_{P_1}^{P_2} V \, dP$$

$$= V(P_2 - P_1)$$

or

$$G_2 = G_1 + V(P_2 - P_1)$$

That is, the volume V may be taken outside the integral. In general, the Gibbs free energies of liquids and solids are much less dependent of pressure.

Gibbs Free Energy and Phase Equilibria 7.8

To see how the Gibbs free energy can be applied to a practical system, we shall study phase equilibria in this section. Phase equilibria are concerned with physical processes such as freezing and boiling. In Chapter 10 we shall apply the Gibbs free energy to the study of chemical equilibria. Our discussion is restricted to only one component systems in this section.

Consider that at some temperature and pressure, two phases, say solid and liquid, of a one-component system are in equilibrium. How shall we formulate the condition for equilibrium between these two phases? We might be tempted to equate the Gibbs free energies as follows:

$$G_{\text{solid}} = G_{\text{liquid}}$$

But this formulation will not hold, for it is possible to have a small ice cube floating in an ocean of water at 0°C, and yet the free energy of water is obviously much larger than that of the ice cube. Instead, we must insist that the Gibbs free energy *per mole* (or the molar Gibbs free energy) of the substance must be the same in both phases at equilibrium:

$$\overline{G}_{\text{solid}} = \overline{G}_{\text{liquid}}$$

If external conditions (temperature or pressure) are altered so that $\overline{G}_{\text{solid}} > \overline{G}_{\text{liquid}}$, then some solid would melt because

$$\Delta \overline{G} = \overline{G}_{\text{liquid}} - \overline{G}_{\text{solid}} < 0$$

On the other hand, if $\overline{G}_{\text{solid}} < \overline{G}_{\text{liquid}}$, then some liquid would freeze spontaneously.

Our next step is to see how the molar Gibbs free energies of solid, liquid, and vapor depend on temperature and pressure. Equation (7.20a) expressed in molar quantities becomes

$$\left(\frac{\partial \overline{G}}{\partial T}\right)_P = -\overline{S}$$

Since the entropy of a substance in any phase is always positive, it follows that a plot of $\overline{G}$ versus T at constant pressure gives us a curve with a negative slope. For the three phases of a single substance, we have

$$\left(\frac{\partial \overline{G}_{\text{solid}}}{\partial T}\right)_P = -\overline{S}_{\text{solid}} \qquad \left(\frac{\partial \overline{G}_{\text{liquid}}}{\partial T}\right)_P = -\overline{S}_{\text{liq}} \qquad \left(\frac{\partial \overline{G}_{\text{vap}}}{\partial T}\right)_P = -\overline{S}_{\text{vap}}$$

At any temperature, the molar entropies decrease in the order*

$$\overline{S}_{\text{vap}} \gg \overline{S}_{\text{liquid}} > \overline{S}_{\text{solid}}$$

These differences are reflected in the slopes of lines shown in Figure 7.7. At high temperatures, gas is the most stable phase because it has the lowest molar Gibbs free energy. As temperature decreases, however, liquid becomes the stable phase and, finally, at even lower temperatures solid becomes the most stable phase. The intercept between the gas and liquid lines is the point at which these two phases are in equilibrium, that is, $\overline{G}_{\text{gas}} = \overline{G}_{\text{liq}}$. The corresponding temperature is T_b, the boiling point. Similarly, solid and liquid coexist in equilibrium at the temperature T_f, the melting (or fusion) point.

Figure 7.7 is that obtained for a particular pressure. How would an increase in pressure affect the phase equilibria? It was pointed out in the last section that the Gibbs free energy of a substance always increases with pressure [see Eq. (7.22)]. Further, for a given change in pressure the increase is greatest for gases, much less for liquids and solids. This follows from the fact that, from Eq. (7.22),

$$\left(\frac{\partial \overline{G}}{\partial P}\right)_T = \overline{V}$$

and molar volume of a gas is normally about a thousand times greater than

Figure 7.7

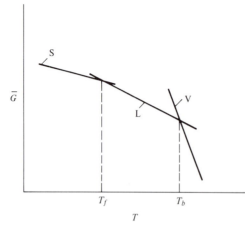

Dependence of the molar Gibbs free energy on temperature for the gas, liquid, and solid phases of a substance. The phase with the lowest G is the most stable phase at that temperature. The pressure is held constant.

---

* The relative magnitudes of S_{solid}, S_{liquid}, and S_{vap} may be seen from Figure 7.5.

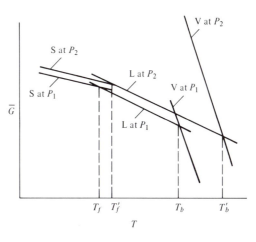

Figure 7.8

Pressure dependence of the molar Gibbs free energy. For the majority of substances, an increase in pressure leads to an increase in both the melting point and the boiling point. (Here we have $P_2 > P_1$.)

that for a liquid or a solid. Figure 7.8 shows the increases of $\overline{G}$ for the three phases as the pressure increases from P_1 to P_2. We see that both T_f and T_b are shifted to higher values, but T_b more so than T_f because of the larger increase in $\overline{G}$ for the gas. Thus in general an increase in external pressure will raise both the melting point and boiling point of a substance. Although it is not shown in Figure 7.8, the reverse also holds true; that is, decreasing the pressure will lower both the melting point and boiling point. It is important to keep in mind that our conclusion about the effect of pressure on melting point is based on the assumption that the molar volume of liquid is greater than that of solid. This is true of most, but not all, substances. An important exception is water. Here the molar volume of ice is actually greater than that of liquid water. For water, then, an increase in pressure will *lower* the melting point. More will be said about this later.

THE CLAPEYRON AND THE CLAUSIUS–CLAPEYRON EQUATIONS

We shall now derive some useful general relations for the quantitative understanding of phase equilibria. Consider a substance that exists in two phases, α and β. The condition for equilibrium at constant temperature and pressure is that

$$\overline{G}_\alpha = \overline{G}_\beta$$

so that

$$d\overline{G}_\alpha = d\overline{G}_\beta$$

To establish the relation of dT to dP in the change that links these two equilibrium states, we have, from Eq. (7.19),

$$d\overline{G}_\alpha = \overline{V}_\alpha\, dP - \overline{S}_\alpha\, dT = d\overline{G}_\beta = \overline{V}_\beta\, dP - \overline{S}_\beta\, dT$$

$$(\overline{S}_\beta - \overline{S}_\alpha)\, dT = (\overline{V}_\beta - \overline{V}_\alpha)\, dP$$

or

$$\frac{dP}{dT} = \frac{\Delta\overline{S}}{\Delta\overline{V}}$$

153

where $\Delta\bar{V}$ and $\Delta\bar{S}$ are the change in molar volume, and molar entropies for the $\alpha \to \beta$ phase transition, respectively. Since at equilibrium

$$\Delta\bar{S} = \frac{\Delta\bar{H}}{T}$$

the equation becomes

$$\frac{dP}{dT} = \frac{\Delta\bar{H}}{T \,\Delta\bar{V}} \tag{7.25}$$

where T is the phase transition temperature (it may be the melting point or the boiling point or any other temperature at which the two phases can coexist in equilibrium). Equation (7.25) is known as the Clapeyron equation. This simple relation gives us the ratio of the change in pressure to the change in temperature in terms of some readily measurable quantities such as the volume and enthalpy change for the process; it applies to fusion, vaporization, and sublimation as well as equilibria between two allotropic forms like graphite and diamond.

The Clapeyron equation can be expressed in a convenient approximate form when dealing with vaporization and sublimation equilibria. In these cases, the molar volume of the gas is much greater than that for the condensed phase, so that

$$\Delta\bar{V} = \bar{V}_{\text{gas}} - \bar{V}_{\text{condensed}} \simeq \bar{V}_{\text{gas}}$$

Further, if we assume ideal gas behavior, then

$$\Delta\bar{V} = \bar{V}_{\text{gas}} = \frac{RT}{P}$$

Substitution for $\Delta\bar{V}$ in Eq. (7.25) yields

$$\frac{dP}{dT} = \frac{P \,\Delta\bar{H}}{RT^2}$$

or

$$\frac{dP}{P} = d \ln P = \frac{\Delta\bar{H} \, dT}{RT^2} \tag{7.26}$$

Equation (7.26) is known as the Clausius–Clapeyron equation. Integrating Eq. (7.26) between limits of P_1, T_1 and P_2, T_2, we obtain

$$\int_{P_1}^{P_2} d \ln P = \ln \frac{P_2}{P_1} = \frac{\Delta\bar{H}}{R} \int_{T_1}^{T_2} \frac{dT}{T^2} = -\frac{\Delta\bar{H}}{R}\left(\frac{1}{T_2} - \frac{1}{T_1}\right)$$

or

$$\ln \frac{P_2}{P_1} = \frac{\Delta\bar{H}}{R} \frac{(T_2 - T_1)}{T_1 T_2} \tag{7.27}$$

We assume that $\Delta\bar{H}$ is independent of temperature. Alternately, we can express $\ln P$ as a function of temperature as follows:

$$\ln P = -\frac{\Delta\bar{H}}{RT} + \text{constant} \tag{7.28}$$

Thus a plot of ln P versus $1/T$ gives a straight line whose slope (which is negative) is equal to $-\Delta\bar{H}/R$.

We shall now apply the foregoing equations to some practical systems.

PHASE DIAGRAMS

At this point we are ready to examine the phase equilibria of some familiar systems. In particular, we shall consider water and carbon dioxide.

Water. Figure 7.9 shows the phase diagram of water, where S, L, and V represent regions where only one phase can exist. Along any one curve, however, the corresponding two phases can coexist. The slope of any curve is given by dP/dT. Finally, there is only one point at which all three phases can coexist. This is called the *triple point*; for water, the triple point is at $T = 273.16$ K and $P = 0.006$ atm.

EXAMPLE 7.8 Calculate the slope of the S–L curve at 273.15 K. $\Delta\bar{H}_{fus} = 6.01$ kJ mol^{-1}, $\bar{V}_L = 0.0180$ liter mol^{-1}, and $\bar{V}_S = 0.0196$ liter mol^{-1}.

Answer: From Eq. (7.25),

$$\frac{dP}{dT} = \frac{\Delta\bar{H}_{fus}}{T_f \, \Delta\bar{V}_{fus}}$$

Using the conversion factor 1 J = 9.87×10^{-3} liter atm, we obtain

$$\frac{dP}{dT} = \frac{(6010 \text{ J mol}^{-1})(9.87 \times 10^{-3} \text{ liter atm J}^{-1})}{(273.15 \text{ K})(0.0180 - 0.0196) \text{ liter mol}^{-1}}$$

$$= -136 \text{ atm K}^{-1}$$

Comments: (1) Because the molar volume of liquid water is smaller than that for ice, the slope is negative. Further, since the quantity $(\bar{V}_L - \bar{V}_S)$ is small, the slope is also quite steep. (2) An interesting result is obtained by

Phase diagram of water. Note that the solid–liquid curve has a negative slope. The liquid–vapor curve stops at x, the critical point (647.6 K and 219.5 atm).

Figure 7.9

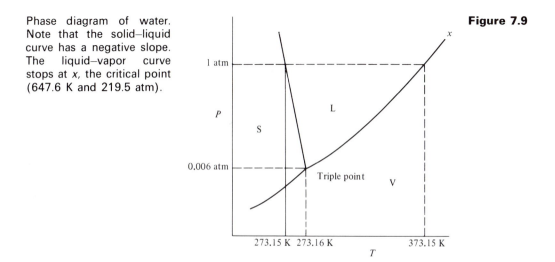

calculating the quantity dT/dP, which gives the change (decrease) in melting point as a function of pressure. We find that $dT/dP = -7.35 \times 10^{-3}$ K atm^{-1}, or that the melting point of ice decreases by 7.35×10^{-3} K whenever the pressure increases by 1 atm. This is what makes ice skating possible. The pressure exerted by the weight of a skater over the ice can be quite appreciable (of the order of 1000 atm) because of the small areas of the blades. On melting, the film of water formed between skates and ice acts as a lubricant to facilitate movement over ice. However, as many people have found out, skating becomes difficult if the temperature is below $-15°C$.

EXAMPLE 7.9 The following data show the variation of the vapor pressure of water as a function of temperature:

P (mm Hg)	17.54	31.82	55.32	92.51	149.38	233.7	355.1
T (°C)	20	30	40	50	60	70	80

Determine the heat of vaporization for water.

Answer: We need Eq. (7.28). The first step is to convert the data into a suitable form for plotting:

ln P	2.865	3.460	4.013	4.527
$1/T$ (K^{-1})	3.41×10^{-3}	3.30×10^{-3}	3.19×10^{-3}	3.10×10^{-3}

	5.006	5.454	5.872
	3.00×10^{-3}	2.91×10^{-3}	2.83×10^{-3}

Figure 7.10 shows the plot of ln P versus $1/T$. From the measured slope, we have

$$-5090 \text{ K} = -\frac{\Delta \bar{H}_{vap}}{R}$$

or

$$\Delta \bar{H}_{vap} = (8.314 \text{ J K}^{-1} \text{ mol}^{-1})(5090 \text{ K})$$

$$= 42,300 \text{ J mol}^{-1}$$

$$= 42.3 \text{ kJ mol}^{-1}$$

Comments: (1) The molar heat of vaporization for water measured at its normal boiling point is 40.79 kJ mol^{-1}. However, since $\Delta \bar{H}$ does depend on temperature to a certain extent, our graphically determined value is taken to be the average value between 20°C and 80°C. (2) In plotting the data, the quantity ln P is of course dimensionless. It is interesting to note that the same slope is obtained whether we express the pressure as mm Hg or as atm. The reason is as follows. The relation between atm and mm Hg is

$$P(\text{atm}) = CP'(\text{mm Hg})$$

Plot of ln P versus $1/T$. The slope of the line is given by $-\Delta H_{vap}/R$.

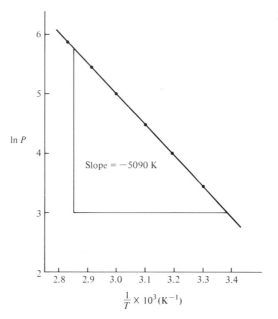

Figure 7.10

Slope = -5090 K

$\frac{1}{T} \times 10^3 (\mathrm{K}^{-1})$

where C is a conversion constant. Suppose that we have originally expressed pressures in atm unit. The slope of the line is given by

$$\text{slope} = \frac{\ln P_2 - \ln P_1}{1/T_2 - 1/T_1} = \frac{\ln CP'_2 - \ln CP'_1}{1/T_2 - 1/T_1}$$

$$= \frac{\ln P'_2 - \ln P'_1}{1/T_2 - 1/T_1}$$

Thus whether the pressure is in atm (P_1 and P_2) or in mm Hg (P'_1 and P'_2), the slope is the same.

Carbon Dioxide. Figure 7.11 shows the phase diagram for carbon dioxide. The main difference between this diagram and that for water is that the S–L curve for CO_2 has a positive slope. This follows from the fact that

Phase diagram of carbon dioxide. Note that the solid–liquid curve has a positive slope. This is true of most substances. The liquid–vapor curve stops at *x*, which is the critical point (304.2 K and 73.0 atm).

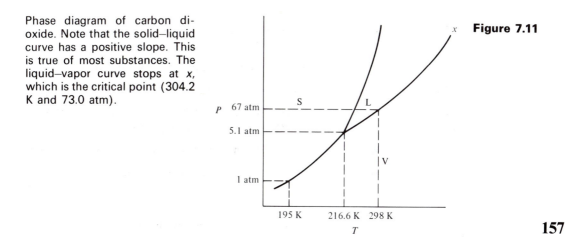

Figure 7.11

P 67 atm

5.1 atm

1 atm

195 K   216.6 K   298 K

T

since $\overline{V}_{liq} > \overline{V}_{solid}$, the quantity on the right-hand side of Eq. (7.25) is positive and therefore so is dP/dT. Note that liquid CO_2 is not stable at pressures below 5 atm. This is the reason that solid CO_2 is called "Dry Ice"—under atmospheric conditions it does not melt. Liquid CO_2 does exist at room temperature, of course, but it is normally confined to a cylinder under a pressure of 67 atm!

THE PHASE RULE

Useful information regarding phase equilibria can be predicted by a simple rule that was originally derived by Gibbs*:

$$f = c - p + 2$$

where c is the number of components and p is the number of phases present in a system. The degree of freedom f gives the number of variables (pressure, temperature, and composition) that must be fixed to completely describe the system. For example, for a pure gas we have $c = 1$ and $p = 1$ so that $f = 2$. This means that to completely describe the system, we only have to know two of the three variables P, V, and T. The third variable can be calculated from the equations of state described in Chapter 2. Now consider the phase diagram of water shown in Figure 7.9. In the pure phase region (S, L, or V) we again have $f = 2$, meaning that the pressure can be varied independently of temperature. Along the S–L, L–V, or S–V boundary, however, $p = 2$, and $f = 1$. Thus for every value of P, there can be only one specific value of T. Finally, at the triple point, $p = 3$ and $f = 0$. Under these conditions the system is totally fixed, and no variation of either the pressure or the temperature is possible.

Suggestions for Further Reading

See the texts listed in Chapter 6.

Reading Assignments

SECTION A

"Why Does Methane Burn?" R. T. Sanderson, *J. Chem. Educ.* **45**, 423 (1968).

"How Can You Tell Whether a Reaction Will Occur?" G. E. MacWood and F. H. Verhoek, *J. Chem. Educ.* **38**, 334 (1961).

"The Vapor Pressure of Water: A Good Reference System?" J. Pupezin, G. Jancso, and W. A. van Hook, *J. Chem. Educ.* **48**, 114 (1971).

"The Triple Point of Water," F. L. Swinton, *J. Chem. Educ.* **44**, 541 (1967).

"The Second Law of Thermodynamics," H. A. Bent, *J. Chem. Educ.* **39**, 491 (1962).

"The Laws of Disorder," G. Porter, *Chemistry* **41**(5), 23, **41**(6), 21, **41**(7), 36, **41**(8), 16, **41**(9), 28, **41**(10), 24, **41**(11), 28 (1968); **42**(1), 21, **42**(2), 19 (1969).

"Haste Makes Waste: Pollution and Entropy," H. A. Bent, *Chemistry* **44**(9), 6 (1971).

"Some Observations Concerning the van't Hoff Equation," J. T. MacQueen, *J. Chem. Educ.* **44**, 755 (1967).

---

* For derivation, see any of the standard physical chemistry texts listed in Chapter 1.

"Introduction to the Thermodynamics of Biopolymer Growth," C. Kittel, *Am. J. Phys.* **40**, 60 (1972).

"Entropy and Chemical Reactions," J. N. Spencer, O. Gordon, and H. D. Schreiber, *Chemistry* **47**(11), 12 (1974).

SECTION B

"The Use and Misuse of the Laws of Thermodynamics," M. L. McGlashan, *J. Chem. Educ.* **43**, 226 (1966).

"The Scope and Limitations of Thermodynamics," K. G. Denbigh, *Chem. Brit.* **4**, 339 (1968).

"Thermodynamics of Hard Molecules," L. K. Runnels, *J. Chem. Educ.* **47**, 742 (1970).

"Recent Improvements in Explaining the Periodicity of Oxygen Chemistry," R. T. Sanderson, *J. Chem. Educ.* **46**, 635 (1969).

"Principles of Chemical Reaction," R. T. Sanderson, *J. Chem. Educ.* **41**, 13 (1964).

"Maxwell's Demon," W. Ehrenberg, *Sci. Am.*, Nov. 1967.

"Temperature–Entropy Diagrams," A. Wood, *J. Chem. Educ.* **47**, 285 (1970).

"Energy Conservation: Better Living Through Thermodynamics," W. D. Metz, *Science* **188**, 820 (1975).

"The Thermodynamic Transformation of Organic Chemistry," D. E. Stull, *Am. Sci.* **54**, 734 (1971).

"High Pressure Synthetic Chemistry," A. P. Hagen, *J. Chem. Educ.* **55**, 620 (1978).

"Reversibility and Returnability," J. A. Campbell, *J. Chem. Educ.* **57**, 345 (1980).

"The Synthesis of Diamond," H. Hall, *J. Chem. Educ.* **38**, 484 (1961).

Problems

7.1 One of the many statements of the second law of thermodynamics is: Heat cannot flow from a colder body to a warmer one without external aid. Assume two systems 1 and 2 at T_1 and $T_2 (T_2 > T_1)$. Show that if a quantity of heat Q did flow spontaneously from 1 to 2, the process would result in a decrease in the entropy of the universe. (You may assume that the heat flows very slowly so that the process can be regarded as reversible. Assume also that the loss of heat by 1 and the gain of heat by 2 do not affect T_1 and T_2.)

7.2 What is the probability that all the molecules of a gas will be found in one half of a container when the gas consists of (a) 1 molecule, (b) 2 molecules, (c) 20 molecules, and (d) 6×10^{23} molecules?

7.3 A quantity of 35.0 g of water at 25.0°C (called A) is mixed with 160.0 g of water at 86.0°C (called B). (a) Calculate the final temperature of the system, assuming that the mixing is carried out adiabatically. (b) Calculate the entropy change of A, B, and the entire system.

7.4 A quantity of ice at 0°C is allowed to melt in a large body of water also at 0°C. Calculate the molar entropy change for this process. (The molar heat of fusion of ice is 6.01 kJ mol^{-1}.)

7.5 The heat capacity of chlorine gas is given by

$$C_P = (31.0 + 0.008T) \text{ J K}^{-1} \text{ mol}^{-1}$$

Calculate the entropy change when 2 mol of the gas is heated from 300 K to 400 K at constant pressure.

7.6 The vapor pressure of mercury at various temperatures has been determined as follows:

T (K)	P (mm Hg)
323.15	0.0127
353.15	0.0888
393.15	0.7457
413.15	1.845
433.15	4.189

Calculate $\Delta \overline{H}_{vap}$ for mercury.

7.7 In the old days the domestic gas for cooking was called "water gas" and was prepared as follows:

$$H_2O(g) + C(\text{graphite}) \longrightarrow CO(g) + H_2(g)$$

From the thermodynamic quantities listed in Appendix 3, predict whether this reaction will occur at 298 K. If not, at what temperature will the reaction occur? Assume $\Delta H°$ and $\Delta S°$ to be temperature-independent.

7.8 The second law of thermodynamics says that heat cannot flow spontaneously from a cold body to a hot one. Describe a practical system in which this process is made to occur by an external driving force. (*Hint*: Consider the refrigerator.)

7.9 According to Trouton's rule, the molar entropy of vaporization is approximately a constant (about 88 J K^{-1} mol^{-1}) for most liquids. Provide a molecular interpretation for this rule. The $\Delta \overline{S}_{vap}$ for water is unusually high (109.3 J K^{-1} mol^{-1}). Explain.

7.10 List two important differences between a steady state and an equilibrium state.

7.11 The molar enthalpy of fusion of ice at 273.15 K and 1 atm is 6.01 kJ mol^{-1}. (a) Calculate the molar entropy of fusion and compare its value with the molar entropy of vaporization of water (see Problem 7.9). Explain the difference. (b) Calculate ΔG for the melting of 1 mol of ice at 297 K.

*7.12 (a) The molar enthalpy of fusion of ice at 0°C is 6.01 kJ mol^{-1} and the molar heat capacity of water between 0°C and -10°C is 75.3 J K^{-1} mol^{-1}. One mole of supercooled water at -10°C is induced to crystallize in an insulated vessel. The result is a mixture of ice and water at 0°C. What fraction of this mixture is ice? (b) Changes in entropy have often been described in terms of order and disorder. For example, a process that leads to an increase in order has a negative ΔS. For the freezing of the supercooled water discussed in (a), two things are obvious. First, the process is spontaneous. Second, the freezing is accompanied by an increase in order. How would you account for the apparent contradiction that in a spontaneous process such as this, ΔS is negative?

*7.13 Suppose that your friend told you of the following extraordinary event. A block of metal weighing 500 g was seen rising spontaneously from the table on which it was resting to a height of 1.00 cm above the table. He stated that the metal had absorbed thermal energy from the table which was then used to raise itself against gravitational pull. (a) Does this process violate the first law of thermodynamics? (b) How about the second law? Assume that the room temperature was 298 K and that the table was large enough so that its temperature was unaffected by this transfer of energy. (*Hint*: First calculate the decrease in entropy as a result of this process and then estimate the probability for the occurrence of such a process.)

7.14 A quantity of 0.35 mol of an ideal gas initially at 15.6°C is expanded from 1.2 liters to 7.4 liters. Calculate W, Q, ΔU, ΔH, ΔS, and ΔG if the process is carried out (a) isothermally and reversibly, and (b) isothermally and irreversibly against an external pressure of 1.0 atm.

7.15 One mole of an ideal gas is isothermally expanded from 5.0 liters to 10 liters at 300 K. Compare the entropy changes for the system, surroundings, and the universe if the process is carried out (a) reversibly, and (b) irreversibly against an external pressure of 2.0 atm.

7.16 The heat capacity of hydrogen may be represented by

$$C_P = (1.554 + 0.0022T) \text{ J K}^{-1} \text{ mol}^{-1}$$

Calculate the entropy changes for the system, surrounding, and the universe for the (a) reversible heating, and (b) irreversible heating of 1 mol of hydrogen from 300 K to 600 K. [*Hint:* In (b) assume the surroundings to be at 600 K.]

7.17 Show a specific example for each of the following spontaneous processes: (a) the process is largely driven by entropy change, (b) the process is largely driven by enthalpy change, (c) an exothermic reaction with an increase in entropy, and (d) an exothermic reaction with a decrease in entropy.

7.18 Calculate ΔU, ΔH, and ΔS for the following process:

$$\text{1 mol of liquid water at 25°C and 1 atm} \longrightarrow \text{1 mol of steam at 100°C and 1 atm}$$

The molar heat of vaporization of water at 373 K is 40.79 kJ mol^{-1} and the molar heat capacity of water is 75.30 J K^{-1} mol^{-1}. Assume the molar heat capacity to be temperature-independent and ideal gas behavior.

7.19 Describe as many ways as you can think of for measuring the ΔG of a process.

7.20 Use the values listed in Appendix 3 to calculate $\Delta G°$ for the following alcohol fermentation:

$$\alpha\text{-D-glucose(aq)} \longrightarrow 2C_2H_5OH(l) + 2CO_2(g)$$

7.21 Molecules of a gas at any temperature T above the absolute zero are always in constant motion. Does this "perpetual motion" violate the laws of thermodynamics?

7.22 Without referring to any table, predict whether the entropy change is positive, nearly zero, or negative for each of the following reactions:
(a) $N_2(g) + O_2(g) \longrightarrow 2NO(g)$
(b) $2Mg(s) + O_2(g) \longrightarrow 2MgO(s)$
(c) $2H_2O_2(l) \longrightarrow 2H_2O(l) + O_2(g)$
(d) $H_2(g) + CO_2(g) \longrightarrow H_2O(g) + CO(g)$
(e) $Ag^+(aq) + Cl^-(aq) \longrightarrow AgCl(s)$

7.23 Suppose that the standard state is defined as $P = 2$ atm instead of $P = 1$ atm. Calculate the new $\Delta \overline{G}_f^\circ$ value of each of the following substances at 298 K: (a) $H_2(g)$, (b) $HBr(g)$, and (c) $NO(g)$.

7.24 According to the second law of thermodynamics, the entropy of an irreversible process in an isolated system must always increase. On the other hand, it is a well-known fact that the entropy of living systems remains small. (For example, the synthesis of highly complex protein molecules from individual amino acids is a process that leads to a decrease in entropy.) Is the second law invalid for living systems? Explain.

7.25 As an approximation, we can assume that proteins exist either in the native (or physiologically functioning) state and the denatured state. The standard molar enthalpy and entropy of the denaturation of a certain protein are 512 kJ mol^{-1} and 1.60 kJ K^{-1} mol^{-1}, respectively. Comment on the signs and magnitudes of these quantities and calculate the temperature at which the denaturation becomes spontaneous.

7.26 Certain bacteria in the soil obtain the necessary energy for growth by oxidizing nitrite to nitrate:

$$2NO_2^-(aq) + O_2(g) \longrightarrow 2NO_3^-(aq)$$

Given that the standard free energies of formation of NO_2^- and NO_3^- are -34.6 kJ mol^{-1} and -110.5 kJ mol^{-1}, respectively, calculate the amount of free energy released when 1 mol of NO_2^- is oxidized to 1 mol of NO_3^-.

7.27 Consider the synthesis of urea according to the equation

$$CO_2(g) + 2NH_3(g) \longrightarrow (NH_2)_2CO(s) + H_2O(l)$$

From the data listed in Appendix 3, calculate ΔG° for the reaction at 298 K. Assuming ΔG° to be temperature-independent, what is the ΔG° for the reaction at a pressure of 10.0 atm? The $\Delta \overline{G}_f^\circ$ of urea is -197.15 kJ mol^{-1}.

7.28 The pressure exerted on ice by a 60.0-kg skater is about 800 atm. Calculate the depression in freezing point.

7.29 Use the change in free energies to predict the direction for the following changes: (a) at the triple point of water, temperature is lowered at constant pressure, and (b) somewhere along the S–L curve of water, pressure is increased at constant temperature.

7.30 Consider the reaction

$$N_2(g) + 3H_2(g) \longrightarrow 2NH_3(g)$$

Calculate ΔS° for the reaction mixture, surroundings, and the universe at 298 K.

★7.31 The ΔH° for the conversion

$$C(graphite) \longrightarrow C(diamond)$$

was obtained by solving Problem 6.35. (a) Calculate the ΔS° for the reaction. Will the conversion occur spontaneously at 25°C or any other temperature? (b) From density measurements, the molar volume of graphite is found to be 2.1 cm^3 greater than that of diamond. Can the conversion of graphite to diamond be brought about at 25°C by applying pressure on graphite? If so, estimate the pressure at which the process becomes spontaneous. [*Hint*: Starting from Eq. (7.22), derive the equation $\Delta \overline{G} = (\overline{V}_{diamond} - \overline{V}_{graphite}) \Delta P$ for a constant-temperature process. Next, calculate the ΔP that would lead to the necessary decrease in Gibbs free energy.]

*7.32 Protein molecules are polypeptide chains made up of amino acids. In their physiologically functioning or native state, these chains fold up in an unique manner such that the nonpolar groups of the amino acids are usually buried in the interior region of the proteins, where there is little or no contact with water. When a protein denatures, the chain unfolds so that these nonpolar groups are exposed to water. A useful estimate of the changes of the thermodynamic quantities as a result of denaturation is to consider the transfer of a hydrocarbon such as methane (a nonpolar substance) from an inert solvent (such as benzene or carbon tetrachloride) to the aqueous environment:

(a) CH_4(inert solvent) $\longrightarrow$ CH_4(g)

(b) CH_4(g) $\longrightarrow$ CH_4(aq)

If ΔH° and ΔG° for (a) and (b) are approximately 2.0 kJ mol^{-1} and -14.5 kJ mol^{-1} and -13.5 kJ mol^{-1} and 26.5 kJ mol^{-1}, respectively, calculate ΔH° and ΔG° for the transfer of 1 mol of CH_4 according to the equation

$$CH_4\text{(inert solvent)} \longrightarrow CH_4\text{(aq)}$$

Comment on your results. Assume T = 298K.

7.33 In the reversible adiabatic expansion of an ideal gas, there are two contributions to entropy changes: the expansion of the gas and the cooling of the gas. Show that these two contributions are equal in magnitude but opposite in sign. Show also that for an irreversible adiabatic gas expansion, these two contributions are no longer equal in magnitude. Predict the sign of ΔS.

*7.34 Calculate the entropy change when neon at 25°C and 1 atm in a container of volume 0.780 liter is allowed to expand to 1.25 liters and is simultaneously heated to 85°C. Assume ideal behavior. (*Hint*: Since S is a state function, you can first calculate ΔS for expansion and then calculate ΔS for heating at constant final volume.)

7.35 When ammonium nitrate is dissolved in water, the solution becomes colder. What conclusion can you draw about ΔS° for the process?

8 Nonelectrolyte Solutions

If a solution of iodine in benzene is cooled, the red color deepens, while if it is warmed, the color approaches the violet of iodine vapor, indicating that the solvation decreases with rising temperature, as would be expected.
J. H. Hildebrand and C. A. Jenks*

The study of solutions is of great importance because most of the interesting and useful chemical and biological processes occur in liquid solutions. Generally, a solution is defined as a homogeneous mixture of two or more components that form a single phase. Gas solutions (for example, air) and solid solutions (for example, solder) are possible, as are liquid solutions.

This chapter is devoted to the thermodynamic study of ideal and nonideal solutions of nonelectrolytes, solutions that do not contain ionic species, and the colligative properties of these solutions.

8.1 Concentration Units

Any quantitative study of solutions requires that we know the amount of solute dissolved in a solvent or the concentration of the solution. Chemists employ several different concentration units in their work, each one having its advantages as well as its limitations. The particular way of expressing the concentration of a solution is generally determined by the use of the solution. This section will consider the following four concentration units: percent by weight, mole fraction, molarity, and molality.

PERCENT BY WEIGHT

The percent by weight of a solute in a solution is defined as

$$\text{percent by weight} = \frac{\text{weight of solute}}{\text{weight of solute} + \text{weight of solvent}} \times 100$$

$$= \frac{\text{weight of solute}}{\text{weight of solution}} \times 100$$

* J. H. Hildebrand and C. A. Jenks, *J. Am. Chem. Soc.* **42**, 2180 (1920).

MOLE FRACTION (*X*)

The concept of mole fraction was first introduced in Section 2.7. The mole fraction of a component *i* of a solution, X_i, is defined as

$$X_i = \frac{\text{number of moles of component } i}{\text{number of moles of all components}}$$

The mole fraction has no units.

MOLARITY (*M*)

Molarity is defined as the number of moles of solute dissolved in 1 liter of solution, that is,

$$\text{molarity} = \frac{\text{number of moles of solute}}{\text{number of liters of solution}}$$

Thus molarity has the units moles per liter. By convention, we use square brackets [ ] to represent molarity.

MOLALITY (*m*)

Molality is defined as the number of moles of solute dissolved in 1 kg (1000 g) of solvent, that is,

$$\text{molality} = \frac{\text{number of moles of solute}}{\text{weight of solvent in kg}}$$

Thus molality has the units of moles per kg of solvent.

Having introduced these four concentration terms, we shall now compare their usefulness here. The percent by weight unit has the advantage that we do not need to know the molar mass of the solute. This fact is useful to biochemists, who frequently work with macromolecules either of unknown molar mass or of unknown purity. Furthermore, the percent by weight of a solute in a solution is independent of temperature, since it is defined in terms of weights. The mole fraction term is not normally used to express concentration of solutions. It is useful, however, for calculating partial pressures of gases (see Section 2.7) and in the discussion of vapor pressures of solutions (to be introduced later). Molarity is one of the most commonly employed concentration units. The advantage of using molarity is that it is generally easier to measure the volume of a solution using precisely calibrated volumetric flasks than to weigh the solvent. Its main drawback is that it is temperature-dependent, since the volume of a solution usually increases with increasing temperature. Another drawback is that molarity does not tell us the amount of solvent present. Molality, on the other hand, is temperature-independent, since it is defined as a ratio of number of moles of solute and weight of solvent. For this reason, molality is the preferred concentration unit in studies that involve changes in temperature, as in those of the colligative properties of solution (see Section 8.7).

8.2 Partial Molar Quantities

The extensive properties of a one-component system at a constant temperature and pressure depend only on the amount of the system present. For example, the volume of water depends on the quantity of water present. Now if the volume is expressed as a molar quantity, it becomes an intensive property. Thus the molar volume of water at 1 atm and 0°C is 0.018 liter, no matter how little or how much water is present. The situation is different when we study solutions. A solution, by definition, contains at least two components. In this case, we find that the extensive properties of a solution depend on temperature, pressure, and the composition of the solution. In discussing the properties of any solution, we can no longer employ molar quantities; instead, we must use *partial molar quantities*. Perhaps the easiest partial molar quantity to understand is the *partial molar volume*, described below.

Let us imagine that there is a very large amount of water present. The addition of 1 mol of water to this quantity of water will increase its volume by 0.018 liter. As mentioned earlier, this increase corresponds to the molar volume of water. Now suppose that we add 1 mol of water to a very large amount of ethanol. We find that the increase in volume is only 0.014 liter. The reason is that water and ethanol molecules interact with each other in different ways than the interaction among water molecules themselves. The volume 0.014 liter, then, is the partial molar volume of water in ethanol at a particular composition, that is, 1 mol of water in a very large quantity of ethanol. What becomes clear, therefore, is the fact that in general the total volume (V) of a solution containing two components *cannot* be written as a sum of the individual volumes of components 1 and 2; that is,

$$V \neq V_1 + V_2$$

Instead, we must write*

$$V = n_1 \overline{V}_1 + n_2 \overline{V}_2 \tag{8.1}$$

where n_1 and n_2 are the numbers of moles of components 1 and 2 present and $\overline{V}_1$ and $\overline{V}_2$ are their partial molar volumes, respectively. The partial molar volumes are defined as

$$\overline{V}_1 = \left(\frac{\partial V}{\partial n_1} \right)_{T, P, n_2} \qquad \overline{V}_2 = \left(\frac{\partial V}{\partial n_2} \right)_{T, P, n_1} \tag{8.2}$$

Thus $\overline{V}_1$ can be viewed as the coefficient that gives the change in the volume of solution upon the addition of 1 mol of component 1 at constant temperature and pressure to a very large (actually infinite) amount of solution containing components 1 and 2 in the mole ratio of n_1/n_2. A similar interpretation can be given to $\overline{V}_2$.

Partial molar volumes are experimentally determinable quantities. Referring to our water–ethanol example above, we can determine the partial molar volume of ethanol as follows. A number of solutions can be prepared at some desired temperature and pressure, all containing the same fixed amount of

* For proof, see Van Holde, K. E. *Physical Biochemistry*. Prentice-Hall, Inc., Englewood Cliffs, N.J., 1971, p. 27.

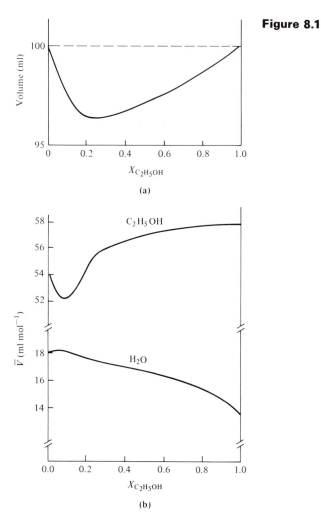

(a) Total volume of ethanol—water solution as a function of mole fraction of ethanol. The sum of separate water and ethanol volumes is 100 ml. (b) Partial molar volumes in water—ethanol solutions as a function of mole fraction of ethanol.

Figure 8.1

water but different amounts of ethanol. The sum of volumes of pure components equals 100 ml in each case. The volumes of the solutions are then plotted against their compositions (Figure 8.1). The slope of the curve at any point gives the partial molar volume of ethanol at that particular composition. The partial molar volume of water can be similarly determined by graphical means or evaluated by an equation relating the two partial molar volumes [Eq. (8.1)].

Partial molar quantities permit us to express the total extensive properties, such as volume, energy, enthalpy, and Gibbs free energy, of a solution of any composition. The partial molar Gibbs free energy of the ith component in solution $\overline{G}_i$ is given by

$$\overline{G}_i = \left(\frac{\partial G}{\partial n_i}\right)_{T, P, n_j} \tag{8.3}$$

where n_j represents the number of moles of all other components present. Again, $\overline{G}_i$ can be viewed as the coefficient that gives the increase in the Gibbs free energy of the solution upon the addition of 1 mol of component i at constant temperature and pressure to an infinite amount of solution of

167

specified concentration. Partial molar Gibbs free energy is also called the *chemical potential* (μ) so that we can write

$$\bar{G}_i = \mu_i \tag{8.4}$$

The chemical potential can be used as a criterion for equilibrium and spontaneity for a multicomponent system just as molar Gibbs free energy is used for a single-component system. Consider the transfer of dn_i moles of component i from some initial state A, where its chemical potential is μ_i^A, to some final state B, where its chemical potential is μ_i^B. If the process is carried out at constant temperature and pressure, then the change in the Gibbs free energy, dG, is given by

$$dG = \mu_i^B \, dn_i - \mu_i^A \, dn_i$$
$$= (\mu_i^B - \mu_i^A) \, dn_i$$

If $\mu_i^B < \mu_i^A$, $dG < 0$, and transfer of i from A to B will be a spontaneous process, if $\mu_i^B > \mu_i^A$, $dG > 0$ and the process would be spontaneous from B to A. At equilibrium, $dG = 0$ and $\mu_i^B = \mu_i^A$.

As we shall see later in many examples, the transfer we speak of here could be from one phase to another or from one state of chemical combination to another. The transfer can be transport by diffusion, evaporation, sublimation, condensation, crystallization, solution formation, or chemical reaction. Regardless of the nature of the process, in each case the transfer proceeds from a higher μ_i to a lower μ_i. This characteristic also explains the special name: chemical potential. In mechanics the direction of spontaneous change is always such that the system goes from a higher potential-energy state to a lower one. In thermodynamics the system is not quite so simple because we have to consider both energy and entropy factors. Nevertheless, we know that at constant temperature and pressure, the direction of a spontaneous change is always toward a decrease in the Gibbs free energy of the system. Thus the Gibbs free energy plays an analogous role in thermodynamics as potential energy does in mechanics. For this reason, molar Gibbs free energy or more frequently, partial molar Gibbs free energy, is called the chemical potential.

8.3 The Thermodynamics of Mixing

The formation of solutions is governed by the principles of thermodynamics. In this section we shall consider the changes in thermodynamic quantities as a result of mixing. In particular, we shall concentrate on gases.

In Section 7.7 we obtained an expression for the molar Gibbs free energy of an ideal gas as [see Eq. (7.24)]

$$\bar{G} = \bar{G}^\circ + RT \ln P$$

In a mixture of ideal gases, the chemical potential of the ith component is given by

$$\mu_i = \mu_i^\circ + RT \ln P_i \tag{8.5}$$

where P_i is the partial pressure of component i in the mixture and μ_i° is the standard chemical potential of component i when its partial pressure is 1 atm.

Now consider the mixing of n_1 moles of gas 1 at pressure P with n_2 moles of gas 2 also at pressure P. Before mixing, the total Gibbs free energy of the system is given by employing the generalized form of Eq. (8.1) as

$$G_{\text{initial}} = n_1\{\mu_1^\circ + RT \ln P\} + n_2\{\mu_2^\circ + RT \ln P\}$$

After mixing, the gases exert partial pressures P_1 and P_2, where $P_1 + P_2 = P$, and the Gibbs free energy is*

$$G_{\text{final}} = n_1\{\mu_1^\circ + RT \ln P_1\} + n_2\{\mu_2^\circ + RT \ln P_2\}$$

The Gibbs free energy of mixing, ΔG_{mix}, is given by

$$\Delta G_{\text{mix}} = G_{\text{final}} - G_{\text{initial}} = n_1 RT \ln \frac{P_1}{P} + n_2 RT \ln \frac{P_2}{P}$$

$$= n_1 RT \ln X_1 + n_2 RT \ln X_2$$

where $P_1 = X_1 P$ and $P_2 = X_2 P$, and X_1 and X_2 are the mole fractions of 1 and 2, respectively. Further, from the relations

$$\frac{n_1}{n_1 + n_2} = \frac{n_1}{n} = X_1 \qquad \frac{n_2}{n_1 + n_2} = \frac{n_2}{n} = X_2$$

we have

$$\Delta G_{\text{mix}} = nRT\{X_1 \ln X_1 + X_2 \ln X_2\} \tag{8.6}$$

One thing is immediately obvious from Eq. (8.6). Since both X_1 and X_2 are less than unity, $\ln X_1$ and $\ln X_2$ are negative quantities and hence so is ΔG_{mix}. It follows that the mixing of gases is a spontaneous process at constant T and P.

Next, we calculate the entropy of mixing, ΔS_{mix}. From Eq. (7.20a), we have

$$\left(\frac{\partial G}{\partial T}\right)_P = -S$$

Thus the entropy of mixing is given by differentiating Eq. (8.6) with respect to T:

$$\left(\frac{\partial \Delta G_{\text{mix}}}{\partial T}\right)_P = nR\{X_1 \ln X_1 + X_2 \ln X_2\}$$

$$= -\Delta S_{\text{mix}} \tag{8.7}$$

The minus sign in Eq. (8.7) now makes ΔS_{mix} a positive quantity, in accord with a spontaneous process. The enthalpy of mixing is given by

$$\Delta H_{\text{mix}} = \Delta G_{\text{mix}} + T \Delta S_{\text{mix}}$$

$$= 0$$

This is not an unexpected result, since molecules of ideal gases do not interact with one another, so no heat is absorbed or produced as a result of mixing. Figure 8.2 shows the plots of ΔG_{mix}, $T\Delta S_{\text{mix}}$, and ΔH_{mix} for a two-component system as a function of composition.

* Note that $P_1 + P_2 = P$ only if there is no change in volume as a result of mixing, that is, $\Delta V_{\text{mix}} = 0$. This condition always holds for ideal gas solutions and, as we shall see later, for ideal liquid solutions as well.

Figure 8.2

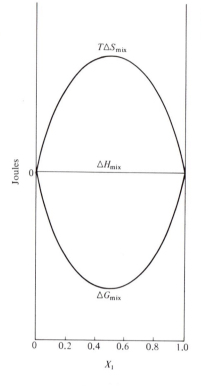

Mixing of a two-component ideal solution. Plots of $T \Delta S_{mix}$, ΔH_{mix} and ΔG_{mix} as a function of mole fraction X_1.

The unmixing process for a two-component solution of equal mole fractions leads to an increase in free energy and a decrease in entropy of the system so that energy must be supplied to the system from the surroundings. Initially, at $X_1 \simeq X_2$, the curves are fairly flat, and separation can be carried out easily. However, as the solution becomes progressively richer in one component, say 1, the curves become very steep. Then a considerable amount of energy input is needed to separate component 2 from 1. This difficulty is encountered when one tries to clean up a lake contaminated by small amounts of undesirable chemicals. The same consideration applies to the purification of compounds. It is relatively easy to prepare most compounds in 95% purity, but much more effort is needed to attain 99% or higher purity (as the silicon crystals used in solid-state electronics).

As another example, let us explore the possibility of mining gold from the oceans. It is estimated that there is about 4×10^{-12} g of gold/ml of seawater. This may not seem like much, but when we multiply it by the total volume of ocean water, 1.5×10^{21} liters, we find the amount of gold present to be 6×10^{12} g or 7 million tons, which should satisfy anybody's needs. Unfortunately, in addition to gold's very low concentration, it is one of some 60 different elements present in the ocean. The problem of separating one pure component initially present in a very low concentration in seawater (that is, starting at the steep portions of the curves in Figure 8.2) would be very formidable indeed.

EXAMPLE 8.1 Calculate the entropy and Gibbs free energy of mixing of 1.6 mol of argon at 1 atm and 25°C with 2.6 mol of nitrogen at 1 atm and 25°C. Assume ideal behavior.

Answer: The mole fractions of argon and neon are

$$X_{Ar} = \frac{1.6}{1.6 + 2.6} = 0.38 \qquad X_{N_2} = \frac{2.6}{1.6 + 2.6} = 0.62$$

From Eq. (8.7), we have

$$\Delta S_{mix} = -(1.6 + 2.6)\ \text{mol}\ (8.314\ \text{J K}^{-1}\ \text{mol}^{-1})$$

$$\times \{0.38 \ln 0.38 + 0.62 \ln 0.62\}$$

$$= 23\ \text{J K}^{-1}$$

Since $\Delta H_{mix} = 0$, we have $\Delta G_{mix} = -T\ \Delta S_{mix}$, or

$$\Delta G_{mix} = -(298\ \text{K})(23\ \text{J K}^{-1})$$

$$= -6.9\ \text{kJ}$$

Binary Mixtures of Volatile Liquids **8.4**

The results obtained in Section 8.3 for the mixing of gases also apply to ideal liquid solutions. In the study of solutions we need to know how to express the chemical potential of each component. We shall consider a solution containing two volatile liquids, that is, liquids with easily measurable vapor pressures.

We start by considering a liquid in equilibrium with its vapor in a closed container. Since the system is at equilibrium, the chemical potential of the substance in the liquid and the vapor phase must be the same, that is,

$$\mu(l) = \mu(g)$$

Further, from the expression for $\mu(g)$, we can write

$$\mu^*(l) = \mu(g) = \mu^\circ + RT \ln P^*$$

where the asterisk denotes a pure component. For a two-component solution at equilibrium with its vapor, the chemical potential of each component is still the same in the two phases. Thus for component 1 we write

$$\mu_1(l) = \mu_1(g) = \mu_1^\circ + RT \ln P_1$$

Since $\mu^\circ = \mu_1^\circ$, we can combine the last two equations to get

$$\mu_1(l) = \mu_1^*(l) + RT \ln \frac{P_1}{P_1^*} \tag{8.8}$$

Thus the chemical potential of component 1 in solution is expressed in terms of the chemical potential of the liquid in the pure state and the vapor pressures of the pure liquid and the liquid in solution.

According to Raoult, the ratio P_1/P_1^* in Eq. (8.8) is equal to the mole fraction of component 1, that is,

$$\frac{P_1}{P_1^*} = X_1$$

or

$$P_1 = X_1 P_1^* \tag{8.9}$$

Figure 8.3

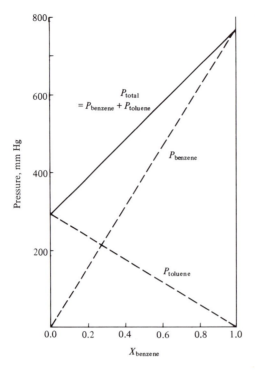

Total vapor pressure of the benzene–toluene mixture as a function of the benzene mole fraction at 80.1 °C. The dashed lines represent the partial pressures of the two components.

Equation (8.9) is known as *Raoult's law*, which states that the vapor pressure of a component of a solution is equal to the product of its mole fraction and the vapor pressure of the pure liquid. Substituting Eq. (8.9) into Eq. (8.8), we obtain

$$\mu_1(l) = \mu_1^*(l) + RT \ln X_1 \tag{8.10}$$

We see that in a pure liquid ($X_1 = 1$), $\mu_1(l) = \mu_1^*(l)$. Solutions that obey Raoult's law are called *ideal solutions*. An example of an ideal solution is the benzene–toluene system. Figure 8.3 shows a plot of the vapor pressures versus composition of the solution.

EXAMPLE 8.2 Liquids A and B form an ideal solution. At 45°C the vapor pressure of pure A and pure B are 66 mm Hg and 88 mm Hg, respectively. Calculate the composition of vapor in equilibrium with a solution containing 36 mol % A at this temperature.

Answer: Since $X_A = 0.36$ and $X_B = 1 - 0.36 = 0.64$, we have

$$P_A = X_A P_A^* = 0.36(66) = 23.8 \text{ mm Hg}$$

$$P_B = X_B P_B^* = 0.64(88) = 56.3 \text{ mm Hg}$$

The total vapor pressure P_T is given by

$$P_T = P_A + P_B = 23.8 + 56.3 = 80.1 \text{ mm Hg}$$

Finally, the compositions of A and B in the vapor phase, X_A^v and X_B^v, are given by

$$X_A^v = \frac{23.8}{80.1} = 0.297$$

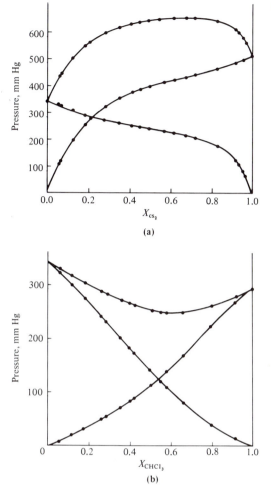

Nonideal solutions. (a) Positive deviation: carbon disulfide–acetone system at 35.2°C. (b) Negative deviation: chloroform–acetone system at 35.2°C. (From *The Solubility of Nonelectrolytes* by J. Hildebrand and R. Scott. © 1950 by Litton Educational Publishing, Inc. Reprinted by permission of Van Nostrand Reinhold Company.)

Figure 8.4

and

$$X_B^v = \frac{56.3}{80.1} = 0.703$$

In an ideal solution, all intermolecular forces are equal, whether the molecules are alike or not. The benzene–toluene system satisfies this requirement because benzene and toluene molecules have similar shapes and electronic structure. Most solutions behave nonideally, however. Figure 8.4 shows the positive and negative deviations from Raoult's law. In the region where one component is in excess (this component is called the solvent), we see that its vapor pressure is quite accurately described by Eq. (8.9). The regions where Raoult's law is applicable are shown for the carbon disulfide–acetone system in Figure 8.5. On the other hand, the vapor pressure of the component present in small amount (this component is called the solute) does not vary with composition of the solution according to Eq. (8.9). Still, the vapor pressure of the solute varies with concentration in a linear manner*:

$$P_2 = kX_2 \tag{8.11}$$

* There is, of course, no sharp distinction between solvent and solute. In cases where applicable, we shall call component 1 solvent and component 2 solute.

Figure 8.5

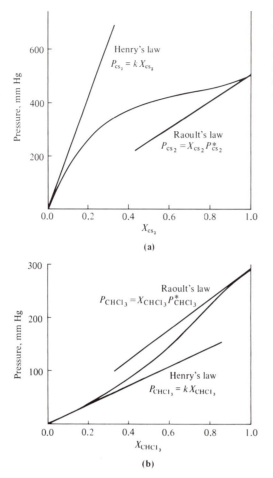

Diagrams showing regions over which Raoult's law and Henry's law are applicable for a two-component system (see Figure 8.4). The Henry's law constants can be obtained from the intercepts on the y (pressure) axis.

Equation (8.11) is known as *Henry's law*, where k, the Henry's law constant, has the unit of atm or torr. Henry's law relates the mole fraction of the solute to its partial (vapor) pressure. Alternatively, Henry's law can be expressed as

$$P_2 = k'm \tag{8.12}$$

where m is the molality of the solution and the constant k' now has the units atm mol^{-1} kg of solvent. Table 8.1 lists the values of k and k' for several gases in water at 298 K.

Table 8.1

Henry's Law Constants for Some Gases in Water at 298 K

Gas	k (torr)	k' (atm mol^{-1} kg H_2O)
H_2	5.54×10^7	1311
He	1.12×10^8	2649
Ar	2.80×10^7	662
N_2	6.80×10^7	1610
O_2	3.27×10^7	773
CO_2	1.24×10^6	29.3
H_2S	4.27×10^5	10.1

EXAMPLE 8.3 Calculate the solubility of carbon dioxide in water at 298 K at a partial pressure of CO_2 over the solution of 0.030 atm.

175

Sec. 8.4

Binary Mixtures of Volatile Liquids

Answer: The mole fraction of solute (carbon dioxide) is given by Eq. (8.11):

$$X_{CO_2} = \frac{P_{CO_2}}{k}$$

Since the number of moles of carbon dioxide dissolved in 1000 g of water is small, we can approximate the mole fraction as follows:

$$X_{CO_2} = \frac{n_{CO_2}}{n_{CO_2} + n_{H_2O}} \simeq \frac{n_{CO_2}}{n_{H_2O}}$$

so that

$$n_{CO_2} = \frac{P_{CO_2} n_{H_2O}}{k}$$

Finally, we write

$$n_{CO_2} = (0.030 \times 760) \text{ mm Hg} \times \frac{1000 \text{ g}}{18 \text{ g mol}^{-1}} \times \frac{1}{1.24 \times 10^6 \text{ mm Hg}}$$

$$= 1.02 \times 10^{-3} \text{ mol (in 1 kg } H_2O)$$

Comment: If we had used the value for k' in Table 8.1, then we would have proceeded as follows:

$$m = \frac{P_{CO_2}}{k'}$$

$$= \frac{0.030 \text{ atm}}{29.3 \text{ atm mol}^{-1} \text{ kg } H_2O} = 1.02 \times 10^{-3} \text{ mol (kg } H_2O)^{-1}$$

which is the same as above.

Henry's law is normally associated with solutions of gases in liquids,* although it is equally applicable to solutions containing nongaseous volatile solutes. It is of great practical importance in chemical and biological systems and merits further discussion. The effervescence observed upon opening a champagne bottle is a nice demonstration of the decrease in gas—mostly CO_2—solubility as its partial pressure is lowered. The emboli (gas bubbles in the bloodstream) suffered by deep-sea divers also illustrate Henry's law. At a point some 40 m below the surface of seawater, the total pressure is about 6 atm. The solubility of nitrogen in the blood plasma is then about 4.8/1610, or 3.0×10^{-3} mol (kg $H_2O)^{-1}$, six times that at sea level. If the diver swims upward too rapidly, dissolved nitrogen gas will start boiling off. The mildest result is dizziness; the most serious, death.† Because helium is less soluble in the blood plasma, it is the preferred gas for diluting oxygen gas for deep-sea divers.

---

* The difference between gas and vapor is as follows. Vapor is the gaseous form of any substance which is usually a liquid or solid. Thus under room temperature and 1 atm conditions, we speak of water vapor and oxygen gas.
† For other interesting illustrations of Henry's law, see T. C. Loose, *J. Chem. Educ.* **48**, 154 (1971); W. J. Ebel, *J. Chem. Educ.* **50**, 559 (1973); and E. D. Cook, *J. Chem. Educ.* **50**, 425 (1973).

There are several different types of deviations from Henry's law. First, as mentioned earlier, the law holds only for dilute solutions. Second, if the dissolved gas interacts chemically with the solvent, then the solubility can be greatly enhanced. Gases such as CO_2, H_2S, NH_3, SO_2, and HCl all possess high solubilities in water because of their reactions with the solvent. The third type of deviation is shown by the dissolution of oxygen in blood. Normally, oxygen is only sparingly soluble in water (see Table 8.1), but its solubility increases drastically if the solution contains hemoglobin or myoglobin. The nature of oxygen binding the heme group in these molecules will be discussed further in later chapters.

8.5 Real Solutions

Most solutions, as was pointed out in Section 8.4, do not behave ideally. One problem that immediately arises in dealing with nonideal solutions is how to write the chemical potentials for the solvent and solute components.

We consider first the solvent component. For an ideal solution, the chemical potential of the solvent is given by

$$\mu_1(l) = \mu_1^*(l) + RT \ln X_1$$

where $X_1 = P_1/P_1^*$. The standard state is the pure liquid and is attained when $X_1 = 1$. Now for a nonideal solution we write

$$\mu_1(l) = \mu_1^*(l) + RT \ln a_1 \qquad (8.13)$$

where a_1 is the *activity* of the solvent. Nonideality arises as a result of the unequal intermolecular forces between solvent–solvent and solvent–solute molecules. Consequently, the extent of nonideality depends on the composition of solution. Thus the activity of the solvent, which is defined by Eq. (8.13), plays the role of "effective" concentration. The activity can be expressed in terms of vapor pressures as

$$a_1 = \frac{P_1}{P_1^*} \qquad (8.14)$$

where P_1 is the partial vapor pressure of component 1 over the solution. The activity is related to concentration (mole fraction) as follows:

$$a_1 = \gamma_1 X_1 \qquad (8.15)$$

where γ_1 is the *activity coefficient*. The value of γ_1 is a measure of the deviation from ideality. In the limiting case, where $X_1 \rightarrow 1$, $\gamma_1 \rightarrow 1$ and activity and the mole fraction are one and the same. This also holds, of course, for an ideal solution at all concentrations.

Equation (8.14) provides a way of obtaining the activity of the solvent. By measuring P_1 of the solvent vapor over a range of concentrations, the value of a_1 can be calculated at each concentration if P_1^* is known.* Table 8.2 shows the calculation of the activity of water in water–urea solutions.

---

* To obtain P_1, we need to measure the total equilibrium pressure P and also analyze the composition of the mixture. The partial pressure P_1 can then be calculated from $P_1 = X_1^v P$, where X_1^v is the mole fraction of the solvent in the vapor phase.

Molality of Urea, m_2	Mole Fraction of Water, X_1	Vapor Pressure of Water, P_1 (atm)	Activity of Water, a_1	Activity Coefficient of Water, γ_1
0	1.000	6.025×10^{-3}	1.000	1.000
1	0.943	5.933×10^{-3}	0.985	1.045
2	0.833	5.846×10^{-3}	0.970	1.165
4	0.806	5.672×10^{-3}	0.942	1.169
6	0.735	5.501×10^{-3}	0.913	1.242
10	0.625	5.163×10^{-3}	0.857	1.371

[a] Data from *International Critical Tables of Numerical Data: Physics, Chemistry, and Technology*, Vol. 3. Prepared by the National Research Council. Copyright 1928 by McGraw-Hill Book Company. Used with permission of McGraw-Hill Book Company.

We now come to the solute. An ideal solution is one in which both components obey Raoult's law over the entire concentration range. In all dilute nonideal solutions in which there is no chemical interaction, the solvent obeys Raoult's law and the solute obeys Henry's law.* Such solutions are sometimes called "ideally dilute solutions" or "ideal dilute solutions." At equilibrium, the chemical potentials of the solute are equal in the liquid and vapor phase so that

$$\mu_2(l) = \mu_2(g) = \mu_2^\circ + RT \ln P_2$$

Since $P_2 = kX_2$, we have

$$\mu_2(l) = \mu_2^\circ + RT \ln kX_2$$
$$= \mu_2^\circ + RT \ln k + RT \ln X_2$$
$$= \mu_2^\ominus + RT \ln X_2 \tag{8.16}$$

where $\mu_2^\ominus = \mu_2^\circ + RT \ln k$. Although Eq. (8.16) seems to take the same form as Eq. (8.10), there is an important difference. The difference lies in the choice of standard state. According to Eq. (8.16), the standard state is defined to be the pure solute, attained by setting $X_2 = 1$. However, Eq. (8.16) holds only for dilute solutions. How can these two conditions be met simultaneously? There is a simple way out of this dilemma; that is, we must realize that standard states are often hypothetical states, not physically realizable. Thus the standard state of solute defined by Eq. (8.16) is the hypothetical pure component 2 with a vapor pressure equal to k (when $X_2 = 1$, $P_2 = k$). This is in a sense an "infinite dilution state of unit mole fraction"; that is, it is infinitely dilute with respect to component 1, the solvent, with the solute at unit mole fraction. For nonideal solutions, Eq. (8.16) is written as

$$\mu_2 = \mu_2^\ominus + RT \ln a_2 \tag{8.17}$$

where a_2 is the activity of the solute. As in the case of the solvent, we have $a_2 = \gamma_2 X_2$, where γ_2 is the activity coefficient of the solute. Here we have $a_2 \to X_2$ or $\gamma_2 \to 1$ as $X_2 \to 0$. Henry's law is now given by

$$P_2 = a_2 k \tag{8.18}$$

---

* For ideal solutions, Raoult's law and Henry's law become identical, that is, $P_2 = kX_2 = P_2^* X_2$.

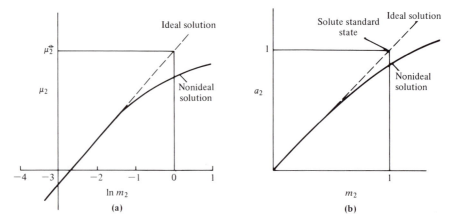

Figure 8.6 (a) Chemical potential of solute plotted against logarithm of molality for a nonideal solution. (b) Activity of a solute as a function of molality for a nonideal solution.

Compositions are usually expressed in molalities or molarities instead of mole fractions. In molality, Eq. (8.16) takes the form

$$\mu_2 = \mu_2^\ominus + RT \ln m_2 \qquad (8.19)$$

Here the standard state is defined as a state at unit molality but in which the solution is behaving ideally. Again this standard state is a hypothetical state, not realizable in practice (Figure 8.6). For nonideal solutions, Eq. (8.19) is rewritten as

$$\mu_2 = \mu_2^\ominus + RT \ln a_2 \qquad (8.20)$$

where $a_2 = \gamma_2 m_2$. In the limiting case of $m_2 \rightarrow 0$, we have $a_2 \rightarrow m_2$ or $\gamma_2 \rightarrow 1$. (See Figure 8.6b.)

It is important to keep in mind that although Eqs. (8.16) and (8.19) were derived by considering Henry's law, they are applicable to any solute, whether or not it is volatile. The expressions of chemical potentials are useful in discussing the colligative properties of solutions and in deriving the equilibrium constant (see Chapter 10). But first let us consider an aspect of solutions of considerable practical importance—distillation.

8.6 Distillation

PRESSURE-COMPOSITION DIAGRAM

The separation of two liquid components is usually accomplished by fractional distillation, the success of which depends on the difference of vapor pressure of the components at a given temperature. It is useful, therefore, to construct diagrams that show the vapor pressure of a solution as a function of mole fractions, as well as the composition of the vapor in equilibrium with the solution. Assuming ideal behavior, we can express the vapor pressures of both components in terms of Raoult's law:

$$P_1 = X_1 P_1^* \qquad \text{and} \qquad P_2 = X_2 P_2^*$$

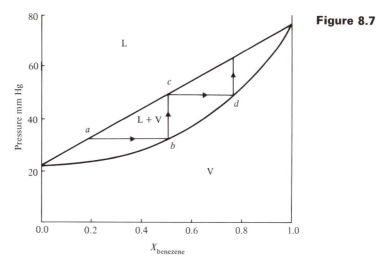

Pressure – composition diagram of the benzene–toluene system at 23°C.

Figure 8.7

Furthermore, using Dalton's law, we can derive the following relations:

$$X_1^v = \frac{P_1}{P_1 + P_2} \tag{8.21a}$$

$$X_2^v = \frac{P_2}{P_1 + P_2} \tag{8.21b}$$

where X_1^v and X_2^v are the mole fractions of components 1 and 2 in the vapor phase, respectively. These mole fractions can be rewritten in the form

$$X_1^v = \frac{X_1 P_1^*}{P_1 + P_2}$$

$$X_2^v = \frac{X_2 P_2^*}{P_1 + P_2}$$

and their ratio is

$$\frac{X_1^v}{X_2^v} = \frac{X_1}{X_2} \frac{P_1^*}{P_2^*}$$

A vapor–liquid diagram for the benzene–toluene system is shown in Figure 8.7. At point a on the liquid curve the mole fractions are $X_{\text{benzene}} = 0.2$ and $X_{\text{toluene}} = 1 - 0.2 = 0.8$, respectively. The composition of the vapor in equilibrium with the solution (point b) has the following values: $X_{\text{benzene}}^v = 0.5$ and $X_{\text{toluene}}^v = 0.5$. Thus the vapor phase is richer in benzene than is the liquid phase. Now, if we condense the vapor $(b \rightarrow c)$ and reevaporate the liquid $(c \rightarrow d)$, the mole fraction of benzene will be even higher in the vapor phase. Repeated processes at constant temperature will eventually lead to a quantitative separation of these two components.

TEMPERATURE-COMPOSITION DIAGRAM

In practice, distillations are usually carried out at constant pressure rather than at constant temperature; therefore, we need to examine the *temperature-composition* or *boiling-point diagram*. The relation between temperature and composition is complex and is usually determined experimentally.

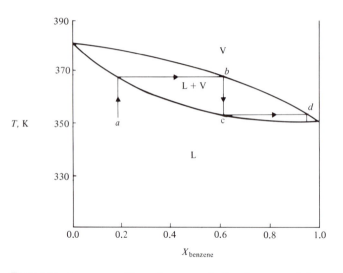

Figure 8.8 Temperature–composition diagram of the benzene–toluene system at 1 atm.

Let us again consider the benzene-toluene system shown in Figure 8.8. Comparing this plot with Figure 8.7, we see that the liquid and vapor regions are inverted and that neither the liquid nor the vapor curve is a straight line. The more volatile component, benzene, has a higher vapor pressure and hence a lower boiling point. Figure 8.7 shows that at a constant temperature, liquid is the stable phase at high pressures. Similarly, Figure 8.8 tells us that at a constant pressure, the stable phase at low temperatures is also the liquid. The fractional distillation process can be imagined as follows. At point a the solution is allowed to evaporate ($a \rightarrow b$). The vapor, which is richer in benzene, is first condensed ($b \rightarrow c$) and then reevaporated ($c \rightarrow d$). By repeating the process, these two components can eventually be completely separated.

AZEOTROPES

Since most solutions are nonideal, their temperature-composition diagrams determined experimentally are more complicated than that shown in Figure 8.8. If the system exhibits a positive deviation from Raoult's law, the curve will show a minimum boiling point. Conversely, a negative deviation from Raoult's law will result in a maximum boiling point (Figure 8.9). Examples of the former are acetone–carbon disulfide, ethanol–water, and n-propanol–water. Systems that show a maximum boiling point are less common; among the known examples are acetone–chloroform and hydrochloric acid–water. In every case the mixture cannot be *completely* separated into pure components by simple fractional distillation.

Consider the following steps, shown in Figure 8.9a. The solution is boiled at a certain composition (below T_2) denoted by the point a. The condensed vapor ($b \rightarrow c$) becomes richer in component 1, while the solution remaining in the pot becomes richer in component 2. Consequently, the point representing the composition of the solution in the pot on the liquid curve will move toward the left as the distillation proceeds and the boiling point rises. Boiling the solution, condensing the vapor, and again boiling the condensed vapor results in a distillate having the composition of the *azeotrope* (to boil unchanged). The boiling point of the solution remaining in the pot will eventually

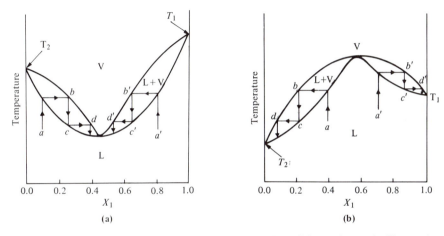

Azeotropes: (a) minimum boiling point; (b) maximum boiling point. **Figure 8.9**

**Variation of Boiling Point and Composition of
the Ethanol–Water Azeotrope with Pressure**

Table 8.3

Pressure (mm Hg)	Boiling Point (K)	Composition (percent by weight)	
		C_2H_5OH	H_2O
760	343.6	91.6	8.4
500	332.5	92.5	7.5
200	310.7	94.2	5.8

reach that of pure component 2 (that is, T_2). Once the azeotrope distillate is produced, further distillation will result in no further separation and it will boil at a constant temperature. The situation with the maximum boiling system can be similarly explained except that the pure components appear in the distillate and the azeotrope appears in the pot.

Although an azeotrope behaves in distillation as if it were a single component, this can be easily shown not to be true. As the data in Table 8.3 show, the composition of the azeotrope depends on the pressure.

Colligative Properties 8.7

This section will examine some general properties of solutions: vapor-pressure lowering, boiling-point elevation, freezing-point depression, and osmotic pressure. These are commonly referred to as *colligative*, or *collective*, *properties* because they are bound together through their common origin. Each of the properties depends only on the number of solute molecules present, not on the size or molar mass of the molecules. Our derivations will be based on three important assumptions: (1) The solutions are ideal; (2) the solutions are dilute; and (3) we consider only nonelectrolyte solutions. As usual, we shall only consider a two-component system.

VAPOR-PRESSURE LOWERING

Consider a solution that contains a solvent 1 and a *nonvolatile* solute 2, for example, a solution of sucrose in water. Since the solution is ideal, Raoult's law applies:

$$P_1 = X_1 P_1^*$$

Since $X_1 = 1 - X_2$, the equation above becomes

$$P_1 = (1 - X_2)P_1^*$$

Rearrangement of this equation gives

$$P_1^* - P_1 = \Delta P = X_2 P_1^* \tag{8.22}$$

where ΔP, the lowering of vapor pressure from pure solvent, is directly proportional to the mole fraction of the solute.

Why is the vapor pressure of a solution lowered by the presence of solute? It is tempting to suggest that it is because of the modification of intermolecular forces. But this cannot be so, since vapor pressure lowering is observed even for ideal solutions where there is no difference between solute–solvent and solvent–solvent interactions. A more convincing explanation is provided by considering the entropy effect. When a solvent evaporates, the entropy of the universe increases because the entropy of any substance in the gaseous state is greater than that in the liquid state (at the same temperature). As we saw in Section 8.3, the solution process is accompanied by an increase in entropy. What this means is that in a solution there is an extra amount of randomness or disorder present that was not present in the pure solvent. Therefore, the evaporation of solvent from a solution will result in a *smaller* increase in entropy. Consequently, the solvent has a smaller tendency to leave the solution, and the solution will have a lower vapor pressure than the pure solvent (see Problem 8.32).

BOILING-POINT ELEVATION

The boiling point of a solution is the temperature at which its vapor pressure is equal to the external pressure. The previous discussion would lead one to expect that since the addition of a nonvolatile solute lowers the vapor pressure, this should result in a rise in the boiling point of solution. This is indeed the case.

We consider a solution containing a *nonvolatile* solute. The boiling-point elevation has its origin in the modification of the chemical potential of the solvent by the presence of the solute. From Eq. (8.10) we see that the chemical potential of the solvent in a solution is less than the chemical potential of the pure solvent by an amount equal to $RT \ln X_1$. How this change affects the boiling point of the solution can be seen from Figure 8.10. The solid lines refer to the pure solvent component. Since the solute is nonvolatile, it does not appear in the vapor phase; therefore, the curve for the gas is the same as that for the pure gas. On the other hand, because the liquid contains a solute, the chemical potential of the solvent is lowered (see the dashed curve). The intersection points between the curve for the vapor and the curves for the liquids (pure and solution) correspond to the boiling points of the pure solvent and the solution, respectively. We see that the boiling point of the solution is higher than that of the pure solvent.

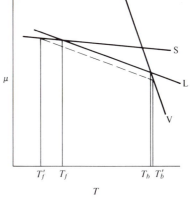

Plot of chemical potential versus temperature to illustrate colligative properties.

Figure 8.10

We now present a quantitative treatment of the elevation of the boiling point. At the boiling point the solvent vapor is in equilibrium with the solvent in solution so that

$$\mu_1(g) = \mu_1(l) = \mu_1^*(l) + RT \ln X_1$$

or

$$\Delta\mu_1 = \mu_1(g) - \mu_1^*(l) = RT \ln X_1$$

where $\Delta\mu_1$ is the free energy change associated with the evaporation of 1 mol of solvent from the solution at temperature T, the boiling point. Thus we can write $\Delta\mu_1 = \Delta\bar{G}_{vap}$. Dividing the preceding equation by T, we obtain

$$\frac{\Delta\bar{G}_{vap}}{T} = \frac{\mu_1(g) - \mu_1^*(l)}{T} = R \ln X_1$$

We saw earlier that the Gibbs–Helmoltz equation [Eq. (7.20b)] can be expressed as

$$\frac{d(\Delta G/T)}{dT} = -\frac{\Delta H}{T^2} \qquad \text{(at constant } P\text{)}$$

or

$$\frac{d(\Delta\bar{G}_{vap}/T)}{dT} = \frac{-\Delta\bar{H}_{vap}}{T^2} = R\frac{d(\ln X_1)}{dT}$$

where $\Delta\bar{H}_{vap}$ is the molar enthalpy of vaporization of the solvent from the solution. Since the solution is assumed to be dilute, $\Delta\bar{H}_{vap}$ is taken to be the same as the molar enthalpy of vaporization of the pure solvent. Rearrangement of the last equation gives

$$d \ln X_1 = \frac{-\Delta\bar{H}_{vap}}{RT^2} dT \tag{8.23}$$

To obtain a relation between X_1 and T, we integrate Eq. (8.23) between the limits of T and T_0, the boiling points of the solution and pure solvent, respectively. It is clear that the mole fraction of the solvent is X_1 at T and 1 at T_0 so that

$$\int_{\ln 1}^{\ln X_1} d \ln X_1 = \int_{T_0}^{T} \frac{-\Delta\bar{H}_{vap}}{RT^2} dT$$

183

or

$$\ln X_1 = \frac{\Delta \bar{H}_{vap}}{R}\left(\frac{1}{T} - \frac{1}{T_0}\right)$$

$$= \frac{-\Delta \bar{H}_{vap}}{R}\left(\frac{T - T_0}{TT_0}\right)$$

$$= \frac{-\Delta \bar{H}_{vap}}{R}\frac{\Delta T}{T_0^2} \tag{8.24}$$

where $\Delta T = T - T_0$. Two assumptions have been employed in obtaining Eq. (8.24), both of which are based on the fact that T and T_0 differ only by a small amount (a few degrees). First, $\Delta \bar{H}_{vap}$ is treated as a constant. Second, since $T \simeq T_0$, we have $TT_0 \simeq T_0^2$.

Equation (8.24) gives the elevation of boiling point ΔT in terms of the concentration of the solvent. However, it is customary to express the concentration in terms of the amount of solute present, we write

$$\ln X_1 = \ln (1 - X_2) = \frac{-\Delta \bar{H}_{vap}}{R}\frac{\Delta T}{T_0^2}$$

where*

$$\ln (1 - X_2) = -X_2 - \frac{X_2^2}{2} - \frac{X_2^3}{3} - \cdots$$

$$= -X_2 \qquad \text{if } X_2 \ll 1$$

We now have

$$\Delta T = \frac{RT_0^2}{\Delta \bar{H}_{vap}} X_2$$

To convert the mole fraction X_2 into a more practical concentration unit, such as molality (m_2), we write

$$X_2 = \frac{n_2}{n_1 + n_2} \simeq \frac{n_2}{n_1} = \frac{w_2/M_2}{w_1/M_1} \qquad (n_1 \gg n_2)$$

where w_1 and w_2 are the masses of the solvent and the solute and M_1 and M_2 the molar masses of the solvent and solute, respectively. Thus

$$\Delta T = \frac{RT_0^2 w_2 M_1}{\Delta \bar{H}_{vap} w_1 M_2}$$

This equation may also be written in the form

$$\Delta T = \frac{RT_0^2 M_1}{1000\,\Delta \bar{H}_{vap}}\frac{1000 w_2}{w_1 M_2}$$

Since all the quantities in the first term on the right are constants for a given system, it is represented by the symbol K_b, the boiling-point-elevation constant. The ratio in the second term on the right gives the molality m_2 of the solution. Thus

$$\Delta T = K_b m_2 \tag{8.25}$$

---

* This series expansion is known as *Maclaurin's theorem*. The reader can verify this relation by employing a small numerical value for X_2 ($\lesssim 0.2$).

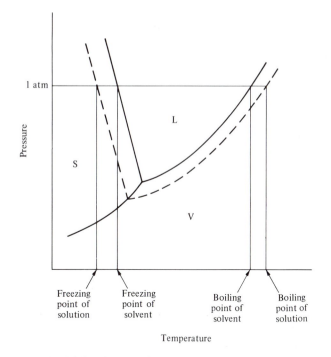

Phase diagrams of pure water (solid lines) and of water in an aqueous solution containing a nonvolatile solute (dashed lines).

Figure 8.11

The advantage of using molality, as mentioned in Section 8.1, is that it is independent of temperature. Figure 8.11 shows the phase diagrams of pure water and an aqueous solution. Upon the addition of a nonvolatile solute, the vapor pressure of the solution is lowered at every temperature. Consequently, the boiling point of the solution at 1 atm will be greater than 373.15 K.

FREEZING-POINT DEPRESSION

A layman may be forever unaware of the boiling-point-elevation phenomenon, but any casual observer living in a cold climate witnesses an illustration of the freezing-point depression. The ice on frozen roads and sidewalks melts readily when sprinkled with salt.* This method of thawing employs a depression of the freezing point.

The thermodynamic analysis of freezing-point depression parallels that for boiling-point elevation. If we assume that when a solution freezes, the solid that separates from the solution contains only the solvent, then the curve for the chemical potential of the solid is unchanged (see Figure 8.10). Consequently, the solid curve for the solid and the dashed curve for the solvent in solution now intersect at a point *below* the freezing point of the pure solvent. By following exactly the same procedure as that for the boiling-point elevation, we can show that the lowering in freezing point ΔT is given by

$$\Delta T = K_f m_2 \qquad (8.26)$$

* The salt employed is usually sodium chloride, which attacks cement and is harmful to many plants. Also see "Freezing Ice Cream and Making Caramel Topping," J. O. Olson and L. H. Bowman, *J. Chem. Educ.* **53**, 49 (1976).

Table 8.4

Boiling-Point-Elevation and Freezing-Point-Depression Constants of Some Common Solvents

	K_b (K m^{-1})	K_f (K m^{-1})
H_2O	0.51	1.86
C_2H_5OH	1.22	—
C_6H_6	2.53	5.12
$CHCl_3$	3.63	—
CH_3COOH	2.93	3.90
CCl_4	5.03	—

[a] Some of the K_f values have not been determined as a result of experimental difficulties.

where $\Delta T = T_0 - T$ and K_f is the freezing-point-depression constant given by

$$K_f = \frac{RT_0^2 M_1}{1000 \, \Delta\overline{H}_{fus}} \qquad (8.27)$$

where T_0 is the freezing point of the pure solvent.

The freezing-point-depression phenomenon can also be understood by studying Figure 8.11. At 1 atm, the freezing point of solution is given by the intersection point made by the dashed curve (between the solid and liquid phases) and the horizontal line at 1 atm. It is interesting to note that whereas in the boiling-point-elevation case the solute must be nonvolatile,* no such restriction applies to lowering the freezing point. A proof of this statement is the use of ethanol (bp = 351.65 K) as an antifreeze.

Both Eqs. (8.25) and (8.26) can be used to determine the molar mass of a solute. In general, the freezing-point-depression experiment is much easier to carry out. It is commonly employed for measuring the molar mass of organic compounds in nonpolar solvents and that of compounds such as urea and sugars in water. Table 8.4 lists the K_b and K_f values for several common solvents.

EXAMPLE 8.4 The quantity 45.20 g of sucrose is dissolved in 316.0 g of water. Calculate (a) the boiling point, and (b) the freezing point of the solution.

Answer
(a) Boiling point: Since

$$K_b = 0.51 \text{ K } m^{-1}$$

$$M_2 = 342.3 \text{ g mol}^{-1}$$

and

$$m_2 = \frac{(45.20 \text{ g})(1000 \text{ g})}{(316.0 \text{ g})(342.3 \text{ g mol}^{-1})} = 0.418 \text{ mol (kg } H_2O)^{-1}$$

$$= 0.418 \, m$$

* A volatile solute component can actually lower the boiling point of the solution. For example, 95% by volume ethanol in water (an azeotrope) boils at 351.3 K.

From Eq. (8.32), we write

$$\Delta T = (0.51 \text{ K } m^{-1})(0.418 \text{ } m)$$

$$= 0.21 \text{ K}$$

Thus the solution will boil at $(373.15 + 0.21)$ K, or 373.36 K.
(b) Freezing point: Here we have, from Eq. (8.24),

$$\Delta T = (1.86 \text{ K } m^{-1})(0.418 \text{ } m)$$

$$= 0.78 \text{ K}$$

Thus the solution will freeze at $(273.15 - 0.78)$ K, or 272.37 K.

Comments: (1) For aqueous solutions of equal concentrations, the depression in freezing point is almost always greater than the corresponding elevation in boiling point. The reason can be seen by comparing the following two equations:

$$\Delta T = \frac{RT_0^2}{\Delta \overline{H}_{vap}} X_2 \qquad \Delta T = \frac{RT_0^2}{\Delta \overline{H}_{fus}} X_2$$

Although $(T_0)_{bp} > (T_0)_{mp}$, $\Delta \overline{H}_{vap}$ for water is 40.79 kJ mol^{-1} while $\Delta \overline{H}_{fus}$ is only 6.01 kJ mol^{-1}. It is this large quantity $(\Delta \overline{H}_{vap})$ in the denominator that results in a small value of ΔT. (2) Of the four colligative properties, *only* the osmotic pressure provides a reliable method of determining the molar masses of macromolecules. Consider a solution that contains 200 g of hemoglobin in 1000 g of water. Here, the freezing point would be depressed by only 0.006 K, a quantity too small to measure accurately. Keep in mind that hemoglobin is one of the most soluble proteins known.

OSMOTIC PRESSURE

The phenomenon of osmotic pressure can be understood by considering Figure 8.12. The left compartment of the apparatus contains pure solvent; the right compartment contains a solution. The two compartments are separated by a *semipermeable membrane* (for example, a cellophane membrane), one that permits the solvent molecules to pass through but does

Apparatus demonstrating the osmotic pressure phenomenon.

Figure 8.12

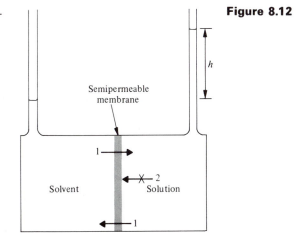

not permit the movement of solute molecules from right to left. Practically speaking, then, this system has two different phases. At equilibrium the height of the solution in the tube is greater than that of the pure solvent by h. We can now derive an expression for the osmotic pressure as follows.

Let μ_1^L and μ_1^R be the chemical potential of the solvent in the left and right compartments, respectively. Initially, before equilibrium is established, we have

$$\mu_1^L = \mu_1^* + RT \ln X_1$$

$$= \mu_1^* \qquad (X_1 = 1)$$

and

$$\mu_1^R = \mu_1^* + RT \ln X_1 \qquad (X_1 < 1)$$

Thus

$$\mu_1^L = \mu_1^* > \mu_1^R = \mu_1^* + RT \ln X_1$$

Note that μ_1^L is the same as the standard chemical potential for the pure solvent μ_1^*, and the inequality sign denotes that $RT \ln X_1$ is a negative quantity. Consequently, more solvent molecules, on the average, will move from left to right across the membrane. The process is spontaneous because the dilution of the solution in the right compartment by solvent leads to a decrease in the Gibbs free energy and an increase in entropy. Equilibrium is finally reached when the flow of solvent is exactly balanced by the hydrostatic pressure difference in the two side arms. This extra pressure increases μ_1^R. From Eq. (7.22), we saw that

$$\left(\frac{\partial G}{\partial P}\right)_T = V$$

A similar equation can be written for the variation of the chemical potential with pressure at constant temperature. Thus for the solvent component in the right compartment, we have

$$\left(\frac{\partial \mu_1^R}{\partial P}\right)_T = \overline{V}_1 \qquad (8.28)$$

where $\overline{V}_1$ is the partial molar volume of the solvent. For a dilute solution, $\overline{V}_1$ is approximately equal to $\overline{V}_1^*$, the molar volume of the pure solvent. The increase in chemical potential $\Delta\mu_1^R$ when the pressure increases from P, the external (atmospheric) pressure, to $(P + \pi)$ is given by

$$\Delta\mu_1^R = \int_P^{P+\pi} \overline{V}_1^* \, dP = \overline{V}_1^* \pi$$

$\overline{V}_1^*$ is treated as a constant, since the volume of a liquid changes little with pressure. The quantity π is called the *osmotic pressure*; it is equal to the excess hydrostatic pressure exerted on the solution side. The term *osmotic pressure of a solution* indicates the pressure that must be applied to the solution to increase the chemical potential of the solvent to the value of its pure liquid at 1 atm.

At equilibrium the following relation must hold:

$$\mu_1^L = \mu_1^R = \mu_1^* + RT \ln X_1 + \pi\overline{V}_1^*$$

Since $\mu_1^L = \mu_1^*$, we have

$$\pi \overline{V}_1^* = -RT \ln X_1 \tag{8.29}$$

We want to relate π to the concentration of solute. This is accomplished in the following steps. From the procedure employed in the boiling-point elevation:

$$-\ln X_1 = -\ln (1 - X_2)$$
$$= X_2 \qquad (X_2 \ll 1)$$

Furthermore,

$$X_2 = \frac{n_2}{n_1 + n_2} = \frac{n_2}{n_1} \qquad (n_1 \gg n_2)$$

where n_1 and n_2 are the number of moles of solvent and solute, respectively. Equation (8.29) now becomes

$$\pi \overline{V}_1^* = RT X_2$$
$$= RT \left(\frac{n_2}{n_1} \right) \tag{8.30}$$

The total volume of the solution V is given by [see Eq. (8.1)]

$$V = n_1 \overline{V}_1^* + n_2 \overline{V}_2^*$$
$$= n_1 \overline{V}_1^* \qquad \text{(for a dilute solution)}$$

or

$$\overline{V}_1^* = \frac{V}{n_1}$$

Substituting the preceding expression into Eq. (8.30), we get

$$\pi V = n_2 RT \tag{8.31}$$

If V is in liters, then

$$\pi = \frac{n_2}{V} RT$$
$$= MRT \tag{8.32}$$

where M is the molarity of the solution. Note that molarity is a convenient concentration unit here, since osmotic pressure measurements are normally carried out at a constant temperature. Furthermore, for dilute solutions, molarity and molality are practically the same quantities. Alternatively, we can write

$$\pi = \frac{c_2}{M_2} RT \tag{8.33a}$$

or

$$\frac{\pi}{c_2} = \frac{RT}{M_2} \tag{8.33b}$$

where c_2 is the concentration of the solute in grams per liter of solution and M_2 is the molar mass of the solute. Equation (8.33), then, provides a way

Figure 8.13

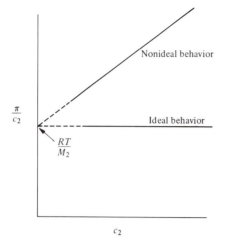

Nonideal behavior

Ideal behavior

$\dfrac{\pi}{c_2}$

$\dfrac{RT}{M_2}$

c_2

Determination of molar mass of a solute by osmotic pressure measurement for an ideal and a nonideal solution.

for determining molar masses of compounds from osmotic pressure measurements.

Equation (8.33) is derived assuming ideal behavior, so it is desirable to measure π at several different concentrations and extrapolate to zero concentration for molar mass determination (Figure 8.13). For a nonideal solution the osmotic pressure at *any* concentration c_2 is given by

$$\frac{\pi}{c_2} = RT\left(\frac{1}{M_2} + Bc_2 + Cc_2^2 + Dc_2^3 + \cdots\right) \tag{8.34}$$

where B, C, and D are called the second, third, and fourth virial coefficients, respectively.* The values of the virial coefficients are such that $B \gg C \gg D$. In dilute solutions we need only be concerned with the second virial coefficient. For an ideal solution, the virial coefficients are all equal to zero, so that Eq. (8.34) reduces to Eq. (8.33).

It is interesting to note that although osmosis is a well-studied phenomenon, the mechanism involved is not always clearly understood. In some cases a semipermeable membrane may act like a molecular sieve, allowing smaller solvent molecules to pass through while blocking larger solute molecules. In other cases, osmosis may be caused by the higher solubility of the solvent in the membrane than the solute. Each system must be studied individually. The foregoing discussion illustrates both the usefulness and limitation of thermodynamics. We have derived a convenient equation relating the molar mass of the solute to an experimentally measurable quantity—the osmotic pressure—simply in terms of the chemical potential difference. However, since thermodynamics is not based on any specific model, Eq. (8.33) tells us nothing about the mechanism of osmosis.

EXAMPLE 8.5 Consider the following arrangement, in which a solution containing 20 g of hemoglobin in 1 liter of the solution is placed in the right compartment and pure water is placed in the left compartment (see Figure 8.12). At equilibrium, the height of the water in the right-hand column is 77.8 mm in excess of that in the left-hand column. What is the molar mass of hemoglobin? The temperature of the system is maintained at 298 K.

* The reader should compare Eq. (8.34) with Eq. (2.13).

CGS units: To calculate the molar mass of hemoglobin, we first need to calculate the osmotic pressure of the solution. Since

$$\text{pressure} = \frac{\text{force}}{\text{area}}$$

$$= \frac{Ah\rho g}{A} = h\rho g$$

where A is the area of the cross section of the side arm, h the excess height in the right-hand column, ρ the density of the solution, and g the acceleration due to gravity, we have

$$h = 7.78 \text{ cm}$$

$$g = 980.7 \text{ cm s}^{-2}$$

$$\rho \simeq 1.0 \text{ g cm}^{-3}$$

We assume here that the density of the solution is close to that of pure water. Thus

$$\text{osmotic pressure} = \pi = 7.78 \text{ cm} \times 1 \text{ g cm}^{-3} \times 980.7 \text{ cm s}^{-2}$$

$$= 7629.8 \text{ g cm}^{-1} \text{ s}^{-2}$$

$$= 7629.8 \text{ dyn cm}^{-2}$$

Since $1 \text{ atm} = 1.0133 \times 10^6 \text{ dyn cm}^{-2}$ (see Section 2.6), we have

$$\pi = \frac{7629.8}{1.0133 \times 10^6} = 7.53 \times 10^{-3} \text{ atm}$$

From Eq. (8.33),

$$\pi = 7.53 \times 10^{-3} \text{ atm}$$

$$c = 20 \text{ g liter}^{-1}$$

$$R = 0.08206 \text{ liter atm K}^{-1} \text{ mol}^{-1}$$

$$T = 298 \text{ K}$$

so that

$$M_2 = \frac{c_2 RT}{\pi}$$

$$= \frac{(20 \text{ g liter}^{-1})(0.08206 \text{ liter atm K}^{-1} \text{ mol}^{-1})(298 \text{ K})}{7.53 \times 10^{-3} \text{ atm}}$$

$$= 65{,}000 \text{ g mol}^{-1}$$

SI units: Here we have

$$h = 0.0778 \text{ m}$$

$$g = 9.807 \text{ m s}^{-2}$$

$$\rho \simeq 1 \times 10^3 \text{ kg m}^{-3}$$

Then

$$\pi = 0.0778 \text{ m} \times 1 \times 10^3 \text{ kg m}^{-3} \times 9.807 \text{ m s}^{-2}$$

$$= 762.98 \text{ kg m}^{-1} \text{ s}^{-2}$$

$$= 762.98 \text{ N m}^{-2}$$

From Eq. (8.31),

$$\pi = 762.98 \text{ N m}^{-2}$$

$$c = 20 \text{ kg m}^{-3}$$

$$R = 8.314 \text{ J K}^{-1} \text{ mol}^{-1}$$

$$T = 298 \text{ K}$$

so that

$$M_2 = \frac{(20 \text{ kg m}^{-3})(8.314 \text{ J K}^{-1} \text{ mol}^{-1})(298 \text{ K})}{762.98 \text{ N m}^{-2}}$$

$$= 65 \text{ kg mol}^{-1}$$

Example 8.5 shows that osmotic pressure measurement is a more sensitive method for determining molar mass than the freezing-point-depression technique, since 7.8 cm is an easily measurable height. Most proteins are less soluble than hemoglobin. Nevertheless, their molar masses can often be determined by osmotic pressure measurements. In Chapter 21 we shall discuss other useful techniques for determining the molar mass of macromolecules.

The osmotic-pressure phenomenon is important to many chemical and biological systems, so we shall consider a few examples here. If two solutions are of equal concentration, and hence the same osmotic pressure, they are said to be *isotonic*. For two solutions of unequal osmotic pressures, the more concentrated solution is said to be *hypertonic* and the less concentrated solution is said to be *hypotonic*. To study the contents in red blood cells, which are protected from the outside environment by a semipermeable membrane, biochemists employ a technique called *hemolysis*. Placing the red blood cells in a hypotonic solution causes water to move into the cell. The cells will swell and eventually burst, releasing hemoglobin and other protein molecules. When a cell is placed in a hypertonic solution, on the other hand, the intracellular water tends to move out of the cell to the more concentrated solution outside by osmosis. This process, known as *crenation*, causes the cell to shrink and eventually cease to function.

The mammalian kidney is a particularly effective osmotic device. Its main function is to remove metabolic waste products and other impurities from the bloodstream by osmosis to the more concentrated urine outside through a semipermeable membrane. The biologically important ions, such as Na^+ and Cl^-, lost in this manner are then actively pumped back into the blood through the same membrane (see Section 9.6). The loss of water through the kidney is controlled by the antidiuretic hormone (ADH), which is secreted into the blood by the hypothalamus and posterior pituitary gland. When little or no ADH is secreted, large amounts of water (perhaps 10 times normal) passes into the urine each day. On the other hand, when large quantities of

193

Sec. 8.8

Colligative
Properties
of Real
Solutions

ADH are present in the blood, the permeability of water through the membrane deceases so that the volume of urine formed may be as little as one-half normal. Thus the kidney–ADH combination controls both the rate of loss of water and the loss of other small waste molecules. The chemical potential of water within the body fluids of freshwater fishes is lower than that in their environment, so they are able to draw in water by osmosis through their gill membranes. Surplus water is excreted as urine. An opposite process occurs for the marine teleost fishes. They lose body water to the more concentrated environment by osmosis across the gill membranes. To balance the loss, they drink seawater.

In Chapter 5 we mentioned that the capillary effect is only partially responsible for water rising upward in plants—osmotic pressure is the major mechanism. The leaves of trees constantly lose water to their surroundings, a process called *transpiration*, so the solute concentrations in leaf fluids increases. Water is then pushed up through the trunks and branches by osmotic pressure, which, to reach the tops of the tallest trees, may be as high as 10 to 15 atm.* Leaf movement is an interesting phenomenon that may also be related to osmotic pressure. It is believed that some unknown process in the presence of light can increase salt concentration in leaf cells. Osmotic pressure is increased and cells become enlarged and turgid, causing the leaves to move toward light.

We shall now briefly discuss a related phenomenon called *reverse osmosis*. To the solution compartment shown in Figure 8.12, if we apply pressure greater than the equilibrium osmotic pressure, pure solvent will flow from the solution to the solvent compartment. This reversal of the osmotic process results in an unmixing of the solution. An important application of reverse osmosis is the desalination of water. Several techniques discussed in this chapter are suitable, at least in principle, for obtaining pure water from the sea. For example, either distilling or freezing seawater would achieve the goal. However, these processes involve a phase change from liquid to vapor or liquid to solid, requiring considerable energy input for maintaining the operation. Reverse osmosis is more appealing, for it does not involve a phase change and is economically sound for large amounts of water. Seawater, which is approximately 0.7 M in NaCl, has an estimated osmotic pressure of 30 atm. For a 50% recovery of pure water from the sea, an additional 60 atm would have to be applied on the seawater-side compartment to cause reverse osmosis. The main practical obstacle to desalination is the selection of a suitable membrane permeable to water but not to other dissolved salts and able to withstand the high pressure over long periods of time.†

Colligative Properties of Real Solutions **8.8**

The equations that describe the colligative properties of solutions were derived assuming ideal behavior. These equations hold quite well for dilute solutions (concentrations less than 0.2 m). For more concentrated solutions or in cases where deviation from ideality are known to occur, we need to replace

---

* Entropy Makes Water Run Uphill—in Trees," P. E. Stevenson, *J. Chem. Educ.* **48**, 837 (1971).
† "Desalination of Water by Reverse Osmosis," C. E. Hecht, *J. Chem. Educ.* **44**, 53 (1967).

concentrations with activities. The colligative properties would then be expressed as

$$\Delta P = a_2 P_1^* \qquad (a_2 = \gamma_2 X_2) \qquad (8.35)$$

$$\Delta T = K_b a_2 \qquad (a_2 = \gamma_2 m_2) \qquad (8.36)$$

$$\Delta T = K_f a_2 \qquad (a_2 = \gamma_2 m_2) \qquad (8.37)$$

$$\pi = a_2 RT \qquad (a_2 = \gamma M) \qquad (8.38)$$

Thus by measuring the colligative properties we can determine the activities of the solutes in the solutions.

Suggestions for Further Reading

The topics discussed in this chapter are covered in any of the standard texts of physical chemistry listed in Chapter 1. The following are more specialized texts.

HILDEBRAND, J. H., and R. L. SCOTT, *Regular Solutions*. Prentice-Hall, Inc., Englewood Cliffs, N.J., 1962.

HILDEBRAND, J. H., and R. L. SCOTT. *The Solubility of Nonelectrolytes*. Van Nostrand Reinhold Company, New York, 1950.

Reading Assignments

SECTION A

"Colligative Properties," F. Rioux, *J. Chem. Educ.* **50**, 490 (1973).

"A Biological Antifreeze," R. E. Feeney, *Am. Sci.* **62**, 712 (1974).

"The Osmotic Pump," O. Levenspiel and N. de Nevers, *Science* **183**, 157 (1974).

"Desalination of Water by Reverse Osmosis," C. E. Hecht, *J. Chem. Educ.* **44**, 53 (1967).

"Reverse Osmosis," M. J. Suess, *J. Chem. Educ.* **48**, 190 (1971).

"Reverse Osmosis," K. W. Boddeker, *Angew. Chem. Int. Ed.* **16**, 607 (1977).

"Deviations from Raoult's Law," M. L. McGlashan, *J. Chem. Educ.* **40**, 516 (1963).

SECTION B

"The Mechanism of Vapor Pressure Lowering," K. J. Mysels, *J. Chem. Educ.* **32**, 179 (1955).

"Ideal Solutions," W. A. Oates, *J. Chem. Educ.* **46,** 501 (1969).

"Colligative Properties of a Solution," H. T. Hammel, *Science* **192**, 748 (1976).

"Osmotic Pressure," D. W. Kupke, *Advan. Protein Chem.* **15**, 57 (1960).

"Osmotic Pressure in the Physics Course for Students of the Life Sciences," R. K. Hobbie, *Am. J. Phys.* **42**, 188 (1974).

"Demonstrating Osmotic and Hydrostatic Pressures in Blood Capillaries," J. W. Ledbetter, Jr., and H. D. Jones, *J. Chem. Educ.* **44**, 362 (1967).

"Desalting Water by Freezing," A. E. Snyder, *Sci. Am.*, Dec. 1962.

"The Kidney," H. W. Smith, *Sci. Am.*, Jan. 1953.

"Removal of an Assumption in Deriving the Phase Change Formula $\Delta T = Km$," F. E. Schubert, *J. Chem. Educ.* **56**, 259 (1979).

8.1 How many grams of water must be added to 20.0 g of urea to prepare a 5.00% aqueous urea solution?

8.2 What is the molarity of a 2.12 m aqueous sulfuric acid solution? The density of this solution is 1.30 g cm^{-3}.

8.3 Calculate the molality of a 1.50 M aqueous ethanol solution. The density of the solution is 0.980 g cm^{-3}.

8.4 The concentrated sulfuric acid we use in the laboratory is 98.0% sulfuric acid by weight. Calculate the molality and molarity of concentrated sulfuric acid if the density of the solution is 1.83 g cm^{-3}.

8.5 The blood sugar (glucose) level of a diabetic woman patient is about 0.140 g of glucose/100 ml of blood. Every time the patient ingests 40 g of glucose, her blood glucose level rises to about 0.240 g/100 ml of blood. Calculate the number of moles of glucose per ml of blood and the total number of moles and grams of glucose in the blood before and after consumption of glucose. (Assume that the total volume of blood in her body is 5.0 liters.)

8.6 The strength of alcoholic beverages is usually described in terms of "proof," which is defined as twice the percentage by volume of ethanol. Calculate the number of grams of alcohol in 2 quarts of 75% proof gin. What is the molality of the gin? (The density of ethanol = 0.80 g cm^{-3}; 1 quart = 0.946 liter.)

8.7 Convert a 0.25 m sucrose solution into percent by weight. The density of the solution is 1.2 g cm^{-3}.

8.8 A mixture of ethanol and n-propanol behaves ideally at 36.4°C. (a) Determine graphically the mole fraction of n-propanol in a mixture of ethanol and n-propanol that boils at 36.4°C and 72 mm Hg. (b) What is the total vapor pressure over the mixture at 36.4°C when the mole fraction of n-propanol is 0.60? (c) Calculate the composition of the vapor in (b). (The equilibrium vapor pressures of ethanol and n-propanol at 36.4°C are 108 mm Hg and 40.0 mm Hg, respectively.)

8.9 Two beakers, 1 and 2, containing 50 ml of 0.10 M urea and 50 ml of 0.20 M urea, respectively, are placed under a tightly sealed bell jar at 298 K. Calculate the mole fraction of urea in the solutions at equilibrium. Assume ideal behavior. (*Hint*: Use Raoult's law and note that at equilibrium, the mole fraction of urea is the same in both solutions.)

8.10 Consider a binary liquid mixture A and B, where A is volatile and B is nonvolatile. The composition of the solution in terms of mole fraction is $X_A = 0.045$ and $X_B = 0.955$. The vapor pressure of A is 5.60 mm Hg while that of pure A is 196.4 mm Hg at the same temperature. Calculate the activity coefficient of A at this concentration.

8.11 At 298 K, the vapor pressure of pure water is 23.76 mm Hg and that of seawater is 22.98 mm Hg. Assuming the seawater to contain only NaCl, estimate its concentration. (*Hint*: Sodium chloride is a strong electrolyte.)

8.12 Calculate the changes in entropy for the following processes: (a) mixing of 1 mol of nitrogen and 1 mol of oxygen, and (b) mixing of 2 mol of argon, 1 mol of helium, and 3 mol of hydrogen. Both (a) and (b) are carried out under conditions of constant temperature and constant pressure. Assume ideal behavior.

8.13 The Henry's law constant of oxygen in water at 25°C is 773 atm mol^{-1} kg of water. Calculate the molality of oxygen in water under a partial pressure of 0.20 atm. Assuming that the solubility of oxygen in blood at 37°C is roughly the same as that in water at 25°C, comment on the prospect for our survival without hemoglobin molecules. (The total volume of blood in the human body is about 5 liters.)

8.14 Prove the statement that an alternative way to express Henry's law of gas solubility is to say that the volume of gas that dissolves in a fixed volume of solution is independent of pressure at a given temperature.

8.15 A miner working 900 ft below the surface had a bottle of soft drink during the lunch break. To his surprise, the drink seemed very flat (that is, not much effervescence was observed upon removing the cap). Shortly after lunch he took the elevator up to the surface. During the ascent, he felt a great urge to eructate. Explain.

8.16 At 67°C, a solution containing 45 mol % *n*-hexane in *n*-heptane behaves ideally. What is the change in chemical potential at 67°C for the transfer of 1 mol of *n*-hexane from solution to the vapor phase?

8.17 Trees in cold-climate countries may be subjected to temperatures as low as −60°C. Estimate the concentration of an aqueous solution in the body of the tree that would remain unfrozen at this temperature. Is this a reasonable concentration? Comment on your result.

8.18 Explain how a pressure cooker reduces cooking time.

8.19 Provide a molecular interpretation for the positive and negative deviations in the boiling-point curves and the formation of azeotropes.

8.20 The freezing-point-depression measurement of benzoic acid in acetone yields a molar mass of 122 g; the same measurement in benzene gives a value of 242 g. Account for this discrepancy. (*Hint*: Consider solvent–solute and solute–solute interactions.)

8.21 A common antifreeze for car radiators is ethylene glycol, $CH_2(OH)CH_2(OH)$. How many milliliters of this substance would you add to 6.5 liters of water in the radiator if the coldest day in winter is −20°C? Would you keep this substance in the radiator in the summer to prevent the water from boiling? (The density and boiling point of ethylene glycol are 1.11 g cm^{-3} and 470 K, respectively.)

8.22 In intravenous injections, great care is taken to ensure that the concentration of solutions to be injected must be comparable to that of blood plasma. Why?

8.23 The tallest trees known are the redwoods in California. Assuming the height of a redwood to be 105 m (ca. 350 ft), estimate the osmotic pressure required to push water up from the roots to the treetop.

8.24 Calculate the change in the Gibbs free energy at 37°C for the kidneys to secrete 0.275 mol of urea/kg of water from blood plasma to urine if the concentrations of urea in blood plasma and urine are 0.005 m and 0.326 m, respectively.

8.25 Calculate ΔS for the mixing of 5.00 g of Ar at 95.0°C and 0.780 atm with 8.50 g of N_2 at 105°C and 1.20 atm.

8.26 A mixture of liquids *A* and *B* exhibits ideal behavior. At 84°C, the total vapor pressure of a solution containing 1.2 mol of *A* and 2.3 mol of *B* is 331 mm Hg. Upon the addition of 1 more mol of *B* to the solution, the vapor pressure increases to 347 mm Hg. Calculate the vapor pressures of pure *A* and *B* at 84°C.

8.27 At 45°C, the vapor pressure of water from a glucose solution in which the mole fraction of glucose is 0.080 is 65.76 mm Hg. Calculate the activity and activity coefficient of the water in the solution. The vapor pressure of pure water at 45°C is 71.88 mm Hg.

8.28 Fish breathe the dissolved air in water through their gills. Assuming the partial pressures of oxygen and nitrogen in air to be 0.20 atm and 0.80 atm, respectively, calculate the mole fractions of oxygen and nitrogen in water at 298 K. Comment on your results.

8.29 Liquids A (molar mass 100 g mol^{-1}) and B (molar mass 110 g mol^{-1}) form an ideal solution. At 55°C, A has a vapor pressure of 95 mm Hg and B a vapor pressure of 42 mm Hg. A solution is prepared by mixing equal weights of A and B. (a) Calculate the mole fraction of each component in the solution. (b) Calculate the partial pressures of A and B over the solution at 55°C. (c) Suppose that some of the vapor described in (b) is condensed to a liquid. Calculate the mole fraction of each component in this liquid and the vapor pressure of each component above this liquid at 55°C.

8.30 Lysozyme extracted from chicken egg white has a molar mass of 13,930 g mol^{-1}. Exactly 0.1 g of this protein is dissolved in 50 g of water at 298 K. Calculate the vapor pressure lowering, the depression in freezing point, the elevation of boiling point, and the osmotic pressure of this solution. The vapor pressure of pure water at 298 K is 23.76 mm Hg.

8.31 The osmotic pressure of poly(methyl methacrylate) in toluene has been measured at a series of concentrations at 298 K:

π (atm)	8.40×10^{-4}	1.72×10^{-3}	2.52×10^{-3}	
c (g liter^{-1})	8.10	12.31	15.00	

		3.23×10^{-3}	7.75×10^{-3}
		18.17	28.05

Determine graphically the molar mass of the polymer.

8.32 The fact that the vapor pressure of the solvent is lower over a solution than over the pure solvent and that lowering is proportional to the concentration is frequently explained as follows. A dynamic equilibrium exists in both cases, so that the rate at which molecules of solvent evaporate from the liquid is always equal to that at which they condense. The rate of condensation is proportional to the partial pressure of the vapor, while that of evaporation is unimpaired in the pure solvent but is impaired by solute molecules in the surface of the solution. Hence the rate of escape is reduced in proportion to the concentration of the solute and maintenance of equilibrium requires a corresponding lowering of the rate of condensation and therefore of the partial pressure of the vapor phase. List arguments showing the incorrectness of this explanation. [*Source*: K. J. Mysels, *J. Chem. Educ.* **32**, 179 (1955).]

8.33 At 25°C and 1 atm pressure, the absolute third-law entropies of methane and ethane are 186.19 J K^{-1} mol^{-1} and 229.49 J K^{-1} mol^{-1} in the gas phase. Calculate the absolute third-law entropy of a "solution" containing 1.0 mol of each gas. Assume ideal behavior.

8.34 A compound weighing 0.458 g is dissolved in 30.0 g of acetic acid. The freezing point of the solution is found to be 1.50 K below that of the pure solvent. Calculate the molar mass of the compound.

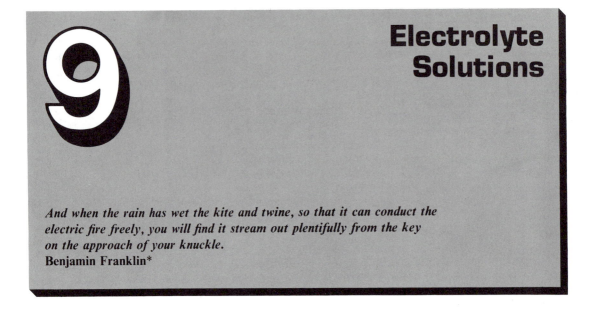

9 Electrolyte Solutions

And when the rain has wet the kite and twine, so that it can conduct the electric fire freely, you will find it stream out plentifully from the key on the approach of your knuckle.
Benjamin Franklin*

Chapter 8 dealt with solutions containing nonelectrolytes. However, all biological and many chemical systems are aqueous solutions containing various ions. The stability of biomacromolecules and the rate of many biochemical reactions are very much dependent on the type and concentration of ions present. It is important to have at least a qualitative understanding of the behavior of ions in solution.

An electrolyte is a substance whose dissolving, usually in water, results in an electrically conducting solution. An electrolyte can be an acid, a base, or a salt. This chapter will consider ionic conductance, dissociation, the thermodynamics of aqueous electrolyte solutions, and the structure and function of biological membranes.

9.1 Electrical Conduction in Solutions

SOME BASIC DEFINITIONS

The ability of an electrolyte to conduct electricity provides us with a simple, direct means of studying ionic behavior in solution. Let us begin by defining a few basic terms.

Ohm's Law. Ohm's law states that the current (I) flowing through a particular medium is directly proportional to the voltage or the electric potential difference (V) across the medium and indirectly proportional to the

* Leonard W. Labaree et al., eds., *The Papers of Benjamin Franklin*, Vol. 4, p. 367, Yale University Press, New Haven, Conn., 1961. Used by permission.

resistance (R) of the medium. Thus

$$I = \frac{V}{R} \qquad (9.1)$$

where I is in amperes, V in volts, and R in ohms.*

Resistance. The *resistance* across a particular medium depends on the geometry of the medium; it is directly proportional to the length (l) and inversely proportional to the cross section of area (A) of the medium. Thus we write

$$R \propto \frac{l}{A}$$

$$= \rho \frac{l}{A} \qquad (9.2)$$

where the proportionality constant ρ is called the *specific resistance* or *resistivity*. Since the units of R are ohms (Ω), l centimeters or meters, and A square centimeters or square meters, the units of ρ are Ω cm or Ω m. Resistivity is a property characteristic of the material comprising the medium.

Conductance (C). *Conductance* is the reciprocal of resistance, that is,

$$C = \frac{1}{R} = \frac{1}{\rho} \frac{A}{l} = \kappa \frac{A}{l} \qquad (9.3)$$

where κ is the *specific conductance* or *conductivity*, equal to $1/\rho$. Conductivity has the units Ω^{-1} cm^{-1} or Ω^{-1} m^{-1}. In the SI system, the symbol for conductance is S (siemens).

A typical conductance cell† is shown in Figure 9.1. The conductance is given by Eq. (9.3). The quantity l/A, called the *cell constant*, is the same for all solutions. Here A is the area and l the distance of separation between the electrodes. In practice, instead of measuring A and l, the cell is calibrated using a standard solution (KCl) of known κ value. Thus l/A can be calculated by measuring the conductance of the solution.

A conductance cell. 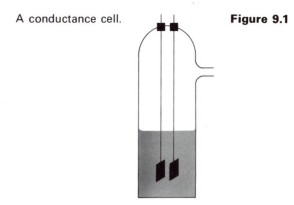 **Figure 9.1**

---

* In the SI system, the symbols for current and resistance are A and Ω, respectively.

† It is customary to measure the conductance rather than the resistance of an electrolyte solution.

Although the specific conductance can be easily measured (from the known cell constant and the experimentally determined conductance), it is not a convenient quantity to use in the discussion of the conduction process in electrolyte solutions. Solutions of different concentrations, for example, will have very different specific conductances simply because a given volume of the different solutions will contain a different number of ions. For this reason, it is preferable to express the conductance as a molar quantity. We define the *molar conductance* (Λ_m) as

$$\Lambda_m = \frac{1000\kappa}{c} \tag{9.4}$$

where c is the concentration of the solution in moles per liter. The factor 1000 is necessary for Λ_m to be expressed in Ω^{-1} mol^{-1} cm^2. Another term, called the *equivalent conductance* (Λ), is also used. In this case the conductance is expressed in terms of the number of individual charges that are being carried. For example, for uni-univalent electrolyte solutions such as KCl or NaCl solution, each ion carries a unit charge, so the equivalent conductance is the same as the molar conductance. In a divalent electrolyte solution such as $MgSO_4$ each ion carries two units of charge. Here the equivalent conductance is half the molar conductance. The equivalent conductance has the units Ω^{-1} equiv^{-1} cm^2.

EXAMPLE 9.1 The resistance of a cell containing an aqueous 0.0560 M KCl solution is found to be 41.8 Ω. When the same cell is filled with an aqueous 0.0836 M NaCl solution, the resistance is 35.1 Ω. Calculate the equivalent conductance of the NaCl at this concentration. Given that the equivalent conductance of KCl at 0.0560 M is 134.5 Ω^{-1} mol^{-1} cm^2.

Answer: We need the cell constant. Our first step is to calculate the specific conductance. From Eq. (9.4),

$$\kappa = \frac{\Lambda_m c}{1000}$$
$$= \frac{(134.5 \ \Omega^{-1} \ \text{mol}^{-1} \ \text{cm}^2)(0.0560 \ \text{mol liter}^{-1})}{1000 \ \text{cm}^3 \ \text{liter}^{-1}}$$
$$= 7.53 \times 10^{-3} \ \Omega^{-1} \ \text{cm}^{-1}$$

Note that the molar conductance is the same as the equivalent conductance in this case. Next, from Eq. (9.3),

$$\frac{l}{A} = \frac{\kappa}{C} = \kappa R$$
$$= (7.53 \times 10^{-3} \ \Omega^{-1} \ \text{cm}^{-1})(41.8 \ \Omega)$$
$$= 0.315 \ \text{cm}^{-1}$$

The specific conductance of the NaCl solution is given by

$$\kappa = \frac{l}{A} C = \frac{l}{A} \frac{1}{R}$$
$$= (0.315 \ \text{cm}^{-1}) \times \frac{1}{35.1 \ \Omega}$$
$$= 8.97 \times 10^{-3} \ \Omega^{-1} \ \text{cm}^{-1}$$

Finally, the equivalence conductance of the NaCl solution is given by

$$\Lambda = \frac{1000 \text{ cm}^3 \text{ liter}^{-1} \times 8.97 \times 10^{-3} \ \Omega^{-1} \text{ cm}^{-1}}{0.0836 \text{ mol liter}^{-1}}$$

$$= 107.3 \ \Omega^{-1} \text{ mol}^{-1} \text{ cm}^2$$

Careful conductance measurements were carried out by Kohlrausch in the nineteenth century; some of his data are shown in Figure 9.2. According to Eq. (9.4), it might appear that Λ or Λ_m would be independent of the concentration of the solution (κ is directly proportional to concentration, but κ/c should be a constant for a given substance). However, this is not the case. Instead, Kohlrausch found the following relation to hold for strong electrolytes:*

$$\Lambda = \Lambda_0 - B\sqrt{c} \qquad (9.5)$$

where B is a constant for a given electrolyte and Λ_0 is the equivalent conductance at infinite dilution; that is, $\Lambda \to \Lambda_0$ as $c \to 0$. Thus Λ_0 can be readily obtained by plotting Λ versus $\sqrt{c}$ and extrapolating to zero concentration. This method is unsatisfactory for weak electrolytes because of the steepness of curves at low concentrations.

Plots of equivalent conductance versus the square root of concentration for several electrolytes.

Figure 9.2

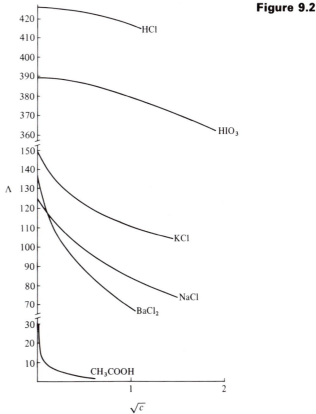

*Strong electrolytes are substances that for all practical purposes are largely or completely dissociated into ionic species in solution.

Table 9.1

Equivalent Conductance at Infinite Dilution for Some Electrolytes in Water at 298 K[a]

Electrolyte	Λ_0 (Ω^{-1} equiv^{-1} cm^2)
HCl	426.16
CH$_3$COOH	390.71
LiCl	115.03
NaCl	126.45
AgCl	137.20
KCl	149.85
LiNO$_3$	110.14
NaNO$_3$	121.56
KNO$_3$	144.96
CuSO$_4$	133.62
CH$_3$COONa	91.00

[a] To express Λ_0 as Ω^{-1} equiv^{-1} m^2, multiply each number by 10^{-4}. Thus Λ_0 for HCl is 426.16 Ω^{-1} equiv^{-1} cm^2 or 426.16×10^{-4} Ω^{-1} equiv^{-1} m^2.

Table 9.1 shows the values of Λ_0 for a number of electrolytes. An interesting pattern emerges when we examine the difference in Λ_0 between two electrolytes containing a common cation or anion. For example,

$$\Lambda_0^{KCl} - \Lambda_0^{NaCl} = 23.4 \ \Omega^{-1} \ \text{equiv}^{-1} \ \text{cm}^2$$

$$\Lambda_0^{KNO_3} - \Lambda_0^{NaNO_3} = 23.4 \ \Omega^{-1} \ \text{equiv}^{-1} \ \text{cm}^2$$

The same difference in these two cases and similar observations led Kohlrausch to suggest the following relation:

$$\Lambda_0 = \lambda_0^+ + \lambda_0^- \tag{9.6}$$

where λ_0^+ and λ_0^- are the equivalent ionic conductances at infinite dilution. Equation (9.6) is known as *Kohlrausch's law of independent migration*. It means that equivalent conductance at infinite dilution is made up of two independent contributions from the cationic and anionic species. We can now see why the same value was obtained in the example above, since

$$\Lambda_0^{KCl} - \Lambda_0^{NaCl} = \lambda_0^{K^+} + \lambda_0^{Cl^-} - \lambda_0^{Na^+} - \lambda_0^{Cl^-} = \lambda_0^{K^+} - \lambda_0^{Na^+}$$

$$\Lambda_0^{KNO_3} - \Lambda_0^{NaNO_3} = \lambda_0^{K^+} + \lambda_0^{NO_3^-} - \lambda_0^{Na^+} - \lambda_0^{NO_3^-} = \lambda_0^{K^+} - \lambda_0^{Na^+}$$

DEGREE OF DISSOCIATION

At a certain concentration, an electrolyte may only be partially dissociated. At infinite dilution, any electrolyte, weak or strong, is completely dissociated. In 1887, Arrhenius suggested that the *degree of dissociation* (α) of an electrolyte can be calculated by the simple relation

$$\alpha = \frac{\Lambda}{\Lambda_0} \tag{9.7}$$

where Λ is the equivalent conductance at a particular concentration to which α refers. Equation (9.7) was used by Ostwald to measure the equilibrium constant of dissociation. Consider the dissociation of a weak acid HA of concentration c (mol liter^{-1}):

$$HA \rightleftharpoons H^+ + A^-$$

$$c(1 - \alpha) \qquad c\alpha \qquad c\alpha$$

By definition, the equilibrium constant is given by

$$K = \frac{[H^+][A^-]}{[HA]} = \frac{c^2\alpha^2}{c(1 - \alpha)}$$

Using the expression for α in Eq. (9.7), we obtain

$$K = \frac{c\Lambda^2}{\Lambda_0(\Lambda_0 - \Lambda)} \tag{9.8}$$

Equation (9.8) can be rearranged to give

$$\frac{1}{\Lambda} = \frac{1}{K\Lambda_0^2}(\Lambda c) + \frac{1}{\Lambda_0} \tag{9.9}$$

Thus K can be obtained either directly from Eq. (9.8) or more accurately from Eq. (9.9) by plotting $1/\Lambda$ versus Λc. Of course, Λ_0 must be known if only Eq. (9.8) is employed.

EXAMPLE 9.2 The equivalent conductance of an aqueous acetic acid solution of concentration 0.10 M is 5.2 Ω^{-1} equiv^{-1} cm^2 at 298 K. Calculate the equilibrium or dissociation constant of acetic acid at this temperature.

Answer: From Table 9.1, $\Lambda_0 = 390.71 \ \Omega^{-1}$ equiv^{-1} cm^2 and from Eq. (9.8), we write

$$K = \frac{(0.10 \text{ equiv liter}^{-1})(5.2 \ \Omega^{-1} \text{ equiv}^{-1} \text{ cm}^2)^2}{(390.71 \ \Omega^{-1} \text{ equiv}^{-1} \text{ cm}^2)(390.71 - 5.2) \ \Omega^{-1} \text{ equiv}^{-1} \text{ cm}^2}$$

$$= 1.8 \times 10^{-5} \text{ equiv liter}^{-1} = 1.8 \times 10^{-5} \text{ mol liter}^{-1}$$

IONIC VELOCITIES

The equivalent conductance of a solution depends on the ease of ionic movement or on the ionic velocity. Imagine two electrodes, each of area 1 cm^2, which are separated by a distance 1 cm. The conductance in this case is given by

$$C = \kappa \frac{A}{l} = \kappa$$

For a dilute solution, we write

$$\Lambda = \frac{\kappa}{c} \simeq \Lambda_0$$

or

$$\kappa = c\Lambda_0 = c(\lambda_0^+ + \lambda_0^-)$$

The current I flowing across the electrodes can be expressed as

$$I = \frac{V}{R} = VC = \frac{V\kappa A}{l} = V\kappa = Vc(\lambda_0^+ + \lambda_0^-)$$

If the voltage across the electrodes is 1 V, then $c(\lambda_0^+ + \lambda_0^-)$ represents the current. Let u_+ and u_- be the ionic velocities of the cation and anion in cm s^{-1}. Consequently, all cations (anions) in a volume of u_+ cm^3 (u_- cm^3) will travel across the plates of the electrodes in 1 s and the total equivalent of cations and anions transported per second is $(u_+ + u_-)c$. According to Faraday's law of electrolysis, the equivalent of any ion carries 96,500 coulombs (C) of electricity.* The amount of electricity carried in this case is therefore $F(u_+ + u_-)c$, a quantity that we call conductance. Thus

$$F(u_+ + u_-)c = C = \kappa = c(\lambda_0^+ + \lambda_0^-)$$

$$F(u_+ + u_-) = \lambda_0^+ + \lambda_0^-$$

Separating the contributions made by the cations and anions, we write

$$u_+ = \frac{\lambda_0^+}{F} \quad \text{and} \quad u_- = \frac{\lambda_0^-}{F}$$

Finally, we obtain the relation

$$\Lambda_0 = \lambda_0^+ + \lambda_0^- = F(u_+ + u_-) \tag{9.10}$$

Ionic velocity is not a constant, since it depends on the voltage across the electrodes; the larger the voltage, the greater is the velocity. However, the quantity ionic velocity per unit electric field strength ($u_+/\mathscr{E}$ or $u_-/\mathscr{E}$) is a constant and does not depend on the applied voltage. It is given a special name, *ionic mobility*, which has the units cm^2 s^{-1} V^{-1}.† Table 9.2 lists the ionic mobilities of various ions at 298 K.

As we can see, the ionic mobilities of H$^+$ and OH$^-$ ions are much higher than those for other ions. These high values are the result of hydrogen bonding. In water, the proton is hydrated and its movement can be represented as follows:

Similarly, the movement of the hydroxide ion is

In each case, the ion can jump along a hydrogen bond, resulting in a very high mobility.

Finally, we note that ionic mobility is utilized in electrophoresis, which is a useful technique for purifying and identifying proteins (see Chapter 21).

---

* Coulomb is ampere × second. The reader should not be confused by the symbols C and C, which are used for coulomb and conductivity, respectively. The charge carried by 1 mol of electrons is called a faraday (F), given by 96,487 coulombs. For convenience, we shall always use 96,500 C in the calculations.

† Ionic velocity is in cm s^{-1} and electric field strength is V cm^{-1} so that ionic mobility is (cm s^{-1})/(V cm^{-1}) or cm^2 s^{-1} V^{-1}.

Equivalent Conductance and Ionic Mobility of Some Common Ions

Table 9.2

Ion	λ_0^a (Ω^{-1} equiv^{-1} cm^2)	Ionic Mobility[b] (cm^2 s^{-1} V^{-1})	Ionic Radius (Å)
H^+	349.81	36.3	
Li^+	38.68	4.01	0.60
Na^+	50.10	5.19	0.95
K^+	73.50	7.62	1.33
Rb^+	77.81	8.06	1.48
Cs^+	77.26	8.01	1.69
NH_4^+	73.5	7.62	
Mg^{2+}	53.05	5.50	0.65
Ca^{2+}	59.50	6.17	0.99
Ba^{2+}	63.63	6.59	1.35
Cu^{2+}	53.6	5.56	0.72
OH^-	198.3	20.06	
F^-	55.4	5.74	1.36
Cl^-	76.35	7.91	1.81
Br^-	78.14	8.10	1.95
I^-	76.88	7.97	2.16
NO_3^-	71.46	7.41	
HCO_3^-	44.50	4.61	
CH_3COO^-	40.90	4.24	
SO_4^{2-}	80.02	8.29	

[a] From R. A. Robinson and R. H. Stokes, *Electrolyte Solutions*, Academic Press, Inc., New York, 1959. Used by permission.
[b] From A. W. Adamson, *A Textbook of Physical Chemistry*, Academic Press, Inc., New York, 1973. Used by permission.

APPLICATIONS OF CONDUCTANCE MEASUREMENTS

Accurate conductance measurements are easy to make and have many different applications. Two examples will be described.

Acid–Base Titration. As mentioned earlier, the conductances for H^+ and OH^- are considerably higher than those for other cations and anions. By following the conductance of a HCl solution as a function of NaOH solution added, we obtain a titration curve such as the one shown in Figure 9.3.

Acid–base titrations monitored by conductance measurements.

Figure 9.3

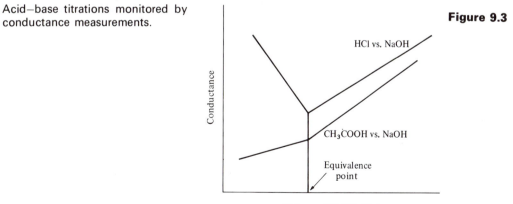

Initially, the conductance of the solution falls, since H^+ ions are replaced by Na^+ ions, which have a lower ionic conductance. This trend continues until the equivalence point is reached. Beyond this point, the conductance begins to rise, as a result of the excess OH^- ions present. If the acid is a weak electrolyte, say acetic acid, the slope of the first part of the curve is much less steep—the conductance actually increases right from the beginning—and there is more uncertainty in determining the equivalence point.

Solubility Determination. We have already seen an example of how the dissociation constant of acetic acid can be obtained from conductance measurements. The same procedure can be applied to determine the solubility of an insoluble salt. Suppose that we are interested in the solubility (mol liter^{-1}), and solubility product of AgCl in water at 298 K. From Eq. (9.4), we write

$$\Lambda = \frac{1000\kappa}{c} = \frac{1000\kappa}{S}$$

where S is the solubility in mol liter^{-1}. Since we are dealing with an insoluble salt, the concentration of the solution is low, so that we can assume that $\Lambda \simeq \Lambda_0$. Thus

$$S = \frac{1000\kappa}{\Lambda} \simeq \frac{1000\kappa}{\Lambda_0}$$

Experimentally, the specific conductance of a saturated AgCl solution is found to be $1.86 \times 10^{-6}\ \Omega^{-1}\ cm^{-1}$. However, since water is a weak electrolyte, we must take out the contribution due to water itself (κ for water is $6.0 \times 10^{-8}\ \Omega^{-1}\ cm^{-1}$). Thus

$$\kappa(AgCl) = 1.86 \times 10^{-6} - 6.0 \times 10^{-8} = 1.8 \times 10^{-6}\ \Omega^{-1}\ cm^{-1}$$

From Table 9.1, we find that $\Lambda_0 = 137.2\ \Omega^{-1}\ equiv^{-1}\ cm^2$ for AgCl. Finally, we have

$$S = \frac{1000\ cm^3\ liter^{-1} \times 1.8 \times 10^{-6}\ \Omega^{-1}\ cm^{-1}}{137.2\ \Omega^{-1}\ equiv^{-1}\ cm^2} = 1.3 \times 10^{-5}\ equiv\ liter^{-1}$$

$$= 1.3 \times 10^{-5}\ mol\ liter^{-1}$$

Since AgCl is a $1:1$ electrolyte, the number of equivalents per liter is the same as the number of moles per liter. The solubility product K_{sp} for AgCl is given by

$$K_{sp} = [Ag^+][Cl^-] = (1.3 \times 10^{-5})(1.3 \times 10^{-5})$$

$$= 1.7 \times 10^{-10}$$

9.2 Ions in Aqueous Solution

When an electrolyte (MX) is dissolved in water, two questions are relevant. First, how do the ions interact with each other? Second, how do the ions interact with the solvent molecules? This section will examine these questions.

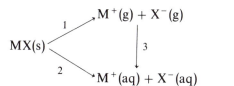

The enthalpy change for 1 corresponds to the energy required to separate the ions from the crystal lattice to infinite distance. This energy is called the *lattice energy* (U). The enthalpy change for 2 is the heat of solution, ΔH_{soln}, the heat absorbed or released when solid MX dissolves in a large amount of water. The heat of hydration, ΔH_{hydr}, for 3 is given by Hess's law:

$$\Delta H_{hydr} = \Delta H_{soln} - U$$

The quantity ΔH_{soln} is experimentally measurable; U can either be estimated if the structure of the lattice is known or directly measured. For NaCl, we have $U = 765 \text{ kJ mol}^{-1}$ from calculation and $U = 778 \text{ kJ mol}^{-1}$ from experiment. Further, $\Delta H_{soln} = 3.8 \text{ kJ mol}^{-1}$, so that

$$\Delta H_{hydr} = 3.8 - 778 = -774 \text{ kJ mol}^{-1}$$

Thus the hydration of Na^+ and Cl^- ions by water releases a large amount of heat.

The heat of hydration obtained above comes from both ions together. It is often desirable to know the value of individual ions. The value can be determined as follows. The hydration energy for the process

$$H^+(g) \longrightarrow H^+(aq)$$

has been reliably estimated to be about 1089 kJ mol^{-1}. Using this value as a

Thermodynamic Values for the Hydration of Gaseous Ions at 298.15 K[a]

Table 9.3

Ion	$-\Delta H^\circ_{hydr}$ (kJ mol^{-1})	$-\Delta S^\circ_{hydr}$ (J mol^{-1} K^{-1})	Ionic Radius (Å)
H^+	1089	109	—
Li^+	506	119	0.60
Na^+	398	87	0.95
K^+	314	52	1.33
Ag^+	468	92	1.26
Mg^{2+}	1908	268	0.65
Ca^{2+}	1577	209	0.99
Ba^{2+}	1288	159	1.35
Mn^{2+}	1832	243	0.80
Fe^{2+}	1908	272	0.76
Cu^{2+}	2092	259	0.72
Fe^{3+}	4355	460	0.64
F^-	506	151	1.36
Cl^-	377	98	1.81
Br^-	343	83	1.95
I^-	297	60	2.16

[a] Reproduced from *Metal Ions in Aqueous Solution*, written by John P. Hunt, with permission of publishers W. A. Benjamin Inc., Advanced Book Program, Reading, Mass., 1963.

Figure 9.4

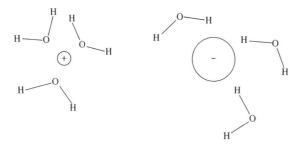

Hydrations of a cation and
an anion.

starting point, it is possible to calculate the ΔH_{hydr} values for individual anions such as F^-, Cl^-, Br^-, and I^- (from data on HF, HCl, HBr, and HI), which in turn enable us to obtain ΔH_{hydr} values for Li^+, Na^+, K^+, and other cations (from data on alkali metal halides). The other quantity of interest is the entropy of hydration, ΔS_{hydr}. Table 9.3 summarizes the ΔH_{hydr} and ΔS_{hydr} for a number of ions.

Generally, smaller ions have larger heats of hydration. A small ion contains a more concentrated charge, leading to a greater electrostatic interaction between the ion and the polar water molecules.* The negative values for the entropy of hydration indicate that there is an ordered arrangement of H_2O molecules surrounding each individual ion. A schematic diagram of hydration is shown in Figure 9.4. Because a different number of water molecules surround each type of ion, we speak of the *hydration number* of an ion. This number is directly proportional to the charge and inversely proportional to the size of the ion. It should be noted that water in the "hydration sphere" and bulk water molecules have different properties,† which can be distinguished by spectroscopic techniques such as nuclear magnetic resonance. There is a dynamic equilibrium between the two types of molecules. Depending on the ion, the mean lifetime of a H_2O molecule in the hydration sphere can vary tremendously. For example, consider the mean lifetime of H_2O in the hydration sphere for the following ions: Br^- (10^{-11} s), Na^+ (10^{-9} s), Cu^{2+} (10^{-7} s), Fe^{3+} (10^{-5} s), Al^{3+} (7 s), and Cr^{3+} (1.5×10^5 s or 42 h).

The ion–dipole interaction (see Chapter 17) between dissolved ions and water molecules can affect a number of bulk properties of water. Small and/or multicharged ions such as Li^+, Na^+, Mg^{2+}, Al^{3+}, Er^{3+}, OH^-, and F^- are often called *structure-making ions*. The high electric fields exerted by these ions can polarize the water molecules, producing additional order beyond the first hydration layer. This interaction leads to an increase in the solution's viscosity. On the other hand, large monovalent ions such as K^+, Rb^+, Cs^+, NH_4^+, Cl^-, NO_3^-, and ClO_4^- are *structure-breaking ions*. As a result of their diffuse surface charges and hence weak electric fields, these ions are unable to polarize water molecules beyond the first layer of hydration. Consequently, the viscosities of solutions containing these ions are often *lower* than that of pure water.

The fact that ions are hydrated in solution means that their effective radii can be appreciably greater than their crystal or ionic radii. For example, the radii of the hydrated Li^+, Na^+, and K^+ ions are estimated to be 3.66 Å,

---

* According to electrostatic theory, the electric field at the surface of a charged sphere of radius R is proportional to ze/R^2, where z is the number of charges and e is the electronic charge.
† Hydrated water molecules do not possess individual translational motion. They move with the ion as a whole.

2.80 Å, and 1.87 Å, respectively, although the ionic radii actually increase from Li^+ to K^+. We might expect that the mobility of an ion is inversely proportional to its hydrated radius. Table 9.2 nicely demonstrates the truth of this expectation. Although the proton is very small and we expect it to be strongly hydrated, it has a high ionic mobility as a result of the rapid movement of the H^+ ions via hydrogen bonds.

We now turn to the other question raised earlier in this section: How do ions interact with one another? To maintain electrical neutrality in solution, an anion must be in the vicinity of a cation, and vice versa. Depending on the proximity of these two ions, we can think of them either as "free" ions or as "ion pairs." Each free ion is surrounded by at least one, and perhaps several, layers of water molecules. In an ion pair, the cation and anion are close to each other, with little or no solvent molecules between them. Generally, free ions and ion pairs are thermodynamically distinguishable species, having quite different chemical reactivities. For dilute 1 : 1 aqueous electrolyte solutions, ions are believed to be in the free-ion form. In higher valent electrolytes such as $CaCl_2$ or Na_2SO_4, the formation of ion pairs is indicated by conductance measurements, for a neutral ion pair cannot conduct electricity.

Two opposing factors determine whether we have free ions or ion pairs in solution: the potential energy of attraction between the cation and anion and the kinetic or thermal energy, of the order of kT, for individual ions. In addition, we know that ion-pair formation is favored in mixed solvents such as water–dioxane or water–alcohol. Water has a high *dielectric constant* (ε), so the electrostatic force between two ions of opposite charge is considerably reduced, as shown by the equation (see Appendix 9.1 for a discussion of dielectric constant)

$$F = \frac{Q_1 Q_2}{\varepsilon r^2} \tag{9.11}$$

Ions would then prefer to be in the free form. Adding an organic solvent decreases the dielectric constant of the solution, resulting in a greater attractive force and hence a larger amount of ion pairs. Table 9.4 lists the dielectric

Dielectric Constant of Some Pure Liquids at 298.15 K

Table 9.4

Liquid	Dielectric Constant, ε^a
H_2SO_4	101
H_2O	78.54
$(CH_3)_2SO$ (dimethylsulfoxide)	49
$C_3H_8O_3$ (glycerol)	42.5
CH_3NO_2 (nitromethane)	38.6
$HOCH_2CH_2OH$ (ethylene glycol)	37.7
CH_3CN (acetonitrile)	36.2
CH_3OH	32.6
C_2H_5OH	24.3
CH_3COCH_3	20.7
CH_3COOH	6.2
C_6H_6	4.6
$C_2H_5OC_2H_5$	4.3
CS_2	2.6

[a] The dielectric constant is a dimensionless quantity.

constant of a number of solvents. It should be kept in mind that ε always decreases with increasing temperature. For example, at 343 K the dielectric constant of water is reduced to about 64.

9.3 Ionic Activities

The intermolecular forces between uncharged species generally depend on $1/r^7$, r being the distance of separation. On the other hand, Coulomb's law has a $1/r^2$ dependence. Here lies the main difference between nonelectrolyte and electrolyte solutions. The $1/r^2$ dependence means that even in quite dilute solutions (for example, at 0.01 M), the electrostatic forces exerted by ions on one another are appreciable enough to cause deviation from ideal behavior. On the other hand, a 0.01 M nonelectrolyte solution is considered ideal for most practical purposes.

As in the case of nonelectrolytes, we again use the concept of activity to represent concentration. Since it is not possible to separate the cation from the anion, we are forced to express the activity of the electrolyte in terms of the ionic activities of both ions. Consider the dissociation of the following salt:

$$M_{v_+}X_{v_-} \rightleftharpoons v_+M^{z+} + v_-X^{z-}$$

where v_+ and v_- are the number of cations and anions per unit and z_+ and z_- are the number of charges on the cation and anion, respectively. We define the *mean* activity of the electrolyte $a_\pm$ to be

$$a_\pm{}^v = a_+{}^{v_+}a_-{}^{v_-}$$

or

$$a_\pm = (a_+{}^{v_+}a_-{}^{v_-})^{1/v} \tag{9.12}$$

where a_+ and a_- are the activities for the cation and anion and $v = v_+ + v_-$. Note that Eq. (9.12) gives us the *geometric mean* of the individual ionic activities. Two examples will illustrate the use of this equation. For NaCl, we have $v_+ = v_- = 1$, so that

$$v = 1 + 1 = 2$$

and

$$a_\pm = (a_+ a_-)^{1/2}$$

For $Mg_3(PO_4)_2$, we have $v_+ = 3$ and $v_- = 2$. Thus

$$v = 3 + 2 = 5$$

and

$$a_\pm = (a_+^3 a_-^2)^{1/5}$$

The activity coefficient for the cation (γ_+) and anion (γ_-) can be defined as

$$a_+ = \gamma_+ m_+ \quad \text{and} \quad a_- = \gamma_- m_- \tag{9.13}$$

where m_+ and m_- are the molalities of the cation and anion. It follows that if the concentration of the electrolyte is expressed in molality m, then

$$m_+ = v_+ m \quad \text{and} \quad m_- = v_- m$$

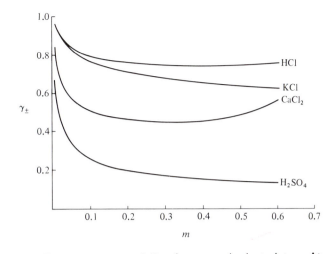

Plots of mean activity coefficient versus molality for several electrolytes. At infinite dilution ($m \to 0$), the mean activity coefficient approaches unity.

Figure 9.5

The mean activity coefficient $\gamma_{\pm}$ is given by

$$\gamma_{\pm}^{\,v} = \gamma_+^{\,v_+}\gamma_-^{\,v_-}$$

or

$$\gamma_{\pm} = \left(\gamma_+^{\,v_+}\gamma_-^{\,v_-}\right)^{1/v} \qquad (9.14)$$

Like $a_{\pm}$, $\gamma_{\pm}$ is the geometric mean of the individual ionic activity coefficients.

Finally, the (geometric) mean molality $m_{\pm}$ can be defined in a similar manner, as follows:

$$m_{\pm} = \left(m_+^{\,v_+}m_-^{\,v_-}\right)^{1/v}$$

Thus we write

$$\gamma_{\pm} = \frac{a_{\pm}}{m_{\pm}} = \frac{a_{\pm}}{\left(m_+^{\,v_+}m_-^{\,v_-}\right)^{1/v}} \qquad (9.15)$$

Again taking $Mg_3(PO_4)_2$ as an example, we have, for a 1 m solution, $m_+ = 3m$ and $m_- = 2m$ so that

$$m_{\pm} = [(3m)^3(2m)^2]^{1/5}$$
$$= m(108)^{1/5} = 2.55m$$

Thus

$$\gamma_{\pm} = \frac{a_{\pm}}{2.55m} = \frac{(a_+^3 \, a_-^2)^{1/5}}{2.55m}$$

Experimental values of $\gamma_{\pm}$ can be obtained from freezing-point-depression and osmotic-pressure measurements* or electrochemical measurements (see Chapter 11). Hence $a_{\pm}$ can be calculated from Eq. (9.15). In the limiting case of infinite dilution ($m \to 0$), we have

$$\lim_{m \to 0} \gamma_{\pm} = 1$$

Figure 9.5 shows the plots of $\gamma_{\pm}$ versus m for several electrolytes. We see that

* Interested readers should consult any of the standard physical chemistry texts listed in Chapter 1.

at very low concentrations $\gamma_\pm$ approaches unity for all types of electrolytes. As the concentrations of electrolytes increase, drastic deviations from ideality occur. The variation of $\gamma_\pm$ with concentration, at least in dilute solutions, can be explained by the Debye–Hückel theory, to be discussed next.

9.4 Debye–Hückel Theory of Electrolytes

THE DEBYE–HÜCKEL LIMITING LAW

Deviations from ideality has thus far been treated on an empirical basis. We replace the concentration of an electrolyte with activity whose value can be obtained from the experimentally determined activity coefficient and the known concentration. Once ionic activities are known, we can then proceed to calculate chemical potential, equilibrium constant, and other properties in terms of these quantities. However, this approach does not provide us with any physical interpretation of ionic behavior in solution. In 1923, Debye and Hückel put forward a quantitative theory of electrolyte solution which has greatly advanced our knowledge in this field. The theory is based on a rather simple model and allows us to calculate the quantity $\gamma_\pm$ from the basic properties of the solution.

The mathematical details of Debye–Hückel's treatment are too difficult to present here. (The interested reader should consult the standard physical chemistry texts listed in Chapter 1.) We shall simply discuss the underlying assumptions and final results. Debye and Hückel began by assuming the following: (1) electrolytes are completely dissociated into ions in solution; (2) the solutions are dilute, that is, the concentration is 0.01 m or lower; and (3) on the average each ion is surrounded by ions of opposite charge, forming the *ionic atmosphere* shown in Figure 9.6a. Working from these assumptions, Debye and Hückel calculated the average electric potential at each ion caused by the presence of other ions in the ionic atmosphere. The Gibbs free energy of the ions was then related to the activity coefficient of the individual ion. Since neither γ_+ nor γ_- could be measured directly, the final result is expressed in terms of the mean ionic activity coefficient of the electrolyte as follows:

$$\log \gamma_\pm = -\frac{1.824 \times 10^6}{(\varepsilon T)^{3/2}} \left| z_+ z_- \right| \sqrt{I} \qquad (9.16)$$

Thus for $CuSO_4$, we have $z_+ = 2$ and $z_- = -2$, but $\left| z_+ z_- \right| = 4$. The quantity I, called the *ionic strength*, is defined as follows:

$$I = \tfrac{1}{2} \sum_i m_i z_i^2 \qquad (9.17)$$

Figure 9.6

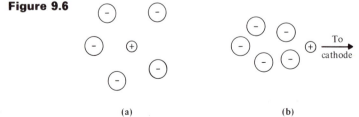

(a)　　　　　　　　(b)

(a) Simplified presentation of an ionic atmosphere. (b) In a conductance measurement, the movement of a cation toward the cathode is retarded by the electric field exerted by the ionic atmosphere left behind.

where m_i and z_i are the molality and the charge of the ith ion in the electrolyte. This quantity was first introduced by Lewis, who pointed out that nonideality observed in electrolyte solutions owes primarily to the *total* concentration of ions present rather than the chemical nature of the individual ionic species. Equation (9.17) allows us to express the ionic concentrations for all types of electrolytes on a common basis. Since most studies are carried out in water at 298 K (we have $\varepsilon = 78.54$ and $T = 298$ K), Eq. (9.16) becomes

$$\log \gamma_\pm = -0.509 |z_+ z_-| \sqrt{I} \tag{9.18}$$

Either Eq. (9.16) or Eq. (9.18) is known as the *Debye–Hückel limiting law*.

EXAMPLE 9.3 Calculate the mean activity coefficient of a 0.01 m aqueous solution of $CuSO_4$ at 298 K.

Answer: The ionic strength of the solution is given by

$$I = \tfrac{1}{2}[(0.01) \times 2^2 + (0.01) \times 2^2]$$

$$= 0.04$$

From Eq. (9.18),

$$\log \gamma_\pm = -0.509(2 \times 2)\sqrt{0.04}$$

$$= -0.407$$

or

$$\gamma_\pm = 0.392$$

Experimentally, $\gamma_\pm$ is found to be 0.41 at the same concentration.

Figure 9.7 shows comparisons of calculated and measured values of $\log \gamma_\pm$ at various ionic strengths. As can be seen, Eq. (9.18) holds quite well for dilute solutions, but drastic deviations occur at high concentrations of the electrolytes. A number of improvements and modifications have been applied to this equation for treating more concentrated solutions.

The good agreement generally obtained between experiments and the Debye–Hückel theory provides strong support for the existence of ionic atmosphere in solution. The model can be tested by taking a conductance measurement as follows. In reality, the ions do not move toward the electrode in a straight line, but along a zigzag path. From a microscopic point of view, the solvent is not a continuous medium so that each ion actually jumps from one hole to another. As the ion moves across the solution, its ionic atmosphere is constantly being destroyed and formed again. The formation process does not occur instantaneously but requires a finite amount of time, called the *relaxation time*, which is about 10^{-7} s in a 0.01 m solution. Under normal conditions of conductance measurement, the velocity of an ion is sufficiently slow so that the electrostatic force exerted by the atmosphere on the ion tends to retard its motion and hence decrease the conductance (see Figure 9.6b). If the conductance measurement were carried out at a very high electric field (about 2×10^5 V cm^{-1}), the ionic velocity would be about 10 cm s^{-1}. The radius of the ionic atmosphere in a 0.01 m solution is about 5 Å or 5×10^{-8} cm so that the time required for the ion to move out of the atmosphere is $5 \times 10^{-8}/10$ or

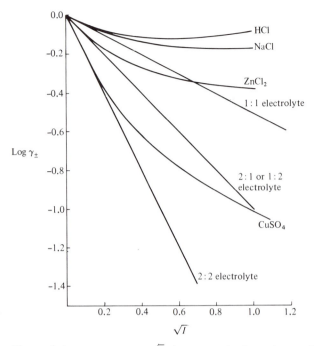

Figure 9.7 Plots of log $\gamma_{\pm}$ versus $\sqrt{I}$ for several electrolytes. Straight lines are those predicted by Eq. (9.18).

5×10^{-9} s, which is considerably shorter than the relaxation time. Consequently, the ion can move through the solution free of the influence of the ionic atmosphere. The free movement leads to a marked increase in conductance. This phenomenon is called the *Wien effect*, after the scientist who first performed the experiment in 1927. The Wien effect is one of the strongest pieces of evidence for the existence of an ionic atmosphere.

THE SALTING-IN AND THE SALTING-OUT EFFECTS

The Debye–Hückel limiting law can be applied to study the solubility of proteins. How soluble a protein is in an aqueous solution depends on the temperature, pH, dielectric constant, ionic strength, and other characteristics of the medium. However, we shall only be concerned with the influence of ionic strength in this section.

Let us first investigate the effect of ionic strength on the solubility of AgCl. The dissolution process is

$$AgCl(s) \rightleftharpoons Ag^+(aq) + Cl^-(aq)$$

The *thermodynamic* solubility product for the process, K_{sp}°, is

$$K_{sp}^{\circ} = a_{Ag^+} a_{Cl^-}$$

Using the relation in Eq. (9.13), we write

$$K_{sp}^{\circ} = m_{Ag^+} \gamma_{Ag^+} m_{Cl^-} \gamma_{Cl^-}$$
$$= \gamma_{Ag^+} \gamma_{Cl^-} K_{sp}$$

214

where $K_{sp} = m_{Ag^+} m_{Cl^-}$ is the *apparent* solubility product. The difference between the thermodynamic and apparent solubility products is as follows. As we can see, the apparent solubility product is expressed in molalities (or some other concentration unit). This quantity can be readily calculated knowing the amount of AgCl dissolved in a known amount of water to produce a saturated solution. However, because of electrostatic forces, the dissolved ions are under the influence of their immediate neighbors. The result of such interionic interactions is that the actual or effective number of ions is no longer the same as that calculated based on the concentration of the solution. For example, if a cation forms a tight ion pair with an anion, then the actual number of species in solution, as far as thermodynamic quantities are concerned, is one and not two as we would expect. This is the reason why we find it necessary to replace concentration with activity, which is the effective concentration. Thus the thermodynamic solubility product represents the true value of the solubility product, which in general is different from the apparent solubility product. Since

$$\gamma_{Ag^+} \gamma_{Cl^-} = \gamma_{\pm}^2$$

we write

$$K_{sp}^\circ = \gamma_{\pm}^2 K_{sp}$$

or

$$\gamma_{\pm}^2 = \frac{K_{sp}^\circ}{K_{sp}}$$

Taking the logarithm of both sides and rearranging, we obtain

$$-\log \gamma_{\pm} = \log \left(\frac{K_{sp}}{K_{sp}^\circ} \right)^{1/2} = 0.509 |z_+ z_-| \sqrt{I}$$

The last equality is the Debye–Hückel limiting law. The solubility product can be directly related to the solubility (S) itself; for a 1:1 electrolyte

$$(K_{sp})^{1/2} = S \qquad \text{and} \qquad (K_{sp}^\circ)^{1/2} = S^\circ$$

where S and S° are the apparent and thermodynamic solubilities in mol liter^{-1}. Finally, we obtain the following equation relating the solubility of an electrolyte to the ionic strength of the solution:

$$\log \frac{S}{S^\circ} = 0.509 |z_+ z_-| \sqrt{I} \tag{9.19}$$

Note that S° can be determined by plotting $\log S$ versus $\sqrt{I}$. The intercept on the $\log S$ axis ($I = 0$) gives $\log S^\circ$ and hence S°.

If AgCl is dissolved in pure water, its solubility (S) is 1.3×10^{-5} mol liter^{-1}. If it is dissolved in a KNO_3 solution, according to Eq. (9.19), its solubility is greater because of the solution's increase in ionic strength. In a KNO_3 solution, the ionic strength is a sum of two concentrations, one from AgCl and the other from KNO_3. The increase in solubility, caused by the increase in ionic strength, is called the *salting-in effect*.

As the ionic strength of a solution increases further, Eq. (9.19) no longer holds, and we replace it with the following expression:

$$\log \frac{S}{S^\circ} = -K'I \tag{9.20}$$

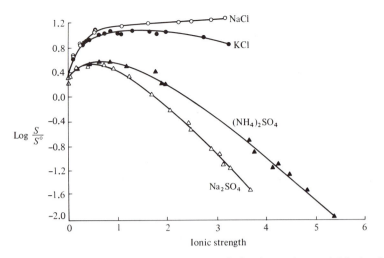

Figure 9.8 Plots of log ($S/S°$) versus ionic strength for horse hemoglobin in the presence of various inorganic salts. Note that when $I = 0$, all the curves converge to the same point on the log ($S/S°$) axis at zero and we have $S = S°$. (From *Proteins, Amino Acids and Peptides* by E. Cohen and J. Edsall, © 1943 by Litton Educational Publishing, Inc. Reprinted by permission of Van Nostrand Reinhold Company.)

where K' is a positive constant whose value depends on the nature of the solute and the electrolyte present. The larger the solute molecule, the greater is K'. Equation (9.20) tells us that the ratio of the solubilities in the region of high ionic strength actually decreases with I (note the negative sign). The decrease in solubility with increasing ionic strength of the solution is called the *salting-out effect*. Combining Eqs. (9.19) and (9.20), we have

$$\log \frac{S}{S°} = 0.509 \, |z_+ z_-| \sqrt{I} - K'I \qquad (9.21)$$

Equation (9.21) is applicable over a wide range of ionic strengths.

Figure 9.8 shows how the ionic strength of various inorganic salts affects the solubility of horse hemoglobin. As we can see, the protein exhibits an

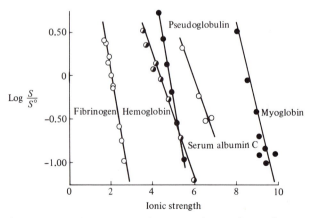

Figure 9.9 Salting-out phenomenon for several proteins using ammonium sulfate. [From E. J. Cohn, *Chem. Rev.* **19**, 241 (1936). Used with permission of The Williams & Wilkins Co., Baltimore.]

initial salting-in region at low ionic strengths.* As I increases, the curve goes through a maximum and eventually the slope becomes negative, indicating that the solubility now decreases with increasing ionic strength. In this region, the second term in Eq. (9.21) predominates. This trend is most pronounced for Na_2SO_4 and $(NH_4)_2SO_4$.

The salting-out effect has great value in precipitating proteins from solutions. In addition, it can also be used to purify proteins. Figure 9.9 shows the range of the salting-out phenomenon for several proteins using ammonium sulfate. By carefully selecting the proper ionic strength, it is possible to quantitatively precipitate out a single protein while leaving others in solution. The point is that higher ionic strengths are needed to salt out proteins, but precipitation occurs over a small range of ionic strength, providing sharp separations.

Colligative Properties of Electrolyte Solutions **9.5**

The colligative properties of an electrolyte solution are influenced by the number of ions present in solution. For example, we expect the aqueous freezing-point depression caused by a 0.01 m solution of NaCl to be twice that for a 0.01 m sucrose solution, assuming complete dissociation of the former. The situation becomes more complicated for incompletely dissociated salts, but its understanding provides us with another way to measure an electrolyte's degree of dissociation.

Let us consider a factor i, called the van't Hoff factor, as follows:

$$i = \frac{\text{actual number of particles in solution}}{\text{number of particles in solution before dissociation}}$$

If a solution contains N units of an electrolyte and if α is the degree of dissociation,

$$M_{v_+}X_{v_-} \rightleftharpoons v_+M^{z+} + v_-X^{z-}$$

there will be $N(1 - \alpha)$ undissociated units and $(Nv_+\alpha + Nv_-\alpha)$ or $Nv\alpha$ ions in solution at equilibrium, where $v = v_+ + v_-$. We can now write the van't Hoff factor as

$$i = \frac{N(1 - \alpha) + Nv\alpha}{N} = 1 - \alpha + v\alpha$$

and

$$\alpha = \frac{i - 1}{v - 1}$$

Our treatment here assumes there are no ion pairs present.

EXAMPLE 9.4 The osmotic pressure of a 0.01 m solution of $CaCl_2$ and a 0.01 m sucrose solution at 298 K are 0.605 atm and 0.224 atm, respectively. Calculate the van't Hoff factor and the degree dissociation for $CaCl_2$. Assume ideal behavior.

---

* When I is less than unity, $\sqrt{I} > I$. Thus at low ionic strengths the first term in Eq. (9.21) predominates.

Answer: As far as osmotic pressure measurements are concerned, the main difference between calcium chloride and sucrose is that only the former can dissociate into ions (Ca^{2+} and Cl^-). Otherwise, equal concentrations of $CaCl_2$ and sucrose solutions would have the same osmotic pressure. Since the osmotic pressure of a solution is directly proportional to the number of particles present, we can calculate the van't Hoff factor of the $CaCl_2$ solution as follows:

$$i = \frac{0.605 \text{ atm}}{0.224 \text{ atm}} = 2.70$$

Since for $CaCl_2$, $v_+ = 1$, and $v_- = 2$, we have $v = 3$ and

$$\alpha = \frac{2.70 - 1}{3 - 1} = 0.85$$

Finally, we note that the equations for the colligative properties of nonelectrolyte solutions [Eqs. (8.25), (8.26), and (8.33)] must be modified for electrolyte solutions as follows:

$$\Delta T = K_b(im_2)$$

$$\Delta T = K_f(im_2)$$

$$\pi = \frac{RT(ic_2)}{M_2}$$

Ideal behavior is assumed for the electrolyte solutions.

THE DONNAN EFFECT

The Donnan effect has its starting point in the treatment of osmotic pressure. It describes the distribution of small diffusible ions on the two sides of a membrane, freely permeable to these ions but impermeable to macromolecular ions, in the presence of a macromolecular electrolyte on one side of the membrane. Suppose that a cell is separated into two components by a semipermeable membrane that allows the diffusion of water and small ions but not protein molecules. Let us consider the following three cases.

Case 1. The protein solution is placed in the left compartment; water is placed in the right compartment. We assume that the protein molecules are neutral species.* Let the concentration of the protein solution be c (mol $liter^{-1}$) so that the osmotic pressure of the solution, according to Eq. (8.32), is given by

$$\pi_1 = cRT$$

Thus from a measurement of the osmotic pressure we can readily determine the molar mass of the protein molecule.

Case 2. In this case the protein is present as the anion of the sodium salt, Na^+P^-, which we assume to be a strong electrolyte. Again, the protein solution of concentration c is placed in the left compartment and pure water is in

---

* Proteins are ampholytes, that is, they possess both acidic and basic properties. Depending on the pH of a solution, a protein can exist as an anion, a cation, or a neutral species. For more details, see Chapter 12.

the right compartment. To maintain electrical neutrality, all of the Na^+ ions remain in the left compartment; the osmotic pressure of the solution now becomes

$$\pi_2 = (c + c)RT = 2cRT$$

Since $\pi_2 = 2\pi_1$, it follows that the molar mass determined in this case will only be half that of the true molar mass. [From Eq. (8.33), we have $M_2 = c_2 RT/\pi$, so that doubling π would decrease the value of M_2 by half.] In practice the situation is actually much worse, since the protein ion may bear as many as 20 or 30 net negative (or positive) charges. In the early days of protein-molar-mass determination by osmotic pressure, disastrously poor results were obtained when the dissociation process was not recognized and efforts made to correct it (see Appendix 9.2).

Case 3. We start with an arrangement similar to that discussed in Case 2 and then add NaCl (of concentration b in mol liter^{-1}) to the right compartment (Figure 9.10). At equilibrium, a certain amount, x (mol liter^{-1}), of Na^+ and Cl^- ions have diffused through the membrane from right to left, creating a final state shown in Figure 9.10b. Both sides of the membrane must be electrically neutral: In each compartment the number of cations equals the number of anions. The condition of equilibrium allows us to equate the chemical potentials of NaCl in the two compartments as follows:

$$(\mu_{NaCl})^L = (\mu_{NaCl})^R$$

or

$$(\mu^\circ + RT \ln a_\pm)^L_{NaCl} = (\mu^\circ + RT \ln a_\pm)^R_{NaCl}$$

Since μ°, the standard chemical potential, is the same on both sides, we obtain

$$(a_\pm)^L_{NaCl} = (a_\pm)^R_{NaCl}$$

From Eq. (9.12),

$$(a_{Na^+} a_{Cl^-})^L = (a_{Na^+} a_{Cl^-})^R$$

If the solutions are dilute, the ionic activities may be replaced by the corresponding concentrations, that is, $a_{Na^+} = [Na^+]$ and $a_{Cl^-} = [Cl^-]$. Hence

$$([Na^+][Cl^-])^L = ([Na^+][Cl^-])^R$$

or

$$(c + x)x = (b - x)(b - x)$$

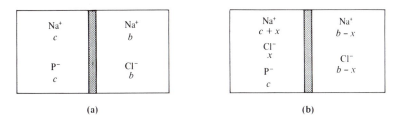

(a)   (b)

Donnan equilibrium: (a) before diffusion has begun; (b) at equilibrium. The membrane separating the left and right compartments is permeable to all but the P^- ions.

Figure 9.10

Initial Concentration		Equilibrium Concentration		
Left Compartment	Right Compartment	Left Compartment		
$c = [Na^+] = [P^-]$	$b = [Na^+] = [Cl^-]$	$(c + x) = [Na^+]$	$c = [P^-]$	$x = [Cl^-]$
0.1	0	0.1	0.1	0
0.1	0.01	0.1008	0.1	0.00083
0.1	0.1	0.1333	0.1	0.0333
0.1	1.0	0.576	0.1	0.476
0.1	10.0	5.075	0.1	4.975
$[P] = 0.1$	0	0	$[P] = 0.1$	0

Solving for x, we obtain

$$x = \frac{b^2}{c + 2b} \tag{9.22}$$

Equation (9.22) says that the amount of NaCl, x, diffused from right to left is inversely proportional to the concentration of the nondiffusible ion (P^-), c, in the left compartment. This unequal distribution of the diffusible ions (Na^+ and Cl^-) in the two compartments is the result of the Donnan effect.

The osmotic pressure of the protein solution is now determined by the *difference* between the number of particles in the left compartment and that in the right compartment. We write

$$\pi_3 = \{(c + c + x + x) - 2(b - x)\}RT = (2c + 4x - 2b)RT$$
$$\underset{\text{compartment}}{\text{left}} \qquad \underset{\text{compartment}}{\text{right}}$$

From Eq. (9.22),

$$\pi_3 = \left(2c + \frac{4b^2}{c + 2b} - 2b\right)RT = \left(\frac{2c^2 + 2cb}{c + 2b}\right)RT$$

Two limiting cases may be applied to the equation above. If $b \ll c$, $\pi_3 = 2cRT$, which gives the same result as Case 2. On the other hand, if $b \gg c$, $\pi_3 = cRT$, which is identical to Case 1. The important conclusion we arrive at is that the presence of NaCl in the right compartment decreases the osmotic pressure of the protein solution compared to Case 2 and therefore minimizes the Donnan effect. When a very large amount of NaCl is present, the Donnan effect can be completely eliminated. In general, we have $\pi_1 \le \pi_3 \le \pi_2$. Since proteins are usually studied in buffer solutions that contain ionic species, the osmotic pressure measured will be less than the case of using pure water as the solvent. Table 9.5 shows the Donnan effect for the NaCl example at several concentrations and the corresponding osmotic pressure at 298 K.

An alternative approach to eliminate the Donnan effect is to choose a pH at which the protein has no net charge, called the *isoelectric point* (see Chapter 12). At this pH, the distribution of any diffusible ion will always be equal in both compartments. This method is difficult in practice because most proteins are least soluble at their isoelectric points.

Right Compartment		Percent NaCl Transferred from Right to Left	Osmotic Pressure (atm)
$(b - x) = [Na^+]$	$(b - x) = [Cl^-]$		
0	0	0	4.90
0.00917	0.00917	8.3	4.48
0.0667	0.0667	33.3	3.26
0.524	0.524	47.6	2.56
5.025	5.025	49.75	2.46
0	0	0	2.45

We have simplified the discussion of the Donnan effect by assuming no change in either pH or volume of the solution. Also, for the sake of simplicity, we have used a common diffusible ion, Na^+, in deriving Eq. (9.22). A Donnan effect does not depend on simple conditions. It would still be observed if we started with K^+P^- and NaCl, although the treatment would become more complicated.

The Donnan effect is essential to understand the distribution of ions across the membranes of living organisms and membrane potentials (see Chapter 11). A particularly important case is the distribution of bicarbonate and chloride ions between plasma water and red-blood-cell water, discussed in Chapter 12.

Biological Membranes 9.6

In this last section we shall consider the structure and function of biological membranes. In particular, we shall concentrate on the transport of ions across these membranes.

Cell membranes are basically composed of two kinds of molecules: proteins and lipids. Examples of lipids are fats and waxes, insoluble in water but soluble in many organic solvents. There are three types of membrane lipids: *phospholipids, cholesterol,* and *glycolipids*. For our purpose here we shall consider only the phospholipids. One of the most common types of phospholipid found in cellular membranes is phosphatidic acid, shown in Figure 9.11. Membrane lipids are unique in that one end of the molecule contains a polar group that is hydrophilic (water-liking), while the other end is a long hydrocarbon chain that is hydrophobic (water-fearing). The lipids form a bilayer (about 70 Å in thickness), arranged so their polar groups constitute the top and bottom surfaces of the membrane while nonpolar groups are buried in the interior region. Figure 9.12 shows a widely accepted model of cell membrane structure, proposed by Singer, called the *fluid mosaic model*. Protein molecules may lie at or near the inner or outer membrane surface or they may penetrate partially or totally through the membrane. The extent of interaction between the proteins and the lipids would depend upon the types of intermolecular forces and thermodynamic considerations. Generally, cell membranes have great physical strength and high electrical insulating properties.

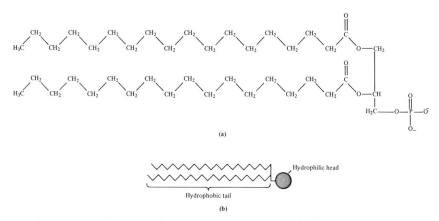

(a)

Hydrophilic head

Hydrophobic tail

(b)

Figure 9.11 (a) Structure of phosphatidic acid (a phospholipid). (b) Simplified structure.

However, it should be realized that biological membranes are not rigid structures. On the contrary, many membrane proteins and lipids are constantly in motion. Experiments using fluorescence probes have shown that membrane proteins can normally diffuse a distance of about several hundred angstroms in 1 min. The phospholipid molecules, because of their smaller size, can diffuse much more quickly.

The function of proteins is threefold: to maintain the overall structural integrity of the membranes*; to act as enzymes in, for example, the synthesis of important molecules such as adenosine triphosphate (ATP) at the mitochondrial membrane, or in initiation of photosynthetic steps at the chloroplast membrane; and to act as carriers for ions and other molecules across the membrane.

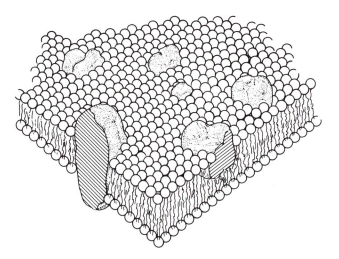

Figure 9.12 The fluid mosaic model of membrane structure. The large bodies are protein molecules that are embedded to varying degrees in a lipid bilayer. [From S. J. Singer and G. L. Nicolson, *Science* **175**, 723 (1972). Copyright © 1972 by The American Association for the Advancement of Science.]

* Recent studies have cast doubt on the role of membrane proteins in providing order to the lipid bilayer. Furthermore, it is a well-known fact that artificial lipid bilayers can be constructed in the absence of proteins.

One of the most important functions of cell membranes is that of a permeability barrier. In general membranes serve as selective barriers that control the passage of substances from one region to another. The movement of substances across membranes is known as membrane transport, and it involves two aspects: membrane permeability and transport mechanism.

Cell membranes are freely permeable to water, carbon dioxide, and oxygen, but less so to other substances. It is possible to make some generalizations about the passage of substances through the membrane. Small molecules pass through the membrane more readily than larger ones and most membranes are impermeable to large molecules such as proteins. Because the interior region of the lipid bilayer is hydrophobic in nature, we would expect the membrane to be more permeable to uncharged species and nonpolar molecules than to ions and polar molecules. This is indeed the case. Because ions are hydrated in solution and the extent of hydration depends on the charge and size of the ion, di- and trivalent ions move more slowly across the membrane than do monovalent ions. Anions such as Cl^- and HPO_4^{2-} pass through the membrane more readily than cations such as H^+, Na^+, and K^+.

To carry out many essential biological processes, the ionic and molecular composition of the internal aqueous phase of the cell must be appreciably different from its external environment. For example, the concentration of potassium ions is some 35 times higher in the red blood cells than in the extracellular blood plasma. The reverse is true for the sodium ions—the extracellular sodium-ion concentration is about 15 times that of the intracellular fluid. Although the red blood cell membrane (and indeed other types of membranes) is more permeable to K^+ ions than Na^+ ions, we would expect that, given enough time, the concentrations of the same types of ions would eventually become equal in the intra- and extracellular fluids. That this is not the case means that processes other than normal diffusion must be at play to maintain the differences in ionic concentrations. We shall now briefly consider the three types of membrane transport mechanisms.

Simple Diffusion.   A number of substances are able to pass through the membrane by simple diffusion, that is, from one region to another containing a lower concentration of the substance. It is believed that there may be discrete pores (thought to be formed by the membrane-bound proteins) in the membrane structure through which small molecules pass, thereby circumventing the hydrophobic lipid bilayer. This mechanism not only accounts for the rapid rate of movement of oxygen and carbon dioxide through the membrane, but also explains why polar molecules such as water can penetrate the membrane with ease. The pores can be envisioned as protein-lined channels that penetrate the lipid bilayer. Because of the fluidity of the membrane structure, these pores are continually being destroyed and created. The rate of movement of the molecules across the membrane is directly proportional to the concentration gradient as described by Fick's first law [Eq. (5.9)]:

$$J = -D\left(\frac{\partial c}{\partial x}\right)_t$$

The majority of substances that pass through membranes do so by processes other than simple diffusion. The transfer invariably involves the interaction of the transported molecules with a specific carrier system.

Figure 9.13

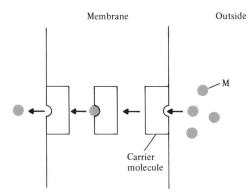

Inside   Membrane   Outside

M

Carrier
molecule

Facilitated diffusion. A metabolite (M) binds to a carrier molecule C at the outer face of the membrane. The complex (CM) diffuses through the membrane and dissociates at the inner surface, releasing M into the cell interior.

Facilitated Diffusion. Facilitated diffusion also involves the movement of molecules from a higher concentration region to a lower one, but it differs from simple diffusion in several important respects. It is a carrier-assisted transport process. The substance to be transported is first complexed with a carrier molecule (probably a protein molecule). The complex formed is readily soluble in the lipid bilayer, so that the rate of facilitated diffusion is normally much higher than that of the uncomplexed molecule by simple diffusion. An example of a facilitated diffusion is the transport of sugar molecules (for example, glucose) to the interior of the cells. Sugar molecules by themselves are quite insoluble in the lipid bilayer so that the rate of simple diffusion of these molecules would be too slow to maintain metabolic processes. A glucose–carrier complex, on the other hand, is freely soluble in the membrane and can therefore diffuse through it with ease. It is interesting to note that the overall rate of glucose transport from the blood plasma into the red blood cell is greatly enhanced by the presence of the hormone insulin. Whether the action of insulin is to increase the concentration of carrier molecules or to promote the glucose–carrier molecule complex formation is still not well understood.

Figure 9.13 shows a schematic diagram for facilitated diffusion. The carrier molecule is free to move back and forth across the membrane. Each carrier has one or several specific binding sites for the material to be transported. At the other face of the membrane (that is, the surface in contact with extracellular fluid), the carrier can bind to the material, and the complex diffuses across the membrane from a high to low concentration region, where it dissociates into the carrier and the free material. The carrier molecule will eventually diffuse back to the other side of the membrane and the process will be repeated. Facilitated diffusion will continue as long as there is a concentration gradient across the membrane for the transported material.

Active Transport. Unlike simple diffusion and facilitated diffusion, active transport involves the movement of substances across the membrane *against* a concentration gradient. Thermodynamics tells us that this is a nonspontaneous process; therefore, energy must be supplied from an external source to carry out active transport. Earlier we mentioned the unequal concentrations of K^+ and Na^+ ions in the intra- and extracellular fluids. The fact that these differences in ionic concentration can be maintained is the result of active transport. On one hand, we have the normal diffusion processes involving the movement of Na^+ ions from the exterior to the interior of the cell and the movement of K^+ ions in the opposite direction. On the other hand, Na^+ ions are constantly "pumped" out of the cell while K^+ ions are pumped into the cell. The word *pump* is used to describe the movement of ions against concentration gradients. A steady state is reached when the flow of the ions of one type in one direction by active transport is balanced by the "leaking" (that is, diffusion) of the ions of the same type in the opposite direction.

As a result of intensive study over the years, the mechanism of active transport is fairly well understood. We now know, for example, that the active transport of Na^+ and the active transport of K^+ ions across the membrane do not occur independently but are linked together. Thus the active transport system is often referred to as the *sodium-potassium pump*. As in facilitated diffusion, we need a carrier molecule for complexing the ions. Furthermore, the molecule must be such that an energy supply can be directly coupled to it for carrying out active transport. This carrier molecule turns out to be an enzyme called *sodium–potassium ATPase* (where ATPase stands for adenosine-triphosphatase). The energy supplier is the ATP molecule, which, upon hydrolysis, yields adenosine diphosphate (ADP) and inorganic phosphate (HPO_4^{2-})*:

$$ATP^{4-} + H_2O \longrightarrow ADP^{3-} + HPO_4^{2-} + H^+ \qquad \Delta G° \simeq -30 \text{ kJ}$$

This spontaneous process (note the large decrease in the standard Gibbs free energy) can then be coupled to sodium–potassium ATPase to drive the energetically unfavorable process of moving ions against a concentration gradient. Figure 9.14 shows the schematic diagram for active transport.

An estimate of the increase in Gibbs free energy can be made for transporting against a concentration gradient. Let us suppose that K^+ ions are transported from the blood plasma into the red blood cells. The difference in concentration in these two media means that

$$(\mu_{K^+})_C > (\mu_{K^+})_P$$

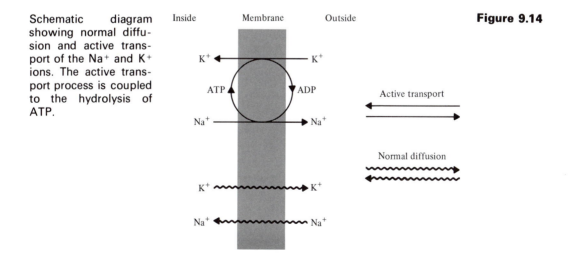

Schematic diagram showing normal diffusion and active transport of the Na^+ and K^+ ions. The active transport process is coupled to the hydrolysis of ATP.

Figure 9.14

* We shall discuss the structure and function of ATP in more detail in Chapter 10.

where C and P denote red blood cell and blood plasma, respectively. Since the chemical potential of the K^+ ion is

$$\mu_{K^+} = \mu_{K^+}^{\circ} + RT \ln a_{K^+}$$

where μ_{K^+} and $\mu_{K^+}^{\circ}$ are the chemical potential and standard chemical potential of the K^+ ions, and a_{K^+} the activity of the K^+ ions, we write

$$(\mu_{K^+})_C - (\mu_{K^+})_P = \Delta\mu = RT \ln \frac{(a_{K^+})_C}{(a_{K^+})_P}$$

(Note that $\mu_{K^+}^{\circ}$ is a constant and therefore cancels in the subtraction.) Approximating concentrations for activities and using a ratio of 35/1 for the K^+-ion concentration in the intra- and extracellular fluids, we obtain at the physiological temperature of 310 K (37°C)*:

$$\Delta\mu = RT \ln \frac{[K^+]_C}{[K^+]_P}$$

$$= (8.314 \text{ J K}^{-1} \text{ mol}^{-1})(310 \text{ K}) \ln \tfrac{35}{1}$$

$$= 9.2 \text{ kJ mol}^{-1}$$

Figure 9.15

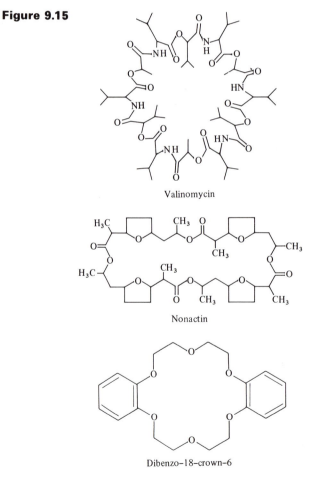

Valinomycin

Nonactin

Dibenzo-18-crown-6

Structures of valinomycin, nonactin, and dibenzo-18-crown-6.

* This calculation neglects the existence of membrane potential, a topic we shall discuss in Chapter 11.

Finally, we note that the ionic permeability across cell membranes can be affected by the naturally occurring antibiotics such as valinomycin and nonactin. These molecules possess the macrocyclic structure, having a cavity in the center and a hydrophobic exterior (Figure 9.15). The antibiotics can form complexes with alkali metal ions (usually 1:1 complex), which, because of their external hydrophobic nature, can pass through the membrane. Moreover, the complex formation is quite specific for individual ions, since the ion must have the appropriate size to fit into the cavity. For example, the formation constant for the valinomycin-K^+ complex is about 1000 times greater than that of the valinomycin-Na^+ complex. The increase in K^+ ion permeability disturbs the delicate intra- and extracellular concentration balance and can lead to eventual cell death.

During the last 10 years or so, a number of model compounds for cyclic antibiotics have been prepared. These molecules are known as "crown ethers," so called because of their appearance when viewed in three dimensions. Initial studies on these compounds have shown that these molecules can affect ionic transport across cell membranes in a manner that is analogous to that of the antibiotics. Furthermore, these crown ethers offer a unique way of carrying out many ionic reactions in nonaqueous and nonpolar solvents because of the central hydrophilic cavities and the hydrophobic exteriors. A dramatic demonstration is the dissolution of NaOH in benzene in the presence of the crown ether dibenzo-18-crown-6.

Appendix 9.1
Notes on Electrostatics

The force between two charges Q_1 and Q_2 in a vacuum separated by distance r is given by Coulomb's law:

$$F = \frac{Q_1 Q_2}{r^2}$$

If Q_1 and Q_2 have the same sign ($+$ and $+$ or $-$ and $-$), F is a positive quantity, meaning that the force between the two charges is repulsive; if Q_1 and Q_2 have opposite signs ($+$ and $-$ or $-$ and $+$), F is a negative quantity, meaning that the force is attractive. The equation above is written in the CGS–esu system. Since electric charge is in esu, force in dyn, and distance in cm, we have, after rearrangement,

$$Q_1 Q_2 = Fr^2$$

or

$$(\text{charge})^2 = \text{force} \times (\text{distance})^2$$

Hence

$$(\text{esu})^2 = \text{dyn cm}^2$$
$$= \text{g cm s}^{-2} \text{ cm}^2$$
$$= \text{g cm}^3 \text{ s}^{-2}$$

Thus

$$1 \text{ esu} = 1 \text{ g}^{1/2} \text{ cm}^{3/2} \text{ s}^{-1}$$

If the two charges are in a medium of dielectric constant ε, Coulomb's law is modified as follows:

$$F = \frac{Q_1 Q_2}{\varepsilon r^2}$$

The dielectric constant is a dimensionless quantity, since it is defined as the ratio of the capacitance of a capacitor filled with a certain medium (C) and the capacitance of the same capacitor with a vacuum between the plates (C_0):

$$\varepsilon = \frac{C}{C_0}$$

The capacitance of a capacitor is a measure of the capacitor's ability to hold charges, given a certain potential difference across the plates. It is defined as

$$C = \frac{Q}{V}$$

where Q and V represent the charge on the plates and potential difference between the plates of a capacitor, respectively. Figure 9.16 shows an arrangement for comparing the capacitances of two geometrically similar capacitors.

In the SI system, the unit of charge is not defined by Coulomb's law. The unit of current is taken to be the fundamental quantity. The SI unit of current is the *ampere*, which is defined to be that current in each of two infinitely long parallel wires 1 m apart that causes an electromagnetic force of 2×10^{-7} N per meter of its length to act on each wire. One *coulomb* (C) is the amount of charge when 1 ampere flows for 1 second. Thus

$$1 \, C = 1 \, A \times 1 \, s$$

Defining charge in the foregoing manner means that Coulomb's law must now be written as

$$F = k \frac{Q_1 Q_2}{r^2}$$

where k is the proportionality constant. The value of k is given by the equation

$$k = \frac{1}{4\pi\varepsilon_0}$$

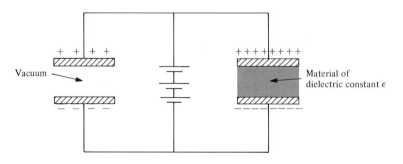

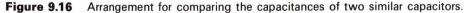

Figure 9.16 Arrangement for comparing the capacitances of two similar capacitors.

where ε_0 is the *permittivity* of free (empty) space, equal to 8.8542×10^{-12} C^2 N^{-1} m^{-2}. In the SI system, Coulomb' law becomes

$$F = \frac{Q_1 Q_2}{r^2} \frac{1}{4\pi\varepsilon_0}$$

If the charges are in a medium of dielectric constant ε, we have*

$$F = \frac{Q_1 Q_2}{\varepsilon r^2} \frac{1}{4\pi\varepsilon_0}$$

EXAMPLE 9.5   Calculate the electrostatic force of attraction between a Na^+ and a Cl^- ion separated by 2.0 Å in water.

Answer

CGS–esu units

$$e = 4.80 \times 10^{-10} \text{ esu}$$

$$r = 2.0 \text{ Å} = 2.0 \times 10^{-8} \text{ cm}$$

$$\varepsilon = 78.54$$

Thus

$$F = \frac{(4.80 \times 10^{-10} \text{ esu})(-4.80 \times 10^{-10} \text{ esu})}{(78.54)(2.0 \times 10^{-8} \text{ cm})^2}$$

$$= -7.3 \times 10^{-6} \text{ esu}^2 \text{ cm}^{-2}$$

$$= -7.3 \times 10^{-6} \text{ dyn}$$

SI units

$$e = 1.602 \times 10^{-19} \text{ C}$$

$$r = 2.0 \text{ Å} = 2.0 \times 10^{-10} \text{ m}$$

$$\varepsilon = 78.54$$

Thus

$$F = \frac{(1.602 \times 10^{-19} \text{ C})(-1.602 \times 10^{-19} \text{ C})}{(78.54)(2.0 \times 10^{-10} \text{ m})^2}$$

$$\times \frac{1}{4\pi \times 8.8542 \times 10^{-12} \text{ C}^2 \text{ N}^{-1} \text{ m}^{-2}}$$

$$= -7.3 \times 10^{-11} \text{ N}$$

Comments: (1) The negative sign indicates that the force between the two ions is attractive. (2) Strictly speaking, the calculation here does not provide us with the exact value. The reason is that dielectric constant is a bulk property. On the other hand, for two ions separated by only 2 Å, there can be but a few water molecules present in between the ions. Thus using the

* For further details, see T. I. Quickenden and R. C. Marshall, *J. Chem. Educ.* **49**, 114 (1972); and T. Cvitas and N. Kallay, *J. Chem. Educ.* **54**, 530 (1977).

value of 78.54 in the denominator is not correct, but it does serve as an estimate of the magnitude of the force. A rigorous calculation would require that we know the exact structure of the hydrated ions.

Appendix 9.2 The Donnan Equilibrium Involving Proteins Bearing Multiple Charges

For a protein at a pH other than its isoelectric point where it will possess either a net positive or net negative charge, an additional factor must be considered; that is, the counterions needed to maintain charge neutrality. In the chapter we considered the simple case where the protein bears only one negative charge. Here we deal with the situation in which the protein bears a number of negative charges (z). We assume that the protein, $Na_z^+ P^{z-}$, is a strong electrolyte so that

$$Na_z^+ P^{z-} \longrightarrow zNa^+ + P^{z-}$$

Referring to Figure 9.10, we shall consider two following cases.

Case 1. The protein solution is placed in the left compartment and water in the right compartment. The osmotic pressure of the solution (π_1) is given by

$$\pi_1 = (z + 1)cRT$$

where c is the concentration (molarity) of the protein solution. Since z is typically of the order of 30, determination of the molar mass of protein using this arrangement yields a value that is only $\frac{1}{30}$ that of the true value.

Case 2. Again the protein solution is placed in the left compartment while a NaCl solution is now placed in the right compartment. Now the requirement that the chemical potential of a component be the same throughout the system applies to the NaCl as well as to the water. To attain equilibrium, NaCl will move from the right to the left compartment and the actual amount transported can be calculated. The initial molar concentration of $Na_z^+ P^{z-}$ is c and that of NaCl is b. At equilibrium, the concentrations are

$$[P^{z-}]^L = c \qquad [Na^+]^L = (zc + x) \qquad [Cl^-]^L = x$$

and

$$[Na^+]^R = (b - x) \qquad [Cl^-]^R = (b - x)$$

where x is the decrease in concentration after transport of NaCl from right to left.

Since $(\mu_{NaCl})^L = (\mu_{NaCl})^R$ at equilibrium and for dilute solutions, we replace activities with concentrations so that

$$([Na^+][Cl^-])^L = ([Na^+][Cl^-])^R$$

or

$$(zc + x)(x) = (b - x)(b - x)$$

$$x = \frac{b^2}{zc + 2b}$$

The osmotic pressure (π_2), which is proportional to the difference in solute concentration between the two sides, is now given by

$$\pi_2 = \{(c + zc + x + x) - (b - x + b - x)\}RT$$
$$\underset{\text{left}}{} \qquad \underset{\text{right}}{}$$
$$\underset{\text{compartment}}{} \qquad \underset{\text{compartment}}{}$$

or

$$\pi_2 = (c + zc - 2b + 4x)RT$$

Substituting for x, we obtain

$$\pi_2 = \left(c + zc - 2b + \frac{4b^2}{zc + 2b}\right)RT$$

$$= \frac{zc^2 + 2cb + z^2c^2}{zc + 2b}RT$$

The equation above was derived assuming no change in either pH or volume of the solutions.

Two limiting cases follow.

If $b \ll zc$ (the salt concentration is much less than the protein concentration), then

$$\pi_2 = \frac{zc^2 + z^2c^2}{zc}RT = (zc + c)RT$$

$$= (z + 1)cRT$$

$$= \pi_1$$

If $b \gg z^2c$ (the salt concentration is much greater than the protein concentration),* then

$$\pi_2 = \frac{2cb}{2b}RT = cRT$$

In this limiting case, the osmotic pressure approaches that of the pure isoelectric protein. In effect, the added salt reduces (and at high enough salt concentrations, eliminates) the Donnan effect. Under these conditions, the molar mass determined by osmotic pressure measurement would correspond closely to the true value.

INTRODUCTORY–INTERMEDIATE

HUNT, J. P. *Metal Ions in Aqueous Solution*. W. A. Benjamin, Inc., Menlo Park, Calif., 1963.

PASS, G. *Ions in Solution*. Clarendon Press, Oxford, 1973.

These two texts provide a descriptive treatment of metal ions in solution. The main emphasis is on thermodynamic and chemical properties.

---

* In practice, $c \lesssim 1 \times 10^{-4}\ M$, $z \lesssim 30$, so that $z^2c \simeq 0.1\ M$. Thus, in order for this limiting case to hold, the concentration of added salt should be about $1\ M$.

INTERMEDIATE–ADVANCED

DAVIES, C. W. *Ion Association*. Butterworth, Washington, D.C., 1962.

HARNED, H., and B. B. OWEN. *The Physical Chemistry of Electrolyte Solutions*, 3rd ed. Van Nostrand Reinhold Company, New York, 1958.

ROBINSON, R. A., and R. H. STOKES. *Electrolyte Solutions*, 2nd ed. Academic Press, Inc., New York, 1959.

These three texts present a rather comprehensive treatment of electrolyte solutions. Recommended for advanced students only.

HARRISON, R., and G. G. LUNT. *Biological Membranes*. Halsted Press, New York, 1976.

TOMBS, M. P., and A. R. PEACOCKE. *The Osmotic Pressure of Biological Macro-molecules*. Clarendon Press, New York, 1975.

These two texts provide many useful examples of biological systems.

Reading Assignments

SECTION A

"The Motion of Ions in Solution Under the Influence of an Electric Field," C. A. Vincent, *J. Chem. Educ.* **53**, 490 (1976).

"Calcium Carbonate Equilibria in Lakes," S. D. Morton and G. F. Lee, *J. Chem. Educ.* **45**, 511 (1968).

"Calcium Carbonate Equilibria in the Oceans—Ion Pair Formation," S. D. Morton and G. F. Lee, *J. Chem. Educ.* **45**, 513 (1968).

"How Things Get into Cells," H. Holter, *Sci. Am.*, Sept. 1961.

"The Structure of Cell Membranes," C. F. Fox, *Sci. Am.*, Feb. 1972.

"A Dynamic Model of Cell Membranes," R. A. Capaldi, *Sci. Am.*, Mar. 1974.

"Living Membranes," A. Katchalsky, *Sci. Technol.*, Dec. 1967.

"The Membrane of the Living Cell," J. R. Robertson, *Sci. Am.*, Apr. 1962.

"The Membrane of the Mitochondrion," E. Racker, *Sci. Am.*, Feb. 1968.

"The Kidney," H. W. Smith, *Sci. Am.*, Jan. 1953.

"The Donnan Equilibrium and Osmotic Pressure," R. Chang and L. J. Kaplan, *J. Chem. Educ.* **54**, 218 (1977).

"Ion Channels in the Nerve Cell Membrane," R. D. Keynes, *Sci. Am.*, Mar. 1979.

"Steps Toward Building a Living Cell," G. Baumann, *Chemistry* **51**(9), 16, **51**(10), 11 (1978); **52**(1), 12 (1979).

"Crown Ethers," A. C. Knipe, *J. Chem. Educ.* **53**, 618 (1976).

SECTION B

"Thermodynamics of Ion Solvation and Its Significance in Various Systems," C. M. Criss and M. Salomon, *J. Chem. Educ.* **53**, 763 (1976).

"Ionic Hydration Enthalpies," D. W. Smith, *J. Chem. Educ.* **54**, 540 (1977).

"Electrolyte Theory and SI Units," R. I. Holliday, *J. Chem. Educ.* **53**, 21 (1976).

"Ion Pairs and Complexes: Free Energies, Enthalpies, and Entropies," J. E. Prue, *J. Chem. Educ.* **46**, 12 (1969).

"Are Solubilities and Solubility Products Related?" L. Meites, J. S. F. Pode, and H. C. Thomas, *J. Chem. Educ.* **43**, 667 (1966).

"How Have Sea Water and Air Got Their Present Compositions?" L. G. Sillén, *Chem. Brit.* **3**, 291 (1967).

"On Squid Axons, Frog Skins, and the Amazing Uses of Thermodynamics," W. H. Cropper, *J. Chem. Educ.* **48**, 182 (1971).

"Osmotic Pressure in the Physics Course for Students of the Life Sciences," R. K. Hobbie, *Am. J. Phys.* **42**, 188 (1974).

"The State of Water in Red Cells," A. K. Solomon, *Sci. Am.*, Feb. 1971.

"Pumps in the Living Cells," A. K. Solomon, *Sci. Am.*, Aug. 1962.

"The Fluid Mosaic Model of the Structure of Cell Membranes," S. J. Singer and G. L. Nicolson, *Science* **175**, 720 (1972).

"Membrane Structure: Some General Principles," M. S. Bretscher, *Science* **181**, 622 (1973).

"The Chemistry of Cell Membranes," L. Hokin and M. Hokin, *Sci. Am.*, Oct. 1965.

"Biochemical Aspects of Active Transport," R. S. Albers, *Ann. Rev. Biochem.* **36**, 727 (1967).

"Ionic Binding by Synthetic Macrocyclic Compounds," J. J. Christensen, J. O. Hill, and R. M. Izatt, *Science* **174**, 459 (1971).

"Vectorial Biochemistry," S. W. Browne, Jr., and G. D. Browne, *J. Chem. Educ.* **52**, 565 (1975).

"Colicins and the Energetics of Cell Membrane," S. E. Luria, *Sci. Am.*, Dec. 1975.

"Across the Living Barrier," D. E. Fenton, *Quart. Rev.* **6**(3), 325 (1977).

"The Assembly of Cell Membranes," H. F. Lodish and J. E. Rothman, *Sci. Am.*, Jan. 1979.

"Ion Fluxes Through Membranes," M. E. Starzak, *J. Chem. Educ.* **54**, 200 (1977).

9.1 The cell constant (l/A) of a conductance cell is 388.1 m^{-1}. The resistances of a 4.8×10^{-4} mol dm^{-3} aqueous solution of sodium chloride and a sample of water are 6.4×10^4 Ω and 7.4×10^6 Ω, respectively, at 25°C. Calculate the molar conductivity of the NaCl solution at this concentration.

9.2 It was pointed out in the text of the chapter that the measurement of Λ_0 for weak electrolytes is generally difficult. How would you deduce the value of Λ_0 for CH_3COOH from the data listed in Table 9.1? (*Hint*: Consider CH_3COONa, HCl, and NaCl.)

9.3 A simple way to determine the salinity of water is to measure its conductivity and assume that the conductivity is entirely due to sodium chloride. In a particular experiment, the resistance of a sample solution is found to be 254 Ω. The resistance of a 0.05 *M* KCl solution measured in the same cell is 467 Ω. Estimate the concentration of NaCl in the solution. (*Hint*: First derive an equation relating R to Λ and c and then use Λ_0 values for Λ.)

Problems

9.4 A conductivity cell consists of two electrodes of equal areas ($4.2 \times 10^{-4} \ m^2$) and separated by 0.020 m. The resistance of the cell when filled with a 6.3×10^{-4} M KNO_3 solution is 26.7 Ω. What is the molar conductivity of the solution?

9.5 From the following data, calculate the heat of solution for KI:

	NaCl	NaI	KCl	KI
Lattice energy ($kJ \ mol^{-1}$)	778	700	716	643
Heat of solution ($kJ \ mol^{-1}$)	3.8	−5.1	17.1	?

9.6 Referring to Figure 9.3, explain why the slope of the conductance versus volume of NaOH added plot rises right at the start if the acid employed in the titration is weak.

9.7 Express the mean activity, mean activity coefficient, and mean molality in terms of the individual ionic quantities (a_+, a_-, γ_+, γ_-, m_+, and m_-) for the following electrolytes: RbI, $SrSO_4$, $CaCl_2$, Li_2CO_3, $K_3Fe(CN)_6$, and $K_4Fe(CN)_6$.

9.8 Calculate the ionic strength and the mean activity coefficient for the following solutions at 298 K: (a) 0.10 m NaCl, (b) 0.01 m $MgCl_2$, and (c) 0.10 m $K_4Fe(CN)_6$.

9.9 The mean activity coefficient of a 0.01 m H_2SO_4 solution is 0.544. What is its mean ionic activity?

9.10 The freezing-point depression of a 0.010 m acetic acid solution is 0.0193 K. Calculate the degree of dissociation for acetic acid at this concentration.

9.11 A 0.010 m aqueous solution of the ionic compound $Co(NH_3)_5Cl_3$ has a freezing-point depression of 0.0558 K. What can you conclude about its structure? Assume the compound to be a strong electrolyte.

9.12 The osmotic pressure of blood plasma is about 7.5 atm at 37°C. Estimate the total concentration of dissolved species and the freezing point of blood plasma.

9.13 Calculate the ionic strength of a 0.0020 m aqueous solution of $MgCl_2$ at 298 K. Use the Debye–Hückel limiting law to estimate (a) the activity coefficients of the Mg^{2+} and Cl^- ions in this solution, and (b) the mean ionic activity coefficients of these ions.

9.14 Calculate the solubility of $BaSO_4$ (in g liter^{-1}) in (a) water, and (b) a 6.5×10^{-5} M $MgSO_4$ solution. The solubility product of $BaSO_4$ is 1.1×10^{-10}. Assume ideal behavior.

9.15 The thermodynamic solubility product of AgCl is 1.6×10^{-10}. What is [Ag^+] in (a) a 0.020 M KNO_3 solution, and (b) a 0.020 M KCl solution?

9.16 Referring to Problem 9.15, calculate $\Delta G°$ for the process

$$AgCl(s) \longrightarrow Ag^+(aq) + Cl^-(aq)$$

to yield a saturated solution at 298 K.

9.17 The apparent solubility products of CdS and CaF_2 at 25°C are 3.8×10^{-29} and 4.0×10^{-11}, respectively. Calculate the solubility (g/100 g of solution) of these compounds.

★**9.18** Oxalic acid, $(COOH)_2$, is a poisonous compound present in many plants and vegetables, including spinach. Calcium oxalate is only slightly soluble in water ($K_{sp} = 3.0 \times 10^{-9}$ at 25°C) and can be deposited in renal calculi. Calculate (a) the apparent and thermodynamic solubility of calcium oxalate in water, and (b) the concentrations of calcium and oxalate ions in a 0.010 M $Ca(NO_3)_2$ solution. Assume ideal behavior in (b).

9.19 Referring to Figure 9.10, calculate the osmotic pressure for the following cases at 298 K: (a) The left compartment contains 200 g of hemoglobin in 1 liter of solution; the right compartment contains pure water. (b) The left compartment contains the same hemoglobin solution; the right compartment initially contains 6.0 g of NaCl in 1 liter of solution. Assume that the pH of the solution is such that the hemoglobin molecules are in the Na^+Hb^- form (the molar mass of hemoglobin is 65,000 g mol^{-1}).

9.20 The concentrations of K^+ and Na^+ ions in the intracellular fluid of a nerve cell are about 400 mM and 50 mM, while those of the same ions in the extracellular fluid are 20 mM and 440 mM, repectively. Calculate the free energy change for the transfer of 1 mol of each type of ions against the concentration gradient at 37°C.

9.21 The Debye–Hückel limiting law is more reliable for 1:1 electrolytes than 2:2 electrolytes. Explain.

9.22 Theory shows that the size of the ionic atmosphere is called the Debye radius $(1/\kappa)$, where κ is given by (see the physical chemistry texts listed in Chapter 1):

$$\kappa = \left(\frac{8\pi N_0 e^2 \rho}{1000 \, \varepsilon kT}\right)^{1/2} \sqrt{I}$$

where N_0 is Avogadro's number, e the electronic charge, ρ the density of the solvent, ε the dielectric constant of the solvent, k the Boltzmann constant, T the absolute temperature, and I the ionic strength. Calculate the Debye radius in a 0.01 m Na_2SO_4 solution at 25°C.

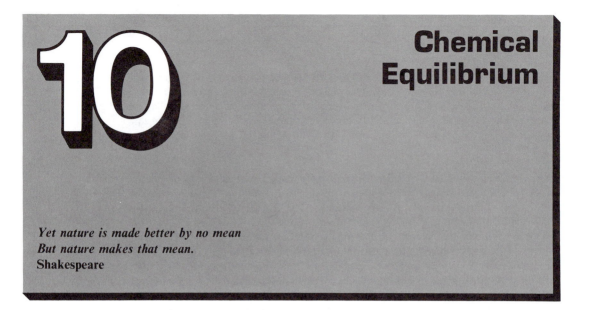

Chemical Equilibrium

Yet nature is made better by no mean
But nature makes that mean.
Shakespeare

Chapters 8 and 9 dealt with the application of Gibbs free energy to nonelectrolyte and electrolyte solutions and physical equilibria. In this chapter we concentrate on chemical equilibrium in gaseous and condensed phases and take a look at bioenergetics.

10.1 Chemical Equilibrium in Gaseous Systems

IDEAL GASES

In this section we derive an expression relating the Gibbs free energy change for a reaction in the gas phase to the concentrations of the reacting species and temperature. We consider first the case in which all gases exhibit ideal behavior.

In Chapter 7 we saw that the molar Gibbs free energy of a pure gas at temperature T and pressure P is given by Eq. (7.24):

$$\overline{G} = \overline{G}^\circ + RT \ln P$$

where $\overline{G}^\circ$, the standard Gibbs free energy, is a function of temperature only. In a mixture of gases, the chemical potential of the ith component μ_i is given by [see Eq. (8.5)]:

$$\mu_i = \mu_i^\circ + RT \ln P_i$$

where P_i is the partial pressure of component i in the mixture and μ_i° is the standard chemical potential of component i when its partial pressure is 1 atm. Consider a hypothetical reaction at temperature T:

$$aA + bB \longrightarrow cC + dD$$

236

where a, b, c, and d are the stoichiometric coefficients. The chemical potentials of the components are

$$\mu_A = \mu_A^\circ + RT \ln P_A \qquad (10.1a)$$

$$\mu_B = \mu_B^\circ + RT \ln P_B \qquad (10.1b)$$

$$\mu_C = \mu_C^\circ + RT \ln P_C \qquad (10.1c)$$

$$\mu_D = \mu_D^\circ + RT \ln P_D \qquad (10.1d)$$

The Gibbs free energy change for the reaction, ΔG, is given by

$$\Delta G = \sum G(\text{products}) - \sum G(\text{reactants})$$

$$= c\mu_C + d\mu_D - a\mu_A - b\mu_B \qquad (10.2)$$

Substituting the expressions in Eq. (10.1) into Eq. (10.2), we get

$$\Delta G = c\mu_C^\circ + d\mu_D^\circ - a\mu_A^\circ - b\mu_B^\circ + cRT \ln P_C$$

$$+ dRT \ln P_D - aRT \ln P_A - bRT \ln P_B \quad (10.3)$$

The standard Gibbs free energy change of the reaction, ΔG°, is just the difference between the standard free energies of products and reactants, that is,

$$\Delta G^\circ = c\mu_C^\circ + d\mu_D^\circ - a\mu_A^\circ - b\mu_B^\circ$$

Therefore, we can write Eq. (10.3) as

$$\Delta G = \Delta G^\circ + RT \ln \frac{P_C^c P_D^d}{P_A^a P_B^b} \qquad (10.4)$$

Equation (10.4) gives the Gibbs free energy change that accompanies the conversion of reactants at pressures P_A, P_B to products at pressures P_C, P_D. By definition, $\Delta G = 0$ at equilibrium so that Eq. (10.4) becomes

$$0 = \Delta G^\circ + RT \ln \frac{P_C^c P_D^d}{P_A^a P_B^b}$$

$$0 = \Delta G^\circ + RT \ln K_P$$

or

$$\Delta G^\circ = -RT \ln K_P \qquad (10.5)$$

where K_P, the equilibrium constant (where the subscript P denotes that concentrations are expressed in pressures) is given by

$$K_P = \frac{P_C^c P_D^d}{P_A^a P_B^b} \qquad (10.6)$$

The equilibrium constant K_P and the standard Gibbs free energy change of the reaction are related in a remarkably simple fashion. Keep in mind that ΔG° is a constant at a given temperature whose value depends only on the nature of the reactants and products. In our example ΔG° is the standard Gibbs free energy change when all reactants at 1 atm pressure and temperature T are converted to products at 1 atm pressure and the same temperature. Equation (10.5) tells us that if we know the value of ΔG°, we can calculate the equilibrium constant K_P, and vice versa. We have already seen how to obtain the standard Gibbs free energy of formation of compounds (Section 7.6). The standard Gibbs free energy of a reaction is just the difference between the standard Gibbs free energies of formation of the products and the standard Gibbs free energies of formation of the reactants. Thus once we have defined a reaction, we can frequently calculate the equilibrium constant from the listed $\Delta\overline{G}_f^\circ$ values in Appendix 3 and Eq. (10.5). Note that these values all refer to 298 K so that if we are interested in a reaction at some other temperature, we shall need to first calculate ΔG° at that temperature before evaluating K_P. This calculation can be accomplished by using the Gibbs–Helmholz equation [Eq. (7.21)].

The change in standard Gibbs free energy ΔG° is a number generally different from zero. If ΔG° is negative, the conversion of reactants in their standard states to products in their standard states is spontaneous, and the reaction is said to be *exergonic*. If ΔG° is negative, the equilibrium constant must be greater than unity; in fact, the more negative ΔG° is, the larger is K_P. The reverse holds true if ΔG° is a positive number. In this case the conversion of reactants in their standard states to products in their standard states is not a spontaneous process, and the reaction is said to be *endergonic*. Here the equilibrium constant is less than unity. It should not be concluded that just because ΔG° is positive no reaction will take place. Far from it. If we have $\Delta G^\circ = 10$ kJ, $T = 298$ K, we find from Eq. (10.5) that $K_P = 0.018$. While 0.018 is a small number compared to unity, an appreciable amount of products can still be obtained if we employ large quantities of reactants for the reaction. There is also the special case in which $\Delta G^\circ = 0$, which corresponds to an equilibrium constant of unity.

EXAMPLE 10.1 From the thermodynamic data listed in Appendix 3, calculate the equilibrium constant for the reaction at 298 K:

$$N_2(g) + 3H_2(g) \rightleftharpoons 2NH_3(g)$$

Answer: The equilibrium constant for the reaction is given by

$$K_P = \frac{P_{NH_3}^2}{P_{N_2} P_{H_2}^3}$$

To calculate K_P, we need Eq. (10.5) and ΔG°. From Appendix 3 we have

$$\Delta G^\circ = 2\Delta\overline{G}_f^\circ(NH_3) - \Delta\overline{G}_f^\circ(N_2) - 3\Delta\overline{G}_f^\circ(H_2)$$

$$= (2 \text{ mol})(-16.64 \text{ kJ mol}^{-1}) - (1 \text{ mol})(0) - (3 \text{ mol})(0)$$

$$= -33.28 \text{ kJ}$$

Finally,

$$-33,280 \text{ J} = -(8.314 \text{ J K}^{-1} \text{ mol}^{-1})(298 \text{ K}) \ln K_P$$

$$\ln K_P = 13.43$$

or

$$K_P = 6.80 \times 10^5$$

Comments: (1) It should be understood that if the reaction had been written as

$$\tfrac{1}{2}N_2(g) + \tfrac{3}{2}H_2(g) \rightleftharpoons NH_3(g)$$

the value of $\Delta G°$ would have been -16.64 kJ and the equilibrium constant would have been calculated as

$$K_P = \frac{P_{NH_3}}{P_{N_2}^{1/2} P_{H_2}^{3/2}} = 825$$

Thus whenever we multiply a balanced equation throughout by a factor n, we change the equilibrium constant K_P to K_P^n. Here we have changed K_P to $K_P^{1/2}$. (2) The equilibrium constant K_P has no units. That it must be so can be seen from Eq. (10.5), since K_P appears as $\ln K_P$ and we cannot take the log of units. In deriving Eq. (7.24), it was shown that the equation should strictly be written as

$$\overline{G} = \overline{G°} + RT \ln \frac{P}{1 \text{ atm}}$$

where 1 atm corresponds to the standard state. Similarly, the chemical potential of the ith component should strictly be written as

$$\mu_i = \mu_i° + RT \ln \frac{P_i}{1 \text{ atm}}$$

so that the equilibrium constant K_P in our example becomes

$$K_P = \frac{\left(\dfrac{P_{NH_3}}{1 \text{ atm}} \right)^2}{\left(\dfrac{P_{N_2}}{1 \text{ atm}} \right) \left(\dfrac{P_{H_2}}{1 \text{ atm}} \right)^3}$$

Each partial pressure is divided by its standard-state value. Numerically, K_P is unchanged but all the units cancel so that K_P is dimensionless. This is a most important conclusion and, as we shall see later, it applies to any equilibrium constant regardless of what concentration units (pressure, molality, molarity, ...) we choose for the reacting species.

So far we have concentrated on the change in standard Gibbs free energy $\Delta G°$. Since reactions cannot always be carried out under standard-state conditions, we must consider the more general case involving ΔG. Equation (10.4) enables us to calculate ΔG for a reaction under any condition, standard or nonstandard. The value of ΔG is determined by two terms: $\Delta G°$ and a concentration-dependent term. At a given temperature the value of $\Delta G°$ is fixed, but we can still change the value of ΔG by properly adjusting the partial pressures of A, B, C, and D. Although the quotient composed of the pressures of reactants and products has the form of an equilibrium constant, it is *not* equal to the equilibrium constant unless P_A, P_B, P_C, and P_D are the partial pressures at equilibrium. In general, we may rewrite Eq. (10.4) as

$$\Delta G = \Delta G° + RT \ln Q_P \qquad (10.7)$$

where Q is the reaction quotient and $Q_P \neq K_P$ unless $\Delta G = 0$. The usefulness of Eq. (10.4) or Eq. (10.7) is that it tells us the direction of a spontaneous change if the concentrations of the reacting species are known. If $\Delta G°$ is a large positive or a large negative number (say 20 kJ or more), then the direction of the reaction (or the sign of ΔG) is primarily determined by $\Delta G°$ alone, unless either the reactants or the products are present in a much larger amount so that the $RT \ln Q_P$ term in Eq. (10.7) becomes comparable to $\Delta G°$

in magnitude but opposite in sign. If $\Delta G°$ is a small number, either positive or negative (say, 10 kJ or less), then the reaction can go either way.

EXAMPLE 10.2 At 298 K the partial pressures of gases in a reaction mixture described in Example 10.1 are $P_{N_2} = 0.250$ atm, $P_{H_2} = 0.550$ atm, and $P_{NH_3} = 0.950$ atm. Calculate ΔG for the reaction.

Answer: From Eq. (10.7), we write

$$\Delta G = -33280 \text{ J} + (8.314 \text{ J K}^{-1} \text{ mol}^{-1})(298 \text{ K}) \times \ln \frac{0.95^2}{(0.25)(0.55)^3}$$

$$= -25.7 \text{ kJ}$$

Since ΔG is negative, the reaction is spontaneous in the forward direction. The increase in NH_3 and decrease in N_2 and H_2 concentrations will continue until the reaction quotient becomes equal to the equilibrium constant K_P.

Comment: Note that in both Examples 10.1 and 10.2 we have expressed $\Delta G°$ and ΔG as kJ rather than kJ mol^{-1}. On the other hand, R has the units J K^{-1} mol^{-1}. Thus we simply ignore this "mol^{-1}" unit in calculating ΔG, $\Delta G°$, or K_P.

REAL GASES

What form would the equilibrium constant take if we are dealing with real gases? As we saw in Chapter 2, the behavior of real gases cannot be described by the ideal gas equation and must be treated by a more accurate equation of state such as the van der Waals equation. However, if we tried to calculate P using the van der Waals equation for every gas, say, and substitute this quantity in the equilibrium constant expression, the final form would look very unwieldy. Instead, we adopt a simpler procedure which is analogous to the discussion of activity in Chapter 8. We define for real gases a new variable called the *fugacity* (f) such that in every equation in which a partial pressure P_i appears, the fugacity f_i is used instead. Equation (8.5) applies only to an ideal gas; for a real gas we must write

$$\mu_i = \mu_i° + RT \ln f_i \tag{10.8}$$

Note that f_i is a dimensionless quantity in Eq. (10.8). For an ideal gas, the standard state is chosen at unit pressure. For a real gas, the standard state is chosen at unit fugacity. When the real gas behaves ideally, the fugacity and pressure are identical. But as the real gas becomes less and less ideal (for example, if the gas is being compressed), the observed pressure gets further and further from fugacity. In general, we have

$$\lim_{P_i \to 0} f_i = P_i$$

The fugacity coefficient γ_i of the ith component is given by

$$\gamma_i = \frac{f_i}{P_i} \tag{10.9}$$

and

$$\lim_{P_i \to 0} \gamma_i = 1$$

It is important to realize that pressure is a directly measurable quantity, but fugacity can only be calculated. The relationship between pressure and fugacity is given in Appendix 10.1.

Starting with Eq. (10.8), we can derive an equilibrium constant K_f for the hypothetical reaction discussed earlier:

$$K_f = \frac{f_C^c f_D^d}{f_A^a f_B^b} \tag{10.10}$$

Since $f = \gamma P$, Eq. (10.10) can be rewritten as

$$K_f = \frac{\gamma_C^c \gamma_D^d}{\gamma_A^a \gamma_B^b} \times \frac{P_C^c P_D^d}{P_A^a P_B^b} = K_\gamma K_P \tag{10.11}$$

K_f as defined by Eq. (10.10) or Eq. (10.11) is called the *thermodynamic equilibrium constant*, since it gives the exact result, while K_P is called the *apparent equilibrium constant*. For an ideal gas reaction, K_f is equal to K_P.

Reactions in Solution 10.2

The treatment of chemical equilibrium in solution is quite analogous to that in the gas phase. The concentrations of gases are usually most conveniently expressed in pressures; in a solution, the concentrations of reacting species are normally expressed in molality or molarity. We again start with the hypothetical reaction in solution:

$$aA + bB \rightleftharpoons cC + dD$$

where A, B, C, and D are the solutes. Assuming ideal behavior and expressing the solute concentrations in molalities, we have, from Eq. (8.19),

$$\mu_A = \mu_A^\ominus + RT \ln m_A \tag{10.12a}$$

$$\mu_B = \mu_B^\ominus + RT \ln m_B \tag{10.12b}$$

$$\mu_C = \mu_C^\ominus + RT \ln m_C \tag{10.12c}$$

$$\mu_D = \mu_D^\ominus + RT \ln m_D \tag{10.12d}$$

Following the same procedure as that employed for ideal gases in Section 10.1, we arrive at the following result:

$$\Delta G^\circ = -RT \ln K_m \tag{10.13}$$

where

$$K_m = \frac{m_C^c m_D^d}{m_A^a m_B^b} \tag{10.14}$$

If we express the solute concentrations in molarities, the equilibrium constant would then take the form

$$K_c = \frac{[C]^c [D]^d}{[A]^a [B]^b} \tag{10.15}$$

where the [ ] sign means mol liter^{-1}. Again, both K_m and K_c are dimensionless quantities, since each concentration term is divided by its standard-state value (1 m or 1 M). For reactions not at equilibrium, the Gibbs free energy change is given by

$$\Delta G = \Delta G° + RT \ln Q_c$$

where Q_c is the reaction quotient.

For nonideal solutions we must replace concentrations with activities. From Eq. (8.20), we write the chemical potential of the ith component as

$$\mu_i = \mu_i^{\ominus} + RT \ln a_i \qquad (10.16)$$

The substitution of activity for concentration is analogous to the substitution of fugacity for pressure. Starting with Eq. (10.16), we obtain the thermodynamic equilibrium constant K_a as

$$K_a = \frac{a_C^c a_D^d}{a_A^a a_B^b} \qquad (10.17)$$

Since $a = \gamma m$, Eq. (10.17) can be written as

$$K_a = \frac{\gamma_C^c \gamma_D^d}{\gamma_A^a \gamma_B^b} \times \frac{m_C^c m_D^d}{m_A^a m_B^b}$$

$$= K_\gamma K_m \qquad (10.18)$$

where K_m is the apparent equilibrium constant for a real solution reaction.

10.3 Heterogeneous Equilibria

So far we have concentrated on homogeneous equilibria, that is, reactions that occur in a single phase. Here we shall discuss chemical equilibria in heterogeneous systems.

Consider the thermal decomposition of calcium carbonate in a closed system:

$$CaCO_3(s) \rightleftharpoons CaO(s) + CO_2(g)$$

The two solids and one gas constitute three separate phases. The equilibrium constant of this reaction in molarities is given by

$$K_c' = \frac{[CaO][CO_2]}{[CaCO_3]}$$

However, it is a common practice not to include the concentrations of solids in the equilibrium-constant expression. The concentration of any pure solid is the ratio of the total number of moles present in the solid divided by the volume of the solid. If part of the solid is removed, the number of moles of the solid will decrease, but so will its volume. The same holds true for the addition of more solid substance—the increase in the number of moles of the solid is accompanied by an increase in volume. For this reason, the ratio of moles to volume always remains unchanged. Thus the amount of CO_2 and CaO produced is always the same, regardless of the amount of $CaCO_3$ employed in the beginning, as long as some of the solid is present at equilibrium. The equilibrium-constant expression given above can now be arranged as follows:

$$\frac{[CaCO_3]}{[CaO]} K_c' = [CO_2]$$

Since both $[CaCO_3]$ and $[CaO]$ are constants, every term on the left-hand side is a constant and we write

$$K_c = [CO_2]$$

where K_c, the "new" equilibrium constant, is given by $[CaCO_3]K_c'/[CaO]$. More conveniently, we can measure the pressure of CO_2 and obtain

$$K_P = P_{CO_2}$$

The equilibrium constants K_c and K_P are related to each other in a simple manner (see Problem 10.5).

The situation becomes easier to handle if we write the thermodynamic equilibrium constant instead of the apparent equilibrium constant. Replacing the concentrations with activities, we have

$$K_a = \frac{a_{CaO}a_{CO_2}}{a_{CaCO_3}}$$

By convention, the activities of pure solids (and pure liquids) in their standard states (that is, at 1 atm) are equal to unity; that is,

$$a_{CaO} = 1 \quad \text{and} \quad a_{CaCO_3} = 1$$

so that

$$K_a = f_{CO_2}$$

or

$$K_P = P_{CO_2}$$

if ideal gas behavior is assumed. Figure 10.1 shows the equilibrium pressure of CO_2 over $CaCO_3$ as a function of temperature. Because liquids and solids are normally not influenced by changes in pressure, we shall always assume that their activities are unity in all calculations.

E X A M P L E 1 0 . 3 Calculate the equilibrium constant for the following reaction at 298 K using the data listed in Appendix 3:

$$Na(s) + \tfrac{1}{2}Cl_2(g) \Longrightarrow NaCl(s)$$

Equilibrium pressure of carbon dioxide over calcium carbonate as a function of temperature.

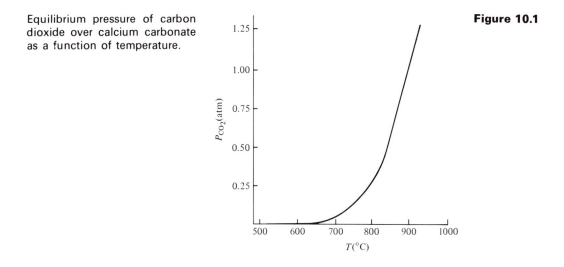

Figure 10.1

Answer: The thermodynamic equilibrium constant is given by

$$K = \frac{a_{NaCl}}{a_{Na} f_{Cl_2}^{1/2}} = \frac{1}{f_{Cl_2}^{1/2}} \simeq \frac{1}{P_{Cl_2}^{1/2}}$$

Since gases usually behave ideally at pressures equal to or smaller than 10 atm, we have replaced fugacity with pressure. From Appendix 3 we have

$$\Delta \overline{G}_f^\circ(Na) = 0 \qquad \Delta \overline{G}_f^\circ(Cl_2) = 0 \qquad \Delta \overline{G}_f^\circ(NaCl) = -384.03 \text{ kJ mol}^{-1}$$

Thus from Eq. (10.5), we write

$$-384,030 \text{ J} = -(8.314 \text{ J K}^{-1} \text{ mol}^{-1})(298 \text{ K}) \ln K_P$$

$$\ln K_P = 155$$

or

$$K_P = 2.02 \times 10^{67}$$

Comment: Note that the measurement of the pressure of chlorine at equilibrium gives K_P and hence also $\Delta \overline{G}_f^\circ(NaCl)$. In the thermal decomposition of $CaCO_3$ example, the measurement of the pressure of carbon dioxide gives K_P and hence also $\Delta \overline{G}_f^\circ(CO_2)$. In simple systems such as these we can often obtain the $\Delta \overline{G}_f^\circ$ values of compounds from equilibrium-constant measurements. For more complex systems, $\Delta \overline{G}_f^\circ$ can be obtained from $\Delta \overline{S}^\circ$ and $\Delta \overline{H}_f^\circ$ values.

10.4 Influence of Temperature, Pressure, and Catalyst on the Equilibrium Constant

EFFECT OF TEMPERATURE

Although Eq. (10.5) relates the change in the standard Gibbs free energy to the equilibrium constant at *any* temperature, normally we find it most convenient to calculate K_P at 298 K. In practice, however, a reaction may be run at temperatures other than 298 K so that we either have to know the ΔG° values at that particular temperature or find some other way to calculate K_P. The question we ask is: If we know the equilibrium constant of a reaction K_1 at temperature T_1, can we calculate the equilibrium constant of the same reaction K_2 at temperature T_2? The answer is yes.

A very useful equation relating the equilibrium constant to temperature can be derived as follows. Since

$$\Delta G^\circ = -RT \ln K$$

and

$$\Delta G^\circ = \Delta H^\circ - T \Delta S^\circ$$

combination of the two equations above gives

$$\ln K = -\frac{\Delta H^\circ}{RT} + \frac{\Delta S^\circ}{R} \tag{10.19}$$

Equation (10.19) is known as *van't Hoff's equation*. Thus from a plot of $\ln K$ versus $1/T$, we obtain a straight line with a slope (which is negative) equal to

$-\Delta H^\circ/R$, and the intercept on the ordinate gives $\Delta S^\circ/R$. This is a convenient way of obtaining ΔH° and ΔS°, the standard enthalpy and standard entropy changes of the reaction. Note that this plot will yield a straight line only if both ΔH° and ΔS° are independent of temperature. For small temperature ranges (say 50 K or less), this is a fairly good approximation. If we consider only two temperatures T_1 and T_2, then Eq. (10.19) can be written as

245

Sec. 10.4

Influence of
Temperature,
Pressure, and
Catalyst on the
Equilibrium Constant

$$\ln K_1 = -\frac{\Delta H^\circ}{RT_1} + \frac{\Delta S^\circ}{R} \tag{10.20a}$$

$$\ln K_2 = -\frac{\Delta H^\circ}{RT_2} + \frac{\Delta S^\circ}{R} \tag{10.20b}$$

Equation (10.20a) minus Eq. (10.20b) gives

$$\ln \frac{K_1}{K_2} = \frac{\Delta H^\circ}{R}\left(\frac{1}{T_2} - \frac{1}{T_1}\right)$$

$$= \frac{\Delta H^\circ}{R}\left(\frac{T_1 - T_2}{T_1 T_2}\right) \tag{10.21}$$

Equation (10.21) now enables us to calculate K_2 if we know the values of K_1, T_1, T_2, and ΔH°.

Some interesting observations can be made from Eq. (10.21). If the reaction is endothermic in the forward direction (that is, ΔH° is a positive quantity) and assuming that $T_2 > T_1$, we see that the quantity in the right-hand side of Eq. (10.21) is negative, which must mean that $K_2 > K_1$. On the other hand, if the reaction is exothermic in the forward direction (that is, ΔH° is a negative quantity), then the quantity on the right-hand side of Eq. (10.21) is positive, or $K_2 < K_1$. Thus we can conclude that an increase in temperature shifts the equilibrium from left to right (favors the formation of products) in an endothermic reaction and shifts the equilibrium from right to left (favors the formation of reactants) in an exothermic reaction. These results, which may be familiar to you, were probably stated in a nonquantitative manner as Le Châtelier's principle in an introductory chemistry course.

EFFECT OF PRESSURE

The question of whether the equilibrium constant can be changed by altering the pressure can be dealt with by examining Eq. (10.5). We see that the equilibrium constant K_P is related to ΔG°. However, since ΔG° is a property that is defined for the reacting species at a specific pressure (1 atm), it does not vary when the pressure of the experiment is changed. The apparent equilibrium constant K_P is independent of pressure if the reaction involves ideal gases. For a reaction involving real gases, the apparent equilibrium constant does depend on pressure. On the other hand, the thermodynamic equilibrium constant K_f is independent of pressure for reactions involving either ideal or real gases. At constant temperature T we write, for a reaction involving ideal gases:

$$\left(\frac{\partial K_P}{\partial P}\right)_T = 0$$

Just because K_P is not affected by pressure does not mean that the amounts of various gases at equilibrium do not change with pressure. To illustrate this point, let us consider the gas-phase reaction at equilibrium:

$$A(g) \rightleftharpoons 2B(g)$$
$$n(1 - \alpha) \qquad 2n\alpha$$

where n is the number of moles of A originally present and α is the fraction of the A molecules dissociated. The total number of moles of molecules present is $n(1 + \alpha)$ and the mole fraction of A and B are

$$X_A = \frac{n(1 - \alpha)}{n(1 + \alpha)} = \frac{1 - \alpha}{1 + \alpha}$$

$$X_B = \frac{2n\alpha}{n(1 + \alpha)} = \frac{2\alpha}{1 + \alpha}$$

and the partial pressures of A and B are (assuming ideal behavior)

$$P_A = \left(\frac{1 - \alpha}{1 + \alpha}\right)P$$

$$P_B = \left(\frac{2\alpha}{1 + \alpha}\right)P$$

where P is the total pressure of the system. The equilibrium constant is given by

$$K_P = \frac{P_B^2}{P_A} = \left(\frac{4\alpha^2}{1 - \alpha^2}\right)P$$

Rearranging the last equation, we obtain

$$\alpha = \sqrt{\frac{K_P}{K_P + 4P}}$$

Because K_P is a constant, the value of α depends only on P. If P is large, α is small; and if P is small, α is large. These predictions again remind us of Le Châtelier's principle. If the "stress" applied to the system is an increase in pressure, then the equilibrium will shift to the side that produces fewer molecules, which is right to left in our example, and hence α decreases. The reverse holds true for a decrease in pressure.

EFFECT OF A CATALYST

By definition, a catalyst is a substance that can speed up the rate of a reaction without itself being used up. Will the addition of a catalyst to a reacting system at equilibrium shift the equilibrium in a particular direction? To answer this question, let us again consider the gas-phase equilibrium

$$A(g) \rightleftharpoons 2B(g)$$

Suppose that there exists a catalyst which favors the reverse reaction $2B \rightarrow A$ but not the forward reaction ($A \rightarrow 2B$). We can then construct an apparatus such as that shown in Figure 10.2. A small box is placed inside a cylinder fitted with a movable piston. The cover of the box is connected to the piston

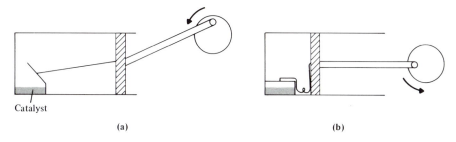

Figure 10.2

Catalyst

(a)   (b)

Perpetual-motion machine operating with a hypothetical catalyst capable of shifting the equilibrium position of a gaseous reaction in only one direction.

by a piece of string, so that the cover can be closed and opened by the movement of the piston. We start with the equilibrium mixture of gases A and B in the cylinder and then add the catalyst to the box (Figure 10.2a). Immediately, the equilibrium shifts toward the formation of A. Since two molecules of B are consumed to produce one molecule of A, this step decreases the total number of molecules present in the cylinder and hence the internal gas pressure. Consequently, the piston will be pushed inward by the external pressure until the internal and external pressures are again balanced. At this stage, the box lid drops (Figure 10.2b). Without the catalyst, the gases gradually return to their original concentrations and the increase in the number of molecules as a result of the formation of B pushes the piston from left to right until the cover is lifted. The whole process will then repeat itself. There is, as you may have noticed, something strange about the whole arrangement. The piston can be made to perform work even though there is no energy input or net consumption of chemicals. Such a device is known as a perpetual-motion machine, which is impossible in practice because it violates the law of conservation of energy.

The reason that such a machine can never be realized is that there are no known catalysts which can speed up the rate of a reaction in one direction without affecting the rate of the reverse reaction in a similar manner. The important conclusion from this simple illustration is that a catalyst cannot shift the position of an equilibrium. For a reaction mixture not at equilibrium, the presence of a catalyst will speed up both the forward and reverse rate processes so that we will reach the equilibrium state sooner, but the same equilibrium state will eventually be attained even if no catalyst has been added.

Binding of Ligands and Metal Ions to Macromolecules 10.5

In this section we apply the equilibrium treatment to the study of binding of ligands and metal ions to macromolecules in solution. We concentrate on the two cases in which the macromolecule possesses one binding site per molecule and *n equivalent* binding sites per molecule. In Chapter 14 we shall deal with the more complicated systems in which each macromolecule possesses *n* binding sites that are not equivalent. Because our approach here is strictly thermodynamic, we need not be concerned with details such as the structure of macromolecules and the nature of covalent and other intermolecular forces responsible for the binding.

This is the simplest case in which one site of a macromolecule P binds one molecule of ligand (or one ion) L. The reaction can be represented as

$$P + L \rightleftharpoons PL$$

The equilibrium constant for this association reaction is*

$$K_a = \frac{[PL]}{[P][L]}$$

Frequently, we find it more convenient to refer to such an equilibrium process in terms of the dissociation constant K_d:

$$K_d = \frac{[P][L]}{[PL]} \tag{10.22}$$

Thus the smaller the value of K_d, the "tighter" is the complex PL. These two equilibrium constants are related by the simple relation $K_a K_d = 1$. We define a quantity Y such that

$$Y = \frac{\text{concentration of L bound to P}}{\text{total concentration of all forms of P}}$$

$$= \frac{[PL]}{[P] + [PL]} \tag{10.23}$$

From Eq. (10.22), we have

$$[PL] = \frac{[P][L]}{K_d}$$

so that

$$Y = \frac{[P][L]/K_d}{[P] + [P][L]/K_d}$$

$$= \frac{[L]}{[L] + K_d} \tag{10.24}$$

The quantity Y is sometimes called the "fractional saturation of sites"; its value ranges from zero (when $[PL] = 0$) to 1 (when $[P] = 0$). For example, when $Y = 0.5$, half of the P molecules are complexed with L and half of the P molecules are in the free form and $[L] = K_d$.

The quantity of interest is, of course, K_d. By taking the reciprocal of Eq. (10.24), we obtain

$$\frac{1}{Y} = 1 + \frac{K_d}{[L]} \tag{10.25}$$

A plot of $1/Y$ versus $1/[L]$ gives a straight line of slope K_d. Alternatively, Eq. (10.24) can be rearranged to give

$$\frac{Y}{[L]} = \frac{1}{K_d} - \frac{Y}{K_d} \tag{10.26}$$

In this case a plot of $Y/[L]$ versus Y gives a straight line whose slope (which is

---

* We assume ideal behavior and employ concentrations instead of activities.

negative) is $-1/K_d$. Thus if we can determine the amount of L bound to P (at given total concentrations of P and L), we can evaluate the dissociation constant. The experimental procedure for this determination will be discussed shortly.

n EQUIVALENT BINDING SITES PER MACROMOLECULE

We now consider the case in which a macromolecule has n equivalent sites; that is, each binding site has the same K_d value, regardless of whether other sites on the same molecule are occupied. We shall first take $n = 2$ and then generalize the result to cases where $n > 2$.

If a macromolecule has two equivalent binding sites, there will be two binding equilibria:

$$P + L \rightleftharpoons PL \qquad K_1 = \frac{[P][L]}{[PL]}$$

$$PL + L \rightleftharpoons PL_2 \qquad K_2 = \frac{[PL][L]}{[PL_2]}$$

where K_1 and K_2 are the dissociation constants. We define a quantity Y such that

$$Y = \frac{\text{concentration of L bound to P}}{\text{total concentration of all forms of P}}$$

$$= \frac{[PL] + 2[PL_2]}{[P] + [PL] + [PL_2]} \tag{10.27}$$

Note that since PL_2 has two molecules of L bound to one molecule of P, its concentration must be multiplied by 2. Since

$$[PL] = \frac{[P][L]}{K_1} \qquad \text{and} \qquad [PL_2] = \frac{[PL][L]}{K_2}$$

Equation (10.27) can now be written as

$$Y = \frac{[P][L]/K_1 + 2[P][L]^2/K_1 K_2}{[P] + [P][L]/K_1 + [P][L]^2/K_1 K_2}$$

$$= \frac{[L]/K_1 + 2[L]^2/K_1 K_2}{1 + [L]/K_1 + [L]^2/K_1 K_2} \tag{10.28}$$

Equation (10.28) can be simplified by realizing that since the two sites are independent and equivalent, K_1 and K_2 are related to each other by a statistical factor. The ligand L can dissociate from the PL_2 complex in two ways (that is, L can come off from either of the two sites), but L can associate with the vacant site in the PL complex in only one way. The general relationship is that the ith dissociation constant K_i is given by

$$K_i = \left(\frac{i}{n - i + 1}\right) K \tag{10.29}$$

where K is an *intrinsic dissociation constant*. For our case here $i = 1, 2$ and $n = 2$ so that

$$K_1 = \frac{K}{2} \qquad \text{and} \qquad K_2 = 2K$$

Thus purely on the basis of statistical consideration, we find that the first dissociation constant is four times as small as the second dissociation constant, that is, $K_2 = 4K_1$. Note that K as defined takes into account the statistical relationship between K_1 and K_2. It is actually the geometric mean of K_1 and K_2, that is,

$$K = \sqrt{K_1 K_2} \qquad (10.30)$$

Equation (10.28) can now be written as

$$
\begin{aligned}
Y &= \frac{2[L]/K + 2[L]^2/K^2}{1 + 2[L]/K + [L]^2/K^2} \\
&= \frac{2[L]/K(1 + [L]/K)}{(1 + [L]/K)^2} = \frac{2[L]}{[L] + K}
\end{aligned} \qquad (10.31)
$$

Equation (10.31) is the result obtained for two equivalent sites. In general, for n equivalent sites we have

$$Y = \frac{n[L]}{[L] + K} \qquad (10.32)$$

Equation (10.32) can be rearranged into several forms suitable for graphical treatment. Three of the most common procedures will now be discussed.

1. The Direct Plot. Figure 10.3 shows a plot of Y versus the ligand concentration. A direct plot of this type yields a hyperbolic curve which is characteristic of simple binding (that is, the binding sites are all equivalent and noninteracting). When $[L] = K$, we have $Y = n/2$ and at very high ligand concentrations, we can assume that $[L] \gg K$, so that $Y = n$. However, the direct plot is generally not very useful in determining n and K, since it is often difficult to determine the asymptotic value n at very high ligand concentrations. (Note that we need to know n to determine K.)

2. The Double Reciprocal Plot. By taking the reciprocal of each side of Eq. (10.32) we obtain

$$\frac{1}{Y} = \frac{1}{n} + \frac{K}{n[L]} \qquad (10.33)$$

Figure 10.3

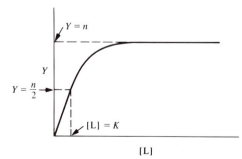

Plot of fractional saturation versus ligand concentration.

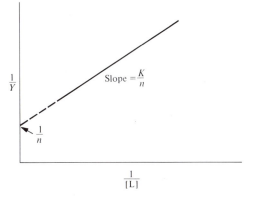

Plot of $1/Y$ versus $1/[L]$. This plot is also known as the Hughes–Klotz plot.

Figure 10.4

Thus a plot of $1/Y$ versus $1/[L]$ gives a straight line of slope K/n and an intercept on the y axis of $1/n$ (Figure 10.4). This plot is also known as the Hughes–Klotz plot.

3. The Scatchard Plot. Starting with Eq. (10.32), we have

$$Y[L] + KY = n[L]$$

$$\frac{Y[L]}{K} + Y = \frac{n[L]}{K}$$

$$Y = \frac{n[L]}{K} - \frac{Y[L]}{K}$$

or

$$\frac{Y}{[L]} = \frac{n}{K} - \frac{Y}{K} \tag{10.34}$$

Equation (10.34) is known as the Scatchard equation. Thus a plot of $Y/[L]$ versus Y gives a straight line of slope $-1/K$ and an intercept on the x axis of n (Figure 10.5).

It is useful to keep in mind that the treatment presented here is quite general and applies to all types of binding phenomena. In Chapter 14 we shall see applications of some of these equations to enzyme kinetics.

Plot of $Y/[L]$ versus Y. This plot is also known as the Scatchard plot.

Figure 10.5

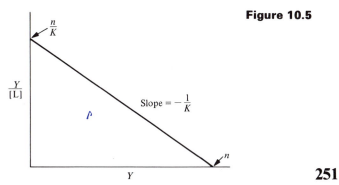

Having discussed the theoretical aspects of binding of ligands to macromolecules, we now describe one particular experimental arrangement for the determination of n and K—*equilibrium dialysis*.

Dialysis is the process of removing small ions and other solute molecules from a protein solution. Suppose that in an experiment we have precipitated hemoglobin from a solution by the salting-out technique using ammonium sulfate (see Section 9.4). The protein can be freed of the $(NH_4)_2SO_4$ salt as follows. First, the precipitate is dissolved in water or, more frequently, in a buffer solution. The protein solution is then placed in a cellophane bag, which, in turn, is immersed in a beaker containing the same buffer (Figure 10.6). Since both the NH_4^+ and SO_4^{2-} ions are small enough to diffuse through the membrane but not the protein molecules, the ions in the bag will begin to enter into the outside solution as a result of the unequal chemical potentials:

$$(\mu_{NH_4^+})_{inside} > (\mu_{NH_4^+})_{outside}$$

$$(\mu_{SO_4^{2-}})_{inside} > (\mu_{SO_4^{2-}})_{outside}$$

This process continues until the chemical potentials of each type of ions become equal inside and outside of the bag and an equilibrium is established. It is possible to remove all the $(NH_4)_2SO_4$ by continually changing the buffer solution in the beaker.

The procedure described above may be reversed to study the binding of ions or small ligands to proteins. In this case we begin by placing the *pure* protein solution in the cellophane bag (called phase 1), which is then immersed in a buffer solution (called phase 2) that contains the ligands (L) of known concentration. At equilibrium, the chemical potentials of the free (unbound) ligands in both phases must be the same (Figure 10.7) so that

$$(\mu_L)_1^{unbound} = (\mu_L)_2^{unbound}$$

or

$$(\mu^\ominus + RT \ln a_L)_1^{unbound} = (\mu^\ominus + RT \ln a_L)_2^{unbound}$$

Since the standard chemical potential $\mu^\ominus$ is the same, we have

$$(a_L)_1^{unbound} = (a_L)_2^{unbound}$$

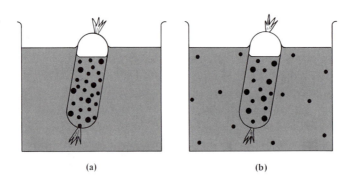

(a)   (b)

Figure 10.6 Arrangement for dialysis. The small dots denote ions and the large dots proteins. (a) At the start of dialysis. (b) At equilibrium, most of the small ions have diffused out of the cellophane bag. By repeatedly replacing the buffer solution in the beaker, it is possible to remove all of the small ions not bound to the protein.

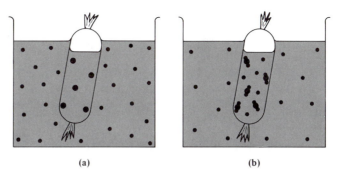

(a)   (b)

Arrangement for studying equilibrium dialysis. The small dots denote ligands and the large dots proteins. (a) Initially, a cellophane bag containing a protein solution is immersed in a buffer solution containing the ligand molecules. (b) At equilibrium, some of the ligand molecules have diffused into the bag and complexed with the protein molecules.

Figure 10.7

If the solutions are dilute, we can replace activities with concentrations so that

$$[L]_1^{unbound} = [L]_2^{unbound}$$

However, the total concentration of the ligands inside the bag is given by

$$[L]_1^{total} = [L]_1^{bound} + [L]_1^{unbound}$$

Therefore, the concentration of ligands bound to protein molecules is

$$[L]_1^{bound} = [L]_1^{total} - [L]_1^{unbound}$$

The first quantity on the right-hand side of the last equation can be determined by analyzing the solution in the bag *after* its removal from the beaker; the second quantity is obtained by measuring the concentration of the ligands in the solution remaining in the beaker. (Remember that the concentrations of the unbound ligands are equal in phases 1 and 2.) We can now see how both the intrinsic dissociation constant K and the number of binding sites n can be obtained by using the equilibrium dialysis technique. Suppose that we start with a protein solution of known concentration in phase 1 and a ligand solution of known concentration in phase 2. At equilibrium, the quantity Y is given by [see Eq. (10.27)]

$$Y = \frac{[L]_1^{bound}}{[P]^{total}}$$

where $[P]^{total}$ is the concentration of the original protein solution. The experiment can be repeated by using different concentrations for the protein and ligand solutions and the values of K and n can be determined from either the Hughes–Klotz plot or the Scatchard plot. Keep in mind that the quantity [L] in Eqs. (10.33) and (10.34) refers to the concentration of unbound ligands, which is $[L]_1^{unbound}$ in our case.

The equilibrium dialysis technique has been successfully employed in the binding of drugs, hormones, and other small molecules to proteins and nucleic acids. It is useful to keep in mind that our discussion implicitly

253

assumes that the ligand is a nonelectrolyte so that the concentrations of unbound ligands are equal in phases 1 and 2 at equilibrium. If the ligand is an electrolyte, we must apply the Donnan equilibrium condition for treating the dialysis data.

10.6 Bioenergetics

Bioenergetics is the study of energy transformation in living systems. As scientists are beginning to understand many biological and biochemical phenomena at the molecular level, they are learning to apply thermodynamics to the study of living systems. This section will examine the energetics of biochemical reactions, using glycolysis as an example. In particular, we shall be concerned with the role of adenosine-5'-triphosphate (ATP) in these processes.

BIOCHEMISTS' STANDARD STATE

Before we begin discussing thermodynamics, it is important to first distinguish between the *standard states* employed by physical chemists and biochemists. In physical chemistry, the standard state refers to the situation in which all the reactants and products are at unit molar (or molal) concentration. In biochemistry, we follow the same procedure except that we define the hydrogen-ion concentration for the standard state to be 10^{-7} M because the physiological pH is about 7. Consequently, the change in the standard Gibbs free energy according to these two conventions will be different for reactions involving uptake or liberation of hydrogen ions. We shall therefore replace ΔG° with $\Delta G^{\circ\prime}$ in discussing biochemical processes. Consider the reaction

$$A + B \longrightarrow C + xH^+$$

The Gibbs free energy change (ΔG) for the process is given by

$$\Delta G = \Delta G^\circ + RT \ln \frac{\dfrac{[C]}{1\ M}\left(\dfrac{[H^+]}{1\ M}\right)^x}{\dfrac{[A]}{1\ M}\dfrac{[B]}{1\ M}}$$

where 1 M represents the physical chemists' standard state of solute in solution.* Since the biochemists' standard state for H^+ ions is 10^{-7} M, the Gibbs free energy change for the same process is now given by

$$\Delta G = \Delta G^{\circ\prime} + RT \ln \frac{\dfrac{[C]}{1\ M}\left(\dfrac{[H^+]}{10^{-7}\ M}\right)^x}{\dfrac{[A]}{1\ M}\dfrac{[B]}{1\ M}}$$

Note that regardless of the convention for the standard state, ΔG must remain unchanged. From the last two equations we obtain

$$\Delta G^\circ = \Delta G^{\circ\prime} + xRT \ln \frac{1}{10^{-7}}$$

* We assume ideal behavior and use concentrations instead of activities.

If $x = 1$ and $T = 298$ K, then

$$\Delta G° = \Delta G°' + 40.0 \text{ kJ}$$

This means that for reactions producing H^+ ions, $\Delta G°$ is greater than $\Delta G°'$ by 40.0 kJ per mole of H^+ ions released. Hence the reaction is more spontaneous at pH 7 than at pH 0. On the other hand, if H^+ ion appears as the reactant,

$$C + xH^+ \longrightarrow A + B$$

we can show that

$$\Delta G° = \Delta G°' - 40.0 \text{ kJ}$$

and this reaction will be more spontaneous at pH 0 than at pH 7. For reactions not involving H^+ ions, $\Delta G°$ is equal to $\Delta G°'$.

EXAMPLE 10.4 NAD^+ and NADH are the oxidized and reduced forms of nicotinamide adenine dinucleotide. The $\Delta G°$ for the oxidation of NADH,

$$NADH + H^+ \longrightarrow NAD^+ + H_2$$

is -21.8 kJ at 298 K. Calculate $\Delta G°'$, K, and K' for the reaction, where $\Delta G°' = -RT \ln K'$. Also calculate the Gibbs free energy change using both the physical chemical standard state and biochemical standard state for the reaction when $[NADH] = 1.5 \times 10^{-2}$ M, $[H^+] = 3.0 \times 10^{-5}$ M, $[NAD^+] = 4.6 \times 10^{-3}$ M, and $p_{H_2} = 0.010$ atm.

Answer: Since H^+ ion appears as the reactant, we write

$$\Delta G° = \Delta G°' - 40.0 \text{ kJ}$$

From the relation $\Delta G° = -RT \ln K$,

$$-21,800 \text{ J mol}^{-1} = -(8.314 \text{ J K}^{-1} \text{ mol}^{-1})(298 \text{ K})(2.303) \log K$$

Hence

$$\log K = 3.82$$

$$K = 6600$$

Similarly, from $\Delta G°' = -RT \ln K'$,

$$18,200 \text{ J mol}^{-1} = -(8.314 \text{ J K}^{-1} \text{ mol}^{-1})(298 \text{ K})(2.303) \log K'$$

$$\log K' = -3.19$$

$$K' = 6.46 \times 10^{-4}$$

Thus*

$$\frac{K}{K'} = 10^7$$

This ratio corresponds to the difference in the two standard states for $[H^+]$.

As mentioned earlier, ΔG for the reaction should be the same regardless of what standard state we employ.

* This ratio would be much closer to 10^7 if we had used -39.93 kJ as the difference between $\Delta G°$ and $\Delta G°'$.

Physical Chemists' Standard State

$$\Delta G = \Delta G^\circ + RT \ln \frac{([NAD^+]/1\ M)(p_{H_2}/1\ atm)}{([NADH]/1\ M)([H^+]/1\ M)}$$

$$= -21{,}800\ J + (8.314\ J\ K^{-1}\ mol^{-1})(298\ K)(2.303)$$

$$\times \log \frac{(4.6 \times 10^{-3})(0.010)}{(1.5 \times 10^{-2})(3 \times 10^{-5})}$$

$$= -21{,}800 + 11{,}500 = -10.3\ kJ$$

Biochemists' Standard State

$$\Delta G = \Delta G^{\circ\prime} + RT \ln \frac{([NAD^+]/1\ M)(p_{H_2}/1\ atm)}{([NADH]/1\ M)([H^+]/10^{-7}\ M)}$$

$$= 18{,}200\ J + (8.314\ J\ K^{-1}\ mol^{-1})(298\ K)(2.303)$$

$$\times \log \frac{(4.6 \times 10^{-3})(0.010)}{(1.5 \times 10^{-2})(3 \times 10^{-5})/10^{-7}}$$

$$= 18{,}200 - 28{,}500 = -10.3\ kJ$$

ATP—THE CURRENCY OF ENERGY

Adenosine-5'-triphosphate (Figure 10.8) is the primary energy source for numerous biological reactions, ranging from protein synthesis and ion transport to muscle contraction and electrical activities in nerve cells. The energy required to carry out these processes is derived from the hydrolysis reaction at pH 7:

$$ATP^{4-} + H_2O \longrightarrow ADP^{3-} + P_i + H^+$$

where P_i denotes the inorganic phosphate HPO_4^{2-}. This reaction is accompanied by a decrease in the standard Gibbs free energy of as much as 25 to 40 kJ per mole of ATP hydrolyzed. The exact value depends on the pH, temperature, and the counter metal ions present. One of the most accurately studied systems is the Mg–ATP complex,* which, at pH = 7 and $T = 310$ K, has the value $\Delta G^\circ = -30.5$ kJ. The hydrolysis reaction may proceed further as follows:

$$ADP^{3-} + H_2O \longrightarrow AMP^{2-} + P_i + H^+ \qquad \Delta G^{\circ\prime} \simeq -30\ kJ$$

$$AMP^{2-} + H_2O \longrightarrow adenosine + P_i \qquad \Delta G^{\circ\prime} \simeq -14\ kJ$$

However, for most cases we are only concerned with the hydrolysis of ATP to ADP.

The large decrease in the Gibbs free energy as a result of the hydrolysis of ATP and ADP prompted some biochemists to use the term *high-energy phosphate bond* to describe these compounds. The usage of this term was unfortunate because it implied that the P—O bond in these molecules is somehow different from the normal covalent bond, which is untrue. Then why such a large negative $\Delta G^{\circ\prime}$? To answer this question, we must examine the structures

---

* R. A. Alberty, *J. Biol. Chem.* **243**, 1337 (1968). This work is presented at a lower level by the same author in *J. Chem. Educ.* **46**, 713 (1969).

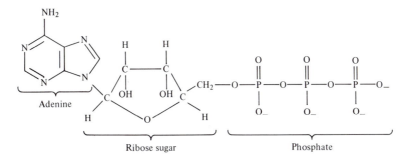

Structure of adenosine-5′-triphosphate (ATP). Upon hydrolysis, ATP loses the end phosphate group to form adenosine-5′-diphosphate (ADP). ADP may undergo further hydrolysis to form adenosine-5′-monophosphate (AMP).

Figure 10.8

of ATP and its hydrolysis products, ADP and P_i, since $\Delta G^{\circ\prime}$ depends on the *difference* in the standard Gibbs free energies of the products and reactants. At least two factors must be taken into consideration: electrostatic repulsion and resonance stabilization. At pH 7 the triphosphate unit of ATP carries four negative charges:

$$\text{adenine}-\text{ribose}-\text{O}-\overset{\overset{\text{O}}{\|}}{\underset{\underset{\text{O}_-}{|}}{\text{P}}}-\text{O}-\overset{\overset{\text{O}}{\|}}{\underset{\underset{\text{O}_-}{|}}{\text{P}}}-\text{O}-\overset{\overset{\text{O}}{\|}}{\underset{\underset{\text{O}_-}{|}}{\text{P}}}-\text{O}^-$$

As a result of the proximity of the charges there is considerable electrostatic repulsion. This repulsion is reduced when ATP is hydrolyzed to ADP and P_i (there are only three negative charges on ADP). The other factor contributing to the large $\Delta G^{\circ\prime}$ value is that ADP and P_i possess more resonance structures than ATP. For example, the P_i group has a number of resonance structures of similar energy:

$$\overset{\overset{\text{O}^-}{|}}{\underset{\underset{\text{O}_-}{|}}{\text{HO}-\text{P}^+}}-\text{O}^- \longleftrightarrow \overset{\overset{\text{O}}{\|}}{\underset{\underset{\text{O}_-}{|}}{\text{HO}-\text{P}}}-\text{O}^- \longleftrightarrow \overset{\overset{\text{O}^-}{|}}{\underset{\underset{\text{O}_-}{|}}{\text{HO}-\text{P}}}=\text{O} \longleftrightarrow \overset{\overset{\text{O}^-}{|}}{\underset{\underset{\text{O}}{\|}}{\text{HO}-\text{P}}}-\text{O}^-$$

On the other hand, the terminal portion of ATP has fewer significant resonance structures per phosphate group. The following resonance structure for ATP is improbable because one of the oxygen atoms has three bonds and the fact that there is a positive charge on the oxygen atom adjacent to a positively charged phosphorus atom:

$$\text{adenine}-\text{ribose}-\text{O}-\overset{\overset{\text{O}}{\|}}{\underset{\underset{\text{O}_-}{|}}{\text{P}}}-\text{O}-\overset{\overset{\text{O}^-}{|}}{\underset{\underset{\text{O}_-}{|}}{\text{P}^+}}-\text{O}^+=\overset{\overset{\text{O}^-}{|}}{\underset{\underset{\text{O}_-}{|}}{\text{P}}}-\text{O}^-$$

Finally, to a less extent, the release of the steric crowding among oxygen atoms on adjacent phosphate groups in ATP upon hydrolysis can also contribute to the large decrease in the standard Gibbs free energy.

As Table 10.1 shows, the hydrolysis of ATP is by no means the most exergonic. However, the significance of its relative position among other phosphates will be mentioned shortly.

**Standard Free Energies
of Hydrolysis of Some Phosphate
Compounds at pH 7[a]**

Table 10.1

Phosphates	$\Delta G^{\circ\prime}$ (kJ mol^{-1})
Phosphoenolpyruvate	−61.9
Acetyl phosphate	−43.1
Creatine phosphate	−43.1
Pyrophosphate	−33.5
ATP	−30.5
Glucose 1-phosphate	−20.9
Glucose 6-phosphate	−13.8
Glycerol 1-phosphate	−9.2

[a] From H. A. Sober, ed., *Handbook of Biochemistry*, © The Chemical Rubber Co., 1968. Used by permission of The Chemical Rubber Co.

Figure 10.9

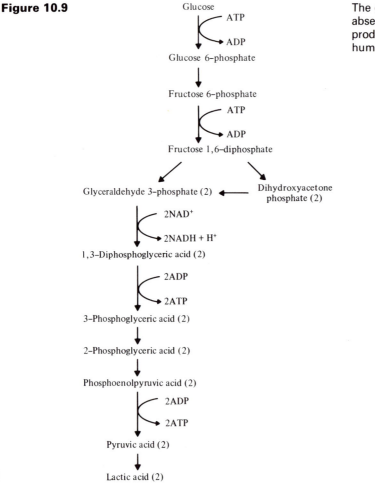

The glycolytic pathway. In the absence of oxygen, the end product is lactic acid in the human body.

All living organisms require energy for growth and function. Plants derive their energy from the sun through the process of photosynthesis. They are called *autotrophs* because they contain self-feeding cells. We, as *heterotrophs*, survive by feeding on others. The food we eat consists mostly of carbohydrates, proteins, and fats. It is through stepwise oxidation that we obtain the necessary energy stored in these molecules. The carbohydrates, for example, serve two purposes: They provide the building blocks for biosynthesis and they produce energy through oxidation. We now turn our attention to energy production.

Figure 10.9 shows the metabolic pathway from glucose to pyruvic acid. A total of nine steps are involved in breaking a six-carbon atom molecule, glucose, to two three-carbon atom molecules, pyruvic acid.* Each of the steps is enzymatically catalyzed. We see that ATP is utilized in some steps and synthesized in others. However, there is a net gain of 2 mol of ATP per mole of glucose metabolized to pyruvic acid.

The first step along the glycolytic pathway involves the conversion of glucose to glucose 6-phosphate:

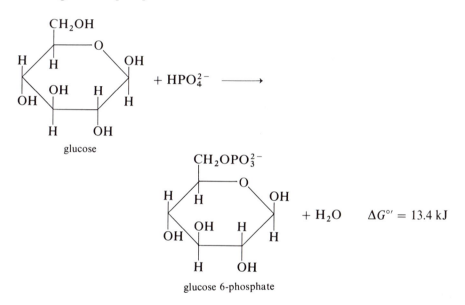

Since this is an endergonic process, the reaction will not be spontaneous if the reactants and products are in their standard states. However, this reaction can be driven by coupling it with a second reaction, which is exergonic, that is, the hydrolysis of ATP. Thus we can write a *coupled* reaction as follows:

$$\text{glucose} + \text{ATP} \longrightarrow \text{glucose-6-phosphate} + \text{ADP} \qquad \Delta G^{\circ\prime} = -17.2 \text{ kJ}$$

This reaction will now take place spontaneously because $\Delta G^{\circ\prime}$ is a large, negative quantity.

The preceding reaction serves to illustrate the importance of coupled reactions in biological systems. Essentially, we have a reaction that is energetically unfavorable but which is necessary for metabolism. It can be made to proceed by connecting it with another reaction, which releases a large amount of Gibbs

---

* For a detailed discussion of glycolysis, the reader is referred to standard texts in biochemistry.

free energy. It must be realized that these two reactions do not take place separately; it is the presence of the enzyme glucokinase (or hexokinase) that makes the coupling possible.

Another coupled reaction takes place from fructose 6-phosphate to fructose 1,6-diphosphate. Again, 1 mol of ATP is hydrolyzed for each mole of fructose 6-phosphate phosphorylated.

The investment of ATP for these reactions is more than compensated for by the subsequent steps, resulting in a total synthesis of 4 mol of ATP; hence there is a net gain of 2 mol of ATP. At pH 7.5 and 298 K, the synthesis of ATP from ADP and P_i is accompanied by an increase in Gibbs free energy, $\Delta G^{\circ\prime} = 31.4$ kJ mol^{-1}. Thus this reaction must be driven by linking it to an exergonic reaction. This occurs in the oxidation of glyceraldehyde 3-phosphate to 1,3-diphosphoglycerate:

$$
\begin{array}{c}
O \diagup H \\
\diagdown C \\
| \\
H\!-\!C\!-\!OH \\
| \\
H\!-\!C\!-\!OPO_3^{2-} \\
| \\
H
\end{array}
\quad + \ NAD^+ + HPO_4^{2-} \quad \longrightarrow
$$

glyceraldehyde 3-phosphate

$$
\begin{array}{c}
O \diagup OPO_3^{2-} \\
\diagdown C \\
| \\
H\!-\!C\!-\!OH \\
| \\
H\!-\!C\!-\!OPO_3^{2-} \\
| \\
H
\end{array}
\quad + \ NADH + H^+ \qquad \Delta G^{\circ\prime} = 6.3 \text{ kJ}
$$

1,3-diphosphoglycerate

where NAD^+ and NADH are the oxidized and reduced forms of nicotinamide adenine dinucleotide (Figure 10.10). NAD^+ is another important biological molecule, which functions as an electron carrier (from glyceraldehyde to pyruvate). The preceding reaction is slightly endergonic and therefore cannot be used to drive any other energetically unfavorable processes. What is important, however, is the fact that the hydrolysis of the product, 1,3-diphosphoglycerate, is highly exergonic so that in the presence of the enzyme phosphoglycerate kinase, we have

1,3-diphosphoglycerate + ADP $\longrightarrow$

3-phosphoglycerate + ATP $\qquad \Delta G^{\circ\prime} = -18.8$ kJ

The large decrease in the standard Gibbs free energy for this process ensures the direction of the reaction from left to right. Stoichiometrically, for every mole of glucose decomposed, 2 mol of 1,3-diphosphoglycerate is produced and hence 2 mol of ATP is formed. The conversion of phosphoenopyruvate to pyruvate also takes place via a coupling process, resulting in an additional 2 mol of ATP.

So far we have considered the energetics for each individual step. Next, we would like to know how the equilibrium constant can be applied to a process like glycolysis. For a sequence of reactions the important quantity is not the equilibrium constant for an isolated reaction; rather, we must consider the

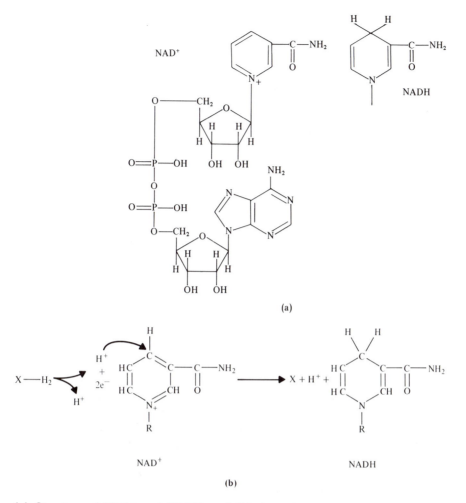

(a) Structure of NAD$^+$ and NADH, and (b) the reduction of NAD$^+$ to NADH. **Figure 10.10**
R designates the rest of the NAD$^+$ molecule and XH$_2$ designates the substrate
molecule being oxidized as NAD$^+$ is being reduced. The substrate, XH$_2$, loses
two hydrogen ions plus two electrons. One of the hydrogen ions is released
to the solution. The other hydrogen ion, accompanied by two electrons, combines with the NAD$^+$ to form NADH. Note that the reaction can be represented
as NAD$^+$ + 2H$^+$ + 2e^- → NADH + H$^+$ or simply as NAD$^+$ + H$_2$ → NADH + H$^+$

equilibrium constant for the overall process. Consider the following reactions:

$$A + B \rightleftharpoons C + D \qquad K_1 = \frac{[C][D]}{[A][B]}$$

$$C + D \rightleftharpoons E + F \qquad K_2 = \frac{[E][F]}{[C][D]}$$

$$E + F \rightleftharpoons G + H \qquad K_3 = \frac{[G][H]}{[E][F]}$$

The equilibrium constant for the overall process

$$A + B \rightleftharpoons G + H$$

261

is given by

$$K_4 = \frac{[G][H]}{[A][B]} = K_1 K_2 K_3$$

The important point is that even if one of the intermediate steps has a small equilibrium constant, appreciable amount of the products can still be formed if the other steps are favored from left to right. For example, if $K_1 = 10^5$, $K_2 = 10^{-4}$, and $K_3 = 10^2$, we still have $K_4 = 10^3$. Furthermore, since

$$RT \ln K_4 = RT \ln K_1 + RT \ln K_2 + RT \ln K_3$$

it follows that

$$\Delta G_4^{\circ\prime} = \Delta G_1^{\circ\prime} + \Delta G_2^{\circ\prime} + \Delta G_3^{\circ\prime}$$

The glycolytic pathway discussed so far is an *anaerobic* process; that is, the reactions are carried out in the absence of molecular oxygen. If oxygen is lacking, the reaction does not stop at pyruvic acid but proceeds one more step as follows:

$$CH_3COCOOH + NADH + H^+ \longrightarrow CH_3CH(OH)COOH + NAD^+$$
$$\text{pyruvic} \qquad\qquad\qquad\qquad\qquad\qquad \text{lactic}$$
$$\text{acid} \qquad\qquad\qquad\qquad\qquad\qquad\qquad \text{acid}$$

For anaerobic cells this is the end of the line—the lactic acid formed is finally discharged from the cell.* Obviously, this is not a terribly efficient process, since much of the free energy can still be extracted from a molecule as complex as lactic acid. The conversion of glucose to lactic acid via the glycolytic pathway probably arose very early in evolutionary development when there was little or no molecular oxygen present on earth. More recently, perhaps a couple of billion years ago, as a result of the change in the earth's atmosphere, *aerobic* cells were developed. In these cells, molecular oxygen is used for the further degradation of pyruvic acid to carbon dioxide and water via the Krebs cycle and the terminal respiratory chain (see Chapter 11). It is sufficient to mention that an additional 36 mol of ATP is synthesized so that the complete degradation of 1 mol of glucose yields a total of 38 mol of ATP:

$$C_6H_{12}O_6 + 38H^+ + 38ADP^{3-} + 38P_i + 6O_2 \longrightarrow$$
$$38ATP^{4-} + 6CO_2 + 44H_2O$$

Keep in mind that each P_i bears two negative charges.

It is interesting to estimate the efficiency of a biological process such as the one discussed here. As we saw earlier (Example 7.5), the complete combustion of glucose in air gives

$$C_6H_{12}O_6(s) + 6O_2(g) \longrightarrow 6CO_2(g) + 6H_2O(l) \qquad \Delta G^\circ = -2879 \text{ kJ}$$

On the other hand, the synthesis of ATP from ADP and P_i is

$$ADP^{3-} + H^+ + P_i \longrightarrow ATP^{4-} + H_2O \qquad \Delta G^{\circ\prime} = 31.4 \text{ kJ}$$

---

* This is the reaction in man. The lactic acid causes muscle cramps, or "charley horse." This painful effect is felt when muscles are exerted too suddenly with an inadequate supply of oxygen. In other organisms, such as yeast, ethanol is formed instead of lactic acid. No doubt, nature has been criticized for this unfair discrimination.

Thus the efficiency (η) for the degradation of glucose to carbon dioxide and water via glycolysis is given by

$$\eta = \frac{\text{energy stored in ATP molecules}}{\text{total energy released}}$$

$$= \frac{38 \times 31.4 \text{ kJ}}{2879 \text{ kJ}} \times 100 = 42\%$$

This is a lower limit value. When the physiological concentrations of various components are taken into account, the efficiency is believed to be above 60%.

This impressively high efficiency is, of course, the result of several billion years' work. In our earlier discussion of gas expansion (Section 6.1), it was shown that a reversible process can perform much more work than an irreversible one. The combustion of glucose in air is a highly irreversible reaction; consequently, the energy stored in the parent molecule appears in the much less useful form, that is, heat. By breaking the reaction down into a number of steps with the aid of enzymes, it is possible to store much of the energy via the synthesis of ATP.

It is important to keep in mind that although high efficiencies are preferred in general, we must also consider the *rate* of a process. As mentioned earlier, the maximum efficiency is obtained only for a truly reversible process, but it will take an infinite amount of time to complete. Thus, if glycolysis were lengthened by 10 more steps, the efficiency would undoubtedly go up because the process would become more reversible. However, the rate would then be slower, perhaps dangerously so for our survival. It seems that a compromise between efficiency and rate will finally be reached through evolution.

One reason why ATP is unique in so many biological reactions is undoubtedly its intermediate value of $\Delta G^{o\prime}$ for hydrolysis (see Table 10.1). A larger $\Delta G^{o\prime}$ value would also mean that more energy is required for its synthesis, which is undesirable. On the other hand, a smaller $\Delta G^{o\prime}$ value for hydrolysis would make ATP a much less useful compound in coupled reactions.

SOME LIMITATIONS OF THERMODYNAMICS

So far we have presented a rather general and qualitative description of how some of the thermodynamic concepts can be applied to help us understand the nature of biochemical reactions. In this section we consider a more fundamental question: Under what conditions is thermodynamics applicable to biology? The answer to this question is not a trivial one, and at present there is much debate on this subject.

To begin with, we note that biochemical reactions are usually discussed in terms of standard free energies ($\Delta G^{o\prime}$). Strictly, the direction of a reaction such as

$$A + B \rightleftharpoons C + D$$

is given by ΔG, where

$$\Delta G = \Delta G^{o\prime} + RT \ln \frac{[C][D]}{[A][B]}$$

Only in the case where the reactants and products are in their standard states can we employ $\Delta G^{o\prime}$. Thus the fact that the hydrolysis of ATP at 298 K and pH 7 results in a decrease of the standard free energy equal to

Figure 10.11

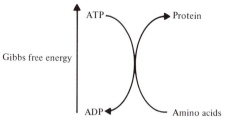

Gibbs free energy

Schematic representation of the Gibbs free energy changes that occur in protein synthesis.

-30.5 kJ mol^{-1} does not necessarily mean that this is also the value for ΔG. In living cells, the physiological temperature, pH, concentration of reactants and products, as well as metal ions can vary from system to system, and these factors must affect the value of ΔG. It is difficult and in some cases impossible to accurately measure the concentration of various species in solution. Nevertheless, reasonable estimates can be made of the concentrations in many cases and hence of the sign of ΔG.

Another point that has often been raised is that thermodynamics deals only with closed systems at equilibrium. Living systems are open systems and they are maintained at steady state rather than at equilibrium.* In fact, any cell at true equilibrium is a dead cell. The treatment of open systems requires *irreversible thermodynamics*, which is beyond the scope of this book. The interested reader is referred to some important papers on the application of thermodynamics to biology listed at the end of the chapter.

Finally, reactions in a living cell, like any other reactions, can be classified into two categories—those that are thermodynamically controlled and those that are kinetically controlled. As an example of the former, let us consider the synthesis of a dipeptide from two amino acids:

$$\text{alanine} + \text{glycine} \longrightarrow \text{alanylglycine} \qquad \Delta G^{\circ\prime} = 17.2 \text{ kJ}$$

The corresponding equilibrium constant for this reaction is about 1×10^{-3} at 298 K. Clearly, such a process will not proceed by an appreciable extent if left on its own. However, if this process could be coupled to the hydrolysis of ATP by the action of an enzyme, the reaction would proceed from left to right (Figure 10.11). Such a reaction is said to be thermodynamically controlled because the reaction itself is not spontaneous, and the energy (or free energy) of reaction must be supplied from an outside source.

A kinetically controlled reaction is one in which the overall ΔG is negative (and hence thermodynamically favorable), but the rate is negligibly small in the absence of appropriate enzyme catalysis. The phosphorylation of glucose to glucose 6-phosphate (coupled to ATP hydrolysis) is certainly an exergonic process, but the reaction occurs at a very slow rate in the absence of the enzyme hexokinase.

Appendix 10.1 Relationship Between Fugacity and Pressure

Consider a pure gas. Expressing Eq. (7.22) in molar quantities, we write

$$\left(\frac{\partial \overline{G}}{\partial P}\right)_T = \overline{V}$$

---

* A steady state bears some superficial resemblance to equilibrium in that the concentration of species remains unchanged with time. However, to achieve this, we need to constantly supply and remove materials from the system.

or

$$d\bar{G} = \bar{V}\,dP$$

This equation applies to all gases, real or ideal. We have

$$d\bar{G}_{\text{ideal}} = \bar{V}_{\text{ideal}}\,dP$$

$$= \left(\frac{RT}{P}\right)dP$$

since $P\bar{V}_{\text{ideal}} = RT$ and

$$d\bar{G}_{\text{real}} = \bar{V}_{\text{real}}\,dP$$

The quantity $\bar{V}_{\text{real}}$, the molar volume of a real gas, can be measured directly or calculated using an equation of state for real gases. Thus

$$d(\bar{G}_{\text{real}} - \bar{G}_{\text{ideal}}) = \left(\bar{V}_{\text{real}} - \frac{RT}{P}\right)dP$$

Integrating between $P = 0$, where $\bar{G}_{\text{real}} = \bar{G}_{\text{ideal}}$ (when pressure is equal to fugacity) and P, the pressure of interest, we obtain

$$\bar{G}_{\text{real}} - \bar{G}_{\text{ideal}} = \int_0^P \left[\bar{V}_{\text{real}} - \left(\frac{RT}{P}\right)\right]dP$$

Since

$$\bar{G}_{\text{real}} = \bar{G}^\circ + RT\ln f$$

and

$$\bar{G}_{\text{ideal}} = \bar{G}^\circ + RT\ln P$$

we have

$$\bar{G}_{\text{real}} - \bar{G}_{\text{ideal}} = RT\ln\frac{f}{P} = \int_0^P \left[\bar{V}_{\text{real}} - \left(\frac{RT}{P}\right)\right]dP$$

We assume here that $\bar{G}^\circ$ is the same for real and ideal gases. At 1 atm, most gases behave ideally so that $f = P = 1$ and $\bar{G}_{\text{real}} = \bar{G}_{\text{ideal}} = \bar{G}^\circ$. Thus

$$\ln\frac{f}{P} = \int_0^P \left(\frac{\bar{V}_{\text{real}}}{RT} - \frac{1}{P}\right)dP \qquad (A10.1)$$

Equation (A10.1) relates the pressure of a gas to its fugacity of the gas at temperature T.

EXAMPLE 10.5 A satisfactory equation of state for ammonia is $P(\bar{V} - b) = RT$, where b is 0.037 liter mol^{-1}. Calculate the fugacity of the gas when its pressure is 50 atm at 298 K.

Answer: Since $\bar{V}_{\text{real}} = (RT/P) + b$, we have

$$\int_0^P \left(\frac{\bar{V}_{\text{real}}}{RT} - \frac{1}{P}\right)dP = \int_0^P \left(\frac{b}{RT}\right)dP$$

$$= \int_0^{50} \left(\frac{0.037 \text{ liter mol}^{-1}}{RT}\right)dP$$

$$= \frac{(0.037 \text{ liter mol}^{-1})(50 \text{ atm})}{(0.08206 \text{ liter atm K}^{-1}\text{ mol}^{-1})(298 \text{ K})} = 0.0757$$

Finally,

$$f = Pe^{0.0757}$$

$$= (50 \text{ atm})(1.079)$$

$$= 54 \text{ atm}$$

Suggestions for Further Reading

See the texts listed in Chapter 6.

Reading Assignments

SECTION A

"Effect of Ionic Strength on Equilibrium Constants," M. D. Seymour and Q. Fernando, *J. Chem. Educ.* **54**, 225 (1977).

"Equilibrium Dialysis," S. A. Katz, C. Parfitt, and R. Purdy, *J. Chem. Educ.* **47**, 721 (1970).

"High Altitude Acclimatization," R. C. Plumb, *J. Chem. Educ.* **48**, 75 (1971).

"Thermodynamics and Biology," B. E. C. Banks, *Chem. Brit.* **5**, 514 (1969).

"Structure of High-Energy Molecules," L. Pauling, *Chem. Brit.* **6**, 468 (1970).

"Thermodynamics and Biology," D. Wilkie, *Chem. Brit.* **6**, 473 (1970).

"Energetics of Muscle," A. F. Huxley, *Chem. Brit.* **6**, 477 (1970).

The preceding four articles represent some of the current views on the applicability of thermodynamics to biology.

"Biological Oxidation and Energy Conservation," J. Kirschbaum, *J. Chem. Educ.* **45**, 28 (1968).

"How Cells Make ATP," P. C. Hinkle and R. E. McCarty, *Sci. Am.*, Mar. 1978.

SECTION B

"The Sources of Muscular Energy," R. Margaria, *Sci. Am.*, Mar. 1972.

"Glucogenesis: A Teaching Pathway," F. Gabrielli, *J. Chem. Educ.* **53**, 86 (1976).

"The Production of Heat by Fat," M. J. R. Dawkins and D. Hull, *Sci. Am.*, Aug. 1965.

"Bioenergetics and the Problem of Tumor Growth," E. Racker, *Am. Sci.* **60**, 56 (1972).

"Composite Formulated Biochemical Equilibria," R. R. Richards, *J. Chem. Educ.* **56**, 514 (1979).

Problems

10.1 Use the data listed in Appendix 3 to calculate the equilibrium constant K_P for the synthesis of HCl at 298 K:

$$H_2(g) + Cl_2(g) \rightleftharpoons 2HCl(g)$$

What is the value of K_P if the equilibrium is expressed as

$$\tfrac{1}{2}H_2(g) + \tfrac{1}{2}Cl_2(g) \rightleftharpoons HCl(g)?$$

10.2 At 720 K, the equilibrium partial pressures are $P_{NH_3} = 321.6$ atm, $P_{N_2} = 69.6$ atm, and $P_{H_2} = 208.8$ atm, respectively. (a) Calculate K_P for the reaction described in Example 10.1. (b) Calculate the thermodynamic equilibrium constant if $\gamma_{NH_3} = 0.782$, $\gamma_{N_2} = 1.266$, and $\gamma_{H_2} = 1.243$.

10.3 Consider the following reaction:

$$CO_2(g) + H_2(g) \rightleftharpoons CO(g) + H_2O(g)$$

The equilibrium constant is 0.534 at 960 K and 1.571 at 1260 K. What is the enthalpy of the reaction?

10.4 Nitric oxide from car exhaust is a primary air pollutant. Calculate the equilibrium constant for the reaction

$$N_2(g) + O_2(g) \rightleftharpoons 2NO(g)$$

at 25°C using the data listed in Appendix 3. Assume both $\Delta H°$ and $\Delta S°$ to be temperature-independent. Calculate the equilibrium constant at 1500°C, which is the typical temperature inside the cylinders of a car's engine after it has been running for some time.

10.5 Equilibrium constant of gaseous reactions can be expressed in terms of pressures only (K_P), concentrations only (K_c), or mole fractions only (K_X). Consider the hypothetical reaction

$$aA(g) \rightleftharpoons bB(g)$$

Derive the following relationships: (a) $K_P = K_c(RT)^{\Delta n}$ and (b) $K_P = K_X P^{\Delta n}$, where Δn is the difference in the number of moles between products and reactants and P is the total pressure of the system. Assume ideal gas behavior.

10.6 Consider the decomposition of calcium carbonate:

$$CaCO_3(s) \rightleftharpoons CaO(s) + CO_2(g)$$

(a) Write an equilibrium constant expression (K_P) for the reaction. (b) The rate of decomposition is slow until the partial pressure of carbon dioxide is equal to the atmospheric pressure. Calculate the temperature at which the decomposition becomes spontaneous. Assume that $\Delta H°$ and $\Delta S°$ are temperature-independent. Use the data in Appendix 3 for your calculation.

10.7 The vapor pressure of dry ice (solid CO_2) is 672.2 mm Hg at -80°C and 1486 mm Hg at -70°C. Calculate the molar heat of sublimation of CO_2.

10.8 Consider the reaction

$$2NO_2(g) \rightleftharpoons N_2O_4(g) \qquad \Delta H° = -58.04 \text{ kJ}$$

Predict what happens to the system at equilibrium if (a) the temperature is raised, (b) the pressure on the system is increased, (c) an inert gas is added to the system at constant pressure, (d) an inert gas is added to the system at constant volume, and (e) a catalyst is added to the system.

10.9 Calculate the equilibrium constant K_P for the following reaction at 298 K:

$$H_2(g) + \tfrac{1}{2}O_2(g) \rightleftharpoons H_2O(g)$$

In spite of the very large value of K_P, a mixture of H_2 and O_2 gases is stable at room temperature in the absence of external disturbances such as heat or catalyst. Explain. Use the data in Appendix 3 for your calculations.

10.10 The dissociation of N_2O_4 into NO_2 is 16.7% complete at 298 K and 1 atm:

$$N_2O_4(g) \rightleftharpoons 2NO_2(g)$$

Calculate the equilibrium constant and the standard free energy change for the reaction. [*Hint:* Let α be the degree of dissociation and show that $K_P = 4\alpha^2 P/(1 - \alpha^2)$, where P is the total pressure.]

10.11 Referring to Problem 10.10, calculate the degree of dissociation of N_2O_4 if the total pressure is 10 atm. Comment on your result.

10.12 Calculate $\Delta G°$ for each of the following equilibrium constants: 1.0×10^{-4}, 1.0×10^{-2}, 1.0, 1.0×10^2, 1.0×10^4 at 298 K.

10.13 The standard Gibbs free energies of formation of gaseous *cis*- and *trans*-2-butene are 67.15 kJ mol^{-1} and 64.10 kJ mol^{-1}, respectively. Calculate the ratio of equilibrium pressures of the gaseous isomers at 298 K.

10.14 The reaction

$$\text{L-glutamate + pyruvate} \longrightarrow \alpha\text{-ketoglutarate + L-alanine}$$

is catalyzed by the enzyme L-glutamate-pyruvate aminotransferase. At 300 K, the equilibrium constant for the reaction is 1.11. Predict whether the forward reaction (left to right) will occur spontaneously if the concentrations of the reactants and products are [L-glutamate] = 3.0×10^{-5} M, [pyruvate] = 3.3×10^{-4} M, [α-ketoglutarate] = 1.6×10^{-2} M, and [L-alanine] = 6.25×10^{-3} M.

10.15 As mentioned in the text of the chapter, the standard Gibbs free energy for the hydrolysis of ATP to ADP at 310 K is about -30.5 kJ. Calculate the $\Delta G^{°\prime}$ for the reaction in the muscle of a polar sea fish at $-1.5°C$. [*Hint:* $\Delta H^{°\prime} = -20.1$ kJ.]

10.16 Under standard-state conditions one of the steps in glycolysis does not occur spontaneously:

$$\text{glucose} + HPO_4^{2-} \longrightarrow \text{glucose-6-phosphate} + H_2O \qquad \Delta G^{°\prime} = 13.4 \text{ kJ}$$

Can the reaction take place in the cytoplasm of a cell where the concentrations are [glucose] = 4.5×10^{-2} M, [HPO_4^{2-}] = 2.7×10^{-3} M, and [glucose-6-phosphate] = 1.6×10^{-4} M and the temperature is 310 K?

10.17 The formation of a dipeptide is the first step toward the synthesis of a protein molecule. Consider the following reaction:

$$\text{glycine + glycine} \longrightarrow \text{glycylglycine} + H_2O$$

Use the data in Appendix 3 to calculate the $\Delta G^{°\prime}$ and the equilibrium constant at 298 K, keeping in mind that the reaction is carried out in an aqueous buffer solution. Assume that $\Delta G^{°\prime}$ is essentially the same at 310 K. What conclusion can you draw about your result?

10.18 From the following reactions:

$$\text{fumarate}^{2-} + NH_4^+ \longrightarrow \text{aspartate}^- \qquad \Delta G^{°\prime} = -36.7 \text{ kJ}$$

$$\text{fumarate}^{2-} + H_2O \longrightarrow \text{malate}^{2-} \qquad \Delta G^{°\prime} = -2.9 \text{ kJ}$$

Calculate the standard Gibbs free energy change and the equilibrium constant for the following reaction at 37°C:

$$\text{malate}^{2-} + NH_4^+ \longrightarrow \text{aspartate}^- + H_2O$$

10.19 In this chapter we introduced the quantity $\Delta G^{\circ\prime}$, which is the standard Gibbs free energy change for a reaction in which the reactants and products are in their biochemical standard states. The discussion was concentrated on the uptake or liberation of H^+ ions. The $\Delta G^{\circ\prime}$ can also be applied to reactions involving the uptake and liberation of gases such as O_2 and CO_2. In these cases the biochemical standard states are $P_{O_2} = 0.2$ atm and $P_{CO_2} = 0.03$ atm, where 0.2 atm and 0.03 atm are the partial pressures of O_2 and CO_2 in air, respectively. Consider the reaction

$$A(aq) + B(aq) \longrightarrow C(aq) + CO_2(g)$$

where A, B, and C are molecular species. Derive a relation between ΔG° and $\Delta G^{\circ\prime}$ for this reaction.

10.20 The binding of oxygen to hemoglobin (Hb) is quite complex, but for our purpose here we can represent the reaction as

$$Hb(aq) + O_2(g) \rightleftharpoons HbO_2(aq)$$

If ΔG° for the reaction is -11.2 kJ at 20°C, calculate $\Delta G^{\circ\prime}$ for the reaction. (*Hint*: The biochemical standard state for gaseous oxygen is $P_{O_2} = 0.2$ atm, where 0.2 atm is oxygen's partial pressure in air.)

10.21 The calcium ion binds to a certain protein to form a $1:1$ complex. The following data were obtained in an experiment:

Total Ca^{2+} (μM)	24	60	120	180	240	480
Ca^{2+} bound to protein (μM)	13.9	31.2	51.2	63.4	70.8	83.4

Determine by graphical method the dissociation constant of the Ca^{2+}-protein complex. The protein concentration was kept at 96 μM for each run. ($1 \mu M = 1 \times 10^{-6}$ M.)

10.22 A polypeptide can exist in either the helical or random coil form. The equilibrium constant for the helix $\rightleftharpoons$ random coil transition is 0.86 at 40°C and 0.35 at 60°C. Calculate ΔH° and ΔS° for the reaction.

10.23 Based on the material covered so far in the text, describe as many ways as you can think of for measuring ΔG of a process.

10.24 In an equilibrium dialysis experiment it is found that the concentrations of the free ligand, bound ligand, and protein are 1.2×10^{-5} M, 5.4×10^{-6} M, and 4.9×10^{-6} M, respectively. Calculate the dissociation constant for the reaction $PL \rightleftharpoons P + L$. Assume that there is one binding site per protein molecule.

10.25 Derive Eq. (10.26) from Eq. (10.24).

10.26 Use the data in Appendix 3 to calculate the following equilibrium constant (K_P) at 25°C:

$$SO_2(g) + \tfrac{1}{2}O_2(g) \rightleftharpoons SO_3(g)$$

Calculate K_P for the reaction at 60°C using the following methods: (a) use the van't Hoff equation, that is, Eq. (10.21); (b) use the Gibbs–Helmholtz equation, that is, Eq. (7.21), to find ΔG° at 60°C and hence K_P at the same temperature; and (c) use $\Delta G^{\circ} = \Delta H^{\circ} - T \Delta S^{\circ}$ to find ΔG° at 60°C and hence K_P at the same temperature. State the approximations employed in each case and compare your results.

10.27 Derive the van't Hoff equation using the Gibbs-Helmholtz equation [Eq. (7.21)] and Eq. (10.5).

★**10.28** The solubility of *n*-heptane in water is 0.050 g per liter of solution at 25°C. What is the Gibbs free energy change for the hypothetical process of dissolving *n*-heptane in water at a concentration of 2.0 g liter^{-1} at the same temperature? [*Hint*: First calculate $\Delta G°$ from the equilibrium process and then ΔG using Eq. (10.7).]

Electrochemistry

The extraordinary power of matter which we have thus traced exhibiting itself as the cause of chemical, electric and calorific effects exhibits another set of phenomena unlike any of these but equal to any of them in the irregularity and generality of its actions. We cannot say that any one is the cause of the others, but only that all are connected and due to a common cause. . . .
Michael Faraday*

Electrochemical reactions reverse the action of electrolysis. While electrolysis converts electrical energy into chemical energy, electrochemical reactions convert chemical energy directly into electrical energy. There is a convenient difference from normal chemical reactions: The Gibbs free energy change for an electrochemical reaction can be readily measured from the maximum electrical work done.

In this chapter we consider the basic principles of electrochemistry, their applications to chemical and biological systems, and some aspects of membrane potentials.

Electrochemical Cells 11.1

When a piece of zinc metal is placed in a $CuSO_4$ solution, two things happen. Some of the zinc metal dissolves as Zn^{2+} ions and, more obviously, some of the Cu^{2+} ions are converted into metallic copper. This spontaneous reaction is represented by

$$Zn(s) + Cu^{2+}(aq) \longrightarrow Zn^{2+}(aq) + Cu(s)$$

Similarly, if a piece of copper wire is placed in a $AgNO_3$ solution, silver metal is deposited on the copper wire. In each case, nothing will happen if we exchange the roles of the metals involved.

Let us now consider the arrangement shown in Figure 11.1. Zinc and copper metals are placed in two separate compartments containing $ZnSO_4$ and $CuSO_4$, respectively. The two solutions are connected by a *salt bridge*, a tube that contains an electrolyte solution such as NH_4NO_3 or KCl. This

* Joseph Agassi, *Faraday as a Natural Philosopher*, The University of Chicago Press, Chicago, 1971, pp. 210–211. Used by permission.

Figure 11.1

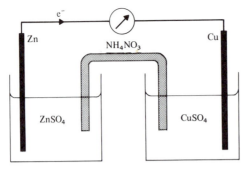

Schematic diagram of a galvanic cell.

solution is kept from flowing into the compartments by either a sintered disc on each end of the tube or mixing the electrolyte solution with a gelatinous material, for example, agar-agar.* When the two electrodes are connected by a piece of metal wire, electrons will flow from the zinc electrode to the copper electrode through the external wire. At the same time, zinc will dissolve in the left compartment as Zn^{2+} ions and Cu^{2+} ions will be converted into metallic copper at the copper electrode. The purpose of the salt bridge is to complete the electrical circuit between the two solutions.

The setup described above is known as the *Daniell cell*, which is an example of *galvanic* or *voltaic cells*. The operation of galvanic cells is based on oxidation–reduction or *redox* reactions. In our case the redox reactions can be expressed in terms of two *half-cell reactions*:

$$Zn(s) \longrightarrow Zn^{2+}(aq) + 2e^-$$
$$Cu^{2+}(aq) + 2e^- \longrightarrow Cu(s)$$

The zinc electrode is called the *anode*, where oxidation (loss of electrons) takes place; the copper electrode is called the *cathode*, where reduction (gain of electrons) takes place. By convention, the *cell diagram* for the Daniell cell is given by

$$Zn(s)\,|\,ZnSO_4(aq)\,|\,NH_4NO_3(aq)\,|\,CuSO_4(aq)\,|\,Cu(s)$$

where the vertical lines indicate phase boundaries. The anode is on the left and the cathode is on the right.

The fact that external current (that is, the electrons) flows from the anode to the cathode means that there is a potential difference between the electrodes, which we shall call the *electromotive force*, or emf (ε), of the cell. For the Daniell cell, a value of $\varepsilon = 1.10$ V is measured at 298 K and equal concentrations of $CuSO_4$ and $ZnSO_4$.

The emf of a cell is usually measured using a potentiometer (Figure 11.2). A cell S, which has a larger emf than that of any other cell to be measured, is connected to the ends of a uniform wire (AB) of high resistance. The cell under study, X, is connected to A and a galvanometer G. In a typical experiment, the sliding contact—the arrow—is moved along AB until a point (C) is reached when the galvanometer registers zero current flow. At this point the potential from cell S across AC is exactly balanced by the emf of cell X, ε_X. The same procedure is repeated by replacing cell X with cell W, whose emf ε_W

* Agar-agar is a polysaccharide.

Measurement of the emf of a cell using a potentiometer.

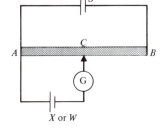

Figure 11.2

is accurately known. If the new balance point is C', we write

$$\frac{\mathcal{E}_X}{\mathcal{E}_W} = \frac{AC}{AC'}$$

Normally, W is a Weston cell ($\mathcal{E}_W = 1.018$ V at 298 K), so that

$$\mathcal{E}_X = 1.018 \frac{AC}{AC'} \text{ V}$$

Thermodynamics of Electrochemical Cells 11.2

In order for us to relate the electrochemical energy associated with a cell to various thermodynamic parameters, the cell must behave reversibly in the following manner. If we apply an external potential that is exactly equal but opposite to that of the cell, no reaction occurs within the cell. An infinitesimal decrease or increase in the external potential would lead either to normal or reverse cell reactions. Any cell that satisfies these conditions is called a *reversible cell*. The situation described here is quite analogous to the reversible expansion of a gas discussed in Chapter 6. Of course, under normal conditions, cells never operate reversibly; if they do, no current would ever flow through them. However, the condition of reversibility is attained when we measure the emf of a cell. Indeed, at the balance point, the net current flow is zero.

Consider an electrochemical cell reaction in which n moles of electrons are transferred from one electrode to the other (that is, from the anode to the cathode). The quantity of electricity Q associated with n moles of electrons is given by

$$Q = nF$$

where F is the faraday. The faraday is the charge carried by 1 mol of electrons, that is,

$$\text{faraday} = \text{charge of electron} \times \text{number of electrons per mole}$$
$$= 1.602 \times 10^{-19} \text{ C} \times 6.022 \times 10^{23} \text{ mol}^{-1}$$
$$= 96{,}472 \text{ C mol}^{-1}$$

Except for very accurate work, for simplicity the value of the faraday is taken to be 96,500 C mol^{-1}. The resulting electric current can in principle be used totally for work. The amount of work* is characterized as $-nF\mathcal{E}$. The negative sign appears because work is done by the cell rather than on the cell, in keeping with the convention established in Section 6.1. For a reversible cell at a given temperature and pressure, this is the maximum work done, which is equal to the decrease in the Gibbs free energy of the system (see Appendix 11.1):

$$\Delta G = -nF\mathcal{E} \tag{11.1}$$

where ΔG is the difference between the Gibbs free energy of the products and that of the reactants, at some particular concentrations. Equation (11.1) may be rewritten as

$$\mathcal{E} = \frac{-\Delta G}{nF} \tag{11.2}$$

A special case of this equation is that of a cell with all reactants and products in their standard states. In this case the emf of the cell is called the standard emf $\mathcal{E}°$, which is related to the change in the standard Gibbs free energy as follows:

$$\mathcal{E}° = \frac{-\Delta G°}{nF} \tag{11.3}$$

Recall that the criterion for a reaction to proceed spontaneously in the direction written is that ΔG is negative. According to Eq. (11.2), this corresponds to a positive emf $\mathcal{E}$. Thus, when $\mathcal{E}$ for an electrochemical cell as written is positive, the cell is galvanic and the cell reaction proceeds as written. When $\mathcal{E}$ for an electrochemical cell as written is negative, the cell is electrolytic. External voltage larger than $\mathcal{E}$ must be supplied to the cell to sustain the nonspontaneous process.

THE NERNST EQUATION

We now derive an equation relating the emf of a cell to variables such as temperature and concentrations of reacting species. The Gibbs free energy change for the cell reaction

$$a\mathrm{A} + b\mathrm{B} \longrightarrow c\mathrm{C} + d\mathrm{D}$$

is given by [see Eq. (10.7)]

$$\Delta G = \Delta G° + RT \ln Q$$

$$= \Delta G° + RT \ln \frac{a_C^c a_D^d}{a_A^a a_B^b}$$

where a denotes activity. Dividing this equation throughout by $-nF$ and using both Eqs. (11.2) and (11.3), we obtain

$$\mathcal{E} = \mathcal{E}° - \frac{RT}{nF} \ln \frac{a_C^c a_D^d}{a_A^a a_B^b} \tag{11.4}$$

Equation (11.4) is known as the *Nernst equation*. Here $\mathcal{E}$ is the observed emf of the cell and $\mathcal{E}°$ is the emf the cell would have if all the reactants and products

---

* The electrical work done is given by the product of potential and charge. Here we have $V = \mathcal{E}$ and $Q = nF$.

were in their unit activity standard states. At equilibrium $\mathcal{E} = 0$ so that

$$\mathcal{E}° = \frac{RT}{nF} \ln K = \frac{-\Delta G°}{nF} \tag{11.5}$$

Since most electrochemical cells operate at or near room temperature, we can evaluate the quantity RT/F by taking $R = 8.314$ J K^{-1} mol^{-1}, $T = 298$ K, and $F = 96,500$ C mol^{-1} so that

$$\frac{(8.314 \text{ J K}^{-1} \text{ mol}^{-1})(298 \text{ K})}{96,500 \text{ C mol}^{-1}} = 0.02567 \text{ J C}^{-1}$$

Since

$$1 \text{ J} = 1 \text{ C} \times 1 \text{ V}$$

we have

$$\frac{RT}{F} = 0.02567 \text{ V}$$

Finally, Eq. (11.4) can be expressed as

$$\mathcal{E} = \mathcal{E}° - \frac{0.0591}{n} \log \frac{a_C^c a_D^d}{a_A^a a_B^b} \tag{11.6}$$

Note that we have changed the natural log to the log of base 10 and that Eq. (11.6) is applicable for $T = 298$ K.

Applying Eq. (11.6) to the cell reaction

$$\text{Zn(s)} + \text{Cu}^{2+}(\text{aq}) \longrightarrow \text{Zn}^{2+}(\text{aq}) + \text{Cu(s)}$$

we write

$$\mathcal{E} = \mathcal{E}° - \frac{0.0591}{n} \log \frac{a_{\text{Zn}^{2+}} a_{\text{Cu}}}{a_{\text{Zn}} a_{\text{Cu}^{2+}}}$$

Since $n = 2$ (two electrons are transferred per redox reaction) and by definition $a_{\text{Cu}} = 1$ and $a_{\text{Zn}} = 1$, we have

$$\mathcal{E} = \mathcal{E}° - 0.0296 \log \frac{a_{\text{Zn}^{2+}}}{a_{\text{Cu}^{2+}}}$$

If solutions are dilute enough so that activities can be replaced by concentrations, the emf of the cell is given by

$$\mathcal{E} = \mathcal{E}° - 0.0296 \log \frac{[\text{Zn}^{2+}]}{[\text{Cu}^{2+}]}$$

Thus from a knowledge of the concentrations of the solutions, we can calculate $\mathcal{E}$ if we know the value of $\mathcal{E}°$. We shall return to this point in Section 11.4.

Types of Electrodes 11.3

Before we consider the determination of the standard emf of a cell ($\mathcal{E}°$), it is useful to first examine the types of electrodes commonly employed in constructing electrochemical cells. The following are a few examples.

Figure 11.3

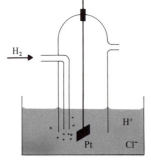

METAL ELECTRODE

A metal electrode consists of a piece of metal in contact with a solution that contains ions of the same metal, as in the Daniell cell.

METAL-INSOLUBLE SALT ELECTRODE

A metal-insoluble salt electrode can be prepared by coating a piece of metal with an insoluble salt of the same metal. An example is the silver–silver chloride (Ag|AgCl) electrode, which must be placed in a solution containing chloride ions to operate. The half-cell reduction reaction is

$$\text{AgCl(s)} + e^- \longrightarrow \text{Ag(s)} + \text{Cl}^-(\text{aq})$$

GAS ELECTRODE

A hydrogen electrode, a type of gas electrode, is shown in Figure 11.3. In this arrangement, hydrogen gas is bubbled through the inlet into an acid solution. The half-cell reduction reaction is

$$\text{H}^+(\text{aq}) + e^- \longrightarrow \tfrac{1}{2}\text{H}_2(\text{g})$$

The functions of the platinum metal are to provide a surface on which the foregoing reaction takes place and to serve as an electrical conductor to the external circuit.*

GLASS ELECTRODE

One of the most widely used electrodes, the glass electrode is an example of an *ion-selective electrode*, since it is specific for H^+ ions. The essential features of a glass electrode are shown in Figure 11.4a. The electrode consists of a very thin bulb or membrane made of a special type of glass that is permeable to H^+ ions. An Ag|AgCl electrode is immersed in a buffer solution (constant pH) containing Cl^- ions. When the electrode is placed in a solution whose pH is different from that of the buffer solution, a potential difference develops between the two sides which is a measure of the difference in the two pH values.†

---

* The platinum metal acts as a catalyst in the decomposition of H_2 into atomic hydrogen, that is, $\tfrac{1}{2}\text{H}_2 \rightarrow \text{H}$, which is followed by the electrode reaction, $\text{H} \rightarrow \text{H}^+ + e^-$.
† See R. A. Durst, *J. Chem. Educ.* **44**, 175 (1967), for a detailed discussion of the glass electrodes.

As we shall see in Section 11.4, the potential of electrodes is measured with reference to the hydrogen electrode. However, a gas electrode is inconvenient to use in practice, so other reference electrodes are of value. One of the most versatile reference electrodes is the calomel electrode, shown in Figure 11.4b. The side arm containing a saturated KCl solution in agar-agar—the salt bridge—is placed in another solution for emf measurements. The electrode system is represented by

$$Hg(l) \,|\, Hg_2Cl_2(s) \,|\, KCl(aq, satd)$$

and the half-cell reduction reaction is

$$\tfrac{1}{2}Hg_2Cl_2(s) + e^- \longrightarrow Hg(l) + Cl^-(aq)$$

ION-SELECTIVE ELECTRODE

In recent years, a variety of electrodes that are specific for cations such as Na^+, K^+, Li^+, Ca^{2+}, NH_4^+, Ag^+, and Cu^{2+} and anions such as the halides, S^{2-}, and CN^- have been developed. Since these specific electrodes lend them-

(a) Glass electrode.
(b) Calomel electrode.

Figure 11.4

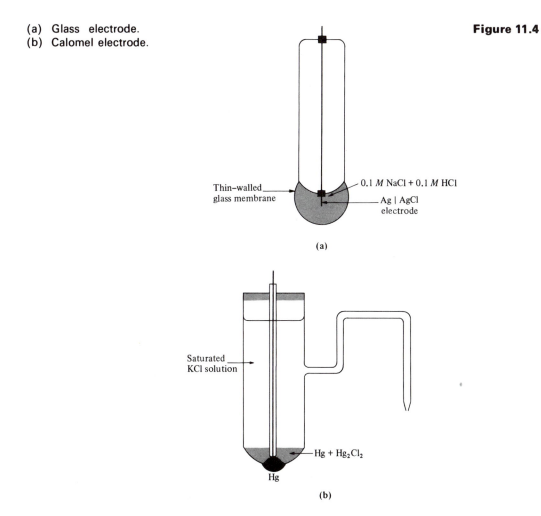

Thin–walled glass membrane

0.1 M NaCl + 0.1 M HCl

Ag | AgCl electrode

(a)

Saturated KCl solution

Hg + Hg$_2$Cl$_2$

Hg

(b)

selves to easy and accurate emf measurements, they have become very useful analytical tools in areas ranging from medicine to environmental studies. As mentioned earlier, the glass electrode was the first ion-selective electrode known. The operational details of these electrodes will not be given here. The interested reader should consult the references listed at the end of the chapter.

11.4 Single-Electrode Potential

Just as in ionic activities, where it is impossible to monitor a single ion, so it is not possible to measure the potential of a single electrode. Any complete circuit must by necessity contain two electrodes. In practice, we measure the potential of all electrodes with reference to a common electrode. By convention, this is chosen to be the *standard hydrogen electrode*, whose potential at 298 K, 1 atm H_2 pressure, and 1 M H^+ concentration (more correctly, unit activity) is arbitrarily set to be zero. The measured emf is then taken to be the potential of the other electrode. Note that it is not necessary to employ the standard hydrogen electrode for all measurements. For example, we can first calibrate the calomel electrode against the standard hydrogen electrode and then use the former in conjunction with other electrodes.

Table 11.1 lists some common half-cell reactions together with their *standard reduction potentials*. The values given apply for the reaction from left to right. The more positive the reduction potential, the greater strength is the oxidizing agent. Thus F_2 is the strongest oxidizing agent, since it has the highest tendency to pick up electrons. On the other hand, Li^+ is the weakest oxidizing agent, which makes lithium metal the most powerful reducing agent. The numerical values apply to aqueous solutions at 298 K, in which the concentration of each dissolved species is at 1 M.

The standard electrode potential for any electrochemical cell can be readily obtained from Table 11.1. Again referring to the Daniell cell, we write

$$\mathscr{E}°$$

At anode:	$Zn(s) \longrightarrow Zn^{2+}(aq) + 2e^-$	0.763 V
At cathode:	$Cu^{2+}(aq) + 2e^- \longrightarrow Cu(s)$	0.337 V
Overall:	$Zn(s) + Cu^{2+}(aq) \longrightarrow Zn^{2+}(aq) + Cu(s)$	1.10 V

Note that the sign of the reduction potential must be reversed when the half-reaction takes place in the opposite direction. A positive value of $\mathscr{E}°$ means that the reaction proceeds spontaneously from left to right as shown when reactants and products are in their standard states.

EXAMPLE 11.1 Predict whether the following reaction would occur spontaneously under standard state conditions. Assume both ions to be at unit activity.

$$Mg(s) + 2Ag^+ \longrightarrow Mg^{2+} + 2Ag(s)$$

Answer: From Table 11.1, we have

$$\mathscr{E}°$$

At anode:	$Mg(s) \longrightarrow Mg^{2+} + 2e^-$	2.52 V
At cathode:	$2Ag^+ + 2e^- \longrightarrow 2Ag(s)$	0.799 V
Overall:	$Mg(s) + 2Ag^+ \longrightarrow Mg^{2+} + 2Ag(s)$	3.32 V

Standard Reduction Potentials $\mathcal{E}°$ for Half-Cells at 298.15 K (pH = 0)[a]

Table 11.1

Electrode	Electrode Reaction	$\mathcal{E}°$ (V)
$Pt \mid F_2 \mid F^-$	$F_2(g) + 2e^- \longrightarrow 2F^-$	+2.87
$Pt \mid Co^{3+}, Co^{2+}$	$Co^{3+} + e^- \longrightarrow Co^{2+}$	+1.81
$Pt \mid Ce^{4+}, Ce^{3+}$	$Ce^{4+} + e^- \longrightarrow Ce^{3+}$	+1.61
$Pt \mid MnO_4^-, Mn^{2+}$	$MnO_4^- + 8H^+ + 5e^- \longrightarrow Mn^{2+} + 4H_2O$	+1.51
$Pt \mid Mn^{3+}, Mn^{2+}$	$Mn^{3+} + e^- \longrightarrow Mn^{2+}$	+1.51
$Pt \mid Cl_2 \mid Cl^-$	$Cl_2(g) + 2e^- \longrightarrow 2Cl^-$	+1.36
$Pt \mid Cr_2O_7^-, Cr^{3+}$	$Cr_2O_7^{2-} + 14H^+ + 6e^- \longrightarrow 2Cr^{3+} + 7H_2O$	+1.33
$Pt \mid Tl^{3+}, Tl^+$	$Tl^{3+} + 2e^- \longrightarrow Tl^+$	+1.25
$Pt \mid O_2, H_2O$	$O_2(g) + 4H^+ + 4e^- \longrightarrow 2H_2O$	+1.229
$Pt \mid Br_2, Br^-$	$Br_2 + 2e^- \longrightarrow 2Br^-$	+1.065
$Pt \mid Hg^{2+}, Hg_2^{2+}$	$2Hg^{2+} + 2e^- \longrightarrow Hg_2^{2+}$	+0.92
$Hg \mid Hg^{2+}$	$Hg^{2+} + 2e^- \longrightarrow Hg$	+0.854
$Ag \mid Ag^+$	$Ag^+ + e^- \longrightarrow Ag$	+0.799
$Pt \mid Fe^{3+}, Fe^{2+}$	$Fe^{3+} + e^- \longrightarrow Fe^{2+}$	+0.771
$Pt \mid I_2, I^-$	$I_2 + 2e^- \longrightarrow 2I^-$	+0.536
$Pt \mid O_2, OH^-$	$O_2(g) + 2H_2O + 4e^- \longrightarrow 4OH^-$	+0.401
$Pt \mid Fe(CN)_6^{3-}, Fe(CN)_6^{4-}$	$Fe(CN)_6^{3-} + e^- \longrightarrow Fe(CN)_6^{4-}$	+0.36
$Cu \mid Cu^{2+}$	$Cu^{2+} + 2e^- \longrightarrow Cu$	+0.337
$Pt \mid Hg \mid Hg_2Cl_2 \mid Cl^-$	$Hg_2Cl_2 + 2e^- \longrightarrow 2Hg + 2Cl^-$	+0.268
$Ag \mid AgCl \mid Cl^-$	$AgCl + e^- \longrightarrow Ag + Cl^-$	+0.223
$Pt \mid Sn^{4+}, Sn^{2+}$	$Sn^{4+} + 2e^- \longrightarrow Sn^{2+}$	+0.154
$Pt \mid Cu^{2+}, Cu^+$	$Cu^{2+} + e^- \longrightarrow Cu^+$	+0.153
$Ag \mid AgBr \mid Br^-$	$AgBr + e^- \longrightarrow Ag + Br^-$	+0.07
$Pt \mid H_2 \mid H^+$	$2H^+ + 2e^- \longrightarrow H_2$	0.0
$Pb \mid Pb^{2+}$	$Pb^{2+} + 2e^- \longrightarrow Pb$	−0.126
$Sn \mid Sn^{2+}$	$Sn^{2+} + 2e^- \longrightarrow Sn$	−0.136
$Co \mid Co^{2+}$	$Co^{2+} + 2e^- \longrightarrow Co$	−0.277
$Tl \mid Tl^+$	$Tl^+ + e^- \longrightarrow Tl$	−0.336
$Pb \mid PbSO_4 \mid SO_4^{2-}$	$PbSO_4 + 2e^- \longrightarrow Pb + SO_4^{2-}$	−0.359
$Cd \mid Cd^{2+}$	$Cd^{2+} + 2e^- \longrightarrow Cd$	−0.403
$Pt \mid Cr^{3+}, Cr^{2+}$	$Cr^{3+} + e^- \longrightarrow Cr^{2+}$	−0.41
$Fe \mid Fe^{2+}$	$Fe^{2+} + 2e^- \longrightarrow Fe$	−0.440
$Zn \mid Zn^{2+}$	$Zn^{2+} + 2e^- \longrightarrow Zn$	−0.763
$Mn \mid Mn^{2+}$	$Mn^{2+} + 2e^- \longrightarrow Mn$	−1.180
$Al \mid Al^{3+}$	$Al^{3+} + 3e^- \longrightarrow Al$	−1.662
$Mg \mid Mg^{2+}$	$Mg^{2+} + 2e^- \longrightarrow Mg$	−2.52
$Na \mid Na^+$	$Na^+ + e^- \longrightarrow Na$	−2.714
$Ca \mid Ca^{2+}$	$Ca^{2+} + 2e^- \longrightarrow Ca$	−2.866
$Sr \mid Sr^{2+}$	$Sr^{2+} + 2e^- \longrightarrow Sr$	−2.888
$Ba \mid Ba^{2+}$	$Ba^{2+} + 2e^- \longrightarrow Ba$	−2.905
$K \mid K^+$	$K^+ + e^- \longrightarrow K$	−2.925
$Li \mid Li^+$	$Li^+ + e^- \longrightarrow Li$	−3.05

[a] From M. J. Sienko and R. A. Plane, *Chemical Principles and Properties*, 2nd ed. Copyright 1974 by McGraw-Hill Book Company. Used with permission of McGraw-Hill Book Company.

Since $\mathcal{E}° = +3.32$ V, the reaction takes place spontaneously from left to right.

An alternative way to show the spontaneity of the reaction is to write

$$\Delta G° = -nF\mathcal{E}°$$

$$= -2(96,500 \text{ C mol}^{-1})(3.32 \text{ V})$$

$$= -641,000 \text{ C V mol}^{-1}$$

$$= -641 \text{ kJ mol}^{-1}$$

279

Comment: Although this example is a straightforward one, it brings out an important feature, that is, the addition of half-cell voltages. We see that in order to obtain the overall reaction, we had to multiply the equation representing the reduction of silver ions by a factor 2. The cell voltage then is given as the sum of the two half-cell voltages. But the basis of the addition of voltages is the additivity property of the Gibbs free energy. We write

$$Mg(s) \longrightarrow Mg^{2+} + 2e^- \qquad \Delta G_1^\circ = -n_1 F \mathcal{E}_1^\circ$$

$$2Ag^+ + 2e^- \longrightarrow 2Ag(s) \qquad \Delta G_2^\circ = -n_2 F \mathcal{E}_2^\circ$$

and for the overall reaction

$$Mg(s) + 2Ag^+ \longrightarrow Mg^{2+} + 2Ag(s) \qquad \Delta G_3^\circ = -n_3 F \mathcal{E}_3^\circ$$

Since G is a state function, we have

$$\Delta G_3^\circ = \Delta G_1^\circ + \Delta G_2^\circ$$

or

$$-n_3 F \mathcal{E}_3^\circ = -n_1 F \mathcal{E}_1^\circ - n_2 F \mathcal{E}_2^\circ$$

In our case $n_1 = n_2 = n_3$ so that

$$\mathcal{E}_3^\circ = \mathcal{E}_1^\circ + \mathcal{E}_2^\circ$$

When the number of moles of electrons in each of the half-reactions is not identical, n must be included, as follows:

$$\mathcal{E}_3^\circ = \frac{n_1 \mathcal{E}_1^\circ + n_2 \mathcal{E}_2^\circ}{n_3}$$

For example, from Table 11.1 we find that

$$Fe(s) \longrightarrow Fe^{2+} + 2e^- \qquad \mathcal{E}_1^\circ = 0.440 \text{ V}$$

and

$$Fe^{2+} \longrightarrow Fe^{3+} + e^- \qquad \mathcal{E}_2^\circ = -0.771 \text{ V}$$

The sum of these two half-reactions is

$$Fe(s) \longrightarrow Fe^{3+} + 3e^- \qquad \mathcal{E}_3^\circ = ?$$

However, $\mathcal{E}_3^\circ$ is not $(0.440 - 0.771) = -0.331$ V; instead, it is given by

$$\mathcal{E}_3^\circ = \frac{(2)(0.440 \text{ V}) + (1)(-0.771 \text{ V})}{3}$$

$$= 0.036 \text{ V}$$

As mentioned earlier, the emf of a cell, $\mathcal{E}$, depends on both $\mathcal{E}^\circ$ and the concentrations. If we are given two electrodes and the concentrations of the ions, how can we tell which electrode acts as the anode and which electrode acts as the cathode? One way to deal with this problem is to *arbitrarily* assign one electrode as the anode and the other as the cathode and calculate for the assumed cell reaction using Eq. (11.6). The cell emf if given by

$$\mathcal{E} = \mathcal{E}_{\text{cathode}} - \mathcal{E}_{\text{anode}}$$

If $\mathcal{E}$ turns out to be a positive quantity, then we guessed right; otherwise, the cell reaction must be the reverse of our reaction. For example, suppose that we have a galvanic cell made up of $Sn|Sn^{2+}$ and $Pb|Pb^{2+}$ electrodes where the concentrations are $[Sn^{2+}] = 1.6\ M$ and $[Pb^{2+}] = 0.20\ M$. Let us assume that the cell reaction proceeds as follows:

$$Sn(s) + Pb^{2+} \longrightarrow Sn^{2+} + Pb(s)$$

First, we determine the $\mathcal{E}°$ for the reaction:

$$\mathcal{E}°$$

			$\mathcal{E}°$
At anode:	$Sn(s)$	$\longrightarrow Sn^{2+} + 2e^-$	0.136 V
At cathode:	$Pb^{2+} + 2e^-$	$\longrightarrow Pb(s)$	-0.126 V
Overall:	$Sn(s) + Pb^{2+}$	$\longrightarrow Sn^{2+} + Pb(s)$	0.010 V

Thus we have

$$\mathcal{E} = \mathcal{E}° - \frac{0.0591}{n} \log \frac{[Sn^{2+}]}{[Pb^{2+}]}$$

$$= 0.010 - 0.027 = -0.017\ V$$

Since $\mathcal{E}$ is negative, the actual cell reaction must be

$$Pb(s) + Sn^{2+} \longrightarrow Pb^{2+} + Sn(s)$$

Temperature Dependence of EMF **11.5**

Useful thermodynamic quantities for the cell reaction can be easily measured from the temperature dependence of the emf. We start with

$$\Delta G° = -nF\mathcal{E}°$$

Differentiating $\Delta G°$ with respect to temperature at constant pressure, we obtain

$$\left(\frac{\partial \Delta G°}{\partial T}\right)_P = -nF\left(\frac{\partial \mathcal{E}°}{\partial T}\right)_P$$

Equation (7.20), when expressed in terms of changes in G and S, is given by

$$\left(\frac{\partial \Delta G°}{\partial T}\right)_P = -\Delta S°$$

so that

$$\Delta S° = nF\left(\frac{\partial \mathcal{E}°}{\partial T}\right)_P \tag{11.7}$$

Thus, from the variation of $\mathcal{E}°$ with temperature,* we can measure the standard entropy change of the cell reaction. Suppose that we wish to determine $(\partial \mathcal{E}°/\partial T)_P$ for the Daniell cell. The simplest way is to set $[Zn^{2+}] = 1.00\ M$ and

---

* The temperature dependence of the emf of most car batteries is generally quite small, of the order of 5×10^{-4} V K^{-1}, insufficient to explain why cars will not start on a cold morning. For an interesting explanation of the real cause, see L. K. Nash, *J. Chem. Educ.* **47**, 382 (1970).

$[Cu^{2+}] = 1.00\ M$ (the standard states) and then measure the emf of the cell at several different temperatures. Once $\Delta S°$ and $\Delta G°$ are known, we can then calculate $\Delta H°$ at some temperature T as follows:

$$\Delta G° = \Delta H° - T\ \Delta S°$$

or

$$\Delta H° = \Delta G° + T\ \Delta S°$$

$$= -nF\mathscr{E}° + nFT\left(\frac{\partial \mathscr{E}°}{\partial T}\right)_P \tag{11.8}$$

In general, $\Delta H°$ and $\Delta S°$ are approximately temperature-independent, while $\Delta G°$ varies with temperature. Note that Eq (11.8) provides us with a purely non-calorimetric method of determining the enthalpy change of a reaction.

EXAMPLE 11.2 Consider an electrochemical cell represented by the following cell diagram:

$$Zn\,|\,ZnSO_4(0.10\ M)\,|\,KCl(1.0\ M)\,|\,AgCl(s)\,|\,Ag$$

The emf of the cell is 1.015 V at 298 K and 0.9953 V at 338 K. Calculate the values of ΔG, ΔS, and ΔH at 298 K.

Answer: The Gibbs free energy change for the reaction at 298 K is

$$\Delta G = -nF\mathscr{E}$$

First we have to determine the value of n. The half-reactions at:

Anode: $\qquad Zn(s) \longrightarrow Zn^{2+}(0.10\ M) + 2e^-$

Cathode: $\quad 2AgCl(s) + 2e^- \longrightarrow 2Ag(s) + 2Cl^-(1.0\ M)$

Overall: $\;Zn(s) + 2AgCl(s) \longrightarrow 2Ag(s) + Zn^{2+}(0.10\ M) + 2Cl^-(1.0\ M)$

Since $n = 2$, we have, at 298 K,

$$\Delta G = -(2)(96{,}500\ \text{C mol}^{-1})(1.015\ \text{V})$$

$$= -195{,}900\ \text{C V mol}^{-1}$$

$$= -195.9\ \text{kJ mol}^{-1}$$

The temperature coefficient of emf is given by

$$\left(\frac{\partial \mathscr{E}}{\partial T}\right)_P = \frac{(0.9953 - 1.015)\ \text{V}}{(338 - 298)\ \text{K}} = -4.93 \times 10^{-4}\ \text{V K}^{-1}$$

Thus

$$\Delta S = (2)(96{,}500\ \text{C mol}^{-1})(-4.93 \times 10^{-4}\ \text{V K}^{-1})$$

$$= -95.2\ \text{J K}^{-1}\ \text{mol}^{-1}$$

Finally,

$$\Delta H = -195.9\ \text{kJ mol}^{-1} + (298)\left(\frac{-95.2}{1000}\ \text{kJ K}^{-1}\ \text{mol}^{-1}\right)$$

$$= -224.3\ \text{kJ mol}^{-1}$$

In addition to the galvanic cell discussed earlier, there are other types of electrochemical cells in use. Two examples will now be briefly described.

CONCENTRATION CELLS

Consider the arrangement

$$Pt\,|\,H_2(P_1)\,|\,HCl\,|\,H_2(P_2)\,|\,Pt$$

The two hydrogen electrodes are placed in the same HCl solution. If pressure P_1 is greater than P_2, the following reactions will occur:

At anode: $\qquad H_2(P_1) \longrightarrow 2H^+ + 2e^-$

At cathode: $\quad 2H^+ + 2e^- \longrightarrow H_2(P_2)$

Thus the overall reaction

$$H_2(P_1) \longrightarrow H_2(P_2)$$

is simply a transfer of the hydrogen gas at P_1 to P_2. The HCl solution serves as a sink and source for H^+ ions as well as a medium for conducting electricity. Its concentration remains constant. The emf of the cell is purely due to a "dilution" process of H_2 gases. At 298 K, we have

$$\mathcal{E} = \mathcal{E}^\circ - \frac{RT}{2F} \ln \frac{P_2}{P_1}$$

$$= -0.0296 \log \frac{P_2}{P_1}$$

since by definition $\mathcal{E}^\circ$ is zero for the hydrogen electrode.

The case described above is a rather simple example of a series of cells known as *concentration cells*. The operation of a concentration cell depends only on the difference in concentration of the two compartments of the cell and not on the identities of the substances taking part in the electrochemical reaction. Generally, most concentration cells involve solution species only.

FUEL CELLS

Fossil fuel is a major source of energy supply at present. Unfortunately, the combustion of fossil fuel is a highly irreversible process and so of low thermodynamic efficiency. On the other hand, fuel cells can make combustion largely reversible, converting a greater amount of chemical energy into useful work.

In its simplest form, a fuel cell consists of an electrolyte solution such as sulfuric acid or sodium hydroxide and two inert electrodes. Hydrogen and oxygen gases are bubbled through the anode and cathode compartments, where the following reactions take place:

At anode: $\quad H_2(g) \longrightarrow 2H^+ + 2e^-$

At cathode: $\quad \frac{1}{2}O_2(g) + H_2O + 2e^- \longrightarrow 2OH^-$

Overall: $\qquad H_2(g) + \frac{1}{2}O_2(g) \longrightarrow H_2O$

283

Figure 11.5

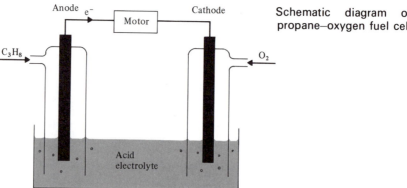

A potential difference is established between the two electrodes and current flows through the wire connecting the two electrodes from the anode to the cathode.

This reaction exactly reverses electrolysis of water. The function of the electrodes is twofold. First, the anode acts as a source, the cathode as a sink, of electrons. Second, the electrodes provide the necessary surface for the initial decomposition of the molecules into atomic species. They are *electrocatalysts*. Metals such as platinum, iridium, and rhodium are good electrocatalysts.

In addition to the H_2–O_2 system, a number of other fuel cells have been developed. Among these is the propane–oxygen fuel cell shown in Figure 11.5. The half-cell reactions are

$$\text{At anode:} \qquad C_3H_8 + 6H_2O \longrightarrow 3CO_2 + 20H^+ + 20e^-$$

$$\text{At cathode:} \quad 5O_2 + 20H^+ + 20e^- \longrightarrow 10H_2O$$

The overall reaction,

$$C_3H_8 + 5O_2 \longrightarrow 3CO_2 + 4H_2O$$

is identical to burning propane in oxygen. With proper design, an efficiency as high as 70% may be attained, about twice that for an internal combustion engine. In addition, fuel cells generating electricity are free of the noise, vibration, heat transfer, and other problems normally associated with conventional power plants. The fuel cells are so attractive that they are likely to become operational on a large scale within the next decade or so. At present, much effort is being spent in search of suitable electrocatalysts for various gases.

11.7 Applications of EMF Measurements

We shall now discuss two important applications of emf measurements.

DETERMINATION OF ACTIVITY COEFFICIENTS

Electromotive force measurement provides one of the most convenient and accurate methods for determining the activity coefficient of ions. As an example, let us consider the following cell arrangement:

$$Pt\,|\,H_2(1\text{ atm})\,|\,HCl(aq, m)\,|\,AgCl(s)\,|\,Ag$$

The overall reaction for the cell is

$$\tfrac{1}{2}H_2 + AgCl(s) \longrightarrow Ag(s) + H^+ + Cl^-$$

and the emf of the cell at 298 K is given by

$$\varepsilon = \varepsilon^\circ - 0.0591 \log \frac{a_{H^+}\, a_{Cl^-}\, a_{Ag}}{f_{H_2}^{1/2}\, a_{AgCl}}$$

Since both Ag and AgCl are solids, their activities are unity; the fugacity of hydrogen gas at 1 atm is also approximately unity, so that the preceding equation reduces to

$$\varepsilon = \varepsilon^\circ - 0.0591 \log a_{H^+}\, a_{Cl^-}$$

From Eq. (9.15) we find that

$$a_{H^+}\, a_{Cl^-} = \gamma_\pm^2\, m_\pm^2$$
$$= \gamma_\pm^2\, m^2$$

For HCl $m_\pm = m$. The emf of the cell can now be expressed as

$$\varepsilon = \varepsilon^\circ - 0.0591 \log (\gamma_\pm m)^2$$
$$= \varepsilon^\circ - 0.1182 \log m - 0.1182 \log \gamma_\pm$$

Thus, by measuring ε at a particular concentration m, $\gamma_\pm$ can be calculated since ε° is known. (From Table 11.1, $\varepsilon^\circ = 0.223$ V.)

pH MEASUREMENTS

Determining pH from emf measurements has now become a standard technique. As mentioned earlier, using the hydrogen electrode itself is impractical. Instead, the combination of glass electrode–calomel electrode provides the most suitable arrangement for this purpose. Consider the cell (also see Figure 11.4)

$$Ag(s)\,|\,AgCl(s)\,|\,HCl(aq),\ NaCl(aq)\ \begin{vmatrix} H^+(aq) \\ X^-(aq) \end{vmatrix}\ KCl(satd)\,|\,Hg_2Cl_2(s)\,|\,Hg(l)$$

glass electrode   solution of   calomel electrode
unknown pH

The overall emf ε for this arrangement is given by

$$\varepsilon = \varepsilon_{ref} - \frac{RT}{F} \ln a_{H^+}$$

$$= \varepsilon_{ref} + \frac{2.303 RT}{F}\, pH$$

where pH is defined in terms of activity, that is,

$$pH = -\log a_{H^+}$$

In practice, a_{H^+} can be replaced by $[H^+]$, except in very precise work. Rearranging the equation for ε and setting $T = 298$ K, we obtain

$$pH = \frac{\varepsilon - \varepsilon_{ref}}{0.0591}$$

Measuring $\mathcal{E}$ for a number of solutions of accurately known pH determines the constant $\mathcal{E}_{ref}$ term. Once $\mathcal{E}_{ref}$ is known, the combination of the glass electrode and the calomel electrode can then be used to determine the pH of other solutions from $\mathcal{E}$ values. The practical arrangement of this combination is called a pH meter.

11.8 Potentiometric Titration of Redox Reactions

The oxidation of a reducing agent by an oxidizing agent can be followed "potentiometrically" by placing an inert platinum electrode in the solution containing the reducing agent. When this half-cell is coupled with a reference electrode (for example, a calomel electrode), the progress of the redox titration can be followed by measuring the emf of the completed cell during the addition of a solution containing the oxidizing agent.

Suppose that a platinum electrode is immersed in a solution containing both the Fe^{2+} and Fe^{3+} ions. The reaction taking place at the electrode is

$$Fe^{3+} + e^- \rightleftharpoons Fe^{2+}$$

The reaction does not proceed appreciably in either direction unless the electrode is connected with another electrode. Nevertheless, this redox system has a certain electrode potential $\mathcal{E}$, which at 298 K is given by

$$\mathcal{E} = \mathcal{E}° - \frac{RT}{nF} \log \frac{[Fe^{2+}]}{[Fe^{3+}]}$$

$$= 0.771 + 0.0591 \log \frac{[Fe^{3+}]}{[Fe^{2+}]}$$

The value of $\mathcal{E}$ depends on the concentration of Fe^{3+} relative to Fe^{2+}. For example, if $[Fe^{3+}]/[Fe^{2+}] = 10$, $\mathcal{E} = 0.830$ V; if the ratio is 0.1, $\mathcal{E} = 0.712$ V. At an equal concentration of Fe^{3+} and Fe^{2+}, $\mathcal{E} = 0.771$ V. A plot of $\mathcal{E}$ versus the percent of the reduced form is shown in Figure 11.6.

Now let us consider a more complicated system, one that consists of two redox couples. The reaction can be represented by

$$Rd_1 + Ox_2 \rightleftharpoons Ox_1 + Rd_2$$

Figure 11.6

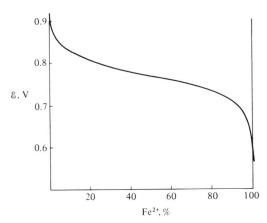

Potential of the Fe^{2+}/Fe^{3+} couple as a function of the Fe^{2+} concentration.

where Rd and Ox are the reduced and oxidized forms of species 1 and 2. It is convenient to write the two half-reactions as reductions:

$$Ox_1 + ne^- \longrightarrow Rd_1$$

$$Ox_2 + ne^- \longrightarrow Rd_2$$

With a suitable electrode such as the platinum electrode, it is possible to measure the electrode potential for each redox couple $(Ox_1 | Rd_1)$ or $(Ox_2 | Rd_2)$ with reference to the calomel electrode. The Nernst equation for each couple at 298 K is

$$\mathcal{E}_1 = \mathcal{E}_1^\circ + \frac{0.0591}{n} \log \frac{[Ox_1]}{[Rd_1]} \qquad (11.9a)$$

$$\mathcal{E}_2 = \mathcal{E}_2^\circ + \frac{0.0591}{n} \log \frac{[Ox_2]}{[Rd_2]} \qquad (11.9b)$$

$\mathcal{E}_1$ and $\mathcal{E}_2$ must be the same, since the platinum electrode is in equilibrium with both redox couples, yet there can be only one potential. Thus we have $\mathcal{E}_1 = \mathcal{E}_2 = \mathcal{E}$.

The dependence of $\mathcal{E}$ on the concentrations of Rd and Ox suggests that the oxidation of a reducing agent by an oxidizing agent can be followed potentiometrically. A well-studied system is

$$Fe^{2+} + Ce^{4+} \longrightarrow Fe^{3+} + Ce^{3+}$$

$$Rd_1 + Ox_2 \longrightarrow Ox_1 + Rd_2$$

Ceric sulfate solution (oxidizing agent) is added to the ferrous sulfate solution (reducing agent) to carry out a titration. The progress of the reaction is monitored by measuring the emf of the cell employing the calomel reference electrode. The standard reduction potentials are (see Table 11.1):

$$Fe^{3+} + e^- \longrightarrow Fe^{2+} \qquad \mathcal{E}^\circ = 0.771 \text{ V}$$

$$Ce^{4+} + e^- \longrightarrow Ce^{3+} \qquad \mathcal{E}^\circ = 1.61 \text{ V}$$

Before the addition of ceric sulfate, the solution consists mostly of ferrous ions, plus a minute amount of ferric ions. Consequently, the system does not have a definite electrode potential, although its value must be less than 0.771 V. However, as soon as a few drops of the ceric solution have been added, the ratio $[Fe^{3+}]/[Fe^{2+}]$ assumes a definite value and a redox potential as well. In the initial stages of the titration, $[Ce^{4+}] \simeq 0$, so that the electrode potential is determined by the Fe^{3+}/Fe^{2+} couple; the potential only rises gradually in this region. Approaching the equivalence point brings a sharp rise in the potential as a result of the large change in the concentration ratio. At the equivalence point, we have $[Ox_1] = [Rd_2]$ and $[Ox_2] = [Rd_1]$. From Eq. (11.9), we write

$$\mathcal{E}_{ep} = \mathcal{E}_1^\circ + \frac{0.0591}{n} \log \frac{[Ox_1]}{[Rd_1]}$$

and

$$\mathcal{E}_{ep} = \mathcal{E}_2^\circ + \frac{0.0591}{n} \log \frac{[Ox_2]}{[Rd_2]}$$

where $\mathcal{E}_{ep}$ is the potential at the equivalence point. Adding the equations and substituting the equalities, we get

$$2\mathcal{E}_{ep} = \mathcal{E}_1^\circ + \mathcal{E}_2^\circ + \frac{0.0591}{n} \log \frac{[Ox_1][Ox_2]}{[Rd_1][Rd_2]}$$

$$= \mathcal{E}_1^\circ + \mathcal{E}_2^\circ + \frac{0.0591}{n} \log(1)$$

or

$$\mathcal{E}_{ep} = \frac{\mathcal{E}_1^\circ + \mathcal{E}_2^\circ}{2} = 1.19 \text{ V}$$

Beyond the equivalence point $[Fe^{2+}] \simeq 0$ so that $\mathcal{E}$ is mainly determined by the Ce^{4+}/Ce^{3+} couple. Table 11.2 summarizes the results for the potentiometric titration of 50 ml of 0.1 M $FeSO_4$ with a solution of 0.1 M $Ce(SO_4)_2$. The corresponding titration curve is shown in Figure 11.7. The curve is quite similar to the acid–base titration curve to be discussed in Chapter 12.

Example 11.3

(a) From the standard emfs of the Fe^{2+}/Fe^{3+} and Ce^{3+}/Ce^{4+} couples, calculate the equilibrium constant for the following reaction at 298 K:

$$Ce^{4+} + Fe^{2+} \longrightarrow Ce^{3+} + Fe^{3+}$$

(b) The reaction in part (a) is employed in a redox titration. Calculate the emf of the cell after the addition of 10.0 ml of a 0.1 M Ce^{4+} solution to 50.0 ml of a 0.1 M Fe^{2+} solution.

Table 11.2

Potentiometric Titration of a 50-ml 0.1 M Fe²⁺ Solution Against a 0.1 M Ce⁴⁺ Solution

Amount of Ce^{4+} Solution Added (ml)	$\dfrac{[Fe^{3+}]}{[Fe^{2+}]}$	$\mathcal{E}_{Fe^{3+}/Fe^{2+}}$ (V)
0.2	0.02/4.98	0.63
2.0	0.20/4.80	0.69
10.0	1.00/4.00	0.74
25.0	2.50/2.50	0.77
40.0	4.00/1.00	0.81
49.0	4.90/0.1	0.87
49.9	4.99/0.01	0.93
50.0	$5.00/4 \times 10^{-7}$	1.19

	$\dfrac{[Ce^{4+}]}{[Ce^{3+}]}$	$\mathcal{E}_{Ce^{4+}/Ce^{3+}}$ (V)
50.2	0.02/5.00	1.51
52.0	0.20/5.00	1.57
60.0	1.00/5.00	1.61
75.0	2.50/5.00	1.63
100	5.00/5.00	1.65

Profile of the potentiometric titration of 50 ml of 0.1 M Fe²⁺ solution against a 0.1 M Ce⁴⁺ solution.

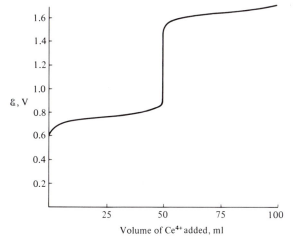

Figure 11.7

&, V (y-axis)

Volume of Ce⁴⁺ added, ml (x-axis)

Answer

(a) The redox reaction can be written as two half-reactions:

$$\mathcal{E}°$$

		$\mathcal{E}°$
Oxidation:	$Fe^{2+} \longrightarrow Fe^{3+} + e^{-}$	-0.771 V
Reduction:	$Ce^{4+} + e^{-} \longrightarrow Ce^{3+}$	1.61 V
Overall:	$Ce^{4+} + Fe^{2+} \longrightarrow Ce^{3+} + Fe^{3+}$	0.839 V

From Eq. (11.5),

$$\ln K = \frac{nF\mathcal{E}°}{RT}$$

$$= \frac{(1)(96{,}500 \text{ C mol}^{-1})(0.839 \text{ V})}{(8.314 \text{ J K}^{-1} \text{ mol}^{-1})(298 \text{ K})}$$

$$= 32.7$$

or

$$K = \frac{[Ce^{3+}][Fe^{3+}]}{[Ce^{4+}][Fe^{2+}]} = 1.59 \times 10^{14}$$

(b) The number of moles of Fe²⁺ initially present is

$$\frac{50 \text{ ml}}{1000 \text{ ml}} \times 0.1 \text{ mol} = 5.00 \times 10^{-3} \text{ mol}$$

The number of moles of Ce⁴⁺ present in 10.0 ml is

$$\frac{10 \text{ ml}}{1000 \text{ ml}} \times 0.1 \text{ mol} = 1.00 \times 10^{-3} \text{ mol}$$

Since the equilibrium constant is very large, the reaction above proceeds almost to completion. Let the number of moles of Ce^{4+} at equilibrium be x; then

$$[Ce^{4+}] = \frac{x}{0.06} \qquad\qquad [Fe^{3+}] = \frac{1.00 \times 10^{-3} - x}{0.06}$$

$$[Ce^{3+}] = \frac{1.00 \times 10^{-3} - x}{0.06} \qquad [Fe^{2+}] = \frac{5.00 \times 10^{-3} - 1.00 \times 10^{-3} + x}{0.06}$$

Note that the total volume is now 60 ml or 0.06 liter. Thus

$$K = \frac{\left(\dfrac{1.00 \times 10^{-3} - x}{0.06}\right)\left(\dfrac{1.00 \times 10^{-3} - x}{0.06}\right)}{\left(\dfrac{x}{0.06}\right)\left(\dfrac{4.00 \times 10^{-3} + x}{0.06}\right)} = 1.59 \times 10^{14}$$

It turns out that x is an exceedingly small number so that for all practical purposes we have

$$[Ce^{3+}] = [Fe^{3+}] = \frac{1.00 \times 10^{-3}}{0.06}$$

and

$$[Fe^{2+}] = \frac{4.00 \times 10^{-3}}{0.06}$$

so that $x = 1.57 \times 10^{-18}$. Finally, the emf of the cell is given by

$$\varepsilon = 0.771 \text{ V} + \frac{0.0591}{1} \log \frac{(1.00 \times 10^{-3})/0.06}{(4.00 \times 10^{-3})/0.06}$$

$$= 0.74 \text{ V}$$

Comment: In general, we shall employ the approximation shown above in calculating ε. This procedure is used to obtain ε values shown in Table 11.2.

Without following the foregoing titration potentiometrically, the equivalence point may be determined by using a redox indicator. A redox indicator is a compound whose color in the oxidized form differs from that in the reduced form. For the titration described, the o-phenanthroline iron complex would be a suitable indicator, for the standard electrode potential for the reaction

$$\text{Fe}(o\text{-phenanthroline})_3^{3+} + e^- \longrightarrow \text{Fe}(o\text{-phenanthroline})_3^{2+}$$

$$\text{blue} \qquad\qquad\qquad\qquad\qquad \text{red}$$

is 1.05 V, which lies on the steep portion of the titration curve. Note that in this case, the redox indicator contains the reacting species themselves, Fe^{2+} and Fe^{3+}. Usually, this is not the case.

In some biological oxidation–reduction reactions, or simply *biological oxidations*, hydrogen ions are transferred along with electrons. Thus we may have a reaction of the form

$$AH_2 + B \longrightarrow [A + 2H^+ + 2e^- + B] \longrightarrow A + BH_2$$

In other instances the substance being oxidized may lose hydrogen ions, while transferring only its electrons to the substance being reduced

$$AH_2 + B \longrightarrow [A + 2H^+ + 2e^- + B] \longrightarrow A + B^{2-} + 2H^+$$

The third type of biological oxidation involves only the transfer of electrons

$$A^{2-} + B \longrightarrow [A + 2e^- + B] \longrightarrow A + B^{2-}$$

Biological oxidations, seldom, if ever, take place in a simple, direct manner. Generally, the mechanism is quite complex, involving a number of enzymes.

In this section we briefly describe how a knowledge of redox potentials provides useful information relevant to some biological processes. An example is the terminal oxidation chain, or respiratory chain, located in *mitochondria.** The net or overall reaction is the transfer of electrons from "fuel" molecules such as glucose to reduce molecular oxygen to water. The other product is carbon dioxide.

We saw in Chapter 10 that glycolysis is not an efficient process. If the system lacks oxygen, the end product of glycolysis, pyruvic acid, is reduced to lactic acid with NADH as the reducing agent. In an aerobic process, the degradation of glucose proceeds through two more steps called the *Krebs cycle* and *terminal respiratory chain* (Figure 11.8), with carbon dioxide and water as the end products.

The "primer" step before the Krebs cycle is the combination of pyruvic acid with a molecule called reduced coenzyme A (CoA—SH) to form acetyl coenzyme A:

$$CH_3COCOOH + NAD^+ + CoA—SH$$

$$\longrightarrow CH_3CO—S—CoA + CO_2 + NADH + H^+$$

acetyl coenzyme A

During the cycle, a pyruvic acid is transformed into three CO_2 molecules that may ultimately be used as raw material for continued photosynthesis (Figure 11.9). More important, the cycle also generates reduced forms of the carrier molecules NADH and $FADH_2$,† which will be used in the synthesis of ATP in the terminal respiratory chain. Since one glucose molecule yields two

The three main stages involved in the breakdown of glucose to carbon dioxide and water.

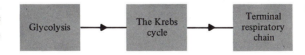

Figure 11.8

---

* Mitochondria are subcellular organelles in the cytoplasm of eucaryotic cells (that is, cells with nuclei), bounded by a double membrane and responsible for the reactions of respiratory metabolism, among other things.

† FAD and $FADH_2$ are the oxidized and reduced forms of flavin adenine dinucleotide, another oxidation–reduction carrier molecule of the same general type as NAD^+ and NADH.

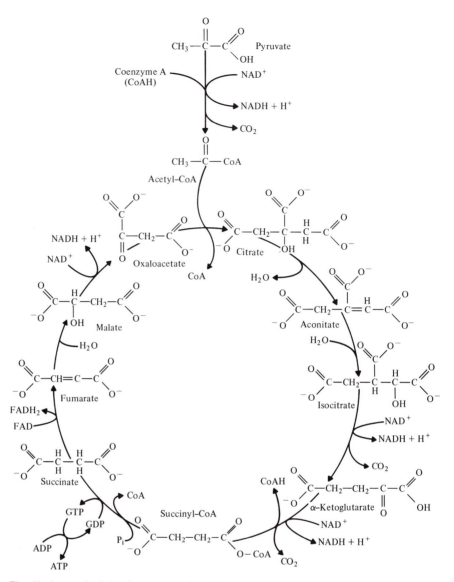

Figure 11.9   The Krebs cycle (also known as the citric acid cycle).

molecules of pyruvic acid in glycolysis (see Figure 10.9), we need to take two complete trips around the Krebs cycle for every glucose molecule degraded.

We now come to the last stage in the metabolic process—the terminal respiratory chain (Figure 11.10). The electrons donated by glucose and carried by NADH are transferred from one electron carrier molecule to another. Eventually, they are passed on to a molecule of oxygen, which thereby becomes converted to water. The cytochromes are electron-carrying proteins containing the heme group. The letters b, c, a, and a_3 denote the various forms of cytochromes. The iron atom of each cytochrome molecule can exist in the oxidized (Fe^{3+}) and reduced (Fe^{2+}) form. Cytochromes a and a_3 together are called *cytochrome oxidase* or the *respiratory enzyme*, which can transfer its electron directly to molecular oxygen. The sequence of the electron carriers in the respiratory chain is determined by their relative redox potentials. Table 11.3 lists the redox potentials for a number of important biological systems.

293

Sec. 11.9

Biological Oxidation

It should be kept in mind that the standard reduction potentials ($\mathcal{E}^{\circ\prime}$) are based on the hydrogen electrode scale at pH 7, rather than pH 0 as for the values listed in Table 11.1. A similar case was discussed in Section 10.6 for ΔG° and $\Delta G^{\circ\prime}$. Again we consider the reaction that produces H^+ ions:

$$A + B \longrightarrow C + xH^+$$

for which we write (at $T = 298$ K and $x = 1$)

$$\Delta G^\circ = \Delta G^{\circ\prime} + 40.0 \text{ kJ mol}^{-1}$$

Dividing the last equation by $-nF$, we obtain

$$\mathcal{E}^\circ = \mathcal{E}^{\circ\prime} - \frac{40,000 \text{ J mol}^{-1}}{n(96,500 \text{ C mol}^{-1})}$$

$$= \mathcal{E}^{\circ\prime} - \frac{0.415}{n} \text{ V}$$

This means that for reactions producing H^+ ions, $\mathcal{E}^\circ$ is less than $\mathcal{E}^{\circ\prime}$ by $0.415/n$ per mole of H^+ ions produced. Hence the reaction is more spontaneous at pH 7 than at pH 0. On the other hand, if H^+ ions appear as the reactant,

$$C + xH^+ \longrightarrow A + B$$

Standard Reduction Potentials $\mathcal{E}^{\circ\prime}$ for Some Biological Half-Reactions at 298.15 K (pH 7)[a]   **Table 11.3**

System	Half-Cell Reaction	$\mathcal{E}^{\circ\prime}$ (V)
O_2/H_2O	$O_2(g) + 4H^+ + 4e^- \longrightarrow 2H_2O$	+0.816
Cu^{2+}/Cu^+ hemocyanin	$Cu^{2+} + e^- \longrightarrow Cu^+$	+0.540
Cyt f^{3+}/Cyt f^{2+}	$Fe^{3+} + e^- \longrightarrow Fe^{2+}$	+0.365
Cyt a^{3+}/Cyt a^{2+}	$Fe^{3+} + e^- \longrightarrow Fe^{2+}$	+0.29
Cyt c^{3+}/Cyt c^{2+}	$Fe^{3+} + e^- \longrightarrow Fe^{2+}$	+0.254
Fe^{3+}/Fe^{2+} hemoglobin	$Fe^{3+} + e^- \longrightarrow Fe^{2+}$	+0.17
Fe^{3+}/Fe^{2+} myoglobin	$Fe^{3+} + e^- \longrightarrow Fe^{2+}$	+0.046
Fumarate/succinate	$^-OOCCH{=}CHCOO^- + 2H^+ + 2e^- \longrightarrow {}^-OOCCH_2CH_2COO^-$	+0.031
MB/MBH$_2$[b]	$MB + 2H^+ + 2e^- \longrightarrow MBH_2$	+0.011
Oxaloacetate/malate	$^-OOC{-}COCH_2COO^- + 2H^+ + 2e^- \longrightarrow {}^-OOCCHOHCH_2COO^-$	−0.166
Pyruvate/lactate	$CH_3COCOO^- + 2H^+ + 2e^- \longrightarrow CH_3CHOHCOO^-$	−0.185
Acetaldehyde/ethanol	$CH_3CHO + 2H^+ + 2e^- \longrightarrow CH_3CH_2OH$	−0.197
FAD/FADH$_2$	$FAD + 2H^+ + 2e^- \longrightarrow FADH_2$	−0.219
NAD^+/NADH	$NAD^+ + 2H^+ + 2e^- \longrightarrow NADH + H^+$	−0.320
$NADP^+$/NADPH	$NADP^+ + 2H^+ + 2e^- \longrightarrow NADPH + H^+$	−0.324
CO_2/formate	$CO_2 + H^+ + 2e^- \longrightarrow HCOO^-$	−0.414
H^+/H_2	$2H^+ + 2e^- \longrightarrow H_2$	−0.421
Fe^{3+}/Fe^{2+} ferredoxin	$Fe^{3+} + e^- \longrightarrow Fe^{2+}$	−0.432
Acetic acid/acetaldehyde	$CH_3COOH + 2H^+ + 2e^- \longrightarrow CH_3CHO + H_2O$	−0.581
Acetate/pyruvate	$CH_3COOH + CO_2 + 2H^+ + 2e^- \longrightarrow CH_3COCOOH + H_2O$	−0.70

[a] From *Handbook of Biochemistry*, H. A. Sober (ed.), © The Chemical Rubber Co., 1968. Used by permission of The Chemical Rubber Co.
[b] The symbols MB and MBH$_2$ represent the oxidized and reduced forms of methylene blue, which is used as a redox indicator.

Figure 11.10

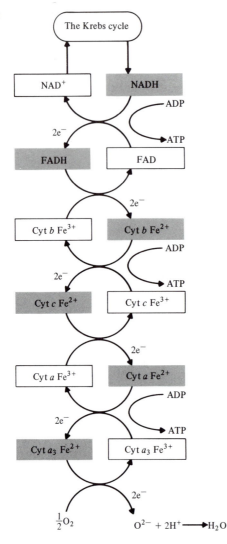

The terminal respiratory chain. This chain consists of two separable sets of reactions: electron transport and phosphorylation. Electrons derived from reactions of the Krebs cycle are passed in sequence from one carrier to another. Each carrier alternates between the reduced state and the oxidized state. The final electron acceptor is molecular oxygen.

the following reaction can be readily derived:

$$\mathcal{E}^\circ = \mathcal{E}^{\circ\prime} + \frac{0.415}{n}\,\text{V}$$

Hence the reaction is more spontaneous at pH 0 than at pH 7. For reactions not involving hydrogen ions, $\mathcal{E}^\circ = \mathcal{E}^{\circ\prime}$.

We can calculate the Gibbs free energy change as a pair of electrons move through the terminal respiratory chain. The conversion of NADH to NAD$^+$ releases two electrons (the first step) which are used to reduce molecular oxygen to water. From Table 11.3 the half-reactions are

$$\text{NADH} + \text{H}^+ \longrightarrow \text{NAD}^+ + 2\text{H}^+ + 2e^- \qquad \mathcal{E}^{\circ\prime} = 0.32 \text{ V}$$

and

$$\tfrac{1}{2}\text{O}_2 + 2\text{H}^+ + 2e^- \longrightarrow \text{H}_2\text{O} \qquad \mathcal{E}^{\circ\prime} = 0.816 \text{ V}$$

The overall reaction is

$$\text{NADH} + \text{H}^+ + \tfrac{1}{2}\text{O}_2 \longrightarrow \text{NAD}^+ + \text{H}_2\text{O} \qquad \mathcal{E}^{\circ\prime} = 1.136 \text{ V}$$

Since $\Delta G^{\circ\prime} = -nF\mathcal{E}^{\circ\prime}$, we write

$$\Delta G^{\circ\prime} = -(2)(96{,}500 \text{ C mol}^{-1})(1.136 \text{ V})$$
$$= -219{,}000 \text{ C V mol}^{-1}$$
$$= -219 \text{ kJ mol}^{-1}$$

The free energy released here is considerably larger than that can be stored in the synthesis of ATP molecules ($\Delta G^{\circ\prime} = 31.4 \text{ kJ mol}^{-1}$). The solution to this dilemma is a series of smaller steps in releasing free energy along the chain shown in Figure 11.10. There are actually three sites of ATP synthesis: that between NADH and FAD, that between cytochrome b and cytochrome c, and that between cytochrome a and cytochrome a_3.

Finally, let us calculate the total number of moles of ATP synthesized. A total of 12 electron pairs are passed down the respiratory chain for every molecule of glucose oxidized to carbon dioxide and water. The total decrease in the Gibbs free energy is $-12 \times 219 = -2628 \text{ kJ mol}^{-1}$. Since the complete combustion of glucose in air results in a Gibbs free energy decrease of $-2879 \text{ kJ mol}^{-1}$ (see Section 10.6), we see that most of the Gibbs free energy available in the oxidation of glucose is released during the respiratory chain portion of the metabolic pathway. Much of this Gibbs free energy is used to synthesize 34 mol of ATP. Including the 2 mol of ATP synthesized during glycolysis and 2 mol produced in the Krebs cycle (remember that each trip around the cycle produces 1 mol of ATP and there are two trips), a total of 38 mol of ATP is obtained from the conversion of 1 mol of glucose to carbon dioxide and water (Table 11.4).

Products from the Biological Degradation of 1 Mol of Glucose to Carbon Dioxide and Water

Table 11.4

Process	Products (mol)		
	NADH	FADH	ATP
Glycolysis			
ATP needed for coupled reactions			-2
ATP produced			$+4$
NADH produced	$+2$		
Krebs cycle			
ATP produced			$+2$
NADH produced	$+8$		
FADH produced		$+2$	
Terminal respiratory chain			
ATP produced from 2 mol of NADH from glycolysis (2×3)			$+6$
ATP produced from 8 mol of NADH from the Krebs cycle (8×3)			$+24$
ATP produced from 2 mol of FADH from the Krebs cycle (2×2)			$+4$
ATP from glycolysis			2 (5%)
ATP from the Krebs cycle			2 (5%)
ATP from the terminal respiratory chain			34 (90%)

THE CHEMIOSMOTIC THEORY OF OXIDATIVE PHOSPHORYLATION

So far we have considered a series of metabolic reactions in which reduced coenzymes, formed during the series of reactions in the Krebs cycle, pass their electrons to an assembly of electron acceptors called the electron transport chain. As the electrons travel down this chain, much of the energy they release is coupled to the phosphorylation of ADP to ATP. This process is known as *oxidative phosphorylation*. The question of central importance is: How is the transfer of electrons through the series of carrier molecules coupled to the synthesis of ATP? Several theories have been put forward to explain the mechanism of this coupling. Here we shall discuss only the *chemiosmotic theory* proposed by the British biochemist Peter Mitchell because of its general acceptance by scientists working in bioenergetics.

Figure 11.11a shows a schematic diagram of a mitochondrion, which consists of an outer membrane and an inner membrane. The inner membrane has embedded in it the enzymes and other components of the respiratory chain. According to Mitchell, the energy released during transport of the electrons along the carrier chain is conserved in a hydrogen-ion gradient and an electrical gradient, which then drive the oxidative phosphorylation.

As a pair of electrons flows down the electron transport chain, a total of six hydrogen ions are expelled from the inner compartment to the outer compartment. This expulsion results in a rise in pH on the inside and a fall in pH on the outside of the inner membrane. Since the inner membrane is impermeable to these ions, a pH gradient is maintained (Figure 11.11b). Consequently, there is also a rise in the electric potential across the membrane, since there are more positive ions (H^+) on the outside and more negative ions (OH^-) on the inside. Protons at the outer surface will seek to move back to the inside, down the potential gradient; this proton gradient, analogous to the electric current produced by a battery, can be drawn upon to do work. The molecular mechanism of the step involving the flow of protons to the synthesis of ATP has not yet been clearly elucidated, but scientists believe that it probably involves the enzyme ATPase located in the inner membrane (Figure 11.12). Mitchell has postulated that the transport of two H^+ ions from the external medium through the ATPase complex into the mitochondrial matrix provides the necessary energy for the formation of one molecule of ATP from ADP and P_i.

Finally, we note that the theory is called the chemiosmotic theory because

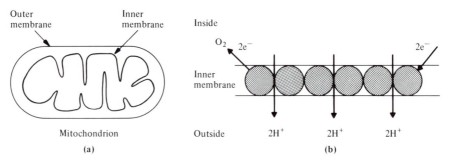

Figure 11.11 (a) Schematic diagram of a mitochondrion. (b) Transfer of energy during electron transport leads to a higher outside H^+ ion concentration. The gray spheres represent the carrier molecules.

The energy stored in the pH gradient drives the oxidative phosphorylation of ADP to form ATP. The ATPase (sphere) is embedded in the membrane. When it combines ADP and P_i to form ATP, H^+ ions are preferentially released to the inside of the membrane while OH^- ions are preferentially lost to the outside. Consequently, there is a drop in the pH gradient across the membrane.

Figure 11.12

Inner membrane

Inside   Outside

$ADP + P_i$

OH^- H^+ H^+

OH^- OH^- OH^- + H^+

OH^- + H^+ H^+ H_2O

H_2O H^+ H^+

OH^- ATP

the primary event involves the osmotic work needed to accumulate ions. Once a pH gradient has been established, the energy stored can be used to perform chemical work, which is the phosphorylation of ADP to ATP.

Membrane Potential 11.10

Electrical potentials exist across the membranes of various kinds of cells. Some cells, such as the nerve cell and muscle cell, are said to be excitable because they are capable of transmitting a change of potential along their membranes. In this section we briefly discuss some aspects of membrane potentials.

A human nerve cell consists of a cell body and a single long fiber extension about 10^{-5} to 10^{-3} cm in diameter, called the *axon* (Figure 11.13). The walls

Schematic diagram of a neuron (a nerve cell), made up of a cell body and an axon.

Figure 11.13

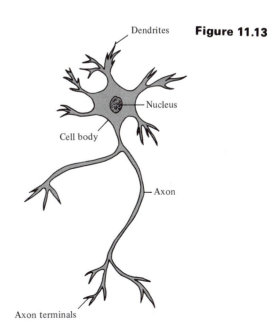

Dendrites

Nucleus

Cell body

Axon

Axon terminals

Table 11.5

Typical Composition of a Nerve Cell[a]

	Interior	Exterior
K^+	400	15
Na^+	20	440
Cl^-	120	550
Nondiffusible organic ions	350	—

[a] Concentrations in mM.

of the cell and the axon are made of membranes whose structure was discussed in Section 9.6. Table 11.5 shows the typical composition of the interior and exterior of a nerve cell. The fluid in the axon is similar in composition to the fluid in the cell body.

The difference in ionic concentrations across the membrane establishes a *membrane potential*. To understand the nature of membrane potentials, let us first consider several simple chemical systems shown in Figure 11.14. Figure 11.14a depicts two NaCl solutions at concentrations of 0.1 M and 0.01 M separated by a membrane permeable to both Na^+ and Cl^- ions. The ions will therefore diffuse from the right compartment to the left compartment. However, because the ionic mobility of the Cl^- ions is greater than that of the Na^+ ions (see Table 9.2), Cl^- ions will move across the membrane

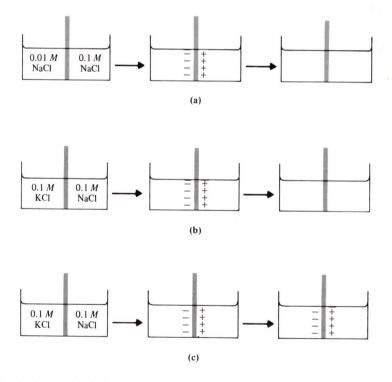

(a)

(b)

(c)

Figure 11.14   In (a) and (b) the diffusion potential established as a result of unequal concentration or unequal ionic mobility disappears with time. (c) The diffusion potential across a membrane permeable to the K+ ions but not to Na+ ions will not disappear with time.

faster than the Na$^+$ ions. Initially, then, the left compartment will become more negative with respect to the right compartment. This electric field gradient is called a *diffusion potential* because it arises as a result of the difference in the diffusion rates of the ions in solution. The diffusion potential gradually disappears over time as the concentrations of Na$^+$ and Cl$^-$ ions in the two compartments become equal. A somewhat similar situation is shown in Figure 11.14b. Here a 0.1 M NaCl solution is separated from a 0.1 M KCl solution by a membrane that is permeable to all the ions. Because the K$^+$ ion has a greater ionic mobility than the Na$^+$ ion, a diffusion potential is established initially such that the left compartment will become negative with respect to the right compartment. Again, this diffusion potential will disappear over time as the concentrations of NaCl and KCl in the two compartments become equal. A different situation is shown in Figure 11.14c. Here a 0.1 M KCl solution is separated from a 0.1 M NaCl solution by a membrane that is permeable to the K$^+$ ions but not to Na$^+$ and Cl$^-$ ions. Initially, the K$^+$ ions will diffuse down the concentration gradient from left to right so that the left compartment will become negative with respect to the right compartment. This process continues until the driving force for diffusion of the K$^+$ ions is exactly balanced by the electrostatic repulsive force preventing the entry of K$^+$ ions into the right compartment. Under these circumstances, the potential established across the membrane will not disappear with time and we call this potential in the equilibrium state the *membrane potential.*

The value of the membrane potential for any ion depends on the concentration gradient of that ion across the membrane—the larger the concentration gradient, the larger is the membrane potential. If the membrane is impermeable to certain types of ions, then no such ions can cross the membrane and contribute to a diffusion potential, regardless of the concentration gradient (or electric field gradient) that may exist. We can now estimate the membrane potential as follows. For a particular diffusible ion X, the potential difference across the membrane can be calculated by applying the Nernst equation

$$\mathcal{E}_x = \frac{RT}{nF} \ln \frac{[X]_{ex}}{[X]_{in}}$$

where the subscripts denote the exterior and interior of the cell, respectively. Using the values for K$^+$ listed in Table 11.5 and setting $T = 298$ K and $n = 1$, and converting to base-10 logs, we find that the membrane potential due to the K$^+$ ions, $\mathcal{E}_{K^+}$, is given by

$$\mathcal{E}_{K^+} = 0.0591 \log \frac{15}{400} = -0.084 \text{ V}$$

$$= -84 \text{ mV}$$

The negative sign means that the interior of the cell wall is more negative than the exterior of the cell. Using an arrangement such as that shown in Figure 11.15, we find that the membrane potential of a nerve cell is only about -75 mV.* The discrepancy in the calculated and measured values is due to the distribution of Na$^+$ ions across the membrane. Because the

---

* Because the membrane is not at the equilibrium potassium membrane potential of -84 mV, there is a continual diffusion of K$^+$ ions out of the cell. To maintain proper concentrations, some of the K$^+$ ions are then pumped back into the cell.

Figure 11.15

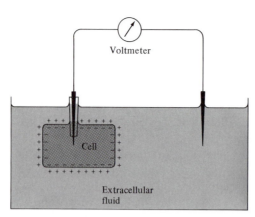

Na^+-ion concentration is much higher in the extracellular fluid, the movement of the Na^+ ions into the cell tends to make the membrane potential more positive. As mentioned earlier, the value of a membrane potential depends not only on the concentration gradient, but also on the permeability of the membrane to ions. Nerve cell membranes in the resting (unperturbed) state are about 100 times more permeable to K^+ ions than to Na^+ ions. Therefore, the Na^+ ions make a relatively small contribution to the membrane potential. The Cl^- ions also play a minor role in developing the membrane potential. Unlike Na^+ and K^+ ions, the Cl^- ions are not pumped across the membrane by an active transport mechanism (see Section 9.6). Further, the negatively charged interior of the cell repels the normal diffusion of the Cl^- ions. Indeed, experiments on nerve cells have shown that varying the Cl^- or Na^+ concentrations hardly affects the membrane potential (at rest) but that changing K^+ concentration makes a large difference.

When more than one type of ion can diffuse across the membrane, the membrane potential is given by the Goldman–Hodgkin–Katz equation,[*] which for our discussion above takes the form

$$\mathscr{E}_{membrane} = \frac{RT}{nF} \ln \frac{[K^+]_{ex} P_{K^+} + [Na^+]_{ex} P_{Na^+} + [Cl^-]_{ex} P_{Cl^-}}{[K^+]_{in} P_{K^+} + [Na^+]_{in} P_{Na^+} + [Cl^-]_{in} P_{Cl^-}} \quad (11.10)$$

where P is the permeability of the membrane to an ion and $n = 1$. If we neglect the small contribution due to the Cl^- ions, then Eq. (11.10) at 298 K becomes

$$\mathscr{E}_{membrane} = 0.0591 \log \frac{[K^+]_{ex} P_{K^+} + [Na^+]_{ex} P_{Na^+}}{[K^+]_{in} P_{K^+} + [Na^+]_{in} P_{Na^+}}$$

$$= 0.0591 \log \frac{[K^+]_{ex} P_{K^+}/P_{Na^+} + [Na^+]_{ex}}{[K^+]_{in} P_{K^+}/P_{Na^+} + [Na^+]_{in}} \quad (11.11)$$

Since $P_{K^+}/P_{Na^+} \simeq 100$, we write

$$\mathscr{E}_{membrane} = 0.0591 \log \frac{15 \times 100 + 440}{400 \times 100 + 20}$$

$$= -0.078 \text{ V}$$

$$= -78 \text{ mV}$$

This value is now closer to the experimentally determined quantity.

[*] A. L. Hodgkin and B. Katz, *J. Physiol. Lond.* **108**, 37 (1949); T. F. Forland and T. Ostvold, *J. Membrane Biol.* **16**, 101 (1974).

If the nerve cell is stimulated electrically, chemically or mechanically, the cell membrane becomes much more permeable to Na^+ ions than to K^+ ions so that $P_{K^+}/P_{Na^+} \simeq 0.17$. (The permeability of the membrane to K^+ ions does not at first change very much but there is a 600-fold increase in the permeability to Na^+ ions.) Although the molecular mechanism by which the membrane changes its permeability is not yet fully understood, the sequence of events that follows the initial stimulation can be readily monitored. During this change a small fraction of the Na^+ ions rush into the cell, resulting in a change in the membrane potential (the membrane is said to be depolarized). From Eq. (11.10), we write

$$\mathcal{E}_{membrane} = 0.0591 \log \frac{15 \times 0.17 + 440}{400 \times 0.17 + 20}$$

$$= 0.042 \text{ V}$$

$$= 42 \text{ mV}$$

During a very short period of time (less than 1 ms), the membrane potential changes from -75 mV to about 40 mV (inside positive) and then rapidly returns to its original value (Figure 11.16). The sudden rise and fall of the membrane potential is called the *action potential*.

What causes the membrane potential to return so quickly to its resting value? There are two factors involved. First, the increased Na^+ permeability is rapidly turned off after the initial influx of Na^+ ions into the cell. The mechanism for this "sodium inactivation" is not yet understood. Second, the membrane's permeability to K^+ ions is increased over its resting value over a short period of time (about 1 ms). For this reason, the membrane potential actually dips below -75 mV initially before it returns to its normal value (see Figure 11.16), at which point the cell is ready to "fire" again. The small number of excess Na^+ ions present in the cell are eventually pumped out of the cell.

So far we have concentrated on the events that occur in and around a small area of the nerve cell membrane that give rise to an action potential. Now we ask: How is the action potential propagated along a neuron? Looking at Figure 11.13, it is tempting to assume that axons act like cables which can

The rise and fall of an action potential and the changes in membrane permeability to Na^+ and K^+ ions during an action potential.

Figure 11.16

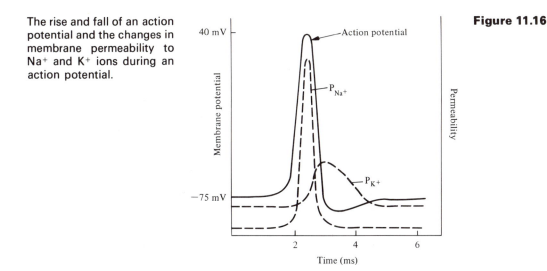

carry electric currents. After all, the axon does have a cablelike structure. It has a core of electrolytic solution and it is surrounded by a membrane that acts as an electrical insulator. However, the resistance of axoplasm (the cytoplasm within the axon) is some 100 million times greater than that of copper. Therefore, the axon would be a poor electrical conductor in this sense. Yet we know that when an action potential is generated at a particular site on a neuron, it moves rapidly and without any decrease in magnitude along the axon. Figure 11.17 shows the mechanism of action-potential propagation. The depolarizing influx of Na^+ ions at the immediate site of the action potential

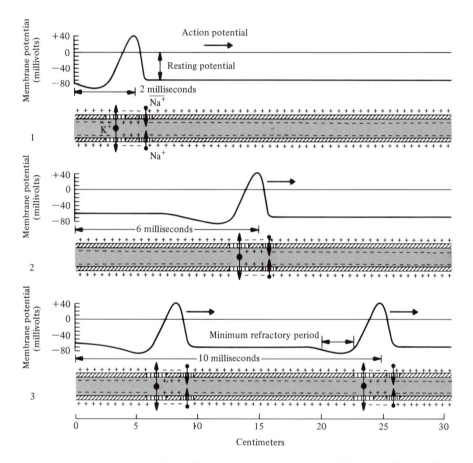

Figure 11.17   Propagation of nerve impulses along the axon coincides with a localized inflow of sodium ions (Na^+) followed by an outflow of potassium ions (K^+) through channels that are "gated," or controlled, by voltage changes across the axon membrane. The electrical event that sends a nerve impulse traveling down the axon normally originates in the cell body. The impulse begins with a slight depolarization, or reduction in the negative potential, across the membrane of the axon, where it leaves the cell body. The slight voltage shift opens some of the sodium channels, shifting the voltage still further. The inflow of sodium ions accelerates until the inner surface of the membrane is locally positive. The voltage reversal closes the sodium channel and opens the potassium channel. The outflow of potassium ions quickly restores the negative potential. The voltage reversal, known as the action potential, propagates itself down the axon (1, 2). After a brief refractory period a second impulse can follow (3). The impulse-propagation speed is that measured in the giant axon of the squid. (From "The Neuron," Charles S. Stevens. Copyright © 1979 by Scientific American, Inc. All rights reserved.)

causes the membrane potential in adjacent regions to depolarize slowly. When this slow depolarization has pushed the potential of the adjacent membrane beyond a certain value, called the *threshold potential*,* the membrane's permeability to Na^+ ions increases drastically and there is an influx of Na^+ ions into the cell that is greater than the efflux of K^+ ions. Consequently, the potential becomes more positive and an action potential is generated at this site. This event, in turn, causes a slow depolarization at an adjacent site farther down from the original site, and so on. In this manner the action potential moves down the neuron without any decrease in magnitude. The speed at which the fastest action potentials in human nerves move along their axon is about 30 m s^{-1}.

The action potential travels along the axon until it reaches either a *synaptic junction* (the connection between nerve cells) or a neuromuscular junction. The arrival of an action potential at a synapse triggers the release of a *neurotransmitter*, which is a small, diffusible molecule such as acetylcholine present in the synaptic vesicles. The acetylcholine molecules then diffuse to the postsynatic membranes, where they produce a large change in the permeability of the membranes. The conductance of both Na^+ and K^+ ions increases markedly, resulting in a large inward current of Na^+ ions and a small outward current of K^+ ions. The inward flow of Na^+ ions again depolarizes the postsynaptic membrane and triggers an action potential in the adjacent axon. Finally, acetylcholine is hydrolyzed to acetate and choline by the enzyme acetylcholinesterase as follows:

$$\underset{\text{acetylcholine}}{H_3C-\overset{\overset{\textstyle O}{\|}}{C}-O-CH_2-CH_2-\overset{+}{N}-(CH_3)_3} + H_2O \longrightarrow$$

$$\underset{\text{choline}}{HO-CH_2-CH_2-\overset{+}{N}-(CH_3)_3} + CH_3COOH$$

In a similar manner, an action potential generated in a nerve cell can be transmitted to a muscle cell. It is interesting to note that a large action potential is generated in the muscle cell of the heart during each heartbeat. This potential produces enough current to be detected by placing electrodes on the chest. After amplification, the signals can be recorded either on a moving chart or displayed on an oscilloscope. The record, called an *electrocardiogram* (EKG), is of great value in diagnosing heart diseases.

Appendix 11.1 Relationship Between ΔG and Maximum Work (Other Than *P–V* Work)

We start with

$$G = H - TS$$

At constant T, the change in the Gibbs free energy is given by

$$dG = dH - T \, dS$$

---

* The threshold potential is the potential at which gradual depolarization is replaced by explosive depolarization. It is about 20 to 40 mV more positive than the resting membrane potential; that is, it lies somewhere between -55 and -35 mV.

Now

$$dH = dU + d(PV)$$

For a reversible process,

$$dU = dQ_{rev} + dW_{rev}$$

so that

$$dH = dQ_{rev} + dW_{rev} + d(PV)$$

and

$$dG = dQ_{rev} + dW_{rev} + V\,dP + P\,dV - T\,dS$$
$$= dW_{rev} + V\,dP + P\,dV$$

The quantity dW_{rev} represents the maximum work of the system, and it consists of expansion work $(P\,dV)$ and possibly some other kind of work (for example, electrical work). Therefore, we write

$$dW_{rev} = -P\,dV + dW_e$$

where dW_e represents electrical work. Now if the pressure is also held constant, so that $V\,dP = 0$,

$$dG = -P\,dV + dW_e + P\,dV$$
$$= dW_e$$

Thus under conditions of constant temperature and pressure, the change in the Gibbs free energy is equal to the maximum work available for the system other than expansion work. For a finite change we have

$$\Delta G = W_e$$

If a system does electrical work equal to $nF\mathcal{E}$ on the surroundings, the *decrease* in the Gibbs free energy is related to the electrical work as follows:

$$\Delta G = -nF\mathcal{E}$$

Suggestions for Further Readings

INTRODUCTORY–INTERMEDIATE

Most of the topics discussed in this chapter are discussed in greater depth in any of the standard texts on physical chemistry listed in Chapter 1.

Bockris, J. O'M., and Z. Nagy. *Electrochemistry for Ecologists.* Plenum Publishing Corporation, New York, 1974.
    A very interesting and nonmathematical treatment of many of the important topics in electrochemistry. Highly recommended.

Katz, B. *Nerve, Muscle, and Synapse.* McGraw-Hill Book Company, New York, 1966.
    An excellent introductory text.

INTERMEDIATE–ADVANCED

Dryhurst, G. *Electrochemistry of Biological Molecules.* Academic Press, Inc., New York, 1977.

SECTION A

"On the Relationship Between Cell Potential and Half-Cell Reactions," D. N. Bailey, O. A. Moe, Jr., and J. N. Spencer, *J. Chem. Educ.* **53**, 77 (1976).

"Cation-Sensitive Glass Electrodes," G. A. Rechnitz, *J. Chem. Educ.* **41**, 385 (1964).

"Fuel Cells—Electrochemical Converts of Chemical to Electrical Energy," J. Weissbart, *J. Chem. Educ.* **38**, 267 (1961).

"Electrochemical Principles Involved in a Fuel Cell," A. K. Vijh, *J. Chem. Educ.* **47**, 680 (1970).

"Biogalvanic Cells," W. D. Hobey, *J. Chem. Educ.* **49**, 413 (1972).

"Batteries: The Workhorses of Chemical Energy Conversion," E. Y. Weissman, *Chemistry* **45**(10), 6 (1972).

"Corrosion," W. L. Smith, *Chemistry* **49**(1), 14, **49**(5), 7 (1976).

"Dental Filling Discomforts Illustrates the Electrochemical Potential of Metals," R. E. Treptow, *J. Chem. Educ.* **55**, 189 (1978).

"Thermodynamic Parameters from an Electrochemical Cell," C. A. Vincent, *J. Chem. Educ.* **47**, 365 (1970).

"Biological Oxidations and Energy Conversion," J. Kirschbaum, *J. Chem. Educ.* **45**, 28 (1968).

"Electrochemistry in Organism," T. P. Chirpith, *J. Chem. Educ.* **52**, 99 (1975).

"Neurotransmitters," J. Axelrod, *Sci. Am.*, June 1974.

"The Synapse," Sir J. Eccles, *Sci. Am.*, Jan. 1965.

"The Nerve Axon," P. F. Baker, *Sci. Am.*, Mar. 1966.

"Chemistry and Nerve Conduction," K. A. Rubinson, *J. Chem. Educ.* **54**, 345 (1977).

"How Cells Make ATP," P. C. Hinkle and R. E. McCarty, *Sci. Am.*, Mar. 1978.

"Ion Channels in the Nerve-Cell Membrane," R. D. Keynes, *Sci. Am.*, Mar. 1979.

"The Neuron," C. F. Stevens, *Sci. Am.*, Sept. 1979.

SECTION B

"Energy Recovery from Saline Water by Means of Electrochemical Cells," B. H. Clampitt and F. E. Kiviat, *Science* **194**, 719 (1976).

"Equivalence Point Potential in Redox Titrations," A. H. A. Heyn, *J. Chem. Educ.* **47**, 240 (1970).

"Enzyme Electrodes," D. A. Gough and J. D. Andrade, *Science* **180**, 380 (1973).

"Ion-Selective Electrodes in Science, Medicine and Technology," R. A. Durst, *Am. Sci.* **59**, 353 (1971).

"Fuel Cells—Present and Future," D. P. Gregory, *Chem. Brit.* **5**, 308 (1969).

"Electrochemical Cells for Space Power," R. M. Lawrence and W. H. Bowman, *J. Chem. Educ.* **48**, 359 (1971).

"Membrane Electrode Probes for Biological Systems," G. A. Rechnitz, *Science* **190**, 234 (1975).

"Electrochemical Reactions in Batteries," A. Kozawa and R. A. Powers, *J. Chem. Educ.* **49**, 587 (1972).

Reading Assignments

"Racing Car Batteries," R. E. Combs and J. M. Connelly, *J. Chem. Educ.* **50**, 857 (1973).

"Corrosion," W. H. Slabaugh, *J. Chem. Educ.* **51**, 218 (1974).

"Mechanisms of Oxidation–Reduction Reactions," H. Taube, *J. Chem. Educ.* **45**, 452 (1968).

"Electron Transfer in Chemical and Biological Systems," N. Sutin, *Chem. Brit.* **8**, 148 (1972).

"Keilin's Respiratory Chain Concept and Its Chemiosmotic Consequences," P. Mitchell, *Science* **206**, 1148 (1979).

"The Transport of Substances in Nerve Cells," J. H. Schwartz, *Sci. Am.*, Apr. 1980.

Problems **11.1** Consider the following Daniell cell:

$$Zn\,|\,ZnSO_4(0.01\ M)\,|\,CuSO_4(0.01\ M)\,|\,Cu$$

Predict what would happen to the emf of the cell if the following solutions are added to the copper compartment: (a) 0.1 M $CuSO_4$, (b) 0.1 M K_2SO_4, (c) 0.1 M NH_3, and (d) 0.1 M Na_2S. Assume that there are no dilutions.

11.2 Complete the following table:

$\mathcal{E}$	ΔG	Cell Reaction
+		
0	+	

11.3 Calculate the standard emf for the following reaction:

$$Fe(s) + Tl^{3+} \longrightarrow Fe^{2+} + Tl^{+}$$

11.4 Calculate the emf of the Daniell cell at 298 K when the concentrations of $CuSO_4$ and $ZnSO_4$ are 0.50 M and 0.10 M, respectively. What would the emf be if activities were used instead of concentrations? (The $\gamma_\pm$ values for $CuSO_4$ and $ZnSO_4$ at their respective concentrations are 0.068 and 0.15.)

11.5 The half-reaction at an electrode is

$$Al^{3+}(aq) + 3e^- \longrightarrow Al(s)$$

Calculate the number of grams of aluminum that can be produced by passing 1.00 faraday through the electrode.

11.6 Calculate $\mathcal{E}°$, $\Delta G°$, and K for the following reactions at 25°C:
(a) $Zn + Sn^{4+} \longrightarrow Zn^{2+} + Sn^{2+}$
(b) $Cl_2 + 2I^- \longrightarrow 2Cl^- + I_2$
(c) $5Fe^{2+} + MnO_4^- + 8H^+ \longrightarrow Mn^{2+} + 4H_2O + 5Fe^{3+}$

11.7 The equilibrium constant for the reaction

$$Sr + Mg^{2+} \longrightarrow Sr^{2+} + Mg$$

is 2.69×10^{12} at 25°C. Calculate $\mathcal{E}°$ for a cell made up of the $Sr\,|\,Sr^{2+}$ and $Mg\,|\,Mg^{2+}$ half-cells.

11.8 Look up the $\mathcal{E}°$ values for the following half-cell reactions:

$$Ag^+ + e^- \longrightarrow Ag$$

$$AgBr + e^- \longrightarrow Ag + Br^-$$

Design a cell that would allow you to determine the solubility product of AgBr at 25°C.

11.9 One way to prevent a buried iron pipe from rusting is to connect it with a piece of wire to a magnesium or zinc rod. What is the electrochemical principle for this action?

11.10 Consider a concentration cell consisting of two hydrogen electrodes. At 25°C the cell emf is found to be 0.0267 V. If the pressure of hydrogen gas at the anode is 4.0 atm, what is the pressure of hydrogen gas at the cathode?

11.11 A student is given two beakers in the laboratory. One beaker contains a solution that is 0.15 M in Fe^{3+} and 0.45 M in Fe^{2+}, and the other beaker contains a solution that is 0.27 M in I^- and 0.050 M in I_2. A piece of platinum wire is dipped into each of the two solutions. (a) Calculate the potential of each electrode relative to a standard hydrogen electrode at 25°C. (b) Predict what chemical reaction would occur when these two electrodes are connected and a salt bridge is used to join the two solutions together.

11.12 An electrochemical cell consists of a half-cell in which a piece of platinum wire is dipped into a solution that is 2.0 M in KBr and 0.050 M in Br_2. The other half-cell consists of magnesium metal immersed in a 0.38 M Mg^{2+} solution. (a) Which electrode is the anode and which is the cathode? (b) What is the emf of the cell? (c) What is the spontaneous cell reaction? (d) What is the equilibrium constant of the cell reaction? Assume the temperature to be 25°C.

11.13 At 25°C the standard reduction potentials of $Pt|Ce^{4+}|Ce^{3+}$ and $Pt|Fe^{3+}|Fe^{2+}$ are 1.61 V and 0.771 V, respectively. Calculate the equilibrium constant for the reaction

$$Ce^{4+} + Fe^{2+} \rightleftharpoons Ce^{3+} + Fe^{3+}$$

11.14 From the standard reduction potentials listed in Table 11.1 for $Sn|Sn^{2+}$ and $Pb|Pb^{2+}$, calculate the ratio of $[Sn^{2+}]$ to $[Pb^{2+}]$ at equilibrium at 25°C and the $\Delta G°$ for the reaction.

11.15 From the standard reduction potentials listed in Table 11.1 for $Cu|Cu^{2+}$ and $Pt|Cu^{2+}, Cu^+$, calculate the standard reduction potential for $Cu|Cu^+$.

11.16 Consider the following cell arrangement:

$$Ag(s)|AgCl(s)|NaCl(aq)|Hg_2Cl_2(s)|Hg(l)$$

(a) Write down the half-cell reactions. (b) The standard emfs of the cell at several temperatures are as follows:

T (K)	291	298	303	311
$\mathcal{E}°$ (mV)	43.0	45.4	47.1	50.1

Calculate $\Delta G°$, $\Delta S°$, and $\Delta H°$ for the reaction at 298 K.

$$2Ag(s) + Hg_2Cl_2(s) \longrightarrow 2AgCl(s) + 2Hg(l)$$

*11.17 A well-known organic redox system is the quinone-hydroquinone couple. In an aqueous solution at a pH below 8 we have

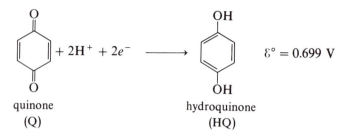

quinone
(Q)

hydroquinone
(HQ)

$\mathcal{E}° = 0.699$ V

This system can be prepared by dissolving quinhydrone, QH (a complex consisting of equimolar of Q and HQ), in water. A quinhydrone electrode can be constructed by immersing a piece of platinum wire in a quinhydrone solution. (a) Derive an expression for the electrode potential of this couple in terms of $\mathcal{E}°$ and the hydrogen-ion concentration. (b) When the quinone–hydroquinone couple is joined to a saturated calomel electrode, the emf of the cell is found to be 0.18 V. In this arrangement the saturated calomel electrode acts as the anode. Draw a cell diagram and calculate the pH of the quinhydrone solution. Assume the temperature to be 25°C.

11.18 A quantity 25 ml of a 0.10 M Fe^{2+} solution is titrated against a 0.10 M Ce^{4+} solution. What is the emf at the equivalence point? The temperature is at 298 K.

11.19 Aluminum has a more negative standard reduction potential than iron. Yet aluminum does not form "rust" or corrode as easily as iron. Explain.

11.20 The $\mathcal{E}°'$ for the reaction

$$NAD^+ + H^+ + 2e^- \longrightarrow NADH$$

is -0.320 V at 25°C. Calculate the value at pH $= 1$. Assume that both NAD^+ and NADH are at unimolar concentration.

11.21 Look up the $\mathcal{E}°'$ values in Table 11.3 for the reactions

$$\text{acetaldehyde} + 2H^+ + 2e^- \longrightarrow \text{ethanol}$$

$$NAD^+ + H^+ + 2e^- \longrightarrow NADH$$

Calculate the equilibrium constant for the reaction

$$\text{acetaldehyde} + NADH + H^+ \rightleftharpoons \text{ethanol} + NAD^+$$

at 298 K.

11.22 The following reaction, which takes place just prior to the Krebs cycle, is catalyzed by the enzyme lactate dehydrogenase:

$$CH_3COCOO^- + NADH + H^+ \rightleftharpoons CH_3CH(OH)COO^- + NAD^+$$

pyruvate   lactate

From the data listed in Table 11.3, calculate $\Delta G°'$ and the equilibrium constant for the reaction at 298 K.

11.23 Calculate the number of moles of cytochrome c^{3+} formed from cytochrome c^{2+} with the Gibbs free energy derived from the oxidation of 1 mol of glucose. ($\Delta G° = 2879$ kJ for the degradation of 1 mol of glucose to CO_2 and H_2O.)

11.24 Use Table 11.3 to construct a diagram that shows the $\mathcal{E}°'$ values on a vertical scale.

11.25 In the terminal respiratory chain, the following redox couples are involved: $NAD^+|NADH$ and $FAD|FADH_2$. Calculate the $\Delta G^{\circ\prime}$ for the reaction at 298 K:

$$NADH + FAD + H^+ \longrightarrow NAD^+ + FADH_2$$

Is this free energy change sufficient to synthesize ATP from ADP and inorganic phosphate? Draw a diagram showing the experimental arrangement for measuring the emf of a cell consisting of these two couples.

11.26 The oxidation of malate to oxaloacetate is a key reaction in the Krebs cycle:

$$malate + NAD^+ \longrightarrow oxaloacetate + NADH + H^+$$

Calculate $\Delta G^{\circ\prime}$ and the equilibrium constant for the reaction at pH 7 and 298 K.

11.27 Calculate $\Delta G^{\circ\prime}$ for the oxidation of succinate to fumarate by cytochrome c at 298 K.

11.28 Flavin adenine dinucleotide (FAD) participates in a number of biological redox reactions according to the half-reaction

$$FAD + 2H^+ + 2e^- \longrightarrow FADH_2$$

If $\mathcal{E}^{\circ\prime}$ of this couple is -0.219 V at 298 K and pH 7, calculate its reduction potential at this temperature and pH when the solution contains (a) 85% of the oxidized form, and (b) 15% of the oxidized form.

*11.29 According to the chemiosmotic theory, the synthesis of ATP is coupled to the movement of $2H^+$ from the low-pH side of the membrane to the high-pH side. (a) Derive an expression for ΔG for this movement of $2H^+$. (b) Calculate the change in pH across the membrane that is required at 25°C to synthesize ATP from ADP and P_i under standard-state conditions. Given that $\Delta G^{\circ\prime} = 31.4$ kJ for the synthesis of 1 mol of ATP.

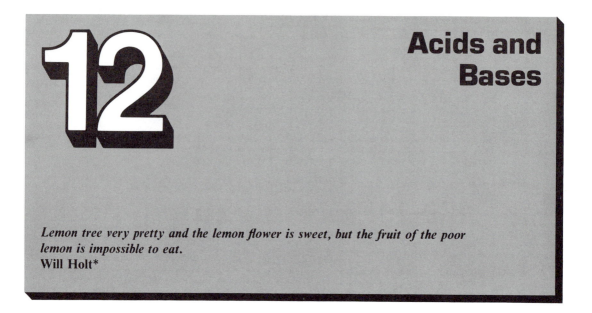

Lemon tree very pretty and the lemon flower is sweet, but the fruit of the poor lemon is impossible to eat.
Will Holt*

Acids and bases form a particularly important class of electrolytes. No chemical equilibria are as widespread as those involving acids and bases. The precise balance of their concentrations or pH in our bodies is necessary for the proper function of enzymes, maintenance of osmotic pressure, and so on. Often a deviation from the normal pH value by as small an amount as one-tenth of a unit leads to disease or even death.

A proper understanding of acid–base balance in chemical and biological systems requires a clear comprehension of the behavior of weak acids and weak bases, and of the hydrogen ion. In this chapter we discuss the general acid–base reactions, the amino acids, and buffers.

12.1 Definitions of Acids and Bases

According to the concept of Brönsted and Lowry (1923), acids are substances that can donate protons, and bases are substances that can accept protons. Thus HCl is an acid because of the reaction

$$HCl \longrightarrow H^+ + Cl^-$$

acid   conjugate base

HCl and Cl^- are said to be *conjugate* or joined in a pair to each other. The chloride ion is a base because of its ability to accept protons. A strong acid has a weak conjugate base, as in this case, while a weak acid has a strong conjugate base. For example,

$$CH_3COOH \longrightarrow H^+ + CH_3COO^-$$

weak acid   strong conjugate base

* From "Lemon Tree" by Will Holt. Used by permission of Lemon Tree Music, Inc., New York.

Ammonia is a base, since it can accept a proton,*

$$NH_3 + H^+ \longrightarrow NH_4^+$$

base   conjugate acid

and the ammonium ion is the conjugate acid. On the other hand, metallic hydroxides such as NaOH or KOH are not themselves bases in Brönsted–Lowry's definition. However, these compounds are for all practical purposes completely dissociated in water as Na^+, K^+, and OH^- ions. The hydroxide ion itself is, of course, a base because it can accept a proton.

A more general definition of acids and bases was provided by Lewis in 1923. According to Lewis, an acid is any substance that can accept an electron pair and a base is any substance that can donate an electron pair. This definition not only covers the examples discussed above, but it also describes acid–base reactions not involving protons. Some examples are

$$BF_3 + F^- \longrightarrow BF_4^-$$

acid   base

$$BCl_3 + (CH_3)_3N \longrightarrow (CH_3)_3NBCl_3$$

acid   base

$$Ag^+ + 2CN^- \longrightarrow Ag(CN)_2^-$$

acid   base

Lewis's definition applies to a greater range of substances and is more fundamental from a theoretical viewpoint. In this chapter and largely throughout this book, however, we shall only be concerned with the proton concept.

Dissociation of Acids **12.2**

The dissociation of an acid in water can be represented by

$$HA + H_2O \rightleftharpoons H_3O^+ + A^-$$

and the equilibrium constant is

$$K_a = \frac{a_{H_3O^+} a_{A^-}}{a_{HA} a_{H_2O}}$$

where the a's are the activities of the species. Since the concentrations of water in most aqueous solutions is very high (1 liter of pure water contains $1000/(16+2) = 55.6$ mol), it is essentially unchanged by the dissociation process. This means that H_2O can effectively be considered to be in its standard state (that is, pure liquid) and hence its activity is unity. As a good approximation, we can usually ignore the correction due to nonideality and replace all activities with concentrations, so that (see Appendix 12.1)

$$K_a = \frac{[H_3O^+][A^-]}{[HA]}$$

$$= \frac{[H^+][A^-]}{[HA]} \tag{12.1}$$

* In pure form NH_3 also qualifies as an acid because of the reaction $NH_3 \rightarrow H^+ + NH_2^-$. However, we shall only be concerned with aqueous solutions.

We can now write H^+ instead of H_3O^+, since water is no longer treated as a reactant.

In water the proton does not exist as H^+ but is a hydrated species. In addition to the hydronium ion H_3O^+, evidence suggests that other species* such as $H_9O_4^+$ also exist in solution.

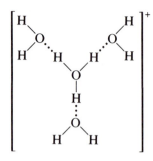

The thermodynamic treatment is unaffected by whichever species we employ. For simplicity we shall use H^+ in all cases.

The strength of the acid is indicated by the magnitude of K_a; that is, the larger K_a, the stronger the acid. Strong acids such as HCl and HNO_3 are assumed to be completely dissociated and we shall not discuss them further. Instead, we shall concentrate on weak acids ($K_a \lesssim 10^{-2}$). Table 12.1 listed the dissociation constant of a number of common acids.

EXAMPLE 12.1 Calculate the pH and percent dissociation of a 1.0×10^{-2} M HCN solution.

Answer: Let x be the concentrations of H^+ and CN^- at equilibrium. Thus we have

$$HCN \rightleftharpoons H^+ + CN^-$$

$$(0.010 - x)\, M \qquad x\, M \quad x\, M$$

From the K_a value in Table 12.1, we write

$$4.9 \times 10^{-10} = \frac{x^2}{0.010 - x}$$

We can either solve the quadratic equation in x or apply the approximation that $0.010 \gg x$. The second case gives

$$x^2 = 4.9 \times 10^{-12}$$

or

$$x = 2.2 \times 10^{-6}\, M$$

Since $[H^+] = 2.2 \times 10^{-6}$ M, the pH of the solution is 5.7. Finally, the percent dissociation is given by

$$\frac{2.2 \times 10^{-6}}{0.010} \times 100 = 0.022\%$$

* See H. L. Clever, *J. Chem. Educ.* **40**, 637 (1963); and P. A. Giguère, *J. Chem. Educ.* **56**, 571 (1979).

Dissociation Constants of Some Common Weak Acids at 298 K[a]

Table 12.1

Acid	K_a	pK_a
HF	1×10^{-3}	3.00
HCN	4.9×10^{-10}	9.31
HNO_2	4×10^{-4}	3.40
H_2S	$5.7 \times 10^{-8} (K'_a)$	7.24 (pK'_a)
	$1.2 \times 10^{-15} (K''_a)$	14.92 (pK''_a)
H_2CO_3	$4.2 \times 10^{-7} (K'_a)$	6.38 (pK'_a)
	$4.8 \times 10^{-11} (K''_a)$	10.32 (pK''_a)
H_3BO_3	$7.3 \times 10^{-10} (K'_a)$	9.14 (pK'_a)
	$1.8 \times 10^{-13} (K''_a)$	12.74 (pK''_a)
	$1.6 \times 10^{-14} (K'''_a)$	13.80 (pK'''_a)
H_3PO_4	$7.5 \times 10^{-3} (K'_a)$	2.13 (pK'_a)
	$6.2 \times 10^{-8} (K''_a)$	7.21 (pK''_a)
	$4.8 \times 10^{-13} (K'''_a)$	12.32 (pK'''_a)
CH_3COOH	1.75×10^{-5}	4.76
C_6H_5COOH	6.30×10^{-5}	4.20
HCOOH	1.77×10^{-4}	3.75
$ClCH_2COOH$	1.36×10^{-3}	2.87
C_6H_5OH	1.30×10^{-10}	9.89
HOOCCOOH	$6.5 \times 10^{-2} (K'_a)$	1.19 (pK'_a)
(oxalic acid)	$6.1 \times 10^{-5} (K''_a)$	4.21 (pK''_a)
$(CH_2COOH)_2$	$6.4 \times 10^{-5} (K'_a)$	4.19 (pK'_a)
(succinic acid)	$2.7 \times 10^{-6} (K''_a)$	5.57 (pK''_a)
$C_6H_8O_6$	$8 \times 10^{-5} (K'_a)$	4.1 (pK'_a)
(ascorbic acid)	$1.6 \times 10^{-12} (K''_a)$	11.79 (pK''_a)
$(CHCOOH)_2$	$9.3 \times 10^{-4} (K'_a)$	3.03 (pK'_a)
(fumaric acid)	$3.4 \times 10^{-5} (K''_a)$	4.47 (pK''_a)
$CH_3CH(OH)COOH$	1.39×10^{-4}	3.86
(lactic acid)		
$HOOCCH(OH)CH_2COOH$	$4 \times 10^{-4} (K'_a)$	3.40 (pK'_a)
(malic acid)	$9 \times 10^{-6} (K''_a)$	5.50 (pK''_a)
$HOOCCH_2C(OH)COOHCH_2COOH$	$8.7 \times 10^{-4} (K'_a)$	3.06 (pK'_a)
(citric acid)	$1.8 \times 10^{-5} (K''_a)$	4.74 (pK''_a)
	$4.0 \times 10^{-6} (K'''_a)$	5.40 (pK'''_a)

$$
\begin{array}{c}
\overset{\displaystyle O}{\overset{\displaystyle \|}{\text{O}-\text{C}-\text{CH}_3}} \\
\text{COOH}
\end{array}
\qquad 3 \times 10^{-4} \qquad 3.5
$$

(acetylsalicylic acid
or aspirin)

[a] From I. H. Segel, *Biochemical Calculations,* John Wiley & Sons, Inc., New York, 1968. Used by permission.

Comment: Note that in this calculation we have neglected the contribution to the hydrogen-ion concentration from water molecules. This assumption always holds unless we are dealing with very dilute solutions. A more rigorous treatment is given in Appendix 12.2. Further, the small value of x justifies our approximation.

The same treatment can be applied to bases. For example, when ammonia dissolves in water, we have*

$$NH_3 + H_2O \rightleftharpoons NH_4^+ + OH^-$$

* Ammonium hydroxide or NH_4OH does not exist.

By analogy with the acid-dissociation constant, we can write the base-dissociation constant, K_b, as

$$K_b = \frac{a_{NH_4^+} \, a_{OH^-}}{a_{NH_3} \, a_{H_2O}}$$

Setting $a_{H_2O} = 1$ and replacing activities with concentrations, we have

$$K_b = \frac{[NH_4^+][OH^-]}{[NH_3]}$$

Further, the reaction of ammonium ion with water (for example, when NH_4Cl is dissolved in water) is

$$NH_4^+ + H_2O \rightleftharpoons NH_3 + H_3O^+$$

or simply

$$NH_4^+ \rightleftharpoons NH_3 + H^+$$

The dissociation constant of the acid, NH_4^+, is given by

$$K_a = \frac{[NH_3][H^+]}{[NH_4^+]}$$

Table 12.2 lists the dissociation constants of a number of bases and their conjugate acids.

The dissociation of water is a particularly important case, since most acid-base reactions occur in aqueous solutions. Water behaves both as an acid as well as a base:

$$H_2O \rightleftharpoons H^+ + OH^-$$

The dissociation constant or the *ionic product* of water is

$$K_w = [H^+][OH^-] \tag{12.2}$$

Table 12.2

Dissociation Constants of Some Weak Bases at 298 K[a]

Base	K_b	pK_b	$pK_a{}^b$
Ammonia	1.8×10^{-5}	4.75	9.25
Aniline	3.80×10^{-10}	9.42	4.58
Caffeine	4.1×10^{-4}	3.39	10.61
Cocaine	2.57×10^{-6}	5.59	8.41
Creatine	1.92×10^{-11}	10.72	3.28
Ethylamine	5.6×10^{-4}	3.25	10.75
Methylamine	4.38×10^{-4}	3.36	10.64
Morphine	7.4×10^{-7}	6.13	7.87
Nicotine	7×10^{-7}	6.2	7.8
Novocaine	7×10^{-6}	5.16	8.84
Pyridine	1.71×10^{-9}	8.77	5.23
Quinine	1.1×10^{-6} (K_b')	5.96 (pK_b')	8.04 (pK_a')
	1.35×10^{-10} (K_b'')	9.87 (pK_b'')	4.13 (pK_a'')
Strychnine	1×10^{-6} (K_b')	6.0 (pK_b')	8.0 (pK_a')
	2×10^{-12} (K_b'')	11.7 (pK_b'')	2.3 (pK_a'')
Urea	1.5×10^{-14}	13.82	0.18

[a] From I. H. Segel, *Biochemical Calculations*, John Wiley & Sons, Inc., New York, 1968. Used by permission.
[b] The pK_a values refer to the conjugate acids of the bases. By definition, $pK_a + pK_b = 14$.

Like any equilibrium constant, K_w is a function of temperature:

$T(K)$	273	297	313	373
K_w	0.05×10^{-14}	1.0×10^{-14}	3.8×10^{-14}	4.8×10^{-14}

As the values of K_w indicate, water is an extremely weak acid as well as an extremely weak base. In a neutral solution, $[H^+] = [OH^-] = 10^{-7}$ at about 298 K. If the solution is acidic, $[H^+] > [OH^-]$. In a basic solution we have $[OH^-] > [H^+]$. The important fact is that the product of the concentration of H^+ and OH^- must be a constant at a given temperature regardless of the nature of solutes present.* Because $[H^+]$ and $[OH^-]$ are normally very small numbers and therefore inconvenient to work with, Sorensen proposed in 1909 that the acidity of a solution may be more conveniently measured on the pH scale, defined as

$$pH = -\log [H^+] \qquad (12.3)$$

Strictly speaking, we should define pH in terms of activity instead of concentration, that is,

$$pH = -\log a_{H^+}$$

where $a_{H^+} = [H^+]\gamma_{H^+}$. As we saw in Chapter 9, only mean ionic activity coefficients may be determined experimentally so that we can only estimate γ_{H^+} using Eq. (9.18). For example, for a 0.10 M HCl solution, we find that

$$pH = -\log [H^+] = 1.0$$

However, the ionic strength of the solution is 0.10 so that

$$\log \gamma_{\pm} = -0.509(1 \times 1)\sqrt{0.10}$$
$$= -0.16$$

or

$$\gamma_{\pm} = 0.69$$

Finally,

$$pH = -\log (0.1)(0.69) = 1.2$$

Generally, for relatively dilute solutions ($[H^+] \lesssim 0.5\ M$) we can calculate the pH using concentration instead of activity. Of course, in pure water the pH is 7 whether we use concentration or activity, since the ionic concentrations are very low.

Taking the negative logarithm of both sides of Eq. (12.2), we obtain

$$-\log K_w = -\log [H^+] - \log [OH^-]$$

or

$$pK_w = pH + pOH \qquad (12.4)$$

where pK_w and pOH are defined in the same manner as pH. At 298 K, Eq. (12.4) becomes

$$pH + pOH = pK_w = 14$$

Thus a knowledge of pH enables us to calculate pOH and hence $[OH^-]$.

---

* This is not strictly true, although for simplicity it will always be assumed to be the case. See T. P. Dirkse, *J. Chem. Educ.* **38**, 261 (1961).

Although the practical pH range is between 1 and 14, negative pH values exist, as do pHs greater than 14. For example, a 10 M HCl solution has pH = -1, whereas 10 M NaOH solution has pH = 15. Because of solubility restrictions, the practical concentration limit of acids and bases is below 50 M. It is difficult to work with very concentrated acid and base solutions, since they tend to attack electrodes (for pH measurements) and Pyrex glassware. Another complication is that in dealing with such concentrated solutions, we must also use activities rather than concentrations for the hydrogen ions.

Finally, we note that the pK values listed in Tables 12.1 and 12.2 are defined as the negative logarithm of the equilibrium constant:

$$pK = -\log K \tag{12.5}$$

Remember that the larger the pK value, the weaker is the acid or the base.

12.3 Salt Hydrolysis

Dissolving NaCl or K_2SO_4 in water creates an essentially neutral solution. A sodium acetate or an ammonium chloride solution, on the other hand, is anything but neutral. Depending on the concentration, the pH of a sodium acetate solution can be appreciably higher than 7, whereas the pH of an ammonium chloride solution is appreciably lower than 7. The departures from pH 7 result from *salt hydrolysis*, or the reaction between the salt and water. Consider the hydrolysis of sodium acetate:

$$CH_3COONa \longrightarrow CH_3COO^- + Na^+$$

$$CH_3COO^- + H_2O \rightleftharpoons CH_3COOH + OH^-$$

Sodium acetate, being a strong electrolyte, is completely dissociated. Acetic acid is a weak acid, so the equilibrium for the second step is shifted to the right, causing a surplus of hydroxide ions for a basic solution. The equilibrium or hydrolysis constant is the same as the base-dissociation constant:

$$K_b = \frac{[CH_3COOH][OH^-]}{[CH_3COO^-]}$$

Multiplying this equation by $[H^+]/[H^+]$, we get

$$K_b = \frac{[CH_3COOH][H^+][OH^-]}{[H^+][CH_3COO^-]}$$

$$= \frac{K_w}{K_a} \tag{12.6}$$

where K_a is the dissociation constant of acetic acid. Since both K_w and K_a are known, we write

$$K_b = \frac{1 \times 10^{-14}}{1.75 \times 10^{-5}} = 5.7 \times 10^{-10}$$

EXAMPLE 12.2 What is the pH of a 0.10 M sodium acetate solution?

Answer: We start with the following complete dissociation:

$$CH_3COONa \longrightarrow CH_3COO^- + Na^+$$

The acetate ion reacts with water to produce CH_3COOH and OH^- ions. Let the hydroxide-ion concentration be x M. Since every reaction of an acetate ion with water produces one hydroxide ion and one undissociated CH_3COOH molecule, the concentration of each of the latter two species must be x M. Thus we have

$$CH_3COO^- + H_2O \rightleftharpoons CH_3COOH + OH^-$$
$$(0.10 - x)\,M \qquad\qquad x\,M \qquad x\,M$$

and arrive at the familiar expression

$$K_b = \frac{x^2}{0.10 - x} = 5.7 \times 10^{-10}$$

We could either solve the quadratic equation in x or apply the approximation that $0.10 \gg x$. The second case gives

$$x^2 = 5.7 \times 10^{-11}$$

or

$$x = 7.6 \times 10^{-6}\,M$$

The small value of x certainly justifies our approximation. Since $[OH^-] = 7.6 \times 10^{-6}$ M, we write

$$[H^+] = \frac{1.0 \times 10^{-14}}{7.6 \times 10^{-6}} = 1.3 \times 10^{-9}\,M$$

and

$$pH = 8.9$$

Comment: Note that in this calculation we have neglected the contribution to the hydroxide-ion concentration from water molecules. This assumption always holds unless we are dealing with very dilute solutions. A more rigorous treatment is given in Appendix 12.2.

Acid–Base Titrations **12.4**

This section will deal with the most common and important techniques in the arsenal of analytical chemistry. Acid–base titration is a simple procedure. A base is added to an acid solution until the equivalence point is reached. If the concentration of one of the solutions is known, the concentration of the other solution can be easily calculated. A pH meter is most convenient for following the titration, but many other devices are employed as well.

The results of a titration experiment are most clear-cut when one uses a strong acid and a strong base; for example, HCl versus NaOH:

$$HCl + NaOH \longrightarrow NaCl + H_2O$$

Since the sodium and chloride ions do not hydrolyze, the solution at the

Figure 12.1

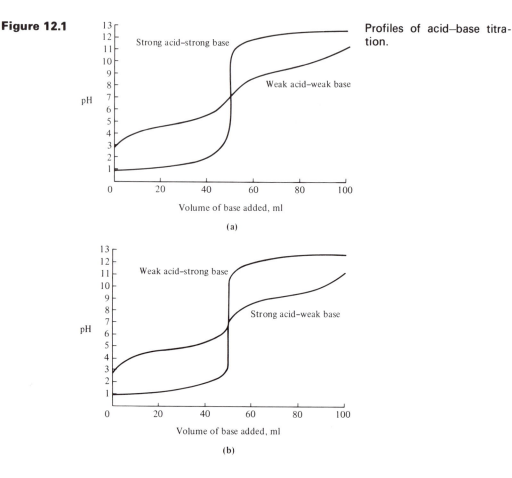

Profiles of acid–base titration.

equivalence point should be neutral, with a pH of 7. Now consider the titration between a weak acid such as acetic acid with NaOH. We have

$$CH_3COOH + NaOH \longrightarrow CH_3COONa + H_2O$$

As mentioned earlier, the acetate ion does hydrolyze to an appreciable extent. Consequently, at the point when the acetic acid is completely neutralized by NaOH, the resulting solution is basic instead of being neutral. Figure 12.1 shows the four possible titration profiles.

Another way to follow titration is to use an indicator whose color depends on the pH of the solution. An indicator is itself an acid or a base. Let us imagine a compound whose acid (HIn) form is distinctly different in color than its conjugate base (In$^-$) form. In solution, the following equilibrium is established:

$$HIn \rightleftharpoons H^+ + In^-$$

for which we write

$$K_{In} = \frac{[H^+][In^-]}{[HIn]}$$

Rearrangement of this equation yields

$$\frac{[HIn]}{[In^-]} = \frac{[H^+]}{K_{In}}$$

Table 12.3

Indicator	Color		pK_{In}	pH Range[b]
	Acid	Base		
Thymol blue	Red	Yellow	1.51	1.2–2.8
Bromophenol blue	Yellow	Blue	3.98	3.0–4.6
Chlorophenol blue	Yellow	Red	5.98	4.8–6.4
Bromothymol blue	Yellow	Blue	7.0	6.0–7.6
Cresol red	Yellow	Red	8.3	7.2–8.8
Methyl orange	Orange	Yellow	3.7	3.1–4.4
Methyl red	Red	Yellow	5.1	4.2–6.3
Phenolphthalein	Colorless	Pink	9.4	8.3–10.0

[a] From *Handbook of Biochemistry*, H. A. Sober (ed.), © The Chemical Rubber Co., 1968. Used by permission of The Chemical Rubber Co.
[b] These values are determined experimentally, rather than calculated using Eq. (12.7).

Since K_{In} is a constant, the color of the indicator depends on $[H^+]$ or the pH of the solution. To obtain a rough estimate for color dependence on relative concentrations of HIn and In^-, we proceed with the assumption that

$$\frac{[HIn]}{[In^-]} \gtrsim 10 \qquad \text{acid color}$$

$$\frac{[HIn]}{[In^-]} \lesssim \frac{1}{10} \qquad \text{base color}$$

Thus for the solution to show the acid color, we must have

$$[H^+] = K_{In}\frac{[HIn]}{[In^-]} = 10K_{In}$$

or

$$pH = -1 + pK_{In}$$

Similarly, for the solution to show the base color,

$$[H^+] = K_{In}\frac{[HIn]}{[In^-]} = 0.1K_{In}$$

or

$$pH = 1 + pK_{In}$$

The range of pH over which the color change occurs is given by the combined equations,

$$pH = pK_{In} \pm 1 \qquad (12.7)$$

EXAMPLE 12.3 The K_{In} for phenolphthalein is 3.98×10^{-10}. Calculate the range of pH values over which its color changes.

Answer: From Eq. (12.7), we have

$$pH = pK_{In} \pm 1$$

$$= 9.40 \pm 1$$

According to this estimate, the color of phenolphthalein begins to change from acid form (colorless) at pH 8.4 to the base form (reddish pink) at pH 10.4. Thus we may conclude that phenolphthalein is a suitable indicator for the strong acid–strong base and weak acid–strong base titrations (see Figure 12.1).

A number of commonly employed acid–base indicators are listed in Table 12.3.

12.5 Diprotic and Polyprotic Acids

So far we have discussed only monoprotic acids. The treatment becomes more involved when we study acids having two or more ionizable protons. In this section we concentrate on two acids of considerable importance in living systems—carbonic acid and phosphoric acid.

Carbon dioxide readily dissolves in water, but only a small percentage (about 0.25%) of the dissolved CO_2 forms what we might call the hydrated form, H_2CO_3. The equilibrium constant for the reaction

$$CO_2(aq) + H_2O \rightleftharpoons H_2CO_3(aq)$$

is only 0.00258. Since we cannot distinguish experimentally between dissolved CO_2 gas and H_2CO_3, the first dissociation of "carbonic acid" can be written either as

$$CO_2(aq) + H_2O \rightleftharpoons H^+(aq) + HCO_3^-(aq)$$

or as

$$H_2CO_3(aq) \rightleftharpoons H^+(aq) + HCO_3^-(aq)$$

The form of the equilibrium constant, that is, the first acid-dissociation constant would be the same as long as we used the total concentration of CO_2 in water as the acid concentration. The common practice is to use the last equation in describing the dissociation and to assume that all the dissolved CO_2 is in the form of H_2CO_3. On this basis, the first acid-dissociation constant has the value $K_a' = 4.2 \times 10^{-7}$ (see Table 12.1). The conjugate base, HCO_3^-, for the first dissociation becomes the acid in the second dissociation step:

$$HCO_3^- \rightleftharpoons H^+ + CO_3^{2-}$$

and we have $K_a'' = 4.8 \times 10^{-11}$. Thus K_a' is greater than K_a'' by some four orders of magnitude.

> **EXAMPLE 12.4** Calculate the solubility of carbon dioxide in equilibrium with water at 298 K and a partial pressure of 0.030 atm. What are the concentrations of all the species in this solution?
>
> **Answer:** From Example 8.3, we find that the solubility of CO_2 is 1.0×10^{-3} per mole kg of H_2O. Since the solution is dilute, we can equate molality to molarity. Thus initially the concentration of H_2CO_3 is 1.0×10^{-3} M.

In this system, there are three equilibria to be considered:

$$H_2CO_3 \rightleftharpoons H^+ + HCO_3^-$$

$$HCO_3^- \rightleftharpoons H^+ + CO_3^{2-}$$

and

$$H_2O \rightleftharpoons H^+ + OH^-$$

There are altogether five unknowns: $[H^+]$, $[OH^-]$, $[H_2CO_3]$, $[HCO_3^-]$, and $[CO_3^{2-}]$. From the mass balance of species containing the carbonate group,

$$1.0 \times 10^{-3} \, M = [H_2CO_3] + [HCO_3^-] + [CO_3^{2-}] \tag{1}$$

Moreover, electrical neutrality requires that

$$[H^+] = [HCO_3^-] + 2[CO_3^{2-}] + [OH^-] \tag{2}$$

Since each carbonate ion carries two negative charges, we need to multiply its concentration by a factor of 2. These five unknowns can be determined from five independent equations (1), (2), K_a', K_a'', and K_w. Certain assumptions will simplify the procedure. Since $K_a' \gg K_a''$ and K_a' itself is a small number, we have

$$[H_2CO_3] \gg [HCO_3^-] \gg [CO_3^{2-}]$$

From the first stage of acid dissociation,

$$H_2CO_3 \rightleftharpoons H^+ + HCO_3^-$$
$$(1.0 \times 10^{-3} - x) \, M \qquad x \, M \quad x \, M$$

we write

$$4.2 \times 10^{-7} = \frac{x^2}{1.0 \times 10^{-3} - x} \simeq \frac{x^2}{1.0 \times 10^{-3}}$$

$$x = 2.1 \times 10^{-5} \, M$$

Since $[H^+] = 2.1 \times 10^{-5} \, M$, we have $[OH^-] = 1.0 \times 10^{-14}/2.1 \times 10^{-5} = 4.8 \times 10^{-10} \, M$. We have assumed that the contributions to $[H^+]$ due to dissociations of HCO_3^- and H_2O are both negligible compared to H_2CO_3. The second dissociation is given by

$$HCO_3^- \rightleftharpoons H^+ + CO_3^{2-}$$
$$(2.1 \times 10^{-5} - y) \, M \qquad (2.1 \times 10^{-5} + y) \, M \quad y \, M$$

Here we have

$$4.8 \times 10^{-11} = \frac{(2.1 \times 10^{-5} + y)y}{2.1 \times 10^{-5} - y}$$

Since $2.1 \times 10^{-5} \gg y$, we find that

$$y = 4.8 \times 10^{-11} \, M$$

We summarize the concentrations of all the species below:

$$[H^+] = 2.1 \times 10^{-5} \, M$$

$$[OH^-] = 4.8 \times 10^{-10} \, M$$

$$[H_2CO_3] = 1.0 \times 10^{-3} \, M$$

$$[HCO_3^-] = 2.1 \times 10^{-5} \, M$$

$$[CO_3^{2-}] = 4.8 \times 10^{-11} \, M$$

Figure 12.2

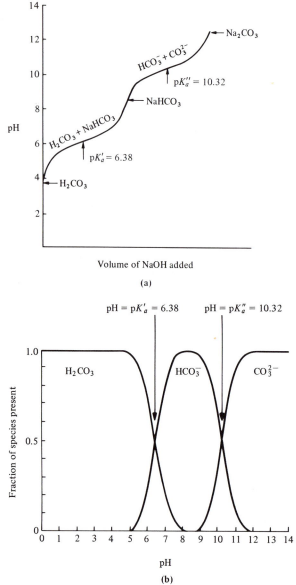

(a) Titration curve of 0.1 M carbonic acid with 0.1 M sodium hydroxide of equivalent strength. (b) Distribution diagram for the carbonic acid system as a function of pH.

Comments: (1) The proper approximations save us from solving five simultaneous equations. Acid–base calculations are frequently simplified by approximations, which should always be justifiable. (2) This calculation shows that if freshly distilled water is left standing in air, it will slowly turn acidic. The pH of the "pure water" at equilibrium will be about 4.7.

Figure 12.2a shows the titration curve of H_2CO_3 versus NaOH. The curve has two inflection points* characteristic of diprotic acid (Appendix 12.2 gives a general treatment of the titration of a weak diprotic acid with a strong base). Often, it is useful to have a *distribution diagram* from which we can determine the predominant species present at a given pH. For carbonic acid, the variations of $[H_2CO_3]$, $[HCO_3^-]$, and $[CO_3^{2-}]$ with pH are shown in Figure 12.2b.

* The *inflection point* of a curve is the point at which the second derivative is zero.

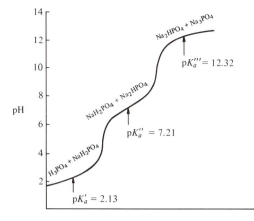

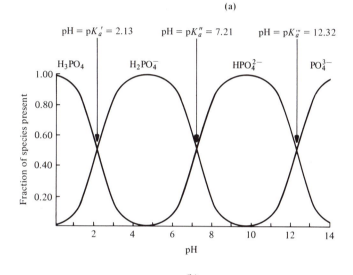

(a) Titration curve of 0.1 M phosphoric acid with 0.1 M sodium hydroxide. **Figure 12.3**
(b) Distribution diagram for the phosphoric acid system as a function of pH.

A particularly important polyprotic acid is H_3PO_4. The multistage equilibria are

$$H_3PO_4 \rightleftharpoons H^+ + H_2PO_4^- \qquad K_a' = 7.5 \times 10^{-3}$$

$$H_2PO_4^- \rightleftharpoons H^+ + HPO_4^{2-} \qquad K_a'' = 6.2 \times 10^{-8}$$

$$HPO_4^{2-} \rightleftharpoons H^+ + PO_4^{3-} \qquad K_a''' = 4.8 \times 10^{-13}$$

Using the procedure shown in Example 12.4, we can calculate the concentration of all the species present. Figure 12.3 shows the titration curve of H_3PO_4 versus NaOH as well as the distribution diagrams.

Amino Acids **12.6**

Amino acids are the building blocks of proteins. To understand the structure and function of proteins, we must first investigate the properties of individual

		pK_a' —COOH	pK_a'' —$\overset{+}{N}H_3$	pK_a''' R	pI	Abbreviation

Table 12.4   **Amino Acids Isolated from Proteins**[a] $\overset{\overset{+}{N}H_3}{\underset{|}{R\text{—CH—COOH}}}$

Name	R	pK_a' —COOH	pK_a'' —$\overset{+}{N}H_3$	pK_a''' R	pI	Abbreviation
Alanine	CH_3—	2.35	9.69		6.02	Ala
Arginine	$H_2N{-}\overset{\overset{+}{N}H_2}{\overset{\|}{C}}{-}NH{-}$	2.17	9.04	12.48	10.76	Arg
Asparagine	$H_2N{-}\overset{\overset{O}{\|}}{C}{-}CH_2{-}$	2.02	8.8		5.41	Asn
Aspartic acid	$HOOC{-}CH_2{-}$	2.09	9.82	3.86	2.98	Asp
Cysteine	$HS{-}CH_2{-}$	1.71	8.9	8.5	5.02	Cys
Cystine	$\overset{S{-}CH_2{-}}{\underset{S{-}CH_2{-}}{\|}}$	1.65 / 2.26	7.86 / 9.85		5.06	Cys⌐Cys
Glycine	H—	2.34	9.60		5.97	Gly
Glutamic acid	$HOOC{-}CH_2{-}CH_2{-}$	2.19	9.67	4.25	3.22	Glu
Glutamine	$H_2N{-}\overset{\overset{O}{\|}}{C}{-}CH_2{-}CH_2{-}$	2.17	9.13		5.70	Gln
Histidine	$\overset{CH_2{-}}{\underset{HN{\diagdown}\underset{\underset{H}{\|}}{C}{\diagup}\overset{+}{N}H}{}}$	1.82	9.17	6.00	7.59	His
Isoleucine	$\overset{CH_3{-}CH_2{\diagdown}}{\underset{CH_3{-}CH_2{\diagup}}{}}CH{-}$	2.36	9.68		6.02	Ile

amino acids. By definition, an amino acid contains at least one amino group and at least one carboxyl group.

Of the many naturally occurring amino acids, only 20 are found in proteins. All except glycine are optically active, having the L-configuration. Table 12.4 shows the structure and the pK values of the 20 amino acids.

DISSOCIATION OF AMINO ACIDS

Amino acids, like water, are ampholytes; that is, they behave both as acids and as bases. For many years it was uncertain whether these substances exist in solution as $NH_2CHRCOOH$, or as $\overset{+}{N}H_3CHRCOO^-$, which is called a *dipolar ion* or a *zwitterion*, meaning hybrid ion. Now much evidence suggests that zwitterions are predominant in solution. Among the facts of evidence are high dipole moments and high solubility in polar solvents.

Let us start with the simplest amino acid, glycine. In solution glycine exists as the zwitterion, $\overset{+}{N}H_3CH_2COO^-$. It behaves as a base when titrated with hydrochloric acid:

$$\overset{+}{N}H_3CH_2COO^- + HCl \longrightarrow \overset{+}{N}H_3CH_2COOH + Cl^-$$

Table 12.4 (continued)

Name	R	pK'_a —COOH	pK''_a $-\overset{+}{N}H_3$	pK'''_a R	p/	Abbreviation
Leucine	CH₃>CH—CH₂— (CH₃)	2.36	9.60		5.98	Leu
Lysine	$H_3\overset{+}{N}(CH_2)_3CH_2$—	2.18	8.95	10.53	9.74	Lys
Methionine	CH_3S—CH_2—CH_2—	2.28	9.21		5.75	Met
Phenylalanine	⬡—CH_2—	1.83	9.13		5.48	Phe
Serine	HO—CH_2—	2.21	9.15		5.68	Ser
Threonine	CH_3—$\overset{OH}{\underset{\mid}{CH}}$—	2.09	9.10		5.60	Thr
Tryptophan	(indole)—CH_2—	2.38	9.39		5.88	Trp
Tyrosine	HO—⬡—CH_2—	2.20	9.11	10.07	5.67	Tyr
Valine	CH₃>CH— (CH₃)	2.32	9.62		5.97	Val

[a] From Robert Barker, *Organic Chemistry of Biological Compounds*, © 1971, pp. 56–57. Adapted by permission of Prentice-Hall, Inc., Englewood Cliffs, N.J.

and behaves as an acid when titrated with sodium hydroxide:

$$\overset{+}{N}H_3CH_2COO^- + NaOH \longrightarrow NH_2CH_2COO^- + Na^+ + H_2O$$

Figure 12.4 shows the titration curve of glycine versus HCl and NaOH. The pH at the first half-equivalence point is equal to pK'_a and the pH at the second half-equivalence point is equal to pK''_a (see Appendix 12.2 for details). At the equivalence point (the first inflection point), the pH is given by

$$pH = \frac{pK'_a + pK''_a}{2} = \frac{2.34 + 9.60}{2}$$
$$= 5.97$$

At this pH the zwitterion predominates. The dissociation of glycine can be summarized as follows:

$$
\begin{array}{ccccc}
\overset{+}{N}H_3 & & \overset{+}{N}H_3 & & NH_2 \\
\mid & pK'_a = 2.34 & \mid & pK''_a = 9.60 & \mid \\
CH_2 & \rightleftharpoons & CH_2 & \rightleftharpoons & CH_2 \\
\mid & & \mid & & \mid \\
COOH & & COO^- & & COO^- \\
\text{cation} & & \text{zwitterion} & & \text{anion} \\
& & pH = 5.97 & &
\end{array}
$$

Figure 12.4

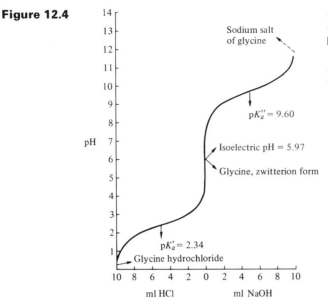

Calculated titration curve of 0.1 *M* glycine with equivalent amounts of hydrochloric acid and sodium hydroxide. (From E. S. West, *Textbook of Biophysical Chemistry*, 2nd ed., Macmillan Publishing Co., Inc., New York, 1956.)

ISOELECTRIC POINT

When the net charge of a molecule is zero, for example, the dipolar ion of glycine, the molecule is electrically neutral. In this condition the molecule is said to be *isoelectric*. The pH at which the dipolar ion does not migrate in an electric field is called the *isoelectric point*, denoted by p*I*. As we saw earlier, the isoelectric point of glycine is equal to 5.97.

The situation becomes more involved for acids containing more than two ionizable protons. Consider aspartic acid, whose dissociations are as follows:

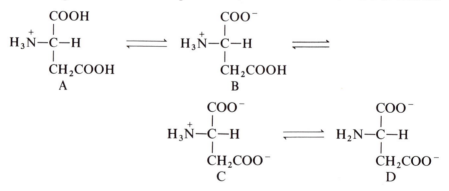

the pK_a's for the A → B, B → C, and C → D dissociations are 2.09, 3.86, and 9.82, respectively. Since only B has an equal number of positive and negative charges, the isoelectric point is given by the mean of the first two pK_a's, that is,

$$pI = \frac{2.09 + 3.86}{2} = 2.98$$

As another example of a polyprotic amino acid, let us consider histidine. This amino acid is important because the dissociation of a proton from the imidazole ring is primarily responsible for the buffering action of proteins—

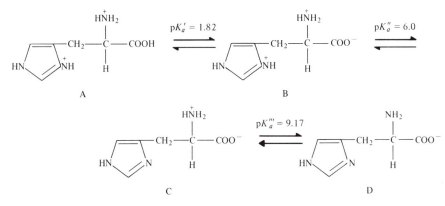

Stepwise dissociation of a fully protonated histidine molecule. The dissociated H+ ions are not shown.

Figure 12.5

particularly hemoglobin—in blood (see Section 12.7). Figure 12.5 shows the stepwise dissociation of a fully protonated histidine molecule. Since only C has an equal number of positive and negative charges, the isoelectric point is given by the mean of the second and third pK_a's, that is,

$$pI = \frac{6.00 + 9.17}{2} = 7.59$$

Table 12.4 lists the pI values of all 20 amino acids.

In Chapter 21 we shall see how the different values of pI's are used to separate a mixture of proteins by a technique called *isoelectric focusing*.

TITRATION OF PROTEINS

It may seem that proteins are far too complex to handle by titration. Actually, the situation is more manageable than we might at first expect. Although a protein has numerous ionizable protons, most of them can be characterized according to the pK values using Table 12.4. However, because of the spread in pK's, the titration curve is broad and much care is needed for accurate assignments.

The primary reason for titrating proteins is to count the number of ionizable protons, identify them, and compare the result with amino acid analysis. Although this is valuable information by itself, the technique often gives additional information regarding the conformation of the macromolecule.

Figure 12.6 shows the titration curve of the enzyme ribonuclease. The shape of the curve is somewhat dependent on the ionic strength of the solution. This curve can be divided into three regions: between pH 1 and 5, eleven protons are dissociated; between pH 5 and 8, five protons are dissociated; between pH 8 and 12, seventeen protons are dissociated. The assignments of the protons are shown in Table 12.5. With the exception of phenolic groups, agreement is remarkably good. This exception is easily explained if we assume that the three nontitratable phenolic groups are located in the interior region of the protein molecule, inaccessible to acid–base reaction. Similar instances are also found in other systems. For example, only 6 of the 12 imidazole groups in myoglobin are titratable. In each case, the buried group(s) can be brought to the surface of the molecule by denaturation, and can then be accounted for by titration.

327

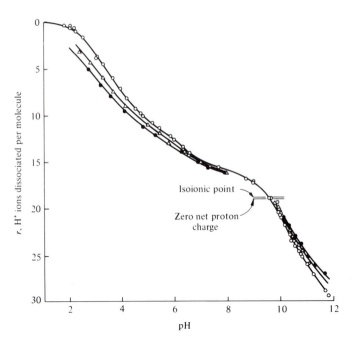

Figure 12.6 Titration curve of ribonuclease. The three curves correspond to three different ionic strengths of the solution. [Reprinted with permission from C. Tanford and J. D. Hauenstein, *J. Am. Chem. Soc.* **78**, 5288 (1956). Copyright by the American Chemical Society.]

Titration of Ribonuclease[a]

Table 12.5

Acid Group	Number of Protons		pK	
	Titration	Amino Acid Analysis	Observed	Normal
α-COOH	1	1	3.75	3.75
β,γ-COOH	10	10		4.6
Imidazole	4	4	6.5	6.5–7
α-$\overset{+}{N}H_3$	1	1	7.8	7.8
ε-$\overset{+}{N}H_3$	10	10	10.2	10.1–10.6
Phenolic	3	6	9.5	9.6
Guanidyl	4	4	$\geqslant 12$	>12

[a] Reprinted with permission from C. Tanford and J. D. Hauenstein, *J. Am. Chem. Soc.* **78**, 5290 (1956). Copyright by the American Chemical Society.

12.7 Buffers

A *buffer solution* is a solution whose pH remains essentially constant despite the addition of a *small* amount of acid or base. Buffers are enormously important to chemical and biological systems. The pH of human body fluids varies greatly, depending on the location. For example, the pH of blood plasma is about 7.4, whereas that of gastric juice, a fluid produced by glands in the mucosa membrane lining the stomach, is about 1. Yet these pH's must be maintained by buffers for the proper functioning of enzymes, the balance of osmotic pressure, and so on. Table 12.6 lists the pH of a number of fluids.

Material	pH Value
Body fluids and tissues	
Blood serum	7.35–7.45
Cerebrospinal fluid	7.35–7.45
Aqueous humor of eye	7.4
Saliva	6.35–6.85
Pure gastric juice	About 0.9
Pancreatic juice	7.5–8.0
Intestinal juice	7.0–8.0
Hepatic duct bile	7.4–8.5
Gallbladder bile	5.4–6.9
Urine	4.8–7.5
Feces	7.0–7.5
Tears	7.4
Milk	6.6–6.9
Skin (intracellular, various layers)	6.2–7.5
Liver (intracellular)	
Kupffer cells	6.4–6.5
Peripheral cells	7.1–7.4
Central cells	6.7–6.9
Miscellaneous	
Distilled water, exposed to air from which it absorbs some CO_2	About 5.5
Sea water	8.0
Vinegar	3.0
Orange juice	2.6–4.4
Grapefruit juice	3.2
Tomatoes (ripe)	4.3
Egg white (fresh)	8.0

[a] From E. S. West, W. R. Todd, H. S. Mason, and J. T. Van Bruggen, *Textbook of Biochemistry*, 4th ed., Macmillan Publishing Co., Inc., New York, 1966. Used by permission.

One of the simplest buffers is the acetic acid–sodium acetate system. The dissociation constant of acetic acid is given by

$$K_a = \frac{[H^+][CH_3COO^-]}{[CH_3COOH]} = 1.75 \times 10^{-5}$$

Rearrangement of this equation leads to

$$[H^+] = K_a \frac{[CH_3COOH]}{[CH_3COO^-]}$$

Taking the negative logarithm of both sides, we obtain

$$-\log [H^+] = -\log K_a + \log \frac{[CH_3COO^-]}{[CH_3COOH]}$$

or

$$pH = pK_a + \log \frac{[CH_3COO^-]}{[CH_3COOH]}$$

In general, this equation can be written as

$$pH = pK_a + \log \frac{[\text{conjugate base}]}{[\text{acid}]} \qquad (12.8)$$

Equation (12.8) is called the *Henderson–Hasselbalch equation.***** It is quite useful in buffer and acid–base titration calculations, as we shall see in the following example.

> **EXAMPLE 12.5** Calculate the pH of the 1.0 M CH_3COOH/1.0 M CH_3COONa buffer system. What is the pH after the addition of 0.10 mol of HCl to 1 liter of this buffer?

Answer: In the buffer we have

$$CH_3COONa \longrightarrow CH_3COO^- + Na^+$$

and

$$CH_3COOH \rightleftharpoons CH_3COO^- + H^+$$

Since acetic acid is a weak acid, as a good approximation we can write

$$[CH_3COOH] \simeq 1.0 \ M$$

and

$$[CH_3COO^-] \simeq 1.0 \ M$$

From Eq. (12.8),

$$pH = 4.76 + \log \frac{1.0}{1.0} = 4.76$$

or

$$[H^+] = 1.75 \times 10^{-5} \ M$$

After the addition of 0.10 mol of HCl, the following reaction takes place:

$$CH_3COO^- + H^+ \longrightarrow CH_3COOH$$
$$0.10 \ M \quad 0.10 \ M \qquad \quad 0.10 \ M$$

Again we have neglected the small contribution to the total hydrogen-ion concentration which results from the ionization of acetic acid. The new concentrations are

$$[CH_3COO^-] = 1.0 - 0.10 = 0.90 \ M$$

and

$$[CH_3COOH] = 1.0 + 0.1 = 1.10 \ M$$

Hence the pH of the buffer is

$$pH = 4.76 + \log \frac{0.90}{1.10} = 4.67$$

or

$$[H^+] = 2.13 \times 10^{-5} \ M$$

---

* Note the interesting similarity between Eqs. (12.8) and (11.9). Acid–base and redox reactions are quite analogous in many respects.

Thus there is a change of about 0.1 unit on the pH scale, equivalent to an increase of $[H^+]$ by a factor of $2.13 \times 10^{-5}/1.75 \times 10^{-5}$, or 1.22.

To see how effective this buffer is against the acid, let us compare the pH change when 0.10 mol of HCl is added to 1 liter of water. The pH decreases from 7 to 1, amounting to a millionfold increase in $[H^+]$!

Often it is necessary to know the effectiveness of a buffer on a quantitative basis. To do so, we employ the term *buffer capacity* (β), first introduced by van Slyke in 1922. Buffer capacity is defined as the amount of acid or base that must be added to the buffer to produce a unit change of pH. Hence

$$\beta = \frac{d[B]}{d\text{pH}} \tag{12.9}$$

where $d[B]$ is the increase (in mol liter^{-1}) of strong base B. If a strong acid is added, Eq. (12.9) becomes

$$\beta = \frac{-d[B]}{-d\text{pH}}$$

The buffer capacity always has a positive value, however, since addition of base increases the pH and addition of acid decreases the pH. Thus $d[B]$ and $d\text{pH}$ always have the same signs. The value of β depends not only on the nature of the buffer, but also on the pH, which is determined by the relative concentrations of the acid and its conjugate base. Figure 12.7 shows plots of buffer capacity versus pH for the CH_3COOH–CH_3COONa system. We see that the buffer functions best around its pK_a value of 4.76. This is not surprising, for according to Eq. (12.8), pH = pK_a when $[HA] = [A^-]$, and there are equal amounts of acid and conjugate base to react with the added base or acid.

All that has been said so far applies equally well to the buffer system of a weak base (B) and its conjugate acid (BH^+), for which the reader should be able to derive the following Henderson–Hasselbalch equation:

$$\text{pH} = pK_a + \log \frac{[B]}{[BH^+]} \tag{12.10}$$

A number of common buffer systems are given in Table 12.7.

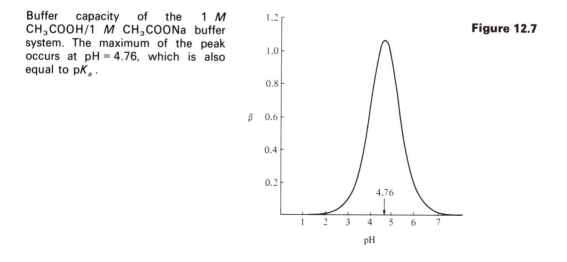

Buffer capacity of the 1 M CH_3COOH/1 M CH_3COONa buffer system. The maximum of the peak occurs at pH = 4.76, which is also equal to pK_a.

Figure 12.7

Table 12.7

Buffer	pH Range[a]
KH phthalate/phthalic acid	2.1–4.1
CH_3COONa/CH_3COOH	3.8–5.8
KNa phalate/KH phthalate	4.4–6.4
Na_2HPO_4/KH_2PO_4	6.2–8.2
Tris/HCl[b]	7.1–9.1
Na borate/boric acid	8.1–10.1
$Na_2CO_3/NaHCO_3$	9.3–11.3
Na_3PO_4/Na_2HPO_4	11.3–13.3

[a] Defined as pH = $pK_a \pm 1$.
[b] Tris is an abbreviation for tris(hydroxymethyl)aminomethane.

MAINTAINING THE pH IN BLOOD

The pH of most intracellular fluids varies between 6.8 and 7.8. Among the large number of buffer systems needed to maintain the proper pH are $HCO_3^- - H_2CO_3$ and $HPO_4^{2-} - H_2PO_4^-$. As can be seen from Figures 12.2 and 12.3, around pH 7.3 the predominant buffers are bicarbonate and carbonic acid and HPO_4^{2-} and $H_2PO_4^-$. They react with acids and bases as follows:

$$HA + HCO_3^- \rightleftharpoons A^- + H_2CO_3$$

$$B + H_2CO_3 \rightleftharpoons BH^+ + HCO_3^-$$

and

$$HA + HPO_4^{2-} \rightleftharpoons A^- + H_2PO_4^-$$

$$B + H_2PO_4^- \rightleftharpoons BH^+ + HPO_4^{2-}$$

In an adult weighing 70 kg, it is estimated that about 0.1 mol of H^+ ions and 12 mol of CO_2 are produced everyday as a result of metabolism. The body has two mechanisms for handling the acid produced by metabolism to prevent a lowering of pH: buffering and excreting H^+. We shall concentrate first on the buffering action of blood. Blood consists essentially of two components: blood plasma, a complex solution containing many biochemically important compounds (carbohydrates, amino acids, proteins, enzymes, hormones, vitamins, and inorganic ions), and erythrocytes or red blood cells. The blood of the average adult contains about 5 million erythrocytes per milliliter of blood. Hemoglobin, the "respiratory" protein, is present inside the erythrocyte. There are about 2×10^8 hemoglobin molecules per erythrocyte. Blood plasma is maintained at pH 7.4 largely by the HCO_3^-/H_2CO_3 and $HPO_4^{2-}/H_2PO_4^-$ buffers, and by various plasma proteins. Proteins are polyamino acids which can themselves act as buffers. A particularly important plasma protein, called albumin, contains 16 histidine residues per molecule. As we saw in Figure 12.5, the imidazole group of the histidine molecule in proteins can combine with and buffer H^+. Equation (12.8) can be used to estimate the relative concentrations of H_2CO_3 and HCO_3^- as follows. At pH 7.4 we write*

$$7.4 = 6.1 + \log \frac{[HCO_3^-]}{[H_2CO_3]}$$

---

* From Table 12.1 we find that $pK_a' = 6.38$ for H_2CO_3/HCO_3^-. However, at the ionic strength of blood and at the physiological temperature of 37°C, pK_a' has a value of 6.1 (see Appendix 12.1).

or $[HCO_3^-]/[H_2CO_3] = 20$. In the normal person, the concentration of CO_2 is about 1.2×10^{-3} M and HCO_3^- is about 0.024 M, giving a ratio of $24/1.2$ or 20, as we calculated above. In the erythrocytes the buffers are HCO_3^-/H_2CO_3 and hemoglobins. (Hemoglobin molecules also possess histidine residues.) There the pH is about 7.25. Again from Eq. (12.8) the $[HCO_3^-]/[H_2CO_3]$ is found to be about 14. The membrane of the red blood cell is more permeable to anions such as HCO_3^-, OH^-, and Cl^- than to K^+ and Na^+ cations.

Oxyhemoglobin, formed by the combination of oxygen with hemoglobin in the lungs, is carried in the arterial blood to the capillary beds, where the oxygen is unloaded to the tissues via myoglobin.* Both hemoglobin and oxyhemoglobin are weak acids, although the latter is considerably stronger than the former:

$$HHb \rightleftharpoons H^+ + Hb^- \qquad pK_a = 8.2$$

$$HHbO_2 \rightleftharpoons H^+ + HbO_2^- \qquad pK_a = 6.95$$

where HHb and $HHbO_2$ represent "monoprotic" hemoglobin and oxyhemoglobin, respectively. This means that at pH = 7.25, about 65% of $HHbO_2$ is in the dissociated form, while only 10% of HHb is dissociated. The release of oxygen by $HHbO_2$ is strongly influenced by the presence of carbon dioxide. In metabolizing tissues, the partial pressure of CO_2, P_{CO_2}, is higher in the interstitial fluid (that is, fluid within the tissue space) than in the plasma. Thus it diffuses into the blood vessel and then into the erythrocytes. Here, most of the CO_2 is converted to H_2CO_3 by the enzyme *carbonic anhydrase*:

$$CO_2 + H_2O \rightleftharpoons H_2CO_3$$

The presence of H_2CO_3 lowers the pH, which has a direct effect on the release of oxygen. Oxygen may be released from either $HHbO_2$ or HbO_2^- as shown:

$$HHbO_2 \rightleftharpoons HHb + O_2$$

$$HHbO_2 \rightleftharpoons H^+ + HbO_2^-$$

$$HbO_2^- \rightleftharpoons Hb^- + O_2$$

Since $HHbO_2$ releases oxygen more readily than HbO_2^-, a lowering in pH increases the concentration of $HHbO_2$ and promotes the first step. The conjugate base Hb^- of the weaker acid HHb has a greater tendency to react with H_2CO_3 as follows:

$$Hb^- + H_2CO_3 \rightleftharpoons HHb + HCO_3^-$$

The bicarbonate ion formed passes through the membrane and is carried away in the plasma. This is the major mechanism for the elimination of CO_2.† When the venous blood circulates back to the lungs, where P_{CO_2} is low and P_{O_2} is high, hemoglobin recombines with oxygen to form oxyhemoglobin:

$$HHb + O_2 \rightleftharpoons HHbO_2$$

---

* To borrow the analogy used by the noted thermodynamicist Henry A. Bent, this probability is less than that for the production of Shakespeare's complete works 15 quadrillion times in succession without error by a tribe of wild monkeys punching randomly on a set of typewriters.
† Since $W \propto V^N$, we have $W = C^N V^N$, where C is a proportionality constant. Referring to the situation shown in Figure 7.3, this constant is given by $1/V_2$. Thus the probability of finding a He atom in volume V_2 is given by $W = (1/V_2)(V_2) = 1$ and that of finding a He atom in volume V_1 is $W = (1/V_2)(V_1) = (V_1/V_2) = \frac{1}{2}$, since $V_1 = V_2/2$.

The bicarbonate ions in blood plasma now diffuse into the erythrocyte to raise the pH, and we have

$$HHbO_2 + HCO_3^- \rightleftharpoons HbO_2^- + H_2CO_3$$

The H_2CO_3 is then converted to CO_2, catalyzed by carbonic anhydrase:

$$H_2CO_3 \rightleftharpoons CO_2 + H_2O$$

Because of the lower P_{CO_2} in the lungs, the CO_2 formed diffuses out of the erythrocyte and is then exhaled into the atmosphere.

What causes the bicarbonate ions formed in red blood cells to preferentially diffuse into plasma? The Donnan equilibrium, discussed in Chapter 9, gives the answer to this question. The concentration of Hb^- and HbO_2^- is quite high in the erythrocytes; consequently there is an unequal distribution of the diffusible anions in the erythrocytes and in the plasma. Now, according to the Donnan equilibrium, the concentrations of the HCO_3^-, Cl^-, and OH^- ions are greater in the plasma than in the erythrocyte (assuming the proteins to be in the anion form).* Further, for any given salt MX we can write

$$(\mu_{MX})^C = (\mu_{MX})^P$$

where C and P denote red blood cell and blood plasma. Following the same procedure as that used for the Donnan equilibrium, we arrive at the result

$$[M^+]_C[X^-]_C = [M^+]_P[X^-]_P$$

or

$$\frac{[M^+]_P}{[M^+]_C} = \frac{[X^-]_C}{[X^-]_P}$$

Thus for a given cation and varying the anions we can show that

$$\frac{[HCO_3^-]_C}{[HCO_3^-]_P} = \frac{[Cl^-]_C}{[Cl^-]_P} = \frac{[OH^-]_C}{[OH^-]_P}$$

At the capillary beds, the CO_2 diffuses into the red blood cell and is converted into H_2CO_3. The carbonic acid reacts with Hb^- and HbO_2^- to form bicarbonate ion, causing the ratio $[HCO_3^-]_C/[HCO_3^-]_P$ to increase. For proper balance of ionic concentrations, the bicarbonate ions diffuse into the plasma, while the chloride and hydroxide ions diffuse into the cell, maintaining electrical neutrality until the above equalities are restored. There is also a corresponding decrease in pH in the erythrocyte caused by the departure of HCO_3^- ions, but this is balanced by the flow of OH^- ions in the reverse direction. Since

$$[H^+]_C[OH^-]_C = [H^+]_P[OH^-]_P$$

or

$$\frac{[OH^-]_C}{[OH^-]_P} = \frac{[H^+]_P}{[H^+]_C}$$

we see that the difference in pH is always maintained in the cell and in the plasma. Figure 12.8 summarizes this discussion.

The phenomenon described above is sometimes called the bicarbonate–chloride shift. In the lungs, the process is exactly reversed. There the bicarbonate ion reacts with oxygenated hemoglobin, causing the ratio $[HCO_3^-]_C/[HCO_3^-]_P$ to decrease. The HCO_3^- ions diffuse from the plasma

* This can be deduced from Table 9.5, which shows that the concentration of the Cl^- ions is always greater in the right compartment, which corresponds to blood plasma in our discussion.

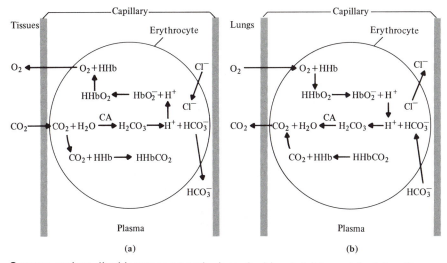

Oxygen–carbon dioxide transport and release by blood. (a) In metabolizing tissues, the partial pressure of CO_2 is higher in the interstitial fluid (fluid in the tissues) than in the plasma. Thus it diffuses into the blood capillaries and then into erythrocytes. There it is converted to carbonic acid by the enzyme carbonic anhydrase (CA). The protons provided by the carbonic acid then combine with the oxyhemoglobin anions to form $HHbO_2$, which eventually dissociates into HHb and O_2. Because the partial pressure of O_2 is higher in the erythrocytes than in the interstitial fluid, oxygen molecules will diffuse out of the erythrocytes and then into the tissues. The bicarbonate ions also diffuse out of the erythrocytes and are carried by the plasma to the lungs. A small portion of the CO_2 also binds to hemoglobin to form carbaminohemoglobin. (b) In the lungs, the processes are exactly reversed.

Figure 12.8

into the erythrocyte, while the Cl^- and OH^- ions diffuse in the opposite direction, until all the ratios of concentrations are again equal.

The foregoing discussion outlines the efficient and fascinating ways our bodies make use of the buffering action. But buffering alone is not enough. The excretion of constantly forming H^+ ions also plays an important role in maintaining a normal blood pH. The kidneys can excrete H^+ ions and return HCO_3^- ions back to the blood, which can then combine with more H^+ ions. Normally, the pH of urine lies between 4.8 and 7.5. But from blood that has a pH of 7.4, the kidneys can produce a urine with a pH as low as 4.5! Two other mechanisms through which the kidneys can excrete H^+ ions are worth noting. The first is the excretion of anions of weak acids, particularly phosphoric acid. At pH 7.4, roughly one-third of the phosphate is present as $H_2PO_4^-$ and two-thirds as HPO_4^-. In an acid urine, however, most of the phosphate will be excreted as $H_2PO_4^-$. The second mechanism involves the formation of NH_4^+ ions. Amino acids are degraded by the kidneys to form ammonia, which combines with H^+ ions as follows:

$$NH_3 + H^+ \longrightarrow NH_4^+$$

The ammonium ions are then excreted in the urine.

Appendix 12.1   Thermodynamic Dissociation Constants of Weak Acids

As was pointed out in the chapter, the dissociation constants of weak acids are usually expressed in terms of concentrations rather than activities. We shall now give a more rigorous treatment of acid dissociations. Consider the following dissociation:

$$HA \rightleftharpoons H^+ + A^-$$

The thermodynamic dissociation constant is given by

$$K_a = \frac{a_{H^+} a_{A^-}}{a_{HA}} = a_{H^+} \frac{[A^-]\gamma_{A^-}}{[HA]\gamma_{HA}}$$

Taking the negative logarithm of this equation, we obtain

$$-\log K_a = -\log a_{H^+} - \log \frac{[A^-]}{[HA]} - \log \frac{\gamma_{A^-}}{\gamma_{HA}}$$

$$pK_a = pH - \log \frac{[A^-]}{[HA]} - \log \frac{\gamma_{A^-}}{\gamma_{HA}}$$

or

$$pH = pK_a + \log \frac{[A^-]}{[HA]} + \log \frac{\gamma_{A^-}}{\gamma_{HA}}$$

Since HA is an uncharged species, a solution containing HA molecules normally behaves ideally so that $\gamma_{HA} \simeq 1$. Hence

$$pH = pK_a + \log \gamma_{A^-} + \log \frac{[A^-]}{[HA]}$$

$$= pK_a' + \log \frac{[A^-]}{[HA]}$$

where

$$pK_a' = pK_a + \log \gamma_{A^-}$$

Since we have no means of knowing the precise value of γ_{A^-}, we can employ Eq. (9.18) as follows:

$$\log \gamma_{A^-} = -0.509 |z_+ z_-| \sqrt{I}$$

so that

$$pK_a' = pK_a - 0.509 |z_+ z_-| \sqrt{I}$$

We see that the dissociation constant term has to be corrected for the ionic strength of the solution. There are two contributions to the ionic strength: one from the dissociation of the acid and the other from dissolved inert salts.

EXAMPLE 12.6 Calculate the dissociation constant of a 0.10 M acetic acid in a 0.20 M KCl solution.

Answer: First we calculate the ionic strength of the solution. Since acetic acid is a weak acid, we may assume that the majority of the ions in the solution came from KCL so that

$$I = \tfrac{1}{2}\{(0.20)(1)^2 + (0.20)(1)^2\}$$

$$= 0.20$$

Hence

$$pK'_a = 4.76 - 0.509(1 \times 1)\sqrt{0.20}$$

$$= 4.53$$

337

App. 12.2

A More Exact
Treatment of
Acid–Base
Equilibria

Comment: In the absence of KCl, the ionic strength is quite small so that $pK'_a \simeq pK_a$. We have assumed that $0.2\ M = 0.2\ m$ in calculating the ionic strength.

Appendix 12.2   A More Exact Treatment of Acid–Base Equilibria

In this appendix we shall carry out an exact derivation of the equilibrium equations for weak acids and their salts.

1. WEAK ACID DISSOCIATION

Consider a weak acid HA whose initial concentration is c_a (mol liter^{-1}). At equilibrium, there are four unknown concentrations: $[H^+]$, $[HA]$, $[A^-]$, and $[OH^-]$ and four equations that relate these unknowns:

$$K_a = \frac{[H^+][A^-]}{[HA]} \tag{A12.1}$$

$$K_w = [H^+][OH^-] \tag{A12.2}$$

Mass balance for the A^- anion:

$$c_a = [HA] + [A^-] \tag{A12.3}$$

Charge balance:

$$[H^+] = [A^-] + [OH^-] \tag{A12.4}$$

Let x be the concentration of hydrogen ions at equilibrium, that is, $[H^+] = x$. Equation (A12.4) can now be written as

$$[A^-] = x - [OH^-]$$

$$= x - \frac{K_w}{x} \tag{A12.5}$$

From Eq. (A12.3), we have

$$[HA] = c_a - [A^-]$$

$$= c_a - x + \frac{K_w}{x} \tag{A12.6}$$

Substituting Eqs. (A12.5) and (A12.6) into Eq. (A12.1), we obtain

$$K_a = \frac{x(x - K_w/x)}{c_a - x + K_w/x} \tag{A12.7}$$

This is a cubic equation in x and in general it is quite tedious to solve. In the majority of cases, however, we can show that $K_w/x \ll x$ so that Eq. (A12.7) is reduced to

$$K_a = \frac{x^2}{c_a - x}$$

This equation is only quadratic in x and is therefore much easier to solve. If the acid is very weak, then $x \ll c_a$ and we have

$$K_a = \frac{x^2}{c_a}$$

This simple equation was employed in Example 12.1.

2. WEAK ACIDS AND THEIR SALTS

Here we consider the case of a solution containing c_a mol liter^{-1} of a weak acid HA and c_s mol liter^{-1} of its salt NaA. We shall first derive a general expression for the acid dissociation constant and then look at some special cases.

In addition to Eqs. (A12.1) and (A12.2), we have

Mass balance for the A$^-$ anion:

$$c_a + c_s = [\text{HA}] + [\text{A}^-] \tag{A12.8}$$

Mass balance for Na$^+$:

$$c_s = [\text{Na}^+] \tag{A12.9}$$

Charge balance:

$$[\text{H}^+] + [\text{Na}^+] = [\text{A}^-] + [\text{OH}^-] \tag{A12.10}$$

For simplicity, let us assume that $[\text{H}^+] = x$ and $[\text{OH}^-] = y$. From Eq. (A12.10), we write

$$[\text{A}^-] = [\text{Na}^+] + x - y$$

$$= c_s + x - y$$

and from Eq. (A12.8),

$$[\text{HA}] = c_a + c_s - [\text{A}^-]$$

$$= c_a + c_s - c_s - x + y$$

$$= c_a - x + y$$

Substituting the expressions for $[\text{A}^-]$ and $[\text{HA}]$ into Eq. (A12.1), we obtain

$$K_a = \frac{x(c_s + x - y)}{c_a - x + y} \tag{A12.11}$$

Equation (A12.11) is in a sense the "master" equation. Let us now consider two special cases.

Case 1. If no salt is present (that is, if we are only dealing with an acid), then $c_s = 0$ and Eq. (A12.11) becomes

$$K_a = \frac{x(x - y)}{c_a - x + y}$$

This equation is the same as Eq. (A12.7), since $y = K_w/x$.

339

App. 12.2

A More Exact
Treatment of
Acid–Base
Equilibria

Case 2. In anionic salt hydrolysis, the hydrogen-ion concentration is usually negligible compared to the hydroxide-ion concentration, that is, $x \ll y$. Thus Eq. (A12.11) now takes the form

$$K_a = \frac{x(c_s - y)}{c_a + y} = \frac{K_w(c_s - y)}{y(c_a + y)}$$

Rearranging this equation, we obtain

$$K_b = \frac{K_w}{K_a} = \frac{y(c_a + y)}{c_s - y}$$

In salt hydrolysis, the initial concentration of the acid is zero, that is, $c_a = 0$ so that

$$K_b = \frac{y^2}{c_s - y}$$

This equation is similar to the expression used in Example 12.2.

3. TITRATION OF A WEAK MONOPROTIC ACID WITH A STRONG BASE

Consider the titration of acetic acid with sodium hydroxide:

$$CH_3COOH + OH^- \longrightarrow H_2O + CH_3COO^-$$

We can follow the pH of the mixture at every stage of the titration by employing Eq. (A12.11). Initially, when no base has been added, $c_s = 0$ and $x \gg y$ so that Eq. (A12.11) can be written as

$$K_a = \frac{x^2}{c_a - x}$$

which is equivalent to Eq. (12.1). After some base has been added, $c_s > 0$; but since we still have $x \gg y$, Eq. (A12.11) takes the form

$$K_a = \frac{x(c_s + x)}{c_a - x}$$

Solving for x then gives us the hydrogen-ion concentration and hence the pH. Note that because of dilution, c_a and c_s must be calculated at each stage by taking the increase in volume into account. As the titration progresses, the concentration of the OH^- ions begins to build up or $x \simeq y$, so that we need to use Eq. (A12.11) or Eq. (A12.7) to solve for x. Because of salt hydrolysis, we predict that the pH at the equivalence point will be greater than 7 so that at and beyond the equivalence point we have $y \gg x$ and Eq. (A12.11) becomes

$$K_a = \frac{x(c_s - y)}{c_a + y} = \frac{x^2 c_s - x K_w}{x c_a + K_w}$$

4. TITRATION OF A WEAK DIPROTIC ACID WITH A STRONG BASE

The rigorous treatment of the titration of a diprotic acid H_2A with a strong base, such as NaOH, is quite involved and will not be presented here. Instead, we shall only describe some qualitative features of the titration curve and indicate how we may obtain the pK_a values.

For a weak diprotic acid H_2A, the dissociations are

$$H_2A \rightleftharpoons H^+ + HA^- \qquad K_a' = \frac{[H^+][HA^-]}{[H_2A]}$$

$$HA^- \rightleftharpoons H^+ + A^{2-} \qquad K_a'' = \frac{[H^+][A^{2-}]}{[HA^-]}$$

The additional equations are

$$K_w = [H^+][OH^-]$$

Mass balance for the A^- anion:

$$c_a = [H_2A] + [HA^-] + [A^{2-}]$$

Mass balance for the Na^+ cation:

$$c_s = [Na^+]$$

Charge balance:

$$[H^+] + [Na^+] = [HA^-] + 2[A^{2-}] + [OH^-]$$

where c_a and c_s are the concentrations of the acid and the base. Note that the coefficient "2" is needed for charge balance so that two negative charges will be counted for each A^{2-} anion. Basically, these six equations tell us all we need to know about the titration. But instead of deriving the necessary mathematical equations, let us divide the titration into five stages and examine some of the salient features.

(i) At the Beginning of the Titration. Here we need only be concerned with the first stage of the dissociation:

$$H_2A \rightleftharpoons H^+ + HA^-$$

$$K_a' = \frac{[H^+][HA^-]}{[H_2A]} \simeq \frac{[H^+]^2}{c_a}$$

or

$$[H^+] = \sqrt{K_a' c_a}$$

(ii) Halfway to the First Inflection Point. At this point, half of the H_2A has been coverted to HA^- so that

$$[HA^-] \simeq [H_2A]$$

giving

$$K_a' = [H^+]$$

or

$$pK_a' = pH$$

(iii) At the First Inflection Point. This is also the first equivalence point. Since the first dissociable protons have been completely neutralized by the base, we have $[H_2A] = [A^{2-}]$. This follows from the fact that equal amounts of H_2A and A^{2-} are formed by the disproportionation: $2HA^- \rightleftharpoons H_2A + A^{2-}$. We neglect the small contribution to A^{2-} from the step: $HA^- \rightleftharpoons H^+ + A^{2-}$. From the acid dissociation constants we can write

341

App. 12.2

A More Exact
Treatment of
Acid–Base
Equilibria

$$K'_a K''_a = \frac{[H^+]^2 [A^{2-}]}{[H_2A]}$$

$$= [H^+]^2$$

$$[H^+] = \sqrt{K'_a K''_a}$$

or

$$pH = \frac{pK'_a + pK''_a}{2}$$

Note that we have neglected salt hydrolysis of HA^- and A^{2-} in arriving at the foregoing result.

(iv) Halfway Between the First and Second Inflection Points.
This condition corresponds to the halfway point in the neutralization of the acid HA^-. Thus we have $[HA^-] = [A^{2-}]$. Since

$$K''_a = \frac{[H^+][A^{2-}]}{[HA^-]}$$

At this point we can write

$$K''_a = [H^+]$$

or

$$pK''_a = pH$$

(v) At the Second Inflection Point. At this point we have a solution of Na_2A for which we can write the hydrolysis equilibrium

$$A^{2-} + H_2O \rightleftharpoons HA^- + OH^-$$

and

$$K_b = \frac{[HA^-][OH^-]}{[A^{2-}]} = \frac{K_w}{K''_a}$$

Also, we may assume that $[HA^-] = [OH^-]$ and $[A^{2-}] = c_a$. Thus

$$[OH^-]^2 = \frac{c_a K_w}{K''_a}$$

Since $[H^+] = K_w/[OH^-]$, we have

$$[OH^-]^2 = \frac{K_w^2}{[H^+]^2} = \frac{c_a K_w}{K''_a}$$

or

$$[H^+] = \sqrt{\frac{K''_a K_w}{c_a}}$$

Figure 12.9

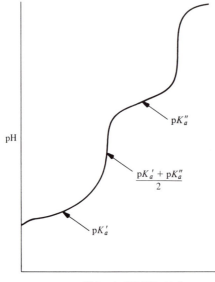

Titration curve of a diprotic acid with sodium hydroxide of equivalent strength.

pH

Volume of NaOH added

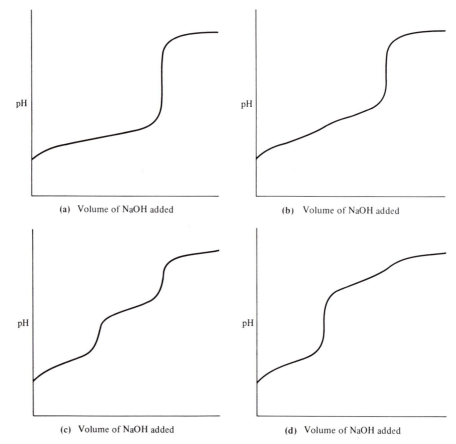

(a) Volume of NaOH added

(b) Volume of NaOH added

(c) Volume of NaOH added

(d) Volume of NaOH added

Figure 12.10 Titration curves of a diprotic acid with sodium hydroxide of equivalent strength. (a) $pK_a' = 4.0$ and $pK_a'' = 4.60$, (b) $pK_a' = 4.0$ and $pK_a'' = 6.0$, (c) $pK_a' = 4.0$ and $pK_a'' = 8.0$, and (d) $pK_a' = 4.0$ and $pK_a'' = 10.0$.

Figure 12.9 shows the titration curve for the H_2A acid and the relationships between pK'_a, pK''_a and pH discussed above.

In arriving at Figure 12.9, we have made the implicit assumption that $K'_a \gg K''_a$. Depending on the nature of the diprotic acid, this condition may or may not be observed in practice. Theory shows that unless K'_a and K''_a differ by about a factor of 1000 or greater, the first inflection point will be difficult if not impossible to detect (Figure 12.10).

INTRODUCTORY–INTERMEDIATE   Suggestions
for Further
Reading

BELL, R. P., *Acids and Bases*. Methuen & Co. Ltd., London (Barnes & Noble, Inc., New York), 1969.
A nice little descriptive book.

CHRISTENSEN, H. N. *Body Fluids and the Acid–Base Balance*. W. B. Saunders Company, Philadelphia, 1964.
A clearly written programmed-learning text.

DAVENPORT, H. W. *The ABC of Acid–Base Chemistry*, 3rd ed. University of Chicago Press, Chicago, 1950.
A useful introductory text that contains many useful data. Strongly recommended.

MASORO, E. J., and P. D. SIEGEL. *Acid–Base Regulations: Its Physiology and Pathophysiology*. W. B. Saunders Company, Philadelphia, 1971.
Provides many examples of acid–base balance in the body.

WEST, E. T., W. R. TODD, H. S. MASON, and J. T. VAN BRUGGEN. *Textbook of Biochemistry*. Macmillan Publishing Co., Inc., New York, 1966, Chapters 15–17.
Rather detailed discussion of acid–base balance and gas transport in the body.

INTERMEDIATE–ADVANCED   Reading
Assignments

EDSALL, J. T., and J. WYMAN. *Biophysical Chemistry*, Vol. 1. Academic Press, Inc., New York, 1958, Chapters 8–10.
Although somewhat outdated, this is still a very useful text.

KING, E. J. *Acid–Base Equilibria*. Pergamon Press, Inc., Elmsford, N.Y., 1965.
A rather detailed treatment of acid–base reactions. Main emphasis is on physiochemical aspects of acids and bases.

TANFORD, C. *Physical Chemistry of Macromolecules*. John Wiley & Sons, Inc., New York, 1961.
Another standard text; strongly recommended for advanced students.

SECTION A

"The Ionic Product Constant of Water," H. L. Clever, *J. Chem. Educ.* **45**, 231 (1968).

"The Ionization Constant of Water," T. P. Dirkse, *J. Chem. Educ.* **38**, 260 (1961).

"Dissociation of Weak Acids and Bases at Infinite Dilution," D. I. Stock, *J. Chem. Educ.* **44**, 764 (1967).

"On Calculating [H^+]," J. D. Burke, *J. Chem. Educ.* **53**, 79 (1976).

"Component Concentration in Solutions of Weak Acids," D. M. Goldish, *J. Chem. Educ.* **47**, 65 (1970).

"Is a Weak Acid Monoprotic?" P. E. Sturrock, *J. Chem. Educ.* **45**, 258 (1968).

"Hydrolysis of Sodium Carbonate," F. S. Nakayama, *J. Chem. Educ.* **47**, 67 (1970). See also *J. Chem. Educ.* **47**, 844, 845 (1970) for additional comments.

"Calcium Carbonate Equilibria in Lakes," S. D. Morton and G. F. Lee, *J. Chem. Educ.* **45**, 511 (1968).

"Effect of Ionic Strength on Equilibrium Constants," M. D. Seymour and Q. Fernando, *J. Chem. Educ.* **54**, 225 (1977).

"Cystinuria: The Relationship of pH to the Origin and Treatment of a Disease," C. Minnier, *J. Chem. Educ.* **50**, 427 (1973).

"Why the Stomach Does Not Digest Itself," H. W. Davenport, *Sci. Am.*, Jan. 1972.

"Stomach Upset Caused by Aspirin," W. D. Hobey, *J. Chem. Educ.* **50**, 212 (1973).

"Physiochemical Properties of Antacids," S. L. Hein, *J. Chem. Educ.* **52**, 383 (1975).

"The Chemistry of Orthophosphoric Acid and Its Sodium Salts," H. A. Neidig, T. G. Teates, and R. T. Yingling, *J. Chem. Educ.* **45**, 57 (1968).

"Actual Effects Controlling the Acidity of Carboxylic Acids," G. V. Calder and T. J. Barton, *J. Chem. Educ.* **48**, 338 (1971).

"Determination of the Microscopic Ionization Constants of Cysteine," G. E. Clement and T. P. Hartz, *J. Chem. Educ.* **48**, 395 (1971).

"Hydrogen Ions in Blood," J. A. Lott, *Chemistry* **51**(4), 6 (1978).

"Pepsin and Antacid Therapy: A Dilemma," W. B. Batson and P. H. Laswick, *J. Chem. Educ.* **56**, 484 (1979).

SECTION B

"Acid–Base Titration and Distribution Curves," J. Waser, *J. Chem. Educ.* **44**, 274 (1967).

"Conjugate Acid–Base and Redox Theory," R. A. Pacer, *J. Chem. Educ.* **50**, 178 (1973).

"A Modern Approach to Acid–Base Chemistry," R. S. Drago, *J. Chem. Educ.* **51**, 300 (1974).

"Development of the pH Concept," F. Szabadvary, *J. Chem. Educ.* **41**, 105 (1964).

"A Laboratory Experiment Using Indicators," C. B. Leonard, Jr., *J. Chem. Educ.* **44**, 363 (1967).

"Thermodynamics of the Ionization of Acetic and Chloroacetic Acids," H. A. Neidig and R. T. Yingling, *J. Chem. Educ.* **42**, 484 (1965).

"The pK_a of a Weak Acid as a Function of Temperature and Ionic Strength," J. L. Bada, *J. Chem. Educ.* **46**, 689 (1969).

"Measurement of pH of Distilled Water," H. L. Youmans, *J. Chem. Educ.* **49**, 429 (1972).

"pK Values for D_2O and H_2O," R. G. Bates, R. A. Robinson, and A. K. Covington, *J. Chem. Educ.* **44**, 635 (1967).

"Standardization of Activity Measurements," R. G. Bates, *Anal. Chem.* **40**(6), 28A (1968).

"The Hydration of Carbon Dioxide," D. W. Kern, *J. Chem. Educ.* **37**, 14 (1960).

"The Hydration of Carbon Dioxide," P. Jones, M. L. Haggett, and J. L. Longridge, *J. Chem. Educ.* **41**, 610 (1964).

"The Great Fallacy of the H^+ ion," P. A. Giguère, *J. Chem. Educ.* **56**, 571 (1979).

"Imidazole—Versatile Today, Prominent Tomorrow," C. A. Matuszak and A. J. Matuszak, *J. Chem. Educ.* **53**, 280 (1976).

"Hydrogen Ion Buffers for Biological Research," N. E. Good et al., *Biochemistry* **5**, 467 (1966).

"The Interpretation of Hydrogen Ion Titration Curves of Proteins," C. Tanford, *Advan. Protein Chem.* **17**, 69 (1962).

Problems

12.1 From the dissociation constants K_a and K_b for an acid and its conjugate base, show that $K_a K_b = K_w$.

12.2 From the dependence of K_w on temperature (see Table 12.1), calculate the enthalpy of ionization of water.

12.3 A 0.040 M solution of a monoprotic acid is 13.5% ionized. What is the dissociation constant of the acid?

12.4 Calculate the pH of the following solutions: (a) 1.0 M HCl, (b) 0.10 M HCl, (c) 1.0×10^{-2} M HCl, (d) 1.0×10^{-4} M HCl, (e) 1.0×10^{-6} M HCl, (f) 1.0×10^{-7} M HCl, (g) 1.0×10^{-8} M HCl, and (h) 1.0×10^{-10} M HCl.

12.5 Calculate the pH of a 0.10 M NH_4Cl solution.

12.6 A quantity of 26.4 ml of a 0.45 M acetic acid solution is added to 31.9 ml of a 0.37 M sodium hydroxide solution. What is the pH of the final solution?

12.7 The dissociation constant of a monoprotic acid at 298 K is 1.47×10^{-3}. Calculate the degree of dissociation assuming (a) ideal behavior, and (b) using a mean activity coefficient $\gamma_{\pm} = 0.93$. The concentration of the acid is 0.010 M.

12.8 Freshly distilled, deionized water has a pH of 7. On standing in air, however, the water gradually becomes acidic. Calculate the pH of the "solution" at equilibrium. (*Hint*: First calculate the solubility of CO_2 in water according to Example 8.3. Assume the partial pressure of CO_2 to be 0.030 atm.)

★**12.9** Show that the acid dissociation constant K_a of a weak monoprotic acid in water is related to its concentration c (mol liter^{-1}) and its degree of dissociation α by $K_a = \alpha^2 c/(1 - \alpha)$ if the self-dissociation of water is ignored. If the latter is taken into account, show that $K_a = \frac{1}{2}\alpha^2 c[1 + (1 + 4K_w \alpha^{-2} c^{-2})^{1/2}]/(1 - \alpha)$.

12.10 To correct for the effect of ionic strength, the dissociation constant of an acid is given by

$$pK_a' = pK_a - \frac{0.509\sqrt{I}}{1 + \sqrt{I}}$$

where K_a is the acid dissociation at zero ionic strength and K_a' the corresponding value for ionic strength I. Calculate the dissociation constant of acetic acid in a 0.15 m KCl solution at 298 K. You may neglect the ionic strength contribution due to the dissociation of the acid itself.

12.11 Calculate the concentration of all the species present in a solution that is 0.12 M in HCN and 0.34 M in NaCN. What is the pH of the solution? Does the solution possess buffer capacity?

*12.12 Depending on the pH of the solution, ferric ions (Fe^{3+}) may exist in the free-ion form or form the insoluble precipitate $Fe(OH)_3$. ($K_{sp} = 1.0 \times 10^{-36}$.) Calculate the pH at which 90% of the Fe^{3+} ions in a 4.5×10^{-5} M Fe^{3+} solution would be precipitated. What conclusion can you draw about the Fe^{3+} ion concentration in blood plasma whose pH is 7.4?

12.13 What is the pH of the buffer 0.10 M Na_2HPO_4|0.10 M KH_2PO_4? Calculate the concentration of all the species in solution.

12.14 A phosphate buffer has a pH equal to 7.3. (a) What is the predominant conjugate pair present in this buffer? (b) If the concentration of this buffer is 0.10 M, what is the new pH after the addition of 5.0 ml of 0.10 M HCl to 20.0 ml of this buffer solution?

12.15 Tris[tris(hydroxymethyl)aminomethane] is a common buffer for studying biological systems. Calculate the pH of the tris buffer by mixing 15.0 ml 0.10 M HCl solution with 25.0 ml 0.10 M tris. (*Hint*: See Table 12.7).

12.16 A 0.020 M aqueous solution of benzoic acid has a freezing point of $-0.0392°C$. Calculate the dissociation constant of benzoic acid. Assume ideal behavior and that molarity is equal to molality at this low concentration.

12.17 The pH of gastric juice is about 1.0, while that in the blood plasma is 7.4. Calculate the free energy required to secrete a mole of H^+ ions from blood plasma to the stomach at 37°C. Assume ideal behavior.

*12.18 Chemical analysis shows that 20.0 ml of a certain sample of blood yields 12.5 ml of CO_2 gas (measured at 25°C and 1 atm) when treated with an acid. Calculate (a) the number of moles of CO_2 originally present in the blood, (b) the concentration of CO_2 and HCO_3^- at equilibrium, and (c) the partial pressure of CO_2 over the blood solution at equilibrium. Assume ideal behavior. The pH of blood is 7.4 and the Henry's law constant of CO_2 in blood is 29.3 atm mol^{-1} (kg H_2O).

*12.19 Calcium oxalate is a major component of kidney stones. From the dissociation constants listed in Table 12.1 and given that the solubility product of CaC_2O_4 is 3.0×10^{-9}, predict whether the formation of kidney stones can be minimized by increasing or decreasing the pH of the fluid present in the kidneys. The pH of normal kidney fluid is about 8.2.

12.20 Derive the Henderson–Hasselbatch equation for the buffer system NH_4^+|NH_3.

12.21 Calculate the pH of the buffer system 0.20 M NH_4Cl|0.10 M NH_3 at 298 K.

12.22 Describe the number of different ways that would allow you to prepare 1 liter of a 0.050 M phosphate buffer at pH 7.8.

12.23 Which of the amino acids listed in Table 12.4 show buffer capacity in the physiological region of pH 7?

*12.24 What is the pH of a 0.050 M glycine solution at 298 K?

12.25 Calculate the ionic strength of a 0.070 M serine buffer at pH 9.2.

*12.26 It was stated in the text of the chapter that amino acids exist in solution as the dipolar ion. Calculate the fraction of the total glycine present as the uncharged species at pH 7.0.

12.27 Sketch the pH versus concentration curve for the titration of 100 ml of 0.1 M aspartic acid hydrogen chloride with sodium hydroxide.

12.28 From the pK_a values listed in Table 12.4, calculate the pI value for each of the 20 amino acids.

12.29 From the dissociation constant of formic acid listed in Table 12.1, calculate the Gibbs free energy and the standard Gibbs free energy for the dissociation of formic acid at 298 K.

12.30 Calculate the pH of a 0.085 M NaCl solution containing 0.020 M Na_2HPO_4 and 0.020 M NaH_2PO_4 assuming (a) ideal behavior, and (b) nonideal behavior. [*Hint*: For part (b), see Example 12.6.]

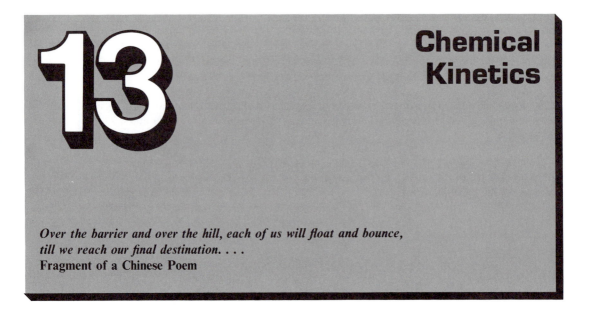

Over the barrier and over the hill, each of us will float and bounce,
till we reach our final destination. . . .
Fragment of a Chinese Poem

The aims of studying chemical kinetics are (1) to determine experimentally the rate of a reaction and its dependence on parameters such as concentration, temperature, and catalysts; and (2) to understand the mechanism of a reaction, that is, the number of steps involved and the nature of intermediates formed.

The subject of chemical kinetics is conceptually easier to understand than some other topics in physical chemistry such as thermodynamics and quantum mechanics, although rigorous, theoretical treatment of the energetics involved is possible only for very simple systems in the gas phase. Nevertheless, the macroscopic, empirical approach to the subject can provide much useful information.

In this chapter we discuss the general topics in chemical kinetics and consider some important examples, such as fast reactions. Enzyme kinetics will be treated in Chapter 14.

13.1 Reaction Rate

The rate of a reaction is expressed as the change in concentration of a reactant molecule with time. Consider the stoichiometrically simple reaction

$$R \longrightarrow P$$

Let the concentrations (in mol liter^{-1}) of R at times t_1 and t_2 $(t_2 > t_1)$ be $[R]_1$ and $[R]_2$. The rate of the reaction over the time interval $(t_2 - t_1)$ is given by

$$\frac{[R]_2 - [R]_1}{t_2 - t_1} = \frac{\Delta[R]}{\Delta t}$$

Since $[R]_2 < [R]_1$, we introduce a minus sign so that the rate will be a positive quantity:

$$\text{rate} = -\frac{\Delta[R]}{\Delta t}$$

The rate could also be expressed in terms of the appearance of a product. In this case we write

$$\text{rate} = \frac{[P]_2 - [P]_1}{t_2 - t_1} = \frac{\Delta[P]}{\Delta t}$$

In this case we have $[P]_2 > [P]_1$. In practice, we find that the quantity of interest is not the rate over a certain time interval (since this is only an average quantity whose value depends on the particular value of Δt); rather, we are interested in the *instantaneous* rate. In the language of calculus, as Δt becomes smaller and smaller and eventually approaches zero, the rate of the foregoing reaction at a specific time t is given by

$$\text{rate} = -\frac{d[R]}{dt} = \frac{d[P]}{dt}$$

The units of reaction rate are usually $M\ s^{-1}$ or $M\ min^{-1}$.

In dealing with stoichiometrically more complicated reactions, the rate must be expressed in an unambiguous manner. Suppose that the reaction of interest is now

$$R \longrightarrow 2P$$

we can no longer express the rate as

$$\text{rate} = -\frac{d[R]}{dt} = \frac{d[P]}{dt}$$

because the product P is appearing twice as fast as the reactant R is disappearing. Instead, the rate should be written as

$$\text{rate} = -\frac{d[R]}{dt} = \frac{1}{2}\frac{d[P]}{dt}$$

In general, for the reaction

$$aA + bB \longrightarrow cC + dD$$

the rate is given by

$$\text{rate} = -\frac{1}{a}\frac{d[A]}{dt} = -\frac{1}{b}\frac{d[B]}{dt} = \frac{1}{c}\frac{d[C]}{dt} = \frac{1}{d}\frac{d[D]}{dt} \tag{13.1}$$

Order of a Reaction **13.2**

The relationship between the rate of a chemical reaction and the concentrations of the reactants is a complex one that must be determined experimentally. Referring to the general equation above, we find that usually (but by no means always) the rate can be expressed as

$$\text{rate} = -\frac{1}{a}\frac{d[A]}{dt} = -\frac{1}{b}\frac{d[B]}{dt} \propto [A]^x[B]^y$$

$$= k[A]^x[B]^y \tag{13.2}$$

What this equation tells us is that the rate of a reaction is not constant; its value at any time t is proportional to the concentrations of A and B raised to some powers. The proportionality constant, k, is called the *rate constant*. The rate constant for a given reaction does not depend on the concentrations of the reactants; it is affected only by temperature, as we shall see later.

Expressing the rate of a reaction as Eq. (13.2) enables us to define the *order* of reaction. We say that the reaction is x order with respect to A and y order with respect to B. Thus the reaction has an overall order of $(x + y)$. It is important to realize that, in general, there is no necessary connection between the order of a reactant in the rate expression and its stoichiometric coefficient in the balanced chemical equation. For example, the rate for the following reaction

$$2N_2O_5(g) \longrightarrow 4NO_2(g) + O_2(g)$$

is given by

$$\text{rate} = k[N_2O_5]$$

The reaction is first-order in N_2O_5—not second-order as we might have inferred from the balanced equation.

The manner in which the rate of a reaction varies with reactant concentrations is called the *rate law*. If we know the order of a reaction and the rate constant, we can always calculate its rate from a knowledge of the reactant concentrations. The order of a reaction is a specification of the empirical dependence of the rate on concentrations. It may be zero, an integer, or even a noninteger. For simplicity, we shall consider only reactions having integral orders. The general reaction of interest is

$$a\text{A} + b\text{B} \longrightarrow \text{products}$$

There is no need to specify the products here, since the order of a reaction is defined in terms of the reactant concentrations only.

ZERO-ORDER REACTIONS

A reaction is said to be *zero-order* if the rate of the reaction is independent of the reactant concentrations. We write

$$\text{rate} = -\frac{1}{a}\frac{d[A]}{dt} = k'[A]^0[B]^0$$

$$= k'$$

Since a is only a coefficient, this equation may be written

$$-\frac{d[A]}{dt} = k \tag{13.3}$$

where $k = ak'$. The quantity k ($M\text{ s}^{-1}$) is the zero-order rate constant. Rearranging Eq. (13.3), we obtain

$$d[A] = -k\ dt$$

Integration between $t = 0$ and $t = t$ gives

$$\int_{[A]_0}^{[A]} d[A] = [A] - [A]_0 = -\int_0^t k\ dt = -kt$$

or

$$[A] = [A]_0 - kt \tag{13.4}$$

Zero-order reaction: (a) plot of rate versus concentration of reactant; (b) plot based on Eq. (13.4).

Figure 13.1

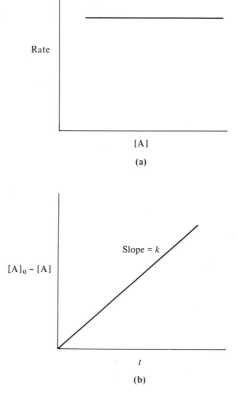

Rate

[A]

(a)

[A]$_0$ - [A]

Slope = k

t

(b)

where $[A]_0$ and $[A]$ are the concentrations at $t = 0$ and $t = t$. A plot of $[A]_0 - [A]$ versus t gives a straight line whose slope is k (Figure 13.1).

Zero-order reactions are relatively rare; among the best-studied reactions are those that occur on metal surfaces. In Chapter 14 we shall see that the zero-order kinetics plays a role in enzyme kinetics.

FIRST-ORDER REACTIONS

Let us assume that the reaction is first-order with respect to A and zero-order with respect to B, so that

$$\text{rate} = -\frac{1}{a}\frac{d[A]}{dt} = k'[A][B]^0$$

$$= k'[A]$$

Again we write

$$-\frac{d[A]}{dt} = k[A] \tag{13.5}$$

where $k = ak'$. The quantity k (s^{-1}) is the first-order rate constant. Upon rearrangement, we have

$$-\frac{d[A]}{[A]} = k\,dt$$

351

Figure 13.2

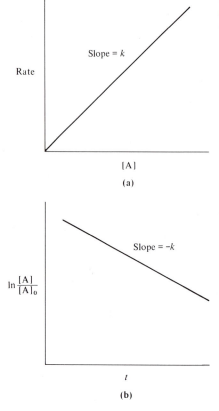

First-order reaction: (a) plot of rate versus concentration of reactant; (b) plot based on Eq. (13.6a).

Integrating between $t = 0$ and $t = t$, we obtain

$$\int_{[A]_0}^{[A]} \frac{d[A]}{[A]} = -\int_0^t k \, dt$$

$$\ln \frac{[A]}{[A]_0} = -kt \qquad (13.6a)$$

or

$$[A] = [A]_0 e^{-kt} \qquad (13.6b)$$

A plot of $\ln ([A]/[A]_0)$ versus t gives a straight line whose slope is given by $-k$ (Figure 13.2).

Many radioactive decays fit first-order kinetics. Examples are

$$^{226}_{88}\text{Ra} \longrightarrow {}^{222}_{86}\text{Rn} + {}^4_2\text{He}$$

$$^{238}_{92}\text{U} \longrightarrow {}^4_2\text{He} + {}^{234}_{90}\text{Th}$$

The thermal decomposition of N_2O_5 mentioned earlier is first-order in N_2O_5. Another example is the rearrangement of methyl isonitrile to acetonitrile:

$$\text{H}_3\text{CNC} \longrightarrow \text{H}_3\text{C}-\text{CN}$$

SECOND-ORDER REACTIONS

We consider two types of second-order reactions here. The first type is represented by

$$a\text{A} + b\text{B} \longrightarrow \text{product}$$

and

$$\text{rate} = -\frac{1}{a}\frac{d[A]}{dt} = k'[A]^2[B]^0$$

$$= k'[A]^2$$

That is, the rate is second-order with respect to A and zero-order with respect to B. We write

$$-\frac{d[A]}{dt} = k[A]^2 \tag{13.7}$$

where $k = ak'$. The quantity $k(M^{-1}\,s^{-1})$ is the second-order rate constant. Separating the variables and integrating, we obtain

$$\int_{[A]_0}^{[A]} \frac{d[A]}{[A]^2} = -\int_0^t k\,dt$$

$$\frac{1}{[A]} - \frac{1}{[A]_0} = kt \tag{13.8}$$

The other type of second-order reaction is represented by

$$aA + bB \longrightarrow \text{product}$$

and

$$\text{rate} = -\frac{1}{a}\frac{d[A]}{dt} = k'[A][B]$$

that is, the reaction is first-order with respect to both A and B. Again we have

$$-\frac{d[A]}{dt} = k[A][B] \tag{13.9}$$

where $k = ak'$. Let

$$[A] = [A]_0 - x$$

$$[B] = [B]_0 - x$$

where x is the mol liter^{-1} of A and B consumed in time t. We write

$$-\frac{d[A]}{dt} = -\frac{d([A]_0 - x)}{dt} = \frac{dx}{dt} = k[A][B]$$

$$= k\{[A]_0 - x\}\{[B]_0 - x\}$$

Rearranging, we obtain

$$\frac{dx}{\{[A]_0 - x\}\{[B]_0 - x\}} = k\,dt$$

By somewhat tedious but straightforward method of integration by parts, we can obtain the final result*:

$$\frac{1}{[B]_0 - [A]_0} \ln \frac{\{[B]_0 - x\}[A]_0}{\{[A]_0 - x\}[B]_0} = kt$$

or

$$\frac{1}{[B]_0 - [A]_0} \ln \frac{[B][A]_0}{[A][B]_0} = kt \tag{13.10}$$

---

* For details, see G. W. Castellan, *Physical Chemistry*, Addison-Wesley Publishing Company, Inc., Reading, Mass., 1971, p. 737.

Figure 13.3

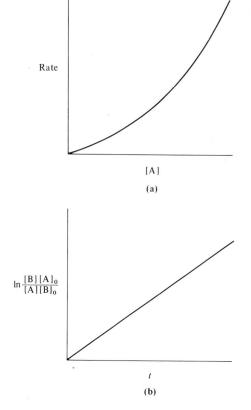

Second-order reaction: (a) plot of rate versus concentration of reactant for the reaction whose rate is second-order in A; (b) plot based on Eq. (13.10).

Equation (13.10) was derived by assuming that (1) $a = b = 1$ and (2) $[A]_0 \neq [B]_0$. A plot of $\ln \left([B][A]_0/[A][B]_0\right)$ versus t is given in Figure 13.3.

Second-order reactions are the most common of all reactions. Some examples are

$$CH_3CHO(g) \longrightarrow CH_4(g) + CO(g)$$

$$CH_3COCH_3(aq) + H^+(aq) \longrightarrow CH_3COHCH_3^+(aq)$$

$$2NO_2(g) \longrightarrow 2NO(g) + O_2(g)$$

When one of the reactants is present in large excess, we have an interesting special case of second-order reactions. For example, consider the hydrolysis of acetyl chloride:

$$CH_3COCl + H_2O \longrightarrow CH_3COOH + HCl$$

Since the concentration of water in the acetyl chloride solution is quite high (about 55.6 M, the concentration of pure water) and the concentration of acetyl chloride is of the order of 1 M or less, the amount of water consumed is negligible. Thus we can express the rate as

$$-\frac{d[CH_3COCl]}{dt} = k[CH_3COCl][H_2O]$$

$$= k'[CH_3COCl]$$

where $k' = k[H_2O]$. The reaction therefore appears to follow first-order kinetics and is called a *pseudo*-first-order reaction.

A quantity of considerable practical importance in kinetic studies is the *half-life* $(t_{1/2})$ of a reaction. The half-life of a reaction is defined as the time it takes for the concentration of the reactant to decrease by half of its original value. Let us consider the first-order reaction as an example. From Eq. (13.6a) we see that as $[A] = [A]_0/2$, $t = t_{1/2}$. Thus

$$\ln \frac{[A]_0/2}{[A]_0} = -kt_{1/2}$$

or

$$t_{1/2} = \frac{\ln 2}{k} = \frac{0.693}{k} \tag{13.11}$$

The half-life for a first-order reaction is independent of the initial concentration. It will take just as much time for A to decrease from 1 M to 0.5 M as it does for A to decrease from 0.001 M to 0.0005 M. The half-lives of other types of reaction do not have this characteristic. We summarize in Table 13.1 the rate equations and half-lives commonly encountered.

DETERMINATION OF REACTION ORDER

Before anything can be said about the mechanism of a reaction, the first task is to determine the order of the reaction. There are a number of methods for determining the order of a reaction; four common approaches follow.

1. Integration Method. An obvious procedure is to measure the concentration of the reactant(s) at various time intervals of a reaction and to substitute the data into the equations listed in Table 13.1. The equation giving the most constant value of the rate constant for a series of time intervals is the one most closely corresponding to the correct order of the reaction.

2. Isolation Method. If a reaction involves more than one type of reactant, we can keep the concentrations of all but one reactant constant and

Summary of Rate Equations for aA → Products **Table 13.1**

Order	Differential Form	Integrated Form	Half-Life	Units of the Rate Constant
0	$-\dfrac{d[A]}{dt} = k$	$[A]_0 - [A] = kt$	$\dfrac{[A]_0}{2k}$	$M\ s^{-1}$
1	$-\dfrac{d[A]}{dt} = k[A]$	$[A] = [A]_0 e^{-kt}$	$\dfrac{\ln 2}{k}$	s^{-1}
2	$-\dfrac{d[A]}{dt} = k[A]^2$	$\dfrac{1}{[A]} - \dfrac{1}{[A]_0} = kt$	$\dfrac{1}{[A]_0 k}$	$M^{-1}\ s^{-1}$
2^a	$-\dfrac{d[A]}{dt} = k[A][B]$	$\dfrac{1}{[B]_0 - [A]_0} \ln \dfrac{[B][A]_0}{[A][B]_0} = kt$	—	$M^{-1}\ s^{-1}$

^{a} For A + B → products.

measure the rate as a function of its concentration. Once the order with respect to this reactant has been determined, the procedure is repeated for the second reactant, and so on.

3. Half-Life Method. Another simple method is to determine the dependence of the half-life of a reaction on the initial concentration, again using the equations in Table 13.1. In general, for a reaction of the nth order, the half-life is found to be

$$t_{1/2} \propto \frac{1}{[A]_0^{n-1}}$$

4. Differential Method. This method was originally suggested by van't Hoff in 1884. Since the rate of an nth-order reaction (v) is proportional to the nth power of the concentration of the reactant, we write

$$\text{rate} = v = k[A]^n$$

Taking common logarithms of both sides, we obtain

$$\log v = n \log [A] + \log k$$

Figure 13.4

(a) Measurement of the initial rates v_0 of a reaction. (b) Plot of log v_0 versus log [A].

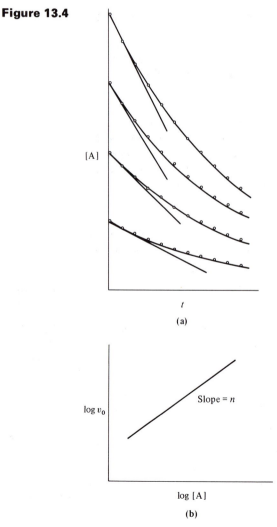

$[A]$

t

(a)

$\log v_0$

Slope = n

$\log [A]$

(b)

Thus, by measuring v at several different concentrations of A, we can obtain the value n from a plot of log v versus log [A]. A satisfactory procedure is to measure the *initial* rates (v_0) of the reaction for several different starting concentrations of A, as shown in Figure 13.4. The advantage is that it avoids possible complications due to the presence of products that might affect the order of the reaction.

It must be realized that the methods described above apply only in ideal cases. In practice the determination of reaction order can be very difficult because of uncertainty in concentration measurements as well as other complicating factors. To a large extent, the procedure is one of trial and error, although intelligent guesses can often be made of the reaction order.

EXAMPLE 13.1 The thermal decomposition of 2,2′-azobisisobutyronitrile (AIBN),

$$N\equiv C-\underset{\underset{CH_3}{|}}{\overset{\overset{CH_3}{|}}{C}}-N=N-\underset{\underset{CH_3}{|}}{\overset{\overset{CH_3}{|}}{C}}-C\equiv N \xrightarrow{\Delta} 2N\equiv C-\underset{\underset{CH_3}{|}}{\overset{\overset{CH_3}{|}}{C}}\cdot \;\; + N_2$$

has been studied in an inert organic solvent at room temperature. The progress of the reaction can be monitored by the optical absorption of AIBN at 350 nm. The following data are obtained:

t (s)	A
0	1.50
2,000	1.26
4,000	1.07
6,000	0.92
8,000	0.81
10,000	0.72
12,000	0.65
∞	0.40

Assume the reaction to be first-order in AIBN; calculate the rate constant.

Answer: From Eq. (13.6), we have

$$\ln \frac{[AIBN]}{[AIBN]_0} = -kt$$

The difference in absorbance at $t = 0$ and at $t = \infty$, $(A_0 - A_\infty)$, is proportional to the concentration of AIBN initially present in the solution. Similarly, the difference $(A - A_\infty)$, where A is the absorbance of AIBN at time t, is proportional to the instantaneous concentration [AIBN].* The rate equation can now be expressed as

$$\ln \frac{A - A_\infty}{A_0 - A_\infty} = -kt$$

---

* These statements hold only if little or no AIBN remains unreacted as t approaches infinity and the absorbance of products does not interfere with that of AIBN at 350 nm.

Figure 13.5

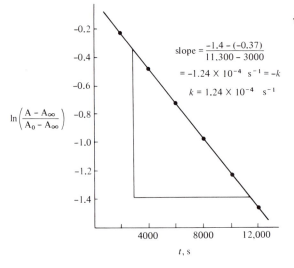

First-order kinetics for the thermal decomposition of AIBN.

$$\text{slope} = \frac{-1.4 - (-0.37)}{11,300 - 3000}$$

$$= -1.24 \times 10^{-4} \text{ s}^{-1} = -k$$

$$k = 1.24 \times 10^{-4} \text{ s}^{-1}$$

Since $A_0 = 1.50$ and $A_\infty = 0.40$, we have

t (s)	$\ln \dfrac{A - A_\infty}{A_0 - A_\infty}$
2,000	−0.246
4,000	−0.496
6,000	−0.749
8,000	−0.987
10,000	−1.240
12,000	−1.482

The first-order rate constant can be obtained from Figure 13.5.

13.3 Molecularity of Reaction

So far we have concentrated on the order of reactions. Knowledge of the order of a reaction is a necessary step toward a detailed understanding of how a reaction occurs, but there is much more to learn about the reaction than just measuring its order. A reaction seldom takes place in the manner suggested by the balanced chemical equation. Frequently, the overall reaction may be the sum of several steps; the sequence of steps by which a reaction occurs is called the *mechanism* of the reaction. To know the mechanism of a reaction is to know how molecules approach one another during a collision, say, and the breaking and making of chemical bonds, charge transfer, and so on when the reactant molecules are in close proximity. The mechanism proposed for a given reaction must be able to account for the overall stoichiometry, for the rate law, and for any other facts that are known. Consider the decomposition of hydrogen peroxide,

$$2H_2O_2 \longrightarrow 2H_2O + O_2$$

358

This reaction is catalyzed by iodide ions in alkaline medium. The rate equation is found to be

$$\text{rate} = -\frac{d[H_2O_2]}{dt} = k[H_2O_2][I^-]$$

Thus the reaction is first-order with respect to both H_2O_2 and I^-.

Whereas the word *order* reflects the overall change, the *molecularity* of a reaction refers to a single, definite kinetic process that may be only one step in the overall reaction. For example, there is evidence to suggest that the decomposition of hydrogen peroxide takes place in two steps, as follows:

$$(1) \qquad H_2O_2 + I^- \longrightarrow H_2O + IO^-$$

$$(2) \quad H_2O_2 + IO^- \longrightarrow H_2O + O_2 + I^-$$

Both *elementary reactions*, that is, reactions that take place on the molecular level, involve the collision or encounter of two molecules and are therefore called *bimolecular reactions*. How do we account for the observed rate dependence in terms of these two steps? Simply by assuming that the rate for the first reaction is much slower than that for the second reaction. The overall rate of decomposition is then completely controlled by the rate for the first step, which is aptly called the *rate-determining* step. Note that the sum of (1) and (2) gives us the overall reaction, since the species IO^- cancels out. Such a species is called an intermediate because it appears in the mechanism of the reaction but is not in the overall balanced chemical equation.

The preceding discussion shows us that our insight into a reaction comes from an understanding of molecularity, not of order. Experimentally, we can determine only the order of a reaction. The experiment serves as a starting point; much more additional work and chemical intuition are needed for a complete understanding of any kinetic process. Although most reactions are kinetically complex, the mechanism for a number of them is sufficiently well understood to be discussed in terms of molecularity. Unlike the order of a reaction, molecularity cannot be zero or a noninteger.

UNIMOLECULAR REACTIONS

Reactions such as cis-trans isomerization, thermal decomposition, ring opening, and racemization are usually unimolecular. For example, the following gas-phase reactions are unimolecular:

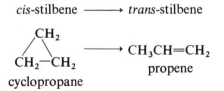

Unimolecular reactions follow a first-order rate law. Since such a reaction presumably occurs as a result of a binary collision through which the reactant molecules acquire the necessary energy to change forms, we would expect it to be a bimolecular process and hence a second-order reaction. How do we account for the difference in the predicted and observed rate laws? To answer this question, let us consider the treatment put forward by Lindermann in

1922. Every now and then a reactant molecule A collides with another A molecule, and one becomes energetically excited at the expense of the other:

$$A + A \xrightarrow{k_1} A + A^*$$

where the asterisk denotes the activated molecule. The activated molecule can then form the desired product according to the elementary step

$$A^* \xrightarrow{k_2} \text{product}$$

Another process that may also be going on is the deactivation of the A* molecule,

$$A^* + A \xrightarrow{k_{-1}} A + A$$

The rate of product formation is given by

$$\frac{d[\text{product}]}{dt} = k_2[A^*]$$

All that remains for us to do is to derive an expression for $[A^*]$. Since A* is an energetically excited species, it has a low stability and a short lifetime. Its concentration in the gas phase is not only low but probably fairly constant as well. Assuming this to be the case, we can apply the *steady-state* approximation as follows. The rate of change of $[A^*]$ is given by the steps leading to the formation of A* minus the steps leading to the removal of A*. Because of the steady-state approximation, however, this rate of change must be zero. Mathematically, we have

$$\frac{d[A^*]}{dt} = 0 = k_1[A]^2 - k_{-1}[A][A^*] - k_2[A^*]$$

Solving for $[A^*]$, we obtain

$$[A^*] = \frac{k_1[A]^2}{k_2 + k_{-1}[A]}$$

The rate of product formation is now given by

$$\frac{d[\text{product}]}{dt} = \frac{k_1 k_2[A]^2}{k_2 + k_{-1}[A]}$$

Two important limiting cases may be applied to this equation. At 1 atm or higher pressures, most A* molecules will be deactivated instead of forming product, and we have

$$k_{-1}[A][A^*] \gg k_2[A^*]$$

or

$$k_{-1}[A] \gg k_2$$

The rate in this case is given by

$$\frac{d[\text{product}]}{dt} = \left(\frac{k_1 k_2}{k_{-1}}\right)[A]$$

which is indeed first-order in A. On the other hand, if the reaction is run at low pressures so that most A* molecules would form the product instead of being deactivated, the following inequality will hold:

$$k_{-1}[A][A^*] \ll k_2[A^*]$$

or

$$k_{-1}[A] \ll k_2$$

The rate now becomes

$$\frac{d[\text{product}]}{dt} = k_1[A]^2$$

which is second-order in A.

Lindemann's theory has been tested for a number of systems and is found to be essentially correct. The analysis for the intermediate case (that is, $k_{-1}[A][A^*] \simeq k_2[A^*]$) is much more complex and will not be dealt with here.

BIMOLECULAR REACTIONS

Any elementary reaction that involves two molecules is a bimolecular reaction. Some of the examples in the gas phase are

$$H + H_2 \longrightarrow H_2 + H$$

$$CO + O_2 \longrightarrow CO_2 + O$$

$$K + HBr \longrightarrow KBr + H$$

$$PH_3 + B_2H_6 \longrightarrow PH_3BH_3 + BH_3$$

In the solution phase we have

$$2CH_3COOH \longrightarrow (CH_3COOH)_2 \qquad \text{(in nonpolar solvents)}$$

$$Fe^{3+} + Fe^{2+} \longrightarrow Fe^{2+} + Fe^{3+}$$

Finally, we note that a termolecular reaction involves the encounter of three molecules in an elementary step. The probability of a three-body collision is usually quite small, and we will not deal with this type of reactions in this text.

More Complex Reactions 13.4

All the reactions discussed so far are simple in the sense that only one reaction is taking place in each case. Unfortunately, this condition is often not satisfied in actual practice. Three examples of more complex reactions will now be discussed.

OPPOSING REACTIONS

Most reactions are reversible to a certain extent, so that we must consider both the forward and reverse rate. Thus for the reversible reaction that proceeds by two *elementary* steps:

$$A \underset{k_{-1}}{\overset{k_1}{\rightleftharpoons}} B$$

we represent the net rate of change of [A] as

$$-\frac{d[A]}{dt} = k_1[A] - k_{-1}[B]$$

At equilibrium, there is no net change in the concentration of A with time, that is, $d[A]/dt = 0$ so that

$$k_1[A] = k_{-1}[B]$$

This leads to

$$\frac{[B]}{[A]} = \frac{k_1}{k_{-1}} = K$$

where K is the equilibrium constant for the reaction.

The discussion of the relationship between reaction rates and equilibria has its roots in a principle of great importance in chemical kinetics. This is the *principle of microscopic reversibility*, which states that at equilibrium the rates of the forward and reverse process are equal for every elementary reaction occurring. This means that the process $A \rightarrow B$ is exactly balanced by $B \rightarrow A$ so that equilibrium cannot be maintained by a cyclic process, with the reaction being $A \rightarrow B$ in one direction and $B \rightarrow C \rightarrow A$ in the opposite direction:

Instead, for every elementary reaction we must write a reverse reaction as follows:

$$A \underset{k_{-1}}{\overset{k_1}{\rightleftharpoons}} C$$

such that

$$k_2[A] = k_{-2}[B]$$

$$k_3[B] = k_{-3}[C]$$

$$k_1[A] = k_{-1}[C]$$

These rate constants are not all independent. By simple algebraic manipulation we can show that $k_{-1}k_2k_3 = k_1k_{-2}k_{-3}$. The usefulness of the principle of microscopic reversibility in chemical kinetics is that it tells us that the reaction pathway for the reverse of a reaction at equilibrium is the exact opposite of the pathway for the forward reaction. This means that the transition states* for the forward and reverse reactions are identical. Consider the base-catalyzed esterification between acetic acid and ethanol:

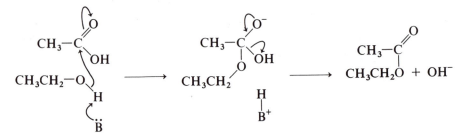

_____

* The transition state of a reaction is the complex between the reactants and products along the reaction pathway (discussed further in Section 13.6).

where B is a base (for example, OH^-). The species formed in the first step is a tetrahedral intermediate. Now according to the principle of microscopic reversibility, the reverse reaction, that is, the hydrolysis of ethyl acetate, must involve the acid-catalyzed expulsion of ethoxide ion from the tetrahedral intermediate:

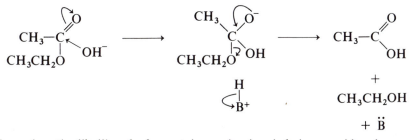

Thus when the likelihood of a certain mechanism is being considered, we can always turn to the principle for guidance. If the reverse mechanism looks ridiculous, then chances are that the proposed mechanism is wrong and we must search for another mechanism.

CONSECUTIVE REACTIONS

A consecutive reaction is represented by

$$A \xrightarrow{k_1} B \xrightarrow{k_2} C$$

The thermal decomposition of acetone is such an example:

$$CH_3COCH_3 \longrightarrow CH_2=CO + CH_4$$

$$2CH_2=CO \longrightarrow 2CO + C_2H_4$$

Many nuclear reactions are also consecutive reactions. For example, upon the capture of a neutron, a uranium-238 isotope is first converted to a uranium-239 isotope, which then decays as follows:

$$^{238}_{92}U + ^{1}_{0}n \longrightarrow ^{239}_{92}U$$

$$^{239}_{92}U \longrightarrow ^{239}_{93}Np + ^{0}_{-1}\beta$$

$$^{239}_{93}Np \longrightarrow ^{239}_{94}Pu + ^{0}_{-1}\beta$$

In the simple case where each step is a first-order, irreversible reaction, we have

$$-\frac{d[A]}{dt} = k_1[A]$$

$$\frac{d[B]}{dt} = k_1[A] - k_2[B]$$

$$\frac{d[C]}{dt} = k_2[B]$$

Plots of concentrations of A, B, and C versus time are shown in Figure 13.6. Since the decrease in A is first-order, we write

$$[A] = [A]_0 e^{-k_1 t}$$

The rate equation for B is quite complex and will not be fully dealt with here. However, this treatment can be greatly simplified if we apply the *steady-state approximation*. We note that B appears as an intermediate in the reaction, and

Figure 13.6

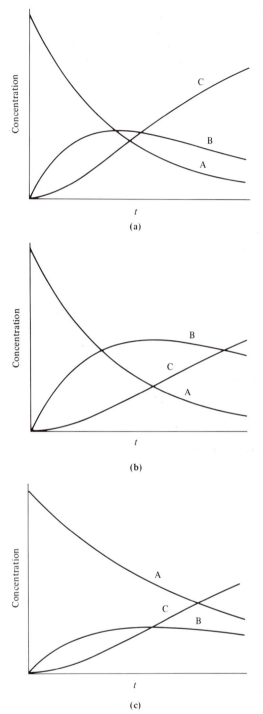

Variation in the concentration of A, B, and C with time for a consecutive reaction $A \rightarrow B \rightarrow C$.
(a) $k_1 = k_2$; (b) $k_1 = 2k_2$;
(c) $k_1 = 0.5k_2$.

its concentration at any time may be both small and fairly constant. Consequently, $d[B]/dt$ can be assumed to be zero. If this is the case, then

$$\frac{d[B]}{dt} = 0 = k_1[A] - k_2[B]$$

$$[B] = \frac{k_1}{k_2}[A] = \frac{k_1}{k_2}[A]_0 e^{-k_1 t} \qquad (13.12)$$

Finally, we have

$$\frac{d[C]}{[B]} = k_2 \, dt$$

Integrating between $t = 0$ and $t = t$ and using Eq. (13.12) for [B], we obtain

$$\frac{1}{[A]_0} \int_0^{[C]} d[C] = \frac{k_1}{k_2} k_2 \int_0^t e^{-k_1 t} \, dt$$

$$[C] = [A]_0(1 - e^{-k_1 t}) \qquad (13.13)$$

Equation (13.13) holds only if the steady-state concentration of B is much smaller than the concentration of A at time t.

It should be emphasized that the steady-state approximation does not always apply to intermediates. Its use must be justified either by experimental evidence or theoretical considerations. In Chapter 14 we shall see how the steady-state approximation is applied to the study of enzyme kinetics.

CHAIN REACTIONS

One of the best-known gas-phase chain reactions involves the formation of hydrogen bromide from molecular hydrogen and bromine. The stoichiometry of the reaction is deceptively simple:

$$H_2 + Br_2 \longrightarrow 2HBr$$

However, the empirical rate equation is given by

$$\frac{d[HBr]}{dt} = \frac{\alpha[H_2][Br_2]^{1/2}}{1 + \beta[HBr]/[Br_2]} \qquad (13.14)$$

where α and β are some constants. Thus the reaction does not have an integral reaction order as described by Eq. (13.2). It has taken many experiments and considerable chemical intuition to come up with Eq. (13.14). We assume a chain of reactions to proceed as follows:

$$Br_2 \xrightarrow{k_1} 2Br \qquad \text{chain initiation}$$

$$Br + H_2 \xrightarrow{k_2} HBr + H \qquad \text{chain propagation}$$

$$H + Br_2 \xrightarrow{k_3} HBr + Br \qquad \text{chain propagation}$$

$$H + HBr \xrightarrow{k_4} H_2 + Br \qquad \text{chain inhibition}$$

$$Br + Br \xrightarrow{k_5} Br_2 \qquad \text{chain termination}$$

Apparently, the following reactions only play a minor role in determining the rate:

$$H_2 \longrightarrow 2H \qquad \text{chain initiation}$$

$$Br + HBr \longrightarrow Br_2 + H \qquad \text{chain inhibition}$$

$$H + H \longrightarrow H_2 \qquad \text{chain termination}$$

$$H + Br \longrightarrow HBr \qquad \text{chain termination}$$

For this reason, they are not included in the kinetic analysis. By applying the steady-state approximation, we can derive Eq. (13.14) using the first five elementary steps (see Problem 13.10).

13.5 Effect of Temperature on Reaction Rates

Figure 13.7 shows four types of temperature dependence for reaction rates. Type (a) represents the normal reactions whose rates increase with increasing temperature. Type (b) shows a rate initially increasing with temperature, reaching a maximum, and finally decreasing with further temperature increase. Type (c) shows a monotonic decrease of rate with temperature. The behavior outlined in (b) and (c) may seem surprising at first. We would expect the rate of a reaction to depend on two quantities: the number of collisions per second and the fraction of collisions that will activate molecules for reaction. Both of these quantities should increase with increasing temperature. The complex nature of reaction mechanism explains this ostensibly startling behavior. For example, in an enzyme-catalyzed reaction, the enzyme molecule must be in a certain specific conformation to react with the substrate molecule. In the *native* state of the enzyme, the rate does increase with temperature. At higher temperatures the molecule may undergo denaturation, thereby losing its effectiveness as a catalyst. Consequently, the rate will decrease with increase in temperature.

The behavior shown in Figure 13.7c is very unusual and is known only for a few systems. Consider the formation of nitrogen dioxide from nitric oxide and oxygen. It is believed that the reaction involves two bimolecular steps:

$$2NO \; \rightleftharpoons \; (NO)_2 \quad \text{rapid} \qquad K = \frac{[(NO)_2]}{[NO]^2}$$

$$(NO)_2 + O_2 \; \xrightarrow{k'} \; 2NO_2 \quad \text{slow, rate-determining}$$

Thus the overall rate is

$$\frac{d[NO_2]}{dt} = k'[(NO)_2][O_2] = k'K[NO]^2[O_2]$$

which is consistent with the experimentally observed rate constant $(k = k'K)$.

Figure 13.7

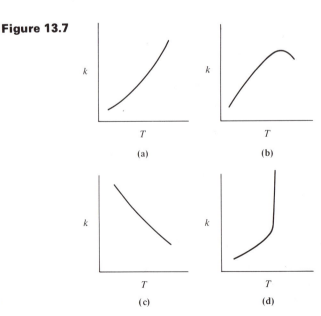

Four types of temperature dependence for reaction rates.

Furthermore, the foregoing equilibrium between NO and $(NO)_2$ is exothermic from left to right. Since the decrease in K with temperature outweighs the increase in k', the overall rate decreases with increasing temperature over a certain range of temperature.

Finally, we note that the behavior shown in Figure 13.7d corresponds to a chain reaction. At first there is a gradual rise in rate with temperature. At a particular temperature, the chain propagation reactions become significant, and the reaction is literally explosive.

Arrhenius discovered in 1889 that the temperature dependence of many reactions could be described by the following equation:

$$k = Ae^{-E_a/RT} \tag{13.15}$$

where k is the rate constant, A is called the *frequency factor* or *preexponential factor*, E_a is the *activation energy*, R is the gas constant, and T is the absolute temperature. The quantity A represents the total frequency of collisions between reactant molecules. The factor $\exp(-E_a/RT)$ resembles the Boltzmann distribution law [see Eq. (3.22)]; it represents the fraction of molecular collisions that have an energy equal to or greater than the activation energy, E_a (Figure 13.8).

Taking the natural logarithm of Eq. (13.15), we obtain

$$\ln k = \ln A - \frac{E_a}{RT} \tag{13.16}$$

Note that the terms $\ln k$ and $\ln A$ pertain to the numerical parts of k and A only. [According to Eq. (13.15), A has the same units as k.] Normally, both A and E_a are approximately constant values over a moderate range of temperature (50 K or so) so that a plot of $\ln k$ versus $1/T$ gives a straight line whose slope (which is negative) is equal to $-E_a/R$ and the intercept on the ordinate gives $\ln A$. Alternatively, if we know the rate constants k_1 and k_2 at T_1 and T_2, we have, from Eq. (13.16),

$$\ln k_1 = \ln A - \frac{E_a}{RT_1}$$

$$\ln k_2 = \ln A - \frac{E_a}{RT_2}$$

Taking the difference between these two equations, we obtain

$$\ln \frac{k_2}{k_1} = -\frac{E_a}{R}\left(\frac{1}{T_2} - \frac{1}{T_1}\right) \tag{13.17}$$

Equation (13.17) provides us with another way of calculating E_a.

Schematic diagram of energy
activation for a reaction.

Figure 13.8

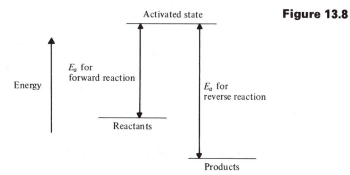

From the standpoint of Arrhenius's rate equation, a complete understanding of the factors determining the rate constant of a reaction requires that we be able to calculate both A and E_a. A great deal of effort has been devoted to this problem, as we shall see below.

13.6 Potential-Energy Surfaces

In order to discuss the activation energy in more detail, we need to learn something about the energetics of a reaction. One of the simplest reactions is the combination of two atoms to form a diatomic molecule, such as

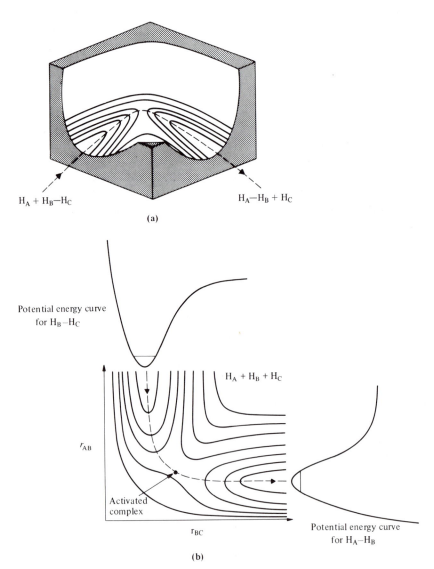

Figure 13.9 $H + H_2$ reaction: (a) potential-energy surface; (b) contour diagram of the potential-energy surface.

H + H → H_2 (see Section 6.6). Basically, we would like to describe more complex reactions in terms of a potential-energy curve such as that shown in Figure 6.8. Unfortunately, potential-energy diagrams are prohibitively complex for all but the simplest systems. One of the simplest and the most studied systems is the exchange reaction between the hydrogen atom and the hydrogen molecule:

$$H + H_2 \longrightarrow H_2 + H$$

Even for a three-atomic system such as this, we need a four-dimensional plot, describing three bond lengths, or two bond lengths and a bond angle, versus energy. The problem is greatly simplified by assuming that the minimum energy configuration to be a linear system so that only two bond lengths need to be specified. Consequently, only a three-dimensional plot is required (Figure 13.9). Labeling the atoms A, B, and C, the reaction can be represented as

$$H_A + H_B{-}H_C \longrightarrow [H_A \cdots H_B \cdots H_C] \longrightarrow H_A{-}H_B + H_C$$
$$\text{activated complex}$$

The plot is a contour map of potential energies corresponding to different values of r_{AB} and r_{BC}. Although the reaction can proceed along any path, the one that requires the *minimum* amount of energy is shown by the dashed line. The system travels along this path through the first valley, over the saddle point, which is the location of the activated complex, and then moves down the second valley. We represent this path in a plot of the potential energy versus the reaction coordinate (Figure 13.10). The plots shown in Figures 13.10b and 13.10c are customarily employed for reactions in general, where the products differ from the reactants. It must be understood that these plots can be used only as a qualitative description for the reaction path because of the complexities involved for large molecules.

Much effort has gone into the calculation of the activation energy for the H + H_2 reaction. The good agreement obtained between the calculated and measured value of E_a (36.8 kJ mol^{-1}) lends support to the correctness of the model. It is interesting to know that if the reaction had taken the path involving the dissociation of H_2 molecule, an activation energy of 436 kJ mol^{-1} would be required.

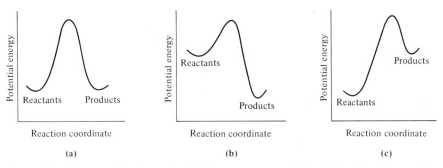

Potential-energy profile along the minimum energy path for (a) a H + H_2 reaction; **Figure 13.10** (b) an exothermic reaction; (c) an endothermic reaction.

13.7 Theories of Reaction Rates

At this point we are ready to consider two important theories of reaction rates: the *collision theory* and the *absolute rate theory*. These theories provide us with a greater insight of the energetic and mechanistic aspects of reactions.

COLLISION THEORY

The collision theory of reaction rates is based on the kinetic theory of gases discussed in Chapter 3. In its simplest form it applies only to bimolecular reactions in the gas phase. Consider the bimolecular elementary reaction

$$A + A \longrightarrow product$$

From Eq. (3.15), the number of binary collisions per cubic centimeters per second is given by

$$Z_{AA} = \frac{\sqrt{2}}{2} \pi d^2 \left(\frac{N_A}{V}\right)^2 \bar{c}$$

$$= \frac{\sqrt{2}}{2} \pi d^2 N_A^2 \bar{c} \qquad (V = 1 \text{ cm}^3)$$

According to Eq. (3.12),

$$\bar{c} = \sqrt{\frac{8kT}{\pi m_A}}$$

we get

$$Z_{AA} = 2N_A^2 d^2 \sqrt{\frac{\pi kT}{m_A}} \tag{13.18}$$

Now for the bimolecular elementary reaction of the type

$$A + B \longrightarrow product$$

the binary collision number becomes

$$Z_{AB} = N_A N_B d_{AB}^2 \sqrt{\frac{8\pi kT}{\mu}} \tag{13.19}$$

where μ, the *reduced mass*, is given by $\mu = m_A m_B / (m_A + m_B)$.

Now if the collisions are 100% effective, that is, a product is formed as a result of every binary collision, the rate of the reaction would be equal to either Z_{AA} or Z_{AB}. But this is not the case. In a gas at a pressure of 1 atm, the collision number is about 10^{28} cm^{-3} s^{-1} at 298 K. If every collision leads to the formation of a product, *all* gas-phase reactions would be complete in about 10^{-9} s, which is contrary to our experience. The additional factor needed for Eqs. (13.18) and (13.19) is of course the term that contains the activation energy. For the A + B → product reaction, we can now write

$$\text{rate} = Z_{AB} \, e^{-E_a/RT}$$

$$= N_A N_B d_{AB}^2 \sqrt{\frac{8\pi kT}{\mu}} \, e^{-E_a/RT} \tag{13.20}$$

The rate constant is then given by

$$k' = \frac{\text{rate}}{N_A N_B} = Ze^{-E_a/RT} \tag{13.21}$$

where

$$Z = d_{AB}^2 \sqrt{\frac{8\pi kT}{\mu}}$$

Note that we have added a "prime" to the rate constant to distinguish it from the Boltzmann constant. Comparison of Eq. (13.15) with Eq. (13.21) gives

$$A = Z = d_{AB}^2 \sqrt{\frac{8\pi kT}{\mu}}$$

We see that the frequency factor A does have a temperature dependence. In practice, we usually treat it as a temperature-independent quantity in the calculation of E_a. This practice does not introduce any serious error, however, because the exponential term [that is, $\exp(-E_a/RT)$] depends on temperature much more strongly than the square-root dependence of temperature.

The collision theory [Eq. (13.21)] predicts the value of the rate constant fairly accurately for reactions that involve atomic species or simple molecules if the activation energy is known. Significant deviations are found, however, when we consider reactions involving complex molecules. The rates tend to be smaller than that predicted by Eq. (13.21), sometimes by a factor of 10^6 or more. To account for the discrepancies observed, we modify Eq. (13.21) as follows:

$$k' = PZe^{-E_a/RT} \tag{13.22}$$

where P, called the *probability* or *steric factor*, takes into account the fact that in a collision complex, molecules must be properly oriented to undergo the reaction (having the proper activation energy is only a *prerequisite* but not a guarantee for a reaction to take place). For reactions involving atomic species as reactants, $P = 1$ (there is no orientation requirement); for reactions involving simple molecules, P varies between 0.5 and 0.2. For complex reactions, P can be a very small number, ranging from 0.1 to 10^{-6} or even smaller. Comparison of Eq. (13.22) with Eq. (13.15) shows that $A = PZ$.

ABSOLUTE RATE THEORY

Although the collision theory is intuitively appealing and does not involve complicated mathematics, it does suffer from some serious drawbacks. Since it is based on the kinetic theory of gases, it takes a "hard-sphere" approach and totally ignores the structures of molecules. For this reason, it cannot satisfactorily account for the probability factor at the molecular level. Furthermore, without quantum mechanics, it cannot calculate the activation energy. A different approach, called the *absolute rate theory*, which was mainly developed by Eyring and others in the 1930s, provides us with a greater insight into the details of mechanism on the molecular scale. It also enables us to calculate the rate constant with considerable accuracy.

The starting point of the absolute rate theory is similar to the collision theory: Reaction is postulated as occurring when molecules collide with each other. An activated complex (also called a transition-state complex) of relatively high energy is formed. Consider the elementary reaction

$$A + B \rightleftharpoons X^{\ddagger} \longrightarrow product$$

where A and B are reactants and $X^{\ddagger}$ represents an activated complex. A fundamental assumption of absolute theory (and one that departs it from the collision theory) is that the reactants are always in equilibrium with $X^{\ddagger}$. The activated complex should not be thought of as an unstable but isolatable intermediate, since it is assumed to be always in the process of decomposing. It is, in fact, neither stable nor isolatable. Thus the equilibrium between the reactants and the activated complex is not of the conventional type. Nevertheless, we can write the equilibrium constant as

$$K^{\ddagger} = \frac{[X^{\ddagger}]}{[A][B]} \tag{13.23}$$

or

$$[X^{\ddagger}] = K^{\ddagger}[A][B]$$

The rate of the reaction is equal to the concentration of the activated complex at the top of the energy barrier, multiplied by the frequency v of crossing the barrier. Hence

rate = number of activated complexes decomposing to form products per cm^3 per second

$$= v[X^{\ddagger}]$$

$$= v[A][B]K^{\ddagger}$$

Since the rate can also be written as

$$rate = k'[A][B]$$

where k' is the rate constant, it follows that

$$k' = vK^{\ddagger}$$

The calculation of k' now depends on our ability to evaluate both v and $K^{\ddagger}$. By the methods of statistical mechanics, we can show that $v = kT/h$,* where k is the Boltzmann constant and h the Planck constant, so that

$$k' = \frac{kT}{h}K^{\ddagger} \tag{13.24}$$

At 298 K, $kT/h = 6.21 \times 10^{12}$ s^{-1}. The equilibrium constant $K^{\ddagger}$ can also be calculated from fundamental physical properties, such as bond length, atomic masses, and vibrational frequencies of the reactants. For this reason, the treatment is called the absolute rate theory.

The rate constant expressed in Eq. (13.24) can now be related to thermodynamic quantities of the reaction. We write†

$$\Delta G^{\ddagger} = -RT \ln K^{\ddagger}$$

Hence

$$K^{\ddagger} = e^{-\Delta G^{\ddagger}/RT} \tag{13.25}$$

---

* The procedure is too lengthy to discuss here. The interested reader is referred to K. J. Laidler, *Chemical Kinetics*, 2nd ed., McGraw-Hill Book Company, New York, 1956, p. 70, or any of the standard physical chemistry texts listed in Chapter 1.

† We omit the superscript ° here for the standard state to avoid cumbersome notations.

Definition of $\Delta G^{\ddagger}$ for a reaction.

Figure 13.11

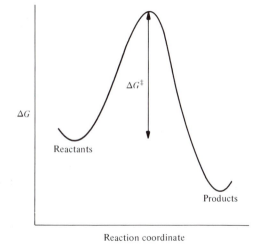

Reactants

Products

ΔG

$\Delta G^{\ddagger}$

Reaction coordinate

where $\Delta G^{\ddagger}$, the free energy of activation (Figure 13.11) is given by

$$\Delta G^{\ddagger} = G^{\circ}(\text{activated complex}) - G^{\circ}(\text{reactants})$$

The rate constant can be written as

$$k' = \frac{kT}{h} \cdot e^{-\Delta G^{\ddagger}/RT} \qquad (13.26)$$

Since kT/h is independent of the nature of A and B, the rate of any reaction at a given temperature is determined by $\Delta G^{\ddagger}$. Furthermore,

$$\Delta G^{\ddagger} = \Delta H^{\ddagger} - T \Delta S^{\ddagger}$$

so that Eq. (13.26) becomes

$$k' = \frac{kT}{h} e^{\Delta S^{\ddagger}/R} e^{-\Delta H^{\ddagger}/RT} \qquad (13.27)$$

where $\Delta S^{\ddagger}$ and $\Delta H^{\ddagger}$ are the entropy and enthalpy of activation. Equation (13.27) is the thermodynamic formulation of the absolute rate theory. A more rigorous approach includes a factor known as the transmission coefficient on the right-hand side of Eq. (13.27), but this is generally close to unity and may be ignored.

It is useful to compare the three expressions for rate constants discussed so far:

$$k' = Ae^{-E_a/RT}$$

$$k' = PZe^{-E_a/RT}$$

$$k' = \frac{kT}{h} e^{\Delta S^{\ddagger}/R} e^{-\Delta H^{\ddagger}/RT}$$

The first equation [Eq. (13.15)] is an empirical equation; both A and E_a must be determined experimentally. The second equation [Eq. (13.22)] is based in part on the collision theory; the value of Z can be calculated from the kinetic theory of gases. On the other hand, it is very difficult in general to estimate accurately the magnitude of P. The last equation [Eq. (13.27)] is based on the absolute rate theory. This equation provides us with the thermodynamic formulation of reaction rate constant and is the most reliable of the three approaches. It is also the most difficult to apply of the three equations.

What is the significance of $\Delta S^\ddagger$ and $\Delta H^\ddagger$? Comparing Eq. (13.27) with Eq. (13.22), we obtain

$$A = PZ = \frac{kT}{h} e^{\Delta S^\ddagger/R}$$

This equation enables us to interpret the probability factor in terms of entropy of activation. If reactants are only atoms or simple molecules, then there is a relatively small amount of rearrangement of energy among the various degrees of freedom in the activated complex. Consequently, $\Delta S^\ddagger$ will be either a small positive or negative number, so that exp $(\Delta S^\ddagger/R)$ or P is close to unity. On the other hand, if complex molecules are involved in a reaction, $\Delta S^\ddagger$ will be either a large positive or negative number. In the former case, the reaction will proceed much faster than predicted by the collision theory; in the latter case, a much slower rate would be observed.

If we compare the coefficients of $1/T$ in Eqs. (13.27) and (13.22), we obtain*

$$\Delta H^\ddagger = E_a$$

that is, the exponential temperature-dependent term now has an enthalpy rather than energy of activation. Furthermore,

$$\Delta H^\ddagger = \Delta U^\ddagger + P\,\Delta V^\ddagger$$

For reactions occurring in solution, the $P\,\Delta V^\ddagger$ term is quite small compared to $\Delta U^\ddagger$ and can usually be neglected. For gas-phase reactions, we have

$$P\,\Delta V^\ddagger = RT\,\Delta n$$

where Δn is the change in moles from reactants to the activated complex. In a unimolecular reaction, $\Delta n = 0$. In general, we have

$$k' = \frac{kT}{h} e^{\Delta S^\ddagger/R} e^{-\Delta H^\ddagger/T}$$

$$\simeq \frac{kT}{h} e^{\Delta S^\ddagger/R} e^{-\Delta U^\ddagger/RT} \qquad \text{in solution}$$

and

$$k' = \frac{kT}{h} e^{\Delta S^\ddagger/R} e^{-\Delta H^\ddagger/RT}$$

$$= \frac{kT}{h} e^{\Delta S^\ddagger/R} e^{-\Delta n} e^{-\Delta U^\ddagger/RT} \qquad \text{in gas phase}$$

Thus from the measurement of A and E_a [Eq. (13.15)], we can obtain $\Delta S^\ddagger$ and $\Delta H^\ddagger$ and, finally, $\Delta G^\ddagger$ for the reaction.

E X A M P L E   **13.2**   The bimolecular dimerization

$$A + A \longrightarrow A_2$$

has been studied in the gas phase. At about 300 K, the experimental rate constant is fitted by the equation $k' = 10^{5.61} e^{-65433/RT}$. Calculate $\Delta S^\ddagger$, $\Delta H^\ddagger$, $\Delta U^\ddagger$, and $\Delta G^\ddagger$ for the reaction.

---

* A more rigorous treatment shows that $\Delta H^\ddagger = E_a - mRT$, where m is the molecularity of the reaction. However, RT is only 5 kJ for a bimolecular reaction and 2.5 kJ for a unimolecular reaction. These values are small compared to $E_a (\gtrsim 50$ kJ) so that we usually write $\Delta H^\ddagger = E_a$.

Answer: From Eq. (13.27), we have

$$k' = \frac{kT}{h} e^{\Delta S^{\ddagger}/R} e^{-\Delta H^{\ddagger}/RT}$$

Since this is a bimolecular reaction,

$$\Delta H^{\ddagger} = E_a - 2RT$$

so that

$$k' = \frac{kT}{h} e^{\Delta S^{\ddagger}/R} e^{-(E_a - 2RT)/RT}$$

$$= \frac{kT}{h} e^{\Delta S^{\ddagger}/R} e^2 e^{-E_a/RT}$$

Since $E_a = 65{,}433$ J mol^{-1}, we have

$$\Delta H^{\ddagger} = 65{,}433 \text{ J mol}^{-1} - 2(8.314 \text{ J K}^{-1} \text{ mol}^{-1})(300 \text{ K})$$

$$= 60{,}445 \text{ J mol}^{-1}$$

$$= 60.5 \text{ kJ mol}^{-1}$$

Comparing the given empirical equation with Eq. (13.27), we obtain

$$\frac{kT}{h} e^{\Delta S^{\ddagger}/R} e^2 = 10^{5.61}$$

or

$$\frac{(1.38 \times 10^{-23} \text{ J K}^{-1})(300 \text{ K})}{6.63 \times 10^{-34} \text{ J s}} \times 7.39 e^{\Delta S^{\ddagger}/R} = 10^{5.61}$$

$$e^{\Delta S^{\ddagger}/R} = 8.81 \times 10^{-9}$$

Taking the natural logarithm on both sides, we get

$$\frac{\Delta S^{\ddagger}}{R} = -18.6$$

or

$$\Delta S^{\ddagger} = -18.6 \times (8.314 \text{ J K}^{-1} \text{ mol}^{-1})$$

$$= -155 \text{ J K}^{-1} \text{ mol}^{-1}$$

Since

$$\Delta H^{\ddagger} = \Delta U^{\ddagger} + RT \, \Delta n$$

and $\Delta n = -1$, we write

$$\Delta U^{\ddagger} = 60.5 \text{ kJ mol}^{-1} - (8.314 \text{ J K}^{-1} \text{ mol}^{-1})(300 \text{ K})(-1)$$

$$= 63.0 \text{ kJ mol}^{-1}$$

Finally,

$$\Delta G^{\ddagger} = \Delta H^{\ddagger} - T \, \Delta S^{\ddagger}$$

$$= 60.5 \text{ kJ mol}^{-1} - (300 \text{ K})\left(-\frac{155}{1000} \text{ kJ K}^{-1} \text{ mol}^{-1}\right)$$

$$= 107 \text{ kJ mol}^{-1}$$

13.8 Kinetic Isotope Effect

The study of kinetic isotope effects has provided much useful information on reaction mechanisms in many branches of chemistry. The theory of kinetic isotope effects is complex, requiring both quantum mechanics and statistical mechanics; therefore, only a qualitative description will be given here.

When an atom in a molecule is replaced by an isotope, there is no change in the electronic structure of the molecule or the potential-energy surface for any reaction the molecule might undergo, yet the rate of the reaction can be profoundly affected by the substitution. To see why this is so, let us consider the H_2, HD, and D_2 molecules, whose zero-point energies are 26.5 kJ mol^{-1}, 21.6 kJ mol^{-1}, and 17.9 kJ mol^{-1}, respectively.* Since D_2 has the lowest

Figure 13.12

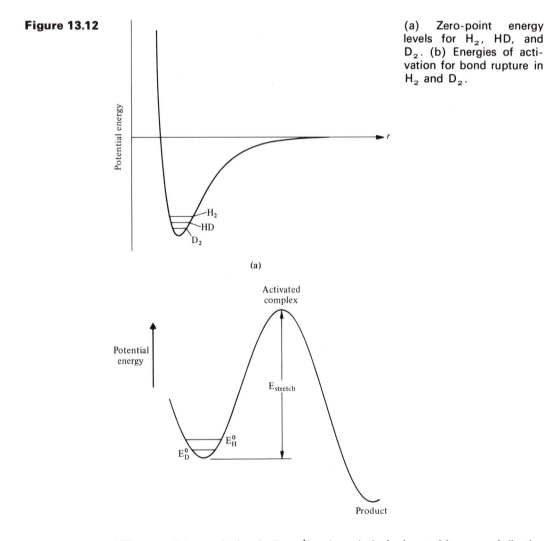

(a) Zero-point energy levels for H_2, HD, and D_2. (b) Energies of activation for bond rupture in H_2 and D_2.

* The zero-point energy is given by $E_{vib} = \frac{1}{2}h\nu$, where ν is the fundamental frequency of vibration. This frequency is given by $\nu = (1/2\pi)\sqrt{k/\mu}$, where k is the force constant of the bond and μ the reduced mass, defined as $1/\mu = 1/m_1 + 1/m_2$ for a diatomic molecule of atomic masses m_1 and m_2. We can readily show that because D_2 has the largest value of μ, it possesses the smallest frequency of vibration and hence E_{vib}. The reverse holds true for H_2. More will be said of molecular vibration in Chapter 18.

zero-point energy (as a result of its largest reduced mass), it will require more energy to dissociate this molecule than H_2 or HD (Figure 13.12a). Consequently, the rate for the reaction $D_2 \rightarrow 2D$ will be the slowest compared to the other two corresponding dissociations. As a rough estimate, we can calculate the ratio of the rate constants for the dissociations of H_2 and D_2, k_H/k_D, as follows. According to Figure 13.12b, the activation energies for these two processes are given by

$$E_H = E_{\text{stretch}} - E_H^\circ$$

$$E_D = E_{\text{stretch}} - E_D^\circ$$

where E_H° and E_D° are the zero-point energies and E_{stretch} is defined as shown in the figure. Using the Arrhenius expression, we write

$$\frac{k_H}{k_D} = \frac{Ae^{-(E_{\text{stretch}} - E_H^\circ)/RT}}{Ae^{-(E_{\text{stretch}} - E_D^\circ)/RT}}$$

$$= e^{(E_H^\circ - E_D^\circ)/RT}$$

$$= e^{(26.5 - 17.9)\,\text{kJ mol}^{-1}/(8.314\,\text{J K}^{-1}\,\text{mol}^{-1})(300\,\text{K})}$$

$$\simeq 31$$

which is quite a large number.

The example above was used to dramatize the difference in the rate constants between D_2 and H_2. In practice, however, we are concerned more with the breaking of a bond between hydrogen and some other atom, such as carbon. Consider, for example, the following reactions:

$$\text{\Large $\gtrless$C}-\text{H} + \text{B} \longrightarrow \text{\Large $\gtrless$C} + \text{H}-\text{B}$$

$$\text{\Large $\gtrless$C}-\text{D} + \text{B} \longrightarrow \text{\Large $\gtrless$C} + \text{D}-\text{B}$$

where B is some group that can take up a hydrogen atom. Again we would expect a kinetic isotope effect, since the fundamental frequencies of vibration are different for the C—H and C—D bonds. But the difference is not as great as that between H_2 and D_2 because the reduced masses are closer to each other in this case (see Problem 13.18). Still, the ratio k_{C-H}/k_{C-D} may be quite appreciable, of the order of 5 or so.

Isotope effects that reflect isotope substitution at an atom involved in a bond-breaking process are called *primary* kinetic isotope effects.* Obviously, such effects are most pronounced for light elements such as H, D, and T. Reactions involving isotopes of mercury (^{199}Hg and ^{201}Hg), for example, would hardly show any detectable difference in rates. A *secondary* kinetic isotope effect occurs whenever the isotope is not directly involved in the bond rupture. We would expect a small change in the reaction rate in this case. This is indeed confirmed by many examples.

Kinetic isotope effects are becoming increasingly important in unraveling the secrets of organic and biological processes. Generally, the primary kinetic isotope effects for light elements (up to about chlorine) are sufficiently large so that many rates can be conveniently and accurately measured. Let us consider briefly one application of the primary kinetic isotope effect for the elucidation of enzyme mechanism. Alcohol dehydrogenase is an enzyme that oxidizes a wide range of aliphatic and aromatic alcohols to their corre-

---

* We assume that this bond-breaking process is also the rate-determining step.

sponding aldehydes and ketones using NAD^+ as coenzyme. The enzyme is found in a variety of sources, including yeast, horse, and human. The reaction is represented by

$$NAD^+ + CH_3 - \overset{\overset{\displaystyle H^*}{|}}{\underset{\underset{\displaystyle H}{|}}{C}} - OH \rightleftharpoons NADH^* + CH_3CHO + H^+$$

The transfer of the hydrogen atom (believed to be a hydride, that is, H^-, transfer) marked * of ethanol and NADH is stereospecific, and does not involve the hydroxyl proton or the solvent. It is found that the deuterium isotope effect of the hydride transfer process from CH_3CD_2OH to NAD^+ is nearly 6 ($k_H/k_D = 6$). It is also possible to study the deuterium isotope effect of the benzaldehyde by deuterated NADH (that is, NADD) for the reverse reaction and again a fairly large deuterium isotope effect ($k_H/k_D = 3$ to 5) is found. These findings show that hydride (or deuteride) transfer occurs in the rate-determining step of the reaction.

How does the kinetic isotope effect affect an equilibrium process? Although the reaction pathway in the forward and reverse direction of an equilibrium process must go via the same transition state, the isotope effect on the two rate constants need not be the same. Consequently, there can be an isotope effect on the equilibrium constant. As a simple example, let us consider the dissociation of a monoprotic acid, such as acetic acid, in H_2O and D_2O. The dissociations are

$$CH_3COOH \rightleftharpoons CH_3COO^- + H^+ \qquad K_H = \frac{[H^+][CH_3COO^-]}{[CH_3COOH]}$$

$$CH_3COOD \rightleftharpoons CH_3COO^- + D^+ \qquad K_D = \frac{[D^+][CH_3COO^-]}{[CH_3COOD]}$$

(In D_2O, all the ionizable protons are replaced by deuterons.) Experimentally, we find that $K_H/K_D = 3.3$.* The greater acid strength of CH_3COOH over CH_3COOD can be explained by noting that the undeuterated molecule has a higher zero-point vibrational level (for the O—H bond) and less energy is required to dissociate the hydrogen than to dissociate the deuterium in CH_3COOD. A useful general rule with regard to the kinetic isotope effect on equilibria is that substitution with a heavier isotope will favor the formation of the stronger bond. As in the case of acetic acid, when we replace H with D, the O—D bond becomes stronger, and hence the molecule has a less tendency to dissociate.

The kinetic isotope effects also have commercial applications. An interesting example is the use of selectively deuterated lubricating oils for watches. These lubricants consist of long-chain hydrocarbons that undergo autoxidation to form solid polymers and eventually lose their effectiveness. One of the initial steps in the autoxidation is a hydrogen abstraction. By replacing the hydrogen atoms with deuteriums at reactive sites, it is possible to substantially slow down the overall oxidation rate. As a result, the lubricant will remain stable during the normal lifetime of the watch.

---

* When a pH meter with a glass electrode and a calomel electrode is calibrated with standard solutions of known $[H^+]$, made up in H_2O, and is then used to measure pD ($-\log [D^+]$) in D_2O, the following correction must be made: pD = pH(meter reading) + 0.4.

The major difference between reactions in the gas phase and in solution lies in the role of solvent. In many cases, the solvent plays a minor role so that rates are not too different in the two phases. In simple kinetic theory, we see that the frequency of collision between reacting molecules depends only on the concentration of the reactants and is not affected by solvent molecules. Actually, molecules in the liquid phase are much closer together and should show less difference in energy between the nonreacting molecules and the activated complexes. As a result, the energy of activation for the same reaction would be somewhat smaller in solution than in the gas phase.

Beside the difference in activation energy, there is also a difference in the encounter of reactant molecules in solution compared to the collision of molecules in the gas phase. If two molecules collide in the gas phase and do not react, they will normally move away from each other. There is very little likelihood that this same pair will collide again. On the other hand, when two solute molecules diffuse together in a solution, they cannot move apart again quickly after the initial encounter because they are surrounded closely by solvent molecules. In this sense the reactants are in a cage of solvent. To be sure, the cage is not a rigid one, as the solvent molecules are constantly in motion and changing positions. Nevertheless, the *cage effect* causes the reactant molecules to remain together for a longer time than they would in the gas phase, and it is not unusual for them to collide hundreds of times with each other before they drift apart.* For reactions having relatively low activation energies, the cage effect virtually ensures reaction during each encounter; the steric factor no longer plays an important role, since the reacting molecules would sooner or later become properly oriented for reaction during the collisions. Under these conditions the rate of the reaction is limited only by how fast the reactants can diffuse together. We shall return to this type of reaction in the next section.

The situation is quite different if the reactants are charged species. The solvation of the ions can be an appreciable factor in determining the sign and magnitude of $\Delta S^{\ddagger}$. In cases where charged species are involved, the value of $\Delta S^{\ddagger}$ depends on the relative net charge of the activated complex. If the activated complex has a greater charge than the reactants, we would expect $\Delta S^{\ddagger}$ to be a negative quantity because of the increase in solvation around the complex:

$$(C_2H_5)_3N + C_2H_5I \longrightarrow (C_2H_5)_4N^+I^- \qquad \Delta S^{\ddagger} = -172 \text{ J K}^{-1} \text{ mol}^{-1}$$

On the other hand, if the formation of an activated complex carries less net charge than the reactants, a positive $\Delta S^{\ddagger}$ is expected:

$$Co(NH_3)_5Br^{2+} + OH^- \longrightarrow Co(NH_3)_4Br(OH)^+ + NH_3$$

$$\Delta S^{\ddagger} = 83.7 \text{ J K}^{-1} \text{ mol}^{-1}$$

We should keep in mind that other factors—steric, for example—also contribute to the large entropy changes.

---

* We speak of molecular collision in the gas phase and molecular encounter in solution. The difference is that after each encounter in solution, the molecules may collide many times before they move away from each other.

Figure 13.13

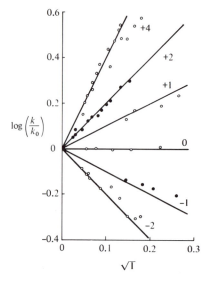

Effect of ionic strength on the rate of reaction between two ions. The slopes and reactions are

+4, $Co(NH_3)_5Br^{2+} + Hg^{2+}$;
+2, $S_2O_8^{2-} + I^-$;
+1, $[NO_2\!=\!N\!-\!CO_2C_2H_5]^- + OH^-$;
0, $CH_3CO_2C_2H_5^+ + OH^-$;
−1, $H_2O_2 + H^+ + Br^-$;
−2, $Co(NH_3)_5Br^{2+} + OH^-$.

[From V. K. LaMer, *Chem. Rev.* **10**, 179 (1932). Used with permission of The Williams & Wilkins Co., Baltimore.]

As we would expect, the rate of a reaction involving ions depends strongly on the ionic strength of the solution. This is known as the kinetic *salt effect*. We state here without proof the following relation*:

$$\log \frac{k}{k_0} = Z_A Z_B A \sqrt{I} \tag{13.28}$$

where A is a constant that depends only on the temperature and the nature of the solvent, k and k_0 are the rate constants at ionic strength I (of an inert salt) and infinitely dilute concentration ($I = 0$), respectively, and Z_A and Z_B the ionic charges of the reactants A and B. Equation (13.28) predicts that (1) if A and B have the same charges, $Z_A Z_B$ is positive and the rate constant k increases with $\sqrt{I}$; (2) if A and B have opposite signs, $Z_A Z_B$ is negative and k decreases with $\sqrt{I}$; and (3) if either A or B is uncharged, $Z_A Z_B = 0$ and k is independent of the ionic strength of the solution.† Figure 13.13 confirms these predictions.

13.10 Fast Reactions in Solution

Roughly speaking, rate constants that lie between 10 and 10^9 for first- and second-order reactions can be described as fast reactions. Recombination of reactive species in the gas phase and solution, acid–base neutralization, and electron and proton exchange reactions are all examples of fast reactions. There has been a great deal of interest in these reactions because of their importance to chemistry and biology. There is also a widespread desire to design experiments for measuring processes whose half-lives are seconds or shorter.

---

* For a derivation, see W. J. Moore, *Physical Chemistry*, 4th ed., Prentice-Hall, Inc., Englewood Cliffs, N.J., 1972, p. 464.
† Note that Eq. (13.28) is derived from the Debye-Hückel limiting law and therefore applies only to dilute solutions.

How fast can a reaction occur in solution? The limit is set by the rate of approach by reacting molecules, which in turn is governed by the rate of diffusion. Thus the fastest reaction is a *diffusion-controlled reaction*. Suppose that we have in a solution two uncharged reactant molecules A and B of radii r_A and r_B. Smoluckowski showed that the number of encounters between A and B per cubic centimeter per second is given by

$$\text{rate of encounter} = 4\pi(D_A + D_B)(r_A + r_B)N_A N_B$$

where N_A and N_B are the numbers of A and B molecules per cubic centimeter and D_A and D_B are the diffusion coefficients. If a reaction takes place upon every encounter, then the rate of reaction is also given by the expression above:

$$\text{rate} = 4\pi(D_A + D_B)(r_A + r_B)N_A N_B$$

or

$$k_D = \frac{\text{rate}}{N_A N_B} = 4\pi(D_A + D_B)(r_A + r_B)$$

where k_D is the diffusion-controlled rate constant in $\text{cm}^3 \text{ molecule}^{-1} \text{s}^{-1}$. If we assume that $D_A = D_B = D$, $r_A = r_B = r$, and use Eq. (5.19) for D, then

$$k_D = 4\pi(2D)(2r)$$

$$= \frac{16\pi k T r}{6\pi\eta r} = \frac{8}{3}\frac{kT}{\eta} \qquad \text{cm}^3 \text{ molecule}^{-1} \text{ s}^{-1}$$

or

$$k_D = \frac{8}{3}\frac{kTN_0}{\eta} = \frac{8RT}{3\eta} \qquad \text{cm}^3 \text{ mol}^{-1} \text{ s}^{-1} \tag{13.29a}$$

A more convenient expression for k_D is

$$k_D = \frac{8RT}{3000\eta} \qquad M^{-1} \text{ s}^{-1} \tag{13.29b}$$

A truly diffusion-controlled reaction has two characteristics. First, such a reaction has a zero activation energy [note the absence of the $\exp(-E_a/RT)$ term in Eq. (13.29)]. Second, the rate is inversely proportional to the viscosity of the medium. The dependence on viscosity is interesting in that η itself depends on temperature as follows:

$$\eta = Be^{E_a'/RT}$$

where E_a' is the "activation energy" of viscosity (note that η *decreases* with increasing temperature) and B is a constant characteristic of the solvent. Thus Eq. (13.29b) can now be written

$$k_D = \frac{8RT}{3000B}e^{-E_a'/RT} \tag{13.30}$$

Eq. (13.30) now takes the form of an Arrhenius equation.

EXAMPLE 13.3 Estimate the rate constant for a diffusion-controlled reaction in water at 298 K.

Answer

CGS units

$$R = 1.987 \text{ cal K}^{-1} \text{ mol}^{-1}$$

$$= 1.987 \times 4.184 \times 10^7 \text{ ergs K}^{-1} \text{ mol}^{-1}$$

$$= 8.314 \times 10^7 \text{ ergs K}^{-1} \text{ mol}^{-1}$$

$$\eta = 0.010 \text{ P}$$

$$= 0.010 \text{ dyn s cm}^{-2}$$

$$= 0.010 \text{ erg s cm}^{-3}$$

From Eq. (13.29a)

$$k_D = \frac{8(8.314 \times 10^7 \text{ ergs K}^{-1} \text{ mol}^{-1})(298 \text{ K})}{3(0.010 \text{ erg s cm}^{-3})}$$

$$= 6.6 \times 10^{12} \text{ ml mol}^{-1} \text{ s}^{-1}$$

$$= 6.6 \times 10^9 \text{ } M^{-1} \text{ s}^{-1}$$

SI units

$$R = 8.314 \text{ J K}^{-1} \text{ mol}^{-1}$$

$$\eta = 0.0010 \text{ P}$$

$$= 0.0010 \text{ N s m}^{-2}$$

$$= 0.0010 \text{ J s m}^{-3}$$

Again,

$$k_D = \frac{8(8.314 \text{ J K}^{-1} \text{ mol}^{-1})(298 \text{ K})}{3(0.0010 \text{ J s m}^{-3})}$$

$$= 6.6 \times 10^6 \text{ m}^3 \text{ mol}^{-1} \text{ s}^{-1}$$

$$= 6.6 \times 10^{12} \text{ cm}^3 \text{ mol}^{-1} \text{ s}^{-1}$$

$$= 6.6 \times 10^9 \text{ } M^{-1} \text{ s}^{-1}$$

Comment: According to Table 13.1, the half-life for a diffusion-controlled reaction, assuming that the starting reactants are identical having concentration equal to 1 M, is

$$t_{1/2} = \frac{1}{1 \text{ } M \times 6.6 \times 10^9 \text{ } M^{-1} \text{ s}^{-1}} = 1.5 \times 10^{-10} \text{ s}$$

which is a very small number.

Many ingenious methods have been devised to study fast reactions. Two examples will be briefly discussed.

FLOW METHODS

In the simplest arrangement, two reactant solutions are first brought together in a mixing chamber and the mixed solution is then passed along an observa-

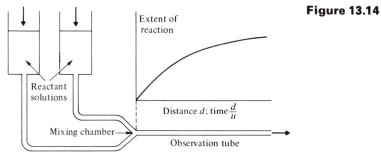

Schematic diagram for a continuous-flow experiment. The velocity of the mixed solution is u. (From E. F. Caldin, *Fast Reactions in Solution*, John Wiley & Sons, Inc., New York, 1964. Used by permission of Blackwell Scientific Publications Ltd., Oxford, England.)

Figure 13.14

tion tube. By monitoring the concentration of either the reactant or product at different points along the tube, we can plot the extent of reaction versus time (Figure 13.14). The limiting factor here is the time required for mixing, which can be as short as 0.001 s. This technique uses large quantities of solutions for every run, a major disadvantage. In an improved arrangement that employs a stop-flow apparatus, the mixed solution is allowed to flow down a tube and is then suddenly stopped. The portion of the solution that comes to rest has been mixed for only a few milliseconds, and the reaction can be monitored spectrophotometrically. The change in absorbance of either the reactant or the product with time can be displayed on an oscilloscope.

RELAXATION METHODS

A system initially at equilibrium is subject to an external perturbation such as a temperature or pressure change. If the change is applied suddenly, there will be a time lag while the system approaches a new equilibrium. This time lag, called the *relaxation time*, can be related to the forward and reverse rate constants. Depending on the systems, reactions with half-lives between 1 s and 10^{-10} s can be studied by the relaxation technique.

Let us illustrate the method with the following equilibrium involving first-order reactions in both directions

$$A \underset{k'_{-1}}{\overset{k'_1}{\rightleftharpoons}} B$$

At equilibrium,

$$k'_1[A]'_e = k'_{-1}[B]'_e$$

where the subscript e denotes the equilibrium concentration. In a temperature-jump experiment, the temperature of the solution can be increased by 10 K in a time as short as 10^{-6} s. This is achieved by discharging a capacitor through the solution. At the elevated temperature, the rate constants are changed to k_1 and k_{-1}. The concentrations of A and B are still, momentarily, at their old equilibrium values, but the system is no longer at equilibrium, so that both A and B will "relax" to the new equilibrium position. At the new equilibrium we have

$$k_1[A]_e = k_{-1}[B]_e$$

Next, we derive an equation that relates the change of either the concentration of A or B to the new position of equilibrium. Let the deviation of [A] from its new equilibrium value, $[A]_e$, be x_t at time t so that at time t after the temperature jump, we have

$$[A]_t = x_t + [A]_e$$

383

At $t = 0$ (immediately after the temperature jump), we have $x_t = x_0$, $[A]_t = [A]_0$, and $[A]_0 = [A]_e'$ so that

$$[A]_0 = x_0 + [A]_e$$

or

$$x_0 = [A]_0 - [A]_e$$
$$= [A]_e' - [A]_e$$

The rate of change of $[A]_t$ is now given by

$$\frac{d[A]_t}{dt} = -k_1[A]_t + k_{-1}[B]_t$$

where $[B]_t$ is defined in the same way as $[A]_t$. Thus we can write

$$\frac{d[A]_t}{dt} = -k_1(x_t + [A]_e) + k_{-1}(-x_t + [B]_e)$$

Note that the x_t terms are the same in magnitude but differ in sign. This is required by the law of conservation of mass—the change in $[A]$ must be equal (but opposite in sign) to the change in $[B]$. Using the $k_{-1}[B]_e = k_1[A]_e$ relationship, we obtain

$$\frac{d[A]_t}{dt} = \frac{d\{x_t + [A]_e\}}{dt} = \frac{dx_t}{dt} = -(k_1 + k_{-1})x_t$$

since $d[A]_e/dt = 0$. Solving this differential equation, we get

$$x_t = x_0 e^{-(k_1 + k_{-1})t}$$
$$= x_0 e^{-t/\tau}$$

where τ, the *relaxation time*, is given by $1/(k_1 + k_{-1})$.

One way to measure the relaxation time is to note that when $t = \tau$, $x_t = x_0/e$ or $x_t = 0.368x_0$. Thus τ is the time required for the concentration of either A or B to decrease from any given value to 0.368 of its original value. By monitoring the decay of $[A]$ after the temperature jump, we can obtain the quantity τ (Figure 13.15). If we can also measure the equilibrium constant $K(k_1/k_{-1})$ by a separate experiment, both k_1 and k_{-1} can be obtained. Note that in a temperature-jump experiment, it is not necessary to know the *actual* concentration change for any species.

Figure 13.15

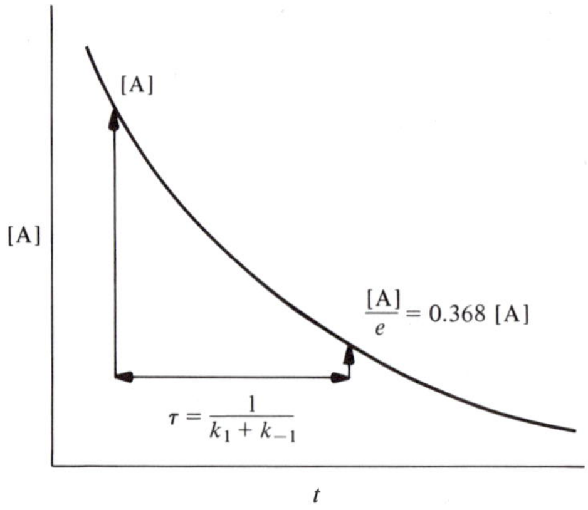

Measurement of relaxation time in a temperature-jump experiment.

Reactions are seldom as simple as the one described above. Often a reaction may have several relaxation times and analysis can be very complex. Nevertheless, relaxation methods are among the most useful and versatile techniques in the study of fast chemical and biochemical reactions.

Suggestions for Further Reading

INTRODUCTORY–INTERMEDIATE

CAMPBELL, J. A. *Why Do Chemical Reactions Occur?* Prentice-Hall, Inc., Englewood Cliffs, N.J., 1965.
A useful introductory text.

EYRING, H., and E. M. EYRING. *Modern Chemical Kinetics.* Reinhold Publishing Corp., New York, 1965.
A rather brief and fairly mathematical treatment.

GARDINER, W. C. *Rates and Mechanisms of Chemical Reactions.* W. A. Benjamin, Inc., Menlo Park, Calif., 1969.
A standard undergraduate text. It does not cover enzyme kinetics.

HAMMES, G. G. *Principles of Chemical Kinetics.* Academic Press, Inc., New York, 1978.
An excellent introductory text.

KING, E. L. *How Chemical Reactions Occur.* W. A. Benjamin, Inc., New York, 1963.
A very readable elementary text.

WESTON, R. E., and H. A. SCHWARZ. *Chemical Kinetics.* Prentice-Hall, Inc., Englewood Cliffs, N.J., 1972.
A brief but well-written text. Presents many up-to-date topics.

INTERMEDIATE–ADVANCED

BENSON, S. W. *The Foundation of Chemical Kinetics.* McGraw-Hill Book Company, New York, 1960.
A comprehensive standard text.

CALDIN, E. F. *Fast Reactions in Solution.* John Wiley & Sons, Inc., New York, 1964.
A very interesting text. Covers many modern techniques and theories.

HAGUE, D. N. *Fast Reactions.* Wiley–Interscience, New York, 1971.
A more up-to-date text than Caldin's text. Also at a lower level.

LAIDLER, K. J. *Chemical Kinetics*, 2nd ed. McGraw-Hill Book Company, New York, 1965.
A very readable standard text.

Reading Assignments

SECTION A

"Concept of Time in Chemistry," O. T. Benfey, *J. Chem. Educ.* **40**, 574 (1963).

"Tables of Conversion Factors for Reaction Rate Constants," D. D. Drysdale and A. C. Lloyd, *J. Chem. Educ.* **46**, 54 (1969).

"On Chemical Kinetics," C. J. Swartz, *J. Chem. Educ.* **46**, 308 (1969).

"Energy Barrier to Chemical Reactions. Why, How, and How Much?" A. A. Zavitsas, *Chem. Tech.*, 434 (July 1974).

"Unconventional Applications of Arrhenius Law," K. J. Laidler, *J. Chem. Educ.* **49**, 343 (1972).

"Method for Determining Order of a Reaction," H. K. Zimmerman, *J. Chem. Educ.* **40**, 356 (1963).

"The Temperature-Jump Method for the Study of Fast Reactions," J. E. Finholt, *J. Chem. Educ.* **45**, 394 (1968).

"Drinking Too Fast Can Cause Sudden Death," C. D. Eskelson, *J. Chem. Educ.* **50**, 365 (1973).

"Drug Receptors," R. E. Rice, *J. Chem. Educ.* **44**, 565 (1967).

"The Time Evolution of Drugs in the Body," G. V. Calder, *J. Chem. Educ.* **51**, 19 (1974).

"Relaxation Methods in Chemistry," L. Faller, *Sci. Am.*, May 1969.

"Kinetics of Solute Permeability in Phospholipid Vesicles," R. Bittman and L. Blau, *J. Chem. Educ.* **53**, 259 (1976).

SECTION B

"Some Aspects of Chemical Dynamics in Solution," J. Halpern, *J. Chem. Educ.* **45**, 372 (1968).

"Along the Reaction Coordinate," W. F. Sheehan, *J. Chem. Educ.* **47**, 853 (1970).

"Unimolecular Gas Reactions at Low Pressures," B. Perlmutter-Hayman, *J. Chem. Educ.* **44**, 605 (1967).

"Mechanistic and Ambiguities of Rate Laws," J. P. Birk, *J. Chem. Educ.* **47**, 805 (1970).

"The Exposition of Isotope Effects on Rates and Equilibria," M. M. Kreevoy, *J. Chem. Educ.* **41**, 636 (1964).

"Application of Isotope Effects," V. Gold, *Chem. Brit.* **6**, 292 (1970).

"The Biology of Heavy Water," J. J. Katz, *Sci. Am.*, July 1960.

"Relaxation Kinetics," J. H. Swinehart, *J. Chem. Educ.* **44**, 524 (1967).

"Steady State and Equilibrium Approximations in Reaction Kinetics," L. Volk, W. Richardson, K. H. Lau, M. Hall, and S. H. Lin, *J. Chem. Educ.* **54**, 95 (1977).

"Some Common Oversimplifications in Teaching Chemical Kinetics," R. K. Boyd, *J. Chem. Educ.* **55**, 84 (1978).

"The Influence of Solvent on Chemical Reactivity," M. R. J. Dack, *J. Chem. Educ.* **51**, 231 (1974).

"Chemistry in a Jiffy," J. Bigeleisen, *Chem. Eng. News*, 26 (Apr. 25, 1977).

"Temperature-Jump Techniques," E. Caldin, *Chem. Brit.* **11**, 4 (1975).

"Ultrashort Phenomena," R. T. Alfano and S. L. Shapiro, *Phys. Today*, 30 (July 1975).

Problems **13.1** Write the rates for the following reactions in terms of the disappearance of reactants and appearance of products:

(a) $3O_2 \longrightarrow 2O_3$.

(b) $C_2H_6 \longrightarrow C_2H_4 + H_2$.

(c) $ClO^- + Br^- \longrightarrow BrO^- + Cl^-$.

(d) $(CH_3)_3CCl + H_2O \longrightarrow (CH_3)_3COH + H^+ + Cl^-$.

(e) $2AsH_3 \longrightarrow 2As + 3H_2$.

13.2 The following reaction is found to be first-order in A: $A \rightarrow B + C$. If half of the starting quantity of A is used up after 56 s, calculate the fraction that will be used up after 6.0 min.

13.3 The half-life of the first-order decay of radioactive ^{14}C is about 5720 years. Calculate the rate constant for the reaction

$$^{14}_{6}C \longrightarrow {}^{14}_{7}N + {}_{-1}^{0}e$$

13.4 The natural abundance of ^{14}C isotope is 1.1×10^{-13} mol % in living matter. Radiochemical analysis of an object obtained in an archeological excavation shows that the ^{14}C isotope content is 0.89×10^{-14} mol %. Calculate the age of the object.

13.5 A certain first-order reaction is 34.5% complete in 4.9 min at 298 K. What is its rate constant?

13.6 The first-order rate constant for the gas-phase decomposition of dimethyl ether,

$$(CH_3)_2O \longrightarrow CH_4 + H_2 + CO$$

is 3.2×10^{-4} s^{-1} at 450°C. The reaction is carried out in a constant-volume container. Initially, there is only dimethyl ether present and the pressure is 0.350 atm. What is the pressure of the system after 8.0 min? Assume ideal gas behavior.

13.7 When the concentration of A in the reaction $A \rightarrow B$ was changed from 1.20 M to 0.60 M, the half-life increased from 2.0 min to 4.0 min at 25°C. Calculate the order of the reaction and the rate constant.

13.8 The progress of a reaction in the aqueous phase was monitored by following the absorbance of a reactant at various times:

Time (s)	0	54	171	390	720	1010	1190
Absorbance	1.67	1.51	1.24	0.847	0.478	0.301	0.216

Determine the order of the reaction and the rate constant.

13.9 The rate constants for the first-order decomposition of an organic compound in solution are measured at several temperatures:

k (s^{-1})	4.92×10^{-3}	0.0216	0.0950	0.326	1.15
T (°C)	5.0	15	25	35	45

Calculate the energy of activation for the reaction.

13.10 Derive Eq. (13.14) using the steady-state approximation for both the H and Br atoms.

13.11 An excited ozone molecule O_3^* in the atmosphere can undergo one of the following reactions:

$$O_3^* \xrightarrow{k_1} O_3 \qquad \text{(1) fluorescence}$$

$$O_3^* \xrightarrow{k_2} O + O_2 \qquad \text{(2) decomposition}$$

$$O_3^* + M \xrightarrow{k_3} O_3 + M \qquad \text{(3) deactivation}$$

where M is an inert molecule. Calculate the fraction of ozone molecules undergoing decomposition.

13.12 Use Eq. (13.15) to calculate the rate constant at 300 K for $E_a = 0$, 2, and 50 kJ mol^{-1}. Assume that $A = 10^{11}$ s^{-1} in each case.

13.13 Many reactions double their rates with every 10° rise in temperature. Assume such a reaction to take place at 305 K and 315 K. What must its activation energy be for this statement to hold?

$\star$**13.14** A certain protein molecule P of molar mass M dimerizes when its solutions are allowed to stand at room temperature. A plausible mechanism is that the protein molecule is first denatured before it dimerizes:

$$P \xrightarrow{k} P^*(\text{denatured}) \qquad \text{slow}$$

$$2P^* \longrightarrow P_2 \qquad \text{fast}$$

This reaction can be followed by viscosity measurements of the average molar mass $\overline{M}$. Derive an expression for $\overline{M}$ in terms of the initial concentration $[P]_0$ and concentration at time t, $[P]$, and M. Write a rate equation consistent with this scheme.

$\star$**13.15** The electron-exchange reaction between naphthalene ($C_{10}H_8$) and its anion radical can be represented by

$$C_{10}H_8^- + C_{10}H_8 \rightleftharpoons C_{10}H_8 + C_{10}H_8^-$$

The reaction is second-order as well as bimolecular. The rate constants are

T (K)	307	299	289	273
k' (M^{-1} s^{-1}) $\times 10^{-9}$	2.71	2.40	1.96	1.43

Calculate $\Delta H^{\ddagger}$, E_a, $\Delta S^{\ddagger}$, and $\Delta G^{\ddagger}$ at 307 K for the reaction. [*Hint*: Use Eq. (13.27) to plot log (k'/T) versus $1/T$.]

13.16 The bromination of acetone is acid-catalyzed:

$$CH_3COCH_3 + Br_2 \xrightarrow{H^+} CH_3COCH_2Br + H^+ + Br^-$$

The progress of the reaction can be conveniently followed by monitoring the optical absorbance of bromine at 450 nm. The simplest way to determine the order and rate constant of the reaction is to use the isolation technique; that is, a series of rates are determined by keeping two of the three reactants constant. The data are as follows:

	$[CH_3COCH_3]$	$[Br_2]$	$[H^+]$	Rate of Disappearance of Br_2 (M s^{-1})
(a)	0.30	0.05	0.05	5.7×10^{-5}
(b)	0.30	0.10	0.05	5.7×10^{-5}
(c)	0.30	0.05	0.10	1.14×10^{-4}
(d)	0.40	0.05	0.20	3.04×10^{-4}
(e)	0.40	0.05	0.05	7.6×10^{-5}

Calculate the rate constant for the reaction.

13.17 A person may die by drinking D_2O instead of H_2O for a few days. Explain.

$\star$**13.18** The rate-determining step of a certain reaction involves breaking a carbon–hydrogen bond. Estimate the ratio of rate constants k_{C-H}/k_{C-D} for the reaction at 300 K. The frequencies of vibration for the particular bonds are $\nu_{C-H} \simeq 3000$ cm^{-1} and $\nu_{C-D} \simeq 2100$ cm^{-1}.

13.19 The equilibrium between dissolved CO_2 and carbonic acid can be represented by

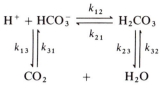

Show that

$$-\frac{d[CO_2]}{dt} = (k_{31} + k_{32})[CO_2] - \left[k_{13} + \frac{k_{23}}{K}\right][H^+][HCO_3^-]$$

where $K = [H^+][HCO_3^-]/[H_2CO_3]$.

* **13.20** The energy of activation for the reaction $2HI \rightleftharpoons H_2 + I_2$ is 180 kJ mol^{-1} at 556 K. Calculate the rate constant using Eq. (13.18). The collision diameter for HI is 3.5×10^{-8} cm. Assume that the pressure is 1 atm.

13.21 Over a range of about $\pm 3°C$ from normal body temperature the metabolic rate, M_T, is given by $M_T = M_{37}(1.1)^{\Delta T}$, where M_{37} is the normal rate and ΔT is the change in T. Discuss this equation in terms of a possible molecular interpretation. [*Source:* "Eco-Chem," J. A. Campbell, *J. Chem. Educ.* **52**, 327 (1975).]

13.22 The rate of bacterial hydrolysis of fish muscle is twice as great at 2.2°C as at $-1.1°C$. Estimate a ΔE_a for this reaction. Is there any relation to the problem of storing fish for food? [*Source:* "Eco-Chem," J. A. Campbell, *J. Chem. Educ.* **52**, 390 (1975)].

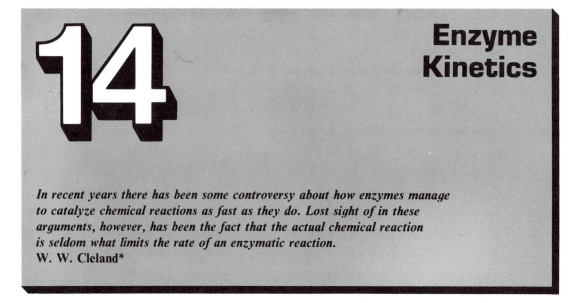

14

Enzyme Kinetics

In recent years there has been some controversy about how enzymes manage to catalyze chemical reactions as fast as they do. Lost sight of in these arguments, however, has been the fact that the actual chemical reaction is seldom what limits the rate of an enzymatic reaction.
W. W. Cleland*

One of the most fascinating studies in chemical kinetics is the investigation of enzyme catalysis. A *catalyst* is a substance that speeds up the rate of a reaction; it is regenerated to its original state after the reactant has been converted to products. Although the mechanisms of catalyzed reaction differ from one another, they all possess a common feature—the lowering of the energy of activation.

An enzyme-catalyzed reaction is usually characterized by a very large increase in the rate (on the order of 10^6 to 10^{12}) and high specificity. By specificity we mean that the enzyme molecule is capable of selectively catalyzing certain reactants, called *substrates*, while discriminating against other molecules.

Since Sumner's work on urease (an enzyme that catalyzes the cleavage of urea to ammonia and carbon dioxide) in 1926, it is now well known that *all* enzymes are protein molecules. Despite the vast amounts of time chemists and biologists have spent working on enzyme catalysis, relatively little is known about the detailed mechanisms involved. An enzyme usually contains one or more *active sites* where reactions with substrates take place. An active site may contain only a few amino acid residues; the rest of the protein molecule is required for maintaining the three-dimensional integrity of the network. Fischer in the 1890s proposed that enzyme specificity may be explained in terms of a *lock-and-key theory*. The active site is assumed to have a rigid structure, similar to a lock. A substrate molecule then has the complementary structure and functions like a key. While appealing in some respects, this theory has been modified to take into account the flexibility of proteins in solution and to explain the phenomenon of cooperativity.

This chapter presents the basic mathematical treatment of enzyme kinetics and discusses the topics of enzyme inhibition, allosterism, and the effect of pH on enzyme kinetics.

* From "What Limits the Rate of an Enzyme-Catalyzed Reaction?" W. W. Cleland, *Acc. Chem. Res.* **8**, 145 (1975). Copyright © 1975 by the American Chemical Society. Used by permission.

In studying enzyme kinetics, it is customary to measure the *initial rate* (v_0) of a reaction. The first reason for this choice is to minimize the rate of reverse reaction, since this rate increases with the concentration of product. Second, during the course of the reaction, there may be appreciable heat and/or pH changes. These changes may alter the conformation of the enzyme and hence the rate of the reaction. Third, in some cases, the product formed may bind to the enzyme in such a way as to inhibit its function. Finally, the initial rate corresponds to a known fixed substrate concentration. As time proceeds, the substrate concentration will drop. Figure 14.1 shows the variation of the initial rate of an enzyme-catalyzed reaction with substrate concentration [S]. (We assume that the enzyme–substrate binding ratio is 1:1.) The rate increases rapidly at low concentrations of the substrate but then gradually levels off at high concentrations of the substrate. Mathematical analysis shows that the relationship between v_0 and [S] can be defined in terms of the equation of a rectangular hyperbola:

$$v_0 = \frac{a[S]}{b + [S]} \tag{14.1}$$

where a and b are constants. Our next step is to develop the necessary equations to account for the experimental data.

MICHAELIS–MENTEN KINETICS

In 1913, Michaelis and Menten proposed a theory to explain the dependence of the initial rate on concentration. They considered the following scheme:

$$E + S \underset{k_{-1}}{\overset{k_1}{\rightleftharpoons}} ES \overset{k_2}{\longrightarrow} P + E$$

where E and S are enzyme and substrate, ES is an enzyme–substrate complex, and P is the product formed. The initial rate of product formation v_0 is given by

$$v_0 = \left(\frac{d[P]}{dt}\right)_0 = k_2[ES] \tag{14.2}$$

The concentrations of the enzyme and substrate at a time shortly after the start of the reaction are

$$[E] = [E]_0 - [ES]$$
$$[S] = [S]_0 - [ES]$$

Plot of the initial rate of an enzyme-catalyzed reaction versus substrate concentration.

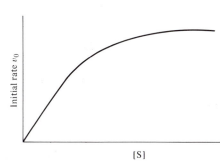

Figure 14.1

However, experimental conditions are usually such that $[S]_0 \gg [E]_0$ so that $[S]_0 \gg [ES]$ or $[S] \simeq [S]_0$. Michaelis and Menten assumed that $k_{-1} \gg k_2$ so that the first step (formation of ES complex) can be treated as a rapid equilibrium process. The dissociation constant K_S is given by

$$K_S = \frac{k_{-1}}{k_1} = \frac{[E][S]}{[ES]}$$

$$= \frac{([E]_0 - [ES])[S]}{[ES]} \tag{14.3}$$

Solving for [ES], we obtain

$$[ES] = \frac{[E]_0[S]}{K_S + [S]} \tag{14.4}$$

Substitution of Eq. (14.4) into Eq. (14.2) yields

$$v_0 = \frac{d[P]}{dt} = \frac{k_2[E]_0[S]}{K_S + [S]} \tag{14.5}$$

We see that Eq. (14.5) has the same form as Eq. (14.1), where $a = k_2[E]_0$ and $b = K_S$. The physical significance of a and b may be understood as follows. At high substrate concentrations, $[S] \gg b$ so that

$$v_0 = \frac{a[S]}{b + [S]} \simeq \frac{a[S]}{[S]} = a = k_2[E]_0$$

Under these conditions all the enzyme molecules are in the complex form, so that the initial rate must be at its maximum value (V_m). Thus we write

$$V_m = k_2[E]_0 \tag{14.6}$$

where V_m is called the maximum rate. Now let us consider what happens when b and $[S]$ are equal. From Eq. (14.1), we find that this condition gives $v_0 = a/2 = V_m/2$. Thus b equals that concentration of S at which the initial rate is half its maximum value.

THE STEADY-STATE APPROXIMATION

Briggs and Haldane in 1925 showed that it is unnecessary to make the assumption that enzyme and substrate are in thermodynamic equilibrium with the enzyme–substrate complex to derive Eq. (14.5). They postulated that a short time after enzyme and substrate are mixed the concentration of the enzyme–substrate complex will reach a constant value so that we can apply the steady-state approximation as follows (Figure 14.2):

$$\frac{d[ES]}{dt} = 0 = k_1[E][S] - k_{-1}[ES] - k_2[ES]$$

$$= k_1([E]_0 - [ES])[S] - (k_{-1} + k_2)[ES]$$

Again, solving for [ES], we get

$$[ES] = \frac{k_1[E]_0[S]}{k_1[S] + k_{-1} + k_2} \tag{14.7}$$

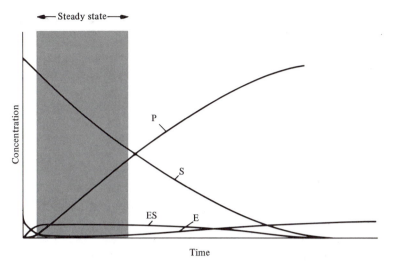

Plot of the concentrations of the various species in an enzyme-catalyzed reaction **Figure 14.2**
$E + S \rightleftharpoons ES \rightarrow P + E$ versus time. We assume that the initial substrate concentration
is much larger than the enzyme concentration and that the rate constants k_1, k_{-1},
and k_2 (see the text) are of comparable magnitude.

Using Eq. (14.7) in Eq. (14.2), we obtain

$$v_0 = \left(\frac{d[P]}{dt}\right)_0 = k_2[ES] = \frac{k_1 k_2[E]_0[S]}{k_1[S] + k_{-1} + k_2}$$

$$= \frac{k_2[E]_0[S]}{[(k_{-1} + k_2)/k_1] + [S]}$$

$$= \frac{k_2[E]_0[S]}{K_m + [S]} \tag{14.8}$$

where K_m, the *Michaelis constant*, is defined as

$$K_m = \frac{k_{-1} + k_2}{k_1} \tag{14.9}$$

Comparing Eq. (14.8) with Eq. (14.5), we see that they both have similar
dependence on substrate concentration; however, $K_m \neq K_S$ in general unless
$k_{-1} \gg k_2$.

The Briggs–Haldane treatment defines the maximum rate exactly the same
as Eq. (14.6). Since $[E]_0 = V_m/k_2$, Eq. (14.8) can also be written as

$$v_0 = \frac{V_m[S]}{K_m + [S]} \tag{14.10}$$

When the initial rate is equal to half the maximum rate, Eq. (14.10) becomes

$$\frac{V_m}{2} = \frac{V_m[S]}{K_m + [S]}$$

or

$$K_m = [S]$$

Thus both V_m and K_m can be determined, at least in principle, from a plot
such as that shown in Figure 14.3. However, in practice we find that the plot

393

Figure 14.3

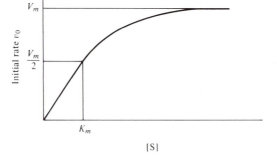

Graphical determination of V_m and K_m.

of v_0 versus [S] is not very useful in determining V_m, since it is often difficult to determine the asymptotic value V_m at very high substrate concentrations. A more satisfactory approach, suggested by Lineweaver and Burk, is to employ the double-reciprocal plot of $1/v_0$ versus $1/[S]$. From Eq. (14.10) we obtain

$$\frac{1}{v_0} = \frac{K_m}{V_m[S]} + \frac{1}{V_m} \qquad (14.11)$$

As Figure 14.4 shows, both K_m and V_m can be obtained from the slope and intercepts of the straight line. Although the Lineweaver–Burk plot is used widely in enzyme kinetic studies, it tends to give too much weight to the least accurate points, that is, those points obtained at low concentrations of substrate. Readers interested in enzyme kinetics should consult the references listed below.*

What is the significance of V_m and K_m? The maximum velocity has a well-defined meaning, both theoretically and empirically. It represents the maximum rate attainable; that is, it is the rate when the total enzyme concentration is present as the enzyme–substrate complex. The kinetic constant k_2 (in units of s^{-1}) in the equation $V_m = k_2[E]_0$ is called the *turnover number*. The turnover number is also known as k_{cat}, the catalytic rate constant. The turnover number of an enzyme is the number of substrate molecules converted into product per unit time, when the enzyme is fully saturated with substrate. Turnover numbers can be measured only with pure enzymes, that is, if $[E]_0$ is known (since V_m can be determined graphically from the Lineweaver–Burk plot). However, it is often difficult in practice to

Figure 14.4

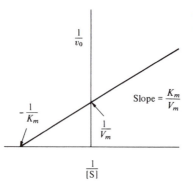

Lineweaver–Burk plot for enzyme-catalyzed reactions obeying Michaelis–Menten kinetics.

* J. E. Dowd and D. S. Riggs, *J. Biol. Chem.* **240**, 863 (1965); R. Eisenthal and A. Cornish-Bowden, *Biochem. J.* **139**, 715, 721 (1974).

obtain highly purified enzymes so that $[E]_0$ is usually an unknown quantity. Partly for this reason the activity of an enzyme is more often given as *units of activity per milligram of protein* (called the *specific activity*). One *International unit* is the amount of enzyme that produces one micromole (1 μmol) of product per minute.* The turnover numbers of most enzymes for their physiological substrates fall in the range 1 to 10^5 s^{-1}. Carbonic anhydrase, an enzyme that catalyses the hydration of carbon dioxide and dehydration of carbonic acid (see Chapter 12),

$$CO_2 + H_2O \rightleftharpoons H_2CO_3$$

has one of the largest known turnover numbers $(k_2 = 1 \times 10^6$ s$^{-1})$ at 25°C. Thus a 10^{-6} M solution of the enzyme can catalyze the formation of 1 M H_2CO_3 from CO_2 and H_2O per second; that is,

$$V_m = 1 \times 10^6 \text{ s}^{-1} \times 1 \times 10^{-6} \text{ M}$$
$$= 1 \text{ M s}^{-1}$$

Without the enzyme, the pseudo first-order rate constant is only about 0.03 s^{-1}.

Unlike the turnover number, K_m lacks a simple interpretation. As mentioned earlier, only in the special case where $k_{-1} \gg k_2$ can K_m be equated to the dissociation constant K_S. When this condition is met, K_m becomes a measure of the strength of the ES complex. A large K_m indicates weak binding, a small K_m indicates strong binding. In general, K_m must be expressed in terms of three rate constants although one of them, k_2, can in principle be obtained from Eq. (14.6). Nevertheless, K_m (in units of M) is customarily reported together with V_m and k_2 for enzyme-catalyzed reactions. The reason is that K_m depends on pH, temperature, substrate, and so on; therefore, its value serves to characterize a particular enzyme–substrate system under certain specified conditions. For most enzymes, K_m lies between 10^{-2} and 10^{-7} M.

α-Chymotrypsin: A Case Study **14.2**

Having developed the basic equations of enzyme kinetics, let us now consider some reactions catalyzed by α-chymotrypsin, which is one of the most studied and best understood enzymes.

α-Chymotrypsin is one of the serine proteases, a family of protein-cutting enzymes that includes trypsin, elastase, and subtilisin. It has a molecular mass of 24,800 and 246 amino acid residues. α-Chymotrypsin is produced in the mammalian pancreas. There it takes the form of an inactive precursor, chymotrypsinogen. Once this precursor enters the intestine, it is activated by another enzyme, trypsin, to become α-chymotrypsin. The reason for the original inactive form is not hard to understand. Certainly, evolution has not spent all these millions of years designing a group of enzymes that will attack one another as well as other essential protein molecules in the pancreas before they can digest food. The enzyme can be prepared in highly purified form by crystallization and has only one active site per molecule.

---

* The International Union of Biochemistry has recommended a new unit, the *katal* (kat), the amount of enzyme that converts one mol s^{-1} of substrate to product. The conversion factor is 1 kat = 6 × 10^7 international units.

One of the reactions catalyzed by α-chymotrypsin is the hydrolysis of *p*-nitrophenylacetate to *p*-nitrophenol and acetate:

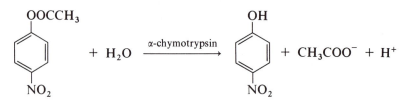

This reaction can be followed spectrophotometrically because *p*-nitrophenol has a substantially different optical absorption spectrum from the substrate molecule. A plot of the absorbance of *p*-nitrophenol (monitored at 400 nm) versus time shows that the reaction takes place in two steps. At first there is a rapid release of *p*-nitrophenol, whose concentration corresponds roughly to the concentration of the enzyme employed. Then a very slow (zero-order in substrate concentration) release of *p*-nitrophenol follows. This behavior suggests that two rate steps are involved in the reaction. It has been suggested that the catalyzed reaction proceeds as follows:

$$E + S \underset{k_{-1}}{\overset{k_1}{\rightleftharpoons}} ES \xrightarrow{k_2} ES' + P_1 \xrightarrow{k_3} E + P_2$$

where P_1 is *p*-nitrophenol and P_2 is acetate ion. The complex ES' is called the acetyl enzyme because the acetyl group (from the substrate) is covalently bound to the serine residue of the active site. Under certain conditions, for example, at pH 5, this complex can be rendered inactive and isolated.

The initial phase of this catalyzed reaction is so rapid that a stopped-flow apparatus must be employed to follow the progress on a millisecond time scale. On the other hand, the catalyzed hydrolysis of *p*-nitrophenyl trimethyl-acetate to *p*-nitrophenol and trimethylacetate has the same characteristics as *p*-nitrophenyl acetate but proceeds much more slowly because of steric hindrance. Consequently, this reaction can be conveniently studied using a conventional spectrometer. Figure 14.5 shows a plot of the absorbance of

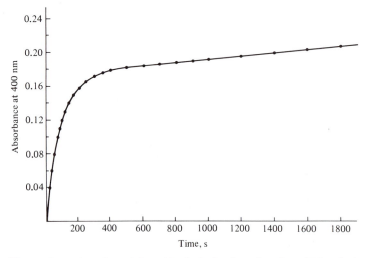

Figure 14.5 The α-chymotrypsin-catalyzed hydrolysis of *p*-nitrophenyl trimethylacetate at 298 K. [From M. L. Bender, F. J. Kézdy, and F. C. Wedler, *J. Chem. Educ.* **44**, 84 (1967).]

p-nitrophenol versus time with *p*-nitrophenyl trimethylacetate as the substrate.

The kinetic analysis, that is, the theoretical fit for the curve in Figure 14.5, starts with the following equations:

$$[E]_0 = [E] + [ES] + [ES']$$

$$\frac{d[P_1]}{dt} = k_2[ES]$$

$$\frac{d[P_2]}{dt} = k_3[ES']$$

$$\frac{d[ES']}{dt} = k_2[ES] - k_3[ES']$$

Since there are five unknowns (k_2, k_3, $[E]_0$, and two of the following three quantities: $[E]$, $[ES]$, and $[ES']$) and only four equations, we need one more equation. This equation is provided by assuming that the first step is a rapid equilibrium, that is, for

$$E + S \underset{k_{-1}}{\overset{k_1}{\rightleftharpoons}} ES$$

we write

$$K_s = \frac{k_{-1}}{k_1} = \frac{[E][S]}{[ES]}$$

From these equations it is possible to fit the curve shown in Figure 14.3 and solve for the pertinent kinetic constants. The results are shown in Table 14.1.* For this mechanism the quantity k_{cat} (catalytic rate constant) is defined by

$$k_{cat} = \frac{k_2 k_3}{k_2 + k_3}$$

since $k_2 \gg k_3$, k_{cat} is essentially equal to k_3. Figure 14.6 shows a proposed mechanism of the hydrolysis of a peptide bond by serine proteases such as trypsin or chymotrypsin.

Kinetic Constants of the α-Chymotrypsin-Catalyzed Hydrolysis of *p*-Nitrophenyl Trimethylacetate at pH 8.2[a,b]

Table 14.1

k_2	0.37 ± 0.11 s^{-1}
k_3	$(1.3 \pm 0.03) \times 10^{-4}$ s^{-1}
K_s	$(1.6 \pm 0.5) \times 10^{-3}$ M
k_{cat}	1.3×10^{-4} s^{-1}
K_m	5.6×10^{-7} M

[a] From M. L. Bender, F. J. Kézdy, and F. C. Wedler, *J. Chem. Educ.* **44**, 84 (1967).
[b] 0.01 M tris–HCl buffer, ionic strength 0.06, 25.6 ± 0.1°C, 1.8% (v/v) acetonitril–water.

---

* The mathematical details involved in the solution are too lengthy (although not too difficult) to present here. The interested reader should study the paper by M. L. Bender, F. J. Kézdy, and F. E. Wedler, *J. Chem. Educ.* **44**, 84 (1967).

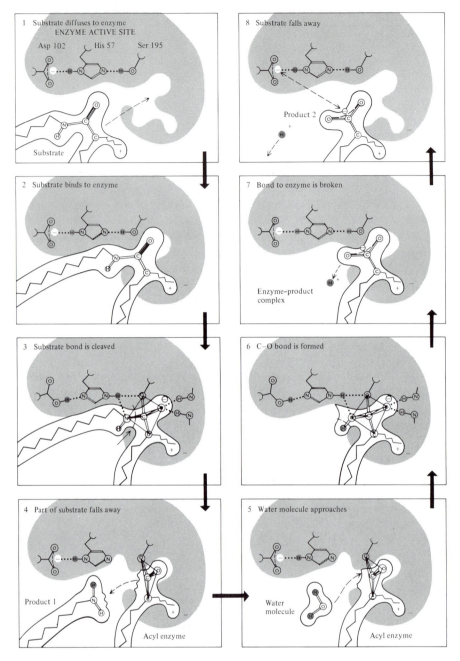

Figure 14.6 The catalytic mechanism of cleavage of a protein chain by serine proteases such as trypsin or chymotrypsin. [Drawing by Irving Geis, from "A Family of Protein-Cutting Proteins," by R. M. Stroud, *Sci. Am.*, July 1974. Copyright © 1974 by Stroud, Dickerson, and Geis.]

So far we have considered enzyme catalysis involving only one substrate. In many instances, two or more substrates may be involved. For example, the reaction

$$C_2H_5OH + NAD^+ \longrightarrow CH_3CHO + NADH + H^+$$

is catalyzed by the enzyme alcohol dehydrogenase, which binds both NAD^+ and the substrate that is to be oxidized. Many of the principles developed for the single-substrate systems may be extended to multisubstrate systems. Although the mathematics involved is too advanced to be presented here, some understanding can be gained by considering the qualitative features below.

The overall picture of a two-substrate reaction can be represented by*

$$A + B \rightleftharpoons P + Q$$

where A and B are the substrates and P and Q the products. The binding of A and B to the enzyme may take place in a number of ways, which can conveniently be organized under two topics.

SEQUENTIAL MECHANISM

In some reactions the binding of both substrates must take place first, before the release of products. A sequential step can be further classified as follows.

1. Ordered sequential mechanism: This takes place according to the scheme

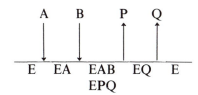

2. Ordered random mechanism: Here the binding of substrates and the release of products do not follow a definite obligatory order. We have

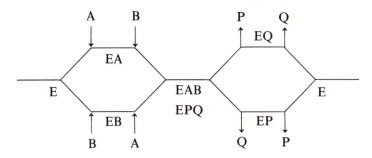

* We follow the scheme suggested by W. W. Cleland. For details, see P. D. Boyer, ed., *The Enzymes*, Vol. 2, Academic Press, Inc., New York, 1970, Chapter 1.

In this mechanism, one product is released before the second substrate is bound to the enzyme. The shuttling of the enzyme back and forth resembles a Ping-Pong game. The reaction is represented by

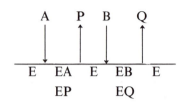

In each of the three cases discussed above, two rate constants must be assigned for each step, one for forward reaction and one for reverse reaction. It is not difficult to see that the exact kinetic analysis will be very complex.

14.4 Enzyme Inhibition

Inhibitors are compounds that decrease the rate of an enzyme-catalyzed reaction. The study of enzyme inhibition has provided much valuable information of the specificity of enzymes and the nature of functional groups at the active site. The activity of certain enzymes may be regulated by a feedback mechanism so that the end products act as specific inhibitors to the enzymes in the initial stages of a chain producing them. The glycolytic pathway is an example of this feedback mechanism. In this manner, the amount of products formed can be controlled (Figure 14.7).

There are two types of enzyme inhibitions: reversible and irreversible. In *reversible inhibitions*, an equilibrium exists between the enzyme and the inhibitor. In *irreversible inhibitions*, inhibition progressively increases with the passage of time. A complete inhibition can be reached if the concentration of the irreversible inhibitor exceeds that of the enzyme. Three important types of reversible inhibition will be discussed.

Figure 14.7  The feedback mechanism. The product *Z* inhibits the enzyme involved in one of the early steps and thereby limits the amount of the final products formed.

In this case both the substrate S and the inhibitor I compete for the same active site. The reactions are

$$E + S \rightleftharpoons ES \longrightarrow E + P$$

$$E + I \rightleftharpoons EI$$

where the complex EI does not form products. This situation can again be treated by Michaelis–Menten kinetics. Applying the steady-state approximation for ES, we obtain*

$$v_0 = \frac{V_m}{1 + (K_m/[S])[1 + ([I]/K_I)]} \qquad (14.12)$$

where

$$K_I = \frac{[E][I]}{[EI]}$$

The Lineweaver–Burk equation is given by

$$\frac{1}{v_0} = \frac{K_m}{V_m}\left(1 + \frac{[I]}{K_I}\right)\frac{1}{[S]} + \frac{1}{V_m} \qquad (14.13)$$

Thus a straight line is obtained by plotting $1/v_0$ versus $1/[S]$ at constant $[I]$ (Figure 14.8a). The difference between Eqs. (14.13) and (14.11) is that in the former the slope is enhanced by a factor $(1 + [I]/K_I)$. The intercept on the $1/v_0$ axis is the same for both plots.

A well-known example of a competitive inhibitor is malonic acid, $CH_2(COOH)_2$, for the dehydrogenation reaction

COOH
|
CH$_2$ succinic   H—C—COOH
| dehydrogenase   ‖
CH$_2$ $\rightleftharpoons$ HOOC—C—H
|
COOH

       succinic acid   fumaric acid

Since malonic acid resembles succinic acid in structure, it can combine with the enzyme succinic dehydrogenase although no product is formed in this reaction.

Dividing Eq. (14.10) by Eq. (14.12), we obtain

$$\frac{v_0}{(v_0)_{\text{inhibition}}} = 1 + \frac{K_m[I]}{[S]K_I + K_m K_I}$$

To overcome this competitive inhibition, we need to increase the substrate concentration relative to the inhibitor; that is, at high substrate concentrations, $[S]K_I \gg K_m K_I$ and

$$\frac{v_0}{(v_0)_{\text{inhibition}}} \simeq 1 + \frac{K_m[I]}{[S]K_I} \simeq 1$$

---

* For this and subsequent derivations, see I. H. Segel, *Enzyme Kinetics*, John Wiley & Sons, Inc., New York, 1975, Chapter 3.

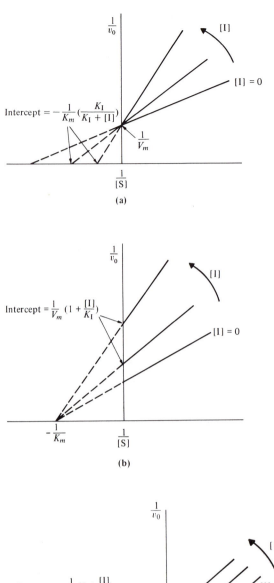

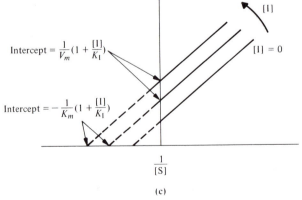

Figure 14.8 Lineweaver–Burk plots: (a) Competitive inhibition; (b) noncompetitive inhibition; (c) uncompetitive inhibition.

A noncompetitive inhibitor usually does not bind at the active site of the enzyme. The reactions are

$$E + S \rightleftharpoons ES \longrightarrow P$$

$$E + I \rightleftharpoons EI$$

$$ES + I \rightleftharpoons ESI$$

Both EI and ESI do not form products. Since I does not interfere with the formation of ES, noncompetitive inhibition cannot be reversed by increasing the substrate concentration. The initial rate is given by*

$$v_0 = \frac{V_m[S]}{(K_m + [S])[1 + ([I]/K_I)]} \tag{14.14}$$

The Lineweaver–Burk equation becomes

$$\frac{1}{v_0} = \frac{K_m}{V_m}\left(1 + \frac{[I]}{K_I}\right)\frac{1}{[S]} + \frac{1}{V_m}\left(1 + \frac{[I]}{K_I}\right) \tag{14.15}$$

From Figure 14.8b we see that a plot of $1/v_0$ versus $1/[S]$ gives a straight line with an increase in slope and intercept compared to Figure 14.4. Dividing Eq. (14.10) by Eq. (14.14), we get

$$\frac{v_0}{(v_0)_{\text{inhibition}}} = 1 + \frac{[I]}{K_I}$$

The result confirms our statement above that the degree of noncompetitive inhibition is independent of $[S]$ and depends only on $[I]$ and K_I.

Examples of noncompetitive inhibitions are the reversible reactions between the sulfhydryl groups of cysteine residues with heavy metal ions.

$$2-SH + Hg^{2+} \rightleftharpoons -S-Hg-S- + 2H^+$$

$$-SH + Ag^+ \rightleftharpoons -S-Ag + H^+$$

UNCOMPETITIVE INHIBITION

An uncompetitive inhibitor does not bind to the enzyme; instead, it binds reversibly to the enzyme–substrate complex to yield an inactive ESI complex. The reactions are

$$E + S \rightleftharpoons ES \longrightarrow P$$

$$ES + I \rightleftharpoons ESI$$

The ESI does not form a product. Again, since I does not interfere with the formation of ES, uncompetitive inhibition cannot be reversed by increasing the substrate concentration. The initial rate is given by

$$v_0 = \frac{V_m[S]}{K_m + [S](1 + [I]/K_I)} \tag{14.16}$$

---

* We assume that $[E][I]/[EI] = [ES][I]/[ESI] = K_I$. This assumption holds if we also have $ESI \rightleftharpoons EI + S$.

where

$$K_1 = \frac{[ES][I]}{[ESI]}$$

The Lineweaver–Burk equation is given by

$$\frac{1}{v_0} = \frac{K_m}{V_m}\frac{1}{[S]} + \frac{1}{V_m}\left(1 + \frac{[I]}{K_1}\right) \tag{14.17}$$

Thus a straight line is obtained by plotting $1/v_0$ versus $1/[S]$ at constant $[I]$ (Figure 14.8c). The difference between Eq. (14.17) and Eq. (14.11) is that the intercept on the $1/v_0$ is increased by $[I]/K_1$, but the slope remains the same. Dividing Eq. (14.10) by Eq. (14.16), we get

$$\frac{v_0}{(v_0)_{\text{inhibition}}} = \frac{K_m + [S](1 + [I]/K_1)}{K_m + [S]}$$

If conditions are such that $[S] \gg K_m$, then the equation above becomes

$$\frac{v_0}{(v_0)_{\text{inhibition}}} \simeq \frac{[S] + [S][I]/K_1}{[S]} = 1 + \frac{[I]}{K_1}$$

Again we see that increasing the substrate concentration cannot overcome the effect of I in uncompetitive inhibition, just as in the case of noncompetitive inhibition.

Uncompetitive inhibition is rarely observed with one-substrate systems. However, multisubstrate enzymes often give parallel line plots with inhibitors.

Michaelis–Menten kinetics cannot be applied to irreversible inhibitions. The inhibitor forms a covalent linkage with the enzyme molecule and cannot be removed by dialysis or similar techniques. The effectiveness of an irreversible inhibitor is not determined by the equilibrium constant but by the rate at which the binding takes place. Iodoacetamides and maleimides act as irreversible inhibitors to the sulfhydryl groups:

$$-SH + ICH_2CONH_2 \longrightarrow -S-CH_2CONH_2 + HI$$

Another example is the action of diisopropylphosphofluoridate (a nerve gas) on the enzyme acetylcholinesterase. When a nerve makes a muscle cell contract, it gives the cell a tiny squirt of acetylcholine molecules. Acetylcholine is called a *neurotransmitter* because it acts as a messenger between the nerve and the final destination (in this case, the muscle cell). Once the proper function has been carried out, the acetylcholine molecules must be destroyed; otherwise, the resulting excess of this substance will hyperstimulate glands and muscle, producing convulsions, choking, and other distressing symptoms. Many victims suffer paralysis or even death. The effective removal of excess acetylcholine is by a hydrolysis reaction (see Section 11.10):

$$CH_3COOCH_2CH_2-\overset{+}{N}(CH_3)_3 + H_2O$$

acetylcholine

$$\longrightarrow HOCH_2CH_2-\overset{+}{N}(CH_3)_3 + CH_3COOH$$

choline

which is catalyzed by acetylcholinesterase. The irreversible inhibition takes place via the formation of a covalent bond between the phosphorus atom and the hydroxyl oxygen atom of the serine residue in the enzyme:

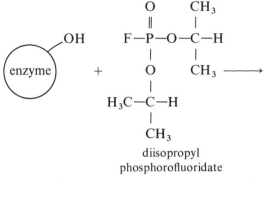

diisopropyl
phosphorofluoridate

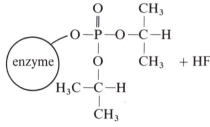

The complex formed is so stable that for practical purposes the restoration of normal nerve function must await the formation of new enzyme molecules by the victim's body.

Allosteric Interactions 14.5

There is a certain class of enzymes whose kinetics do not obey the Michaelis–Menten description. Instead of the usual hyperbolic curve (see Figure 14.1), a sigmoidal or S-shaped curve is obtained. This behavior is exhibited by the *regulatory enzymes*,* which possess multiple binding sites. Sigmoidal curves are characteristic of the *cooperative* binding of ligands to proteins. The term "cooperativity" means that the binding of the first substrate at one site affects the binding of the second substrate at a different site. Enzymes that show cooperativity are called *allosteric* (from the Greek words *allos* = other and *stereos* = solid, which means conformation in our discussion), since the allosteric *effector* (the inhibitor or activator) is generally structurally different from the substrates and binds at its own separate site away from the active site. The term *homotropic* is sometimes used to denote interactions among the identical substrate molecules, and *heterotropic* for the interactions of the allosteric effectors with the substrates. With other proteins, it is the protein itself that can exist in two different structural forms, that is, in other than one solid form (or conformation), hence allosteric forms. A popular analysis of oxygen binding by hemoglobin, to be discussed shortly, is based on the assumption that the protein can exist in either of two such allosteric states.

* Regulatory enzymes are essential for metabolic processes. Their activities are subject to feedback inhibition, or activation.

BINDING OF OXYGEN TO MYOGLOBIN AND HEMOGLOBIN

The phenomenon of cooperativity was first observed for the oxygen–hemoglobin system. Although hemoglobin is not an enzyme, its mode of binding with oxygen is analogous to that for regulatory enzymes. (Note that in this system we have a homotropic interaction, since all the "substrates" are the same, that is, oxygen.) Figure 14.9 shows the percent saturation curves for hemoglobin and myoglobin. The difference between hemoglobin and myoglobin is as follows. A hemoglobin molecule is made up of four polypeptide chains, two α chains of 141 amino acid residues each and two β chains of 146 amino acid residues each. Each chain contains a heme group. These four chains fold up to form similar three-dimensional structures. In an intact hemoglobin molecule, these four chains, or four *subunits*, are joined together to form a tetramer. A myoglobin molecule possesses only one polypeptide of 153 amino acids. It contains one heme group and has a structure that is similar to the β chain of hemoglobin. As we can see, the curve for myoglobin is hyperbolic, indicating its noncooperative nature in binding with oxygen. This observation is consistent with the fact that there is only one heme group and hence only one binding site. On the other hand, the curve for hemoglobin is sigmoidal, indicating that binding is cooperative.

Let us first consider myoglobin, since it is a simpler system. The reaction is

$$Mb + O_2 \rightleftharpoons MbO_2$$

The dissociation constant is given by

$$K = \frac{[Mb][O_2]}{[MbO_2]} \tag{14.18}$$

We define a quantity Y, the fractional saturation, as follows (see p. 248):

$$Y = \frac{[MbO_2]}{[MbO_2] + [Mb]} \tag{14.19}$$

From Eqs. (14.18) and (14.19),

$$Y = \frac{[O_2]}{[O_2] + K} \tag{14.20}$$

Figure 14.9

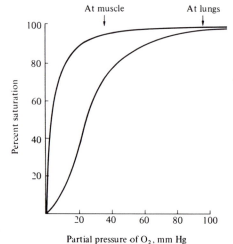

Oxygen saturation curves for myoglobin (left) and hemoglobin (right).

Since oxygen is a gas, its concentration is more conveniently expressed in terms of partial pressure p_{O_2}. Thus we write

$$K = \frac{[Mb]p_{O_2}}{[MbO_2]}$$

Similarly, we replace $[O_2]$ with p_{O_2} in Eq. (14.20). Further, if we represent the oxygen affinity for myoglobin by P_{50}, which is the partial pressure of oxygen at which half or 50% of the binding sites are filled (that is, $[Mb] = [MbO_2]$), it follows that

$$K = \frac{[Mb]p_{O_2}}{[MbO_2]} = p_{O_2} = P_{50}$$

Equation (14.20) now becomes

$$Y = \frac{p_{O_2}}{p_{O_2} + P_{50}} \tag{14.21}$$

Upon rearrangement,

$$\frac{Y}{1 - Y} = \frac{p_{O_2}}{P_{50}}$$

Taking the logarithm of both sides of this equation, we obtain

$$\log \frac{Y}{1 - Y} = \log p_{O_2} - \log P_{50} \tag{14.22}$$

Thus a plot of $\log Y/(1 - Y)$ versus $\log p_{O_2}$ gives a straight line with a slope of unity (Figure 14.10). Equation (14.22) describes the binding of myoglobin with oxygen quite well.

However, Eq. (14.22) does not hold for hemoglobin; instead, it must be modified as follows:

$$\log \frac{Y}{1 - Y} = n \log p_{O_2} - n \log P_{50} \tag{14.23}$$

A similar plot in this case yields a straight line with a slope of 2.8 (that is, $n = 2.8$), also shown in Figure 14.10. The fact that the slope is greater than unity indicates that the binding of hemoglobin with oxygen is cooperative. Note that we cannot explain the binding phenomenon in this case by assuming it to be a higher-order reaction, because n is not integral and it is not identical with the number of sites. If all four sites were equivalent and

Plot of $\log Y/(1 - Y)$ versus $\log p_{O_2}$ for myoglobin and hemoglobin.

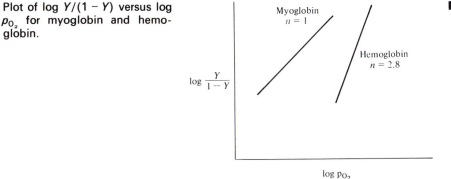

Figure 14.10

independent of one another, the binding curve would be analyzed by using Eq. (10.32), which is not true in practice. Furthermore, the fact that the slope of the line at very low and very high partial pressures of oxygen tends to unity is inconsistent with a higher-order mechanism, which predicts a constant slope at all partial pressures of oxygen. Equation (14.23) is often referred to as the *Hill equation* and *n* is known as the *Hill coefficient*. The quantity *n* is a convenient measure of cooperativity; if $n = 1$, we have no cooperativity and, if $n > 1$, then the binding is cooperative.

What is the significance of cooperativity? In essence, it enables hemoglobin to be a more efficient oxygen transporter than myoglobin. The partial pressure of oxygen in the lungs is about 100 mm Hg, while that in the capillaries of a muscle is about 20 mm Hg.* Further, the partial pressure for 50% saturation for hemoglobin is about 30 mm Hg (see Figure 14.9). From Eq. (14.23),

$$\frac{Y}{1 - Y} = \left(\frac{p_{O_2}}{P_{50}}\right)^n$$

In the lungs,

$$\frac{Y_{lung}}{1 - Y_{lung}} = \left(\frac{100}{30}\right)^{2.8}$$

or

$$Y_{lung} = 0.97$$

In the muscles,

$$\frac{Y_{muscle}}{1 - Y_{muscle}} = \left(\frac{20}{30}\right)^{2.8}$$

$$Y_{muscle} = 0.24$$

The amount of oxygen delivered is proportional to ΔY, defined as

$$\Delta Y = Y_{lung} - Y_{muscle} = 0.73$$

What would happen if the binding between hemoglobin and oxygen were not cooperative? In this case we have, from Eq. (14.21): In the lungs,

$$Y_{lung} = \frac{100}{100 + 30} = 0.77$$

In the muscles,

$$Y_{muscle} = \frac{20}{20 + 30} = 0.40$$

so that $\Delta Y = 0.37$. It is clear, therefore, that nearly twice as much oxygen is delivered to the tissues when the binding of hemoglobin with oxygen is cooperative.

Because of its great physiological importance, it is worth looking into factors that affect the binding of oxygen to hemoglobin. Hill's coefficient and the oxygen affinity of hemoglobin depend on the concentration of several

---

* A common (non-SI) unit of pressure is *torr*. One torr is equal to the pressure exerted by a column of mercury 1 mm high at 273 K and sea level. Thus 1 torr = 1 mm Hg.

species in the red blood cell: protons, carbon dioxide, chloride ions, and 2,3-diphosphoglycerate (DPG):

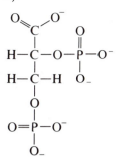

An increase in the concentration of any of these species shifts the oxygen binding curve (Figure 14.9) to the right, which indicates a decrease in the oxygen affinity. This behavior is just what we need. At the tissues, where the partial pressure of carbon dioxide and the concentration of H^+ ions are high, the oxyhemoglobin molecules have a greater tendency to dissociate into hemoglobin and oxygen, the latter is taken up by myoglobin for metabolic processes. About two protons are taken up by the hemoglobin molecule for every four oxygen molecules released. The reverse effect occurs in the alveolar capillaries of the lungs. There the high concentration of oxygen drives off protons and carbon dioxide bound to deoxyhemoglobin. This reciprocal action is known as the *Bohr effect*, first reported in 1904. Figure 14.11a shows the effect of pH on the oxygen affinity of hemoglobin.

The effect of DPG on the oxygen affinity for hemoglobin was only discovered in 1967. It was found that DPG binds only to deoxyhemoglobin and not to oxyhemoglobin,* and that DPG reduces the oxygen affinity by a factor of about 25 (Figure 14.11b). The number of DPG molecules in the red blood cell is roughly the same as the number of hemoglobin molecules, 280 million. A shortage of oxygen, however, causes more DPG to be made, which promotes the release of oxygen. In this respect it is interesting to note that when a person travels quickly from sea level to a high-altitude region, where the partial pressure of oxygen is low, the level of DPG in his or her red blood

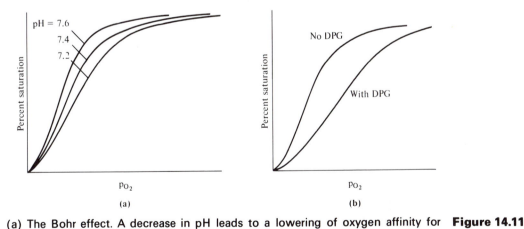

(a) The Bohr effect. A decrease in pH leads to a lowering of oxygen affinity for hemoglobin. (b) The presence of DPG decreases the oxygen affinity for hemoglobin. **Figure 14.11**

* Note that the binding of DPG to hemoglobin is an example of heterotropic interaction, since DPG differs from the normal "substrate," that is, oxygen.

cells increases. This increase lowers the oxygen affinity of hemoglobin and helps to release more oxygen. The fetus has its own kind of hemoglobin, called hemoglobin F, which is made up of two α chains and two γ chains. This hemoglobin differs from adult hemoglobin, called hemoglobin A, which is made up of two α chains and two β chains. Under physiological conditions, hemoglobin F has a higher oxygen affinity than hemoglobin A. This difference in affinity helps the transfer of oxygen from the maternal to the fetal circulation. We now understand that the higher oxygen affinity of hemoglobin F is the result that this molecule binds DPG less strongly than hemoglobin A. The comparison of these systems also shows that DPG only binds to the β chains in hemoglobin A and γ chains in hemoglobin F.

Finally, we note that the binding of oxygen to myoglobin is not affected by any of these factors. It is affected, of course, by temperature. The affinity of oxygen decreases with increasing temperature, which is true for both myoglobin and hemoglobin.

Equation (14.23) is an empirical approach to cooperativity; it says nothing about the mechanism involved. Over the last 50 years a number of theories have been proposed to explain cooperativity. Next, we briefly discuss two recent theories that have played an important role in our understanding of allosteric enzymes.

THE ALLOSTERIC MODEL

In 1965, Monod, Wyman, and Changeux proposed a theory, called the *allosteric model*, to explain cooperativity.* Their theory makes the following

Figure 14.12

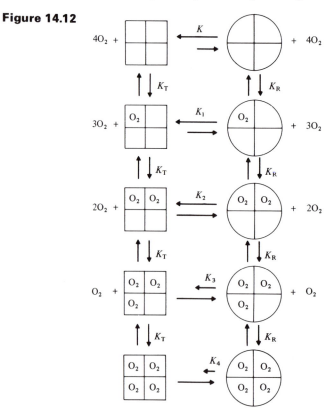

The two-state model of Monod, Wyman, and Changeux for binding of oxygen with hemoglobin. The squares represent the tense state; the quarter circles represent the relaxed state.

---

* J. Monod, J. Wyman, and J.-P. Changeaux, *J. Mol. Biol.* **12**, 88 (1965).

assumptions: (1) The proteins are oligomers, that is, they contain two or more subunits. (2) Each protein molecule can exist in either of two states, called T (tense) and R (relaxed), which are in equilibrium. (3) In the absence of substrate molecules, the T state is favored. When substrate molecules are bound to enzyme, the equilibrium gradually shifts to the R state, which has a higher affinity for the ligand. (4) All binding sites in each state are equivalent and have identical binding constants for ligands. Figure 14.12 shows the two-state model of Monod, Wyman, and Changeux for binding of oxygen with hemoglobin.

The equilibrium constant K for the two states in the absence of oxygen is given by

$$K = \frac{[T]}{[R]}$$

Since K is large, the equilibrium favors the T state and only negligible amounts of the R state are present. When oxygen is present, the equilibrium will shift gradually to the R state, since it has a higher affinity for oxygen. When four molecules of ligand are bound, virtually all of the hemoglobin molecules will be in the R state, which corresponds to the conformation of oxyhemoglobin. The partition between R and T states for the various degrees of ligand saturation is given by the following equations:

$$K_1 = Ka$$
$$K_2 = K_1 a = Ka^2$$
$$K_3 = K_2 a = Ka^3$$
$$K_4 = K_3 a = Ka^4$$

where a represents the relative advantage the R state gains in the R–T equilibrium for each O_2 molecule bound. Thus we write

$$a = \frac{K_R}{K_T}$$

where K_R and K_T are the dissociation constants defined in Figure 14.12. Based on this model the sigmoidal kinetics can be reasonably well accounted for. Figure 14.13 shows that the sigmoidal curve can be analyzed in terms of the binding of the R and T states.

It is important to note that if hemoglobin were always and entirely in the T

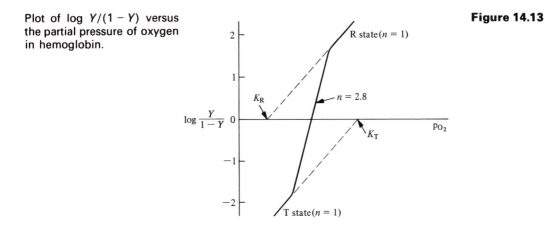

Plot of log $Y/(1 - Y)$ versus the partial pressure of oxygen in hemoglobin.

Figure 14.13

state, its binding of oxygen, although weak, would be completely noncooperative and only characterized by K_T. Conversely, if hemoglobin were always and entirely in the R state, its binding of oxygen, although strong, would also be completely noncooperative and characterized only by K_R. This noncooperativity is a result of the fact that in any given hemoglobin molecule, the allosteric model dictates that all four subunits must either be in the R state or the T state, that is, either R_4 or T_4. Mixed forms such as R_3T or R_2T_2 are considered nonexistent. For this reason the model is sometimes called the "concerted," "all-or-none," or "symmetry-conserved" model. Cooperativity, then, is the result of a free equilibrium between R and T states, governed by the value of K and the concentration of oxygen present. Therefore, as oxygen is added to a solution containing hemoglobin molecules, it preferentially binds to the R state molecules since $K_R < K_T$, thus reducing the concentration of oxygen-free R-state hemoglobin molecules. Since the equilibrium between R- and T-state molecules must be maintained, some T-state molecules shift to the R state, the high oxygen-affinity state.

THE INDUCED-FIT MODEL

An alternative theory, suggested by Koshland, Némethy, and Filmer,* assumes that the affinity of vacant sites for a particular ligand changes progressively as sites are taken up. Again referring to the binding of oxygen to hemoglobin, this means that when an oxygen molecule binds to a vacant site on one of the four-subunits, the interaction causes the site to change its conformation, which in turn affects the binding constants of the other three vacant sites (Figure 14.14). As a result, this model is called the *induced-fit model*. Unlike the concerted mechanism proposed by the allosteric model, the induced-fit model involves a sequential mechanism, in which one can have tetramers that are made up of both R- and T-state subunits such as R_3T or R_2T_2. Again, a sigmoidal curve is predicted by this approach. The binding constant for O_2 molecules increases from left to right in Figure 14.14.

At present both theories are employed by biochemists in the study of enzymes and it is not clear which one is to be preferred in the hemoglobin case. In some cases the induced-fit model has the advantage over the allosteric model in that it can also account for *negative cooperativity*. Negative cooperativity means that the induced conformational changes as a result of the first ligand binding would make the second ligand bind *less* readily. Glyceraldehyde 3-phosphate dehydrogenase, an important enzyme in glycolysis, exhibits this interesting behavior, as do a number of other enzymes. In any case, the two theories have provided biochemists with a deeper insight into the structure and function of many enzymes.

Figure 14.14 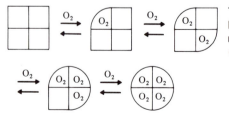 The induced-fit model of Koshland, Némethy, and Filmer. The squares represent the tense state; the quarter circles represent the relaxed state.

---

* D. E. Koshland, Jr., G. Némethy, and D. Filmer, *Biochemistry* **5**, 365 (1966).

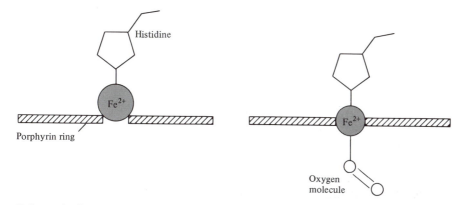

Schematic diagram showing the changes that occur when the heme group in hemo-globin binds an oxygen molecule. *Left*: The heme group in deoxyhemoglobin. It is believed that the radius of the Fe^{2+} ion is too large to fit into the porphyrin ring. *Right*: When O_2 binds to Fe^{2+}, however, the ion shrinks somewhat in size so that it can fit in the plane of the ring. This sinking of the ion into the ring pulls the histidine residue toward the ring and thereby sets off a sequence of structural changes from one subunit to another. Although the details of the changes are not clear, biochemists believe that this is how one heme group learns of the presence of an oxygen molecule at another heme group. The structural changes also drastically affect the affinity of the remaining heme groups for oxygen molecules.

Figure 14.15

Finally, we ask the question: If the four heme groups are well separated from one another in the hemoglobin molecule, as shown by X-ray diffraction work, then how are they able to transmit information regarding binding of oxygen? It is reasonable to assume that the communication among the heme groups, called "heme–heme interaction," probably takes place by means of some kind of conformational change of the molecule. It had long been realized that deoxyhemoglobin and oxyhemoglobin form different crystals, and X-ray crystallographic studies show that there are indeed structural differences between the completely oxygenated and completely deoxygenated hemoglobin. At present much work is being done to understand how the binding of O_2 to one heme group can trigger such extensive structural changes from one subunit to another. It seems that nature has devised a most ingenious mechanism for cooperativity in hemoglobin. The Fe^{2+} ion in deoxyhemoglobin is in the high-spin state,* and its size is too large to fit into the plane of the porphyrin ring of the heme (Figure 14.15). Upon binding to O_2, the Fe^{2+} ion becomes low-spin and its size shrinks sufficiently for it to fit into the plane of the porphyrin ring, an arrangement that is energetically more favorable. This movement of the Fe^{2+} ion pulls the histidine ligand with it and sets off the chain of events that eventually lead to the conformational changes of the entire molecule.

Effects of pH on Enzyme Kinetics **14.6**

A useful way to understand enzyme mechanism is to study the rate of an enzyme-catalyzed reaction as a function of pH.† The activities of many

---

* The electronic structure of iron atoms in the heme group will be discussed in Chapter 16.
† We assume that changes in pH do not lead to denaturation of the enzyme.

Figure 14.16

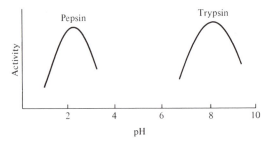

Activites of pepsin and trypsin as a function of pH.

enzymes vary with pH in a manner that can often be explained in terms of the dissociation of simple acids and bases. This is not too surprising, since most active sites contain acidic and basic groups. Figure 14.16 shows the pH-activity profiles of two digestive enzymes, trypsin and pepsin. These enzymes have a characteristic pH at which their activity is maximal (called the *pH optimum*); above or below this pH the activity declines. Pepsin is secreted into the lumen of the stomach, where the pH is around 2. On the other hand, trypsin is secreted into and functions in the alkaline environment of the intestine, where the pH is about 8. Most enzymes that are active within the cells have a pH optimum fairly close to the range of pH that cells normally function. The pH-activity profiles of some enzymes acting on substrates that are electrically neutral, or in which the charge of the substrate plays no role in the reaction, are often quite simple in shape. For example, the activity of papain, a proteolytic enzyme, remains virtually the same over a large pH range from 4 to 8 for certain substrates.

The quantitative treatment of the effect of pH on enzyme-catalyzed reactions is to plot $\log V_m$ versus pH. Let us consider a specific case in which only the acidic form of the dissociating group is catalytically active. Figure 14.17 shows that it is possible to obtain the pK_a of this group by extrapolating the linear portions of the curve (see Appendix 14.1 for the mathematical equations). It is tempting to try to assign this observed pK_a to a particular type of amino acid side chain (see Table 12.4), and hence to identify this amino acid as being responsible for the catalytic action. In practice the situation is usually quite complex, and such simple interpretations can sometimes lead us astray. For example, the pK_a of amino acids in enzymes are influenced by their immediate environment and are not always the same as the pK_a of the free amino acids. A difference of as much as 4 units has been noted for the pK_a of an amino acid in the free form and when it is present in a protein!

Figure 14.17

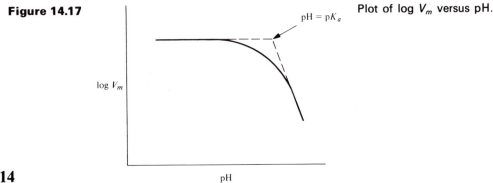

Plot of log V_m versus pH.

Plot of log V_m versus pH for the chymotrypsin-catalyzed hydrolysis of N-acetyl-L-tryptophanamide.

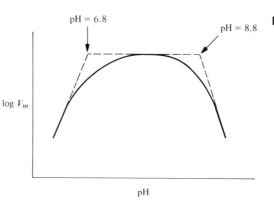

Figure 14.18

(Remember that a difference of 4 units on the log scale is equivalent to a factor of 10^4.)

When there are two dissociating groups present at the active site, the pH-activity profile is usually a bell-shaped curve such as those shown in Figure 14.18. The general scheme for dissociation is as follows:

$$\text{EnH}_2 \underset{}{\overset{\text{p}K_a'}{\rightleftharpoons}} \text{EnH}^- + \text{H}^+ \underset{}{\overset{\text{p}K_a''}{\rightleftharpoons}} \text{En}^{2-} + 2\text{H}^+$$

$$\text{inactive} \qquad \text{active} \qquad\qquad \text{inactive}$$

Figure 14.18 shows the log V_m versus pH plot for the chymotrypsin-catalyzed hydrolysis of N-acetyl-L-tryptophanamide. It has already been established through other studies that the catalytic activity of the enzyme is due to the serine residue at the active site. However, the plot in Figure 14.18 suggests that there are two dissociable groups having pK_a's of 6.8 and 8.8 that affect the activity of chymotrypsin. Since only the imidazole group of histidine has a pK_a of 7, it was suspected that a histidine group might be present at the active site. Further studies indeed confirm this finding. The more basic dissociation ($pK_a = 8.8$) is probably caused by the deprotonation of the α-amino group of aspartic acid.

Appendix 14.1 The Effect of pH on the Rate of a Reaction

We assume that the acid form of an enzyme, EnH, is active while its conjugate base, En$^-$, is inactive. We write

$$\text{EnH} \rightleftharpoons \text{En}^- + \text{H}^+$$

$$\text{active} \qquad \text{inactive}$$

As the pH is lowered, [EnH] is increased and hence so is the rate of the catalyzed reaction. The dissociation constant is given by

$$K_a = \frac{[\text{En}^-][\text{H}^+]}{[\text{EnH}]}$$

or

$$[\text{EnH}] = \frac{[\text{En}^-][\text{H}^+]}{K_a}$$

415

The fraction of the acid, f, is given by

$$f = \frac{[\text{EnH}]}{[\text{EnH}] + [\text{En}^-]}$$

$$= \frac{[\text{En}^-][\text{H}^+]/K_a}{[\text{En}^-][\text{H}^+]/K_a + [\text{En}^-]}$$

$$= \frac{[\text{H}^+]/K_a}{[\text{H}^+]/K_a + 1} = \frac{[\text{H}^+]}{[\text{H}^+] + K_a}$$

The observed rate constant k is given as a product of the intrinsic rate constant k_0 and the fraction of the EnH species in solution:*

$$k = k_0 f$$

$$= k_0 \frac{[\text{H}^+]}{[\text{H}^+] + K_a}$$

Two extreme cases can be considered. (1) If $[\text{H}^+] \gg K_a$ (at low pH's), the equation becomes

$$k = k_0$$

that is, the observed rate is *independent* of pH. The reason is that under this condition the majority of the enzyme molecules are in the active form. (2) If $[\text{H}^+] \ll K_a$ (at high pH's), the equation becomes

$$k = k_0 \frac{[\text{H}^+]}{K_a}$$

or

$$\log k = \log k_0 + \log [\text{H}^+] - \log K_a$$

$$= \log k_0 - \text{pH} + \text{p}K_a$$

Thus a plot of $\log k$ versus pH will give us a straight line with a slope of -1 (Figure 14.19). At the intersection point we have $\text{pH} = \text{p}K_a$ (when $\log k = \log k_0$). Thus such a plot can be used to find the $\text{p}K_a$ of a dissociating group. Since V_m is directly proportional to k, a plot of $\log V_m$ versus pH gives us a similar curve (see Figure 14.17).

Figure 14.19

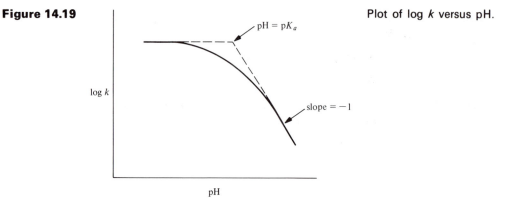

Plot of $\log k$ versus pH.

*The intrinsic rate constant is defined as the rate constant when all the species present in solution are in the active form as EnH.

Suggestions for Further Reading

BENDER, M. L., and L. J. BRAUBACHER. *Catalysis and Enzyme Action*. McGraw-Hill Book Company, New York, 1973.
An excellent introductory text. Emphasis is on mechanism rather than mathematics.

BERNHARD, S. *The Structure and Function of Enzymes*. W. A. Benjamin, Inc., Menlo Park, Calif., 1968.
A very readable introductory text.

HAMMES, G. G. *Principles of Chemical Kinetics*. Academic Press, Inc., New York, 1978.
An excellent introductory text.

BOYER, P. D. (ed.). *The Enzymes*. Academic Press, Inc., New York, 1970.
Includes a detailed discussion of enzyme kinetics.

DIXON, M., and E. C. WEBB. *Enzymes*. Academic Press, Inc., New York, 1964.
Provides a detailed treatment of enzyme kinetics. Suitable for advanced students.

FERSHT, A. *Enzyme Structure and Mechanism*. W. H. Freeman and Company, San Francisco, 1977.
A very readable text. Strongly recommended.

GUTFREUND, H. *An Introduction to the Study of Enzymes*. Blackwell Scientific Publications, Oxford, 1965.
Discusses the kinetic and mechanistic aspects of enzyme-catalyzed reactions in considerable detail with a minimum amount of mathematics. Strongly recommended.

GUTFREUND, H. *Enzymes: Physical Principles*. John Wiley & Sons, Inc., New York, 1972.
An excellent text on the general properties of enzymes.

JOHNSON, F. H., H. EYRING, and B. J. STOVER. *The Theory of Rate Processes in Biology and Medicine*. John Wiley & Sons, Inc., New York, 1974.
A comprehensive and authoritative treatment of many important topics in biological sciences. Strongly recommended for advanced students.

SEGAL, I. H. *Enzyme Kinetics*. John Wiley & Sons, Inc., New York, 1975.
A comprehensive treatment of the subject. Strongly recommended for advanced students.

WALSH, C. *Enzymatic Reaction Mechanisms*. W. H. Freeman and Company, San Francisco, 1979.
A comprehensive treatment of the mechanistic aspects of enzyme-catalyzed reactions. Strongly recommended for advanced students.

Reading Assignments

"An Introduction to Enzyme Kinetics," A. Ault, *J. Chem. Educ.* **51**, 381 (1974).

"Enzymes: Catalysts of the Future?" J. L. Meers, *Chem. Brit.* **12**, 115 (1976).

"The Study of Enzymes," G. K. Radda and R. J. P. Williams, *Chem. Brit.* **12**, 124 (1976).

"Sparkling Pure Water–In a Swimming Pool," R. C. Plumb, *J. Chem. Educ.* **53**, 189 (1976). Discusses enzyme inhibition.

"α-Chymotrypsin: Enzyme Concentration and Kinetics," M. L. Bender, F. J. Kézdy, and F. C. Wedler, *J. Chem. Educ.* **44**, 84 (1967).

"The Kinetics and Inhibition of the Enzyme Methemoglobin Reductase: A Biochemistry Experiment," A. G. Splittgerber, K. Mitchell, G. Dahle, M. Puffer, and K. Bloomquist, *J. Chem. Educ.* **52**, 680 (1975).

"Protein Shape and Biological Control," D. E. Koshland, Jr., *Sci. Am.*, Oct. 1973.

"The Control of Biochemical Reactions," J. P. Changeux, *Sci. Am.*, Apr. 1965.

"K_m as an Apparent Dissociation Constant," J. A. Cohlberg, *J. Chem. Educ.* **56**, 512 (1979).

"Demonstration of Allosteric Behavior," W. H. Sawyer, *J. Chem. Educ.* **49**, 777 (1972).

"Probe-Dependent Cooperativity in Hill-Plots," L. D. Byers, *J. Chem. Educ.* **54**, 352 (1977).

"Hemoglobin Structure and Respiratory Transport," M. F. Perutz, *Sci. Am.*, Dec. 1978.

SECTION B

"Enzyme Action: Views Derived from Metalloenzyme Studies," B. L. Vallee and R. J. P. Williams, *Chem. Brit.* **4**, 397 (1968).

"Molecular Models of Metal Chelates to Illustrate Enzymatic Reactions," H. S. Hendrickson and P. A. Srere, *J. Chem. Educ.* **45**, 539 (1968).

"Enzyme Catalysis and Transition-State Theory," G. E. Linehard, *Science* **180**, 149 (1973).

"Interactions of Enzyme and Inhibitors," B. R. Baker, *J. Chem. Educ.* **44**, 610 (1967).

"Free Energy Diagrams and Concentration Profiles for Enzyme-Catalyzed Reactions," I. M. Klotz, *J. Chem. Educ.* **53**, 159 (1976).

"A Family of Protein-Cutting Proteins," R. M. Stroud, *Sci. Am.*, July 1974.

"Relaxation Spectrometry of Biological Systems," G. G. Hammes, *Advan. Protein Chem.* **23**, 1 (1968).

"What Limits the Rate of an Enzyme-Catalyzed Reaction?" W. W. Cleland, *Acc. Chem. Res.* **8**, 145 (1975).

"Metal Ions in Enzymes Using Ammonia or Amides," N. E. Dixon, C. Gazzola, R. L. Blakeley, and B. Zerner, *Science* **191**, 1144 (1976).

"Collision and Transition State Theory Approaches to Acid–Base Catalysis," H. B. Dunford, *J. Chem. Educ.* **52**, 578 (1975).

"The Three-Dimensional Structures and Chemical Mechanisms of Enzymes," W. N. Lipscomb, *Chem. Soc. Rev.* **1**, 319 (1972).

"An Enzyme-Catalyzed Resolution of Amino Acids," J. R. Mohrig and S. M. Shapiro, *J. Chem. Educ.* **53**, 586 (1976).

"The Activity of Trypsin," S. F. Russo and T. Holzman, *J. Chem. Educ.* **54**, 60 (1977).

"A Kinetic Investigation of an Enzyme-Catalyzed Reaction," W. G. Nigh, *J. Chem. Educ.* **53**, 668 (1976).

"Determination of the Kinetic Constants in a Two-Substrate Enzymatic Reaction," W. T. Yap, B. F. Howell, and R. Schaffer, *J. Chem. Educ.* **54**, 254 (1977).

"Entropy, Binding Energy, and Enzyme Catalysis," M. I. Page, *Angew. Chem., Int. Ed.* **16**, 449 (1977).

"The Active Site of Pepsin," J. S. Fruton, *Acc. Chem. Res.* **7**, 241 (1974).

"Mechanisms of Action of Naturally Occurring Irreversible Enzyme Inhibitors," R. R. Rando, *Acc. Chem. Res.* **8**, 281 (1975).

"Hemocyanin: The Copper Blood," N. M. Senozan, *J. Chem. Educ.* **53**, 684 (1976).

"Ligand Binding to Macromolecules: Allosteric and Sequential Models of Cooperativity," V. L. Hess and A. Szabo, *J. Chem. Educ.* **56**, 289 (1979).

"A Structural Model for the Kinetic Behavior of Hemoglobin," K. Moffat, J. F. Deatherage, and D. W. Seybert, *Science* **206**, 1035 (1979).

Problems

14.1 Explain why a catalyst must affect the rate of a reaction in both directions.

14.2 The hydrolysis of acetylcholine is catalyzed by the enzyme acetylcholinesterase, which has a turnover rate of 25,000 s^{-1}. Calculate how long it will take for the enzyme to cleave one acetylcholine molecule.

14.3 Derive the following equation from Eq. (14.10)

$$\frac{v_0}{[S]} = \frac{V_m}{K_m} - \frac{v_0}{K_m}$$

and show how you would obtain values of K_m and V_m graphically from this equation.

14.4 The hydrolysis of N-glutaryl-L-phenylalanine-p-nitroanilide (GPNA) to p-nitro-aniline and N-glutaryl-L-phenylalanine is catalyzed by α-chymotrypsin. The following data are obtained:

$[S]\ M \times 10^4$	2.5	5.0	10.0	15.0
$v_0\ (M\ \text{min}^{-1}) \times 10^6$	2.2	3.8	5.9	7.1

where $[S]$ = GPNA. Assuming Michaelis–Menten kinetics, calculate V_m, K_m, and k_2 using the Lineweaver–Burk plot. Another way to treat the data is to plot v_0 versus $v_0/[S]$, which is known as the Eadie–Hofstee plot. Calculate V_m, K_m, and k_2 from the Eadie–Hofstee treatment. $[E]_0 = 4.0 \times 10^{-6}\ M$. [*Source*: J. A. Hurlbut, T. N. Ball, H. C. Pound, and J. L. Graves, *J. Chem. Educ.* **50**, 149 (1973).]

14.5 Comment on the following date obtained on an enzyme-catalyzed reaction (no calculations are needed):

T (°C)	10	15	20	25	30	35	40	45
V_m (arbitrary units)	1.0	1.7	2.3	2.6	3.2	4.0	2.6	0.2

14.6 An enzyme having a K_m of $3.9 \times 10^{-5}\ M$ is studied at an initial substrate concentration of 0.035 M. After 1 min, it is found that 6.2 μmol liter^{-1} of product has been produced. Calculate V_m and the amount of product formed at 4.5 min.

14.7 The K_m of lysozome is $6.0 \times 10^{-6}\ M$ with hexa-N-acetylglucosamine as a substrate. It is essayed at the following substrate concentrations: (a) $1.5 \times 10^{-7}\ M$, (b) $6.8 \times 10^{-5}\ M$, (c) $2.4 \times 10^{-4}\ M$, (d) $1.9 \times 10^{-3}\ M$, and (e) 0.061 M. The initial rate measured at 0.061 M was 3.2 μmol liter^{-1} min^{-1}. Calculate the initial rates at the other substrate concentrations.

★14.8 The hydrolysis of urea,

$$H_2N-\overset{\underset{\|}{O}}{C}-NH_2 + H_2O \longrightarrow 2NH_3 + CO_2$$

has been studied by a number of workers. At 100°C, the (pseudo) first-order rate constant is found to be 4.2×10^{-5} s^{-1}. The reaction is catalyzed by the enzyme urease, which at 21°C has a rate constant of 3×10^4 s^{-1}. If the enthalpies of activation for the uncatalyzed and catalyzed reactions are 134 kJ mol^{-1} and 43.9 kJ mol^{-1}, respectively, (a) calculate the temperature that would be required for the nonenzymatic hydrolysis of urea to proceed at the same rate as the enzymatic hydrolysis at 21°C, (b) calculate the lowering of $\Delta G^{\ddagger}$ due to urease, and (c) comment on the sign of $\Delta S^{\ddagger}$. Assume that $\Delta H^{\ddagger} = E_a$ and that $\Delta H^{\ddagger}$ and $\Delta S^{\ddagger}$ are independent of temperature.

14.9 Sketch a potential-energy versus reaction-coordinate plot for a simple enzyme-catalyzed reaction such as

$$E + S \rightleftharpoons ES \rightleftharpoons EP \rightleftharpoons E + P$$

14.10 Silver ions are known to react with the sulfhydryl groups of proteins and can thus inhibit the action of certain enzymes. In a reaction it is found that 0.0075 g of $AgNO_3$ is needed to completely inactivate a 5-ml enzyme solution. Estimate the molar mass of the enzyme. Explain why the molar mass obtained represents the minimum value. The concentration of the enzyme solution is such that 1 ml of the solution contains 75 mg of the enzyme.

14.11 An enzyme has a K_m of 2.8×10^{-5} M and a V_m of 53 μmol liter^{-1} min^{-1}. Calculate v_0 if [S] = 3.7×10^{-4} M and [I] = 4.8×10^{-4} M for (a) a competitive inhibitor, (b) a noncompetitive inhibitor, and (c) an uncompetitive inhibitor. ($K_I = 1.7 \times 10^{-5}$.)

14.12 The degree of inhibition i is given by $i\% = 100(1 - a)$, where $a = (v_0)_{\text{inhibition}}/v_0$. Calculate the percent inhibition in each of the three cases in Problem 14.11.

14.13 The initial rates at various substrate concentrations for an enzyme-catalyzed reaction are as follows:

[S]	v_0 (M min^{-1})
2.50×10^{-5}	38.0
4.00×10^{-5}	53.4
6.00×10^{-5}	68.6
8.00×10^{-5}	80.0
16.00×10^{-5}	106.8
20.00×10^{-5}	114.0

(a) Does this reaction follow Michaelis–Menten kinetics? (b) Calculate the V_m of the reaction. (c) Calculate the K_m of the reaction. (d) Calculate the initial rates at [S] = 5.00×10^{-5} M and [S] = 3.00×10^{-1} M. (e) What is the total amount of product formed during the first 3 min at [S] = 7.2×10^{-5} M? (f) How would an increase in the enzyme concentration by a factor of 2 affect each of the following quantities: K_m, V_m, and v_0 (at [S] = 5.00×10^{-5} M)?

14.14 An enzyme-catalyzed reaction ($K_m = 2.7 \times 10^{-3}$ M) is inhibited by a competitive inhibitor I ($K_I = 3.1 \times 10^{-5}$ M). Suppose that the substrate concentration is 3.6×10^{-4} M. How much of the inhibitor is needed for a 65% inhibition? How much does the substrate concentration have to be increased to reduce the inhibition to 25%?

14.15 Fatality usually results when more than 50% of a human being's hemoglobin is complexed with carbon monoxide. Yet a person whose hemoglobin content is diminished by anemia to half its original content can often function normally. Explain.

14.16 Calculate the concentration of a noncompetitive inhibitor ($K_I = 2.9 \times 10^{-4}$ M) needed to yield a 90% inhibition of an enzyme-catalyzed reaction.

14.17 The presence of small amounts of competitive inhibitors often act as activators to allosteric enzymes. Why?

14.18 The following data were obtained for the variation of V_m with pH for a reaction catalyzed by α-amylase at 24°C. What can you conclude about the pK_a values of the ionizing groups at the active site?

V_m (arbitrary units)	200	501	1584	1778	3300	5248	5250	5251	2818	2510	1584	398	158
pH	3.0	3.5	4.0	4.5	5.0	5.5	6.0	6.5	7.0	7.5	8.0	8.5	9.0

(*Hint*: Plot log V_m versus pH.)

14.19 An enzyme contains a single ionizable group at its active site. It is known that for catalysis to occur, this group must be in the dissociated (that is, negative) form. The substrate bears a net positive charge. The reaction scheme can be represented by

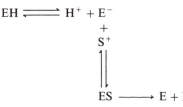

(a) What kind of inhibitor is H^+? (b) Write an expression for the initial rate of the reaction in the presence of the inhibitor.

14.20 What is the advantage of having the heme group in a hydrophobic region in the myoglobin and hemoglobin molecule?

14.21 Although it is possible to carry out X-ray diffraction studies of fully deoxygenated hemoglobin and fully oxygenated hemoglobin, it is much more difficult, if not impossible, to obtain crystals in which each hemoglobin molecule is bound to only one, two, or three oxygen molecules. Explain.

14.22 Compare the allosteric model with the induced-fit model for the binding of oxygen with hemoglobin.

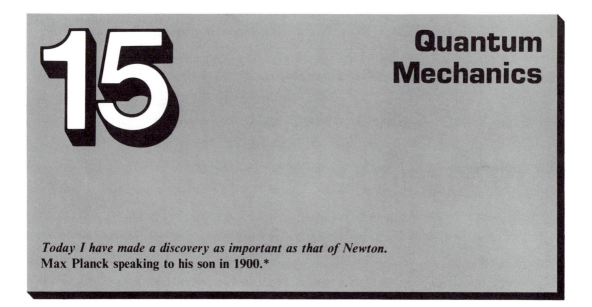

Today I have made a discovery as important as that of Newton.
Max Planck speaking to his son in 1900.*

Up to this point we have been mainly concerned with matter in bulk. We have seen that the methods of thermodynamics and kinetics provide important information regarding chemical processes. Now we shall take a close look at the properties of atoms and molecules. To do so, however, we need to first become familiar with a new topic—quantum mechanics.

In this chapter we give a brief description of the development of the quantum theory, proposed by Planck in 1900. To understand Planck's quantum theory, we must first have some idea about the nature of radiation. Since radiation involves the emission and transmission of energy in the form of waves through space, we shall start at the beginning with a discussion of the properties of waves and the wave theory of light.

15.1 Wave Theory of Light

The first quantitative investigation of the nature of light was carried out by Newton in the seventeenth century. Using a glass prism, he showed that sunlight is composed of seven different colors. The work of Huygens, Fresnel, and Young in the eighteenth and nineteenth centuries firmly established the fact that light has wave properties. Figure 15.1 shows the propagation of a sinusoidal wave along the x direction. The velocity of the wave c is given by

$$c = \lambda v \tag{15.1}$$

where λ is the wavelength and v the frequency of the wave.

The interference phenomenon is a convincing demonstration of the wave theory of light. Consider the interaction of two waves in space shown in

* H. W. Cropper, *The Quantum Physicists*, Oxford University Press, Inc., New York, 1970, p. 7. Used by permission.

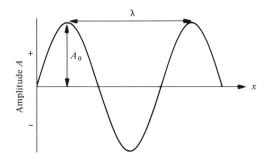

Sinusoidal wave of the form $A = A_0 \sin x$, where A_0 is the amplitude of the wave. **Figure 15.1**

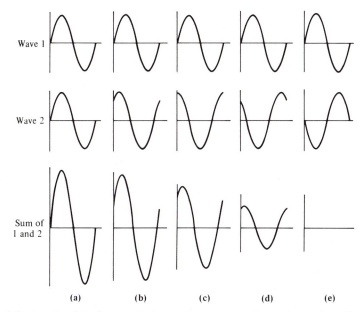

Wave 1

Wave 2

Sum of 1 and 2

(a) (b) (c) (d) (e)

Constructive and destructive interference between two waves of equal wavelength **Figure 15.2** and amplitude: (a) two waves completely in phase; (b)–(d) two waves partially out of phase; and (e) two waves exactly out of phase.

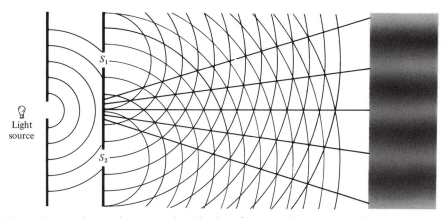

Two-slit experiment demonstrating the interference phenomenon. The pattern of **Figure 15.3** the screen consists of alternating bright and dark bands.

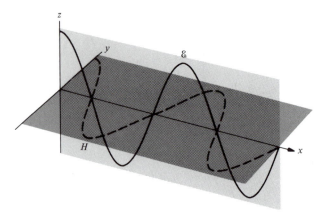

Figure 15.4 Electric field component $\mathscr{E}$ and magnetic field component H of an electromagnetic wave.

Figure 15.2. Depending on the relative displacement or the *phase difference*, the interaction can lead to constructive or destructive interference. Experimentally, this can be observed by the arrangement shown in Figure 15.3. A point source is first passed through a filter, which selects light of approximately one wavelength. Slits S_1 and S_2 are small openings (compared to the distance separating S_1 and S_2), which can act as two separate point sources. Interference occurs between these two waves and constructive and destructive patterns are observed on the screen as alternate bright and dark regions.

A mathematical formulation was put forward by Maxwell in 1873. He showed that light is just one form of electromagnetic radiation. Others are microwaves, infrared, X rays, and so on. An electromagnetic wave consists of an electric field component and a mutually perpendicular magnetic field component oscillating in space with frequency v. The direction of oscillation is perpendicular to the direction of wave propagation (Figure 15.4). For ordinary unpolarized light, the electric and magnetic components can and do rotate about the x axis (the direction of propagation) although they are always perpendicular to each other. When the light becomes polarized, these two components can only oscillate within the two fixed planes (the xy and xz planes). More on this point in Chapter 19.

15.2 Planck's Quantum Theory

Toward the end of the nineteenth century, physicists were in a secure state. The wave theory of light was well established, and Newtonian mechanics, since its formulation in the seventeenth century, was successful in the study of systems ranging from billiard balls to planets. The science of thermodynamics had already become a powerful tool in solving chemical and physical problems. However, this comfortable state of affairs was not to last. In 1899, Lummer and Pringsheim, among others, studied the emission of radiation by solids as a function of temperature and obtained a series of curves that could not be explained by the wave theory or by the laws of thermodynamics. The search for a proper explanation quickly led to a new and exciting era in physics.

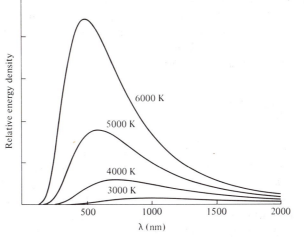

Blackbody radiation curves at various temperatures.

Figure 15.5

All bodies at a temperature above absolute zero emit radiation over a range of wavelengths. The red glow of an electric heater and the bright white light of an electric bulb are familiar examples. If we measure the intensity of radiation versus the wavelength for an emitter at different temperatures, we obtain a series of curves such as those shown in Figure 15.5. These plots are commonly referred to as *blackbody radiation curves*, a blackbody being any object that absorbs all the radiation falling on it. A perfect absorber is also a perfect emitter of radiation.

In 1900, Planck solved the problem with an assumption that departed drastically from classical physics. In classical physics, it was assumed that the radiant energy emitted by a solid could have any energy value within a continuous range. What Planck said was that radiant energy could *not* have any arbitrary value; instead, the energy could only be emitted in small, discrete amounts which he called *quanta*. The energy of the emitted radiation, E, is proportional to the frequency of the emitted radiation:

$$E \propto v$$
$$= hv \tag{15.2}$$

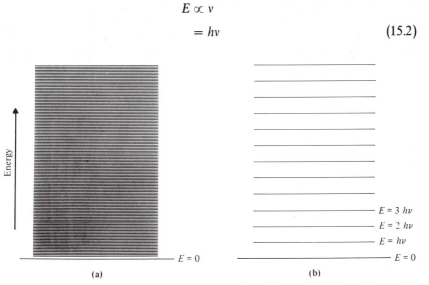

Energy variation for an oscillator: (a) the classical model; and (b) Planck's model. The spacing between successive levels in (a) is so minute that energy can be considered to vary continuously.

Figure 15.6

where h is the Planck constant, equal to 6.626×10^{-27} erg s (CGS units) or 6.626×10^{-34} J s (SI units).* According to Planck's quantum theory, energy is always emitted in multiples of $h\nu$; for example, $h\nu$, $2h\nu$, $3h\nu$, ... but never $1.68h\nu$ or $3.52h\nu$. The difference between the classical model and Planck's model for energy variation is illustrated in Figure 15.6.

15.3 Photoelectric Effect

In the development of science, a single major discovery or the formulation of an important theory can trigger an avalanche of activities. This was the case with the quantum theory. Within a few years of Planck's hypothesis, many previously puzzling observations were also explained by the quantum theory. One of these was the *photoelectric effect*.

When light of a certain frequency shines on a clean metal surface, electrons are ejected. Experimentally, it is found that (1) the number of electrons ejected is proportional to the intensity of light; (2) the kinetic energy of the ejected electrons is proportional to the frequency of incident light; and (3) no electrons can be ejected if the frequency of the light is lower than a certain value, called the *threshold frequency*.

According to the wave theory of light, the energy of radiation is proportional to the square of the amplitude. The energy is related to the intensity, not the frequency, of radiation. This seems to contradict point (2) above. In 1905, Einstein solved this difficulty by using Planck's theory in the following manner. He assumed that light consisted of particles called light quanta or *photons* of energy $h\nu$, where ν is the frequency of light. Shining light onto the metal surface can be viewed as the collision between photons and electrons. If ν is above the threshold frequency, then the Einstein photoelectric equation becomes

$$h\nu = W + \tfrac{1}{2}mv^2 \tag{15.3}$$

where W represents the energy that the photon must possess in order to remove an electron from the metal, and $\tfrac{1}{2}mv^2$ is the kinetic energy of the ejected electron. The quantity W is called the *work function*, which measures how strongly the electrons are held in the metal. A plot of the kinetic energy

Figure 15.7

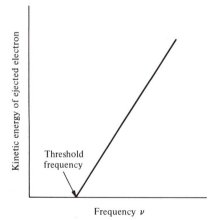

Frequency ν

Plot of the kinetic energy of ejected electrons versus the frequency of incident radiation.

---

* For an interesting discussion of the Planck radiation law, see T. A. Lehman, *J. Chem. Educ.* **49**, 832 (1972).

versus frequency is shown in Figure 15.7. Equation (15.3) is simply another way of stating the law of conservation of energy.

Experimental observations can now be explained. Since the number of photons increases with the intensity of light, more electrons are ejected at higher intensities. Furthermore, the energy of the photons increases with the frequency of light so that electrons ejected at higher frequencies will also possess higher kinetic energies.

The success of Eq. (15.3) immediately poses the following question: What is the nature of light? On the one hand, the wave properties of light have been proved beyond doubt. On the other hand, the photoelectric effect can be explained only in terms of photons that are particles. Can light be both wavelike and particlelike? This explanation seemed strange and unfamiliar at the time the quantum theory was postulated, but scientists were beginning to realize that submicroscopic particles behave very differently from macroscopic objects.

De Broglie's Postulate **15.4**

If we accept the fact that light will, under certain circumstances, exhibit particlelike behavior, the converse might also be true. Particles, under certain circumstances, would also exhibit wave properties. The question of wave–particle duality led de Broglie in 1924 to propose the important relation

$$\lambda = \frac{h}{mv} = \frac{h}{p} \tag{15.4}$$

where p is the momentum. In principle, Eq. (15.4) applies to any system whose wave properties are represented by λ and whose particle properties are represented by mv. The experimental confirmation of Eq. (15.4) was provided by the works of Davisson and Germer in 1927 and G. P. Thomson in 1928. Figure 15.8 shows the diffraction patterns of electron waves by aluminum powder.

Equation (15.4) can also be applied to photons. According to Einstein's relativity theory, energy and mass are related by the equation

$$E = mc^2 \tag{15.5}$$

where m is now the mass of photon and c the velocity of light. The momentum of the photon mc is given by

$$mc = \frac{E}{c} = \frac{h\nu}{c} = \frac{h}{\lambda} \tag{15.6}$$

Equation (15.6) is the *de Broglie equation for photons*.

One practical application of the wavelike behavior of electrons is in the use of the electron microscope. Human eyes are sensitive to the wavelength of light from about 400 nm to 750 nm ($1 \text{ nm} = 1 \times 10^{-7} \text{ cm} = 1 \times 10^{-9} \text{ m}$). The ability to see details of small structures is limited by the resolving power, or resolution, of our optical systems. Resolution refers to the minimum distance at which objects can be distinguished as separate entities. Any two objects separated by less than that distance will blur together as a single object. The lower limit of resolution of the unaided human eye is about 0.02 cm. For this reason, we cannot see *individual* objects whose size lie below 0.02 cm. On the other

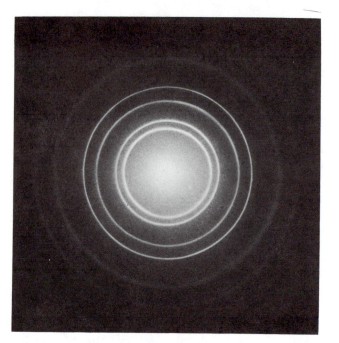

Figure 15.8 Transmission pattern of polycrystalline aluminum obtained with an electron diffraction tube. (By permission of Sargent-Welch Scientific Company, Skokie, Ill.)

hand, the lower limit of resolution of a light microscope is about 200 nm or 2×10^{-5} cm. This means that with the aid of a light microscope, we can see objects whose sizes are about one-half the wavelength of violet light (400 nm) but not smaller. Greater resolution is possible with an electron microscope because a beam of electrons has properties corresponding to wavelengths 100,000 times shorter than visible light. When a beam of electrons is passed through an accelerating electrostatic field (two parallel plates with potential difference of V volts), the potential energy gained by each electron, eV, can be equated to its kinetic energy as follows:

$$eV = \tfrac{1}{2}mv^2$$

or

$$v = \sqrt{\frac{2Ve}{m}} \tag{15.7}$$

where e is the electronic charge. Substituting Eq. (15.7) into Eq. (15.4), we obtain

$$\lambda = \frac{h}{\sqrt{2meV}} \tag{15.8}$$

EXAMPLE 15.1 What is the wavelength of an electron when it is accelerated by 10,000 V?

CGS–esu units: Let us first derive a general expression for λ in terms of V. We have

$$h = 6.626 \times 10^{-27} \text{ erg s}$$

$$m = 9.108 \times 10^{-28} \text{ g}$$

$$e = 4.803 \times 10^{-10} \text{ esu}$$

Equation (15.8) can now be written

$$\lambda = \frac{6.626 \times 10^{-27} \text{ erg s}}{\sqrt{2 \times 9.108 \times 10^{-28} \text{ g} \times 4.803 \times 10^{-10} \text{ esu} \times V}}$$

The quantity V expressed in this equation is the electrostatic volt, or stat volt. To convert it into the practical volt, we use the relation (see Appendix 15.1)

$$300 \text{ stat volts} = 1 \text{ practical volt}$$

Since

$$\text{charge (esu)} \times V = \text{ergs}$$

we have

$$\lambda = \frac{6.626 \times 10^{-27} \text{ erg s}}{\sqrt{2 \times 9.108 \times 10^{-28} \text{ g} \times (4.803 \times 10^{-10}/300)V \text{ ergs}}}$$

$$= \frac{1.23 \times 10^{-7} \text{ erg s}}{\sqrt{V \text{ g}^2 \text{ ergs}}}$$

where V now represents only the numerical part of the volt. Now

$$1 \text{ erg} = 1 \text{ g cm}^2 \text{ s}^{-2}$$

Thus

$$\lambda = \frac{1.23 \times 10^{-7}}{\sqrt{V}} \text{ cm} = \frac{12.3}{\sqrt{V}} \text{ Å}$$

If the electron is accelerated by 10,000 V, the corresponding wavelength is

$$\lambda = \frac{12.3}{\sqrt{10,000}} = 0.123 \text{ Å}$$

$$= 0.0123 \text{ nm}$$

SI units

$$h = 6.626 \times 10^{-34} \text{ J s}$$
$$m = 9.108 \times 10^{-31} \text{ kg}$$
$$e = 1.602 \times 10^{-19} \text{ C}$$

Thus

$$\lambda = \frac{6.626 \times 10^{-34} \text{ J s}}{\sqrt{2 \times 9.108 \times 10^{-31} \text{ kg} \times 1.602 \times 10^{-19} \text{ C} \times V}}$$

Since

$$\text{C} = \text{A s}$$

where A is amperes and

$$\text{A V} = \text{W}$$

where W is watts, we have

$$\text{C V} = \text{A V s} = \text{W s} = \text{J} = \text{N m} = \text{kg m}^2 \text{ s}^{-2}$$

Finally,

$$\lambda = \frac{1.23 \times 10^{-9} \text{ J s}}{\sqrt{V} \text{ kg}^2 \text{ m}^2 \text{ s}^{-2}}$$

$$= \frac{1.23 \times 10^{-9} \text{ m}}{\sqrt{V}} = \frac{12.3}{\sqrt{V}} \text{ Å}$$

Again V pertains only to the numerical part. For 10,000 V, we have

$$\lambda = 1.23 \times 10^{-11} \text{ m} = 0.123 \text{ Å}$$

$$= 0.0123 \text{ nm}$$

Since it is relatively easy to obtain V in the kilo- or even million-volt range, very small wavelengths can be obtained. Thus an electron microscope differs from a light microscope in that visible light is replaced by a beam of electrons. With the much shorter wavelength, better resolution can be obtained. It is now possible to "see" large molecules as well as heavy atoms with this technique. The major advantage of electron microscopy over the X-ray diffraction technique is that electrons are charged particles, so they can be easily focused by electric and magnetic fields, which act as lenses. No condensing lenses are known for X rays.

15.5 Bohr's Theory of Atomic Spectra

It was well known in the nineteenth century that the emission spectrum of atomic hydrogen consisted of a series of sharp, well-defined lines (Figure 15.9). In 1885, Balmer suggested the following equation to account for the regularity of the lines observed:

$$\bar{v} = R_H \left(\frac{1}{n''^2} - \frac{1}{n'^2} \right) \text{ cm}^{-1}$$

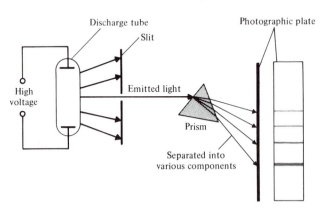

Figure 15.9 Experimental arrangement for studying the emission spectra of atoms and molecules. The gas (hydrogen) under study is placed in a discharge tube containing two electrodes. As electrons flow from the cathode to the anode, they collide with the H_2 molecules, which are then dissociated into atoms. The H atoms are formed in an excited state and quickly decay to the ground state with emission of light. The emitted light is spread into various components. Each component color is focused at a definite position according to its wavelength and forms an image of the slit on the plate. The different-color images of the slit are called spectral lines.

Table 15.1

Series	n''	n'
Lyman	1	2, 3, . . .
Balmer	2	3, 4, . . .
Paschen	3	4, 5, . . .
Brackett	4	5, 6, . . .
Pfund	5	6, 7, . . .

where $\bar{v}$ is the wave number $(1/\lambda)$ of a particular line,* R_H is a constant, called *Rydberg's constant* $(1.09737 \times 10^5 \text{ cm}^{-1})$, and n' and n'' are integers. The series and their discoveries are listed in Table 15.1.

The structure of atoms was fairly well understood in the 1910s through the work of J. J. Thomson, Rutherford, and others. From the α-particle scattering experiment, Rutherford discovered that an atom consisted of a nucleus made up of positively charged particles called protons. The existence of neutrons was postulated for nuclear stability and it was later discovered by Chadwick. Since atoms are electrically neutral species, this means that there must be an equal number of negatively charged particles, called electrons, as there are protons in each atom. Electrons are found outside of the nucleus. The hydrogen atom is the simplest system, since it has only one proton and one electron. Difficulties arose when scientists tried to account for the appearance and position of the emission lines, since the behavior of electrons in atoms was not understood. By using Planck's hypothesis, Bohr was able in 1913 to develop a quantitative treatment to explain the observed spectra.

Bohr's starting point was his assumption that electrons in atoms move in circular orbits of radius r about the nucleus. The Coulombic attraction between the electron and the nucleus is balanced by the outward acceleration due to the circular motion of the electron. For a one-electron atom, we write

$$\frac{Ze^2}{r^2} = \frac{mv^2}{r} \tag{15.9}$$

where Z is the atomic number (the number of protons in the nucleus), e the electronic charge of an electron, and v the instantaneous velocity; that is, at any instant the electron can be thought of as moving tangentially to the circular orbit. The total energy E of the electron can be expressed as the sum of the kinetic energy and the potential energy, where

$$E = \frac{1}{2}mv^2 - \frac{Ze^2}{r} \tag{15.10}$$

The negative sign indicates that the interaction between the electron and the nucleus is attractive. From Eq. (15.9), we have

$$mv^2 = \frac{Ze^2}{r} \tag{15.11}$$

Substituting this expression into Eq. (15.10), we obtain

$$E = -\frac{Ze^2}{2r} \tag{15.12}$$

From Eqs. (15.10) and (15.11),

$$E = -\tfrac{1}{2}mv^2 \tag{15.13}$$

* See Section 18.1 for more discussion of the unit.

Bohr further assumed that the energy of the electron can only be changed in discrete amounts. Thus in a transition from a higher state of energy E_h to a lower one of energy E_l, the energy difference ΔE is given by

$$\Delta E = E_h - E_l = h\nu \tag{15.14}$$

where ν is the frequency of the emitted radiation. Another important assumption is that the circular motion of the electron is restricted so that its angular momentum (see Appendix 2) mvr can have only certain values, that is,

$$mvr = n\frac{h}{2\pi} \qquad n = 1, 2, 3, \ldots \tag{15.15}$$

where n is called a *quantum number*. From Eqs. (15.11) and (15.15), we obtain

$$v = \frac{2\pi Ze^2}{nh} \tag{15.16}$$

Substitution of Eq. (15.16) into Eq. (15.13) yields

$$E = -\frac{2m\pi^2 Z^2 e^4}{n^2 h^2} \tag{15.17}$$

Equation (15.17) gives the possible energies that an electron can have in an atom by occupying different orbits. Again the negative sign says that the electron is attractively held by the nucleus; that is, the energy of the electron is less than that it would have if it were at rest and infinitely far from the nucleus. The most stable state, which we shall call the *ground state*, has the value $n = 1$. Physically, as n increases, the electron moves farther away from the nucleus and becomes less tightly held. In the limit of n approaching infinity, E approaches zero. At this point the electron is completely free of the nucleus and the atom becomes ionized.

The radius of each circular orbit is expressed as

$$r_n = \frac{n^2 h^2}{4\pi^2 m e^2 Z} \tag{15.18}$$

Again we see that, as n increases, so does r_n, indicating a further separation between the electron and the nucleus. The radius of the first orbit ($n = 1$) is 0.529 Å, which is called the *Bohr radius*.

E X A M P L E   15.2 Calculate the energy of an electron in the lowest orbit of a hydrogen atom.

Setting $n = Z = 1$ in Eq. (15.17), we have

$$E_1 = -\frac{2m\pi^2 e^4}{h^2}$$

CGS–esu units

$$e = 4.803 \times 10^{-10} \text{ esu}$$

$$m = 9.1083 \times 10^{-28} \text{ g}$$

$$h = 6.626 \times 10^{-27} \text{ erg s}$$

$$E_1 = -\frac{2(9.1083 \times 10^{-28} \text{ g})\pi^2(4.803 \times 10^{-10} \text{ esu})^4}{(6.626 \times 10^{-27} \text{ erg s})^2}$$

$$= -2.18 \times 10^{-11} \text{ g esu}^4 \text{ erg}^{-2} \text{ s}^{-2}$$

$$E_1 = -2.18 \times 10^{-11} \text{ erg}$$
$$= -2.18 \times 10^{-18} \text{ J}$$

This is the energy required to remove the electron from the lowest level in hydrogen to infinite distance away from the nucleus. Another accepted unit is the *electron volt* (eV). An electron volt is the energy gained by an electron in going through a potential difference of 1 V. The conversion factor is

$$1 \text{ eV} = 1.60 \times 10^{-19} \text{ J}$$

and

$$1 \text{ J} = 6.25 \times 10^{18} \text{ eV}$$

Thus

$$E_1 = -2.18 \times 10^{-18} \times 6.25 \times 10^{18}$$
$$= -13.63 \text{ eV}$$

SI units: Here the expression for energy is modified to give (see Appendix 15.2)

$$E_1 = -\frac{2m\pi^2 e^4}{h^2}\left(\frac{1}{4\pi\varepsilon_0}\right)^2$$

where

$$m = 9.108 \times 10^{-31} \text{ kg}$$
$$e = 4.803 \times 10^{-10} \text{ esu} = 1.602 \times 10^{-19} \text{ C}$$
$$h = 6.626 \times 10^{-34} \text{ J s}$$
$$\varepsilon_0 = 8.8542 \times 10^{-12} \text{ C}^2 \text{ N}^{-1} \text{ m}^{-2}$$

$$E_1 = -\frac{2(9.108 \times 10^{-31} \text{ kg})\pi^2(1.602 \times 10^{-19} \text{ C})^4}{(6.626 \times 10^{-34} \text{ J s})^2}$$

$$\times \left(\frac{1}{4\pi \times 8.8542 \times 10^{-12} \text{ C}^2 \text{ N}^{-1} \text{ m}^{-2}}\right)^2$$

$$= -2.18 \times 10^{-18} \text{ kg m}^2 \text{ s}^{-2}$$
$$= -2.18 \times 10^{-18} \text{ J}$$
$$= -13.63 \text{ eV}$$

Figure 15.10 shows the energy states or levels that an electron can occupy in the hydrogen atom. Returning to Eq. (15.14) and using the result of Eq. (15.17), we can now write

$$\Delta E = E_h - E_l = \frac{2\pi^2 m Z^2 e^4}{h^2}\left(\frac{1}{n_l^2} - \frac{1}{n_h^2}\right) \tag{15.19}$$

and the corresponding wave number as

$$\bar{v} = \frac{1}{\lambda} = \frac{v}{c} = \frac{\Delta E}{hc} = \frac{2\pi^2 m Z^2 e^4}{ch^3}\left(\frac{1}{n_l^2} - \frac{1}{n_h^2}\right) \tag{15.20}$$

Figure 15.10

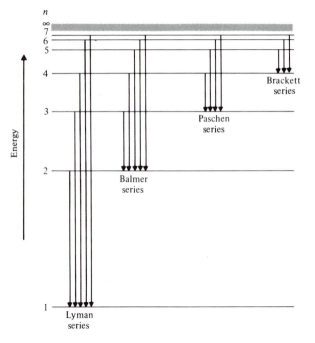

Energy levels and the various series for the atomic hydrogen emission spectrum.

where n_h and n_l are the quantum numbers for the higher and lower state. Thus Bohr's theory was able to account for all the lines observed in the emission spectra of hydrogen as well as other one-electron systems, such as D, T, He^+, and Li^{2+}. This was one of the great triumphs of early quantum theory.

So far we have only considered the process of emission. A reverse process, absorption, also takes place. An electron in a lower energy level can be excited to a higher one by absorbing a photon of energy $h\nu$, which satisfies Eq. (15.14). Figure 15.11 shows a comparison of the absorption and emission processes.

Figure 15.11

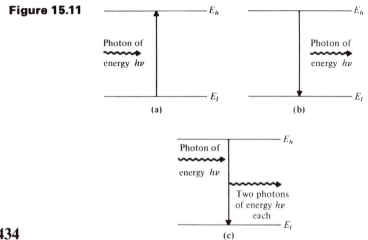

Interaction of electromagnetic radiation with atoms and molecules: (a) absorption; (b) spontaneous emission; and (c) induced emission. Case (c) will be discussed in Chapter 18.

Bohr's theory was soon realized to be inadequate, as it was unable to account for the emission spectra of complex atoms as well as for the behavior of atoms in a magnetic field. A general equation for the submicroscopic systems was needed, one just like Newton's equation for macroscopic bodies. This equation was provided by Schrödinger in 1926.

When expressed in one dimension (say x), the equation is given by

$$-\frac{h^2}{8\pi^2 m}\frac{d^2\psi}{dx^2} + V\psi = E\psi \qquad (15.21)$$

where V is the potential energy and E the total energy of the system with mass m. The wave properties of the particle are represented by the wave function, ψ. Equation (15.21) is known as *Schrödinger's wave equation* in one dimension. In classical mechanics, the total energy E is given by the sum of kinetic energy (T) and potential energy (V):

$$T + V = E \qquad (15.22)$$

The main difference between Eqs. (15.21) and (15.22) is that we have replaced T with the kinetic-energy operator (see Appendix 2, Section A2.1):

$$-\frac{h^2}{8\pi^2 m}\frac{d^2}{dx^2}$$

The reader should keep in mind that Schrödinger's wave equation, like Newton's law of motion, *cannot* be derived from a first principle. Instead, it was deduced by analogy to classical mechanics and optics.

How should we interpret ψ? Being a mathematical wave function, it has no physical meaning by itself. However, Born suggested in 1926 that the product ψ^2 does have physical meaning, since it gives the probability of finding the particle. Earlier we mentioned that the intensity of light is proportional to the square of the amplitude of the wave. In an analogous way, if ψ represents the wave property of the particle, then the probability of locating the particle at some point in space is given by the value of ψ^2 at the same point.*

In order to apply Eq. (15.21) to any system, ψ must be a "well-behaved" wave function, the conditions for which are

1. ψ must be single-valued at any particular point.
2. ψ must be finite at any point.
3. ψ must be a smooth or continuous function of its coordinates.

The Schrödinger wave equation marked the beginning of a new era—what we now know as *wave mechanics* or *quantum mechanics*.

Heisenberg's Uncertainty
Principle **15.7**

In 1927, Heisenberg proposed a principle that has utmost importance in the philosophical groundwork for quantum mechanics. He deduced that when the

---

* In general, ψ is a function of the x, y, and z coordinates.

uncertainties in the simultaneous measurement of momentum and position for a particle are multiplied together, the product is approximately equal to Planck's constant divided by 4π. Mathematically, this can be expressed as

$$\Delta x\, \Delta p \geq \frac{h}{4\pi} \tag{15.23}$$

where Δ means "uncertainty of." Thus Δx is the uncertainty of position and Δp is the uncertainty of momentum. Of course, if the measured uncertainties of position and momentum are large, their product can be substantially greater than $h/4\pi$. The significance of Eq. (15.23), which is the usual statement of the *Heisenberg uncertainty principle*, is that even in the most favorable case of measuring position and momentum, the lower limit of uncertainty is always given by $h/4\pi$. Thus, if we wish to determine the position of an electron with great accuracy (that is, reduce the magnitude of Δx), the corresponding uncertainty in momentum will be large, and vice versa. Clearly, there is no such restriction in classical mechanics, for we can measure the trajectory, or position and momentum, of a tennis ball in flight to whatever degree of accuracy our instruments permit. This is so because the Planck constant is such a small number that it becomes important only when we are dealing with particles on the atomic scale.

It is not difficult to see why the uncertainty relation should exist. Any measurement must, by necessity, result in some disturbance on the system. Thus, when we determine the position of a quantum-mechanical object, say an electron, we have also supplied some energy to it (for example, by shining light on it) so that its velocity or momentum becomes less well defined. However, in dealing with macroscopic bodies, the amount of perturbation is so negligibly small that its momentum can be accurately measured at the same time.

The Heisenberg uncertainty principle can also be expressed in terms of energy and time as follows. Since

$$\text{momentum} = \text{mass} \times \text{velocity}$$

$$= \text{mass} \times \frac{\text{velocity}}{\text{time}} \times \text{time}$$

$$= \text{force} \times \text{time}$$

Thus

$$\text{momentum} \times \text{distance} = \text{force} \times \text{distance} \times \text{time}$$

$$= \text{energy} \times \text{time}$$

or

$$\Delta x\, \Delta p = \Delta E\, \Delta t$$

where ΔE and Δt are the uncertainties in energy and time, respectively. Equation (15.23) can now be written as

$$\Delta E\, \Delta t \geq \frac{h}{4\pi} \tag{15.24}$$

Thus we cannot measure the (kinetic) energy of a particle with absolute precision (that is, to have $\Delta E = 0$) in a finite span of time. Equation (15.24) is particularly useful for estimating spectral line widths (see Section 18.1).

EXAMPLE 15.3 According to Bohr's theory, the electron circles around the nucleus in a hydrogen atom in much the same way that the earth moves around the sun. Suppose that we could locate the position of the electron to within 0.1 Å; what would be the minimum uncertainty in knowing its velocity?

CGS units: Since we are asked to calculate the minimum uncertainty, we use the equal sign in Eq. (15.23).

$$\Delta x \, \Delta p = \frac{h}{4\pi}$$

$$\Delta x = 0.1 \text{ Å} = 1 \times 10^{-9} \text{ cm}$$

Thus

$$\Delta p = \frac{6.626 \times 10^{-27} \text{ erg s}}{4\pi \times 10^{-9} \text{ cm}}$$

$$= 0.53 \times 10^{-18} \text{ erg cm}^{-1} \text{ s}$$

$$= 0.53 \times 10^{-18} \text{ g cm s}^{-1}$$

$$1 \text{ erg} = 1 \text{ g cm}^2 \text{ s}^{-2}$$

Now

$$p = mv$$

so that

$$\Delta p = m \, \Delta v$$

$$m = 9.1083 \times 10^{-28} \text{ g}$$

We have

$$\Delta v = \frac{\Delta p}{m} = \frac{0.53 \times 10^{-18} \text{ g cm s}^{-1}}{9.1083 \times 10^{-28} \text{ g}}$$

$$= 0.60 \times 10^9 \text{ cm s}^{-1}$$

Because the uncertainty in velocity is tremendously large, it is not possible to specify the trajectory of the electron in the hydrogen atom as we can with a macroscopic system. The result therefore shows that Bohr's planetary model is basically incorrect.

SI units

$$\Delta x = 0.1 \text{ Å} = 1 \times 10^{-11} \text{ m}$$

$$\Delta p = \frac{6.626 \times 10^{-34} \text{ J s}}{4\pi \times 10^{-11} \text{ m}}$$

$$= 0.53 \times 10^{-23} \text{ J m}^{-1} \text{ s}$$

$$= 0.53 \times 10^{-23} \text{ kg m s}^{-1}$$

$$1 \text{ J} = 1 \text{ kg m}^2 \text{ s}^{-2}$$

Again,

$$\Delta v = \frac{\Delta p}{m}$$

$$m = 9.1083 \times 10^{-31} \text{ kg}$$

Hence

$$\Delta v = \frac{0.53 \times 10^{-23} \text{ kg m s}^{-1}}{9.1083 \times 10^{-31} \text{ kg}}$$

$$= 0.60 \times 10^7 \text{ m s}^{-1}$$

15.8 Particle in a One-Dimensional Box

The Schrödinger wave equation can now be applied to solve a particularly simple problem, the particle in a one-dimensional box. The situation is only imaginary, but the solution serves to illustrate many of the ideas introduced in this chapter.

Suppose that we have a particle of mass m confined in a one-dimensional box of length L. (Imagine the particle moving along a piece of straight wire.) For simplicity, we assume that the potential energy of the particle is zero inside the box; it has only kinetic energy. At each end of the box there is an infinite potential wall so that there is no probability of finding the particle outside the box (Figure 15.12). Equation (15.21) can now be written as

$$-\frac{h^2}{8\pi^2 m} \frac{d^2\psi}{dx^2} = E\psi \tag{15.25}$$

We are interested in the values of E and ψ that the particle can possess. Equation (15.25) tells us that the wave function ψ is such that when it is differentiated twice with respect to x, the original function is obtained. Examples of such functions are trigonometric and exponential functions. As a trial solution, let ψ be

$$\psi = A \sin ax \tag{15.26}$$

where A and a are constant. We have

$$\frac{d\psi}{dx} = aA \cos ax$$

$$\frac{d^2\psi}{dx^2} = -a^2 A \sin ax$$

$$= -a^2\psi \tag{15.27}$$

From Eqs. (15.25) and (15.27), we get

$$a^2 = \frac{8\pi^2 mE}{h^2}$$

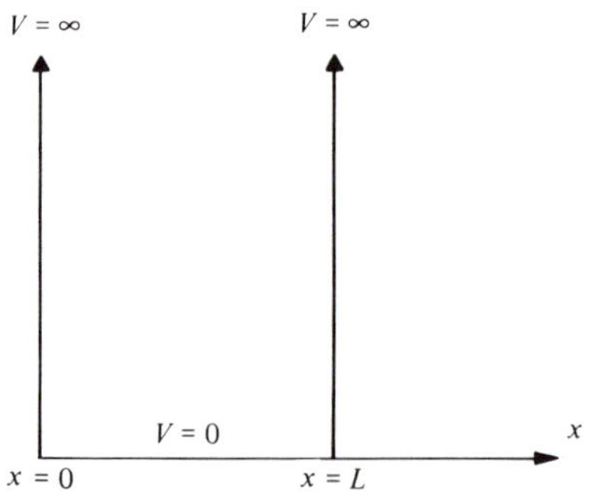

Figure 15.12

One-dimensional box with infinite potential barriers.

or

$$a = \left(\frac{8\pi^2 mE}{h^2}\right)^{1/2} \qquad (15.28)$$

Substitution of Eq. (15.28) into Eq. (15.26) gives

$$\psi = A \sin \left(\frac{8\pi^2 mE}{h^2}\right)^{1/2} x \qquad (15.29)$$

Mathematically, there are an infinite number of solutions that can satisfy Eq. (15.29), since A can have any value. Physically, however, ψ must satisfy the following boundary conditions. Because the probability of finding the particle is zero at either end of the box, we have

$$\text{at } x = 0, \qquad \psi = 0$$

and

$$\text{at } x = L, \qquad \psi = 0$$

The second condition, when applied to Eq. (15.29), gives

$$0 = A \sin \left(\frac{8\pi^2 mE}{h^2}\right)^{1/2} L$$

Since $A = 0$ is a trivial solution, in general we have

$$\sin \left(\frac{8\pi^2 mE}{h^2}\right)^{1/2} L = 0$$

This situation holds if

$$\left(\frac{8\pi^2 mE}{h^2}\right)^{1/2} L = n\pi$$

where $n = 1, 2, \ldots$, since

$$\sin \pi = \sin 2\pi = \sin 3\pi = \cdots = 0$$

Thus $n\pi = (8\pi^2 mE_n/h^2)^{1/2} L$ or

$$E_n = \frac{n^2 h^2}{8mL^2} \qquad (15.30)$$

where E_n is the energy for the nth level. The corresponding wave function is given by

$$\psi_n = A \sin \frac{n\pi}{L} x$$

The next step is to determine A. We start with the knowledge that since the particle must remain inside the box, the total probability of finding the particle between $x = 0$ and $x = L$ must be unity. This *normalization* process gives

$$\int_0^L \psi^2 \, dx = 1$$

where $\psi^2 \, dx$ gives the probability of finding the particle between x and $x + dx$. We write

$$A^2 \int_0^L \sin^2 \frac{n\pi}{L} x \, dx = 1$$

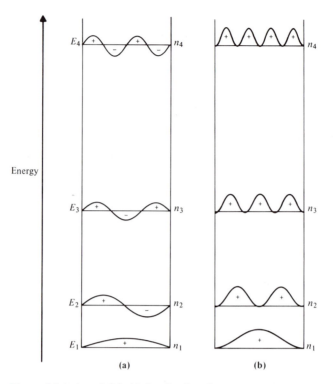

Figure 15.13 Plots of (a) ψ and (b) ψ^2 for the first four energy levels in a one-dimensional box.

The definite integral above gives*

$$A^2 \frac{L}{2} = 1$$

or

$$A = \sqrt{\frac{2}{L}}$$

Finally, we have

$$\psi_n = \sqrt{\frac{2}{L}} \sin \frac{n\pi}{L} x \tag{15.31}$$

Plots of the allowed energy levels as well as ψ and ψ^2 are shown in Figure 15.13.

Several important results must be mentioned:

1. The (kinetic) energy of the particle is quantized according to Eq. (15.30).
2. The lowest energy level is not zero but is equal to $h^2/8mL^2$. This *zero-point energy* can be accounted for by the Heisenberg uncertainty principle. If the particle could possess zero energy, its velocity would also be zero; consequently, there would be no uncertainty in determining its momentum. According to Eq. (15.23), therefore, Δx would be infinite.

---

* This definite integral is evaluated by using the relation

$$\int \sin^2 ax \, dx = \frac{x}{2} - \frac{\sin 2ax}{4a}$$

However, if the box is of finite size, the uncertainty in determining the particle's position cannot exceed ΔL, and this would violate the Heisenberg uncertainty principle.

3. Depending on the value of n, the wave behavior of the particle is described by Eq. (15.31), but the probability is given by ψ_n^2, which is always positive. For $n = 1$, the maximum probability is at $x = L/2$; for $n = 2$, the maxima occur at $x = L/4$ and $x = 3L/4$. In the latter case, the probability is zero at $x = L/2$. The point at which ψ (and hence ψ^2) is zero is called a *node*. Generally, the number of nodes increases with increasing energy. In classical mechanics the probability of finding the particle is the same along the box, irrespective of its energy.

Equation (15.30) can also be derived from de Broglie's relation. From Figure 15.13a we find that the wavelength λ of the particle in the nth level is given by

$$n\frac{\lambda}{2} = L \qquad n = 1, 2, \ldots$$

Since

$$\lambda = \frac{h}{mv}$$

we obtain

$$\frac{nh}{2mv} = L$$

or

$$(mv)^2 = \frac{n^2 h^2}{4L^2}$$

Finally, we have

$$E_n = \frac{1}{2}mv^2 = \frac{(mv)^2}{2m} = \frac{n^2 h^2}{8mL^2}$$

The problem of a particle in a one-dimensional box shows us that when a submicroscopic particle is in a *bound* state, that is, when its movement is restricted by potential barriers, its energy values should be quantized. This is precisely the case for electrons in atoms. Indeed, we can predict a number of atomic properties by considering a particle in a three-dimensional box model. For example, the energies of an electron in a hydrogen atom must be quantized. Further, the electron should possess three quantum numbers. We shall discuss this system in Chapter 16.

Quantum-Mechanical Tunneling **15.9**

What would happen if the potential walls surrounding the particle in the one-dimensional box were not infinitely high? Two things. The particle would escape when its kinetic energy is equal to or greater than the potential energy of the barrier. What is more surprising, however, is the fact that we may find the particle outside the box even if its kinetic energy is not sufficient to reach the top of the barrier! This phenomenon, known as *quantum-mechanical tunneling*, has many profound consequences in chemistry and biology.*

The probability of the particle tunneling through the barrier is proportional to the quantity†

$$\exp\left\{-\frac{4\pi a}{h}[2m(V-E)]^{1/2}\right\} \qquad V > E$$

where exp means exponential and a is the thickness of the barrier. Clearly, unless $V = \infty$ or $a = \infty$, there is always a probability that the particle will escape. This is especially the case for light particles such as electrons and protons (small masses).

The energy profile for a chemical reaction is usually described in terms of reactant molecules acquiring sufficient activation energy to get over the energy barrier to form more stable products. However, in a number of cases (for example, some electron exchange reactions) reactions are known to proceed even when an insufficient amount of activation energy is available. Such results have been attributed to quantum-mechanical tunneling.

Appendix 15.1 Relation Between Stat Volt and Practical Volt

The potential difference V between two points A and B is equal to the work required to move a charge Q between these two points. Thus

$$V = \frac{W}{Q}$$

If 1 erg of work is required to move a charge of 1 stat C (1 stat C = 4.803×10^{-10} esu) between A and B, then the potential difference between these two points is 1 stat V. On the other hand, if 1 joule of work is required to move a charge of 1 C from A to B, the potential difference between these two points is 1 V. We have

$$\text{CGS–esu units:} \quad 1 \text{ stat V} = \frac{1 \text{ erg}}{1 \text{ stat C}}$$

$$\text{SI units:} \quad 1 \text{ V} = \frac{1 \text{ J}}{1 \text{ C}}$$

Now 1 J = 10^7 ergs and 1 C = 3×10^9 stat C (4.803×10^{-10} esu/1.602×10^{-19} C). Thus

$$1 \text{ stat V} = 300 \text{ V}$$

* See W. T. Scott, *J. Chem. Educ.* **48**, 524 (1971), for an interesting illustration of quantum-mechanical tunneling.
† For derivation, see F. L. Pilar, *Elementary Quantum Chemistry*, McGraw-Hill Book Company, New York, 1968, p. 93.

Appendix 15.2   Derivation of Equation (15.17) in SI Units

Equation (15.17) was derived in the CGS–esu system. To derive the energy expression for the electron in a hydrogen atom, we must start from Eq. (15.9), as follows. In the SI system, Eq. (15.9) becomes

$$\frac{Ze^2}{r^2}\left(\frac{1}{4\pi\varepsilon_0}\right) = \frac{mv^2}{r} \tag{A15.1}$$

where ε_0 is the permittivity of free space (see Appendix 9.1). The total energy of the system is given by

$$E = \frac{1}{2}mv^2 - \frac{Ze^2}{r}\left(\frac{1}{4\pi\varepsilon_0}\right) \tag{A15.2}$$

From Eq. (A15.1),

$$mv^2 = \frac{Ze^2}{r}\left(\frac{1}{4\pi\varepsilon_0}\right) \tag{A15.3}$$

Substitution of Eq. (A15.3) into Eq. (A15.2) gives

$$E = -\frac{Ze^2}{2r}\left(\frac{1}{4\pi\varepsilon_0}\right) \tag{A15.4}$$

From Eqs. (A15.2) and (A15.3), we get

$$E = -\tfrac{1}{2}mv^2 \tag{A15.5}$$

Since the angular momentum is quantized, we write

$$mvr = n\frac{h}{2\pi} \tag{A15.6}$$

Dividing Eq. (A15.3) by Eq. (A15.6), we obtain

$$v = \frac{2\pi Ze^2}{nh}\left(\frac{1}{4\pi\varepsilon_0}\right) \tag{A15.7}$$

Substitution of Eq. (A15.7) into Eq. (A15.5) gives

$$E = -\frac{2m\pi^2 Z^2 e^4}{n^2 h^2}\left(\frac{1}{4\pi\varepsilon_0}\right)^2 \tag{A15.8}$$

Equation (A15.8) gives the possible energies an electron can have in a hydrogen atom or a hydrogenlike ion in SI units.

The topics discussed in this chapter are covered in most standard physical chemistry texts listed in Chapter 1.

**Suggestions
for Further
Reading**

INTRODUCTORY

HOCHSTRASSER, R. M. *Behavior of Electrons in Atoms.* W. A. Benjamin, Inc., Menlo Park, Calif., 1964.
   A very readable introduction to atomic theory.

CROPPER, W. H. *The Quantum Physicists*. Oxford University Press, Inc., New York, 1970.

Traces the historical development of quantum mechanics. A very interesting book, although some discussions are far beyond the level of this chapter.

HERZBERG, G. *Atomic Spectra and Atomic Structure*. Dover Publications, Inc., New York, 1944.

A standard text, suitable for those interested in more mathematical details.

Reading Assignments

"The Quantum Theory," K. K. Darrow, *Sci. Am.*, Mar. 1952.

"Quantum Theory: Max Planck," A. B. Garrett, *J. Chem. Educ.* **40**, 262 (1963).

"What Is Matter?" E. Schrödinger, *Sci. Am.*, Sept. 1953.

"The Limits of Measurement," R. Furth, *Sci. Am.*, July 1950.

"The Principle of Uncertainty," G. Gamow, *Sci. Am.*, Jan. 1958.

"Demonstration of the Uncertainty Principle," W. Laurita, *J. Chem. Educ.* **45**, 461 (1968).

"The Exclusion Principle," G. Gamow, *Sci. Am.*, July 1959.

"Particles, Waves, and the Interpretation of Quantum Mechanics," N. D. Christoudouleas, *J. Chem. Educ.* **52**, 573 (1975).

"The Bohr Atomic Model: Niels Bohr," A. B. Garrett, *J. Chem. Educ.* **39**, 534 (1962).

Problems

15.1 Calculate the energy associated with a quanta (photon) of light of wavelength 500 nm.

15.2 The threshold frequency from a zinc metal surface is 8.54×10^{14} Hz. Calculate the minimum amount of energy required to remove an electron from the metal.

15.3 Use Eqs. (15.12) and (15.17) to derive an expression for the radius of the orbit. Calculate its value for $n = 1$ for hydrogen.

15.4 What are the wavelengths associated with (a) an electron moving at 1.50×10^8 cm s^{-1}, and (b) a 60-g tennis ball moving at 1500 cm s^{-1}?

15.5 Calculate the frequency and wavelength associated with the transition from the $n = 4$ to the $n = 2$ level in hydrogen.

15.6 The diffraction phenomenon may be observed whenever the wavelength is comparable in magnitude to the size of the slit opening. To be "diffracted," how fast must a person who weighs 84 kg move through a door 1 m in width?

15.7 According to Eq. (15.30), the energy is inversely proportional to the square of the length of the box. How would you account for this dependence in terms of the Heisenberg uncertainty principle?

15.8 Suppose that the uncertainty in determining the position of an electron circling an orbit in an atom is 0.4 Å. What is the uncertainty in its velocity?

15.9 A person weighing 77 kg jogs at 1.5 m s^{-1}. Calculate the momentum and wavelength of this person. What is the uncertainty in determining his position at any given instant if we can measure his momentum to $\pm 0.05\%$? Predict the changes that would take place in this problem if the Planck constant were 1×10^7 ergs s.

15.10 A proton is accelerated through a potential difference of 3.0×10^6 V, starting from rest. Calculate its final wavelength, velocity, and kinetic energy.

$\star$**15.11** What is the probability of locating a particle in a one-dimensional box between $L/4$ and $3L/4$, where L is the length of the box? Assume the particle to be in the lowest level.

15.12 Show that Eq. (15.30) is dimensionally correct.

15.13 When two atoms collide, some of their kinetic energy may be converted into electronic energy of one or both atoms. If the average kinetic energy ($\frac{3}{2}kT$) is about equal to the energy for some allowed electronic transition, an appreciable number of atoms can absorb enough energy through inelastic collision to be raised to an excited electronic state. (a) Calculate the average kinetic energy per atom in a gas sample at 298 K. (b) Calculate the energy difference between the $n = 1$ and $n = 2$ levels in hydrogen. (c) At what temperature will it be possible to excite a hydrogen atom from the $n = 1$ level to $n = 2$ level by collision?

15.14 A student records an emission spectrum of hydrogen and notices that there is one spectral line in the Balmer series that cannot be accounted for by the Bohr theory. Assuming that the gas sample is pure, suggest a species responsible for emitting this line.

15.15 Calculate Rydberg's constant using Eq. (15.20).

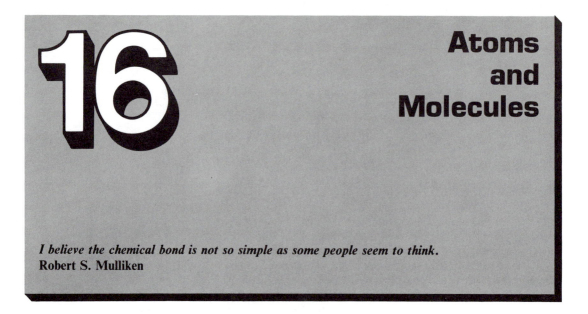

16

Atoms and Molecules

I believe the chemical bond is not so simple as some people seem to think.
Robert S. Mulliken

To understand the behavior of molecules, we must first understand their structure and bonding. How can we explain the fact that two hydrogen atoms will combine to form a stable H_2 molecule, but two helium atoms will not form a stable He_2 molecule? How do we account for the different bond length and bond strength observed for various compounds? Why is the water molecule bent but the carbon dioxide molecule linear in shape? The answers to all these questions, and many more, must come from quantum-mechanical treatment of atoms and molecules.

This chapter will survey some important theories employed by chemists in studying the nature of bonding. Some simple molecules will be discussed and the role of transition metals in biological systems will be examined.

16.1 The Hydrogen Atom

We consider first the simplest atom, hydrogen. This is a three-dimensional problem, so the wave function ψ for the electron will depend on the x, y, and z coordinates. The Schrödinger wave equation is given by

$$\frac{\partial^2 \psi}{\partial x^2} + \frac{\partial^2 \psi}{\partial y^2} + \frac{\partial^2 \psi}{\partial z^2} + \frac{8\pi^2 m}{h^2}(E - V)\psi = 0 \qquad (16.1)$$

where the potential energy V is the Coulombic interaction between the electron and the nucleus, given by $-e^2/r$, and the ∂ sign denotes the partial derivative (Appendix 2). Because the attraction has spherical symmetry (depends only on r), Eq. (16.1) is more conveniently expressed in terms of *polar coordinates*. The relation between Cartesian and polar coordinates is

Relation between Cartesian and polar coordinates.

Figure 16.1

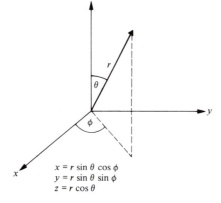

$x = r \sin \theta \cos \phi$
$y = r \sin \theta \sin \phi$
$z = r \cos \theta$

shown in Figure 16.1. With this transformation, we can rewrite Eq. (16.1) as

$$\frac{\partial^2 \psi}{\partial r^2} + \frac{2}{r} \frac{\partial \psi}{\partial r} + \frac{1}{r^2 \sin \theta} \frac{\partial[\sin \theta (\partial \psi / \partial \theta)]}{\partial \theta}$$

$$+ \frac{1}{r^2 \sin \theta} \frac{\partial^2 \psi}{\partial \phi^2} + \frac{8\pi^2 m}{h^2} \left(E + \frac{e^2}{r} \right) \psi = 0 \quad (16.2)$$

Fortunately, this oppressive-looking equation has already been solved, so we need only be concerned with the results. The main point is that the wave function can be expressed as a function of r, θ, and ϕ, as follows:

$$\psi(r, \theta, \phi) = R(r)\Theta(\theta)\Phi(\phi) \quad (16.3)$$

Thus ψ is given as a product of two independent quantities, R and $\Theta\Phi$, which are the *radial* and *angular* parts of the wave function, respectively.

The solution of Eq. (16.2) gives rise to three quantum numbers, n, l, and m_l, which have the following properties. The *principal quantum number*, n, determines the size of the wave function and the energy of the electron. The *azimuthal quantum number* or the *angular momentum quantum number*, l, determines the shape of the wave function. Finally, the *magnetic quantum number*, m_l, determines the orientations of the wave function in space. We shall soon see how these quantum numbers are used to describe the electron in hydrogen atom.

For a given n (where $n = 1, 2, \ldots$), there are n values of l given by $l = 0, 1, 2, \ldots, (n - 1)$; for a given l, there are $(2l + 1)$ values of m_l, given by $m_l = 0$, $\pm 1, \pm 2, \ldots, \pm l$. Thus, if $n = 3$, we have

$$l = 0, 1, \text{ and } 2$$

$$l = 0 \quad m_l = 0$$

$$l = 1 \quad m_l = 0, \pm 1$$

$$l = 2 \quad m_l = 0, \pm 1, \pm 2$$

The names of the wave function are determined by the value of l according to Table 16.1. For example, if $n = 3$ and $l = 2$, we have a 3d wave function. The wave function for a single electron is called an *orbital*. In Bohr's theory, we consider the orbits of an electron around a nucleus. In quantum mechanics, we consider the position of an electron not in terms of its orbits, but in terms of its wave function or orbital ψ.

Table 16.1

l	Wave Function
0	*s*
1	*p*
2	*d*
3	*f*
.	.
.	.
.	.

The energy of the electron in a hydrogen atom is given by

$$E = -\frac{2\pi^2 me^4}{n^2 h^2} \tag{16.4}$$

which is identical to Eq. (15.17). For a nucleus containing Z protons, Eq. (16.4) is modified to

$$E = -\frac{2\pi^2 me^4 Z^2}{n^2 h^2} \tag{16.5}$$

16.2 Atomic Orbitals

Our next step is to investigate the nature of the wave functions. Let us first consider the radial part of ψ given in Table 16.2. Figure 16.2a shows the dependence of $R(r)$ on r for the hydrogen $1s$ orbital. The $R(r)^2$ gives the probability of finding the electron at a point between r and $r + dr$ away from the nucleus. A more informative plot should give the total probability of finding the electron between r and $r + dr$, instead of just at a single point. To do this, we consider two concentric spheres of radii r and $r + dr$. The volume between these two spheres is $4\pi r^2 dr$,* and the probability of finding the electron within this spherical shell is $4\pi r^2 R(r)^2$. The function $4\pi r^2 R(r)^2$ is called the *radial distribution function*. It is interesting to note that for the $1s$ orbital, the maximum value occurs at 0.529 Å, which is the radius of the first orbit in

Radial Part of the Wave Function

Table 16.2

Orbital	$R(r)^a$
$1s$	$2\left(\dfrac{Z}{a_0}\right)^{3/2} e^{-\sigma/2}$
$2s$	$\dfrac{1}{2\sqrt{2}}\left(\dfrac{Z}{2a_0}\right)^{3/2}(2-\sigma)e^{-\sigma/2}$
$2p$	$\dfrac{1}{2\sqrt{6}}\left(\dfrac{Z}{a_0}\right)^{3/2}\sigma e^{-\sigma/2}$

^{a} $\sigma = 2Zr/na_0$; $a_0 = h^2/4\pi^2 me^2$.

* This is obtained by taking the difference between two volumes: $(4\pi/3)(r+dr)^3 - (4\pi/3)r^3$ and neglecting the $(dr)^2$ and $(dr)^3$ terms.

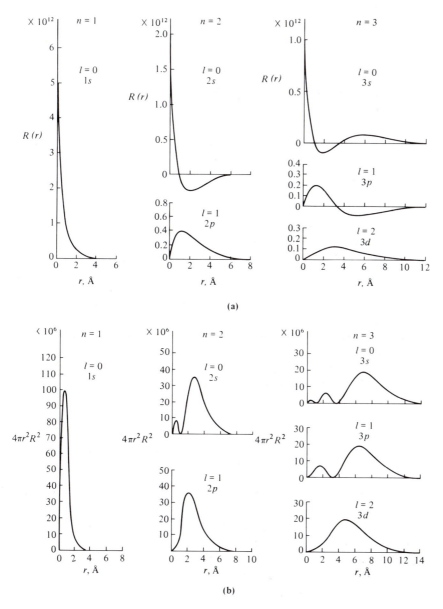

(a) Plots of $R(r)$ for the hydrogen $1s$, $2s$, $2p$, $3s$, $3p$, and $3d$ orbitals. (b) Plots of $4\pi r^2 R(r)^2$ for the same hydrogen orbitals. (From G. Herzberg, *Atomic Spectra and Atomic Structure*, Dover Publications, Inc., New York, 1944. Used by permission.)

Figure 16.2

Bohr's model (see Problem 16.9). Figure 16.2 also shows plots of $R(r)$ and $4\pi r^2 R(r)^2$ for the 2s, 2p, 3s, 3p, and 3d orbitals. From these plots we see that an electron in any particular orbital does not have a well-defined position, and it is therefore more convenient to use the term *electron density* or *electron cloud*.

Mathematically, the probability vanishes only when r approaches infinity. Physically, however, we need only consider each orbital over a relatively small distance (a few angstroms) because the function decreases rapidly with increasing r. The 2s orbital has two maxima. In this case, we can imagine two concentric spheres with a node somewhere in the spherical shell. The radial

449

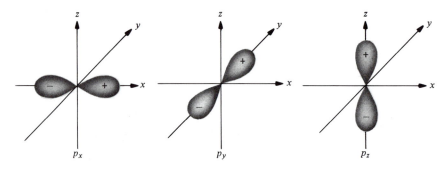

Figure 16.3 The three *p* orbitals.

distribution function plots for the *p* and *d* orbitals are more complex in form, but they can be interpreted in a similar manner.

The shapes of the orbitals are determined by the angular part of the wave functions, which are listed in Table 16.3. The *s* orbitals contain only a constant; therefore, they are spherically symmetrical. The *p* orbitals, on the other hand, depend on θ and ϕ and do not have spherical symmetry. Figure 16.3 shows the plots for the *p* orbitals along the *x*, *y*, and *z* coordinates. Each orbital consists of two regions together, resembling a dumbbell. Furthermore, the wave function is positive in one region and negative in the other, with a nodal plane in between. The wave function is zero in this plane, which also contains the nucleus. These three orbitals are entirely equivalent except for their orientations. Thus all three *p* orbitals have the same energy and are said to be *degenerate*. There is no physical significance in the signs by themselves. The only meaningful quantity is the probability of finding the electron given by the square of the wave function. However, the signs will be useful later when we consider the interaction between orbitals as in chemical-bond formation.

It is important to note that a complete wave function must, according to Eq. (16.3), be given by the product of the radial and angular parts, with the size of the orbital determined by the former and the shape by the latter. An orbital can be represented in several ways. This is illustrated with the 1*s* orbital shown in Figure 16.4. The boundary-surface representation is simplest to use, although it is also least informative. The contour-surface and electron density representations provide a more detailed description, but they are more

Table 16.3

Orbital	Angular Part of the Wave Function $\Theta(\theta)\Phi(\phi)$
s	$\sqrt{\dfrac{1}{4\pi}}$
p_x	$\sqrt{\dfrac{3}{4\pi}}\ \sin\theta\cos\phi$
p_y	$\sqrt{\dfrac{3}{4\pi}}\ \sin\theta\sin\phi$
p_z	$\sqrt{\dfrac{3}{4\pi}}\ \cos\theta$

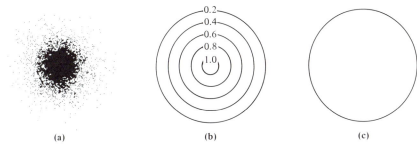

Figure 16.4

Representations for the hydrogen $1s$ orbital: (a) charge cloud, (b) contour surfaces (the numbers represent the relative charge densities), and (c) boundary surface. **Figure 16.4**

The two possible spinning motions of an electron. The ↑ and ↓ arrows are the symbols commonly employed to denote the direction of the spins. The magnetic fields generated by the spinning motion are equivalent to those of two bar magnets.

Figure 16.5

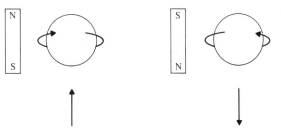

time-consuming to draw. We shall postpone the discussion of d orbitals to Section 16.11.

The solution of the Schrödinger equation for hydrogen atoms provides us with three quantum numbers. However, there is yet a fourth quantum number associated with the electron. We know that each electron has a spinning motion about its own axis, which can be either clockwise or counterclockwise (Figure 16.5). In quantum mechanics we say that the electron has a spin S of $\frac{1}{2}$ and a spin quantum number $m_s = \pm\frac{1}{2}$. The value of m_s gives the orientation of the magnetic moment of the electron* in the presence of an external magnetic field. Thus m_s is analogous to m_l, which determines the orientation of the orbitals in space.

Many-Electron Atoms and the Periodic Table **16.3**

The next simplest atom after hydrogen is helium. Unfortunately, the Schrödinger wave equation cannot be solved exactly for a three-body system of two electrons and one nucleus, because of the mathematical complexities. Instead, a number of approximate methods must be applied. All the approaches use the hydrogen wave functions as a starting point and then try to account for the influence exerted by the extra electron(s) present. We shall not be concerned with any of the theories here. Suffice it to say that the orbitals obtained for *many-electron* atoms are basically similar to those for the hydrogen atom.

Instead, let us examine the *electronic configuration* of atoms, that is, the arrangement of electrons among different orbitals. According to Eq. (16.5),

---

* The spinning of a charged particle generates a magnetic field so that each electron behaves like a small magnet.

Table 16.4 — Electronic Configuration of the Elements[a]

At. No.	Elem.	1s	2s	2p	3s	3p	3d	4s	4p	4d	4f	5s	5p	5d	5f	6s	6p	6d	6f	7s	7p
		K	**L**		**M**			**N**				**O**				**P**				**Q**	
1	H	1																			
2	He	2																			
3	Li	2	1																		
4	Be	2	2																		
5	B	2	2	1																	
6	C	2	2	2																	
7	N	2	2	3																	
8	O	2	2	4																	
9	F	2	2	5																	
10	Ne	2	2	6																	
11	Na	2	2	6	1																
12	Mg	2	2	6	2																
13	Al	2	2	6	2	1															
14	Si	2	2	6	2	2															
15	P	2	2	6	2	3															
16	S	2	2	6	2	4															
17	Cl	2	2	6	2	5															
18	Ar	2	2	6	2	6															
19	K	2	2	6	2	6		1													
20	Ca	2	2	6	2	6		2													
21	Sc	2	2	6	2	6	1	2													
22	Ti	2	2	6	2	6	2	2													
23	V	2	2	6	2	6	3	2													
24	Cr	2	2	6	2	6	5	1													
25	Mn	2	2	6	2	6	5	2													
26	Fe	2	2	6	2	6	6	2													
27	Co	2	2	6	2	6	7	2													
28	Ni	2	2	6	2	6	8	2													
29	Cu	2	2	6	2	6	10	1													
30	Zn	2	2	6	2	6	10	2													
31	Ga	2	2	6	2	6	10	2	1												
32	Ge	2	2	6	2	6	10	2	2												
33	As	2	2	6	2	6	10	2	3												
34	Sc	2	2	6	2	6	10	2	4												
35	Br	2	2	6	2	6	10	2	5												
36	Kr	2	2	6	2	6	10	2	6												
37	Rb	2	2	6	2	6	10	2	6			1									
38	Sr	2	2	6	2	6	10	2	6			2									
39	Y	2	2	6	2	6	10	2	6	1		2									
40	Zr	2	2	6	2	6	10	2	6	2		2									
41	Nb	2	2	6	2	6	10	2	6	4		1									
42	Mo	2	2	6	2	6	10	2	6	5		1									
43	Tc	2	2	6	2	6	10	2	6	6		1									
44	Ru	2	2	6	2	6	10	2	6	7		1									
45	Rh	2	2	6	2	6	10	2	6	8		1									
46	Pd	2	2	6	2	6	10	2	6	10											
47	Ag	2	2	6	2	6	10	2	6	10		1									
48	Cd	2	2	6	2	6	10	2	6	10		2									
49	In	2	2	6	2	6	10	2	6	10		2	1								
50	Sn	2	2	6	2	6	10	2	6	10		2	2								
51	Sb	2	2	6	2	6	10	2	6	10		2	3								
52	Te	2	2	6	2	6	10	2	6	10		2	4								

Table 16.4 (*cont.*)

Shell		K	L		M			N				O				P				Q	
Subshell		1s	2s	2p	3s	3p	3d	4s	4p	4d	4f	5s	5p	5d	5f	6s	6p	6d	6f	7s	7p
At. No.	Elem.																				
53	I	2	2	6	2	6	10	2	6	10		2	5								
54	Xe	2	2	6	2	6	10	2	6	10		2	6								
55	Cs	2	2	6	2	6	10	2	6	10		2	6			1					
56	Ba	2	2	6	2	6	10	2	6	10		2	6			2					
57	La	2	2	6	2	6	10	2	6	10		2	6	1		2					
58	Ce	2	2	6	2	6	10	2	6	10	2	2	6			2					
59	Pr	2	2	6	2	6	10	2	6	10	3	2	6			2					
60	Nd	2	2	6	2	6	10	2	6	10	4	2	6			2					
61	Pm	2	2	6	2	6	10	2	6	10	5	2	6			2					
62	Sm	2	2	6	2	6	10	2	6	10	6	2	6			2					
63	Eu	2	2	6	2	6	10	2	6	10	7	2	6			2					
64	Gd	2	2	6	2	6	10	2	6	10	7	2	6	1		2					
65	Tb	2	2	6	2	6	10	2	6	10	8	2	6	1		2					
66	Dy	2	2	6	2	6	10	2	6	10	9	2	6	1		2					
67	Ho	2	2	6	2	6	10	2	6	10	10	2	6	1		2					
68	Er	2	2	6	2	6	10	2	6	10	11	2	6	1		2					
69	Tm	2	2	6	2	6	10	2	6	10	13	2	6			2					
70	Yb	2	2	6	2	6	10	2	6	10	14	2	6			2					
71	Lu	2	2	6	2	6	10	2	6	10	14	2	6	1		2					
72	Hf	2	2	6	2	6	10	2	6	10	14	2	6	2		2					
73	Ta	2	2	6	2	6	10	2	6	10	14	2	6	3		2					
74	W	2	2	6	2	6	10	2	6	10	14	2	6	4		2					
75	Re	2	2	6	2	6	10	2	6	10	14	2	6	5		2					
76	Os	2	2	6	2	6	10	2	6	10	14	2	6	6		2					
77	Ir	2	2	6	2	6	10	2	6	10	14	2	6	9							
78	Pt	2	2	6	2	6	10	2	6	10	14	2	6	9		1					
79	Au	2	2	6	2	6	10	2	6	10	14	2	6	10		1					
80	Hg	2	2	6	2	6	10	2	6	10	14	2	6	10		2					
81	Tl	2	2	6	2	6	10	2	6	10	14	2	6	10		2	1				
82	Pb	2	2	6	2	6	10	2	6	10	14	2	6	10		2	2				
83	Bi	2	2	6	2	6	10	2	6	10	14	2	6	10		2	3				
84	Po	2	2	6	2	6	10	2	6	10	14	2	6	10		2	4				
85	At	2	2	6	2	6	10	2	6	10	14	2	6	10		2	5				
86	Rn	2	2	6	2	6	10	2	6	10	14	2	6	10		2	6				
87	Fr	2	2	6	2	6	10	2	6	10	14	2	6	10		2	6			1	
88	Ra	2	2	6	2	6	10	2	6	10	14	2	6	10		2	6			2	
89	Ac	2	2	6	2	6	10	2	6	10	14	2	6	10		2	6	1		2	
90	Th	2	2	6	2	6	10	2	6	10	14	2	6	10		2	6	2		2	
91	Pa	2	2	6	2	6	10	2	6	10	14	2	6	10	2	2	6	1		2	
92	U	2	2	6	2	6	10	2	6	10	14	2	6	10	3	2	6	1		2	
93	Np	2	2	6	2	6	10	2	6	10	14	2	6	10	5	2	6			2	
94	Pu	2	2	6	2	6	10	2	6	10	14	2	6	10	6	2	6			2	
95	Am	2	2	6	2	6	10	2	6	10	14	2	6	10	7	2	6			2	
96	Cm	2	2	6	2	6	10	2	6	10	14	2	6	10	7	2	6	1		2	
97	Bk	2	2	6	2	6	10	2	6	10	14	2	6	10	8	2	6	1		2	
98	Cf	2	2	6	2	6	10	2	6	10	14	2	6	10	9	2	6	1		2	
99	Es	2	2	6	2	6	10	2	6	10	14	2	6	10	10	2	6	1		2	
100	Fm	2	2	6	2	6	10	2	6	10	14	2	6	10	11	2	6	1		2	
101	Md	2	2	6	2	6	10	2	6	10	14	2	6	10	12	2	6	1		2	
102	No	2	2	6	2	6	10	2	6	10	14	2	6	10	13	2	6	1		2	
103	Lw	2	2	6	2	6	10	2	6	10	14	2	6	10	14	2	6	1		2	

[a] From S. H. Maron and J. B. Lando, *Fundamentals of Physical Chemistry*, Macmillan Publishing Co., Inc., New York, 1974, pp. 122–124. Used by permission.

the energy of an electron depends only on the principal quantum number, n. Thus the orbitals can be arranged in the following order of increasing energy (decreasing stability) for hydrogen or hydrogenlike ions:

$$1s < 2s = 2p < 3s = 3p = 3d < 4s = 4p = 4d < \cdots$$

However, for many electron atoms, the order of increasing energy is given by

$$1s < 2s < 2p < 3s < 3p < 4s < 3d < 4p < 5s < 4d < \cdots$$

The difference in these two cases can be qualitatively explained as follows. From Figure 16.2 we see that even though the most probable location of a $2p$ electron lies closer to the nucleus than that for a $2s$ electron, the electron density close to the nucleus is actually greater for a $2s$ electron. Put another way, we say that an s electron is more penetrating than a p electron. Thus the $2p$ electron is more shielded from the nucleus by the $2s$ electron than the reverse case, that is, the shielding of the $2s$ electron by the $2p$ electron. Consequently, its energy lies higher than that for a $2s$ electron. In a hydrogen atom there is only one electron; hence no such shielding is present.

The electronic configuration for hydrogen atom is $1s^1$ or simply $1s$, meaning that there is one electron in the s orbital.* The four quantum numbers (n, l, m_l, and m_s) for the electron can be either $(1, 0, 0, +\frac{1}{2})$ or $(1, 0, 0, -\frac{1}{2})$. In the absence of a magnetic field, the energy of the electron is the same whether $m_s = \frac{1}{2}$ or $-\frac{1}{2}$. Helium has two electrons, so its electron configuration is $1s^2$. Furthermore, helium atom is *diamagnetic*, which means that the two electrons must have their spins opposed to each other, with the result that the net magnetic field generated is zero. Thus one electron must have $m_s = \frac{1}{2}$ and the other $m_s = -\frac{1}{2}$, the other three quantum numbers being the same. This is an example of the *Pauli exclusion principle*, which states that no two electrons in an atom (or molecule) can have the same four quantum numbers.† Thus the third electron in lithium atom must enter the $2s$ orbital and we have $1s^2 2s$ for Li.

This procedure of filling orbitals is continued until we reach the carbon atom $(1s^2 2s^2 2p^2)$, and we are faced with three different choices of adding the two electrons to the p orbitals, as shown in Figure 16.6. Although none of the arrangements violates the Pauli exclusion principle, Figure 16.6c is energetically the most favorable. According to *Hund's rule*, when more than one electron enters a set of degenerate levels, the most stable arrangement is one that has the greatest number of parallel spins. The physical interpretation of Hund's rule is that the Pauli exclusion principle forbids two electrons with parallel spins in an atom to approach closely to each other. Consequently, there is a decrease in the electrostatic repulsion between the two electrons, and a more stable arrangement is obtained. Indeed, experimentally we find that the ground-state carbon atom is *paramagnetic*, containing two unpaired electrons. Table 16.4 lists the electronic configuration for various elements.

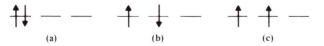

(a)   (b)   (c)

Figure 16.6   Three possible ways of adding two electrons to three equivalent p orbitals.

* We consider only the ground state of the atoms.
† Another way of stating the Pauli exclusion principle is to say that the total wave function of electrons changes its sign when two electrons are interchanged or permuted.

Element	Type of Electron Removed	Ionization Potential[a] (eV)	Electron Affinity (eV)
H	$1s$	13.6	0.8
He	$1s$	24.6	—
Li	$2s$	5.4	0.6
Be	$2s$	9.3	?
B	$2p$	8.3	0.3
C	$2p$	11.3	1.3
N	$2p$	14.5	?
O	$2p$	13.6	1.5
F	$2p$	17.4	3.5
Ne	$2p$	21.6	—

[a] The ionization potential is usually expressed in electron volts (eV), where $1 \text{ eV} = 1.60 \times 10^{-19}$ J or $1 \text{ eV molecule}^{-1} = 96.48 \text{ kJ mol}^{-1}$.

The energy for the orbitals can be obtained from the *ionization-potential* measurements. An ionization potential is defined as the energy required to remove an electron from an atom in the gaseous state to form a positive ion, also in the gaseous state; that is,

$$X(g) \longrightarrow X^+(g) + e^-(g)$$

The terms first, second, ... ionization potential refer to the removal of the outermost electron in an atom, the outermost electron in a unipositive ion, and so on.

Another important quantity is the *electron affinity*, defined as the energy released when a neutral gaseous species accepts an electron:

$$A(g) + e^-(g) \longrightarrow A^-(g)$$

We see that the reverse of the process described above gives the ionization potential of A^-. Generally, the more *electronegative* (to be discussed in Section 16.6) the element, the greater is its electron affinity.

Table 16.5 gives the first ionization potential and electron affinity values for second-period elements.

The Chemical Bond **16.4**

Long before quantum mechanics was applied to molecules, chemists knew a fair amount about the qualitative aspects of chemical bonding. In 1916, Lewis proposed that the atoms in a molecule are held together by chemical bonds formed by the sharing of electrons. He further suggested that this sharing is carried out in such a way that each atom achieves the helium or noble-gas configuration. Consider the following Lewis structures for H_2, H_2O, and NaCl:

$$H \overset{\times}{_\circ} H \qquad \overset{\circ\circ}{\underset{\circ\times}{\circ}} \overset{\circ\circ}{O} \overset{\circ}{_\circ} H \qquad \overset{\circ\circ}{_\circ} Na \overset{\circ}{_\circ}{}^+ \quad \overset{\times\times}{_\circ} Cl \overset{\times}{_\times}{}^-$$
$$H$$

where the cross and circle signs denote the electrons on different atoms. In H_2 each hydrogen atom achieves the helium electronic configuration; in H_2O, the oxygen atom shares two electrons with the two hydrogen atoms and has the Ne configuration of $1s^2 2s^2 2p^6$. There is a complete transfer of an electron from the sodium atom to chlorine in NaCl. In each ion a completely filled shell (that is, filled orbitals for a given principal quantum number n) is obtained, and this is taken to represent stability. This concept is called the *octet rule* and applies mainly to the second-period elements.

Although useful, the octet rule is unsatisfactory in several respects. For example, it fails to account for the difference in bond strength for molecules as well as to explain molecular geometry. A more quantitative description of bonding and structure must therefore be provided by the quantum-mechanical approach, which is based on the concept of atomic orbitals.

In Section 6.6 we discussed the formation of a hydrogen molecule in terms of the potential-energy curve. It is now appropriate to ask how the two hydrogen $1s$ orbitals interact with each other in this process. By analogy with the interference phenomenon discussed in Chapter 15, these two orbitals can interact either constructively or destructively, depending on whether their wave functions add or subtract in the region of "overlap." These interactions then lead to the formation of a bonding molecular orbital called σ (sigma) and an antibonding molecular orbital called σ^* (sigma star) (Figure 16.7). In the bonding σ orbital, there is a buildup of electron density between the two nuclei; in the antibonding σ^* orbital, there is a decrease in electron density. Further, σ and σ^* orbitals correspond to the potential-energy curves with and without a minimum, respectively, as shown in Figure 6.8.

Mathematically, this interaction is treated by taking the *linear combination of the atomic orbitals* (LCAO) as follows*:

$$\psi = N(\psi_A + \psi_B) \tag{16.6}$$

$$\psi^* = N(\psi_A - \psi_B) \tag{16.7}$$

where ψ and ψ^* are the wave functions for σ and σ^* molecular orbitals, respectively, and ψ_A and ψ_B the hydrogen $1s$ orbitals for atoms A and B. Here N is a constant called the *normalization constant* such that

$$\int \psi^2 \, d\tau = \int N^2 (\psi_A + \psi_B)^2 \, d\tau = 1$$

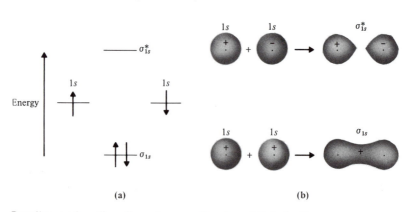

(a)   (b)

Figure 16.7   Bonding and antibonding sigma molecular orbitals in H_2.

* The word *linear* means that both ψ_A and ψ_B are raised to the power of unity.

and

$$\int \psi^{*2} \, d\tau = \int N^2 (\psi_A - \psi_B)^2 \, d\tau = 1$$

where $d\tau$ represents the volume element; that is, $d\tau = dx \, dy \, dz$.

The concept of molecular orbital is a natural extension of atomic orbitals. Each electron in the atomic case is localized in the $1s$ orbital. We can also think of the paired electrons in H_2 as contained in the molecular orbital. Squaring Eq. (16.6), we get

$$\psi^2 = N^2 (\psi_A^2 + \psi_B^2 + 2\psi_A \psi_B) \tag{16.8}$$

Equation (16.8) says that the probability of finding an electron near nucleus A or B is still given by ψ_A^2 or ψ_B^2 (although modified by the N^2 term), and there is also a buildup of electron density between A and B. This is the result of the overlap of the two atomic orbitals, and its magnitude is given by the product $2\psi_A \psi_B$. The main difference between ψ^2 and ψ^{*2} is that for the antibonding molecular orbital, the overlap is given by $-2\psi_A \psi_B$, which actually corresponds to a decrease in electron density.

Finally, an electron configuration of molecules can be understood in the same manner as that for the atomic case. Both the Pauli exclusion principle and Hund's rule must be satisfied. The electronic configuration for H_2 is simply $(\sigma_{1s})^2$, where σ_{1s} denotes the σ molecular orbital formed by the $1s$ orbitals and the superscript 2 denotes that there are two electrons present.

Diatomic Molecules 16.5

This section applies the molecular orbital approach to a few diatomic molecules. Their stability and electronic configurations are of particular interest.

Li₂

The electronic configuration of Li is $1s^2 2s^1$, so in Li_2 the four $1s$ electrons are paired in the σ_{1s} and σ_{1s}^* molecular orbitals. In addition, we also have σ_{2s} and σ_{2s}^* orbitals. Since there are only two $2s$ electrons, the electronic configuration for Li_2 is given by

$$(\sigma_{1s})^2 (\sigma_{1s}^*)^2 (\sigma_{2s})^2$$

B₂

The electronic configuration of B is $1s^2 2s^2 2p^1$. The presence of a p orbital suggests that there are two possible ways of interaction (Figure 16.8). In Figure 16.8a, the overlap of the p orbitals gives rise to a σ molecular orbital; in Figure 16.8b, a π (pi) molecular orbital is formed. The difference between a σ molecular orbital and a π molecular orbital is as follows. A σ molecular orbital has cylindrical symmetry; that is, rotation about the internuclear axis by 180° leaves the molecular orbital in an indistinguishable position from the original one. On the other hand, the same rotation changes the signs of the π

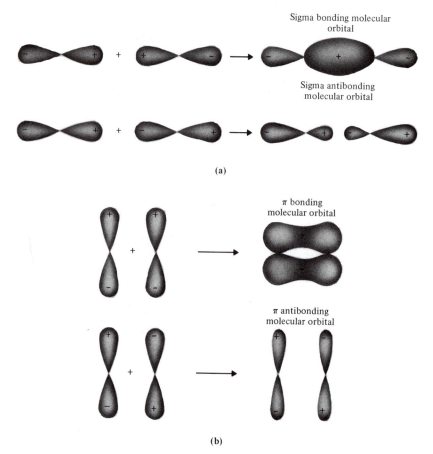

(a)

(b)

Figure 16.8 Formation of (a) σ and (b) π molecular orbitals from two p orbitals.

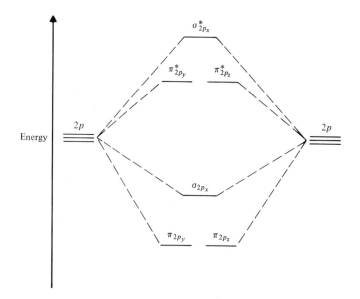

Figure 16.9 Molecular-energy-level diagram for the second-period homonuclear diatomic molecules. The σ_{1s} and σ_{2s} orbitals are not shown.

molecular orbital. Since there are three p orbitals for each atom, by simple symmetry we can see that one σ and two π molecular orbitals will result from the interaction. Although σ orbitals in general have a greater overlap, they do not always lie below the π orbitals on the energy scale because of the influence of the inner electrons ($1s$ and $2s$). A case in point is B_2, which is paramagnetic; thus its electronic configuration is most likely to be

$$(\sigma_{1s})^2(\sigma_{1s}^*)^2(\sigma_{2s})^2(\sigma_{2s}^*)^2(\pi_{2p_y})^1(\pi_{2p_z})^1$$

Figure 16.9 shows the molecular-orbital-energy-level diagram that applies to the homonuclear diatomic molecules of the second-period elements.

O_2

The electronic configuration of O is $1s^22s^22p^4$. Since oxygen molecules are paramagnetic, its electronic configuration is given by

$$(\sigma_{1s})^2(\sigma_{1s}^*)^2(\sigma_{2s})^2(\sigma_{2s}^*)^2(\pi_{2p_y})^2(\pi_{2p_z})^2(\sigma_{2p_x})^2(\pi_{2p_y}^*)^1(\pi_{2p_z}^*)^1$$

Here is an example of the usefulness of the molecular orbital theory—the paramagnetism of the O_2 molecule is readily accounted for by this description. On the other hand, it is not at all clear from the Lewis structure how an oxygen molecule can have two unpaired electrons.

Table 16.6 lists the electronic configuration of all the homonuclear diatomic molecules for the second-period elements. We introduce a term called *bond order*, which is defined as

$$\text{bond order} = \frac{1}{2} \left| \begin{array}{c} \text{number of electrons} \\ \text{in the bonding orbitals} \end{array} - \begin{array}{c} \text{number of electrons in} \\ \text{the antibonding orbitals} \end{array} \right|$$

Qualitatively, bond order gives a measure of the strength of a bond.

Electronic Configuration and Other Properties of the Homonuclear Diatomic Molecules of the First- and Second-Period Elements   **Table 16.6**

Molecule	Electronic Configuration	Bond Energy (kJ mol^{-1})	Bond Length (Å)	Bond Order	Magnetic Properties
H_2	$(\sigma_{1s})^2$	436.0	0.74	1	Diamagnetic
He_2	$(\sigma_{1s})^2(\sigma_{1s}^*)^2$	—	—	0	—
Li_2	$[He_2](\sigma_{2s})^2$	104.6	2.67	1	Diamagnetic
Be_2	$[He_2](\sigma_{2s})^2(\sigma_{2s}^*)^2$	—	—	0	—
B_2	$[Be_2](\pi_{2p_y})^1(\pi_{2p_z})^1$	288.7	1.59	1	Paramagnetic
C_2	$[Be_2](\pi_{2p_y})^2(\pi_{2p_z})^2$	627.6	1.31	2	Diamagnetic
N_2	$[Be_2](\pi_{2p_y})^2(\pi_{2p_z})^2(\sigma_{2p_x})^2$	941.4	1.10	3	Diamagnetic
O_2	$[Be_2](\pi_{2p_y})^2(\pi_{2p_z})^2(\sigma_{2p_x})^2(\pi_{2p_y}^*)^1(\pi_{2p_z}^*)^1$	498.8	1.21	2	Paramagnetic
F_2	$[Be_2](\pi_{2p_y})^2(\pi_{2p_z})^2(\sigma_{2p_x})^2(\pi_{2p_y}^*)^2(\pi_{2p_z}^*)^2$	150.6	1.42	1	Diamagnetic
Ne_2	$[Be_2](\pi_{2p_y})^2(\pi_{2p_z})^2(\sigma_{2p_x})^2(\pi_{2p_y}^*)^2(\pi_{2p_z}^*)^2(\sigma_{2p_x}^*)^2$	—	—	0	—

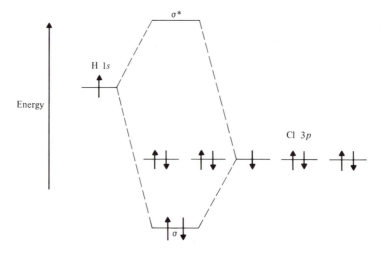

Figure 16.10 Molecular-energy-level diagram for HCl. Note that the hydrogen $1s$ orbital lies higher in energy than the chlorine $2p$ orbitals because an electron in the former orbital has a lower ionization potential. The $1s$ and $2s$ orbitals of chlorine are relatively unaffected by the formation of the HCl molecule.

For heteronuclear diatomic molecules the approach is basically the same, although the interpretation becomes more complicated. Consider the HCl molecule (Figure 16.10). The molecular orbitals are given by

$$\psi = N(\psi_H + \lambda \psi_{Cl}) \tag{16.9}$$

$$\psi^* = N(\psi_H - \lambda \psi_{Cl}) \tag{16.10}$$

where λ is some coefficient and ψ_H and ψ_{Cl} the wave functions for the $1s$ and $3p$ orbitals, respectively. Taking the square of Eq. (16.9), we get

$$\psi^2 = N^2(\psi_H^2 + \lambda^2 \psi_{Cl}^2 + 2\lambda \psi_H \psi_{Cl}) \tag{16.11}$$

The value of λ $(\lambda > 1)$ provides a measure of the *polarity* of the orbital. For example, there is greater electron density nearer the chlorine nucleus than the hydrogen nucleus.

Two other important heteronuclear diatomic molecules are CO and CN$^-$. Carbon monoxide is *isoelectronic* with N_2 (both contain the same number of electrons). Its electronic configuration is

$$(\sigma_{1s})^2(\sigma_{1s}^*)^2(\sigma_{2s})^2(\sigma_{2s}^*)^2(\pi_{2p_y})^2(\pi_{2p_z})^2(\sigma_{2p_x})^2$$

The cyanide group is also isoelectronic with N_2 and has the same electronic configuration as CO. It is important to keep in mind that in both molecules the lowest empty antibonding orbitals are π orbitals $(\pi_{2p_y}^*$ and $\pi_{2p_z}^*)$, a point that we shall return to when we discuss the bonding of these molecules to transition-metal ions in Section 16.11.

16.6 Electronegativity

The unequal distribution of electron density in a heteronuclear molecule results directly from the difference in *electronegativity* (X), which is the tendency of an atom to attract electrons in a molecule. There are several ways to compare the electronegativity of elements. Here we shall discuss a procedure

						H 2.1										
Li 1.0	Be 1.5											B 2.0	C 2.5	N 3.0	O 3.5	F 4.0
Na 0.9	Mg 1.2											Al 1.5	Si 1.8	P 2.1	S 2.5	Cl 3.0

K	Ca	Sc	Ti	V	Cr	Mn	Fe	Co	Ni	Cu	Zn	Ga	Ge	As	Se	Br
0.8	1.0	1.3	1.5	1.6	1.6	1.5	1.8	1.9	1.9	1.9	1.6	1.6	1.8	2.0	2.4	2.8
Rb	Sr	Y	Zr	Nb	Mo	Tc	Ru	Rh	Pd	Ag	Cd	In	Sn	Sb	Te	I
0.8	1.0	1.2	1.4	1.6	1.8	1.9	2.2	2.2	2.2	1.9	1.7	1.7	1.8	1.9	2.1	2.5
Cs	Ba	La–Lu	Hf	Ta	W	Re	Os	Ir	Pt	Au	Hg	Tl	Pb	Bi	Po	At
0.7	0.9	1.0–1.2	1.3	1.5	1.7	1.9	2.2	2.2	2.2	2.4	1.9	1.8	1.9	1.9	2.0	2.2
Fr	Ra	Ac	Th	Pa	U	Np–No										
0.7	0.9	1.1	1.3	1.4	1.4	1.4–1.3										

[a] From *General Chemistry*, 3rd ed., by Linus Pauling, p. 182. W. H. Freeman and Company, San Francisco. Copyright © 1970.

introduced by Pauling in 1932. We define the electronegativity difference between atoms A and B to be

$$|X_A - X_B| = \sqrt{D_{AB} - (D_{A_2}D_{B_2})^{1/2}} \qquad (16.12)$$

where D_{AB}, D_{A_2}, and D_{B_2} are the bond dissociation energies (in kJ mol^{-1}) of molecules AB, A$_2$, and B$_2$ and the vertical lines on the left-hand side of Eq. (16.12) indicate that only the magnitude of the difference is considered. Since only differences are obtained from Eq. (16.12), one element must be assigned a specific electronegativity value, and then the values for the other elements can be calculated readily. By arbitrarily defining $X_F = 4.0$, Pauling was able to set up an electronegativity scale shown in Table 16.7.

Dipole Moment 16.7

A molecule possesses a permanent electric *dipole moment* μ if its center of positive charge does not coincide with its center of negative charge. The dipole moment is defined as

$$\mu = Q \times r \qquad (16.13)$$

where Q is the charge separation and r the distance between the positive and negative centers. In the CGS system, if Q is one electron (4.8×10^{-10} esu) and r is 10^{-8} cm, then μ is given by

$$\mu = 4.8 \times 10^{-10} \times 10^{-8} = 4.8 \times 10^{-18} \text{ esu cm}$$
$$= 4.8 \text{ D}$$

where 1 D $= 1 \times 10^{-18}$ esu cm. The unit of dipole moment is called the *debye* (D), in honor of the pioneer in this field.

In the SI system, $e = 1.602 \times 10^{-19}$ C and $r = 10^{-10}$ m so that

$$\mu = 1.602 \times 10^{-19} \times 10^{-10} = 1.602 \times 10^{-29} \text{ C m} = 4.8 \text{ D}$$

where 1 D $= 3.338 \times 10^{-30}$ C m.

Bond Moments of Some Chemical Bonds

Table 16.8

Bond	Bond (or dipole) Moment (D)	$\mid X_A - X_B \mid$
HF	1.91	1.8
HCl	1.08	1.0
HBr	0.80	0.8
HI	0.42	0.5
H—O	1.5	1.2
H—N	1.3	0.8
H—P	0.4	0.0
C—H	0.4	0.4
C—F	1.4	1.5
C—Cl	1.5	0.5
C—Br	1.4	0.3
C—I	1.2	0.0
C—O	0.7	1.0
C—N	0.2	0.5

The dipole moment of a molecule can be readily measured. However, the value is given as a product of two quantities (that is, charge and distance), so that dipole-moment measurements cannot be used to obtain bond lengths and bond angles. In general, charge separation is difficult if not impossible to estimate accurately. Yet the measurements are extremely useful in determining the *symmetry* of a molecule. An example is the $HgBr_2$ molecule, which can be either linear or bent. By convention, the symbol ↦ indicates that the flow of electron density is from the less electronegative to the more electronegative element. Keeping in mind that the dipole moment is a vector quantity, we find that the molecule is linear because its dipole moment is zero.

$$Br—Hg—Br \qquad Br \overset{Hg}{\diagup} \diagdown Br$$
$$\mu = 0 \qquad\qquad \mu \neq 0$$

The case just discussed shows that it is useful to consider the dipole moment of a molecule in terms of its *bond moments*. The total dipole moment of a molecule can be estimated by vector addition of the individual bond moments. To be sure, the bond moment for a particular bond will vary from molecule to molecule, but usually the variation is not great. In Table 16.8 are given the bond-moment values for some chemical bonds. The table shows a rough correlation between bond moment and difference in electronegativity between the two elements.

16.8 The Ionic Bond

When there is a large difference in electronegativity between atoms in a diatomic molecule, an electron may be completely transferred from one atom to the other. Consider the process

$$Na(g) + Cl(g) \longrightarrow Na^+(g) + Cl^-(g)$$

This process can be broken down into two steps:

$$Na(g) \longrightarrow Na^+(g) + e^-(g) \qquad 5.14 \text{ eV } (8.2 \times 10^{-19} \text{ J})$$

$$Cl(g) + e^-(g) \longrightarrow Cl^-(g) \qquad -3.65 \text{ eV } (-5.8 \times 10^{-19} \text{ J})$$

The net change in energy is $5.14 - 3.65 = 1.49$ eV (2.4×10^{-19} J). This means that the Na^+ and Cl^- ions at infinite separation have 1.49 eV more energy than the Na and Cl atoms at infinite separation.

The electrostatic interaction between two ions is given by *Coulomb's law*:

$$\text{force } (F) \propto \frac{Q_1 Q_2}{r^2}$$

$$= \frac{Q_1 Q_2}{\varepsilon r^2} \tag{16.14}$$

where Q_1 and Q_2 are the electric charges of ions, r the distance of separation, and ε (a dimensionless quantity) the dielectric constant of the medium. If Q_1 and Q_2 are the same kinds of charges (from two anions or two cations), F is positive, indicating a repulsive force. On the other hand, interaction between a positive and a negative charge is attractive and F becomes a negative quantity.

The energy (E) of this electrostatic interaction is obtained by computing the electrical work done in bringing charge Q_1 from an infinite distance to distance r from another charge Q_2. Thus

$$E = \text{force} \times \text{distance}$$

$$= -\int_{\infty}^{r} \frac{Q_1 Q_2}{\varepsilon r^2} \, dr$$

$$= \frac{Q_1 Q_2}{\varepsilon r} \tag{16.15}$$

The negative sign ensures that the final expression, $Q_1 Q_2 / \varepsilon r$, can be interpreted in a manner similar to Eq. (16.14).

EXAMPLE 16.1 Calculate the electrostatic energy of interaction of an isolated pair of Na^+ and Cl^- ions separated by 2 Å in air.

We assume that Na^+ and Cl^- are point charges.

CGS–esu units

$$e = 4.80 \times 10^{-10} \text{ esu}$$

$$r = 2 \text{ Å} = 2 \times 10^{-8} \text{ cm}$$

$$\varepsilon_{air} \simeq 1$$

From Eq. (16.15),

$$E = \frac{(4.80 \times 10^{-10} \text{ esu})(-4.80 \times 10^{-10} \text{ esu})}{1 \times (2 \times 10^{-8} \text{ cm})}$$

$$= -1.15 \times 10^{-11} \text{ erg} = -1.15 \times 10^{-18} \text{ J}$$

since 1 esu = 1 $cm^{3/2}$ $g^{1/2}$ s^{-1} (see Appendix 9.1).

SI units

$$e = 1.602 \times 10^{-19} \text{ C}$$

$$r = 2 \text{ Å} = 2 \times 10^{-10} \text{ m}$$

$$\varepsilon_{air} \simeq 1$$

Equation (16.15) now takes the form (see Appendix 9.1)

$$E = \frac{Q_1 Q_2}{\varepsilon r} \frac{1}{4\pi\varepsilon_0}$$

Hence

$$E = \frac{(1.602 \times 10^{-19} \text{ C})(-1.602 \times 10^{-19} \text{ C})}{1 \times (2 \times 10^{-10} \text{ m})}$$

$$\times \frac{1}{4\pi \times 8.8542 \times 10^{-12} \text{ C}^2 \text{ N}^{-1} \text{ m}^{-2}}$$

$$= -1.15 \times 10^{-18} \text{ J}$$

The negative sign indicates that this is an attractive interaction. Heat is given off when Na^+ and Cl^- ions are brought from infinite separation to 2 Å apart. However, it should not be concluded that the force between Na^+ and Cl^- (or any two oppositely charged ions) is always attractive. As these two ions approach each other to within very short distances ($\lesssim 1$ Å), the repulsive force between electrons and between nuclei becomes dominant.

The situation is more complicated in NaCl and other ionic crystals, for we must now consider *all* the attractive and repulsive interactions between nearest neighbors, next nearest neighbors, and so on. However, if the cell constant and hence the interionic distances can be determined by the X-ray diffraction technique, the total energy per mole for the ions can be calculated with considerable accuracy.*

16.9 Hybridization of Atomic Orbitals

The study of polyatomic molecules involves the problem of accounting for their geometry. One widely used approach to this study is based on the concept of hybridization of atomic orbitals.

Let us consider the carbon atom whose electronic configuration is $1s^2 2s^2 2p^2$. We might reasonably expect that carbon should be divalent, since it contains two unpaired electrons. Indeed, the molecule methylene (or carbene) CH_2 is known although it is a highly reactive species. The stable carbon compounds are best represented by three types of molecules: methane, ethylene, and acetylene.

METHANE

Methane (CH_4) is the simplest hydrocarbon. Both physical and chemical studies show that all four C—H bonds are identical in length and strength,

* See "Ionic Bonding in Solids," J. E. House, Jr., *Chemistry* **43**(2), 18 (1970).

and the molecule has a tetrahedral symmetry. The angle between any pair of C—H bonds is 109°28′. How can we explain the tetravalent state for carbon? A carbon atom in its ground state obviously would not do. Promoting a $2s$ electron into the empty $2p$ orbital would result in four unpaired electrons $(1s^2 2s^1 2p^3)$, which can then form four C—H bonds. However, if this did take place, methane would contain three C—H bonds of one type and a fourth C—H bond of a different type. This configuration is contrary to experimental evidence. The fact that all four bonds are identical suggests that the bonding atomic orbitals of carbon are all equivalent, meaning there is a mixing of the s and p orbitals in such a way that hybridized or *hybrid* orbitals are obtained. Because there are one s and three p orbitals, this process is called sp^3 hybridization. Figure 16.11 shows the changes in energy that take place in the process of hybridization. The state labeled with $2s^1 2p^3$ is real and can be detected by spectroscopic means. The valence state, that is, the state in which the four equivalent hybrids are formed, is not real in the sense that it does not exist for an isolated carbon atom. It is only convenient for us to imagine such a state just prior to the formation of the methane molecule. As Figure 16.11 shows, extra energy is needed to reach this state or to hybridize the atomic orbitals, but the investment is more than compensated for by the release of energy as a result of bond formations.

The mixing of the atomic orbitals to form four hybrid orbitals t_1, t_2, t_3, and t_4 can be represented by

$$t_1 = \tfrac{1}{2}(s + p_x + p_y + p_z)$$
$$t_2 = \tfrac{1}{2}(s + p_x - p_y - p_z)$$
$$t_3 = \tfrac{1}{2}(s - p_x + p_y - p_z) \tag{16.16}$$
$$t_4 = \tfrac{1}{2}(s - p_x - p_y + p_z)$$

where s and p_x, p_y, p_z represent the carbon $2s$ and $2p$ orbitals, and the factor $\tfrac{1}{2}$ is the normalization constant. Each sp^3 hybrid orbital has the shape shown in Figure 16.12; its direction is determined by the relative signs in Eq. (16.16). A C—H σ bond can then be formed by the overlap between a hybrid orbital and a hydrogen s orbital (Figure 16.12b). It is important not to confuse Eq. (16.16) with the LCAO treatment discussed earlier. In the present case, the hybrid orbitals are still atomic orbitals, since they all arise from the same carbon atom.

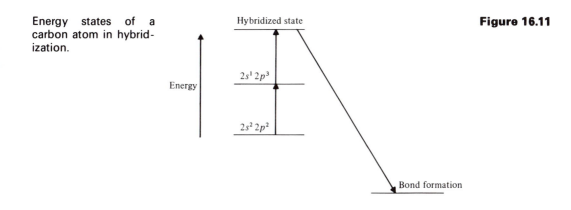

Energy states of a carbon atom in hybridization.

Hybridized state

Energy

$2s^1 2p^3$

$2s^2 2p^2$

Bond formation

Figure 16.11

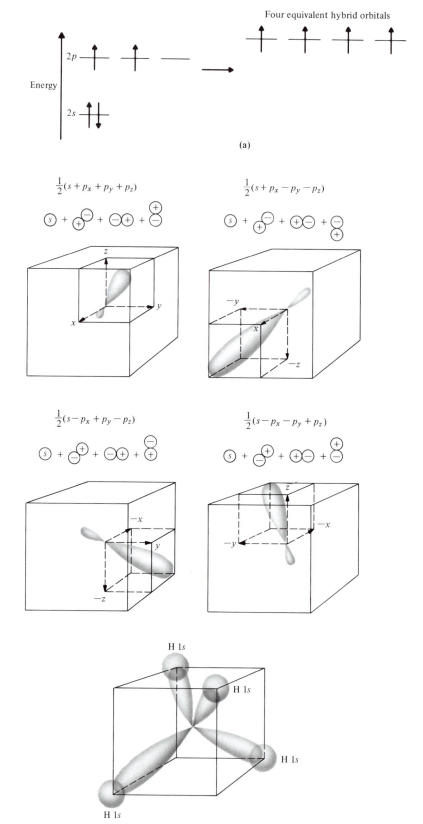

Four equivalent hybrid orbitals

$2p$

Energy

$2s$

(a)

$\frac{1}{2}(s + p_x + p_y + p_z)$

$\frac{1}{2}(s + p_x - p_y - p_z)$

$\frac{1}{2}(s - p_x + p_y - p_z)$

$\frac{1}{2}(s - p_x - p_y + p_z)$

H 1s

H 1s

H 1s

H 1s

(b)

ETHYLENE

Ethylene (C_2H_4) is a planar molecule; the angle between two C—H bonds is 120°. In contrast to methane, each carbon atom is bonded only to three atoms. Both the geometry and bonding can be understood if we assume that each carbon atom is sp^2-hybridized. Consider Figure 16.13. By mixing the s electron with only two p electrons (say p_x and p_y), we obtain three sp^2 hybrid orbitals (which all lie in the same plane), plus a pure p_z orbital. These three hybrid orbitals are then used to form two σ bonds with the two hydrogen atoms and one σ bond with the other carbon atom. The p_z orbitals on the two

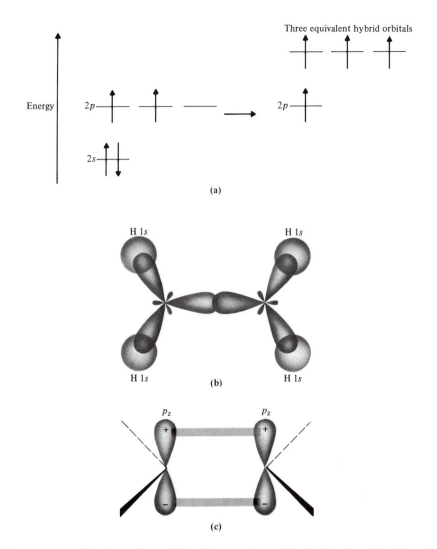

(a) Arrangement of carbon 2s and 2p electrons in sp^2 hybridization. (b) The **Figure 16.13**
three σ bonds between carbon and hydrogen and carbon and carbon atoms in
C_2H_4. (c) π bonds formed between the two p_z orbitals.

carbon atoms can also overlap to form a π bond. These three hybrid orbitals are represented by

$$t_1 = \sqrt{\tfrac{1}{3}}\,s + \sqrt{\tfrac{2}{3}}\,p_x$$
$$t_2 = \sqrt{\tfrac{1}{3}}\,s - \sqrt{\tfrac{1}{6}}\,p_x + \sqrt{\tfrac{1}{2}}\,p_y \qquad (16.17)$$
$$t_3 = \sqrt{\tfrac{1}{3}}\,s - \sqrt{\tfrac{1}{6}}\,p_x - \sqrt{\tfrac{1}{2}}\,p_y$$

ACETYLENE

Acetylene (C_2H_2) is a linear molecule. From Figure 16.14a we see that by mixing the $2s$ electron with only one p orbital (say p_x), we obtain two sp hybrid orbitals plus two pure p orbitals. Consequently, each carbon atom forms two σ bonds (one with hydrogen atom and one with the other carbon atom) and two π bonds (both with the other carbon atom), as shown in Figure 16.14b and c. The two hybrid orbitals are

$$t_1 = \sqrt{\tfrac{1}{2}}\,s + \sqrt{\tfrac{1}{2}}\,p_x$$
$$t_2 = \sqrt{\tfrac{1}{2}}\,s - \sqrt{\tfrac{1}{2}}\,p_x \qquad (16.18)$$

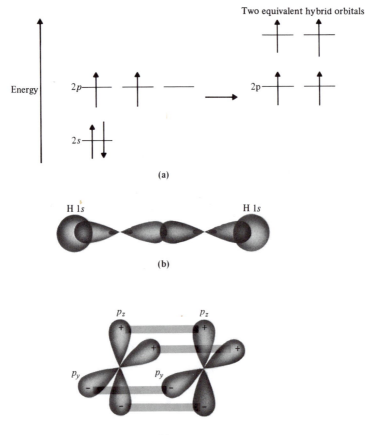

Figure 16.14   (a) Arrangement of carbon $2s$ and $2p$ electrons in sp hybridization. (b) The two σ bonds between the carbon and hydrogen and carbon and carbon atoms in C_2H_2. (c) The two π bonds formed between the two p_y and two p_z orbitals.

Although hybridization has so far been dealt with mathematically in terms of the mixing of atomic orbitals of the same atom, a physical interpretation is also possible. As the hydrogen atoms approach the carbon atom, electrostatic interactions between electrons and between electrons and nuclei cause the s and p orbitals of the carbon atom to become distorted. Consequently, we no longer have pure s and p orbitals, and each atomic orbital is more correctly described as partly resembling an s orbital and partly resembling a p orbital. For example, in methane the three p orbitals are distorted to a different extent from the s orbital, but their final shapes are all identical.

The concept of hybridization applies equally well for other elements. In ammonia, for example, each N—H bond points to the apex of a slightly irregular tetrahedron; the angle between any two N—H bonds is 107°20′. Since the electronic configuration of nitrogen is $1s^2 2s^2 2p^3$, the bonding in NH_3 might be explained by assuming an overlap between the three p orbitals and the hydrogen $1s$ orbitals. In this case, the bond angles would be 90°, because the p orbitals are all mutually perpendicular. It may also be argued that nitrogen is more electronegative than hydrogen, so some charge separation may occur, resulting in a slightly negative nitrogen atom and slightly positive hydrogen atoms. Repulsion between the hydrogen atoms will then increase the bond angles. Although this repulsion undoubtedly occurs, the effect is too small to account for the observed angle. The assumption that nitrogen is sp^3-hybridized in ammonia is more valid. One of the hybrid orbitals contains a pair of nonbonding electrons called a *lone pair*. Repulsion between the lone-pair electrons and those in the bonding orbitals causes a

Structure of (a) ammonia, and (b) water. In both cases the four sp^3 orbitals form a slightly distorted tetrahedron. Ammonia has one lone pair; water has two lone pairs.

Figure 16.15

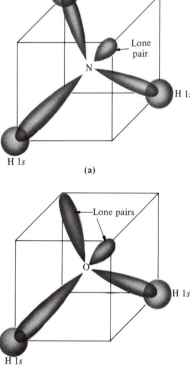

Table 16.9

Covalent Radii for Atoms (Å)[a]

	Single-Bond Radius	Double-Bond Radius	Triple-Bond Radius
H	0.37		
C	0.772	0.667	0.603
N	0.70	0.62	0.55
O	0.66	0.62	
F	0.64		
Si	1.17	1.07	1.00
P	1.10	1.00	0.93
S	1.04	0.94	0.87
Cl	0.99	0.89	
Ge	1.22	1.12	
As	1.21	1.11	
Se	1.17	1.07	
Br	1.14	1.04	
Sn	1.40	1.30	
Sb	1.41	1.31	
Te	1.37	1.27	
I	1.33	1.23	

[a] Reprinted from Linus Pauling, *The Nature of the Chemical Bond*, 3rd ed., p. 224. Copyright 1939 and 1940, third edition © 1960, by Cornell University. Used by permission of Cornell University Press.

decrease in the bond angle from 109°28′ to 107°20′. This lone pair is also responsible for the basicity of ammonia (Figure 16.15). When ammonia is dissolved in water, it readily accepts a proton, to form the ammonium ion, which possesses perfect tetrahedral symmetry.

The water molecule is another example of hybridization. Again, the geometry is best explained by assuming the O atom to be sp^3-hybridized. There are two lone pairs in this case and the bond angle is only 104°31′.

So far we have considered hybridization only in terms of the s and p orbitals. The participation of d orbitals is also possible for the third-period elements and beyond. We shall return to this point in Section 16.11.

Finally, we note that the distance between atoms A and B forming a covalent bond remains fairly constant in molecules. The bond length can be expressed as a sum $r_A + r_B$, where r_A and r_B are the *covalent radii* of A and B. For example, the C—C bond distance in many compounds is about 1.54 Å, so that the covalent radius for a carbon single bond is taken to be $1.54/2 = 0.77$ Å. Now the C—Cl distance in many compounds is about 1.76 Å; therefore, it follows that the covalent radius for Cl is $(1.76 - 0.77) = 0.99$ Å. In this way we can obtain the covalent radius for many atoms shown in Table 16.9.

16.10 Resonance and Electron Delocalization

The advantage of using hybridization is that, in addition to being able to explain the geometry of molecules, it allows us to continue thinking of a chemical bond in terms of pairing of electrons. However, often it is found that the properties of a molecule cannot be completely represented by a single structure. A case in point is the nitrate ion shown in Figure 16.16a. The ion

(a) Lewis structure for the nitrate ion. (b) Resonance among the three equivalent Lewis structures.

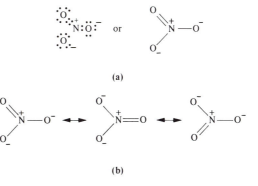

Figure 16.16

(a)

(b)

has a planar structure and the O—N—O bond angle is 120°. This can be readily accounted for if we assume that the nitrogen atom is sp^2-hybridized. Because experimental studies indicate that all three N—O bonds have equal length and equal strength, the structure shown in Figure 16.16a is inadequate to describe the ion. The assignment of the N=O double bond is completely arbitrary, so we must include the other two structures shown in Figure 16.16b. We can imagine that the molecule *resonates* from one structure to another. The point is that the nitrate ion cannot be represented by just one of the *resonance* structures; it must be represented by the three together. Then it is clear that the character of each N—O bond is somewhere between a single and a double bond, in accord with experimental observations. There is no evidence whatsoever that these three structures actually oscillate back and forth, for no one has been able to trap any one of the three structures for study.* The model above merely enables us to solve the dilemma.

The concept of resonance has its roots in *valence bond theory*, which, together with molecular orbital theory, are the two major methods presently employed in the study of chemical bonding. In valence bond theory, the wave function ψ for the nitrate ion is given by

$$\psi = c_A \psi_A + c_B \psi_B + c_C \psi_C \qquad (16.19)$$

where ψ_A, ψ_B, and ψ_C are the wave functions for the three individual resonance structures and c_A, c_B, and c_C are the coefficients that determine the weight or importance of resonance structures. Here we have three equivalent structures so that c_A, c_B, and c_C are all equal.

The concept of resonance is most often applied to aromatic hydrocarbons. In 1865, Kekulé first proposed the ring structure for benzene. Since then, a great deal of progress has been made in the study of these molecules. As in the nitrate ion case, the measured C—C distance in benzene is 1.40 Å, which is somewhere between the single C—C bond (1.54 Å) and C=C bond (1.33 Å). It is more realistic to describe the resonance between two Kekulé structures as follows:

---

* J. D. Roberts of the California Institute of Technology has provided an interesting analogy for resonance. A medieval European traveler returns home from a journey to Africa and describes a rhinoceros as a cross between a griffin and a unicorn. Thus a real animal is described in terms of two familiar (in concept) but imaginary animals. Similarly, the real chemical species, the nitrate ion, is described in terms of the three familiar but nonexistent resonance structures shown in Figure 16.16.

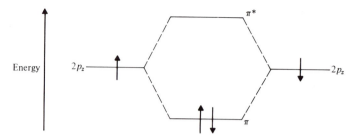

Figure 16.17 Bonding π and antibonding π^* molecular orbitals in ethylene.

Other resonance structures, such as the planar Dewar structures,

may also be included, although they have considerably less importance because of their energetically unfavorable long bonds.

The molecular orbital theory offers an alternative approach to explain the properties discussed above. Instead of the resonance concept, we consider the delocalization of electrons in the molecular orbitals.

First consider the ethylene molecule. An overlap of the two p_z orbitals gives rise to two molecular orbitals—one bonding and one antibonding (Figure 16.17). The two electrons are placed in a bonding molecular orbital which extends over the two carbon atoms. The σ molecular orbitals lie lower on the energy scale as a result of their greater stability. Similarly, for the benzene molecule, the overlap of the six p_z orbitals gives rise to six molecular orbitals, of which three are bonding and three antibonding (Figure 16.18). A pair of electrons is assigned to each of the molecular orbitals. These electrons are free to move within the boundaries denoted by the dashed lines (see Figure 16.18b). For example, the lowest bonding molecule orbital is formed by taking the linear combination of the six p_z orbitals ($\psi_1 + \psi_2 + \psi_3 + \psi_4 + \psi_5 + \psi_6$).* In this molecular orbital the electrons are *delocalized* over the entire molecule. The extent of electron delocalization decreases (and the number of nodes increases) as the energy of the molecular orbitals increases, a situation analogous to the particle-in-a-box case discussed in Chapter 15.

Both the molecular orbital theory and valence bond theory are useful methods for studying chemical bonding; each has its strong and weak points. The former tends to emphasize electron delocalization; the latter emphasizes the electron-pair bond. The concept of resonance is a useful idea, although one needs considerable chemical intuition when dealing with large molecules.† The method of hybridization appears to fit the valence bond theory, dealing as it does with the electron-pair concept rather than electron delocalization. In fact, hybridization can be equally well accounted for by the molecular orbital theory, for hybrid orbitals can also be used to construct molecular orbitals.

---

* Higher molecular orbitals are formed by taking different linear combinations of the p_z orbitals.
† For example, about 450 resonance structures have to be considered for a molecule the size of anthracene ($C_{14}H_{10}$).

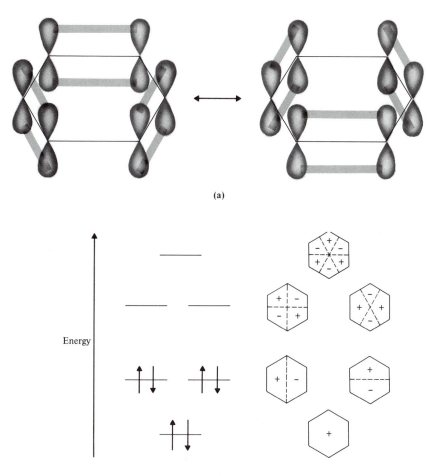

(a)

(b)

(a) The two Kekulé structures represented by the overlap between the p_z orbitals. (b) The three bonding and three antibonding molecular orbitals in benzene together with the π electron wave functions (top view). The dashed lines represent the nodal planes.

Figure 16.18

Energy

Transition-Metal Complexes   **16.11**

The complexes of the transition-metal elements are characterized by their color, paramagnetism, stereochemistry, and other qualities. Because of their importance in biological systems, we shall examine the bonding orbitals involved and discuss a few examples.

Transition elements contain incompletely filled d orbitals either in their neutral or ionic states. As mentioned earlier, when the principal quantum number n is equal to 3, the azimuthal quantum number can be 2, 1, or 0, which correspond to the $3d$, $3p$, and $3s$ orbitals. For $l = 2$, we have $m_l = 0$, ± 1, and ± 2, giving rise to five equivalent d orbitals shown in Figure 16.19. Orbitals d_{xy}, d_{yz}, and d_{xz}, and $d_{x^2-y^2}$ are clearly similar except for their orientations, but the d_{z^2} orbital appears quite different. The reason for this difference is that five, independent atomic wave functions with the same shape are difficult to

473

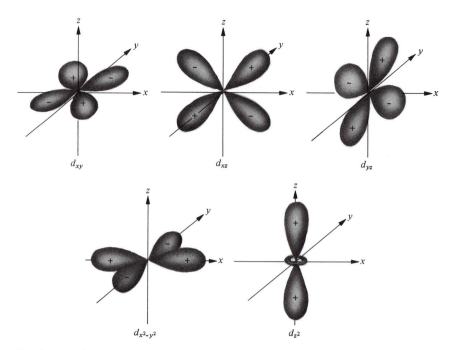

Figure 16.19 The five d orbitals.

express graphically. It is more convenient to show only independent wave functions at the expense of having a differently shaped orbital, d_{z^2}.

Table 16.10 gives the ground-state electronic configuration for the first row of the transition elements. At the beginning of the series, the $4s$ electrons are more stable than the $3d$ electrons, but the trend reverses as we move to the right. Two configurations, d^5 and d^{10}, are preferred because of their half-filled and completely filled arrangements, which result in extra stability. Qualitatively, this can be understood as follows. Electrons in the same subshell have equivalent spatial distributions; consequently, their shielding of one another is relatively small. Thus the effective nuclear charge increases as the actual nuclear charge increases. This is the reason that a completely filled subshell

Table 16.10				**Electronic Configuration and Other Properties of the First-Row Transition Metals**					
	Sc	Ti	V	Cr	Mn	Fe	Co	Ni	Cu
Electronic configuration									
M	$3d^14s^2$	$3d^24s^2$	$3d^34s^2$	$3d^54s^1$	$3d^54s^2$	$3d^64s^2$	$3d^74s^2$	$3d^84s^2$	$3d^{10}4s^1$
M^{2+}	—	$3d^2$	$3d^3$	$3d^4$	$3d^5$	$3d^6$	$3d^7$	$3d^8$	$3d^9$
M^{3+}	[Ne]	$3d^1$	$3d^2$	$3d^3$	$3d^4$	$3d^5$	$3d^6$	—	—
Ionization potential (eV)									
First	6.54	6.82	6.74	6.76	7.43	7.87	7.88	7.63	7.72
Second	12.80	13.57	14.65	16.49	15.64	16.18	17.05	18.15	20.29
Third	24.75	27.47	29.31	30.95	33.69	30.64	33.49	35.17	37.08
Ionic radii (Å)									
M	1.44	1.36	1.22	1.17	1.17	1.16	1.16	1.15	1.17
M^{2+}	—	0.90	0.88	0.85	0.80	0.77	0.75	0.72	0.72
M^{3+}	0.81	0.77	0.74	0.68	0.66	0.63	0.64	—	—

(d^{10}) has high stability. For a half-filled subshell (d^5), each electron goes into a different d orbital. This reduces the electrostatic repulsion between the electrons, since the electrons are as far apart from one another as possible (Hund's rule). Extra stability is again achieved.

The following three theories are useful in studying the bonding of transition-metal ions.

CRYSTAL FIELD THEORY

In an isolated atom or ion, all five d orbitals have the same energy irrespective of their orientations. This is no longer the case when the atom or ion is surrounded by ligands. The perturbations exerted by ligands can greatly affect the energy of the d orbitals. Bethe (1929) and van Vleck (1935) suggested that the metal–ligand interaction can be treated as purely electrostatic by thinking of all species as point charges. One simple case is an atom with six identical ligands arranged along the x, y, and z axes. As a result of this arrangement, the five d orbitals no longer possess the same energy, since the electrostatic field created by the ligands repels the electrons in the $d_{x^2-y^2}$ and d_{z^2} orbitals to a greater extent than those in the d_{xy}, d_{yz}, and d_{xz} orbitals. The orbitals will be split according to Figure 16.20. The symbol Δ, called *crystal field splitting*, measures the magnitude of this electrostatic interaction. The splitting of d orbitals for other arrangements can be interpreted in a similar way.

Spectroscopic technique provides the most direct means for measuring the quantity Δ. For example, the electronic absorption spectrum of $Ti(H_2O)_6^{3+}$ (with only one d electron) has a maximum at about 498 nm, corresponding to an energy of 244 kJ mol^{-1} for Δ. By using a number of different ligands, we can establish a *spectrochemical series* in the decreasing order of Δ:

$$CO > CN^- > NH_3 > CNS^- > H_2O > F^- > OH^- > Cl^- > Br^- > I^-$$

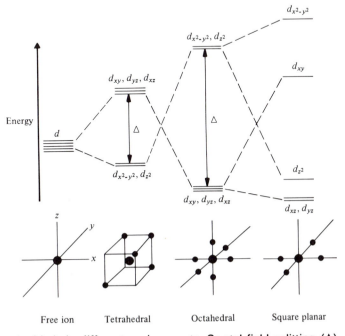

Splitting of the five d orbitals in different environments. Crystal field splitting (Δ) **Figure 16.20** is shown for octahedral and tetrahedral complexes.

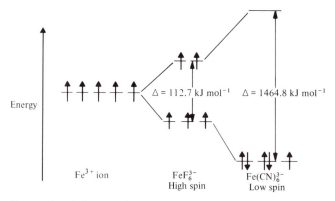

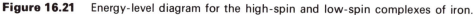

Figure 16.21   Energy-level diagram for the high-spin and low-spin complexes of iron.

The magnitude of Δ can also determine the magnetic properties of a complex. For $Ti(H_2O)_6^{3+}$, the d electron must be in one of the three lower orbitals. However, if there are several d electrons present, as in the Fe^{3+} complexes, two choices can be made. Consider the octahedral complexes FeF_6^{3-} and $Fe(CN)_6^{3-}$. Since F^- has a small Δ or is a weak-field ligand, the five d electrons will enter the five separate d orbitals with their spins parallel according to Hund's rule. This gives rise to a high-spin complex. On the other hand, the cyanide group is a ligand with a very strong field, so energetically it is more preferable to have all five electrons in the lower orbitals and a low-spin complex is formed (Figure 16.21). The actual number of unpaired electrons (or spins) can be obtained by *magnetic susceptibility* measurements, and the good agreement between theory and experiment in general supports the correctness of this scheme.

LIGAND FIELD THEORY

Although crystal field theory is conceptually easy to understand and provides an explanation for the spectral and magnetic properties of some complexes, it has a serious defect. By considering only the electrostatic interaction, it com-

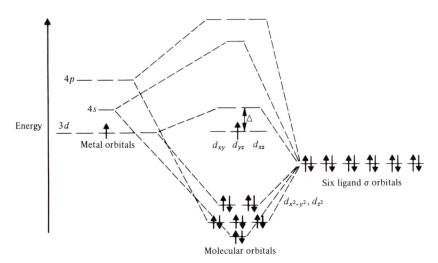

Figure 16.22   Molecular orbitals in $Ti(H_2O)_6^{3+}$.

pletely neglects the covalent character of many metal–ligand bonds. A more satisfactory treatment, which takes this into consideration, is ligand field theory, which is largely based on the molecular orbital approach. The theory can be illustrated by again considering the $Ti(H_2O)_6^{3+}$ complex. Figure 16.22 shows the interaction between the metal and ligand orbitals. Each H_2O ligand provides two electrons from one lone pair on the oxygen atom, amounting to 12 electrons in the six σ-bonding molecular orbitals. The extra electron is placed in one of the three d orbitals, which are not properly oriented for bond formation. These orbitals are appropriately called the *non-bonding orbitals*. The same diagram can be used to describe bonding in other complexes. Thus, for FeF_6^{3-}, we simply add the five d electrons to the three nonbonding orbitals and two lowest antibonding orbitals to give a high-spin complex. Some ligands, such as CO and CN^-, are capable of forming both π bonds and σ bonds with the metal ion. We shall discuss such bonding schemes in Section 16.12.

VALENCE BOND THEORY

A third approach assumes that each ligand donates a pair of electrons to the metal ion to form a *coordinate covalent bond*. In an octahedral complex, the metal ion has six vacant orbitals available for bond formations. We can easily visualize the configuration by assuming the metal ion to be hybridized. In our earlier discussion of hybridization the d orbitals were not considered although their participation was expected for the third-period elements and beyond. The criterion is that the d orbitals must lie close to the s and p orbitals on the energy scale. Thus the trigonal bipyramid shape of PCl_5 can be accounted for if we assume that the P atom is sp^3d-hybridized. The electronic configuration of phosphorus is $[Ne]3s^23p^3$. Promoting an s electron into a low-lying vacant $3d$ orbital followed by mixing generates five hybrid orbitals. Table 16.11 summarizes different geometries in terms of hybridizations.

In octahedral complexes such as $Fe(CN)_6^{3-}$, the metal ion is d^2sp^3-hybridized. This process involves the $3d_{x^2-y^2}$, $3d_{z^2}$, $4s$, and the three $4p$ orbitals. Upon accepting 12 electrons from the six cyanide groups, there are a total of 17 electrons on the metal ion, whose electronic configuration is now $3d^{10}4s^24p^5$. After hybridization, we have 12 electrons in the six hybrid orbitals and 5 more in the unhybridized d_{xy}, d_{xz}, and d_{yz} orbitals. We are not concerned with how the process actually takes place. The important point is that

Hybridization and Geometry of Molecules

Table 16.11

Hybridization of Central Atom	Shape	Bond Angle	Example
sp	Linear	180°	HCN, C_2H_2
sp^2	Planar	120°	BF_3, C_2H_4
sp^3	Tetrahedral	109°28′	CH_4, NH_4^+
dsp^2	Square planar	90°	$Ni(CN)_4^{2-}$, $PtCl_4^{2-}$
dsp^3	Trigonal bipyramid	90° (apical-equatorial)	
		120° (equatorial-equatorial)	PCl_5
d^2sp^3	Octahedral	90°	$Ti(H_2O)_6^{3+}$, SF_6

we are able to account for the geometry as well as many other properties of transition–metal complexes in terms of hybridization.

For the purposes of this book, and in general, we need only be concerned with ligand field treatment, for it has proved far superior to both crystal field theory and valence bond theory. However, the idea of relating the geometry of molecules to hybridization is still very useful, so it will be retained.

16.12 Metal Ions in Biological Systems

It is impossible to cover adequately the role of various metal ions in biological systems in this section. Instead, we shall discuss only a few transition-metal ions and examine their structural aspects in proteins and other biomolecules. Because of the iron's importance to all living organisms, we shall devote most of our discussion to the bioinorganic chemistry of iron.

It is interesting, although not surprising from the evolutionary point of view, that in proteins such as myoglobin, hemoglobin, and cytochromes, and in enzymes such as catalase and peroxidase, iron is situated at the center of a planar porphyrin system as shown in Figure 16.23. Like aromatic hydrocarbons, the porphyrin molecule is extensively delocalized. In both myoglobin and hemoglobin iron is in the $+2$ oxidation state. It forms four σ bonds with the nitrogen atoms in the porphyrin ring. This leaves us two more ligands to account for in an octahedral complex. The fifth ligand in these cases is provided by the histidine group, which is part of the protein chain. In the absence of oxygen, the sixth ligand is a water molecule, which binds the Fe^{2+} ion on the other side of the ring to complete the octahedral complex (Figure 16.24). In oxyhemoglobin, where a water molecule is replaced by molecular oxygen, three different orientations have been proposed for the O_2 molecule (Figure 16.25). Although the exact arrangement is still unknown, indirect evidence strongly suggests that the O_2 molecule assumes a bent configuration, as shown in Figure 16.25c. Although the end-on configuration shown in Figure 16.25b would seem more reasonable, since it allows for a greater extent of orbital overlap, steric hindrance due to neighboring side chains forces the O_2 molecule to tilt at an angle. The nature of binding is essentially the same in oxymyoglobin and oxyhemoglobin, yet their affinity for oxygen is drastically different. The cooperative effect, discussed in Chapter 14, explains this difference.

Beside water and oxygen, a number of other ligands, such as CO and NO, bind to the Fe^{2+} ion. However, CO binds more strongly than O_2 by some 50

Figure 16.23

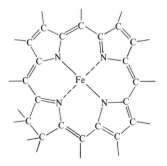

Iron porphyrin system.

The heme group in myoglobin and hemoglobin. In the absence of oxygen, the sixth ligand is water (W).

Figure 16.24

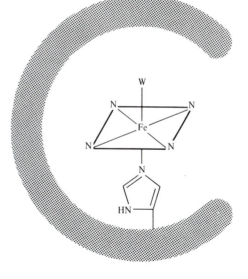

times in myoglobin and 200 times in hemoglobin. The enhanced affinity of CO for Fe^{2+} is the result of the $d_\pi-p_\pi$ interaction. The carbon atom in CO is sp-hybridized. One of the sp hybrid orbitals containing the lone-pair electrons overlaps with the d_{z^2} orbital of iron to form a σ bond. There is also a back donation of electron density from the d_{xz} orbital to an empty π^* orbital on CO (Figure 16.26). As a result of this interaction, the linkage between Fe^{2+} and CO has appreciable double-bond characteristic and is therefore harder to break. The strength of the Fe—C bond depends on the extent of the orbital overlap. It is believed that in carboxyhemoglobin the CO molecule also assumes a bent configuration as the O_2 molecule in oxyhemoglobin. This fact is particularly fortunate for modern man. Were it not true, that is, if the CO molecule should bind the Fe^{2+} ion in an end-on fashion, the CO affinity for Fe^{2+} in myoglobin and hemoglobin would be a thousand times or more greater than that of O_2. A short drive on the Los Angeles freeway during the rush hour, say, could prove fatal to most drivers! Although the cyanide ion is isoelectronic with carbon monoxide, the former does not complex strongly with the Fe^{2+} ion in hemoglobin. The toxicity of the cyanide ion lies in its

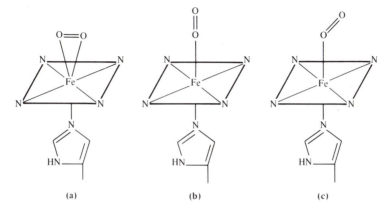

(a) (b) (c)

Three possible ways for oxygen to bind to the heme group in hemoglobin. **Figure 16.25**

Figure 16.26

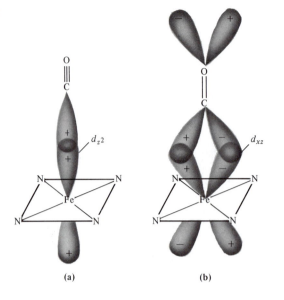

(a)   (b)

The formation of a σ bond (a) and a π bond (b) between the carbon monoxide ligand and Fe^{2+} ion in the heme group. Note that the extent of overlap in both (a) and (b) will be decreased if the CO molecule assumes a bent configuration like that shown in Figure 16.25 (c) for the O_2 molecule.

ability to attack cytochrome oxidase, a respiratory enzyme.* Table 16.12 gives the spin states for some important heme proteins, that is, proteins that contain the heme group.

Another example of a heme protein is cytochrome c. Differing from hemoglobin in that it carries electrons instead of oxygen, this compound is located on the electron transport chains in both photosynthetic and respiratory systems. Again, iron is at the center of the porphyrin ring, with histidine as the fifth ligand. However, in this case the sixth ligand is provided by a methionine segment (Figure 16.27). The iron–sulfur bond is strong enough to prevent the replacement of the methionine ligand by oxygen.

Rubredoxin and ferredoxin are examples of another type of iron protein that does not contain the heme group. These proteins are found in all green plants, algae, and photosynthetic bacteria. They also play an important role in nitrogen fixation. Like the cytochromes, the ferredoxins function as electron carriers. X-ray studies have revealed that the iron atom is tetrahedrally bonded to sulfur atoms. At present, the exact nature of the bonding orbitals involved is still not completely understood.

Spin States for Some Heme Proteins

Table 16.12

Heme Protein	Oxidation State for Fe	Sixth Ligand	Number of Unpaired Electrons	Spin State
Hemoglobin	$+2$	H_2O	4	High
Oxyhemoglobin	$+2$	O_2	0	Low
Myoglobin	$+2$	H_2O	4	High
Cyanohemoglobin	$+2$	CN^-	0	Low
Carboxyhemoglobin	$+2$	CO	0	Low
Methemoglobin	$+3$	F^-	5	High
Cyanomethemoglobin	$+3$	CN^-	1	Low

* For more discussion on the toxicity of CO versus CN^-, see R. Chang and L. E. Vickery, *J. Chem. Educ.* **51**, 800 (1974). Also see D. A. Labianca, *J. Chem. Educ.* **56**, 789 (1979).

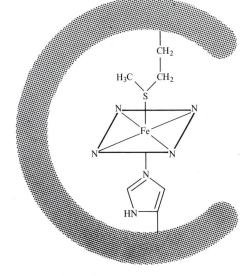

Despite their importance, relatively little is known about the structure and bonding of copper proteins such as cytochrome oxidase, an enzyme in the respiratory chains; hemocyanin, an oxygen-carrying protein; plastocyanin, an electron-carrying protein in photosynthesis; and tyrosinase, an enzyme that catalyzes the conversion of the amino acid tyrosine to melanin pigment in plants and animals. The electronic configuration of Cu^+ is $[Ar]3d^{10}$ so that it is not a transition-metal ion. Most of the Cu(I) complexes are colorless and tetrahedral in shape. Copper in the $+2$ oxidation state usually forms square planar and octahedral complexes that have green or blue color. For example, deoxyhemocyanin is a colorless Cu(I) species, which on combination with oxygen becomes blue, suggestive of Cu(II). There are two copper atoms per molecule of hemocyanin and the protein/oxygen binding ratio is 1 : 1.

Finally, we mention briefly two other biologically important transition metals: cobalt and manganese. Like copper, cobalt has two common oxidation states that exist in solution, cobaltous(II) and cobaltic(III). The biological activity of cobalt is largely confined to its role in the vitamin B_{12} series of coenzymes.* In vitamin B_{12} the cobaltic ion is situated at the center of a conjugated corrin ring that is similar to the porphyrin structure. The corrin system provides four nitrogen atoms whose lone pairs form sigma bonds with cobalt. The fifth and sixth ligands are benzimadazole and a cyanide ion. The cyanide complex itself does not function as a coenzyme. The cyanide ion can be replaced by a number of other ligands, many of which do possess biological activity.

Although there are many known oxidation states of manganese, only Mn(II) and Mn(III) are important in biological systems. Manganese in the $+2$ oxidation state usually forms high-spin octahedral complexes, whereas Mn^{3+} is unstable in aqueous solution unless it is complexed. Photosynthesis cannot occur in many higher plants, such as spinach, without the presence of manganese. There is evidence to suggest that the oxygen evolution in these plants (that is, the oxidation of water to molecular oxygen) is catalyzed by a Mn(III)–enzyme complex, although such a complex has never been isolated.

* A *coenzyme* is a substance that activates enzymes. Most coenzymes are vitamins.

Suggestions for Further Reading

INTRODUCTORY–INTERMEDIATE

COMPANION, A. L. *Chemical Bonding*, 2nd ed. McGraw-Hill Book Company, New York, 1979.
An elementary and descriptive text.

COULSON, C. A. *The Shape and Structure of Molecules*. Clarendon Press, Oxford, 1973.
A very readable introductory text.

COULSON, C. A. *Valence*. Oxford University Press, Inc., New York, 1961.
Although somewhat out of date, it is still one of the best texts on the qualitative aspects of chemical bonding.

GRAY, H. B. *Electrons and Chemical Bonding*. W. A. Benjamin, Inc., New York, 1965.
A good introductory text. Main emphasis is on molecular orbital theory.

HOCHSTRASSER, R. M. *Behavior of Electrons in Atoms*. W. A. Benjamin, Inc., New York, 1965.
A useful introductory text.

WILLIAMS, D. R. *The Metals of Life*. Van Nostrand Reinhold Company, New York, 1971.
Describes the roles of many metal ions in the living system.

INTERMEDIATE–ADVANCED

COTTON, F. A., and G. WILKINSON. *Advanced Inorganic Chemistry*, 3rd ed. Wiley-Interscience, New York, 1972.
An authoritative and comprehensive text.

HANZLIK, R. P. *Inorganic Aspects of Biological and Organic Chemistry*. Academic Press, Inc., New York, 1976.

HUGHES, M. N. *The Inorganic Chemistry of Biological Processes*. John Wiley & Sons, Inc., New York, 1972.
Provides a good survey of the occurrence and role of metal ions in biological systems.

HUHEEY, J. E. *Inorganic Chemistry*. Harper & Row, Publishers, New York, 1978.
A comprehensive text that has a nice chapter on bioinorganic chemistry.

OCHIAI, E. I. *Bioinorganic Chemistry: An Introduction*. Allyn and Bacon, Inc., Boston, 1977.
Another very readable text.

Reading Assignments

SECTION A

"The Stability of the Hydrogen Atom," F. Rioux, *J. Chem. Educ.* **50**, 550 (1973).

"Ionic Bonding in Solids," J. E. House, Jr., *Chemistry* **43**(2), 18 (1970).

"Size and Shape of a Molecule," M. J. Demchik and V. C. Demchik, *J. Chem. Educ.* **48**, 770 (1971).

"The Molecular Orbital Theory," R. Ferreira, *Chemistry* **41**(6), 8 (1968).

"What Is Bond Polarity and What Difference Does It Make?" R. T. Sanderson, *Chemistry* **46**(8), 12 (1973).

"The Shape of Organic Molecules," J. B. Lambert, *Sci. Am.*, Jan. 1970.

"Chemistry by Computer," A. C. Wahl, *Sci. Am.*, Apr. 1970.

"Ligand Field Theory," F. A. Cotton, *J. Chem. Educ.* **41**, 466 (1964).

"Molecular Orbital Theory for Transition Metal Complexes," H. B. Gray, *J. Chem. Educ.* **41**, 2 (1964).

"Bioinorganic Chemistry," G. W. Rayner Canham and A. B. P. Lever, *J. Chem. Educ.* **49**, 656 (1972).

"The Role of Chelation in Iron Metabolism," P. Saltman, *J. Chem. Educ.* **42**, 682 (1965).

"The Role of Metal Ions in Proteins and Other Biological Molecules," E. W. Ainscough and A. M. Brodie, *J. Chem. Educ.* **53**, 156 (1976).

"Environmental Bioinorganic Chemistry," E. I. Ochiai, *J. Chem. Educ.* **51**, 235 (1974).

"Principles in Bioinorganic Chemistry," E. I. Ochiai, *J. Chem. Educ.* **55**, 631 (1978).

"Inorganic Elements in Biology and Medicine," R. J. P. Williams, *Chem. Brit.* **15**, 506 (1979).

"Modeling Coordination Sites in Metallobiomolecules," J. A. Ibers and R. H. Holm, *Science* **209**, 223 (1980).

SECTION B

"The Five Equivalent *d* Orbitals," R. E. Powell, *J. Chem. Educ.* **45**, 45 (1968).

"Five Equivalent Orbitals," L. Pauling and V. McClure, *J. Chem. Educ.* **47**, 15 (1970).

"A Simple, Quantitative Molecular Orbital Theory," W. F. Cooper, G. A. Clark, and C. R. Hare, *J. Chem. Educ.* **48**, 247 (1971).

"Molecular Orbitals and Air Pollution," B. M. Fung, *J. Chem. Educ.* **49**, 26 (1972). Also see p. 654 of the same volume.

"Carbon Monoxide Complexes of Uncharged Metals," J. E. House, Jr., *Chemistry* **48**(5), 12 (1975).

"Metals, Models, Mechanisms, Microbes, and Medicine," H. A. O. Hill, *Chem. Brit.* **12**, 119 (1976).

"Kekulé and Benzene," C. A. Russell, *Chem. Brit.* **1**, 141 (1965).

"Molecular Oxygen Adducts of Transition Metal Complexes: Structure and Mechanism," L. Klevan, J. Peone, Jr., and S. K. Madan, *J. Chem. Educ.* **50**, 670 (1973).

"Hemoglobin: Model Systems Shed Light on Oxygen Binding," T. H. Maugh, II, *Science* **187**, 154 (1975).

"The Chemical Elements of Life," E. Frieden, *Sci. Am.*, June 1972.

"Chemical Toxicology: Part II. Metal Toxicity," D. E. Carter and Q. Fernando, *J. Chem. Educ.* **56**, 491 (1979).

"Lead Poisoning," J. J. Chisolm, Jr., *Sci. Am.*, Feb. 1971.

"Vanadium in the Living World," N. M. Senozan, *J. Chem. Educ.* **51**, 503 (1974).

"Mercury Poisoning," L. E. Strong, *J. Chem. Educ.* **49**, 28 (1972).

"Chelation in Medicine," J. Schubert, *Sci. Am.*, May 1966.

"The Biochemistry of Copper," E. Frieden, *Sci. Am.*, May 1968.

"Iron–Sulfur Coordination Compounds and Proteins," S. J. Lippard, *Acc. Chem. Res.* **6**, 282 (1973).

"Inorganic Oxygen Carriers as Models for Biological Systems," G. McLendon and A. E. Martell, *Coord. Chem. Rev.* **19**, 1 (1976).

"Iron and Susceptibility to Infectious Disease," E. D. Weinberg, *Science* **184**, 952 (1974).

"Therapeutic Chelating Agents," M. M. Jones and T. H. Pratt, *J. Chem. Educ.* **53**, 342 (1976).

"Biochemical Effects of Excited State Molecular Oxygen," J. Bland, *J. Chem. Educ.* **53**, 274 (1976).

"The Cytochromes of Higher Plants and Algae," D. B. Knaff, *Coord. Chem. Rev.* **26**, 47 (1978).

"Hemocyanin: The Copper Blood," N. M. Senozan, *J. Chem. Educ.* **53**, 684 (1977).

"Some Aspects of Bioinorganic Chemistry of Molybdenum," K. B. Swedo and J. H. Enemark, *J. Chem. Educ.* **56**, 70 (1979).

"Cytochrome c and the Evolution of Energy Metabolism," R. E. Dickerson, *Sci. Am.*, March 1980.

Problems   **16.1**   Name two important differences between an s and a p orbital.

16.2   Write down the Schrödinger wave equation for the helium atom in Cartesian coordinates.

16.3   Write down the ground-state electronic configuration for the following ions and determine the number of unpaired electrons in each case: Ti^{3+}, Mn^{3+}, Co^{2+}, Zn^{2+}, Cu^{3+}, Eu^{2+}, Gd^{3+}, and Pb^{2+}.

16.4   In the hydrogen atom the $3s$ and $3p$ orbitals have the same energy, but in the chlorine atom the $3s$ orbital lies at a considerably lower energy than the $3p$ orbital. Account for this difference.

16.5   Give the electronic configuration for the following species: Cl^-, P^+, Tl^{3+}, S^-, and O^+.

16.6   Explain qualitatively the variation in ionization potential listed in Table 16.5.

16.7   Calculate the sixth ionization potential for carbon, that is, the energy required for the process

$$C^{5+}(g) \longrightarrow C^{6+}(g) + e^-(g)$$

16.8   Use the $2s$ wave function given in Table 16.2 to calculate the value of r (other than $r = \infty$) at which this wave function becomes zero.

★**16.9**   Obtain an expression for the most probable radius at which an electron will be found when it occupies the $1s$ orbital. (*Hint*: Differentiate the $1s$ wave function in Table 16.2 with respect to r.)

16.10   Atomic hydrogen is unstable and there is a great tendency for hydrogen atoms to form H_2 molecules. However, the approach of two hydrogen atoms does not always result in the formation of a H_2 molecule. Suggest two reasons for this behavior.

16.11   Describe the bonding scheme in the following species in terms of the molecular orbital theory: H_2^+, H_2, He_2^+, and He_2. List the species in the order of decreasing stability.

16.12 One way to modify Eq. (16.5) for many-electron atoms is to replace Z with $(Z - \sigma)$, where σ is called the screening constant. Calculate the value of σ for helium if the first ionization potential is 24.5 eV.

16.13 The dipole moment of HF is 1.91 D and the equilibrium bond distance is 0.917 Å. Calculate the charge separation in HF. For a completely polar or ionic bond, the charge separation would be one electron. What is the percent ionic character for the H—F bond?

16.14 The dipole moment of the compound CH_2ClCH_2Cl increases from 1.1 D at 298 K to 1.5 D at 550 K. Explain this change in terms of the change in molecular structure.

16.15 Borazine $(B_3N_3H_6)$ is isoelectronic with benzene. Describe qualitatively the bonding in this molecule in terms of (a) resonance, and (b) molecular orbital theory.

16.16 Which molecule would have the longer bond length: F_2 or F_2^+? Explain in terms of molecular orbital theory. (*Hint*: Consider the electronic configuration of F_2 and F_2^+.)

16.17 Which of the following species has the longest bond: CN^+, CN, and CN^-? Why?

16.18 The resonance concept is sometimes described in terms of a mule being a cross between a horse and a donkey. Compare this analogy with that used in the text. (That is, the description of a rhinoceros as a cross between a griffin and a unicorn.) Which of the two descriptions is more appropriate? Why?

16.19 The unstable molecule carbene or methylene (CH_2) has been isolated and studied spectroscopically. Suggest two types of bonding that might be present in this molecule. How would you distinguish between these two possibilities?

16.20 Draw resonance structures for CO_2 and N_2O.

16.21 Disulfide bonds play an important role in the three-dimensional structure of protein molecules. Discuss the nature of the —S—S— bond.

16.22 An s orbital can overlap with a p orbital in one of two possible ways. However, only one of these two ways leads to the formation of a bonding and an antibonding molecular orbital. Explain.

16.23 Draw an energy-level diagram for the CO molecule.

16.24 $Ni(CN)_4^{2-}$ is diamagnetic but $NiCl_4^{2-}$ is paramagnetic. Explain the difference in terms of crystal field theory. Both complexes are square planar.

16.25 The maximum of absorption for the $Ti(H_2O)_6^{3+}$ species occurs at 497.5 nm. Calculate the crystal field splitting in kJ mol^{-1}.

16.26 Draw energy-level diagrams to show the low- and high-spin octahedral complexes of the transition-metal ions having the electronic configurations d^4, d^5, d^6, and d^7.

16.27 Describe the bonding in CO_2 and C_3H_4 (allene) in terms of hybridization.

16.28 Carbon monoxide has a rather small dipole moment $(\mu = 0.12$ D) even though the electronegativity difference between C and O is rather large $(X_C = 2.5$ and $X_O = 3.5)$. How would you explain this in terms of resonance structures?

16.29 Draw energy-level diagrams for Fe^{2+} in deoxyhemoglobin and oxyhemoglobin.

16.30 Suffocation victims usually look purple, but a person poisoned by carbon monoxide often has rosy cheeks. Explain.

16.31 Explain why carbon monoxide's affinity for hemoglobin would be much greater if it could bind to the Fe^{2+} ion in a straight fashion.

16.32 Which of the following two molecules has a greater degree of π-electron delocalization: naphthalene or biphenyl?

16.33 The enzyme carboxypeptidase A contains the Zn^{2+} ion. The Zn^{2+} ion can be replaced by a Co^{2+} ion, which results in a more active enzyme. Given this fact, why do you suppose that Co^{2+} ion is not employed in the natural system?

★**16.34** Silicon is some 146 times more plentiful than carbon. Both elements belong to group 4A of the periodic table and exhibit some of the same properties. Yet only carbon plays a major role in directing the chemistry of living systems. Suggest some reasons for this difference.

16.35 Referring to Figure 16.25, explain why the configuration shown in (a) is considered unlikely.

16.36 Suggest an explanation for the fact that $Fe(CN)_6^{3-}$ is poisonous while $Fe(CN)_6^{4-}$ is not.

16.37 Compare the bonding in FeF_6^{3-} and $Fe(CN)_6^{3-}$ in terms of hybridization.

17

Intermolecular Forces

United we stand.

In Chapter 16 we discussed the covalent bond and related the force holding atoms together to the overlapping of atomic orbitals. The interaction between molecules is best explained by describing several types of intermolecular interactions, interactions responsible, for example, for liquefaction of gases and stability of protein molecules. A special type of interaction is the hydrogen bond, which plays important roles in determining the structure and properties of DNA and water.

Intermolecular Interactions 17.1

DIPOLE–DIPOLE INTERACTION

An intermolecular interaction of the *dipole–dipole* type occurs only between polar molecules or molecules that possess a permanent dipole moment. Consider the electrostatic interaction between the two dipoles μ_A and μ_B separated by distance r. In extreme cases these two dipoles can be aligned as shown in Figure 17.1. In the top arrangement, the energy of interaction is given by

$$E = -\frac{2\mu_A\mu_B}{\varepsilon r^3} \tag{17.1}$$

and in the bottom arrangement we have

$$E = -\frac{\mu_A\mu_B}{\varepsilon r^3} \tag{17.2}$$

where ε is the dielectric constant of the medium (see Section 9.2) and the negative sign indicates that the interaction is attractive; that is, energy is released when these two molecules interact. Reversing the charge signs of one of the dipoles makes E a positive quantity. Then the interaction leads to a repulsion between the two molecules.

487

Figure 17.1

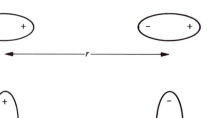

Schematic drawing showing the extreme orientations of two permanent dipoles for attractive interactions.

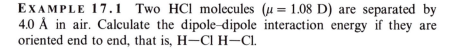

EXAMPLE 17.1 Two HCl molecules ($\mu = 1.08$ D) are separated by 4.0 Å in air. Calculate the dipole–dipole interaction energy if they are oriented end to end, that is, H—Cl H—Cl.

CGS units

$$\mu = 1.08 \text{ D} = 1.08 \times 10^{-18} \text{ esu cm}$$

$$r = 4.0 \text{ Å} = 4.0 \times 10^{-8} \text{ cm}$$

$$\varepsilon_{air} \simeq 1$$

From Eq. (17.1), we write

$$E = -\frac{2(1.08 \times 10^{-18} \text{ esu cm})(1.08 \times 10^{-18} \text{ esu cm})}{1 \times (4.0 \times 10^{-8} \text{ cm})^3}$$

$$= -3.65 \times 10^{-14} \text{ erg}$$

$$= -3.65 \times 10^{-21} \text{ J}$$

$$1 \text{ esu} = 1 \text{ g}^{1/2} \text{ cm}^{3/2} \text{ s}^{-1}$$

SI units

$$\mu = 1.08 \text{ D} = 3.605 \times 10^{-30} \text{ C m (see Section 16.7)}$$

$$r = 4.0 \text{ Å} = 4.0 \times 10^{-10} \text{ m}$$

In SI units, Eq. (17.1) takes the form (see Appendix 9.1)

$$E = -\frac{2\mu_A \mu_B}{\varepsilon r^3} \frac{1}{4\pi\varepsilon_0}$$

where

$$4\pi\varepsilon_0 = 1.1127 \times 10^{-10} \text{ C}^2 \text{ N}^{-1} \text{ m}^{-2}$$

Thus

$$E = -\frac{2(3.605 \times 10^{-30} \text{ C m})(3.605 \times 10^{-30} \text{ C m})}{1 \times (4.0 \times 10^{-10} \text{ m})^3}$$

$$\times \frac{1}{1.1127 \times 10^{-10} \text{ C}^2 \text{ N}^{-1} \text{ m}^{-2}}$$

$$= -3.65 \times 10^{-21} \text{ N m}$$

$$= -3.65 \times 10^{-21} \text{ J}$$

In a macroscopic system where all possible orientations of the dipoles are present, we might expect that the mean value of E would be zero, since there would be as many repulsions as attractions. Even under conditions of free rotation in a liquid or gaseous state, Boltzmann's distribution law (see Chapter 3) tells us that molecules having energetically favorable orientations outweigh those having unfavorable orientations by a factor that is proportional to exp $(-\mu_A\mu_B/\varepsilon r^3 kT)$, where k is Boltzmann's constant and T is absolute temperature. A rather elaborate derivation by Keesom shows that the average or net energy of interaction of permanent dipoles is given by

$$E = -\frac{2}{3}\frac{\mu_A^2\mu_B^2}{\varepsilon r^6}\frac{1}{kT} \qquad (17.3)$$

Note that E is inversely proportional to the sixth power of r, the energy of interaction falls off rapidly with distance. Also, E is inversely proportional to T, because at higher temperatures the average kinetic energy of the molecules is greater, a condition unfavorable to aligning dipoles for attractive interaction.

ION–DIPOLE INTERACTION

The interaction between an ion and polar molecules was first discussed in Chapter 9 in relation to ionic hydration. The energy of interaction between an ion of charge e at a distance r away from a dipole is given by

$$E = -\frac{e\mu}{\varepsilon r^2} \qquad (17.4)$$

Equation (17.4) holds only when the ion and the dipole lie along the same axis. This attractive interaction is mainly responsible for the dissolution of ionic compounds in polar solvents.

ION–INDUCED DIPOLE AND DIPOLE–INDUCED DIPOLE INTERACTIONS

In a neutral nonpolar species such as helium, the electron charge density is spherically symmetric about the nucleus. If an electrically charged object, for example, a positive ion, is brought near the helium atom, electrostatic interaction will cause a redistribution of the charge density (Figure 17.2). The atom will then acquire a dipole moment induced by the charged particle. The

(a) Neutral helium atom. (b) Induced dipole moment in helium due to a cation. (c) Induced dipole moment in helium due to a permanent dipole. The plus and minus signs in helium represent shifts in electron density.

Figure 17.2

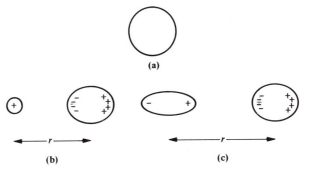

magnitude of the *induced dipole moment* μ_{ind} is directly proportional to the strength of the electric field:

$$\mu_{ind} \propto \mathcal{E}$$
$$= \alpha\mathcal{E} \tag{17.5}$$

where α, the constant of proportionality, is the *polarizability* of helium.* The energy of this ion–induced dipole interaction is given by

$$E = -\frac{1}{2}\frac{\alpha e^2}{\varepsilon r^4} \tag{17.6}$$

Qualitatively, polarizability measures how easily the electron density in an atom or a molecule can be perturbed by an external electric field. It has the units of cm^3 (CGS units) or m^3 (SI units). Unsaturated bonds such as $>\!C\!=\!C\!<$, $>\!C\!=\!N\!-$, $-NO_2$, the phenyl group, the base pairs in DNA, and negative ions are examples of highly polarizable groups. Generally, the larger and more diffuse the electron charge cloud in the molecule, the greater its polarizability.

A permanent dipole can also induce a dipole moment in a nonpolar molecule (Figure 17.2c). In 1920, Debye showed that the energy of interaction for the dipole–induced dipole interaction is given by

$$E = -\frac{2\alpha\mu^2}{\varepsilon r^6} \tag{17.7}$$

Again, the negative sign means attractive interaction. Note that Eq. (17.7), unlike Eq. (17.3), is independent of temperature. This is because the dipole moment can be induced instantaneously so that the value of E is unaffected by the thermal motion of the molecules.

DISPERSION OR LONDON INTERACTIONS

The cases considered thus far consist of at least one charged ion or one permanent dipole among the interacting species, and they can be satisfactorily treated by classical physics. We must now ask: Since nonpolar gases such as helium and nitrogen can be condensed, what kind of attractive interaction exists between neutral and nonpolar molecules?

When we speak of the spherical symmetry of the charge density in helium, we mean that averaged over a certain period of time (for example, a time long enough for us to carry out a physical measurement on the system), the electron density at a fixed distance away from the nucleus is the same in every direction. If we could take snapshots of the *instantaneous* configuration of each individual helium atom, we would most likely find varying degrees of deviation from spherical symmetry, owing to collision among the atoms. Nevertheless, the temporary dipole created at every instant can induce a dipole from its neighboring atom(s), so an attractive interaction will result. We expect that this should be a weak attraction; indeed, the boiling point of helium is 4.2 K, suggesting that very weak forces holding the atoms together in the liquid state. However, for molecules that possess large polarizabilities, this interaction can be comparable to or even greater than dipole–dipole and dipole–induced dipole interactions in many cases.

---

* The *electric field* $\mathcal{E}$ due to a charge q at a distance r from the charge is equal to q/r^2.

A quantum-mechanical treatment for this type of interaction was given in 1930 by London, who showed that

$$E = -\frac{3}{4}\frac{I\alpha^2}{r^6} \tag{17.8}$$

where α is the polarizability and I the first ionization potential of the atom or molecule. For unlike molecules A and B, Eq. (17.8) becomes

$$E = -\frac{3}{2}\frac{I_A I_B}{I_A + I_B}\frac{\alpha_A \alpha_B}{r^6} \tag{17.9}$$

In addition to the various attractive forces discussed so far, atoms and molecules must exert repulsion on one another; otherwise, they would eventually fuse. Fusion is prevented by strong repulsive forces between electron clouds and between nuclei. The potential energy of repulsion is extremely short-ranged, being proportional to $1/r^n$, where n is between 9 and 12. Lennard-Jones proposed the following expression to represent the attractive and repulsive interactions in nonionic systems:

$$E = -\frac{A}{r^6} + \frac{B}{r^{12}} \tag{17.10}$$

where A and B are constants for two interacting atoms or molecules. A more convenient form commonly used is

$$E = -\frac{A}{r^6} + Be^{-ar} \tag{17.11}$$

where the repulsive interaction is represented by an exponential term. Again, A, B, and a are constants characteristic of the interacting systems (Figure 17.3).

The dipole–dipole, ion–induced dipole, dipole–induced dipole, and London interactions make up what physical chemists have referred to as *van der Waals forces*. The London forces or dispersion forces differ from the other types of van der Waals forces in that the former are always attractive whatever the mutual orientation of nonpolar molecules. The term *van der Waals radius* refers to the distance of closest approach between nonbonded atoms. For example, in crystalline iodine the distance between iodine atoms

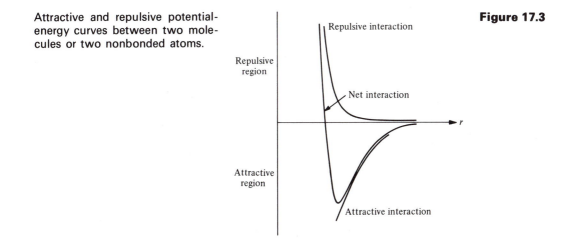

Attractive and repulsive potential-energy curves between two molecules or two nonbonded atoms.

Figure 17.3

Table 17.1

Van der Waals Radii of Atoms and the Methyl Group

Atom	Radius (Å)
H	1.1
C	1.5
N	1.5
O	1.4
P	1.9
S	1.85
F	1.35
Cl	1.8
Br	1.95
I	2.2
—CH_3	2.0

measured from their centers in adjacent molecules is 4.30 Å. The van der Waals radius of iodine is therefore $4.30/2 = 2.15$ Å.* Table 17.1 lists the van der Waals radii of a number of atoms.

Dispersion forces play an important role in protein structures. For instance, the tertiary structure (see Chapter 21) of proteins is determined to a large extent by contact of nonpolar groups. Dispersion forces are partly responsible for maintaining the heme group in the "pocket" formed by the side chains in hemoglobin and myoglobin. In lipoprotein membranes, the interaction between hydrocarbon tails of lipid molecules is mainly caused by dispersion and other van der Waals forces. The term *hydrophobic bond*, or *hydrophobic interaction*, is employed to describe interaction between nonpolar groups of valine, leucine, isoleucine, methionine, phenylalanine, and tryptophan in proteins in aqueous solution. These groups are frequently located in the interior of the protein molecule, where they have little or no contact with water. We shall discuss protein structure in Chapter 21.

As mentioned in Chapter 1, the replacement of glutamic acid by valine in the sixth position in each of the β chains in hemoglobin (called hemoglobin S)

Figure 17.4

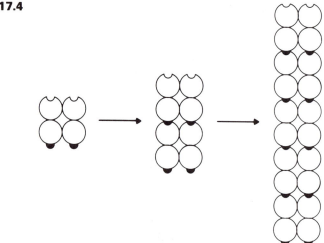

Proposed scheme for the formation of a "super polymer" made up of sickle-cell hemoglobin molecules. Note that this reaction only occurs among deoxyhemoglobin molecules, that is, hemoglobin molecules that do not contain bound oxygen. The black protruding spots represent the valine residues.

* The size of atoms of the space-filling model is proportional to their van der Waals radii.

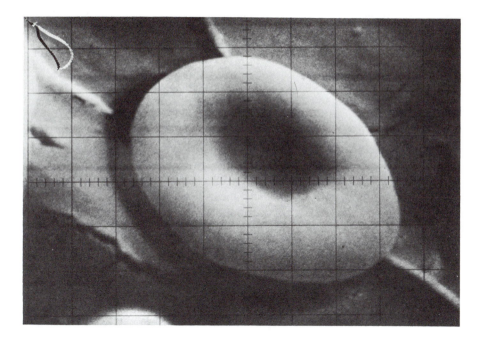

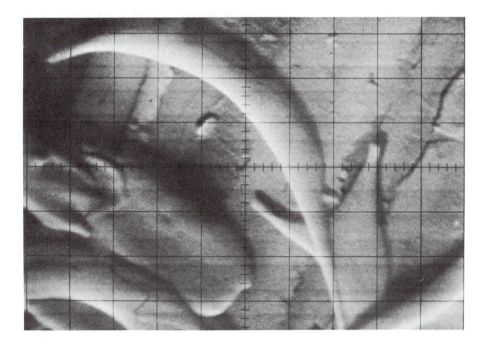

Electron micrograph showing (*top*) a normal red blood cell; and (*bottom*) sickled red blood cells. (Courtesy of Phillips Electronic Instruments, Inc.)   **Figure 17.5**

causes the disease known as sickle-cell anemia. In 1966, Murayama proposed that it is the noncovalent forces between deoxyhemoglobin S molecules that result in the aggregation of the proteins to form a "super polymer." Presumably, the alkyl groups in the valine residues interact with certain non-polar groups in the α chains of another hemoglobin S molecule through dispersion forces. Figure 17.4 shows the "head-to-tail" stacking of hemo-globin S molecules. Eventually, the aggregated hemoglobin S molecules will precipitate out of solution. The precipitate causes the normal disc-shaped red blood cells to assume a warped crescent or sickle shape (Figure 17.5). These deformed cells clog the narrow capillaries, restricting blood flow to vital organs. The usual symptoms of sickle cell anemia are swelling, severe pain, and other complications. Sickle-cell anemia has been termed a molecular disease, because the destructive action of molecules is understood at the molecular level and the disease is, in effect, due to a molecular defect.

Despite an intensive research effort currently underway, no complete cure for this disease is known. The treatment of sickle-cell anemia has been largely based on employing antisickling agents such as urea and the cyanate ion:

$$H_2N-\overset{\underset{\displaystyle O}{\|}}{C}-NH_2 \qquad O=C=N^-$$

urea   cyanate ion

For example, the use of intravenously administered urea in 10% sugar solu-tion has met with limited success. A more effective antisickling agent is the

Table 17.2

Interaction Between Molecules

Type of Interaction	Energy of Interaction	Example	Order of Magnitude (kJ mol^{-1})
Covalent bond[a]	No simple expression	H—H	200–800
Ion–ion	$\dfrac{e^2}{\varepsilon r}$	Na$^+$Cl$^-$	40–400
Ion–dipole	$\dfrac{e\mu}{\varepsilon r^2}$	Na$^+$(H$_2$O)$_n$	4–40
Dipole–dipole	$\dfrac{2}{3}\dfrac{\mu_1^2\mu_2^2}{\varepsilon r^6 kT}$	SO$_2$ SO$_2$	0.4–4
Dipole–induced dipole	$\dfrac{\alpha\mu^2}{\varepsilon r^6}$	HCl C$_6$H$_6$	0.4–4
Ion–induced dipole	$\dfrac{1}{2}\dfrac{\alpha e^2}{\varepsilon r^4}$	Na$^+$C$_6$H$_6$	0.4–4
Dispersion	$\dfrac{3}{4}\dfrac{I\alpha^2}{r^6}$	He He	4–40
Hydrogen bond	No simple expression	H$_2$O $\cdots$ H$_2$O	4–40

[a] This is listed for comparison purposes only.

cyanate ion, which is formed from urea in solution. The function of these molecules is to break up the hydrophobic interactions between different hemoglobin S molecules to reverse the "sickling" of the red blood cells.

Table 17.2 summarizes different types of interactions discussed in this section.

The Hydrogen Bond **17.2**

The hydrogen bond is a special type of interaction between molecules; it forms whenever a polar bond containing the hydrogen atom (for example, O—H or N—H) interacts with an electronegative atom such as oxygen, nitrogen, fluorine, or chlorine. This interaction is represented as A—H···B, where A and B are the electronegative atoms and the dotted line denotes the hydrogen bond. Although hydrogen bonds are relatively weak (about 40 kJ mol^{-1} or less), they play a central role in determining the properties of many compounds. Normally, the boiling points of substances in a homogeneous series increase with increasing polarizability, but this is not the case for the compounds H_2O, H_2S, H_2Se, and H_2Te, whose boiling points are 373 K, 213 K, 231 K, and 271 K, respectively. The abnormally high boiling point of H_2O, despite its low polarizability compared to other analogs in this group, is the result of *intermolecular* hydrogen bonding present in water.

Why is this type of bonding unique to hydrogen? The main reason is that the hydrogen atom has only one electron; when that electron is used to form a covalent bond with an electronegative atom, its nucleus becomes partially unshielded. Consequently, this proton can interact directly with another electronegative atom on a different molecule. Depending on the strength of the interaction, such bonding can exist in the gaseous phases as well as in the solid and liquid phases. In liquids, substances like HF and HCN form polymeric chains as follows:

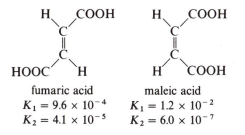

For maximum stability, the donor pair AH and the acceptor B are usually collinear (that is, AHB = 180°), but deviations of up to 30° are known.

A molecule can also form *intramolecular* hydrogen bonds. Consider the first and second acid dissociation constants of fumaric and maleic acids:

fumaric acid
$K_1 = 9.6 \times 10^{-4}$
$K_2 = 4.1 \times 10^{-5}$

maleic acid
$K_1 = 1.2 \times 10^{-2}$
$K_2 = 6.0 \times 10^{-7}$

The first dissociation constant for maleic acid is higher than that for fumaric acid because of steric interaction in the cis isomer, which facilitates the removal of a proton. Although the second dissociation constant of fumaric acid is only

20 times lower than the first dissociation constant, K_2 is smaller than K_1 by a factor of 20,000 for maleic acid. This phenomenon can be explained by assuming a stable intramolecular hydrogen bond formed between the —COOH and —COO⁻ groups in maleic acid as follows:

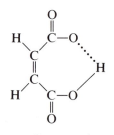

A number of physical techniques are used to detect the presence of hydrogen bonds. For crystals, X-ray diffraction measurements usually provide the most direct evidence. A relatively strong hydrogen bond can shorten the expected distance between AH and B (the sum of the van der Waals radii) by two- to three-tenths of an angstrom. Infrared and nuclear magnetic resonance (nmr) spectroscopy are particularly useful for studying hydrogen bonding in the liquid phase. For example, formation of a hydrogen bond shifts the stretching mode to lower frequencies and the bending modes to higher frequencies.

$$\overset{\leftarrow \quad \rightarrow}{A—H} \qquad \qquad A—H$$

stretching   bending

The width and intensity of the peak due to A—H stretching may also be enhanced by hydrogen-bond formation. In nmr studies we predict and observe that the chemical shift of a proton can be appreciably altered by hydrogen bonding (see Chapter 18).

Hydrogen bonding is largely responsible for the stability of protein conformations. The intramolecular hydrogen bonds formed between the $>C=O$ and $>N—H$ groups of a polypeptide chain result in the α helix. On the other hand, intermolecular hydrogen bonds between two polypeptide chains account for pleated-sheet structures. We shall postpone the discussion of these structures until Chapter 21. Instead, let us consider the importance of hydrogen bonding in deoxyribonucleic acids (DNA).

The DNA molecules are polymers having molecular masses ranging from millions to tens of billions. They consist of three parts: the phosphate groups, the sugar groups (deoxyribose), and purine and pyrimidine bases (adenine, cytosine, guanine, and thymine). Figure 17.6 shows the famous Watson–Crick double-helical structure of DNA. The molecule's backbone contains alternating sugar and phosphate residues. To each sugar residue is attached a purine or pyrimidine base. The double-helical structure results when the two strands are held together by hydrogen bonds formed between bases. The bases are roughly perpendicular to the axis of the helix. Each base can form a strong hydrogen bond with only one among the four bases available. This specificity of the base pairing is responsible for the genetic code.

Energetically, the most favorable pairings in DNA molecules occur between

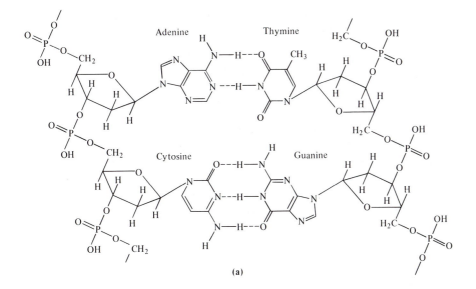

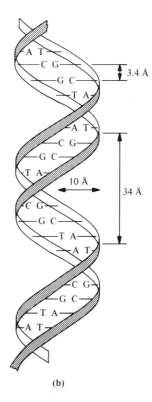

(a) Base-pair formation between adenine (A) and thymine (T) and between cytosine (C) and guanine (G). (b) The double-helical strand of a DNA molecule held together by hydrogen bonds and other intermolecular forces.

Figure 17.6

adenine (A) and thymine (T) and between cytosine (C) and guanine (G). Although the energy per hydrogen bond is rather weak (about 5 kJ mol^{-1}), the double-helical structure of DNA is stable under normal physiological conditions. The stability of the molecule rests on the cooperative nature of hydrogen bond formation. Consider the pairing of two nucleotides, C and G,

497

in solution at room temperature. The ratio of free bases to hydrogen-bonded base pair can be calculated from Boltzmann's distribution law:

$$\frac{(C, G)_{\text{free bases}}}{(C\text{-}G)_{\text{H-bonded base pair}}} = e^{-\Delta E/RT} = \exp\left(\frac{-5000 \text{ J mol}^{-1}}{8.314 \text{ J K}^{-1} \text{ mol}^{-1} \times 300 \text{ K}}\right) = 0.14$$

or there are seven pairs of hydrogen-bonded bases to one pair of free bases. For dinucleotides of the types C_2 and G_2, the complexed form would now be favored over the free bases by a factor of 7×7, or 49. It is not difficult to see that in a polynucleotide containing thousands of bases, the equilibrium would be overwhelmingly in favor of the hydrogen-bonded structure. In practice, base pairing follows strictly the A-T and C-G pattern, and the two strands are said to be complementary. During replication, the two strands part to form two identical DNA molecules.*

17.3 Structure and Properties of Water

Water is so common a substance that we often overlook its unique properties. For example, water should be a gas at room temperature, but because of hydrogen bonding, it has a boiling point of 373.15 K at 1 atm. This section deals with the structure of ice and liquid water and considers some biologically significant aspects of water.

STRUCTURE OF ICE

To understand the behavior of water, we must first investigate the structure of ice. There are nine known crystalline forms of ice; most of them are stable only at high pressures. Ice I, the familiar form, has been studied thoroughly. It has a density of 0.924 g ml^{-1} at 273 K and 1 atm pressure.

The electronic structure of a water molecule was discussed in Chapter 16. There is a significant difference between H_2O,

and other polar molecules, such as NH_3 and HF. In water, the number of protons about each oxygen atom that can form the positive ends of hydrogen bonds is equal to the number of lone pairs on that oxygen atom that can form the negative ends. An extensive three-dimensional network results, each oxygen atom being tetrahedrally bonded to four hydrogen atoms in two covalent bonds and two hydrogen bonds. This equality in number of protons and lone pairs is not characteristic of NH_3 and HF. Consequently, these molecules can form only rings or chains, not a three-dimensional structure.

Figure 17.7 shows the structure of ice I. The distance between adjacent oxygen atoms is 2.76 Å. The O—H distance is between 0.96 and 1.02 Å and the O···H distance between 1.74 and 1.80 Å.

* The role of DNA in protein synthesis is described in any standard text on biochemistry or molecular biology.

Figure 17.7

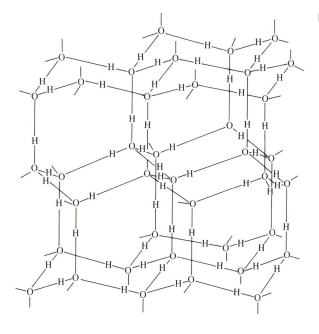

Structure of ice I. The hydrogen bonds are represented by the long lines between O and H. [From G. Némethy and H. A. Scheraga, *J. Chem. Phys.* **36**, 3383 (1962).]

Because of its open lattice, ice has a lower density than water, a fact of profound ecological significance. Were it not for this unique type of hydrogen bonding, ice, like most other solid substances, would be heavier than the corresponding liquid. On freezing, it would sink to the bottom of a lake or pond, causing all the water to freeze gradually. Most living organisms in the body of water would not survive. Fortunately, water reaches its maximum density at 277.15 K, which is 4 degrees above freezing. Cooling below 277.15 K decreases the density of water, allowing it to rise to the surface, where freezing occurs. An ice layer formed on the surface does not sink; just as important, it acts as a thermal insulator for the water below it.

STRUCTURE OF WATER

The structure of liquid water is not as well understood as that of solid ice. Interest in water is great because of its importance in chemical and biological systems, and numerous structural models have been proposed for it. Although none of the existing theories can account satisfactorily for all the observed properties of water, good progress has been made toward a detailed explanation of its structure and properties.

Bernal and Fowler in 1933 suggested that the hydrogen bonds found in ice also exist in water, although we no longer have the regular three-dimensional network of the solid. Instead, some hydrogen bonds are broken and free monomeric water molecules are formed. These monomers can occupy holes in the remaining "icelike" lattice, explaining why the density of water is greater than ice. As temperature goes up, more hydrogen bonds are broken, but at the same time the kinetic energy of molecules increases. Higher temperature increases the density of water, while elevated kinetic energy decreases the density, because each molecule will now occupy a greater volume. The net result is that a maximum density is reached at 277.15 K, above which the density decreases monatonically with increasing temperature.

According to a different theory, proposed by Pople in 1950, very few

499

Figure 17.8

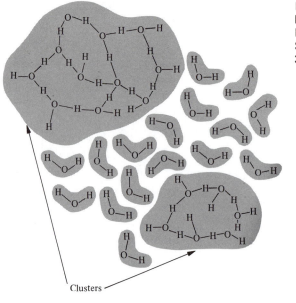

Flickering-cluster model of liquid water. [From G. Némethy and H. A. Scheraga, *J. Chem. Phys.* **36**, 3387 (1962).]

Clusters

hydrogen bonds are actually broken when ice melts. Instead, the bonds are bent and distorted in liquid water. The relatively low heat of fusion for water (6.01 kJ mol^{-1}) lends some support to this model.

In 1952, Pauling showed that the structure of water may be similar to certain hydrocarbon–water complexes, such as $CH_4 \cdot 6H_2O$ and $Cl_2 \cdot 8H_2O$. Water molecules in these complexes form structures that are characteristically open like ice, but the overall arrangement is looser than that of ice. The guest molecules, CH_4 or Cl_2, are trapped inside the cavities of the icelike structure. These complexes are called *clathrate compounds* or simply *clathrates*.* In the water model we would replace the guest molecule by another water molecule.

Pauling's arrangement was considered to be physically too rigid, and a closely similar *flickering-cluster model* was put forward by Frank and Wen. They assumed that the clathrates were constantly breaking up and re-forming, the mean lifetime being of the order of 10^{-10} s. Figure 17.8 shows their model, further modified by Némethy and Scheraga. Here, each water molecule can hydrogen-bond with 4, 3, 2, 1, or no water molecules. In the last case, the molecule becomes a guest in the clathrate. The manner of hydrogen bonding is cooperative: The presence of one hydrogen bond facilitates the formation of additional bonds. When one hydrogen bond is broken, the entire structure tends to disintegrate.

SOME PHYSIOCHEMICAL PROPERTIES OF WATER

Table 17.3 lists some important physiochemical properties of water. Several of the abnormally high values make it a unique solvent, particularly suited to the support of living systems.

Because of its high heat capacity, water acts as a good "thermostat" for proteins and nucleic acids, whose structure and function strongly depend on

---

* It is interesting to note that the formation of methane (or other similar hydrocarbon)–water clathrate can be a cause of trouble for transferring natural gas in pipelines. Since such a clathrate's freezing point is some 20 degrees above that of water, the solid formed either restricts or totally blocks the flow of the gas.

Some Physiochemical Properties of Water[a]

Table 17.3

Melting point	273.15 K
Boiling point	373.15 K
Density of water	0.99987 g ml^{-1} (0.99987 × 10^3 kg m^{-3}) at 273.15 K
	1.00000 g ml^{-1} (1.0000 × 10^3 kg m^{-3}) at 277.15 K
Density of ice	0.9167 g ml^{-1} (0.9167 × 10^3 kg m^{-3}) at 273.15 K
Molar heat capacity	75.3 J K^{-1} mol^{-1}
Molar heat of fusion	6.01 kJ mol^{-1}
Molar heat of vaporization	40.79 kJ mol^{-1} at 373.15 K
Dielectric constant	78.54 at 298.15 K
Dipole moment	1.82 D
	(6.08 × 10^{-30} mC)
Viscosity	0.01 P at 293.15 K
	(0.001 kg m^{-1} s^{-1})
Surface tension	72.75 dyn cm^{-1} at 293.15 K
	(0.07275 N m^{-1})
Diffusion coefficient	2.4 × 10^{-9} m^2 s^{-1} at 298.15 K

[a] When two values are given for the same quantity, the one in parentheses is in SI units.

the temperature of their surroundings. From an environmental point of view, this property has greatly influenced weather. The huge bodies of water in lakes and oceans absorb or give up large amounts of heat with only a small change in temperature. The high heat of vaporization works to reduce the loss of water by evaporation. Further, the cooling effect of evaporation helps to regulate the temperature of the water body.

Although water's heat of fusion is not very high, it is still sizable compared to other solvents. Again, this property helps to protect against freezing. Finally, we note that the high dielectric constant of water is essential for dissolving a great variety of salts, acids, and bases. According to Eq. (16.14), the force between two charges is inversely proportional to the dielectric constant of the medium. When sodium chloride dissolves in water, the attractive force between Na$^+$ and Cl$^-$ ions is greatly diminished, and the ions become hydrated through ion–dipole interaction. On the other hand, when sodium chloride dissolves in benzene (whose dielectric constant is only one-fortieth that of water), a strong attractive force keeps the ions together. Consequently, the amount of free sodium and chloride ions remaining in solution is exceedingly small.

Suggestions for Further Reading

INTRODUCTORY–INTERMEDIATE

Davies, M. *Some Electrical and Optical Aspects of Molecular Behavior.* Pergamon Press, Inc., Elmsford, N.Y., 1965.
A useful introductory text.

Davis, K. S., and J. A. Day. *Water, the Mirror of Science.* Doubleday & Company, Inc., Garden City, N.Y., 1961.
A readable, descriptive book.

Edsall, J. T., and J. Wyman. *Biophysical Chemistry.* Academic Press, Inc., New York, 1958, Chapter 2.
A somewhat dated but still very valuable text.

KAVANAU, J. L. *Water and Water-Solute Interactions.* Holden-Day, Inc., San Francisco, 1964.
> A short, descriptive text that is full of useful information.

PIMENTAL, G. C., and A. L. McCLELLAN. *The Hydrogen Bond.* W. H. Freeman and Company, San Francisco, 1960.
> A standard text of the topic.

INTERMEDIATE–ADVANCED

EISENBERG, D., and W. KAUZMANN. *The Structure and Properties of Water.* Oxford University Press, Inc., New York, 1969.
> Suitable for advanced students.

VINOGRADOR, S. N., and R. H. LINNELL. *Hydrogen Bonding.* Van Nostrand Reinhold Company, New York, 1971.
> Another useful text for advanced students.

Reading Assignments

SECTION A

"Weak Intermolecular Interactions," J. E. House, Jr., *Chemistry* **45**(4), 13 (1972).

"Hydrophobic Interactions," G. Némethy, *Angew. Chem. Int. Ed.* **6**, 195 (1967).

"On Hydrogen Bonds," J. Donohue, *J. Chem. Educ.* **40**, 598 (1963).

"Hydrogen Bond: Special Agent," V. J. Webb, *Chemistry* **41**(6) 17 (1968).

"Teaching Ion–Ion, Ion–Dipole, and Dipole–Dipole Interactions," C. H. Yoder, *J. Chem. Educ.* **54**, 402 (1977).

SECTION B

"Chemical Forces," H. H. Jaffé, *J. Chem. Educ.* **40**, 649 (1963).

"The Structure of Ordinary Water," H. S. Frank, *Science* **169**, 635 (1970).

"A Molecular Theory of General Anesthesia," L. Pauling, *Science* **134**, 15 (1961). (Discusses the structure of clathrate compounds.)

"Inclusion Compounds," J. F. Brown, Jr., *Sci. Am.*, July 1962.

"Clathrates: Compounds in Cages," M. M. Hagan, *J. Chem. Educ.* **40**, 643 (1963).

"The Human Thermostat," T. H. Benzinger, *Sci. Am.*, Jan. 1961.

"Van der Waals Molecules," G. E. Ewing, *Acc. Chem. Res.* **8**, 185 (1975).

"The Role of van der Waals Forces in Surface and Colloid Chemistry," P. C. Hiemenz, *J. Chem. Educ.* **49**, 164 (1972).

"The Antibody Combining Site," J. D. Capra and A. B. Edmondson, *Sci. Am.*, Jan. 1977.

"Sickle-Cell Anemia: Molecular and Cellular Bases of Therapeutic Approaches," J. Dean and A. N. Schechter, *New Engl. J. Med.* **299**, 752, 804, 863 (1978).

Problems

17.1 List all the intermolecular interactions present in each type of the following molecules: Xe, H_2O, C_6H_5F, and LiF.

17.2 Acetic acid is miscible with water, but it also dissolves in nonpolar solvents such as benzene and CCl_4. Explain. (*Hint*: Consider the hydrogen bonds formed between a pair of acetic acid molecules.)

17.3 Which of the following molecules would have a higher melting point: *p*-nitrophenol or *o*-nitrophenol? Explain.

17.4 Which of the following two molecules is a stronger base: $(CH_3)_4N^+OH^-$ or $(CH_3)_3NH^+OH^-$? Explain.

★**17.5** The bases guanine (G) and thymine (T) can exist in the keto and enol forms, while the bases adenine (A) and cytosine (C) can exist in the amino and imino forms. Only the amino form of A can form stable hydrogen bonds with the keto form of T and only the amino form of C can form stable hydrogen bonds with the keto form of G. Draw the structures of the tautomeric forms of these four bases.

17.6 Suggest a chemical analysis to test the A-T and C-G pairing scheme in DNA.

17.7 Assume the energy of hydrogen bonds per base pair to be 10 kJ mol^{-1}. Given two complementary strands of DNA molecule containing 1000 bases each, calculate the ratio of hydrogen-bonded double helix to two separate strands in solution at 300 K.

17.8 Two water molecules are separated by 2.76 Å in air. Use Eq. (17.1) to calculate the dipole–dipole interaction energy.

★**17.9** The internuclear distance between two closest argon atoms in solid argon is about 3.8 Å. The polarizability of argon is 1.66×10^{-24} cm^3, and the first ionization potential is 2.53×10^{-11} erg. Estimate the boiling point of argon. (*Hint*: Calculate the dispersion energy for solid argon and equate this energy with the average kinetic energy of 1 mol of argon gas, which is $\frac{3}{2}RT$.)

17.10 List all the intra- and intermolecular forces that could exist between hemoglobin molecules dissolved in water.

17.11 Diethyl ether has a boiling point of 34.5°C, while 1-butanol has a boiling point of 117°C. These two compounds have the same types and number of atoms. Explain the difference in their boiling points.

17.12 People living in the northeastern United States usually drain the water in their swimming pools before winter each year. Why?

17.13 Butane exists in two structurally different forms called *n*-butane and isobutane. Which one would you expect to have the higher boiling point?

17.14 A small drop of oil in water usually assumes a spherical shape. Explain.

17.15 Coulombic forces are usually referred to as long-range forces (they depend on $1/r^2$) while dispersion forces are called short-range forces (they depend on $1/r^7$). (a) Compare the falloff of these forces with distance by calculating for each the ratio of the force at 2 Å, 3 Å, 4 Å, and 5 Å to that at 1 Å. (b) From your results explain the fact that while a 0.5 *M* nonelectrolyte solution usually behaves ideally, nonideal behavior is quite noticeable in a 0.05 *M* electrolyte solution.

17.16 From Figure 17.6 we see that the average distance between base pairs measured parallel to the axis of a DNA molecule is 3.4 Å. The average molecular mass of a pair of nucleotides is about 650. Estimate the length in cm of a DNA molecule of molecular mass 5.0×10^9. Roughly, how many base pairs are contained in this molecule?

17.17 Look up in a chemistry handbook the melting points and boiling points of the elements or compounds listed. For each series, tabulate the data and account for the trends in terms of the forces involved: (a) He, Ne, Ar, Kr, Xe, and (b) CH_4, CH_3Cl, CH_2Cl_2, $CHCl_3$, and CCl_4.

17.18 The polarizabilities of the noble gases are He: 0.20×10^{-24} cm^3; Ne: 0.40×10^{-24} cm^3; Ar: 1.66×10^{-24} cm^3; Kr: 2.54×10^{-24} cm^3; Xe: 4.15×10^{-24} cm^3. Plot the polarizabilities (y axis) versus the boiling points of the elements. On the same graph also plot the molar masses (y axis) versus the boiling points. Comment on the trends.

$\star$**17.19** Calculate the induced dipole moment of I_2 due to a sodium ion 5 Å away from the center of the iodine molecule. The polarizability of iodine is 1.25×10^{-23} cm^3.

$\star$**17.20** Given the following general properties of water and ammonia, comment on the problems that a biological system (as we know it) would have developing in an ammonia medium.

	H_2O	NH_3
Boiling point	373.15 K	239.65 K
Melting point	273.15 K	195.3 K
Molar heat capacity	75.3 J K^{-1} mol^{-1}	8.53 J K^{-1} mol^{-1}
Molar heat of vaporization	40.79 kJ mol^{-1}	23.3 kJ mol^{-1}
Molar heat of fusion	6.0 kJ mol^{-1}	5.9 kJ mol^{-1}
Dielectric constant	78.54	16.9
Viscosity	0.010 P	0.254 P (at 240 K)
Surface tension	72.75 dyn cm^{-1}	41.2 dyn cm^{-1} (at 244 K)
Dipole moment	1.82 D	1.46 D
Phase at 300 K	Liquid	Gas

18 Spectroscopy

And God said, "Let there be light. . . ."
Genesis 1:3

Spectroscopy is the study of the interaction between electromagnetic radiation and matter: it is a phenomenon of quantum mechanics. Detailed information regarding structure and bonding, as well as various intra- and intermolecular processes, can be obtained from the analysis of atomic and molecular spectra. In this chapter we introduce the vocabulary for spectroscopy and discuss several important spectroscopic techniques.

Vocabulary 18.1

ABSORPTION AND EMISSION

All branches of spectroscopy can be divided into one of two categories: absorption or emission. These processes were mentioned previously, in Chapter 15 (see Figure 15.11). The fundamental equation for both absorption and emission is

$$\Delta E = E_2 - E_1 = h\nu$$

where E_1 and E_2 are the energies of the two quantized levels involved in a transition. Transitions that result from external irradiation are described as *induced* absorption or emission. Further, an excited molecule may lose an amount of energy in the form of radiation without being irradiated. This is called *spontaneous emission*.

BEER'S LAW

Beer's law (or the *Beer–Lambert law*) relates the amount of absorption to the nature of absorbing species. Consider the passage of a monochromatic beam of radiation, that is, radiation of one wavelength, through a homogeneous solution of concentration c (mol liter^{-1}). Let I_0 and I be the intensity of the

Figure 18.1

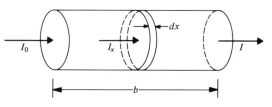

incident and transmitted radiation, and I_x the intensity* of the light at distance x (Figure 18.1). The incremental decrease in intensity, $-dI_x$, is proportional to $I_x \, dx$, that is,

$$-dI_x \propto I_x \, dx$$

$$= kI_x \, dx \tag{18.1}$$

where k is a constant whose value depends on the nature of the absorbing medium. Rearranging Eq. (18.1) and integrating, we obtain

$$\frac{dI_x}{I_x} = -k \, dx$$

$$\ln I_x = -kx + C$$

where C is a constant of integration. At $x = 0$, $I_x = I_0$ so that $C = \ln I_0$. If we consider the entire length of the absorbing medium, I_x can be replaced by I and x by b (cm). Hence

$$\ln I = -kb + \ln I_0$$

$$\ln \frac{I_0}{I} = kb$$

or

$$\log \frac{I_0}{I} = k'b \tag{18.2}$$

where $k = 2.303k'$. Equation (18.2) can be expressed in a more useful form as follows:

$$\log \frac{I_0}{I} = \varepsilon bc \tag{18.3a}$$

$$A = \varepsilon bc \tag{18.3b}$$

where ε is the *molar extinction coefficient* or *molar absorptivity* (liters $mol^{-1} \, cm^{-1}$). Either Eq. (18.3a) or Eq. (18.3b) is known as Beer's law. The quantity $\log(I_0/I)$ is called *absorbance* (A). Theoretically, A can vary from zero to infinity. In experimental designs, however, A is usually measured on a scale between zero and 2. (Note that A is dimensionless.) Beer's law applies to all types of absorption spectroscopy and holds only for relatively low concentrations. At high concentrations ($\gtrsim 0.5 \, M$), deviations occur and A is no longer a linear function of concentration c.

Another useful quantity is the *transmittance* (T), defined as

$$T = \frac{I}{I_0} \tag{18.4}$$

* The intensity of light is determined by the number of photons $cm^{-2} \, s^{-1}$.

Thus

$$-\log T = A = \varepsilon bc$$

The percent transmittance is given by $(I/I_0) \times 100$.

LINE WIDTH AND RESOLUTION

Every spectral line has a finite, nonzero width which is usually defined as the width at half-height of the peak. If the two levels involved in an absorption have precisely known energy values, then their difference must also be an exactly measurable quantity. In this case we would observe a line of no width, which would be a line of infinite intensity. But atoms and molecules are generally unstable in the excited state and will emit spontaneously. Consequently, the lifetime of an atom or molecule in an excited state is much shorter than it would be in a ground state. The Heisenberg uncertainty principle was discussed in Section 15.7. From Eq. (15.24),

$$\Delta E \, \Delta t \geq \frac{h}{4\pi} \tag{18.5}$$

or

$$\Delta E \geq \frac{h}{4\pi \, \Delta t}$$

where ΔE and Δt are the uncertainties in determining the energy and lifetime of a system. If the lifetime is short, it follows that Δt will also be small so that ΔE will be a large quantity. Thus the width of the absorption line is determined by ΔE. The lifetime for the ground state is usually very long and its level is well defined (Figure 18.2).

Beside "lifetime" broadening, spectral lines can be broadened by collision as well as by other processes, such as chemical exchange reactions. Figure 18.3 shows the electronic absorption spectrum of benzene in the vapor state and in solution. The liquid phase has a higher frequency of collision, so excited molecules are more readily deactivated and hence have a shorter lifetime. Consequently, greater line widths are observed.

Related to line width is a quantity called *resolution*, which measures how

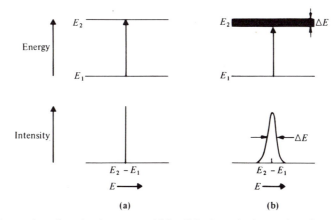

(a) Hypothetical absorption line having no width. (b) Actual absorption line having a width ΔE at half-height.

Figure 18.2

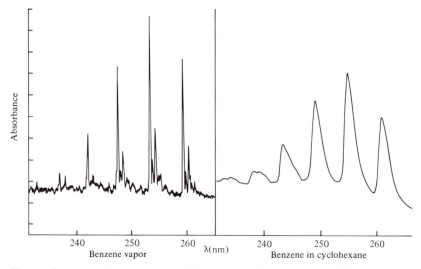

Figure 18.3 Electronic absorption spectrum of benzene: (a) vapor; and (b) in cyclohexane. (By permission of Varian Associates, Palo Alto, Calif.)

well the lines are separated from one another (Figure 18.4). In general, the resolution of any spectrum depends not only on the molecule and its environment but on the spectrometer design as well. We define the *resolving power R* of a spectrometer as its ability to separate radiations at wavelengths λ and $\lambda + \Delta\lambda$, that is,

$$R = \frac{\lambda}{\Delta\lambda} \tag{18.6}$$

Figure 18.4

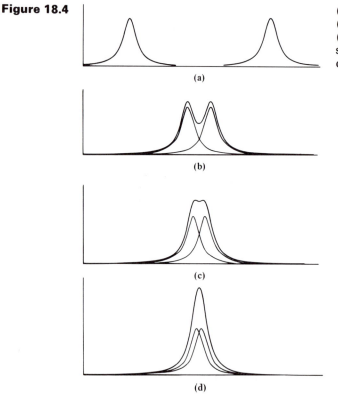

(a) Two well-resolved lines. (b)–(d) Two overlapping lines. From (b) through (d), the observed line shape is the sum of the two overlapping lines.

508

The position of a line corresponds to the difference in energy between two levels involved in a transition. The position can be measured in several different units.

1. *Wavelength.* Wavelength (λ) can be measured in

meter (m)

centimeter (cm)

micrometer (μm)

nanometer (nm)

angstrom (Å)

where

$$1 \text{ m} = 100 \text{ cm}$$

$$1 \text{ } \mu\text{m} = 10^{-4} \text{ cm} = 10^4 \text{ Å}$$

$$1 \text{ nm} = 10^{-3} \text{ } \mu\text{m} = 10 \text{ Å}$$

$$1 \text{ Å} = 10^{-8} \text{ cm} = 0.1 \text{ nm}$$

Both Å and nm are commonly used for wavelength. In this text we usually use the nm unit.

2. *Frequency.* Frequency (v) is given in cps or Hz (Hertz):

$$1 \text{ cycle per second (cps)} = 1 \text{ Hz}$$

3. *Wave Number.* Wave number ($\bar{v}$) is the number of waves per centimeter,

$$\bar{v} = \frac{1}{\lambda} = \frac{v}{c} \tag{18.7}$$

Depending on the particular branch of spectroscopy, any one of these units may be employed for labeling a spectrum. No confusion should arise as long as we remember the fundamental equation, $c = v\lambda$.

INTENSITY

The *intensity* of absorption lines depends on the difference in population (the number of molecules occupying a particular energy level) between the two levels. The ratio of populations in these two levels is given by the Boltzmann expression (see Chapter 3),

$$\frac{n_2}{n_1} = e^{-(E_2 - E_1)/kT} \tag{18.8}$$

where n_1 and n_2 are the populations in the lower and upper levels, k is Boltzmann's constant, and T is the absolute temperature. The greater the difference in population ($n_1 - n_2$), the more intense the absorption line will be.

The reader should keep in mind that any type of spectrum (absorption or emission) is actually a superposition of a very large number of spectra from individual molecules. It is normally impossible to detect the energy absorbed or emitted by a single molecule. Further, the interaction between a photon of electromagnetic radiation and a molecule can give rise to only one transition and hence one line. When we look at any spectrum containing more than one line, as we usually do, we see the statistical sum of all the spectra, each consisting of one line caused by one specific transition.

SELECTION RULES

Transitions may not take place between any two levels in an atom or molecule even if the frequency of radiation appropriate to the resonance condition $(E_2 - E_1 = h\nu)$ is present. Generally, a transition has to obey certain *selection rules*, which are theoretical quantities obtained by quantum-mechanical calculations. Transitions, then, are classified as *allowed* (having a high probability) or *forbidden* (having a low probability), depending on how they occur according to the selection rules.

Theoretically, we predict two types of transitions as forbidden: spin-forbidden transitions and symmetry-forbidden transitions.

Spin-Forbidden Transitions. Spin-forbidden transitions involve a change of *spin multiplicity*. The spin multiplicity is given by the value $(2S + 1)$, as illustrated in Table 18.1. The numerical value of $(2S + 1)$ tells us the number of different ways the unpaired spins can line up in an external magnetic field. The selection rule is that in a transition there must not be a change in the spin multiplicity; that is, we must have $\Delta S = 0$. For example, a transition from a singlet to a triplet, or vice versa, is normally strongly forbidden.

Symmetry-Forbidden Transitions. A quantitative measure for the intensity of a transition is provided by the quantity called *transition dipole moment* μ_{ij}, where

$$\mu_{ij} = \int \psi_i \mu \psi_j \, d\tau \tag{18.9}$$

where ψ_i and ψ_j are the wave functions for states i and j and μ is the dipole moment vector connecting the two states. The integration is taken over all coordinates, and $d\tau$ represents the volume element $(d\tau = dx \, dy \, dz)$. Unless the product $\psi_i \mu \psi_j$ is of a certain symmetry, this integral will be zero and no transition can occur. Consider the electronic transition in a molecule as an

Spin Multiplicity of Atoms and Molecules

Table 18.1

Number of Unpaired Electrons	Electron Spin S	$2S + 1$	Multiplicity
0	0	1	Singlet
1	$\frac{1}{2}$	2	Doublet
2	1	3	Triplet
3	$\frac{3}{2}$	4	Quartet
.	.	.	.
.	.	.	.
.	.	.	.

Figure 18.5 — The various branches of spectroscopy.

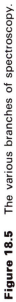

	γ ray	X ray	Ultraviolet	Visible	Infrared	Microwave	Radio frequency
Wavelength, nm	0.0003 0.03	10 30	400	800 1000	3×10^5	3×10^7 3×10^{11}	3×10^{13}
Frequency, Hz	1×10^{21} 1×10^{19}	3×10^{16} 1×10^{16}	8×10^{14}	4×10^{14} 3×10^{14}	1×10^{12} 1×10^{10}	1×10^6	1×10^4
Wave number, cm^{-1}	3×10^{10} 3×10^8	1×10^6 3×10^5	3×10^4	1.3×10^4 1×10^4	33 3	3×10^{-5}	3×10^{-7}
Energy, kJ mol^{-1}	4×10^8 4×10^6	1.2×10^4 4×10^3	330	170 125	0.4 4×10^{-3} 4×10^{-7}		4×10^{-9}
Phenomenon observed	Nuclear transitions	Inner electronic transitions $\sigma \rightarrow \sigma^*$	Outer electronic transitions $\pi \rightarrow \pi^*,\, n \rightarrow \pi^*$		Molecular vibration	Molecular rotation, electron spin resonance	Nuclear magnetic resonance

example. The physical significance of the dipole-moment vector is that it is related to the difference in the electronic dipole moments of the initial and final state. This difference in electronic dipole moment arises as the result of a different electron distribution in the two states and can be thought of as representing charge migration during the transition. When the integral becomes zero, the transition is totally forbidden.

Mechanisms too complex to list here cause various degrees of breakdown in the selection rules. Consequently, some transitions predicted as forbidden may appear as weak lines.

REGIONS OF SPECTRUM

Figure 18.5 summarizes various branches of spectroscopy. As far as biochemistry is concerned, the spectroscopic techniques that are most frequently employed are visible and uv absorption, fluorescence, nuclear magnetic resonance, and infrared. However, to gain a better understanding of molecular spectroscopy, it is useful to have some knowledge of other branches of spectroscopy as well. For this reason, we present a systematic treatment of the various types of interaction between electromagnetic radiation and molecules that give rise to absorption and emission processes. Furthermore, we begin with nuclear magnetic resonance because the interaction between nuclear spins and an external magnetic field lies lowest on the energy scale. Equally important, in the simplest case a nuclear magnetic resonance experiment involves only a two energy-level system which is considerably easier to deal with than the hundreds or even thousands of energy levels that must be considered in infrared, visible, and uv spectroscopies.

18.2 Nuclear Magnetic Resonance Spectroscopy

The fundamental properties of a proton are (1) mass, (2) charge, and (3) spin, represented by the symbol I. Unlike the electron case (where $S = \frac{1}{2}$), I can be zero, an integer, or a half-integer. The actual value of I depends on the particular nucleus.

Since the spinning motion of a charge particle generates a magnetic field, nuclei that have spin $I \neq 0$ will possess a *magnetic dipole moment*, μ_I. When these nuclei are placed between the poles of an external magnet, the energy of interaction is given by

$$E = -\mu_I H \cos \theta \qquad (18.10)$$

where H is the external magnetic field and θ the angle between the vectors μ_I and H. The minus sign indicates that this is an attractive interaction. Classically, θ can have any value so that E varies continuously. Quantum mechanics restricts μ_I in the magnetic field to certain allowed orientations characterized by the nuclear spin quantum number $M_I = \pm\frac{1}{2}$. The energy is now given by

$$E = -\frac{\mu_I}{I} H M_I \qquad (18.11)$$

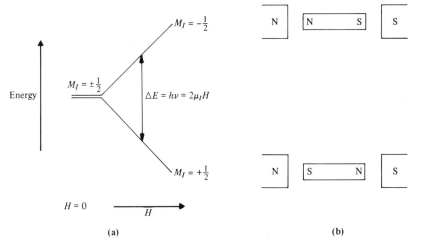

(a)   (b)

Resonance condition for a nuclear spin in the presence of an external magnetic field: (a) quantum-mechanical description; and (b) classical description.

Figure 18.6

In the absence of the magnetic field, that is, when $H = 0$, these two orientations are equivalent in energy, but degeneracy is removed when an external field is turned on. The splitting of the energy levels, together with its classical analog, is shown in Figure 18.6. If the nucleus is now irradiated with the appropriate frequency such that*

$$E_2 - E_1 = \Delta E = h\nu = 2\mu_I H \qquad (18.12)$$

transition from the lower to higher level can be induced. Such a transition gives rise to the phenomenon known as *nuclear magnetic resonance* (*nmr*). Table 18.2 lists some nuclei for which the nmr technique is applicable.

The basic features of an nmr spectrometer are (1) a source of radiation, (2) a receiver coil to detect the absorption of energy, (3) a dc magnetic field, and (4) a recorder or an oscilloscope to display the signals. The sample tube is placed between the poles of the magnet operating at about 14,000 gauss (G),† which for protons corresponds to a frequency of 6×10^7 Hz, or 60 MHz. This falls in the radio-frequency range. According to Eq. (18.12) there are two

Nuclei Suitable for Nmr Study

Table 18.2

Isotope	Nuclear Spin (I)	Natural Abundance (percent)
^{1}H	$\frac{1}{2}$	99.985
^{13}C	$\frac{1}{2}$	1.5×10^{-2}
^{14}N	1	99.63
^{15}N	$\frac{1}{2}$	0.37
^{17}O	$\frac{5}{2}$	3.7×10^{-2}
^{19}F	$\frac{1}{2}$	100
^{31}P	$\frac{1}{2}$	100

* The reader must not be confused by the Δ symbol, which is used for both the uncertainty principle [Eq. (18.5)] and for the change in energy between two levels.
† In SI units, the magnetic field strength is in tesla (T), where 1 T = 10,000 G.

Figure 18.7

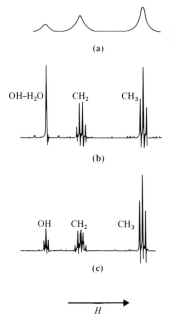

(a) Low-resolution and (b) high-resolution pmr spectrum of ethanol. (c) Pmr spectrum of very pure ethanol. [Parts (b) and (c) from G. Glaros and N. H. Cromwell, *J. Chem. Educ.* **48**, 202 (1971).]

variables: v and H. Experimentally, it is more convenient to hold the frequency constant and vary H until the resonance condition is met. The absorption of energy by the nuclei is then detected, amplified, and displayed on a recorder or oscilloscope.

Since many compounds contain hydrogen atoms, ^1H nmr or *proton magnetic resonance* (*pmr*) is the most studied case. Consider the pmr spectrum of C_2H_5OH shown in Figure 18.7. Following the reasoning above, we expect only one line to result from the protons present. However, three peaks are actually observed (see Figure 18.7a). The ratio of these peaks is $1:2:3$, which suggests that they represent the $-OH$, $-CH_2$, and $-CH_3$ protons.* The fact that three separate peaks are observed implies that the magnetic field H at each type of nuclei is different from the external magnetic field H_0. We write

$$H = H_0(1 - \sigma) \qquad (18.13)$$

where σ, a dimensionless quantity, is called the *screening constant*. In general, σ is a small number (about 10^{-5} for protons) whose magnitude depends on the electronic structure around the nucleus in question. The modified resonance condition is given in Figure 18.8.

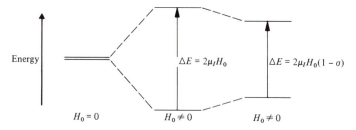

Figure 18.8   Effect of electronic shielding on the nuclear magnetic resonance condition.

* More properly, the area under the peaks, which represents the number of nuclear spins.

Figure 18.9

Relative chemical shift between A and B nuclei.

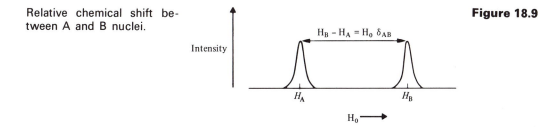

Consider a spectrum consisting of two lines representing two types of protons A and B (Figure 18.9). The magnetic fields at A and B are given by

$$H_A = H_0(1 - \sigma_A)$$

$$H_B = H_0(1 - \sigma_B)$$

where σ_A and σ_B are the screening constants for A and B. The separation
between these two lines is given by the difference

$$H_B - H_A = H_0(1 - \sigma_B) - H_0(1 - \sigma_A)$$

$$= H_0(\sigma_A - \sigma_B) = H_0\delta_{AB} \qquad (18.14)$$

where δ_{AB} is known as the *chemical shift* of A relative to B, or vice versa. The
separation between A and B is directly proportional to H_0. For example, if
these two lines are separated by 100 Hz at 60 MHz, the separation will be
$100 \times 220/60$, or 366.7 Hz for a spectrometer operating at 220 MHz.* The
ethanol case shows how different protons absorb at different fields; therefore,
it is convenient to have a common reference compound for chemical-shift
measurements. The compound chosen for most cases in organic systems is
tetramethylsilane (TMS), $(CH_3)_4Si$, because (1) it contains 12 protons of the
same type, so only a small amount of it is needed as an internal reference; (2)
it is chemically inert; and (3) its protons have a resonance field higher than
most others. The practical chemical-shift scale is defined as

$$\delta = \frac{H_{sample} - H_{TMS}}{H_{TMS}} \times 10^6 \text{ ppm} \qquad (18.15)$$

Since H_{sample} and H_{TMS}, the resonance frequencies of sample and reference,
differ by only an order of 100 Hz or so, while their individual values are of the
order of 60 MHz, the ratio is multiplied by a factor of 10^6 in order to make δ
a convenient number to work with. Generally, δ is expressed as parts per
million, ppm. Figure 18.10 gives the chemical shifts for various types of
protons.

Under high resolutions, the spectrum of ethanol is that shown in Figure
18.7b. The $-CH_2$ and $-CH_3$ peaks actually consist of four and three lines of
relative intensities $1:3:3:1$ and $1:2:1$, respectively. The spacing between each
group of lines is independent of the spectrometer frequency, so it cannot be a
chemical-shift effect. How can we explain this observation? The protons in
each group interact with the external magnetic field, yet they are also affected
by local magnetic fields generated by protons in the adjacent group. With two
possible orientations for each nuclear spin, the methyl peak is split into two
lines by the first proton in the methylene group. Each of the two lines is then

---

* It is customary in nmr work to express the magnetic field in frequency rather than in gauss.
From Eq. (18.12) we can show that 1 G = 2.83×10^6 Hz.

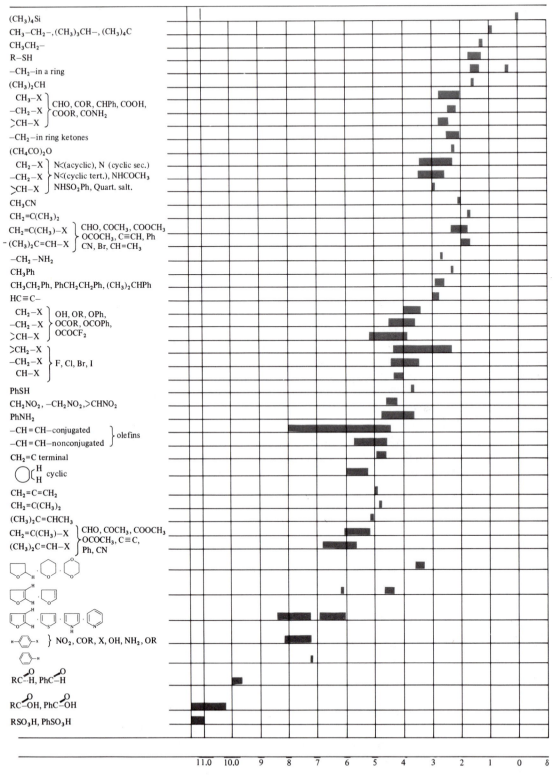

Figure 18.10 Chemical-shift chart for various organic compounds. [From E. Mohacsi, *J. Chem. Educ.* **41**, 38 (1964).]

Spin–spin splitting between —CH$_2$ and —CH$_3$ groups in ethanol. The coupling constant (J) is the same in both cases.

Figure 18.11

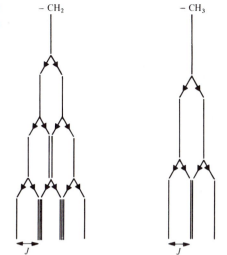

further split by the second proton in the methylene group and a total of four lines are observed. We see only three lines, because two of the lines fall on top of each other. Similarly, four lines are obtained for the —CH$_2$ group (Figure 18.11). The separation between the lines in each group gives the *spin–spin coupling constant* (J), whose magnitude is determined by the extent of this magnetic interaction.*

The ethanol spectrum is as yet incompletely described because the absence of spin–spin interaction between the methylene and hydroxyl group still must be accounted for. Actually, in pure ethanol the hydroxyl peak is indeed split into a 1:2:1 triplet by the methylene group, and each of the four lines in the methylene group is further split into a doublet of equal intensity by the hydroxyl proton. The separation between the —OH and —CH$_3$ groups is too great for us to observe any splittings. If there is a small amount of acid or base present, though, a rapid proton or hydroxide ion-exchange reaction between the —OH group and H$_2$O effectively removes the spin–spin interaction. The exchange reactions are

$$C_2H_5OH + H_3O^+ \rightleftharpoons C_2H_5OH_2^+ + H_2O \qquad \text{acid-catalyzed}$$

$$C_2H_5OH' + C_2H_5OH_2^+ \rightleftharpoons C_2H_5OH + C_2H_5OHH'^+ \qquad \text{exchange}$$

$$C_2H_5OH + OH^- \rightleftharpoons C_2H_5O^- + H_2O \qquad \text{base-catalyzed}$$

$$C_2H_5OH + C_2H_5O^- \rightleftharpoons C_2H_5O^- + C_2H_5OH \qquad \text{exchange}$$

The nmr technique is particularly useful in compound identifications. Measuring the chemical shifts and spin–spin coupling constants makes it possible to deduce the type of protons present as well as the arrangement of atoms in the molecule. Spin–spin interaction makes the task of analyzing spectra of large molecules quite complex. Spectroscopists in the nmr field are fortunate that naturally abundant carbon (^{12}C) and oxygen (^{16}O) both have zero spin. Otherwise, they would have to limit structural studies to rather simple molecules.

* Theory postulates that protons within the same group or protons that have the same chemical shift do not interact with one another in this manner.

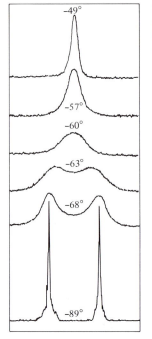

Figure 18.12

Pmr spectra of deuterated cyclohexane ($C_6D_{11}H$) at various temperatures (degrees Celsius). (From F. A. Bovey, *Nuclear Magnetic Resonance Spectroscopy*, Academic Press, Inc., New York, 1969.)

−49°

−57°

−60°

−63°

−68°

−89°

Nuclear magnetic resonance is also convenient for studying a number of chemical rate processes, such as proton exchange reaction, rotation about a single bond, and ring inversion. Consider the conformational change in cyclohexane:

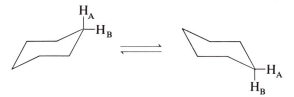

The pmr spectrum of cyclohexane is rather complex because of spin–spin interactions. By a procedure known as *spin decoupling*, these interactions can be removed, leaving only two lines for observation, representing the axial and equatorial protons. At −50°C, the ring inversion frequency is so great that a spectrometer can record only the average signal from the two kinds of protons. On lowering the temperature, the peak first broadens and finally splits into two lines of equal intensity (Figure 18.12). From an analysis of the line width and line shape, the energy of activation for the ring inversion is found to be about 42 kJ mol^{-1}.

Proton magnetic resonance spectra are most frequently encountered, but ^{13}C nmr spectra are increasing in importance to investigations of biological molecules. Because the ^{13}C isotope is not naturally abundant and has a smaller magnetic moment, ^{13}C spectra for most compounds were once difficult to record. A recently developed technique, called *Fourier transform spectroscopy*, now permits us to make many repeated scans of the same spectrum in a short period of time, accumulating signals. As a result of this technique, sensitivity is no longer a major problem. In Figure 18.13 we show the nmr spectra of ^1H, ^{13}C, and ^{31}P of a small biomolecule, adenosine-5′-triphosphate (ATP).

518

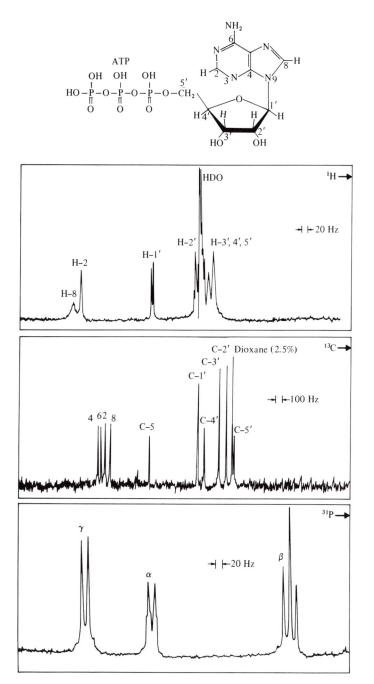

^1H, ^{13}C, and ^{31}P nmr spectra of adenosine-5'-triphosphate. (By permission of Varian Associates, Palo Alto, Calif.)

Figure 18.13

Electron Spin Resonance Spectroscopy **18.3**

Electron spin resonance (esr) is very similar to nmr in theory. The spinning motion of an electron also generates a magnetic field and the orientation of the electron magnetic moment in an external magnetic field is characterized

519

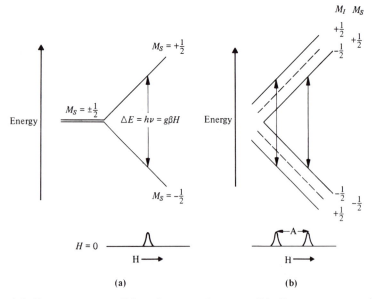

Figure 18.14 (a) Resonance condition for an electron. (b) Resonance condition for the electron in a hydrogen atom.

by the electron spin quantum number $M_S = \pm\frac{1}{2}$. The resonance condition is usually expressed as

$$\Delta E = h\nu = g\beta H \qquad (18.16)$$

where g is a dimensionless constant, called the Landé g factor, equal to 2.0023,* and β is the *Bohr magneton*, given by $eh/2\pi mc$, where e and m are the electronic charge and mass and c is the velocity of light (Figure 18.14).

Because the electron magnetic moment is about 600 times greater than that for a proton, the experimental setup for esr experiments is different from that for the nmr. Normally, an esr measurement is carried out in a magnetic field of about 3400 G and a frequency of 9.5×10^9 Hz or 9.5 GHz, which falls in the microwave region. Spectrometer designs usually require that esr lines be presented as the first derivative of the absorption lines (Figure 18.15).

In most molecules electrons regularly occur in pairs with opposite spins as required by the Pauli exclusion principle; hence esr experiments cannot be performed. A few molecules, such as O_2, NO, NO_2, and ClO_2, do contain

Figure 18.15

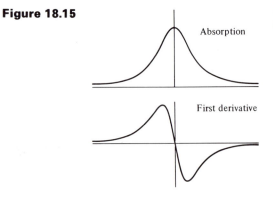

Relation between an absorption line and its first derivative.

* The *g factor* is the ratio of the electron's magnetic moment to its spin angular momentum.

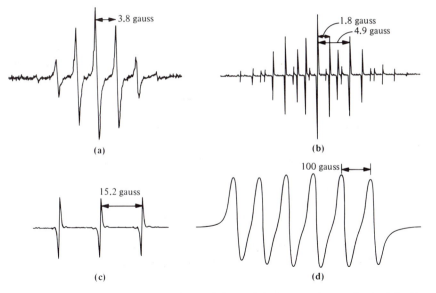

(a) 3.8 gauss

(b) 1.8 gauss 4.9 gauss

(c) 15.2 gauss

(d) 100 gauss

Esr spectra: (a) benzene anion radical; (b) naphthalene anion radical; (c) di-*tert*-butyl nitroxide radical; and (d) Mn^{2+} ion in water. **Figure 18.16**

one or more unpaired electrons in their stable ground state, and the esr spectrum for each molecule has been observed. Many transition-metal ions, such as Fe^{3+}, Mn^{2+}, and Cu^{2+}, also possess unpaired d electrons.* This fact is of particular importance in biochemistry because these paramagnetic ions are required cofactors for many proteins and enzymes.

Although only one transition and therefore one line is observed for isolated electrons, or electrons trapped in a matrix, the esr spectrum of hydrogen atoms consists of two lines of equal intensity, as shown in Figure 18.14b. This *hyperfine splitting* results from the magnetic interaction between the unpaired electron and the nucleus, analogous to the spin–spin interaction discussed earlier for the nmr spectrum of ethanol. Only two transitions are allowed, however, because of the selection rules, $\Delta M_I = 0$ and $\Delta M_S = \pm 1$. An interpretation of the selection rules is that the motion of a nucleus is much slower than that of an electron, so that during the time it takes for an electron to change its orientation, the nuclear spin has no time to reorient. The separation between these two lines gives the *hyperfine splitting constant* (A).

In general, the number of hyperfine lines can be predicted by the quantity $(2nI + 1)$, where n is the number of the same type or equivalent protons and I is the nuclear spin. Figure 18.16 shows the esr spectra of the benzene and naphthalene anion radicals, prepared by the reduction with alkali metals in ether, di-*tert*-butyl nitroxide, a neutral paramagnetic species, and Mn^{2+} ion in water:

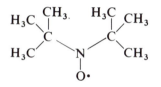

di-*tert*-butyl nitroxide

* Although all transition-metal ions possess unpaired electrons, esr experiments cannot always be performed on these ions. According to a theorem by Kramer, the most suitable ions for esr study are those containing an *odd* number of unpaired electrons.

18.4 Microwave Spectroscopy

Microwave spectroscopy is concerned with the rotational motion of molecules. The molecule must possess a permanent electric dipole moment to enable observation of its pure rotation spectrum. Thus homonuclear diatomic molecules such as O_2 and N_2 are microwave *inactive*. On the other hand, molecules such as HCl and CO can be studied by this technique, as can many polyatomic molecules.

Let us consider a heteronuclear diatomic molecule, which we assume to be a rigid rotor: that is, it behaves like a dumbbell. The moment of inertia (I) of the molecule about the center of gravity is given by

$$I = m_1 r_1^2 + m_2 r_2^2 \tag{18.17}$$

where m_1 and m_2 are the atomic masses and r_1 and r_2 are the separations defined in Figure 18.17. Also, the center of gravity requires that

$$m_1 r_1 = m_2 r_2 \tag{18.18}$$

Therefore, we write

$$r_1 = \frac{m_2}{m_1} r_2 = \frac{m_2}{m_1}(r - r_1) \tag{18.19}$$

From Eqs. (18.18) and (18.19), we obtain

$$r_1 = \frac{m_2}{m_1 + m_2} r \tag{18.20}$$

$$r_2 = \frac{m_1}{m_1 + m_2} r \tag{18.21}$$

Substituting Eqs. (18.20) and (18.21) into (18.17), we get

$$I = m_1 \left(\frac{m_2}{m_1 + m_2}\right)^2 r^2 + m_2 \left(\frac{m_1}{m_1 + m_2}\right)^2 r^2$$

$$= \frac{m_1 m_2}{m_1 + m_2} r^2 = \mu r^2 \tag{18.22}$$

where μ, the reduced mass, is defined as

$$\frac{1}{\mu} = \frac{1}{m_1} + \frac{1}{m_2} \tag{18.23}$$

The reduced mass is physically significant, for now the rotational motion of the diatomic molecule can be treated as a single particle of mass μ describing a circle of radius r.

Figure 18.17 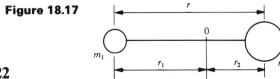 Diatomic molecule as a rigid rotor.

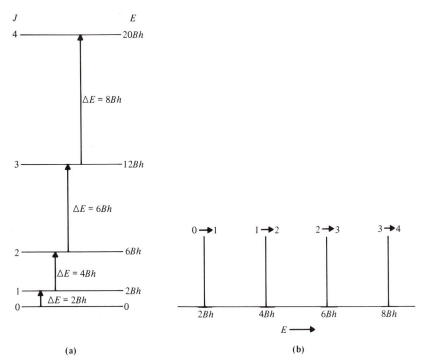

(a) Resonance condition for microwave transitions. (b) Equally spaced rotational lines. In practice, these lines are of unequal intensities. **Figure 18.18**

From the solution of the Schrödinger equation, we obtain the following quantized energies for rotation*:

$$E_{\text{rot}} = \frac{J(J + 1)h^2}{8\pi^2 I} = BJ(J + 1)h \tag{18.24}$$

where B is the *rotational constant*, given by $h/8\pi^2 I$ and J is the rotational quantum number, where

$$J = 0, 1, 2, \ldots$$

Transition from the lower to the upper level can be induced by irradiating the molecule with the appropriate frequency. The reader should note that the interaction here is between the electric field component of the electromagnetic radiation (see Figure 15.4) and the electric dipole of the molecule.† Not all transitions are allowed because of the selection rules‡ $\Delta J = \pm 1$, as shown in Figure 18.18.

Let J' and J'' be the rotational quantum numbers in the upper and lower levels. The energy difference is then given by

$$\Delta E = BJ'(J' + 1)h - BJ''(J'' + 1)h$$
$$= Bh[J'(J' + 1) - J''(J'' + 1)]$$

Since $J' - J'' = 1$, this equation reduces to

$$\Delta E = 2BhJ' \qquad J' = 1, 2, 3, \ldots \tag{18.25}$$

* See any of the standard physical chemistry texts listed in Chapter 1.
† In nmr and esr experiments the interaction is between the magnetic field component of the electromagnetic radiation and the magnetic dipoles.
‡ For a discussion of this selection rule, see C. T. Moynihan, *J. Chem. Educ.* **46**, 431 1969).

Thus absorptions $J = 0 \rightarrow 1$, $1 \rightarrow 2$, $2 \rightarrow 3$, ... will have energy difference of $2Bh$, $4Bh$, $6Bh$, ..., and a set of equally spaced lines with separation of $2Bh$ will be obtained.

Under high resolutions, spacing between adjacent lines in a rotational spectrum is found to decrease with increasing J. In higher levels the molecule rotates faster, so the internuclear bond is stretched somewhat by the centrifugal force. An increase in bond length r increases the moment of inertia I, causing E_{rot} to decrease [see Eq. (18.24)]. This effect can be corrected by adding a term to Eq. (18.24) as follows:

$$E_{rot} = BJ(J + 1)h - D[J(J + 1)]^2 h \qquad (18.26)$$

where D, the centrifugal constant, is about 1000 times smaller than B. Thus the second term can usually be neglected unless J is large in value.

Microwave spectroscopy is an important tool for determining molecular geometry. From the separation between adjacent lines we obtain the rotational constant B and hence the moment of inertia and internuclear distance.

E X A M P L E 18.1 The average spacing between adjacent lines in the rotational spectrum of HCl is 20.8 cm^{-1}. Calculate the moment of inertia and the internuclear distance of HCl.

CGS units: The separation in frequency is given by

$$\Delta v = \frac{c}{\Delta \lambda} = c\Delta \bar{v} = 2.998 \times 10^{10} \text{ cm s}^{-1} \times 20.8 \text{ cm}^{-1}$$

$$= 6.24 \times 10^{11} \text{ s}^{-1} = 6.24 \times 10^{11} \text{ Hz}$$

Since

$$\Delta E = h \, \Delta v = 2Bh$$

$$\Delta v = 2B$$

or

$$B = 3.12 \times 10^{11} \text{ Hz}$$

Now

$$B = \frac{h}{8\pi^2 I}$$

so that

$$I = \frac{h}{8\pi^2 B}$$

$$= \frac{6.626 \times 10^{-27} \text{ erg s}}{8\pi^2 \times 3.12 \times 10^{11} \text{ s}^{-1}}$$

$$= 2.69 \times 10^{-40} \text{ erg s}^2$$

$$= 2.69 \times 10^{-40} \text{ g cm}^2$$

$$1 \text{ erg} = 1 \text{ dyn cm}$$

$$= 1 \text{ g cm}^2 \text{ s}^{-2}$$

Since

$$I = \frac{m_H m_{Cl}}{m_H + m_{Cl}} r^2$$

we write

$$2.69 \times 10^{-40} \text{ g cm}^2 = \frac{1.00797 \times 35.453}{1.00797 + 35.453} \times 1.660 \times 10^{-24} \text{ g } r^2$$

$$r^2 = 1.65 \times 10^{-16} \text{ cm}^2$$

$$r = 1.28 \times 10^{-8} \text{ cm}$$

$$= 1.28 \text{ Å}$$

$$(1 \text{ amu} = 1.660 \times 10^{-24} \text{ g})$$

SI units: Again we have $B = 3.12 \times 10^{11}$ Hz so that

$$I = \frac{h}{8\pi^2 B}$$

$$= \frac{6.626 \times 10^{-34} \text{ J s}}{8\pi^2 \times 3.12 \times 10^{11} \text{ s}^{-1}}$$

$$= 2.69 \times 10^{-47} \text{ J s}^2$$

$$= 2.69 \times 10^{-47} \text{ kg m}^2$$

$$1 \text{ J} = 1 \text{ N m}$$

$$= 1 \text{ kg m}^2 \text{ s}^{-2}$$

Since

$$I = \frac{m_H m_{Cl}}{m_H + m_{Cl}} r^2$$

$$2.69 \times 10^{-47} \text{ kg m}^2 = \frac{1.00797 \times 35.453}{1.00797 + 35.453} \times 1.660 \times 10^{-27} \text{ kg } r^2$$

$$r^2 = 1.65 \times 10^{-20} \text{ m}^2$$

$$r = 1.28 \times 10^{-10} \text{ m}$$

$$= 1.28 \text{ Å}$$

$$(1 \text{ amu} = 1.660 \times 10^{-27} \text{ kg})$$

Infrared Spectroscopy 18.5

Infrared (ir) *spectroscopy* deals with the vibrational motion of molecules. Again, the simplest system is a diatomic molecule that is treated as a *simple harmonic oscillator*. The motion of a simple harmonic oscillator is such that the force acting on it is proportional to its displacement from a position of equilibrium.

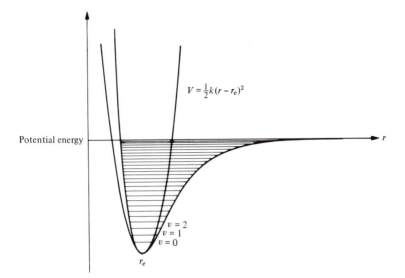

Figure 18.19 Potential-energy curve for a diatomic molecule. The symmetric curve is that given by Eq. (18.28); the other curve represents the actual behavior of the molecule.

Further, the variation of the displacement with time is sinusoidal in nature. The frequency of vibration is given by*

$$v = \frac{1}{2\pi} \sqrt{\frac{k}{\mu}} \tag{18.27}$$

where μ is the reduced mass and k is the force constant determined by the strength of the bond. In CGS units, k is dyn cm^{-1}; in SI units, k is N m^{-1}. The potential energy V of the oscillator as a function of the displacement is represented by the parabolic equation

$$V = \tfrac{1}{2}k(r - r_e)^2 \tag{18.28}$$

where r_e is the equilibrium distance. A plot of V versus r is shown in Figure 18.19.

The quantized vibrational energies are given by the following formula from quantum mechanics:

$$E_{\text{vib}} = (v + \tfrac{1}{2})hv \tag{18.29}$$

where v is the fundamental frequency of vibration as defined by Eq. (18.27) and v is the vibrational quantum number, where

$$v = 0, 1, 2, 3, \ldots$$

This shows that the lowest vibrational energy of a molecule is not zero but is equal to $\tfrac{1}{2}hv$, called the *zero-point energy*.

EXAMPLE 18.2 The fundamental frequency of vibration for HCl is 8.67×10^{13} Hz. Calculate the force constant for the molecule.

---

* For derivation, see H. B. Dunford, *Elements of Diatomic Molecular Spectra*, Addison-Wesley Publishing Company, Inc., Reading, Mass., 1968, p. 26, or any standard text on spectroscopy or quantum mechanics.

$$\frac{1}{\mu} = \frac{1}{m_H} + \frac{1}{m_{Cl}}$$

$$\mu = \frac{m_H m_{Cl}}{m_H + m_{Cl}}$$

$$= \frac{1.00797 \times 35.453}{1.00797 + 35.453} \times 1.660 \times 10^{-24} \text{ g}$$

$$= 1.63 \times 10^{-24} \text{ g}$$

Now

$$v = \frac{1}{2\pi} \sqrt{\frac{k}{\mu}}$$

so that

$$k = 4\pi^2 v^2 \mu$$

$$= 4\pi^2 (8.67 \times 10^{13} \text{ s}^{-1})^2 \times 1.63 \times 10^{-24} \text{ g}$$

$$= 4.84 \times 10^5 \text{ g s}^{-2}$$

$$= 4.84 \times 10^5 \text{ g cm s}^{-2} \text{ cm}^{-1}$$

$$= 4.84 \times 10^5 \text{ dyn cm}^{-1}$$

SI units: Here we have $\mu = 1.63 \times 10^{-27}$ kg so that

$$k = 4\pi^2 v^2 \mu$$

$$= 4\pi^2 (8.67 \times 10^{13} \text{ s}^{-1})^2 \times 1.63 \times 10^{-27} \text{ kg}$$

$$= 4.84 \times 10^2 \text{ kg s}^{-2}$$

$$= 4.84 \times 10^2 \text{ kg m s}^{-2} \text{ m}^{-1}$$

$$= 4.84 \times 10^2 \text{ N m}^{-1}$$

Conversion factor: $1 \text{ N m}^{-1} = 1 \times 10^3 \text{ dyn cm}^{-1}$

Molecules do not behave exactly like simple harmonic oscillators. For example, as r increases, the chemical bond becomes weaker, and eventually dissociation takes place. A more realistic description of molecular vibration is presented by the asymmetric curve in Figure 18.19. Each horizontal line represents a vibrational energy level. Spacing between successive levels decreases with increasing v, as the result of the *anharmonic* character of the vibration. As a correction, we rewrite Eq. (18.29) as

$$E_{vib} = (v + \tfrac{1}{2})hv - x(v + \tfrac{1}{2})^2 hv \qquad (18.30)$$

where x is the anharmonicity constant. Analogous to the rotational spectrum case, x can be neglected except for large values of v.

In order for a particular vibration to be ir-active, that is, for the motion to appear as an absorption line, we must have

$$\frac{d\mu}{dr} \neq 0$$

Figure 18.20

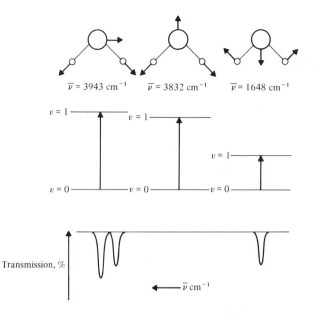

The three fundamental modes of vibration of H_2O. Under high resolution, each of these three peaks shows a number of lines as a result of simultaneous transition between rotational levels.

which means that the electric dipole moment must *change* during a vibration. This requirement immediately rules out all homonuclear diatomic molecules. Further theoretical considerations lead to the following selection rules: $\Delta v = \pm 1$.

A diatomic molecule has only one vibrational degree of freedom and so only one fundamental frequency of vibration. A nonlinear polyatomic molecule such as H_2O or SO_2 has 3 ($3 \times 3 - 6 = 3$) vibrational degrees of freedom (see Section 3.8). We say that it has three different modes of vibration. The apparently complex vibrational motion can be analyzed in terms of these three fundamental frequencies, shown in Figure 18.20. A total of three lines is expected from this scheme if we consider only the $v = 0 \to 1$ transition in each case.

As another example, let us consider a linear molecule such as CO_2, which has four degrees of vibrational freedom (Figure 18.21). Although CO_2 itself does not possess a permanent electric dipole moment, three of the four vibrations are ir-active. In those three vibrations, there is a change of dipole moment with respect to bond distance. Two of the three vibrations are called "degenerate" because they have the same frequencies.

It is possible to obtain pure rotation spectra but not pure vibrational spectra. For any given vibrational state v, there are a set of rotational levels. For example, the rotational levels shown in Figure 18.18 are assumed to associate with the $v = 0$ level. Generally, then, a $v = 0 \to 1$ transition is invariably accompanied by a simultaneous transition between the two rotational levels associated with the lower and upper vibrational states. The

Figure 18.21

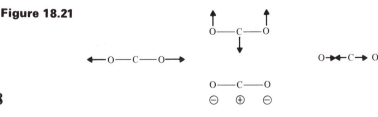

The four fundamental modes of vibration of CO_2. The middle two vibrations have the same frequency. Only the extreme left vibration is ir-inactive.

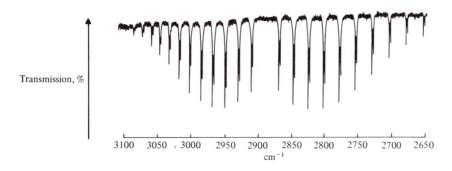

Infrared spectrum of HCl gas. The more intense line of each doublet is due to **Figure 18.22** $H^{35}Cl$ (75%); the weaker line is due to $H^{37}Cl$ (25%). [From J. L. Hollenberg, *J. Chem. Educ.* **47**, 2 (1970).]

selection rules $\Delta J = \pm 1$ still hold in this case. Under high resolution, each line in Figure 18.20 is found to consist of many rotational fine structures. Figure 18.22 shows the experimental spectrum of gaseous HCl.

From Eqs. (18.24) and (18.29), the energy difference for such a transition is

$$\Delta E = (v' + \tfrac{1}{2})h\nu + BJ'(J' + 1)h - (v'' + \tfrac{1}{2})h\nu - BJ''(J'' + 1)h \qquad (18.31)$$

where v' and v'' represent the higher and lower vibrational states, and J' and J'' the rotational levels in the v' and v'' states, respectively. For the $v = 0 \rightarrow 1$ transition, that is, $v' - v'' = 1$, we obtain

$$\Delta E = h\nu + Bh[J'(J' + 1) - J''(J'' + 1)]$$

In most cases, an ir spectrum can be divided into two branches, called P and R, according to the following:

$$\text{P branch:} \quad J' = J'' - 1 \qquad \Delta E = h\nu - 2BJ''$$

$$\text{R branch:} \quad J' = J'' + 1 \qquad \Delta E = h\nu + 2B(J'' + 1)$$

Infrared spectroscopy is a highly useful technique for chemical analysis. The complexity of molecular vibrations virtually assures that two different molecules cannot produce identical ir spectra. Matching the ir spectrum of an unknown with that of a standard compound, a procedure called *fingerprinting*, is the unequivocal method of identification. Over 100,000 reference spectra have been recorded for fingerprinting. The details of an ir spectrum reveals much information about the structure and bonding of the molecule. A group-frequency correlation chart for some common functional groups is given in Figure 18.23. Such a chart is quite analogous to the nmr chemical-shift chart discussed earlier. Figure 18.24 shows the ir spectrum of a relatively simple molecule acrylonitrile, $CH_2=CHCN$, and the assignment of its major peaks.

The ir technique is of great value in the study of intra- and intermolecular hydrogen bonding. Formation of a hydrogen bond,

$$A—H\cdots B$$

will cause the A—H stretching band to broaden and shift to lower frequencies. In addition, new bands might appear as a result of the torsional and bending motions of the system.

529

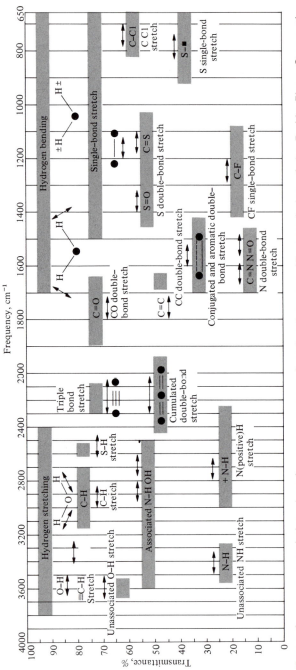

Figure 18.23 Group-frequency correlation chart for some common functional groups. (By permission of the Perkin-Elmer Corporation, Norwalk, Conn.)

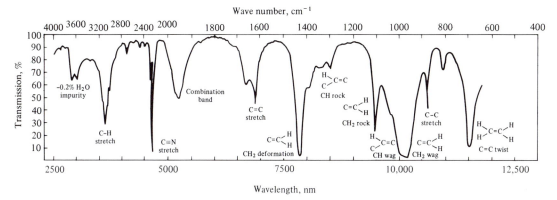

Infrared spectrum of acrylonitrile. (From "Chemical Analysis by Infrared," **Figure 18.24** B. Crawford, Jr. Copyright © 1953 by Scientific American, Inc. All rights reserved.)

In many cases ir can distinguish structural isomers by their characteristic frequencies. For example, in bromocyclohexane, the axial and equatorial C—Br stretching frequencies are

The equilibrium constant for the conformational change is readily obtained from the ratio of the areas under these peaks.

Finally, we note that it is a customary, although unfortunate practice, in ir work to express frequency as cm^{-1} rather than Hz. The reason is that the frequency of molecular vibration is usually very high, ranging from 1×10^{13} Hz to 1×10^{14} Hz. The corresponding wave-number values are between 330 cm^{-1} and 3300 cm^{-1}, which are more convenient to use. Further, the energy of vibration is directly proportional to the wave number, since $E = h\nu = hc/\lambda = hc\bar{\nu}$.

Electronic Spectroscopy **18.6**

Both visible and ultraviolet (uv) spectra involve electronic transitions. Within the framework of molecular orbital theory, electrons can be classified as σ, π, or n (nonbonding), depending on the orbitals in which they reside. A gross selection rule, which is based on symmetry considerations, requires that only transitions of the type $\sigma \to \sigma^*$, $\pi \to \pi^*$, and $n \to \pi^*$ are allowed. Since the energy difference between σ and σ^* orbitals is quite large, these transitions usually fall in the uv region. On the other hand, $\pi \to \pi^*$ and $n \to \pi^*$ transitions may fall in either the uv or visible region.

An electronic transition is always accompanied by simultaneous changes in rotational and vibrational states. Large moments of inertia in polyatomic molecules rule out the observation of rotational fine structures [see Eq. (18.24)], and vibrational transitions usually appear as broad, unresolved bands. Analogous to the group vibrational frequencies in ir, electronic spectra

531

Table 18.3

Some Common Chromophores and Their Approximate Maximum Absorption Wavelengths

Chromophore	λ_{max} (nm)
$>\!C\!=\!C\!<$	190
$>\!C\!=\!C\!-\!C\!=\!C\!<$	210
(benzene ring)	190
	260
$>\!C\!=\!O$	190
	280
$-C\equiv N$	160
$-COOH$	200
$-N\!=\!N\!-$	350
$-NO_2$	270

can often be characterized by functional groups such as $>\!C\!=\!C\!<$, $>\!C\!=\!O$, $-NO_2$, and $-Ph$, which are called *chromophores*. Table 18.3 lists the wavelengths of absorption for some common chromophores. The actual maximum adsorption wavelength for these chromophores depends not only on the molecule involved but also on the environment, being affected by such variables as solvent and temperature.

The visible region is roughly between 400 and 750 nm. An object appears colored because it selectively absorbs certain wavelengths and transmits (or reflects) others in this region. For example, copper sulfate solution, which absorbs in the yellow region, appears as yellow's complementary color, blue. Table 18.4 lists the various wavelengths absorbed in the visible region and the complementary color observed.

Many transition-metal complexes are colored. A particularly simple system is $Ti(H_2O)_6^{3+}$, whose electronic absorption spectrum is shown in Figure 18.25. Since there is only one d electron present (see Figure 16.22) and hence only one transition, analysis is fairly straightforward. $Ti(H_2O)_6^{3+}$ is an example of d-d transition; that is, transition takes place from one d level to another.

Relationship of Wavelength to Color[a]

Table 18.4

Wavelength Absorbed (nm)	Color Observed (transmitted)
400 (violet)	Greenish yellow
425 (dark blue)	Yellow
450 (blue)	Orange
490 (blue-green)	Red
510 (green)	Purple
530 (yellow-green)	Violet
550 (yellow)	Dark blue
590 (orange)	Blue
640 (red)	Bluish green
730 (purple)	Green

[a] From N. J. Juster, *J. Chem. Educ.* **39**, 596 (1962).

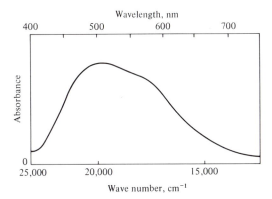

The absorption spectrum of $Ti(H_2O)_6^{3+}$. [From F. A. Cotton, *J. Chem. Educ.* **41**, 474 (1964).]

Figure 18.25

When several *d* electrons are present, analysis can be quite complex. In addition, we may also have charge-transfer bands (see later) as a result of metal–ligand interactions.

In saturated organic molecules, such as the alkanes, the transitions are of the $\sigma \rightarrow \sigma^*$ type. Aromatic molecules or compounds containing $>C=C<$ and $>C=O$ chromophores also have $\pi \rightarrow \pi^*$ and $n \rightarrow \pi^*$ transitions. Diphenylpolyenes $C_6H_5-(CH=CH)_n-C_6H_5$ are an interesting series as the chromophore $>C=C<$ has an important effect on their color. The transition from the highest filled π orbital to the lowest empty π^* orbital falls in the uv for $n = 1$ and 2. As *n* increases, there is a gradual shift toward the visible region, and the color changes from pale yellow for $n = 3$ to greenish black for $n = 15$ (Figure 18.26). Similarly, the highly conjugated porphyrin system (see Figure 16.23) is green in the chlorophyll molecule (complexed with Mg^{2+}) and red in hemoglobin (complexed with Fe^{2+}). In these cases the color of the complex is greatly affected by the extent of electron delocalization and the metal ion present.

The electronic spectra of most amino acids arise from the $\sigma \rightarrow \sigma^*$ transitions that occur in the far uv, below 230 nm. Phenylalanine, tryptophan, and tyrosine are the only exceptions. These molecules contain the —Ph chromophore, which absorbs strongly above 250 nm (Figure 18.27). Absorbance at 280 nm, due mainly to the tryptophan and tyrosyl residues, is used to measure the concentration of protein solutions.

Effect of increased conjugation on the $\pi \rightarrow \pi^*$ transitions in polyenes.

Figure 18.26

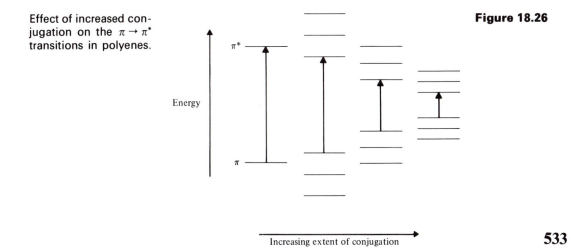

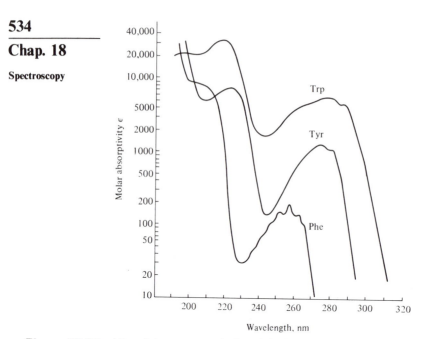

Figure 18.27 Ultraviolet spectra of phenylalanine (Phe), tryptophan, (Trp), and tyrosine (Tyr). [From D. C. Neckers, *J. Chem. Educ.* **50**, 164 (1973).]

The optical properties of DNA and RNA are studied in terms of the purines (adenine and guanine) and pyrimidines (cytosine, thymine, and uracil), whose spectra are shown in Figure 18.28. The concentration of nucleic acid solutions is determined by the absorbance at 260 nm. Both DNA and RNA exhibit an interesting phenomenon called *hypochromism*. In general, the molar absorptivity of intact DNA is some 20 to 40% lower than that which we would expect from a knowledge of the total number of nucleotides present.

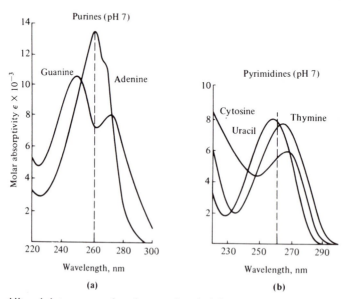

Figure 18.28 Ultraviolet spectra of purines and pyrimidines. (From A. L. Lehninger, *Biochemistry*, 2nd ed., Worth Publishers, Inc., New York, 1975.)

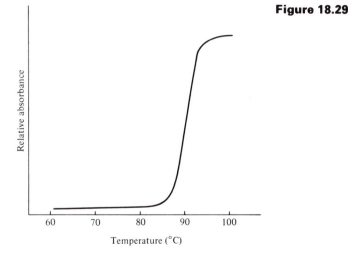

The relative absorbance of DNA at 260 nm as a function of temperature. The melting temperature (Tm) is about 90°C.

Figure 18.29

For example, molar absorptivity of calf thymus DNA at 260 nm increases from about 6500 to 9500 when the polymer undergoes thermal denaturation. Although the theory of hypochromism is beyond the scope of this text, the phenomenon can be explained as Coulombic interactions between electric dipoles induced by light absorption in the base pairs. The extent of this interaction depends on the orientation of the dipoles relative to one another. In a random orientation, there would be little or no interaction and no effect on the absorption spectrum. In the native state, the dipoles are stacked parallel on top of one another, leading to a *decrease* in the absorbance. This property has been successfully employed to follow the helix-coil transition in DNA. Figure 18.29 showing a *melting curve* of a DNA solution. The term *melting* here refers to the unwinding of the double-helical structure. The melting temperature T_m corresponds to the inflection point of the curve; its value depends on the base-pair composition of the DNA.

A special type of electronic spectra arises from *charge-transfer* interaction between a pair of molecules. When tetracyanoethylene, an acceptor, is dissolved in carbon tetrachloride, a colorless solution is obtained. Upon the addition of a small amount of a donor aromatic hydrocarbon, such as benzene or toluene, the solution immediately turns yellow (Figure 18.30). Many similar reactions have been observed. In 1952, Mulliken proposed the following scheme to explain the observed spectra:

$$D + A \rightleftharpoons [(D, A); (D^+, A^-)] \xrightarrow{h\nu} [(D, A); (D^+, A^-)]^*$$

$$\text{ground state} \qquad\qquad \text{excited state}$$

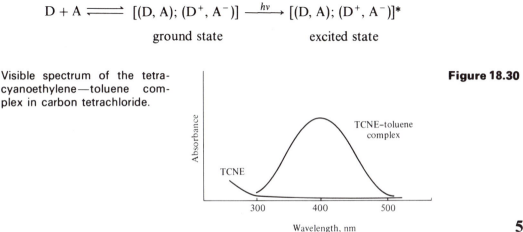

Visible spectrum of the tetracyanoethylene—toluene complex in carbon tetrachloride.

Figure 18.30

where D and A are the donor and acceptor molecules and (D, A) and (D$^+$, A$^-$) represent the covalent and ionic resonance structures of the charge-transfer complex. In the ground state the normal van der Waals forces are responsible for holding the molecules together, and there is little if any actual transfer of charge from D to A. However, when the complex is excited by a suitable wavelength, a large charge transfer takes place, and the ionic structure will now make a major contribution to the excited state. If the exciting wavelength falls in the visible region, the solution will appear colored. There is an interesting difference between this electronic transition and the normal absorption. Here an electron is excited from a lower level (bonding molecular orbital) in the donor molecule to a higher level (antibonding molecular orbital) in the acceptor molecule.

The tendency for charge-transfer formation generally depends on the ionization potential of the donor and the electron affinity of the acceptor. Charge-transfer spectra have also been observed for many transition-metal complexes. There the interaction is between the various ligands and the mental ion, with the former as the donor and the latter the acceptor.

The visible and uv technique is not as reliable for compound identifications as is the ir technique, since an electronic spectrum generally does not possess the fine details of an ir spectrum. However, electronic spectroscopy is a useful tool in quantitative analysis. The concentration of a solution can be readily determined from Beer's law by measuring the absorbance (if the molar absorptivity is known). An interesting situation may arise if a solution contains two absorbing substances in equilibrium whose bands overlap. At some wavelength over the overlapped region, the molar absorptivities of the two species may be equal. If the sum of the concentrations of these two compounds in solution is held constant, there will be *no* change in absorbance at this wavelength as the ratio of these two compounds is varied. This invariant point is called the *isosbestic point*. The existence of one or more isosbestic points in a system is a good indication of chemical equilibrium between two compounds. However, care must be exercised in such analyses; conceivably, a third compound having zero absorptivity at this wavelength could participate

Figure 18.31

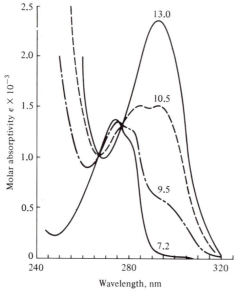

Absorption spectrum of tyrosine at pH values indicated. Note the isosbestic points at 267 nm and 277.5 nm. [From D. Schugar, *Biochem. J.* **52**, 142 (1952).]

in the overall reaction. Figure 18.31 shows the absorption spectra of tyrosine at various pH values. There are two isosbestic points, at 267 nm and 277.5 nm, as a result of the following equilibrium:

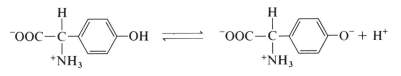

Fluorescence and Phosphorescence 18.7

So far, we have considered only the absorption process. If the excitation of a molecule by light does not lead to a chemical reaction such as dissociation or rearrangement, or energy transfer by collision, then the molecule will eventually return to the ground state through the release of a photon of energy $h\nu$. That release shows up as luminescent emission. In luminescence, there are two paths for the energy depletion in the excited molecule—fluorescence and phosphorescence.

Fluorescence is the emission of radiation that causes molecular transition from an excited state to a ground state without any change in spin multiplicity. Since electrons in molecules are paired according to the Pauli exclusion principle, the initial absorption is from the ground-state singlet S_0 to the first excited singlet S_1 (or some higher singlet level). At first glance, fluorescence might appear to be exactly the reverse of the absorption process. This is true on the atomic level, but comparing the absorption and emission spectra of molecules shows them not to be superimposed on each other. Instead, they form the mirror image of each other, with the emission spectrum displaced toward the longer wavelength (Figure 18.32). Since the time required to execute a vibration (about 10^{-14} to 10^{-13} s) is much shorter than the decay or mean lifetime (about 10^{-9} to 10^{-5} s) of the fluorescent state, most of the excess vibrational energy dissipates to the surroundings as heat, and the excited molecules now decay from their ground vibrational levels.

We define the quantum yield of fluorescence emission ϕ_F as the ratio of photons emitted through fluorescence to the total number of photons absorbed. The maximum value of ϕ_F is 1 although it can be appreciably smaller than 1 if there are other processes present to deactivate the excited molecules. The intensity of radiation emitted after the exciting light is turned off is given by

$$I = I_0 e^{-t/\tau} \tag{18.32}$$

where I_0 is the intensity at $t = 0$, I is the intensity at time t, and τ is the mean lifetime of the fluorescent state. The mean lifetime is equal to the time it takes for the original intensity to decrease to $1/e$, or 0.368 of its original value. Thus when $t = \tau$, $I = I_0/e$.

The fluorescence technique, beside yielding information about the electronic structure of molecules in the excited state, is increasingly important in chemical and biochemical analysis. The technique is also useful in *liquid scintillation counting*, a common method of assaying radioactive tracer-labeled compounds containing tritium (^3H), carbon-14, phosphorus-32, and sulfur-35.

Figure 18.32

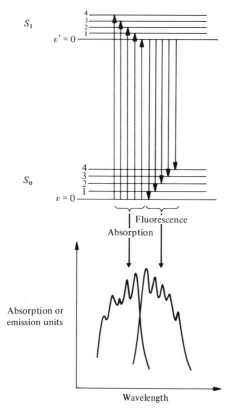

Relation between absorption and fluorescence. (From E. F. H. Brittain, W. O. George, and C. H. J. Wells, *Introduction to Molecular Spectroscopy*, Academic Press, Inc., New York, 1970.)

Consider the example of molecules labeled with tritium dissolved in toluene. The resulting process is expressed by

$$\beta + C_6H_5CH_3 \longrightarrow \beta + C_6H_5CH_3^*$$

and

$$C_6H_5CH_3^* \longrightarrow C_6H_5CH_3 + h\nu \qquad \text{fluorescence}$$

where β represents the beta particle emitted by tritium and the asterisk denotes an electronically excited state. With proper calibrations, the number of β particles and the concentration of tritium are obtained by measuring the intensity of toluene fluorescence. Because the quantum yield of toluene is about 0.2 (quite low), measurements are usually taken in the presence of a *sensitizer*. The sensitizer, characterized by a large ϕ_F value, is excited by receiving energy from toluene via a nonradiative mechanism such as collision.* This singlet–singlet energy transfer is represented by

$$D(S_1) + A(S_0) \longrightarrow D(S_0) + A(S_1)$$

Here the donor molecule is the excited toluene and the acceptor molecule is the sensitizer. The common sensitizer compounds are 2,5-diphenyloxazole (PPO) and 1,4-bis-2-(5-phenyloxazole)benzene (POPOP). Measuring the

---

* Another mechanism, which is responsible for long-range intermolecular transfers of energy, depends on the overlap between the emission band of the donor molecule and the absorption band of the acceptor molecule. The excited donor interacts with the ground-state acceptor through a dipole–dipole mechanism.

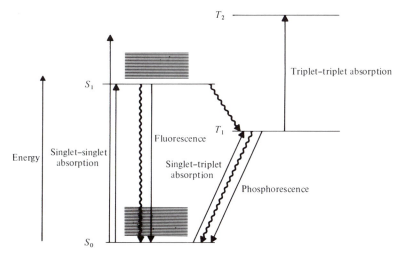

Schematic diagram for absorption, fluorescence, and phosphorescence. The wavy lines indicate radiationless transitions; the closely spaced lines represent the vibrational levels.

Figure 18.33

fluorescence of the sensitizer rather than that of toluene greatly improves the sensitivity of the experiment.

Phosphorescence offers a different path for the return of an excited molecule to the ground state with the emission of light. Phosphorescence can be readily distinguished from fluorescence by two characteristics. First, phosphorescence has a much longer decay period than fluorescence, about 10^{-3} s to several seconds. Second, a molecule in the phosphorescent state is paramagnetic containing two unpaired electrons; that is, it is a triplet state (Figure 18.33). Initially, an electron is promoted from S_0 to S_1. The promotion is followed by a process called *radiationless transition*, in which the electron flips its spin and drops from S_1 to T_1, the lowest triplet level, without the emission of light. In the end, a radiative transition from T_1 to S_0 occurs. This step is called *phosphorescence*. Since the transition involves a change in spin multiplicity (from triplet to singlet), it has a low probability, accounting for the long lifetimes observed. Because the excited state (T_1) is easily deactivated by collision, phosphorescence, unlike fluorescence, cannot be studied in a liquid phase. The most convenient state for studying phosphorescence is in a frozen glass at or below the liquid nitrogen temperature (77 K).

Lasers **18.8**

The principle of *laser** action can be understood by considering the following situation. Suppose that N molecules are irradiated by light at intensity I. The number of induced absorptions per second is given by NPI, where P is the probability of the induced transition and is the same for absorption and emission. In addition, an excited molecule may lose its excess energy by undergoing a spontaneous emission. The rate of this process is AN, where A

---

* Laser is an acronym for "*l*ight *a*mplification by *s*timulated *e*mission of *r*adiation."

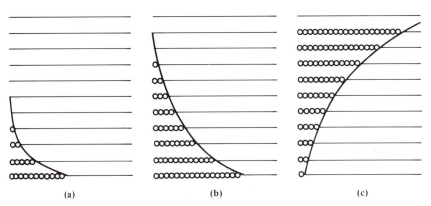

Figure 18.34 (a) and (b) Normal Boltzmann distributions ($T_2 > T_1$). (c) Population inversion. This distribution gives rise to the laser action.

is the probability of spontaneous emission. Under steady-state conditions of illumination, the rates of absorption and emission are equal so that

$$PIN(1 - x) = \underset{\text{induced}}{PINx} + \underset{\text{spontaneous}}{ANx} \qquad (18.33)$$

$$\underset{\text{induced}}{\text{induced}} \quad \underset{\text{induced}}{\text{induced}} \quad \underset{\text{spontaneous}}{\text{spontaneous}}$$
$$\text{absorption} \quad \text{emission} \quad \text{emission}$$

where

$$x = \frac{PI}{2PI + A}$$

is the fraction of molecules in the upper level. It follows, therefore, that the maximum value of x is 0.5, reached as I approached infinity. To put it another way, the population of the upper level can never exceed that of the lower level, a result predicted by Eq. (18.8) (Figure 18.34a and b).

If we can somehow populate the upper level without using the normal radiation process, we might bring about a *population inversion*, causing x to exceed 0.5. In such a state, intense emission can be induced by irradiating the system with a single photon of the appropriate frequency.

The first successful laser experiment was performed by Maiman in 1960. In his arrangement, a ruby rod was constructed by doping synthetic sapphire (Al_2O_3) with about 0.05% Cr_2O_3. Some of the Al^{3+} ions in the crystal lattice

Figure 18.35

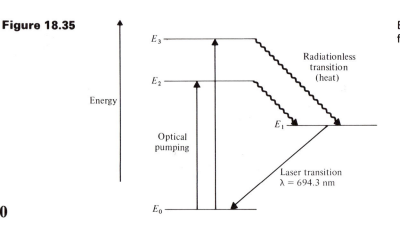

Energy-level diagram for a ruby laser.

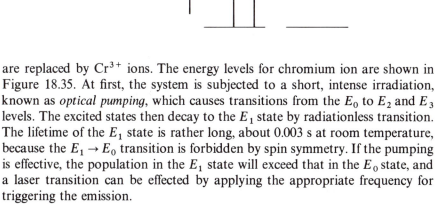

are replaced by Cr^{3+} ions. The energy levels for chromium ion are shown in Figure 18.35. At first, the system is subjected to a short, intense irradiation, known as *optical pumping*, which causes transitions from the E_0 to E_2 and E_3 levels. The excited states then decay to the E_1 state by radiationless transition. The lifetime of the E_1 state is rather long, about 0.003 s at room temperature, because the $E_1 \rightarrow E_0$ transition is forbidden by spin symmetry. If the pumping is effective, the population in the E_1 state will exceed that in the E_0 state, and a laser transition can be effected by applying the appropriate frequency for triggering the emission.

A large number of lasers are now available, operating over frequencies that range from infrared to ultraviolet. We will not survey the different types of lasers here, but will point out that the mechanism for achieving population inversion is very different for different lasers. For example, in the helium–neon laser, the helium atom is first excited by electron impact to the higher electronic state, which is then deactivated by collision with the neon atoms. The population of excited levels in neon builds up to exceed that of the ground level, so laser transition can occur (Figure 18.36). Table 18.5 summarizes the properties of some of the common lasers.

A laser beam is characterized by three unusual properties: high coherence or directionality, high monochromaticity, and high intensity. The high directionality means that a laser beam can be focused by a lens to an area as small

Some Common Laser Systems

Table 18.5

Laser	Emitted Wavelengths (nm)	Power[a]	Mode[b]
Ruby	694.3	5×10^8 W	Pulsed
He–Ne(g)	632.8	100 mW	Continuous wave
	1084	70 mW	
	1152	20 mW	
	3391	10 mW	
Ar(g)	457	50 mW	Continuous wave
	488	10 W	
	514.5	1 W	
N_2(g)	337.1	0.5 mW	Pulsed
CO_2(g)	10600	1 kW	Continuous wave

[a] Approximate values only. 1 mW = 10^{-3} W; 1 kW = 10^3 W.
[b] Lasers may operate in either the continuous-wave (CW) mode, when the light emerges in a continuous flow as long as the pumping operation is in progress, or in the pulsed mode, as in the case of ruby lasers, when the laser light emerges in a short burst as the stimulated emission occurs.

541

as 0.01 nm². Thus lasers have been employed as surgical knives in delicate operations of the vocal cord and the eye. The monochromatic nature of laser light is exemplified by the ruby laser light, centered at 694.3 nm with a width less than 0.01 nm. The precisely well defined energy of such a laser beam finds an important application in the separation of isotopes. The technique involves focusing a laser beam and a uv light onto a heated mixture of $^{235}UF_6$ and $^{238}UF_6$ vapor. The narrow line width associated with the laser makes it possible to selectively excite the electrons of only ^{235}U isotope, enabling the uv radiation to ionize only the $^{235}UF_6$ molecules.

$$^{235}UF_6 \xrightarrow[\text{irradiation}]{\text{laser}} {}^{235}UF_6^*$$

$$^{235}UF_6^* \xrightarrow[\text{irradiation}]{\text{uv}} {}^{235}UF_6^+ + e^-$$

where the asterisk denotes an electronically excited molecule. Consequently, these charged species can be conveniently separated by applying an electrostatic field. Quantitative separations of the two isotopes can be carried out by this technique.

The intensity of a laser beam can be estimated by again considering the ruby laser. A relatively small ruby laser can give rise to about 10^{18} excited Cr^{3+} ions, resulting in 10^{18} photons at 14,403 cm^{-1}. The total energy output is given by

$$E = h\nu = hc\bar{\nu}$$

$$= 6.63 \times 10^{-27} \text{ erg s} \times 3 \times 10^{10} \text{ cm s}^{-1} \times 14{,}403 \text{ cm}^{-1}$$

$$\times 1 \times 10^{18}$$

$$= 2.86 \times 10^6 \text{ ergs} = 0.286 \text{ J} \simeq 0.3 \text{ J}$$

Now 0.3 J may not seem like much; however, this amount of energy can be generated in a very short period of time, about 10^{-9} s. The power associated with such a laser beam is then given by

$$\text{power} = \frac{\text{energy}}{\text{time}} = \frac{0.3 \text{ J}}{10^{-9} \text{ s}} = 3 \times 10^8 \text{ J s}^{-1} = 3 \times 10^8 \text{ W}$$

This is the power output per pulse of the laser action.* When such a laser beam is focused on a small target of area 10^{-4} cm², say, the power flux density is given by

$$\frac{\text{power}}{\text{flux density}} = \frac{\text{power}}{\text{area}} = \frac{3 \times 10^8 \text{ W}}{10^{-4} \text{ cm}^2} = 3 \times 10^{12} \text{ W cm}^{-2}$$

which is a very impressive figure. Such an enormously intensive beam can be used to cut or drill holes in metal. The high intensity can also be a hazard to workers, because even the briefest exposure to the laser radiation can cause severe burns of the skin and damage to the eye. With proper designs, laser pulses of 1000 J in 10^{-10} s can be generated which can be used to produce nuclear fusion reactions.

* An ordinary 100-watt light bulb has a continuous energy output of 100 W or 100 J s^{-1}.

BARROW, G. M. *The Structure of Molecules.* W. A. Benjamin, Inc., Menlo Park, Calif., 1963.

A very readable introductory text. Covers rotation, vibration, and electronic spectra. Strongly recommended.

BRITTAIN, E. F., W. O. GEORGE, and C. H. J. WELLS. *Introduction to Molecular Spectroscopy.* Academic Press, Inc., New York, 1970.

A well-written text that has many worked-out examples.

CHANG, R. *Basic Principles of Spectroscopy.* McGraw-Hill Book Company, New York, 1971.

CLAYTON, R. K. *Light and Living Matter*, Vol. 1. McGraw-Hill Book Company, New York, 1970.

An excellent introductory text. It concentrates mainly on electronic absorption and emission spectra.

DENNY, R. C. *A Dictionary of Spectroscopy.* John Wiley & Sons, Inc., New York, 1973.

DUNFORD, H. B. *Elements of Diatomic Molecular Spectra.* Addison-Wesley Publishing Company, Inc., Reading, Mass., 1968.

LAIDLAW, W. G. *Introduction to Quantum Concepts in Spectroscopy.* McGraw-Hill Book Company, New York, 1970.

A useful text.

SCHENK, G. H. *Absorption of Light and Ultraviolet Radiation.* Allyn and Bacon, Inc., Boston, 1973.

Covers electronic absorption and emission spectra.

SLATER, C. D., D. C. NYBERG, and J. VIKIN. *Infrared Spectroscopy.* Willard Grant Press, Boston, 1974.

A clearly written introductory text.

THOMPSON, C. C. *Ultraviolet-Visible Absorption Spectroscopy.* Willard Grant Press, Boston, 1974.

Another very helpful introductory text.

WHIFFEN, D. H. *Spectroscopy*, 2nd ed. John Wiley & Sons, Inc., New York, 1972.

Covers all the common branches of spectroscopy, briefly but succinctly.

INTERMEDIATE–ADVANCED

BECKER, E. D. *High Resolution NMR: Theory and Chemical Applications.* Academic Press, Inc., New York, 1971.

BOVEY, F. A. *High Resolution NMR of Macromolecules.* Academic Press, Inc., New York, 1972.

CASY, A. F. *PMR Spectroscopy in Medicine and Biological Chemistry.* Academic Press, Inc., New York, 1971.

DWEK, R. A. *Nuclear Magnetic Resonance in Biochemistry.* Clarendon Press, Oxford, 1973.

FOSTER, R. *Organic Charge-Transfer Complexes.* Academic Press, Inc., New York, 1969.

HASCHEMEYER, R. H., and A. E. V. HASCHEMEYER. *Proteins.* John Wiley & Sons, Inc., New York, 1973, Chapters 9 and 12.

JAFFÉ, H. H., and M. ORCHIN. *Theory and Applications of Ultraviolet Spectroscopy.* John Wiley & Sons, Inc., New York, 1962.

JAMES, T. L. *Nuclear Magnetic Resonance in Biochemistry.* Academic Press, Inc., New York, 1975.

JONES, D. W., ed. *Introduction to the Spectroscopy of Biological Polymers.* Academic Press, Inc., New York, 1976.

KNOWLES, P. F., D. MARSH, and H. W. E. RATTLE. *Magnetic Resonance of Biomolecules.* John Wiley & Sons, Inc., New York, 1976.

KONEV, S. V. *Fluorescence and Phosphorescence of Proteins and Nucleic Acids.* Plenum Publishing Corporation, New York, 1967.

PESCE, A. J., C. G. ROSEN, and T. L. PASBY. *Fluorescence Spectroscopy: An Introduction for Biology and Medicine.* Marcel Dekker, Inc., New York, 1971.

RAO, C. N. *Chemical Applications of Infrared Spectroscopy.* Academic Press, Inc., New York, 1963.

SCHAWLOW, A. L., ed. *Laser and Light.* W. H. Freeman and Company, Publishers, San Francisco, 1969.

SWARTZ, H. M., J. R. BOLTON, and D. C. BORG. *Biological Applications of Electron Spin Resonance.* Wiley–Interscience, New York, 1972.

Reading Assignments

SECTION A

"Introduction to Spectroscopy," J. Davis, Jr., *Chemistry* **47**(9), 6 (1974); **48**(1), 11, **48**(5), 19, **48**(7), 15, **48**(11), 5 (1975).

"Experiments in NMR," G. Glarous and N. H. Cromwell, *J. Chem. Educ.* **48**, 202 (1971).

"Nuclear Magnetic Resonance Spectroscopy," F. A. Bovey, *Chem. Eng. News* **43**(35), 98 (1965).

"Microwave Spectroscopy in Chemistry," E. B. Wilson, Jr., *Science* **162**, 59 (1968).

"Microwave Spectroscopy," W. H. Kirchhoff, *Chem. Eng. News* **47**(13), 88 (1969).

"Infrared Spectroscopy," K. Whetsel, *Chem. Eng. News* **46**(6), 82 (1968).

"Vibration–Rotation Spectrum of HCl," F. E. Stafford, C. W. Holt, and G. L. Paulson, *J. Chem. Educ.* **40**, 245 (1963).

"The Ultraviolet Spectra of Aromatic Hydrocarbons," P. E. Steveson, *J. Chem. Educ.* **41**, 234 (1964).

"The Fates of Electronic Excitation Energy," H. H. Jaffe and A. L. Miller, *J. Chem. Educ.* **43**, 469 (1966).

"Color and Chemical Constitution," N. J. Juster, *J. Chem. Educ.* **39**, 596 (1962).

"Color in Nature," P. G. Seybold, *Chemistry* **49**(9), 7 (1976).

"Shedding Light on the Color of Gems and Minerals," B. M. Loeffer and R. G. Burns, *Am. Sci.* **64**, 636 (1976).

"The Chemical Origin of Color," M. V. Orna, *J. Chem. Educ.* **55**, 478 (1978).

"How Light Is Analyzed," P. Connes, *Sci. Am.*, Sept. 1968.

"Application of Absorption Spectroscopy in Biochemistry," G. R. Penzer, *J. Chem. Educ.* **45**, 692 (1968).

"Progress in Our Understanding of the Optical Properties of Nucleic Acids," A. M. Lesk, *J. Chem. Educ.* **46**, 821 (1969).

"Charge-Transfer Complexes and Photochemistry," M. R. J. Dack, *J. Chem. Educ.* **50**, 169 (1973).

"Chemistry and Light Generation," N. Slagg, *J. Chem. Educ.* **45**, 103 (1968).

"Photochemical Reactions of Natural Molecules," D. C. Neckers, *J. Chem. Educ.* **50**, 164 (1973).

"Luminescence," P. G. Seybold, *Chemistry* **42**(2), 6 (1973).

"Fluorescence Spectroscopy of Proteins," L. Stryer, *Science* **162**, 526 (1968).

"Lasers: A Renaissance in Optical Research," N. Bloembergen, *Am. Sci.* **63**(1), 16 (1975).

"Laser Chemistry," D. L. Rousseau, *J. Chem. Educ.* **43**, 566 (1966).

"Liquid Scintillation Counting," W. Yang and E. K. C. Lee, *J. Chem. Educ.* **46**, 277 (1969).

"Energy States of Molecules," J. L. Hollenberg, *J. Chem. Educ.* **47**, 2 (1970).

"Laser Separation of Isotopes," R. N. Zare, *Sci. Am.*, Feb. 1977.

"Laser: The Light Fantastic," *J. Chem. Educ.* **55**, 529 (1978).

"Laser Chemistry," A. M. Ronn, *Sci. Am.*, May 1979.

"The Laser Revolution," W. J. Cromie, *Chemistry* **52**(2), 12 (1979).

SECTION B

"Analysis of Complex NMR Spectra for the Organic Chemist," E. W. Garbisch, *J. Chem. Educ.* **45**, 311, 402, 480 (1968).

"C—H Bond Strengths," G. J. Boobyer and A. P. Cox, *J. Chem. Educ.* **45**, 18 (1968).

"Accurate Molecular Geometry," D. H. Whiffen, *Chem. Brit.* **7**, 57 (1971).

"ESR Study of Electron Transfer Reactions," R. Chang, *J. Chem. Educ.* **47**, 563 (1970).

"Electron Spin Resonance in Transition Metal Chemistry," J. B. Raynor, *Chem. Brit.* **10**, 254 (1974).

"Free Radicals in Biological Systems," W. A. Pryor, *Sci. Am.*, Aug. 1970.

"Ultraviolet Spectra of Proteins and Amino Acids," D. B. Wetlaufer, *Advan. Protein Chem.*, **17**, 303 (1962).

"Ultraviolet Absorption," J. W. Donovan, in *Physical Principles and Techniques of Protein Chemistry*, Part A, S. L. Leach, ed., Academic Press, Inc., New York, 1969.

"Light," G. Feinberg, *Sci. Am.*, Sept. 1968.

"The Chemical Effects of Light," G. Oster, *Sci. Am.*, Sept. 1968.

"The Triplet State," N. J. Turro, *J. Chem. Educ.* **46**, 2 (1969).

"The Spectrum of Atomic Hydrogen," T. W. Hansch, A. L. Schawlow, and G. W. Series, *Sci. Am.*, Mar. 1979.

"The Absorption of Light by Oriented Molecules," R. M. Wilson, E. J. Gardner, and R. H. Squire, *J. Chem. Educ.* **50**, 94 (1973).

"Molecular Charge-Transfer Complexes," W. E. Wentworth, G. W. Drake, W. Hirsch, and E. Chen. *J. Chem. Educ.* **41**, 373 (1964).

"Organic Electron-Donor-Acceptor Complexes," R. Foster, *Chem. Brit.* **12**, 18 (1976).

"Cell Surgery by Laser," M. W. Berns and D. E. Rounds, *Sci. Am.*, Feb. 1970.

"Introduction to the Laser in the Physical Chemistry Course," D. P. Wong, *J. Chem. Educ.* **48**, 654 (1971).

"Organic Lasers," P. Sorokin, *Sci. Am.*, Feb. 1969.

"Liquid Lasers," A. Lempicki and H. Samelson, *Sci. Am.*, June 1967.

"Applications of Lasers to Chemical Research," S. R. Leone, *J. Chem. Educ.* **53**, 13 (1976).

Problems

18.1 Convert 15,000 cm^{-1} into wavelength (nm) and frequency.

18.2 Convert 450 nm into wave number and frequency.

18.3 Convert the following percent transmittance to absorbance: (a) 100%, (b) 50%, and (c) 0%.

18.4 Convert the following absorbance to percent transmittance: (a) 0.0, (b) 2.0, and (c) ∞.

18.5 The absorption of radiation energy by a molecule results in the formation of an excited molecule. It would seem that given enough time all of the molecules in a sample would have been excited and no more absorption would occur. Yet in practice we find that the absorbance of a sample at any wavelength remains unchanged with time. Why?

18.6 The mean lifetime of an electronically excited molecule is 10^{-8} s. If the emission of the radiation occurs at 610 nm, what are the uncertainties in frequency (Δv) and wavelength ($\Delta \lambda$)? (*Hint:* The ratio $\Delta v/v$ is equal to $\Delta \lambda/\lambda$.)

18.7 The familiar yellow D lines of sodium is actually a doublet at 589.0 nm and 589.6 nm. Calculate the difference in energy (in J) between these two lines.

18.8 The typical energy differences for transitions in the microwave, ir, and electronic spectroscopies are 5×10^{-22} J, 0.5×10^{-19} J, and 1×10^{-18} J, respectively. Calculate the ratio of number of molecules in the two adjacent energy levels (for example, the ground level and the first excited level) at 300 K in each case.

18.9 The nmr signal of a compound is found to be 240 Hz downfield from the TMS peak using a spectrometer operating at 60 MHz. Calculate its chemical shift in ppm relative to TMS.

18.10 Analyze the ^{31}P nmr spectrum of ATP shown in Figure 18.13.

18.11 How many lines are there in the pmr spectrum of (a) benzene, (b) toluene, (c) *o*-xylene, (d) *m*-xylene, and (e) *p*-xylene?

18.12 Account for the number of lines observed in the esr spectra of benzene and naphthalene anion radicals shown in Figure 18.16.

★18.13 Both nmr and esr spectroscopy differ from other branches of spectroscopy in one important respect. Explain.

18.14 The resolution of visible and uv spectra can usually be improved by recording the spectra at low temperatures. Why does this work?

18.15 The luminescent first-order decay of a certain organic molecule yields the following data:

t (s)	0	1	2	3	4	5	10
I	100	43.5	18.9	8.2	3.6	1.6	0.02

where I is the relative intensity. Calculate the mean lifetime τ for the process. Is it fluorescence or phosphorescence?

18.16 List some important differences between fluorescence and phosphorescence.

18.17 Many aromatic hydrocarbons are colorless, but their anion and cation radicals are often strongly colored. Give a qualitative explanation for this phenomenon. (*Hint:* Consider only the π molecular orbitals.)

18.18 The fundamental frequency of vibration for $D^{35}Cl$ is given by $\bar{v} = 2081.0 \text{ cm}^{-1}$. Calculate the force constant k and compare this value with the force constant obtained for $H^{35}Cl$ in Example 18.2. Comment on your result.

18.19 The equilibrium bond length in nitric oxide ($^{14}N^{16}O$) is 1.15 Å. Calculate (a) the moment of inertia of NO, and (b) the energy for the $J = 0 \rightarrow 1$ transition. How many times does the molecule rotate per second in the $J = 1$ level?

18.20 The ir spectrum of the carbon monoxide–hemoglobin complex gives a peak at about 1950 cm^{-1}, which is due to the carbonyl stretching frequency. (a) Compare this value with the fundamental frequency of free CO, which is 2143.3 cm^{-1}. Comment on the difference. (b) Convert this frequency to kJ mol^{-1}. (c) What conclusion can you draw from the fact that there is only one band present?

18.21 Calculate the number of vibrational degrees of a protein molecule containing 7947 atoms.

18.22 The lowest triplet in naphthalene molecule is about 11,000 cm^{-1} below the lowest excited singlet electronic level at 77 K. Calculate the ratio of populations in these two states. (*Hint:* The Boltzmann equation is given by $n_2/n_1 = (g_2/g_1)e^{-\Delta E/kT}$, where g_1 and g_2 are the degeneracies for levels 1 and 2.)

18.23 Referring to Figure 18.29, explain why T_m increases as the mole % (G + C) increases.

18.24 How would you employ the particle-in-a-box model to qualitatively explain the difference in the appearance of stilbene (colorless) and $C_6H_5-(CH=CH)_4-C_6H_5$ (brown)?

$\star$**18.25** Assuming that the average carbon–carbon bond length in hexatriene is 1.5 Å, use the particle-in-a-one-dimensional box model to calculate the longest-wavelength peak in its absorption spectrum.

18.26 An aqueous solution contains two species A and B. The absorbance at 300 nm is 0.372 and at 250 nm is 0.478. The molar absorptivities of A and B are:

$$A: \quad \varepsilon_{300} = 3.22 \times 10^4 \text{ liters mol}^{-1} \text{ cm}^{-1}$$
$$\varepsilon_{250} = 4.05 \times 10^4 \text{ liters mol}^{-1} \text{ cm}^{-1}$$
$$B: \quad \varepsilon_{300} = 2.86 \times 10^4 \text{ liters mol}^{-1} \text{ cm}^{-1}$$
$$\varepsilon_{250} = 3.76 \times 10^4 \text{ liters mol}^{-1} \text{ cm}^{-1}$$

If the path length of the cell is 1.00 cm, calculate the concentrations of A and B.

18.27 The fluorescence of a protein is due to tryptophan, tyrosine, and phenylalanine (assuming that the protein does not contain a prosthetic group that is fluorescent). Iodide ions are known to quench the fluorescence of tryptophan. If a protein is known to contain only one tryptophan group and iodide fails to quench its fluorescence, what can you conclude about the location of the tryptophan residue?

18.28 Anthracene is colorless, but tetracene is light orange. Explain.

18.29 Oxyhemoglobin is bright red while deoxyhemoglobin is purple. Explain the difference in color in terms of their electronic configurations.

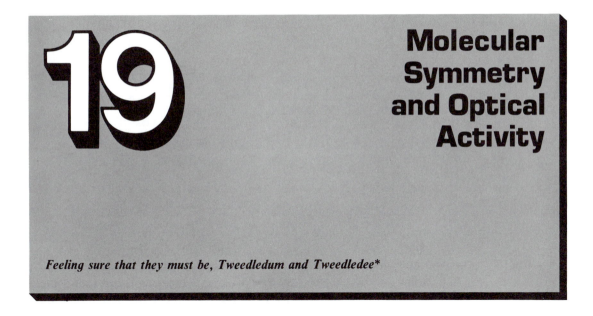

19

Molecular Symmetry and Optical Activity

Feeling sure that they must be, Tweedledum and Tweedledee*

Several branches of chemistry have been enlarged by the study of molecular symmetry. One example of this fruitful application to spectroscopy is the analysis of infrared and electronic spectra, which can be greatly simplified by a procedure based on group theory, involving the symmetry properties of molecules. In this chapter we consider another aspect of molecular symmetry, one concerned with the phenomenon of optical activity.

Symmetry of Molecules 19.1

The symmetry of a molecule relates to the shape or three-dimensional arrangement of its atoms. Tetrahedral methane has a symmetry different from that of a linear molecule such as acetylene. The example of a planar water molecule will help to clarify the concept of symmetry (Figure 19.1). When the molecule is rotated about an axis through the oxygen atom by 180° in either direction as indicated, it reaches a new configuration, indistinguishable from the original one. This rotation is called a *symmetry operation*. The results of a symmetry operation show that the molecule must possess a *symmetry element*. A symmetry element is a geometrical entity such as a point, a plane, or a line with respect to which a symmetry operation is performed. The elements of symmetry are now summarized.

PROPER ROTATION AXIS

A molecule possesses a *proper rotation axis* of order n (C_n) if rotation about the axis by $2\pi/n(360°/n)$ leaves the molecule in an indistinguishable

* Taken from *The Annotated Alice: Alice's Adventures in Wonderland and Through the Looking Glass* by Lewis Carroll, introduction and notes by Martin Gardner. © 1960 by Martin Gardner. Used by permission of Crown Publishers, Inc.

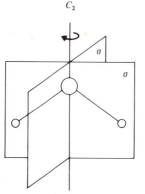

configuration from the original one. Water molecules have a C_2 axis, ammonia molecules a C_3 axis, and so on. All molecules, of course, possess a C_1 axis. A linear molecule also possesses a C_∞ axis, meaning that rotation about the internuclear axis by any amount would leave the molecule unchanged.

PLANE OF SYMMETRY

A molecule possesses a *plane of symmetry* (σ) if a reflection through the plane leaves the molecule in an indistinguishable position from the original one. Thus the water molecule possesses two planes of symmetry. Generally, a planar molecule has at least one plane of symmetry.

CENTER OF SYMMETRY

If the coordinates (x, y, z) of every atom in a molecule are changed into $(-x, -y, -z)$ and the molecule's configuration is indistinguishable from the original one, then the point of origin, that is $(0, 0, 0)$, is the *center of symmetry* (i). The benzene molecule, for example, has a center of symmetry:

A molecule can have only one center of symmetry.

IMPROPER ROTATION AXIS

The *improper rotation axis* (S_n) is a more complicated case than the three preceding elements of symmetry because it involves two operations. A molecule possesses an improper rotation axis of order n if rotation about the axis by $2\pi/n$ followed by reflection in a plane perpendicular to the axis leaves the molecule in a configuration indistinguishable from the original one. From Figure 19.2 we see that the methane molecule has a S_4 axis.

Finally, we have the identity element E, which does not do anything to the molecule and is therefore possessed by all the molecules. These symmetry properties can be represented by a special mathematical technique called *group theory*. The discussion of group theory is beyond the scope of this text; instead, we shall only consider some qualitative aspects of molecular symmetry.

550

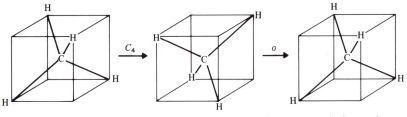

Improper axis in methane. **Figure 19.2**

We can now draw some interesting and important conclusions from our discussion of molecular symmetry. For example, we can deduce the presence of a permanent electric dipole moment from a molecule's symmetry elements. Recalling that dipole moment is a vector quantity and keeping in mind that it remains unchanged by any symmetry operation, we realize that it must lie along the symmetry element. A molecule cannot have a dipole moment if it possesses a center of symmetry, for a vector cannot be a point, or if it possesses more than one C_n axis where $n \geq 2$, for a vector cannot lie along two different axes.

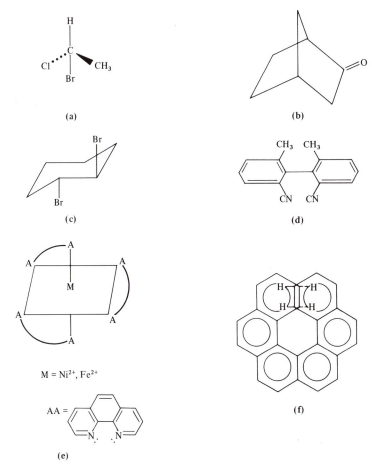

Some optically active (or chiral) molecules: (a) 1,1-bromochloroethane; (b) nor-camphor; (c) 1,2-*trans*-dibromocyclohexane; (d) 2,2'-dimethyl-6,6'-dicyano-biphenyl; (e) a coordination compound with three bidentate ligands (AA here represents 1,10-phenanthroline); and (f) hexahelicene. **Figure 19.3**

Molecular symmetry can also assist in the study of optical activity. A molecule is optically active (or chiral) if it can rotate the plane of polarized light. (Later, we shall discuss the reasons in more detail.) The general criterion for optical activity is that the molecule and its mirror image are not superimposable. However, this rule is difficult to apply on complex molecules without actually constructing three-dimensional models. A more useful criterion for optical activity is the knowledge that a molecule does not possess any one of the following symmetry elements: (1) center of symmetry, (2) plane of symmetry, and (3) improper axis. Since a onefold improper axis is identical with a plane of symmetry ($\sigma \equiv S_1$) and a twofold improper axis is identical with a center of inversion ($i \equiv S_2$), (1), (2), and (3) are reduced to the statement that the molecule must not possess an improper axis.* Some examples of optically active molecules are shown in Figure 19.3.

19.2 Polarized Light and Optical Rotation

As we saw in Chapter 15, light is just one form of electromagnetic radiation, having an electric field component ($\mathcal{E}$) and a mutually perpendicular magnetic field component (H). Both components oscillate in space with the same wavelength and frequency. Light is a *transverse* wave, so the planes containing these components are perpendicular to the direction of propagation. For ordinary unpolarized light, instantaneous polarization, or the direction of $\mathcal{E}$ and H in space, changes rapidly and randomly.

A beam of unpolarized light passed through a polarizer, emerges polarized, meaning that its electric field component is confined to a single plane, as

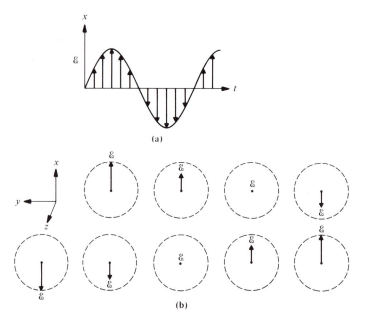

(a)

(b)

Figure 19.4 (a) Electric field component of a linearly polarized light as a function of time. (b) Variation of $\mathcal{E}$ with time as seen by an observer along the z direction. (From C. Djerassi, *Optical Rotatory Dispersion*. Copyright 1960 by McGraw-Hill Book Company. Used with permission of McGraw-Hill Book Company.)

* There are a few exceptions to this general rule.

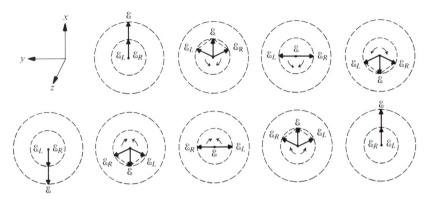

Electric field vector $\mathscr{E}$ as the resultant of two rotating vectors $\mathscr{E}_L$ and $\mathscr{E}_R$. (From C. Djerassi, *Optical Rotatory Dispersion*. Copyright 1960 by McGraw-Hill Book Company. Used with permission of McGraw-Hill Book Company.) **Figure 19.5**

shown in Figure 19.4.* A polarizer can be a properly cut and cemented Nicol prism, a crystal of Iceland spar, $CaCO_3$; or a sheet of Polaroid, polyvinyl alcohol stained with iodine. These polarizers produce what we call *linearly* or *plane-polarized light*.

A plane-polarized light beam can be resolved into two component vectors $\mathscr{E}_L$ and $\mathscr{E}_R$, which correspond to the left and right *circularly* polarized waves. This is not only true mathematically; circularly polarized light can be produced by passing a beam of plane polarized light through an optical device known as a *quarter-wave plate*. Figure 19.5 shows the variation of $\mathscr{E}$ as the resultant of the two rotating vectors (the addition of $\mathscr{E}_L$ and $\mathscr{E}_R$ gives $\mathscr{E}$ at every point).

In an ordinary medium, $\mathscr{E}_L$ and $\mathscr{E}_R$ rotate at the same speed, so $\mathscr{E}$ is confined to the xz plane. In an optically active medium, $\mathscr{E}_L$ and $\mathscr{E}_R$ rotate at different speeds. Although the light still remains polarized, the plane containing $\mathscr{E}$ now makes an angle α with the x axis (Figure 19.6). The plane of polarization rotates because in an optically active medium the refractive index n is different for left and right circularly polarized light and produces different speeds of rotation.† Such a medium is said to be *circularly birefringent*.

In an optical rotation experiment, the polarizer and the analyzer (another polarizer) are first aligned so that no light passes through the analyzer. With an optically active medium present, light is transmitted because of rotation. The angle of rotation, α, can be measured by turning the analyzer until again no light is transmitted. A quantitative treatment shows that the angle of rotation per unit length of the birefringent medium α is given by

$$\alpha = \frac{\pi}{\lambda}\left(n_L - n_R\right) \qquad (19.1)$$

where λ is the wavelength of light employed and n_L and n_R are the refractive indices for the left and right circularly polarized light, respectively. In the CGS system, α has the dimensions of radians per centimeter. To express α in

---

* The magnetic field component shall be omitted in the subsequent discussion, since its interaction with matter is much smaller than the interaction between the electric field component and matter.

† The *refractive index* of a medium is defined as the ratio of the speed of light in vacuum to the speed of light in the medium. It is a dimensionless quantity.

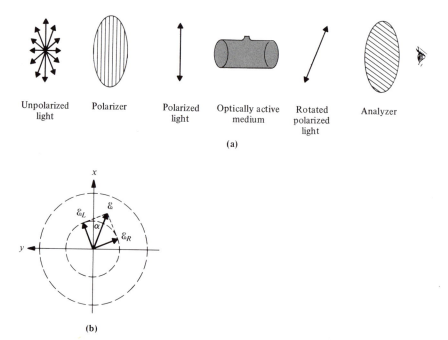

Figure 19.6 (a) Experimental arrangement for optical rotation measurement. (b) Rotation of $\mathscr{E}$ when $\mathscr{E}_L$ and $\mathscr{E}_R$ make unequal angles with the x axis. (From C. Djerassi, *Optical Rotatory Dispersion.* Copyright 1960 by McGraw-Hill Book Company. Used with permission of McGraw-Hill Book Company.)

the more common experimental units of degrees per decimeter, we must multiply the right-hand side of Eq. (19.1) by $1800/\pi$*:

$$\alpha = \frac{1800}{\lambda}(n_L - n_R) \tag{19.2}$$

Another convenient quantity is the *specific rotation* $[\alpha]$, defined as

$$[\alpha]_\lambda^T = \frac{100\alpha}{lc} \tag{19.3}$$

where α is now in degrees per decimeter, l is the pathlength of the cell in decimeter (1 dm = 10 cm), and c is the concentration of the optically active substance in g (100 ml solution)$^{-1}$.

Since the optical activity is a function both of the wavelength of light and of the temperature, it is customary to designate both of these factors in expressing the specific rotation. Note that T is in degrees Celsius. Thus $[\alpha]_\lambda^T$ is in degrees cm^2 g^{-1}. For a pure liquid, Eq. (19.3) becomes

$$[\alpha]_\lambda^T = \frac{\alpha}{l \times d} \tag{19.4}$$

where d is the density (g ml^{-1}) of the liquid. The specific rotation is a function of the wavelength; in 1907, Drude proposed the following relationship:

$$[\alpha]_\lambda^T = \frac{K}{\lambda^2 - \lambda_0^2} \tag{19.5}$$

* See Appendix 2 for the relationship between degrees and radians.

where K and λ_0 are constants whose values depend on the nature of the optically active medium.

Another useful quantity is the *molar rotation*, $[\Phi]_\lambda^T$, given by

$$[\Phi]_\lambda^T = \frac{[\alpha]_\lambda^T M}{100} \tag{19.6}$$

where M is the molar mass of the optically active compound; $[\Phi]_\lambda^T$ is in degrees cm^2 mol^{-1}. Both specific rotation and molar rotation are independent of the concentration of solution and path length.

A final note: If a medium causes the plane of polarization to rotate toward the right, it is called *dextrorotatory*, denoted by $(+)$. If rotation is to the left, the medium is called *levorotatory*, denoted by $(-)$.

EXAMPLE 19.1 The specific rotation of L-lysine is $+13.5°$ at 25°C. Calculate the optical rotation of a lysine solution (concentration: 14.8 g/100 ml H$_2$O) in a 10-cm cell measured at 589.3 nm. What is the difference in refractive index between the left and right circularly polarized light?

Answer: From Eq. (19.3), we have

$$c = 14.8 \text{ g/100 ml H}_2\text{O}$$

$$l = 1 \text{ dm}$$

Hence

$$13.5° = \frac{100\alpha}{l \times 14.8}$$

$$\alpha = 2.00°$$

Substituting this value for α in Eq. (19.2), we obtain

$$2.00 = \frac{1800}{5893 \times 10^{-8}} (n_L - n_R)$$

Hence

$$n_L - n_R = 6.55 \times 10^{-8} \quad \text{at 589.3 nm}$$

which is a very small number.

EXAMPLE 19.2 In an optical rotation measurement, the angle measured for a solution of concentration c in a cell of path length l is $-12.7°$. How can you be certain that this is the correct rotation and not $(-12.7 + 360) = 347.3°$? (Clockwise rotation by 347.3° is equivalent to counterclockwise rotation by 12.7°.)

Answer: The best way to confirm this is to make another measurement at a different concentration, say $c/10$. At $0.1c$, $\alpha = -1.27°$, which is clearly distinguishable from 34.73°. Of course, changing the path length of the cell will also remove this ambiguity.

19.3 Optical Rotatory Dispersion and Circular Dichroism

The refractive index of a medium is not a constant but depends on the wavelength of light employed. In fact, a plot of n versus λ gives a *dispersion curve* (Figure 19.7). Physically, this indicates how a prism made of the material being studied disperses light of various wavelengths. It follows that optical rotation, which depends on refractive index, must also vary with wavelength. Figure 19.8 shows two types of optical rotatory dispersion (ORD) curves. These are called *plain* dispersion curves because they lack maxima and minima. For compounds that do not absorb in the visible region, rotation is quite small—a major experimental difficulty encountered when optical rotation was measured at a single wavelength of sodium D-line.*

Beside playing a part in circular birefringent effect discussed above, optical activity is also manifest in small differences in the molar absorptivity, ε_L and

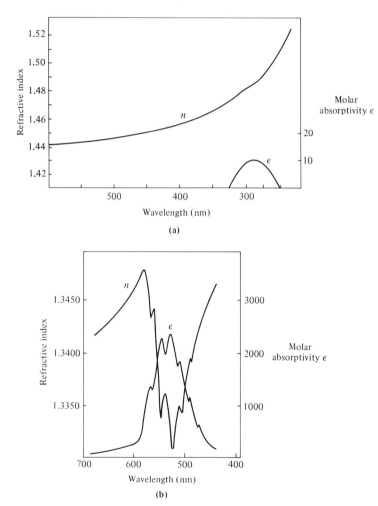

Figure 19.7 Relation between absorption and dispersion: (a) cyclohexanone; (b) potassium permanganate. [From J. G. Foss, *J. Chem. Educ.* **40**, 592 (1963).]

* Actually, the familiar yellow color of the sodium D-line is a closely spaced doublet at 589.0 and 589.6 nm.

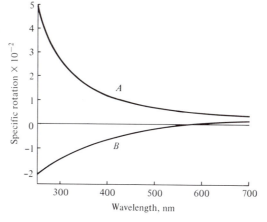

Dispersion curves: (A) plain positive; (B) plain negative. (From C. Djerassi, *Optical Rotatory Dispersion*. Copyright 1960 by McGraw-Hill Book Company. Used with permission of McGraw-Hill Book Company.)

Figure 19.8

ε_R, of circularly polarized components. Then the medium is said to exhibit *circular dichroism* (CD). Since ε_L is not equal to ε_R, $\mathscr{E}$ will no longer oscillate along a single line but will trace out the ellipse shown in Figure 19.9. The difference in ε's is measured according to the equation

$$[\theta] = 3300(\varepsilon_L - \varepsilon_R) \tag{19.7}$$

where $[\theta]$, the molecular ellipticity, is in degrees cm^{-2} per decimole of the optically active compound.

The optical activity of most organic and biological compounds generally increases toward the shorter-wavelength region, because compounds in the ultraviolet possess optically active absorption bands ($\pi \to \pi^*$, $n \to \pi^*$, and $\sigma \to \sigma^*$ transitions). The chromophores for these bands are either intrinsically asymmetric or become asymmetric as a result of the interaction with an asymmetric environment. The hexahelicene molecule (see Figure 19.3) is an example of intrinsic asymmetry. There is no asymmetric carbon atom in the molecule; the entire molecule acts as a chromophore for the $\pi \to \pi^*$ transition. On the other hand, the optically active absorption band in 3-methylcyclohexanone is the $n \to \pi^*$ transition of the symmetric carbonyl group. Its optical activity arises as a result of the asymmetric carbon atom in the 3 position.

Variation of $\mathscr{E}$ for circular dichroism. The $\mathscr{E}$ vector traces out an ellipse. (From C. Djerassi, *Optical Rotatory Dispersion*. Copyright 1960 by McGraw-Hill Book Company. Used with permission of McGraw-Hill Book Company.)

Figure 19.9

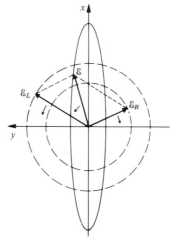

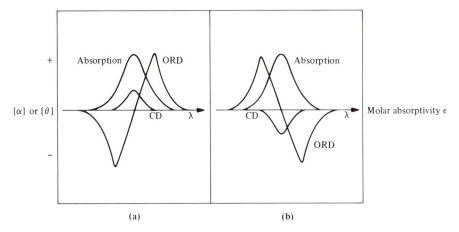

Figure 19.10 The cotton effect: (a) positive; and (b) negative.

As mentioned earlier, optical rotation is a function of wavelength. Within the spectral region of the optically active absorption band, an anomalous behavior is observed. This is known as the *Cotton effect*. A *positive* Cotton effect is characterized by an initial rise of the ORD curve with decreasing wavelength reaching a maximum (called a *peak*) at a longer wavelength than the absorption band. Beyond this point, the curve changes its slope, reaching a minimum (called a *trough*) at a shorter wavelength than the absorption band. Just the opposite holds for a *negative* Cotton effect (Figure 19.10). The vertical distance between the peak and the trough is a measure of the amplitude of the Cotton effect while the horizontal distance between the same two points gives the breadth of the curve.

In some respects, the relationship between CD and ORD is analogous to that between absorption and dispersion. Like absorption, the CD phenomenon is exhibited over a relatively small wavelength region, while both the dispersion and ORD can be observed at wavelengths far removed from the absorption bands. In contrast to an absorption band, however, a CD band is signed, since the difference $(\varepsilon_L - \varepsilon_R)$ can be either positive or negative Although a weak absorption band must necessarily lead to a small anomalous dispersion, the same band may give rise to a large anomalous ORD curve.

Both ORD and CD are important tools for studying the conformation of biomolecules in solution. Much of the early work was carried out on model compounds, synthetic polypeptides such as a poly-γ-benzyl-L-glutamate and poly-L-proline. There are two different contributions to optical activity in these molecules, as well as in proteins: the presence of L-amino acids and the folding of the polypeptide chains, for example, the α helices.* The helix may be either left-handed or right-handed, as shown in Figure 19.11. Protein helices are exclusively right-handed and rotate plane polarized light in opposite direction from the L-amino acids. Studying the optical rotation of many protein molecules allows biochemists to estimate the percent of helical structure. From the change in rotation with temperature, pH, and so on, we can also measure the helix-coil transitions.

The $n \rightarrow \pi^*$ and $\pi \rightarrow \pi^*$ transitions of amino acids occur at about 225 nm and 195 nm. The ORD curve is really a superposition of two Cotton effects

* See Chapter 21 for a detailed discussion of the α helix.

Left-handed and right-handed helixes.

Figure 19.11

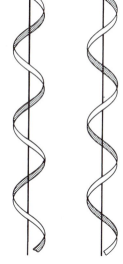

Left–handed   Right–handed

due to these two transitions. Changes in conformation and the like can be more conveniently monitored in this region, since Cotton effects are highly sensitive to small structural changes. The advantage of CD measurements is that the positive and negative CD bands appearing at the wavelengths of optically active transitions are more easily resolved than the corresponding ORD bands would be.

INTRODUCTORY–INTERMEDIATE   **Suggestions for Further Reading**

MISLOW, K. *Introduction to Stereochemistry.* W. A. Benjamin, Inc., Menlo Park, Calif., 1965, Chapter 2.
   A clearly written introductory text.

NATTA, G., and M. FARINA. *Stereochemistry.* Harper & Row, Publishers, New York, 1972, Chapter 4.
   Discusses the symmetry of many optically active molecules.

INTERMEDIATE–ADVANCED

CRABBE, P. *Optical Rotary Dispersion and Circular Dichroism in Organic Chemistry.* Holden-Day, Inc., San Francisco, 1965.

CRABBE, P. *ORD and CD in Chemistry and Biochemistry.* Academic Press, Inc., New York, 1972.

DJERASSI, C. *Optical Rotatory Dispersion.* McGraw-Hill Book Company, New York, 1960.
   These three texts are more suited for advanced students.

HASCHEMEYER, R. H., and A. E. V. HASCHEMEYER. *Proteins.* John Wiley & Sons, Inc., New York, 1973, Chapter 9.
   Provides many useful references.

JIRGENSONS, B. *Optical Rotatory Dispersion of Proteins and Other Macromolecules.* Springer-Verlag, New York, 1969.

An advanced text suitable for graduate students.

VAN HOLDE, K. E. *Physical Biochemistry.* Prentice-Hall, Inc., Englewood Cliffs, N.J., 1971, Chapter 10.

A brief but readable introduction to CD and ORD.

**Reading
Assignments**

SECTION A

"Stereochemistry Since LeBel and van't Hoff," E. L. Eliel, *Chemistry* **49**(1), 6, **49**(3), 8 (1976).

"Criteria for Optical Activity in Organic Molecules," D. F. Mowery, Jr., *J. Chem. Educ.* **46**, 269 (1969).

"Absorption, Dispersion, Circular Dichroism, and Rotatory Dispersion," J. G. Foss, *J. Chem. Educ.* **40**, 593 (1963).

"An Experiment in Optical Rotatory Dispersion," J. P. Schelz and W. C. Purdy, *J. Chem. Educ.* **41**, 645 (1964).

"Proteins," P. Doty, *Sci. Am.*, Sept. 1957.

"The Natural Origin of Optically Active Compounds," W. E. Elias, *J. Chem. Educ.* **49**, 448 (1972).

"Spontaneous Generation of Optical Activity," R. E. Pincock and K. R. Wilson, *J. Chem. Educ.* **50**, 455 (1973).

"Circular Dichroism of Biological Macromolecules," S. Beychok, *Science* **154**, 1288 (1966).

"Optical Rotatory Dispersion and Circular Dichroism," K. P. Wong, *J. Chem. Educ.* **51**, A573 (1974); **52**, A9 (1975).

"Optical Activity and Molecular Dissymmetry," S. F. Mason, *Chem. Brit.* **1**, 245 (1965).

"Optical Rotation and the DNA Helix to Coil Transition," G. L. Baker and M. E. Alden, *J. Chem. Educ.* **51**, 591 (1974).

"Left-Handed and Right-Handed Molecules," G. B. Kauffman, *Chemistry* **50**(3), 14 (1977).

SECTION B

"Symmetry," C. A. Coulson, *Chem, Brit.* **4**, 113 (1968).

"A Brief History of Polarimetry," R. E. Lyle and G. G. Lyle, *J. Chem. Educ.* **41**, 308 (1964).

"A Model for Optical Rotation," L. L. Jones and H. Eyring, *J. Chem. Educ.* **38**, 601 (1961).

"Circular Dichroism—Theory and Instrumentation," A. Abu-Shumays and J. J. Duffield, *Anal. Chem.* **38**, 29A (1966).

"Optical Rotation and the Conformation of Polypeptides and Proteins," P. Urnes and P. Doty, *Advan. Protein Chem.* **16**, 421 (1961).

Problems

19.1 List all the symmetry elements for the following molecules: CH_4, CH_3Cl, CH_2Cl_2, and CHClBrI.

19.2 How many C_n axes does benzene possess?

19.3 Optically active molecules can be classified into two types, asymmetric and dissymmetric. The former does not contain any symmetry elements (except C_1 and E), while the latter contains a single rotation axis $C_n(n > 1)$. Classify the following molecules according to these two definitions:

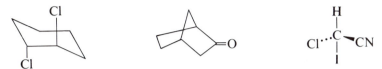

19.4 Is 1,3-dichloroallene ($C_3H_2Cl_2$) optically active?

19.5 The optical rotation of a sucrose solution (concentration: 9.6 g/100 ml H_2O) is 0.34° when measured in a 10-cm cell with the sodium D-line at room temperature. Calculate the specific rotation and molar rotation for sucrose.

19.6 The rotation of a certain solution containing an optically active compound is 2.41°. Calculate the quantity $(n_L - n_R)$. The sodium D-line is used for the measurement.

19.7 Two optical isomers A and B having specific rotations 27.6° and $-19.5°$, respectively, are in equilibrium in solution. If the specific rotation of the mixture is 16.2°, calculate the equilibrium constant for the process: $A \rightleftharpoons B$.

19.8 Two substances A and B have identical absorption spectra and identical CD curves except that one CD curve is positive and the other is negative. What is the structural relationship between A and B?

20 Photochemistry and Photobiology

Green is beautiful.

Chemical reactions can be divided into two types: thermal and photochemical. *Thermal reactions* are those discussed in Chapter 13. They involve atoms and molecules in their electronic *ground* states. By definition a *photochemical reaction* takes place only in the presence of light, which usually ranges from the visible and uv region to high-energy radiation such as X rays and γ rays. Before the photochemical reaction can proceed, the reactant molecules must first be electronically excited by radiation.

A photochemist is interested in the fate of an electronically excited molecule. Depending on the system and the conditions under which photoexcitation is carried out, such a molecule can undergo one of several processes. It can lose its excess energy by collision with other molecules, liberating heat. It can return to the ground state by emitting a photon; that is, it can fluoresce or phosphoresce. Third, it can undergo a chemical reaction—isomerization, dissociation, or ionization, for example. Chapter 18 dealt with fluorescence and phosphorescence in some detail. In this chapter we discuss several important types of photochemical and photobiological reactions.

20.1 Introduction

Before we discuss individual reactions, a few definitions are in order. Photochemical reactions can be classified as *primary* and *secondary processes*. Primary processes lead to vibrational relaxation, or loss of vibrational energy, by collision with other molecules, fluorescence, phosphorescence, isomerization, and dissociation. Dissociation of excited molecules may provide reactive intermediates that can undergo secondary processes of a thermal nature.

Let us illustrate the primary and secondary processes with the decomposition of hydrogen iodide. In the absence of light, the following thermal reaction takes place:

$$2HI \longrightarrow H_2 + I_2$$

When light of the appropriate wavelength is applied, the reactions are

$$HI \xrightarrow{h\nu} H + I \qquad \text{photochemical reaction}$$
$$\text{(primary process)}$$

$$H + HI \longrightarrow H_2 + I \qquad \text{thermal reactions}$$

$$I + I \longrightarrow I_2 \qquad \text{(secondary processes)}$$

$$H + H \longrightarrow H_2$$

where $h\nu$ represents quanta of light absorbed. Although the final products (H_2 and I_2) are the same in both cases, the mechanisms are different for the thermal and photochemical processes.

We define the *photochemical efficiency* or the quantum yield Φ as

$$\Phi = \frac{\text{output}}{\text{input}} = \frac{\text{number of molecules produced}}{\text{number of light quanta absorbed}} \qquad (20.1)$$

Equation (20.1) can be written in molar quantities as

$$\Phi = \frac{\text{number of moles of product formed}}{\text{number of einsteins absorbed}} \qquad (20.2)$$

where an *einstein* is equal to 1 mol of light quanta, or 6.022×10^{23} photons.

Alternatively, a photochemical reaction can be analyzed in terms of product formation. Consider the reactions

$$A \xrightarrow{h\nu} A^*$$
$$A^* \xrightarrow{k_1} A$$
$$A^* \xrightarrow{k_2} \text{product}$$

where A is the reactant and A* is the reactive intermediate. If the steady-state condition holds, then

$$\text{rate of formation of A*} = \text{rate of removal of A*}$$

$$= k_1[A^*] + k_2[A^*]$$

The efficiency of product formation Φ_p is given by

$$\Phi_p = \frac{\text{rate of product formation}}{\text{total rate of removal of A*}}$$

$$= \frac{k_2[A^*]}{k_1[A^*] + k_2[A^*]} = \frac{k_2}{k_1 + k_2} \qquad (20.3)$$

It is important to realize that Φ and Φ_p are not fundamentally related. Two reactions may have very similar photochemical efficiencies but differ greatly in rate constants. Consider the following photochemical decompositions*:

$$C_6H_5COCH_2CH_2CH_3 \xrightarrow{k} C_6H_5COCH_3 + CH_2{=}CH_2$$
$$\Phi = 0.40 \qquad k = 3 \times 10^6 \text{ s}^{-1}$$

$$CH_3COCH_2CH_2CH_2CH_3 \xrightarrow{k} CH_3COCH_3 + CH_2{=}CHCH_3$$
$$\Phi = 0.38 \qquad k = 1 \times 10^9 \text{ s}^{-1}$$

* For a good discussion, see N. J. Turro, *J. Chem. Educ.* **44**, 536 (1967).

Regardless of the mechanism involved, the rate of a photochemical reaction should be proportional to the rate of absorption of light. Thus kinetic studies of photochemical reactions require accurate measurement of the intensity of light employed. The most commonly employed technique is a chemical *actinometer*—a chemical system whose photochemical behavior is quantitatively understood. One of the most useful solution-phase actinometers is the potassium ferrioxalate system. When a sulfuric acid solution of $K_3Fe(C_2O_4)_3$ is irradiated with light in the range 250 to 470 nm, simultaneous reduction of iron from the ferric to the ferrous state and oxidation of the oxalate ion result:

$$Fe(C_2O_4)_3^{3-} \xrightarrow{\ h\nu\ } Fe^{2+} + \text{oxalate} + \text{oxidized oxalate}$$

This reaction has been carefully studied and quantum yields of Fe^{2+} are known with great accuracy at various wavelengths. The amount of Fe^{2+} ions formed can be readily determined by the formation of the red 1,10-phenanthroline–Fe^{2+} complex. In this way the amount of light quanta absorbed, or the intensity of the light, can be calculated using Eq. (20.1) or Eq. (20.2).

> **EXAMPLE 20.1** A 35-ml solution of ferric oxalate is irradiated with monochromatic light at 468 nm for 30 min. The solution is then titrated with 1,10-phenethroline to form the red complex of 1,10-Ph–Fe^{2+}. The absorbance of this complex measured in a 1-cm cuvette at 510 nm is found to be 0.65 ($\varepsilon_{510} = 1.11 \times 10^4$ liters mol^{-1} cm^{-1}). Assume that the quantum yield for the decomposition at 468 nm is 0.93. Calculate the absorption of the monochromatic light in einsteins and total amount of energy absorbed.

Answer: From Eq. (18.3),

$$A = \varepsilon bc$$

$$c = \frac{0.65}{1 \text{ cm} \times 1.11 \times 10^4 \text{ liters mol}^{-1} \text{ cm}^{-1}}$$

$$= 5.86 \times 10^{-5} \ M$$

Number of einsteins absorbed in 30 min is given by

$$\frac{\text{number of moles of } Fe^{2+} \text{ produced}}{\text{quantum yield}} = \frac{5.86 \times 10^{-5} \times 35 \text{ ml}}{0.93 \times 1000 \text{ ml}}$$

$$= 2.2 \times 10^{-6}$$

Finally,

$$\text{total energy absorbed} = \text{number of photons} \times h\nu$$

$$= 2.2 \times 10^{-6} \times 6.022 \times 10^{23} \times 6.626$$

$$\times 10^{-34} \text{ J s} \times \frac{3.00 \times 10^{10} \text{ cm s}^{-1}}{4650 \times 10^{-8} \text{ cm}} = 0.56 \text{ J}$$

The actual rate of photochemical reaction depends on a number of parameters. As an approximation, we write

$$\text{rate of product formation} = If\Phi_p \tag{20.4}$$

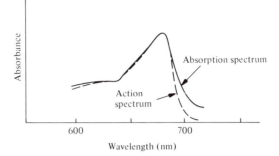

Comparison of the absorption spectrum and the action spectrum for the unicellular alga chlorella. The photosynthetic efficiency (measured by oxygen evolution) of light of different wavelengths (action spectrum) parallels closely the absorption spectrum of chlorophyll molecules. The discrepancy at about 700 nm is known as the "red drop," to be discussed in Section 20.2. This comparison strongly suggests that chlorophyll plays a key role in photosynthesis. (From *Light and Living Matter*, Vol. 2, p. 38, by R. K. Clayton. Copyright 1971 by McGraw-Hill Book Company. Used with permission of McGraw-Hill Book Company.)   **Figure 20.1**

where I is the rate of absorption of light, f the fraction of light that produces the chemically reactive species, and Φ_p the quantum yield of product formation. We see now why it is possible for two reactions to have similar Φ_p's but very different rates if the reactants have very different I and f values.

Finally, we note that often very useful information regarding species responsible for photochemical and photobiological processes can be obtained if we measure the response or the effectiveness of the system as a function of the wavelength of the light employed. This procedure gives rise to the *action spectrum*. In general, if a simple system contains only one type of molecule, the action spectrum should and does resemble the absorption spectrum closely. In a biological system, there are usually several different compounds that absorb strongly the incident radiation over the range of wavelength of interest. The molecules that are responsible for the photochemical reaction may be present in very low concentrations so that their absorption spectra cannot always be easily detected. However, their presence may be revealed by recording the action spectrum instead of the usual absorption spectrum (Figure 20.1).

Photosynthesis $\mathbf{20.2}$

Photosynthesis is the most important photobiological reaction. It is a process by which plants and other organisms store the energy of sunlight. Although green plant photosynthesis involves a number of very complex steps, the overall change can be represented as

$$6CO_2 + 6H_2O \xrightarrow{h\nu} C_6H_{12}O_6 + 6O_2 \qquad \Delta G^\circ = 2861.9 \text{ kJ}$$

or, more generally,

$$CO_2 + H_2O \xrightarrow{h\nu} (CH_2O) + O_2 \qquad \Delta G^\circ = 477.0 \text{ kJ}$$

565

The net results are assimilation of carbon dioxide, oxidation of water to molecular oxygen, and synthesis of carbohydrates. Formation of oxygen as a by-product is not universal among photosynthetic organisms. For example, in photosynthetic bacteria we have

$$2H_2S + CO_2 \xrightarrow{h\nu} (CH_2O) + H_2O + 2S$$

where free sulfur is produced instead of O_2. A more general formula describing the reaction would be

$$2H_2D + CO_2 \xrightarrow{h\nu} (CH_2O) + H_2O + 2D$$

where D is some element (usually O or S) that is oxidized in the reaction. As the changes in standard free energy show, these are highly endergonic reactions—reactions that can only be engendered by a supply of light energy.

The study of photosynthesis encompasses areas ranging from chemical physics to molecular biology. Advances made during the last three decades have given us a fairly complete description of the overall process involved, although some important details are still lacking. This section will discuss only the initial process of light absorption and some of the chemical reactions that immediately follow it. Photosynthesis consists of two main stages. The first stage, called *light reactions*, involves the absorption of photons and initial energy storage that takes place rapidly, perhaps in a period as short as 10^{-12} to 10^{-8} s. This step is then followed by a series of chemical transformations called *dark reactions* because they occur in the absence of light. Dark reactions form the rate-determining steps for the overall change.

CHLOROPHYLL MOLECULES

Chlorophyll molecules are chromophores responsible for the absorption of light. A photosynthetic organism may contain two or more types of chlorophyll molecules. In green plants, these are chlorophyll *a* and chlorophyll *b*, as shown in Figure 20.2. The highly conjugated porphyrin system results in their absorption of light in the visible region. As Figure 20.3 shows, these molecules absorb strongly in the blue and red regions, but reflect light in the green, yellow, and orange regions. Consequently, they all have a characteristic green color.

In plants, the chlorophyll molecules are associated with lipoprotein membranes present in the organelle known as the *chloroplast*. A typical plant cel' contains between 50 and 200 chloroplasts, which are about 10,000 Å in length (Figure 20.4). In addition to the two outer membranes, chloroplasts have interior membranes which form a highly laminated structure consisting of individual sacs piled into cylinders called *grana*. Chlorophyll and other pigment molecules are located in the grana lamellae, and stroma lamellae, and there the photosynthetic process begins.

Upon absorbing a photon, the excited chlorophyll molecule can undergo one of two processes. It can fluoresce or utilize its excess electronic energy to drive some energetically unfavorable reactions. The fluorescence of chlorophyll *a* molecules in chloroplast has been observed although the quantum yield of fluorescence is quite low ($\lesssim 0.05$). Interestingly, the fluorescence of chlorophyll *b* inside the chloroplast has never been observed. This fact led Duysens and others to suggest that chlorophyll *b* molecules do not participate

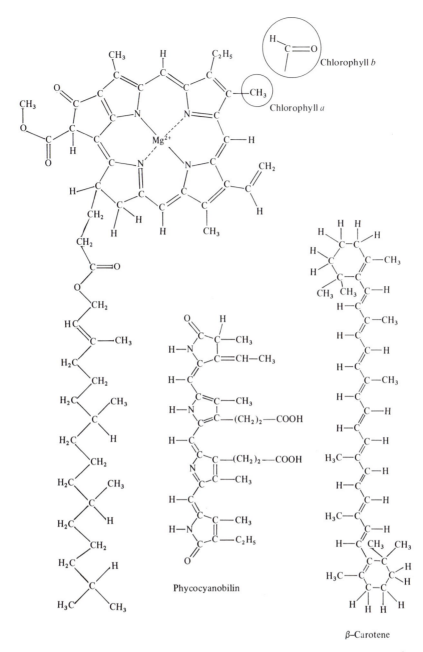

Photosynthetic pigments. Chlorophyll and β-carotene molecules are present in higher plants; phycocyanobilin is found only in certain algaes.

Figure 20.2

directly in photosynthesis. Rather, their role is to act as antennas helping to gather or harvest light energy. Once a chlorophyll b molecule is photoexcited, it rapidly transfers its excess energy to a chlorophyll a molecule, which can take part in photosynthesis. In addition to chlorophyll b, other antenna molecules are the carotenoid pigments and phycobilins. The transfer of excitation energy from chlorophyll b to chlorophyll a takes place via a resonance process, which requires that the molecules be closely spaced together (see Section 18.7). The efficiency for transfer falls rapidly if the molecules are more than 25 Å apart.

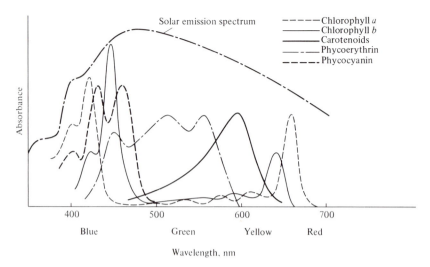

Chlorophyll *a* molecules absorb light by themselves; the role played by the antenna molecules is to improve the supply of energy in regions of the spectrum where chlorophyll *a* does not absorb strongly. Eventually, this excitation energy reaches a *reaction center*, which contains only a few chlorophyll *a* molecules oriented in a special way. It is here that the first step of the chemical reaction begins. The light-harvesting pigments and the reaction center together constitute the photosynthetic unit, which is shown in Figure 20.5. There are about 300 chlorophyll molecules per unit.

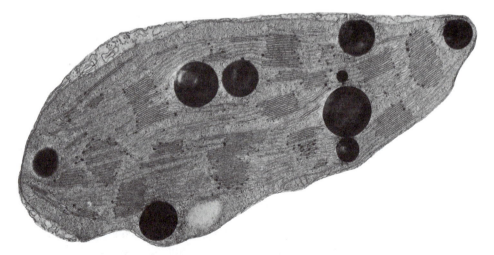

Figure 20.4 Electron micrograph of a chloroplast from a citrus leaf. The closely packed membranes, which are packed into stacks are called grana. They conduct the initial step in photosynthesis—the "light reaction" that traps the photons. The black particles are called plastoglobuli, which consist of lipids. They appear black as a result of the staining technique used in electron microscopy. (Courtesy of W. W. Thomson.)

Schematic diagram of a photosynthetic unit. The wavy arrows represent transfer of the excitation energy by resonance interaction.

Figure 20.5

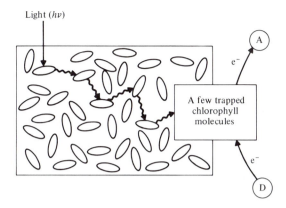

Light ($h\nu$)

A few trapped chlorophyll molecules

A

D

e^-

e^-

PHOTOSYSTEMS I AND II

So far we have implied that photosynthesis takes place upon the absorption of a single photon. However, an important finding made by Emerson and his colleagues in 1956 showed that the actual situation is more complex than that depicted in Figure 20.5. They measured the efficiency of photosynthesis in an isolated chloroplast as a function of wavelength and discovered a curious effect.* Starting from the short wavelength at about 400 nm, the efficiency was found to remain fairly constant and close to the maximum value until 680 nm is reached. Beyond this point the efficiency rapidly drops to zero (called the *red drop*), even though the chlorophyll *a* molecule still absorbs at this wavelength (see Figure 20.1). Efficiency could be restored if the system were simultaneously irradiated with a shorter wavelength, say at 650 nm. Furthermore, the efficiency under simultaneous irradiation is greater than the sum of the efficiencies under separate irradiation at these two wavelengths. This phenomenon, now called the *Emerson enhancement effect*, suggests that there are two separate photochemical systems called *photosystem I* (PSI) and *photosystem II* (PSII). Both of these systems are driven by light of less than 680 nm, but only one of them is driven by light of longer wavelengths. Each photosystem has its own photosynthetic unit, containing 200 to 300 light-harvesting chlorophylls and other pigments. The ratio of chlorophyll *a* to chlorophyll *b* is much greater in PSI than in PSII. The reaction center in PSI contains dimeric chlorophyll *a* molecules whose maximum absorption is at 700 nm, so it is called pigment 700 or P700. The nature of chlorophyll molecules in PSII reaction center is not known, although there is some evidence to suggest that it might be pigment 680 or P680.

BIOCHEMICAL REACTIONS

How are these photosystems related to each other, and what are their roles in the overall process? As mentioned earlier, the conversion of carbon dioxide and water into carbohydrate and oxygen is a highly endergonic process. This energy is obtained photochemically in the following manner. At the reaction

* The efficiency can be defined as the number of oxygen molecules released for each quanta of light absorbed.

center, each chlorophyll molecule is closely associated with an electron donor
D and an electron acceptor molecule A. Upon photoexcitation, the following
electron-transfer reaction occurs:

$$\text{D Chl A} \xrightarrow{h\nu} \text{D Chl*A} \longrightarrow \text{D Chl}^+ \text{ A}^- \longrightarrow \text{D}^+ \text{ Chl A}^-$$

The molecule A accepts an electron from the excited chlorophyll and becomes
A^-. The electron so removed can be replaced by one coming from a donor
molecule D. Since D and A are suitably separated, they do not interact with
each other; instead, the D^+ and A^- species can take part in other redox
processes. The relation between PSI and PSII, as well as other various components, is shown in Figure 20.6. This diagram is known as the Z scheme. Let
us start with PSII. Upon receiving a photon from their pigment molecules, a
strong oxidant D^+ and a weak reductant A^- are formed at the reaction
center. The chemical nature of D^+ is not known, but this species is capable of
oxidizing water to molecular oxygen. The acceptor in this electron-transfer
process is probably a special type of plastoquinone. Note that this is an uphill
reaction; that is, electron is boosted uphill electrochemically from chlorophyll
to Q. On the other hand, the flow of electron from Q to PSI is a downhill
process. Some of the electron carriers along the path are known to be plastoquinone, cytochrome b, cytochrome f, and plastocyanin.

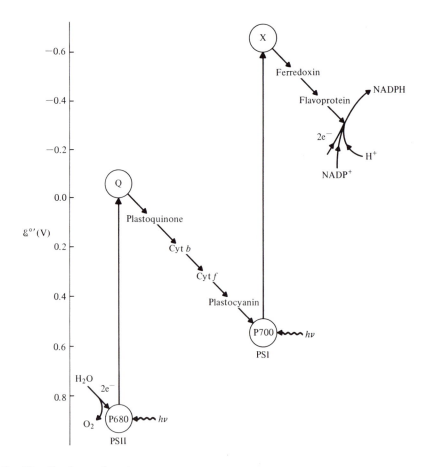

Figure 20.6 The Z scheme for electron transport in higher plants.

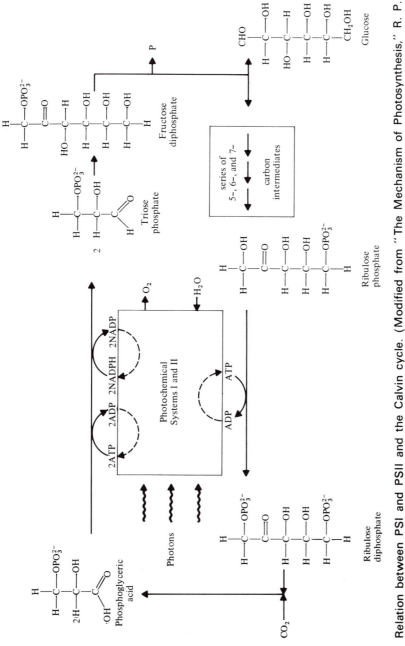

Figure 20.7 Relation between PSI and PSII and the Calvin cycle. (Modified from "The Mechanism of Photosynthesis," R. P. Levine. Copyright © 1969 by Scientific American, Inc. All rights reserved.)

Now we come to PSI. In this case the immediate donor molecule is plasto-cyanin while the acceptor is called "X," probably a membrane-bound iron-sulfur protein (ferredoxin), or perhaps a chlorophyll molecule. In contrast to PSII, the illumination of PSI results in a weak oxidant and a strong reductant. From ferredoxin the electrons are transferred to flavoprotein. Here they are used to reduce the coenzyme nicotinamide adenine dinucleotide phosphate ($NADP^+$) to NADPH.

The red drop and the Emerson enhancement effect can now be readily explained. As mentioned earlier, PSI is driven most efficiently by far-red light (680 nm or longer) alone. In the absence of a short-wavelength radiation to drive PSII, the flow of electrons along the Z scheme soon ceases, since water molecules are not oxidized to molecular oxygen. Irradiation at a wavelength shorter than 680 nm alone does increase the efficiency. But maximum efficiency is obtained only when both wavelengths are employed.

The structure of $NADP^+$ is quite similar to that of NAD^+, shown in Figure 10.11. The redox potential of the $NADP^+$/NADPH couple is about -0.32 V, insufficient to reduce carbon dioxide to carbohydrate (-0.42 V). Reduction can be achieved, however, if an additional source of energy is supplied by ATP. Since photosynthesis is a self-contained process, ATP molecules must also be synthesized during the dark reactions (a process called *photophosphorylation*). The exact mechanism for coupling the synthesis of ATP with photosynthetic electron transport is still not well understood.*

To summarize the chemistry involved, we write

$$2H_2O \longrightarrow O_2 + 4e^- + 4H^+$$

$$2NADP^+ + 4e^- + 2H^+ \longrightarrow 2NADPH$$

$$\underline{2H^+ + 2NADPH + CO_2 \longrightarrow 2NADP^+ + H_2O + CH_2O}$$

$$\text{overall reaction:} \quad CO_2 + H_2O \longrightarrow CH_2O + O_2$$

Figure 20.7 shows the relation between the photosynthetic units and the metabolic pathway, which is known as the Calvin cycle. Like glycolysis, each step is enzyme-catalyzed.

Finally, we mention briefly photosynthesis in bacterial systems. In many photosynthetic bacteria, oxygen is not produced. In contrast to the directional (noncyclic) electron flow from water to $NADP^+$, which is found in green plants and algae, these bacteria utilize a single cyclic photosystem. The basic mechanism of light-induced electron transfer remains the same, but the donor and acceptor chains are connected so that no net oxidation or reduction occurs. Some of the energy of the photon is conserved in ATP formation, which in turn drives NAD^+ reduction, using electrons from H_2S or organic acids. The NADH thus formed along with additional light-generated ATP are used to fix CO_2 using the Calvin cycle.

20.3 Vision

Vision involves absorption of radiation energy and transduction of its stimulation into nerve impulses. Only three of the eleven animal phyla have developed well-formed, image-resolving eyes. These are anthropods, mollusks,

---

* See D. W. Deamer, *J. Chem. Educ.* **46**, 198 (1969), and P. C. Hinkle and R. E. McCarty, *Sci. Am.*, Mar. 1978.

and vertebrates. Although these species have followed widely divergent evolutionary roads, the chemistry of their visual processes is remarkably similar.

As a result of the work carried out by Wald and others, we now understand the basic mechanism involved in vision fairly well. As in photosynthesis, the first step in vision is absorption of light energy by the proper chromophore. The chromophore responsible for absorption of visible light is vitamin A aldehyde or retinal. In the retina of the eye there are some 100 million rod-shaped cells and 5 million cone-shaped cells. Between the cells and the nerve fibers leading to the brain are synapses or junctions (Figure 20.8). Retinal is associated with a protein called *opsin*. Four different types of opsin exist, one in the rod-shaped cells and three in the cone-shaped cells. The chromophore-opsin complexes in these two different cells are called *rhodopsin* and *iodopsin*, respectively. The changes occurring after excitation by light are

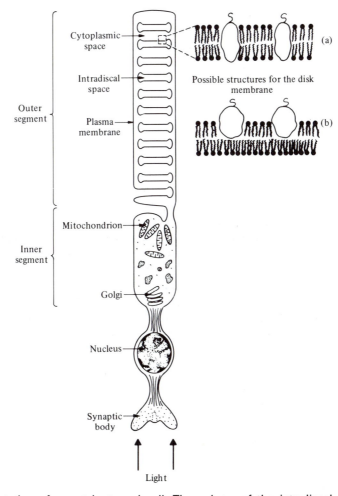

Schematic representation of a vertebrate rod cell. The volume of the intradiscal space is greatly exaggerated, and in the normal bovine outer segment there are approximately 1500 discs within an outer segment 500,000 Å in length. In the magnified views showing possible structures for the disc membrane, the wiggly lines represent the hydrocarbon chains of the phospholipids, and the circles the polar head groups. The "S" on the rhodopsin molecule signifies the carbohydrate moiety. [Reprinted with permission from W. L. Hubbell, *Acc. Chem. Res.* **8**, 85 (1975). Copyright by the American Chemical Society.]

Figure 20.8

basically the same in rhodopsin and iodopsin, namely a cis to trans isomerization of the chromophore. Since isomerization reactions are important not
only in vision, but also in many other organic reactions, we digress a moment
to discuss their basic features.

ROTATION ABOUT C—C AND C=C BONDS

In Chapter 16 we discussed the bonding orbitals involved in the C—C and
C=C bonds. Consider first the rotation about a C—C bond, say in ethane.
Each carbon atom in ethane is sp^3-hybridized. Energetically, the rotation
about the single bond produces a more stable form, called the *staggered
conformation*, and a less stable form, called the *eclipsed conformation*.

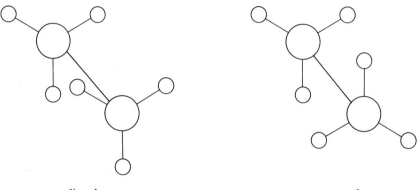

eclipsed   staggered

It is important to realize that the term *conformation* is defined to represent
any one of the infinite number of arrangements of atoms in three-dimensional
space. As long as the molecule can rotate about the single bond, each conformation should only be thought of as momentary. The rotation is restricted to
a certain extent because of steric hindrance between the hydrogen atoms on
adjacent carbon atoms. Figure 20.9 shows a plot of potential energy of ethane
versus some angle θ of rotation. The barrier to rotation is fairly small, about 8
to 10 kJ mol^{-1}. Consequently, the molecule can easily change from one
staggered conformation to another. Even in the absence of the necessary
activation energy (for example, at low temperature), this process can take

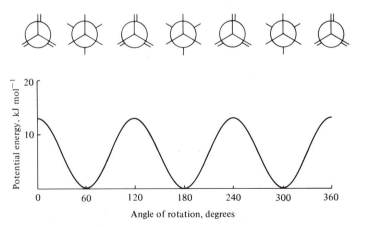

Figure 20.9   Potential energy versus angle of rotation about the C—C bond in ethane.

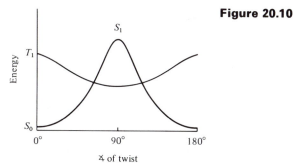

Potential energy versus angle of twist about the C=C bond for the first excited singlet state and the ground triplet state. (Reproduced from *Molecular Photochemistry*, written by N. J. Turro, with permission of publishers W. A. Benjamin, Inc., Advanced Book Program, Reading, Mass. 1965.)

Figure 20.10

place via quantum mechanical tunneling, that is, tunneling of hydrogen atoms through the barrier (see Chapter 15). If we now replace two hydrogen atoms, one on each carbon atom, by two bulky groups such as *tert*-butyl or phenyl, we can increase the barrier height sufficiently to actually "freeze out" or "trap" the various conformational isomers or conformers.

The situation is different for molecules containing the C=C bonds. In addition to the σ bond, there is also a π bond between the two carbon atoms. Rotation about the double bond becomes much more restricted. Consequently, we obtain geometric isomers, called *cis* and *trans isomers*, which generally differ markedly in melting point, boiling point, dipole moment, and other properties.

Much effort has been devoted to the study of cis-trans isomerization. The activation energy required for these reactions is typically of the order of $100 \, kJ \, mol^{-1}$. Heating, irradiation, and chemical catalysis are the usual means to bring about isomerization.

Many of the systems investigated are derivatives of ethylene such as stilbene, as well as compounds containing N=N bonds (for example, azobenzene). Photoexcitation of the C=C bonds results in a $\pi \to \pi^*$ transition. The molecule may be in the first excited singlet or the ground triplet state. Theoretical considerations show that the most stable state is a triplet state which corresponds to a twisted molecule through 90° about the C—C internuclear axis, as shown in Figure 20.10. Consequently, this excess energy is lost, and the molecule returns from the triplet state to the planar singlet ground state. This is the step that leads to isomerization. The same stable perpendicular form of the triplet state is obtained starting from either the cis or the trans isomer.

Photoisomerization can thus occur from cis to trans and trans to cis isomers. If the quantum yields for these processes are equal, then the prolonged irradiation of either isomer produces a photostationary state of constant [cis] to [trans] ratio. The actual ratio depends on the molar extinction coefficients of each isomer over the wavelengths employed. Generally, the trans isomer absorbs light of longer wavelengths and more intensely than does the cis isomer. As an example, consider the stilbene molecules.* For *trans*-stilbene, $\lambda_{max} = 311$ nm and $\varepsilon_{max} = 25{,}600$ liters $mol^{-1} \, cm^{-1}$; for *cis*-stilbene, $\lambda_{max} = 280$ nm and $\varepsilon_{max} = 10{,}000$ liters $mol^{-1} \, cm^{-1}$. As long as the spectra of two isomers do not overlap strongly throughout the entire region, quantitative conversion in either direction may be obtained by using monochromatic light. The isomer that absorbs strongly at a particular wavelength will be converted to the isomer that does not absorb at the same wavelength.

---

* There are several maxima for each isomer. We compare here the long-wavelength absorptions only.

MECHANISM OF VISION

Figure 20.11 shows the structure of 11-*cis* retinal and all-*trans* retinal isomers. Actually, a total of six geometric isomers are possible, but only these two isomers are important in the vision process. The only role played by light is the isomerization of 11-*cis* retinal to all-*trans* retinal. Here is the fundamental difference between the action of light in vision and in photosynthesis. In photosynthesis light energy is used to do the chemical work of boosting electrons against an electrochemical gradient and synthesizing ATP molecules. In vision there is no evidence that chemical reactions are brought about by the light energy. Nerve fibers upon which light acts are ready to discharge because they were previously charged by chemical reactions totally unrelated to excitation of the chromophore. Light is only needed here to trigger their discharges.

Opsin is a protein with a molar mass of about 30,000 g mol^{-1}. Its structure has not been elucidated by X ray although it is known that in rhodopsin the aldehyde group of 11-*cis* retinal forms a Schiff's base with the amino group of a lysyl residue in opsin:

$$\diagdown C{=}O \ + \ -NH_2 \ \longrightarrow \ \diagdown C{=}N{-} \ + \ H_2O$$

There is also evidence to suggest that there are additional sites of binding, probably the sulfhydryl groups. The important point is that the fit between 11-*cis* retinal and opsin is highly stereospecific so that, upon isomerization, the straight all-*trans* retinal is no longer able to bind opsin and dissociation takes place. One or more of the steps, which involve conformation changes and immediately follow the dissociation, somehow induce the signal from the retinal cell to the nerve fiber.

For many years it had been assumed that the cis to trans isomerization is the first step in the vision process. However, recent advances in spectroscopic techniques, particularly in the area of picosecond spectroscopy,* have enabled scientists to probe the fundamental steps that occur in rhodopsin upon receiving a photon in greater detail. It appears that the earliest event in vision involves a proton transfer from an amino acid residue to the Schiff base, which is then followed by the isomerization, although the exact mechanism is still not clear. The interested reader should consult the references listed at the end of this chapter. Figure 20.12 shows the visual cycle in the vertebrate eye

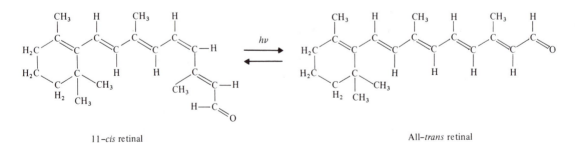

11–*cis* retinal

All-*trans* retinal

Figure 20.11   Structure of 11-*cis* retinal and all-*trans* retinal.

* A picosecond is 10^{-12} s. Picosecond spectroscopy employs light of extremely short duration which is necessary to follow the ultrafast events that occur at the molecular level. For more detail, see "Picosecond Chemical and Biological Events," P. M. Rentzepis, *Science* **202**, 174 (1978).

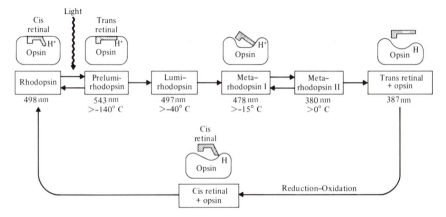

Visual cycle in the vertebrate eye. The wavelengths are in nanometers (nm). (From "How Light Interacts with Living Matter," S. B. Hendricks. Copyright © 1968 by Scientific American, Inc. All rights reserved.)   **Figure 20.12**

starting with the cis-to-trans isomerization step. A number of intermediates (prelumirhodopsin, lumirhodopsin, and metarhodopsins) are formed during the cycle, although not all of them can be isolated at room temperature for characterization.

It is tempting to postulate a triplet-state intermediate for the initial cis-to-trans isomerization although no direct evidence has been obtained to support this speculation. The situation is more complicated in retinal, since in addition to the $\pi \to \pi^*$ transition, we must also consider the $n \to \pi^*$ transition due to the aldehyde group.

The sensitivity of the animal retina is remarkable, for a single quantum of light is enough to stimulate a rod cell. Actual vision, however, requires the absorption of several quanta, perhaps five. It follows that the quantum efficiency for the initial photoisomerization must be close to unity.

Rhodopsin functions when vision takes place under low light intensity, for example, at nighttime. It cannot distinguish different colors, since it has only one pigment. Three types of iodopsin, which contain pigments that absorb light at a maximum of 450 nm (blue), 525 nm (green), and 555 nm (yellow), are responsible for color vision. The pigment with its maximum at 555 nm extends its sensitivity far into the red to allow sensation of red as well. Cone cells are much less sensitive to light than rod cells, so in dim lighting all objects appear in shades of gray. In addition to human beings, primates, bone fishes, and birds possess cone cells and can perceive certain colors at least. The retinas of cats and cattle, on the other hand, contain predominantly rod cells. These animals are therefore colorblind.*

Chemiluminescence and
Bioluminescence **20.4**

Chemiluminescence is the production of light by chemical reaction. During the course of a reaction, one or more of the intermediates is formed in an electronically excited state, so it can emit light by means of fluorescence. Here

* We conclude, therefore, that the matador does not really need to use a red cape in a bullfight.

is an example of chemical energy being converted into light energy—just the opposite of photosynthesis.

A number of compounds are known to be *chemiluminescent*. One of the most studied systems is the oxidation of luminol (5-amino-2,3-dihydro-1,4-phthalazinedione). When luminol is treated with a base, hydrogen peroxide, and potassium ferricyanide, an intense blue light is produced. The steps are as follows:

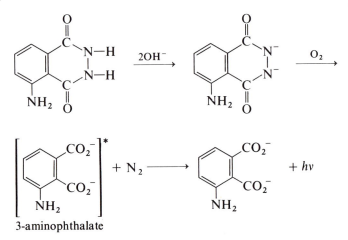

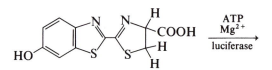

3-aminophthalate

Spectroscopic studies show that the species responsible for light emission is the electronically excited 3-aminophthalate molecules, denoted by the asterisk. The transition is from the first excited singlet state to the ground state.

Bioluminescent reactions generally involve enzyme-catalyzed reactions with oxygen. A large number of organisms, such as bacteria, fungi, corals, clams, and insects, have developed the ability to emit light. Of these, the best known case is undoubtedly the firefly and its larvae, called glowworms. The chemistry of bioluminescence in fireflies has been fairly well clarified by McElroy, Seliger, and their coworkers. The reaction involves luciferin, ATP, oxygen, and the enzyme luciferase (molecular mass about 100,000). The first step is the formation of luciferyl adenylate:

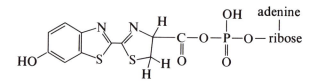

luciferin

luciferyl adenylate

In the presence of molecular oxygen, luciferyl adenylate undergoes bioluminescence; the species responsible for emission is still unknown.* The firefly *Photinus pyralis* emits light at about 560 nm, which is in the yellow-green

---

* It is interesting to note that the combination luciferin–luciferase is routinely used to assay ATP.

region, whereas other species of firefly emit at somewhat longer wavelengths. Since all the species have the same luciferin molecule, the variety of colors observed is probably caused by the difference in the enzyme luciferase, whose structure and/or conformation may change from species to species. *In vitro* studies have shown that the actual wavelength of emission depends on the pH of the medium.

Bioluminescence has often been described as "cold light," meaning that the production of light from a biochemical reaction is complete and no heat is given off. This is an accurate description for fireflies, since their quantum efficiency is indeed close to 100%. On the other hand, a chemiluminescent reaction such as oxidation of luminol has an overall efficiency of about 1%; that is, the ratio of photons emitted to the number of luminol molecules reacted is 0.01.

What is the significance of bioluminescence? A study of animal evolution provides a partial answer to this question. During the gradual appearance of oxygen in the earth's atmosphere, the anaerobic organisms then in existence had to get rid of highly toxic O_2 molecules. One way to eliminate O_2 is to reduce it to water. The energy liberated in such a reaction is sufficient to excite certain molecules or intermediates, which can then emit light. The mechanism is obviously not necessary today, for these organisms have already made a transition from anaerobic to aerobic pathways. Yet this unnecessary by-product, light, evolved a useful purpose, for the flashing light of fireflies now serves as a mating signal.*

UV Irradiation of DNA **20.5**

It has long been known that uv radiation kills bacteria, causes abnormalities in chromosomes, and produces mutations. We are just beginning to understand the mechanisms involved.

The DNA molecules absorb radiation strongly between 200 and 300 nm, with a maximum at about 260 nm (see Figure 18.28). When cytosine is irradiated with uv radiation in solution, it takes up a water molecule as follows:

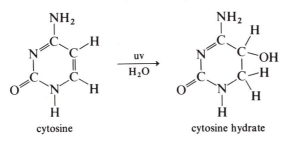

cytosine   cytosine hydrate

However, on treatment with an acid or upon gentle heating, the cytosine hydrate readily loses the water molecule to form cytosine. Because the reaction reverses so easily, the process is not believed to have any biological significance.

Normally, a thymine solution is relatively insensitive to uv light, but when a frozen solution of thymine is irradiated with uv, thymine dimer is formed in

---

* See "Synchronous Fireflies," J. Buck and E. Buck, *Sci. Am.*, May 1976.

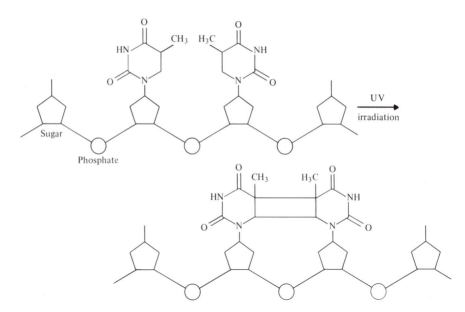

Figure 20.13 Dimerization of two adjacent thymine bases on the same strand of a DNA molecule.

high yield. Unlike the hydrated cytosine, the thymine dimer is a stable molecule, unaffected by acid or heat. It can, however, be decomposed into monomers if irradiated with uv light in solution.

The fact that thymine dimers are formed only in the frozen state shows the reaction to require two thymine molecules not only to be close to each other, but also held in a certain position. Two adjacent thymine base pairs are both close and fixed in position on the same strand of a DNA molecule. We would then expect thymine dimers to form when DNA molecules are exposed to uv radiation, and chemists have confirmed this expectation (Figure 20.13). The reaction can disrupt the normal replication process and may lead to incorrect sequencing in protein synthesis.

Sometimes, a uv-damaged cell, or bacteria apparently killed by uv light, can be repaired by additional uv irradiation. To see how this is accomplished, we need to examine the absorption spectra of thymine monomer and dimer shown in Figure 20.14. As we can see, the dimer does not absorb appreciably

Figure 20.14

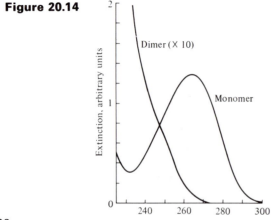

Absorption spectra of thymine monomer and dimer. The molar extinction of the dimer has been exaggerated tenfold. (From *Light and Living Matter*, Vol. 2, p. 184, by R. K. Clayton. Copyright 1971 by McGraw-Hill Company. Used with permission of McGraw-Hill Book Company.)

beyond 280 nm. This means that irradiation at 280 nm favors the formation of dimers, whereas irradiation at a shorter wavelength, say 240 nm, preferentially excites the dimers, resulting in their breakdown into thymine monomers. The latter process then repairs the damaged DNA molecules.

The dimerization of thymine molecules is believed to be the most important photochemical reaction in nucleic acids. Unlike the pyrimidines, the purines (adenine and guanine) are much less sensitive to uv light.

INTRODUCTORY–INTERMEDIATE

Suggestions for Further Reading

CLAYTON, R. K. *Light and Living Matter*, Vols. 1 and 2. McGraw-Hill Book Company, New York, 1971.
An excellent introduction to photochemistry and photobiology. Strongly recommended.

CLAYTON, R. K. *Molecular Physics in Photosynthesis*. Blaisdell Publishing Company, New York, 1965.
Emphasis is on the physical approach.

FOGG, G. E. *Photosynthesis*. American Elsevier Publishing Company, Inc., New York, 1972.
A readable introductory text that deals with many qualitative aspects of photosynthesis.

RABINOWITCH, E., and GOVINDJEE. *Photosynthesis*. John Wiley & Sons, Inc., New York, 1969.
Another useful introductory text.

WAYNE, R. P. *Photochemistry*. American Elsevier Publishing Company, Inc., New York, 1970.

INTERMEDIATE–ADVANCED

CALVERT, J. G., and J. N. PITTS, JR. *Photochemistry*. John Wiley & Sons, Inc., New York, 1966.
This is one of the standard texts in the field. Strongly recommended for advanced students.

FLORKIN, M., and E. H. STORZ (eds.). *Comprehensive Biochemistry:* Vol. 27, *Photobiology, Ionizing Radiations*. Elsevier Publishing Company, New York, 1967.

SMITH, K. C., ed. *The Science of Photobiology*. Plenum Publishing Corporation, New York, 1977.
A collection of many interesting topics on photobiology.

TURRO, N. J. *Molecular Photochemistry*. W. A. Benjamin, Inc., New York, 1965.
Main emphasis is on organic reactions.

SECTION A

Reading Assignments

"Organic Photochemistry and the Excited State," P. A. Leermakers and G. F. Vesley, *J. Chem. Educ.* **41**, 535 (1964).

"The Fates of Electronic Excitation Energy," H. H. Jaffe and A. L. Miller, *J. Chem. Educ.* **43**, 469 (1966).

"Photochemical Reactivity," N. J. Turro, *J. Chem. Educ.* **44**, 536 (1967).

"Life and Light," G. Wald, *Sci. Am.*, Oct. 1959.

"The Chemical Effects of Light," G. Oster, *Sci. Am.*, Sept. 1968.

"The Effects of Light on the Human Body," R. J. Wurtman, *Sci. Am.*, July 1975.

"How Light Interacts with Living Matter," S. B. Hendricks, *Sci. Am.*, Sept. 1968.

"Photochemical Reactions of Natural Macromolecules," D. C. Neckers, *J. Chem. Educ.* **50**, 164 (1973).

"The Path of Carbon in Photosynthesis," J. A. Bassham, *Sci. Am.*, June 1962.

"The Role of Chlorophyll in Photosynthesis," E. I. Rabinowitch and Govindjee, *Sci. Am.*, July 1965.

"The Mechanism of Photosynthesis," R. P. Levine, *Sci. Am.*, Dec. 1969.

"The Absorption of Light in Photosynthesis," Govindjee and R. Govindjee, *Sci. Am.*, Dec. 1974.

"The Photosynthetic Membrane," K. R. Miller, *Sci. Am.*, Oct. 1979.

"How Cells Make ATP," P. C. Hinkle and R. E. McCarty, *Sci. Am.*, Mar. 1978.

"An Apparent Violation of the Second Law of Thermodynamics in Biological Systems," J. J. Klein, *J. Chem. Educ.* **56**, 314 (1979).

"Molecular Isomers in Vision," R. Hubbard and A. Kropf, *Sci. Am.*, June 1967.

"Visual Cells," R. W. Young, *Sci. Am.*, Oct. 1970.

"Visual Pigments in Man," W. A. Rushton, *Sci. Am.*, Nov. 1962.

"Molecular Basis of Visual Excitation," G. Wald, *Science* **162**, 230 (1968).

"Action of Light upon the Visual Pigment Rhodopsin," R. H. Johnson and T. P. Williams, *J. Chem. Educ.* **47**, 736 (1970).

"Ultraviolet Radiation and Nucleic Acid," R. A. Deering, *Sci. Am.*, Dec. 1962.

"Luminescence," P. G. Seybold, *Chemistry* **46**(2), 6 (1973).

"A New, Long-Lasting Luminol Chemiluminescent Cold Light," H. W. Schneider, *J. Chem. Educ.* **47**, 519 (1970).

"Biological Luminescence," W. D. McElroy and H. H. Seliger, *Sci. Am.*, Dec. 1972.

"Radiation Therapy," D. F. Nelsen and P. Rubin, *Chemistry* **50**(6), 6 (1977).

SECTION B

"Photochemistry of Organic Compounds," J. S. Swenton, *J. Chem. Educ.* **46**, 7, 217 (1969).

"Biochemical Effects of Excited State Molecular Oxygen," J. Bland, *J. Chem. Educ.* **53**, 274 (1976).

"Photosynthesis as a Resource for Energy and Materials," M. Calvin, *Am. Sci.* **64**, 270 (1976).

"The Evolution of Photosynthesis," J. M. Olson, *Science* **168**, 438 (1970).

"High Efficiency Photosynthesis," O. Bjorkman and J. Berry, *Sci. Am.*, Oct. 1973.

"The Path of Oxygen from Water to Molecular Oxygen," L. J. Heidt, *J. Chem. Educ.* **43**, 623 (1966).

Acc. Chem. Res., Mar. 1975. (The entire issue is on vision.)

"Three-Pigment Color Vision," E. F. MacNichol, Jr., *Sci. Am.*, Dec. 1964.

"Night Blindness," J. E. Dowling, *Sci. Am.*, Oct. 1966.

"Retinal Processing of Visual Image," C. R. Michael, *Sci. Am.*, May 1969.

"The Origin of Bioluminescence," H. H. Seliger, *Photochem. Photobiol.* **21**, 355 (1975).

"Solar Energy Conversion by Water Photodissociation," V. Balzani, L. Moggi, M. F. Manfrin, F. Bolletta, and M. Gleria, *Science* **189**, 852 (1975).

"Biological Light: α-Peroxylates as Bioluminescent Intermediates," W. Adam, *J. Chem. Educ.* **52**, 97 (1975).

"Sun Tans," W. D. Smith, *SciQuest* **52**(5), 10 (1979).

Problems

20.1 Design an experiment that would enable you to measure the rate of absorption of light by a solution.

20.2 In a photochemical reaction, 428.3 kJ mol^{-1} is required to break a certain bond. What wavelength must be employed for this irradiation?

20.3 Convert 450 nm into kJ einstein^{-1}.

20.4 Consider the photosynthetic reaction

$$H_2O + CO_2 \xrightarrow{\ h\nu\ } (CH_2O) + O_2$$

Suggest an experiment that would enable you to determine whether molecular oxygen is originally derived from water or carbon dioxide.

20.5 Electronically excited 3-aminophthalate has been suggested as the species responsible for the chemiluminescence of luminol. How would you confirm this suggestion?

20.6 The transparency of a new type of sunglasses to light depends on the intensity of the light in the environment. The lenses are clear in dimly lit rooms but darken when the wearer goes outdoors. The material responsible for this change is the AgCl crystals incorporated in the glass. Suggest a photochemical mechanism that would account for this change.

20.7 Suppose that an excited singlet S_1 can be deactivated by three different mechanisms whose rate constants are k_1, k_2, and k_3. The rate of decomposition is given by $-d[S_1]/dt = (k_1 + k_2 + k_3)[S_1]$. (a) If τ is the mean lifetime, that is, the time required for the number of S_1 molecules to decrease to $1/e$ or 0.368 of the original value, show that $(k_1 + k_2 + k_3)\tau = 1$. (b) The overall rate constant k is given by

$$\frac{1}{\tau} = k = k_1 + k_2 + k_3 = \frac{1}{\tau_1} + \frac{1}{\tau_2} + \frac{1}{\tau_3}$$

Show that the quantum yield efficiency Φ_i is given by

$$\Phi_i = \frac{k_i}{\sum k_i} = \frac{\tau}{\tau_i}$$

where i denotes the ith decay mechanism. (c) If $\tau_1 = 10^{-7}$ s, $\tau_2 = 5 \times 10^{-8}$ s, and $\tau_3 = 10^{-8}$ s, calculate the lifetime of the singlet state and the quantum yield for the path having τ_2.

20.8 In the sea, light intensity diminishes with depth. For example, at a depth 20 m below the surface light intensity has diminished to one-half of that at the sea level. In practice, total darkness sets in when 99% of the light is absorbed by the water. With this in mind, explain why green algae are found near the surface, but red algae are located as deep as 100 m.

20.9 Transition metals such as Fe, Cu, Co, and Mn are required in respiration and photosynthesis rather than nontransition metals such as Zn, Ca, and Na. Explain.

20.10 In photosynthesis, the term *quantum requirement* refers to the number of photons required to reduce one CO_2 molecule to (CH_2O):

$$H_2O + CO_2 \xrightarrow{hv} (CH_2O) + O_2$$

The efficiency of this process depends on the wavelength of light employed. Assuming a quantum requirement of 8, calculate the efficiency under standard-state conditions for the synthesis of 1 mol of glucose if the wavelength of light employed is (a) 400 nm, and (b) 700 nm.

20.11 The first-order rate constants for the fluorescence and phosphorescence of naphthalene are $4.5 \times 10^7 \text{ s}^{-1}$ and 0.50 s^{-1}, respectively. Calculate the time for 1.0% of fluorescence and phosphorescence to occur following termination of excitation.

20.12 An organic molecule absorbs at 549.6 nm. If 0.031 mol of the molecule is excited by 1.43 einsteins of light, what is the quantum efficiency for this process? Also calculate the total energy taken up in the absorption.

20.13 Consider the photochemical isomerization $A \rightleftharpoons B$. At 650 nm, the quantum yields for the forward and reverse reactions are 0.73 and 0.44, respectively. If the molar extinction coefficients of A and B are 1.3×10^3 liters $\text{mol}^{-1} \text{ cm}^{-1}$ and 0.47×10^3 liters $\text{mol}^{-1} \text{ cm}^{-1}$, what is the ratio $[B]/[A]$ in the photostationary state?

20.14 At low light intensities, the rate of photosynthesis increases linearly with intensity. At high intensities, however, the rate is constant (saturation rate) independent of variation in light intensity. Suggest an interpretation at the molecular level. The saturation rate varies with T. Suggest an interpretation. [*Source*: "Eco-Chem," J. A. Campbell, *J. Chem. Educ.* **52**, 241 (1975).]

20.15 In the photochemical decomposition of a certain compound, light intensity of 5.4×10^{-8} einstein s^{-1} was employed. Assuming the most favorable conditions, estimate the time needed to decompose 1 mol of the compound.

20.16 The molecules in the retina of the human eye, when completely adapted to darkness, are excited and pass a signal to the optic nerve when the rate of incident radiation from a point source on a black background equals 2×10^{-16} W. Find the minimum number of photons that must reach the retina per second to produce vision. Assume that the wavelength of the light is 550 nm. [*Source*: "Eco-Chem," J. A. Campbell, *J. Chem. Educ.* **49**, 414 (1972).]

20.17 Both green and purple bacteria are known that can oxidize sulfide, but the green can only produce sulfur, whereas the purple can oxidized all the way to sulfate. Suggest a possible interpretation based on the color of the bacteria. [*Source*: "Eco-Chem," J. A. Campbell, *J. Chem. Educ.* **51**, 119 (1974).]

20.18 Calculate the number of moles of ATP that could be synthesized at 80% efficiency by a photosynthetic organism upon the absorption of 2.1 einsteins of photons at 650 nm. (*Hint*: $\Delta G°$ for the synthesis of ATP from ADP and P_i is 31.4 kJ.)

Macromolecules

Polymers, allegedly objects of inordinate complexity, are susceptible to treatment that is simple, yet rigorous and exact.
Paul J. Flory*

A macromolecule is a chemical species distinguished by a high molar mass, ranging from 10^3 to 10^9 or 10^{10} g mol^{-1}. Macromolecules are often called *polymers*. Macromolecular chemistry differs greatly from the chemistry of small, ordinary molecules. Special techniques are required to study the properties of these giant molecules.

Macromolecules are divided into two classes: natural and synthetic. Examples of *natural macromolecules* are proteins, nucleic acids, polysaccharides (cellulose), and polyisoprene (rubber). Most *synthetic macromolecules* are organic polymers, such as polyhexamethylene adipamide (nylon), polyethylene terephthalate (Dacron, Mylar), and polymethylmethacrylate (Lucite, Plexiglas).

In this chapter we examine some of the methods used to characterize macromolecules and discuss their structure, conformation, and stability in solution.

21.1 Methods for Determining the Size, Shape, and Molar Mass of Macromolecules

Molar mass has a different meaning when applied to macromolecules. In a sucrose solution, all solute molecules have the same molar mass; different methods of determining molecular mass yield the same value. Likewise, molecules in a solution containing hemoglobin or other proteins would have identical molar masses, assuming no dissociation of the subunits. Such is not the case, however, for polystyrene, DNA, fibrous proteins, rubber, and other substances composed of polymers. In each of these systems, the molecules are not all identical, and there is a distribution of molar masses. A polymeric

* From "Priestley Medal Address," by P. J. Flory, *Chemical and Engineering News*, April 8, 1974. Copyright © 1974 by the American Chemical Society. Used by permission.

system whose molecules are all of the same molar mass is said to be *monodisperse*; if they are not, the polymer is said to be *polydisperse*.

The molar mass of a polymer can be defined in a number of ways. Two definitions are most frequently encountered.

Number-Average Molar Mass. Consider a sample of N polymer molecules containing n_1 molecules of molar mass M_1, n_2 molecules of molar mass M_2, and so on. The *number-average molar mass* is defined as

$$\overline{M}_n = \frac{n_1 M_1 + n_2 M_2 + \cdots}{n_1 + n_2 + \cdots} = \frac{\sum_i n_i M_i}{\sum_i n_i}$$

$$= \frac{\sum_i n_i M_i}{N} \tag{21.1}$$

where $\sum_i n_i = N$. Thus $\overline{M}_n$ is simply the arithmetic mean of all the molar masses.

Weight-Average Molar Mass. This is defined as

$$\overline{M}_w = \frac{n_1 M_1^2 + n_2 M_2^2 + \cdots}{n_1 M_1 + n_2 M_2 + \cdots} = \frac{\sum_i n_i M_i^2}{\sum_i n_i M_i} \tag{21.2}$$

For a polydisperse system, we have $\overline{M}_w > \overline{M}_n$; for a monodisperse system, $\overline{M}_n = \overline{M}_w$. It follows, therefore, that the determination of molar mass by two different methods can, in principle, test the homogeneity of the system under investigation.

There are a number of techniques for characterizing macromolecules; we shall now briefly discuss three of these techniques.

SEDIMENTATION IN THE ULTRACENTRIFUGE

Sedimentation Velocity. Everyday experience tells us that particles suspended in a solution are pulled downward by the earth's gravitational force. This movement is partially offset by the buoyancy of the particle. Since the earth's gravitational field is weak, a solution containing macromolecules is usually homogeneous, as a result of the random thermal motion of the molecules. The rate of sedimentation of the particles increases with the mass of the particle and with the strength of the gravitational field. Consider a solution under the influence of a strong gravitational field; for example, a solution being spun in a centrifuge tube. The centrifugal force acting on the solute particle of mass m is $m\omega^2 r$, where ω is the angular velocity of the rotor in radians per second (the relation between angle and radian is given in Appendix 2), r the distance from the center of rotation to the particle, and $\omega^2 r$ is the centrifugal acceleration of the rotor (Figure 21.1).

In addition to centrifugal force, we must also consider the particle's buoyancy as a result of the displacement of the solvent molecules by the particle.

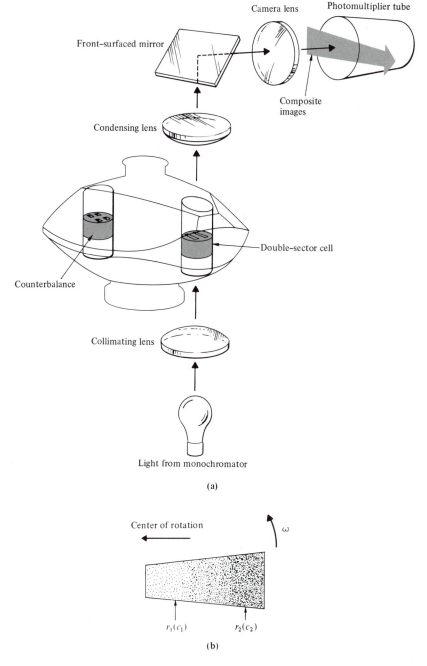

(a) Schematic diagram of an ultracentrifuge. (Courtesy of Beckman Instruments, Inc., Palo Alto, Calif.) (b) Concentration gradient established in the sample cell. **Figure 21.1**

This buoyancy reduces the force on the particle by $\omega^2 r$ times the mass of the displaced solvent. Thus the net force acting on the particle is given by

$$\text{net force} = \text{centrifugal force} - \text{buoyant force}$$
$$= \omega^2 rm - \omega^2 rm_s$$
$$= \omega^2 rm - \omega^2 rv\rho$$

587

The buoyant force is equal to the centrifugal acceleration times the mass of displaced solvent, m_s, and v and ρ are the volume of the particle and the density of solvent, respectively.

According to Newton's second law of motion, a net force acting on a particle would cause it to accelerate. In our case, the initial acceleration lasts for a very short time, of the order 10^{-9} s, after which the particle moves at a constant velocity. This is because the medium exerts a frictional force on the particle, which is proportional to the sedimentation velocity dr/dt. The frictional force is equal to the product of the frictional coefficient f (see Section 5.4) and the sedimentation velocity. It acts in the opposite direction to the net force. At steady state, then, the frictional force is equal to the net force:

$$f \frac{dr}{dt} = \omega^2 rm - \omega^2 rv\rho \qquad (21.3)$$

The volume of the particle is a difficult quantity to measure, so for convenience we use a term called *partial specific volume*, $\bar{v}$, defined as the increase in volume when 1 g of the dry solute is dissolved in a large volume of the solvent. The quantity $m\bar{v}$ is the incremental volume increase when one molecule of mass m is added to the solvent; that is, it is equal to the volume v of the particle.* For most proteins, $\bar{v}$ has a value of about 0.74 ml g^{-1}. Equation (21.3) can now be written as

$$f \frac{dr}{dt} = \omega^2 rm - \omega^2 rm\bar{v}\rho$$

$$= \omega^2 rm(1 - \bar{v}\rho) \qquad (21.4)$$

Rearrangement of Eq. (21.4) gives

$$s = \frac{dr/dt}{\omega^2 r} = \frac{m(1 - \bar{v}\rho)}{f}$$

$$= \frac{M}{N_0} \frac{1 - \bar{v}\rho}{f} \qquad (21.5)$$

where M is the molar mass of the solute and N_0 is Avogadro's number.

The quantity on the left side in Eq. (21.5) is called the *sedimentation coefficient*(s), which has the unit *svedberg*, named after the pioneer in ultracentrifuge studies. One svedberg is equal to 10^{-13} s. For spherical molecules, the frictional coefficient is given by Stokes law [see Eq. (5.18)],

$$f = 6\pi\eta r_s$$

where r_s is the radius of the solute molecule. From Eq. (21.5),

$$M = \frac{sN_0 f}{1 - \bar{v}\rho} = \frac{sN_0(6\pi\eta r_s)}{1 - \bar{v}\rho} \qquad (21.6)$$

Alternatively, according to Eq. (5.17), we can write

$$f = \frac{kT}{D}$$

---

* In practice, the accurate determination of $\bar{v}$ is difficult unless large quantities of protein are available.

Protein	$s_{20\ w}$ (10^{-13} s)[b]	$D \times 10^{7}$[c] ($cm^2\ s^{-1}$)	$\bar{v}$ (ml g^{-1})[d]	M (g mol^{-1})
Cytochrome c_1 (bovine heart)	1.71	11.40	0.728	13,370
Lysozyme (chicken egg white)	1.91	11.20	0.703	13,930
Ribonuclease (bovine pancrease)	2.00	13.10	0.707	12,640
Myoglobin	2.04	11.3	0.74	16,890
Human serum albumin	4.60	6.10	0.733	68,460
Alcohol dehydrogenase (horse liver)	4.88	6.50	0.751	73,050

[a] From *Handbook of Biochemistry*, H. A. Sober (ed.), © The Chemical Rubber Co., 1968. Used by permission of The Chemical Rubber Co.

[b] By convention, sedimentation coefficients are corrected to a reference solvent having the viscosity and density of water at 20°C, hence the subscripts 20, *w*.

[c] To convert to SI units (m^2 s^{-1}), multiply each number by 10^{-2}. For example, D(lysozyme) $= 11.20 \times 10^{-7}$ cm^2 s^{-1} $= 11.20 \times 10^{-9}$ m^2 s^{-1}.

[d] To convert to SI units (m^3 kg^{-1}), multiply each number by 10^{-3}.

and obtain

$$M = \frac{sN_0kT}{D(1 - \bar{v}\rho)} = \frac{sRT}{D(1 - \bar{v}\rho)} \tag{21.7}$$

since $N_0 k = R$.

Since D and $\bar{v}$ can be determined by separate experiments, the only quantity that needs to be measured in determining the molar mass is the sedimentation coefficient. By definition,

$$s = \frac{dr/dt}{\omega^2 r}$$

or

$$s\,dt = \frac{1}{\omega^2}\frac{dr}{r}$$

Integration over distance traveled by the particle from $r = r_0$ ($t = 0$) to $r = r$ ($t = t$) gives

$$\int_0^t s\,dt = \frac{1}{\omega^2}\int_{r_0}^r \frac{dr}{r}$$

or

$$s = \frac{1}{t\omega^2} \ln \frac{r}{r_0} \tag{21.8}$$

Suitable optical means such as refractive-index measurements are used to monitor the movement of sedimenting particles in a given time.* Knowing ω, we are able to calculate s and hence M. The reader should note that the sedimentation coefficient for a given molecule is *independent* of the angular velocity of the rotor. As $\omega^2 r$ increases, so does dr/dt so that the ratio remains a constant. Table 21.1 lists several important properties of proteins for the ultracentrifuge experiment.

---

* For experimental details the reader should consult the texts by Freifelder, Haschemeyer and Haschemeyer, van Holde, and Walton and Blackwell listed at the end of the chapter.

Sedimentation Equilibrium. A concentration gradient is created in a sedimentation velocity experiment. When the rotor speed is great enough, all solute molecules will eventually collect in the bottom of the cell. Now let us suppose that the rotor speed is lowered to about 10,000 rpm (revolutions per minute), instead of the 60,000 rpm or so required for a sedimentation velocity experiment. A perfect balance between sedimentation and diffusion can be achieved. In diffusion, solute molecules move from a higher concentration to a lower one, while sedimentation reverses this process. When an equilibrium is established, no net flow occurs. Referring to the treatment of diffusion in Chapter 5, we see that the amount of solute molecules flowing through per unit area per second is given by Fick's first law*:

$$\frac{dn}{dt} = D\frac{dc}{dr}$$

$$= \frac{kT}{f}\frac{dc}{dr} = \frac{RT}{fN_0}\frac{dc}{dr}$$

Note that the concentration gradient is now expressed as dc/dr instead of dc/dx. According to Eq. (21.4), the sedimentation rate for a solution of concentration c is

$$\frac{dr}{dt} = \frac{\omega^2 rm}{f}(1 - \bar{v}\rho)$$

or

$$c\frac{dr}{dt} = \frac{c\omega^2 rm}{f}(1 - \bar{v}\rho)$$

At equilibrium, the diffusion rate is equal to the sedimentation rate, so that

$$c\frac{dr}{dt} = \frac{RT}{fN_0}\frac{dc}{dr}$$

or

$$c\omega^2 rm(1 - \bar{v}\rho) = \frac{RT}{N_0}\frac{dc}{dr}$$

Rearranging, we obtain

$$\frac{dc}{c} = \frac{M\omega^2 r(1 - \bar{v}\rho)}{RT}dr \qquad M = mN_0$$

Integration between $r_1(c_1)$ and $r_2(c_2)$ yields

$$\int_{c_1}^{c_2}\frac{dc}{c} = \frac{M\omega^2(1 - \bar{v}\rho)}{RT}\int_{r_1}^{r_2}r\,dr$$

$$\ln\frac{c_2}{c_1} = \frac{M(1 - \bar{v}\rho)\omega^2}{2RT}(r_2^2 - r_1^2) \tag{21.9}$$

Again, optical techniques will measure the concentrations of the solute concentrations c_1 and c_2 at r_1 and r_2, and if $\bar{v}$, ρ, and ω are known, M can be calculated. As opposed to the sedimentation velocity approach, this technique

---

* The negative sign [see Eq. (5.9)] is omitted, since for our system here the concentration gradient increases with increasing r.

does not require any knowledge of the shape of the molecule or its diffusion coefficient. It is therefore one of the most accurate methods for molar mass determination.

Density Gradient Sedimentation. One improvement has greatly widened the scope of the ultracentrifuge technique, the cesium chloride sedimentation equilibrium method.

A macromolecular solution mixed with a concentrated CsCl solution, about 6 M, is spun in a celluloid centrifuge tube until equilibrium is reached. At equilibrium, a CsCl density gradient is formed along the tube. At some value of r the density of the solution ρ is equal to the reciprocal of the specific volume of the macromolecule, that is, $\rho = 1/\bar{v}$. This means that $(1 - \bar{v}\rho) = 0$ and, from Eq. (21.4), the sedimentation rate dr/dt is zero. Consequently, a band will be formed for each type of macromolecule at some point along the density gradient. Generally, the bands are narrow and well resolved. Above the band (nearer the center of rotation) we have $\rho < 1/\bar{v}$ and $(1 - \bar{v}\rho)$ is positive. Sedimentation will drive the macromolecules down toward the band. Below the band (toward the bottom of the tube) $\rho > 1/\bar{v}$ and $(1 - \bar{v}\rho)$ is negative. Down here, buoyancy will drive the macromolecules up toward the band. Thus the presence of CsCl density gradient acts as a stabilizing force to prevent intermixing caused by changes in temperature and mechanical disturbances. Using this technique it has been possible to separate the same type of DNA molecules containing ^{14}N and ^{15}N isotopes, respectively.* The ordinary sedimentation equilibrium method cannot obtain well-resolved bands.

Another advantage of the CsCl sedimentation equilibrium method is that once equilibrium is reached, the tube can be removed from the centrifuge, and different bands can be separated from one another by puncturing the bottom of the celluloid tube and collecting each portion into a different tube. Although the CsCl density gradient is particularly suitable for analyzing nucleic acids, other materials, such as sucrose, have been successfully employed for studying proteins.

VISCOSITY

Viscosity measurements are the simplest and cheapest techniques available for characterizing macromolecules. A great deal of theoretical and experimental study has been devoted to the viscosity of macromolecular solutions.

In Chapter 5 an expression was derived for the absolute viscosity of a liquid [see Eq. (5.5)]. Because accurate measurements of absolute viscosities are difficult to make, it is convenient to measure the relative viscosity η_{rel} of a solution defined as

$$\eta_{rel} = \frac{\eta}{\eta_0} = \frac{\rho}{\rho_0}\frac{t}{t_0} \qquad (21.10)$$

where η and η_0 are the viscosities of the solution and the standard, usually the solvent, and t and t_0 the respective flow times, and ρ and ρ_0 the respective densities. Since the presence of the solute molecules normally disturbs the streamline flow of liquids, causing an increase in viscosity, η_{rel} is usually

* M. Meselson and F. Stahl, *Proc. Nat. Acad. Sci, U.S.* **44**, 671 (1958).

greater than unity. Several different definitions of viscosity are commonly in use, as follows:

$$\text{Specific viscosity:} \quad \eta_{sp} = \eta_{rel} - 1 \tag{21.11}$$

$$\text{Reduced viscosity:} \quad \eta_{red} = \frac{\eta_{sp}}{c} \tag{21.12}$$

$$\text{Intrinsic viscosity:} \quad [\eta] = \lim_{c \to 0} \frac{\eta_{sp}}{c} \tag{21.13}$$

where c is the concentration in g ml^{-1} or g 100 ml^{-1}.

Relative viscosity is a measure of the change in the solution's viscosity compared to that of the pure solvent. The specific viscosity measures the increase in viscosity over one. To take the concentration into account, that is, to find out how large the specific viscosity per unit concentration of the solute is, η_{sp} is divided by c to give η_{red}. Further, η_{sp} itself is dependent of concentration, so we still need another quantity, $[\eta]$, the intrinsic viscosity, obtained by the extrapolation shown in Figure 21.2.

A useful equation relating intrinsic viscosity to the molar mass is given by

$$[\eta] = KM^{\alpha} \tag{21.14}$$

where M is the molar mass of the polymer and K and α are empirical constants. The value of α depends on the shape or geometry of the macromolecule: $\alpha = 0$ for a sphere, $\alpha = 0.5$ for a random coil, and $\alpha \simeq 1.8$ for a long rigid rod. For a macromolecule of known K and α values, Eq. (21.14) provides a quick estimate of its molar mass from intrinsic viscosity measurement. On the other hand, if the molar mass of the macromolecule is known, then Eq. (21.14) will enable us to deduce the shape of the molecule.

Intrinsic viscosity has often been used to follow the conformational changes of proteins. For example, when a globular protein (to be discussed later) unfolds to form a random coil, its intrinsic viscosity increases but when a rod-shaped protein such as collagen or myosin unfolds, the intrinsic viscosity decreases because of a decrease in the asymmetry of the molecule. Table 21.2 lists the intrinsic viscosity of a number of macromolecules.

ELECTROPHORESIS

Electrophoresis is the migration of ions under the influence of an applied electric field. It is similar to sedimentation, for in both cases solute molecules move under the influence of an external field. However, electrophoresis

Figure 21.2

Determination of intrinsic viscosity.

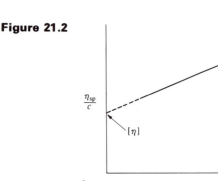

Intrinsic Viscosity of Some Macromolecules

Table 21.2

Molecule	Shape	$[\eta]$ (ml g^{-1}) (CGS units)	$[\eta] \times 10^3$ (m^3 kg^{-1}) (SI units)	Molar Mass (g mol^{-1})
Myoglobin	Globular	3.1	3.1	17,800
Myoglobin[a]	Random coil	21	21	17,800
Hemoglobin	Globular	3.6	3.6	64,450
Hemoglobin[a]	Random coil	19	19	64,450
Ribonuclease	Globular	3.4	3.4	13,683
Bovine serum albumin	Globular	3.7	3.7	67,500
Bovine serum albumin[a]	Random coil	51	51	67,500
Myosin	Rod	217	217	440,000
Tobacco mosaic virus	Rod	37	37	39,000,000
Collagen	Rod	1150	1150	350,000
Polystyrene[b]	Random coil	130	130	500,000

[a] Denatured.
[b] In toluene.

depends primarily on the charge and not on the molar mass of the solute. Electrophoresis is a useful technique for separating proteins in a mixture.

An electric field, $\mathcal{E}$, is applied across a solution. The force acting on the charged solute molecules is given by $ez\mathcal{E}$, where e is the electronic charge and z the number of charges on the molecule. As in the sedimentation case, each ion accelerates for a very short period of time immediately after the field is turned on. Then a steady state is reached as the electrostatic force is balanced by the frictional force exerted by the solvent medium. The ions are now moving at a constant velocity v. Under this condition,

$$ez\mathcal{E} = fv$$

$$= 6\pi\eta rv$$

Hence

$$v = \frac{ez\mathcal{E}}{f}$$

Defining the *electrophoretic mobility* u as the velocity per unit electric field, we write

$$u = \frac{v}{\mathcal{E}} = \frac{ze}{f}$$

$$= \frac{ze}{6\pi\eta r} \tag{21.15}$$

It has the units m^2 V^{-1} s^{-1}.

We see from Eq. (21.15) that a particular ion's ease of movement depends on its charge and is inversely proportional to the size of the ion and the viscosity of the medium. This equation is oversimplified, for it assumes the ion is spherical. The equation also neglects the influence of ionic atmosphere (see Chapter 9) on the movement of the ions. Nevertheless, it allows us to estimate

Figure 21.3

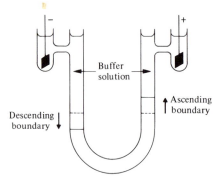

Moving-boundary electrophoresis apparatus.

electrophoretic mobility and therefore suggests a convenient means for separating a mixture of macromolecules into pure components.

One of the simplest measurements of electrophoretic mobility utilizes the *moving-boundary method*, shown in Figure 21.3. The solution under investigation is poured into the bottom of the U tube and buffer solutions are then carefully placed on top to obtain sharp boundaries. Electrodes are inserted in the side arms to prevent products formed from electrolysis falling into the boundary regions. The entire apparatus is then immersed in a thermostat. If the solution contains protein molecules bearing excess charges on their surfaces, the boundaries of the layered buffer solutions will move toward the electrode of the opposite sign. If follows that the direction of movement of protein molecules depends on the pH of the medium. At a pH above its isoelectric point (see Section 12.6), a protein is negatively charged and the boundaries will move toward the anode; at a pH below its isoelectric point, the net charge on the protein is positive and the boundaries will move toward the cathode. At the isoelectric point the net charge on the protein is zero and the boundaries remain stationary (Figure 21.4). The value of the electrophoresis technique is nicely demonstrated in Figure 21.5, which shows the many components of human plasma proteins separated at pH 8.8. Separation is possible because these proteins have rather different isoelectric points. Table 21.3 lists the isoelectric point of a few common proteins.

To minimize diffusion and other disturbances, a gelatinous substance such as starch gel or polyacrylamide gel can be used as the supporting medium for

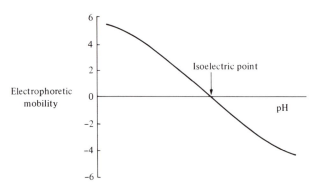

Figure 21.4   Determination of the isoelectric point of a protein by plotting its electrophoretic mobility as a function of pH. At lower pH values than the isoelectric point, the protein moves toward the cathode. At higher pH values the protein moves toward the anode.

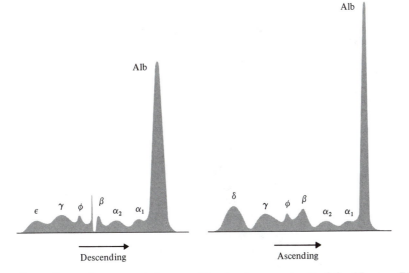

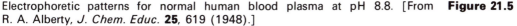

Electrophoretic patterns for normal human blood plasma at pH 8.8. [From **Figure 21.5**
R. A. Alberty, *J. Chem. Educ.* **25**, 619 (1948).]

the protein solution, resulting in a better resolution of the electrophoretic
bands. It is also possible to use moistened filter paper as the supporting
medium.

In the *isoelectric focusing technique*, a pH gradient is first established between the electrodes so that different proteins are brought to form stationary
bands along the gradient at points where the pH is equal to their isoelectric
points. The pH gradient is established in a special way. First, low-molecular-
mass polyampholytes that cover a wide range of isoelectric points are dissolved in water. Before the application of an electric field, the pH of the
solution is the same throughout (the pH is the average value of all the
polyampholytes in solution). After the field is turned on, the polyampholytes
start to migrate toward the electrodes. As a result of their own buffering
capacities, a pH gradient is gradually established between the electrodes.

**Isoelectric Point of Some
Proteins**[a]

Table 21.3

Protein	pI
Bovine serum albumin	4.9
β-Lactoglobulin	5.2
Carboxypeptidase	6.0
Hemoglobin	6.7
Hemoglobin S[b]	6.9
Ribonuclease	9.5
Cytochrome c	10.7
Lysozyme	10.7

[a] The precise value depends on the
temperature and the ionic strength of the
solution.
[b] This is sickle-cell hemoglobin.

Eventually, each type of polyampholyte will come to rest in the self-established gradient at the point corresponding to its own isoelectric point. If a mixture of proteins is introduced to the medium, each type of protein will migrate to the position corresponding to its isoelectric point so that a number of bands will be formed which can then be separated from one another for characterization. It is interesting to note that the isoelectric focusing technique is analogous in principle to the CsCl density equilibrium method discussed earlier.

Finally, we note that the molar mass of a protein can be reliably estimated by the determination of its electrophoretic mobility in a polyacrylamide gel. This method requires treatment of the protein with the denaturing agent sodium dodecyl sulfate (SDS) and β-mercaptoethanol. The denaturing agent SDS binds strongly to most proteins (about 1.4 g of SDS to 1 g of protein) and β-mercaptoethanol ruptures disulfide linkages in proteins. The surface charge on the SDS–protein complex is almost entirely due to the exposed sulfate ions. The complex is not a completely random coil; rather, it assumes the shape of a long rod of constant width. Its length is a function of the protein's molar mass. The important point is that the surface charge per unit area of the complex tends to be a constant regardless of the charge of the polypeptide chain. Thus its electrophoretic mobility depends only on its size (that is, length) rather than the net charge. By carrying out an electrophoresis experiment using polyacrylamide gel as the supporting medium, it is possible to obtain well-resolved bands of various SDS–protein complexes.

From a plot of the logarithm of the molar mass of the proteins versus the electrophoretic mobility of their SDS complexes, a straight line having a negative slope is obtained.* Thus the molar mass of an unknown protein can be readily determined from the electrophoretic mobility of its SDS complex and the standard calibration curve. An additional advantage of this technique is that SDS dissociates oligomeric proteins into their individual polypeptide chains. Thus from the molar-mass determination we can often deduce the number of subunits present. For example, the molar mass of glyceraldehyde 3-phosphate dehydrogenase is found to be 140,000 by the ultracentrifugation technique. However, from the SDS–gel electrophoresis experiment its molar mass is only 36,500. We may conclude that this enzyme has four subunits.

Table 21.4 summarizes various techniques for molar-mass determination.

Summary of Different Methods for Molar-Mass Determination

Table 21.4

Method	Approximate Range	Type of Molar Mass Determined
Freezing-point depression	$\leqslant 500$	$\bar{M}_n$
Osmotic pressure	$\leqslant 100,000$	$\bar{M}_n$
Viscosity	Unrestricted	$\bar{M}_v$
Ultracentrifugation	$\geqslant 5,000$	$\bar{M}_w$
X-ray diffraction	Unrestricted	$\bar{M}_n$
Gel electrophoresis	$\geqslant 5,000$	$\bar{M}_w$
Electron microscopy[a]	$\geqslant 100,000$	$\bar{M}_n$
Light scattering[a]	$\geqslant 5,000$	$\bar{M}_w$

[a] Not discussed in this text.

* See K. Weber and M. Osborn, *J. Biol. Chem.* **244**, 4406 (1969).

While determining the molar mass and overall shape of a macromolecule is relatively easy, the problem of understanding the three-dimensional arrangement of individual atoms in such a system is much more complex. As mentioned in Chapter 4, only the X-ray diffraction technique can yield the complete structural details of a macromolecule. However, even in favorable cases where the complex data can be analyzed, an endeavor that may take time to perform, the information obtained pertains only to the solid-state structure. In chemical processes, we are interested in the solution conformation of the macromolecule, which may be appreciably different from the crystalline state.* The situation is not as hopeless as it seems, for recently a number of powerful techniques have been developed, enabling us to probe different aspects of the molecule. Pieces of information are then put together like parts of a jigsaw puzzle, creating an overall picture. This section will briefly consider the stereochemistry of macromolecules, factors that govern the folding of proteins, and protein denaturation.

THE RANDOM-WALK MODEL

Imagine a long polymer chain made up of identical units, say $-CH_2-$, dissolved in some solvent. For simplicity, ignore all the solute–solvent and solute–solute interactions for the moment. The question to ask is: What shape does the polymer assume? In one extreme, the chain would be completely stretched out; in the other extreme, the chain would be wound up on itself like a ball of string. Generally, the situation lies somewhere in between these two cases.

If the repeating units have no preferred orientations, the problem can be treated by the random-walk model. Place a drunken person in the center of a large room and instruct him to take a succession of steps of equal length. No restriction is placed on his direction, so each step is completely uncorrelated with the previous one. It turns out that the lines joining the successive steps quite accurately represent the arrangement of a polymer chain. The analogy is not exact, however, since polymers have three-dimensional structure, while the random walk occurs on a two-dimensional floor.

Figure 21.6 shows a two-dimensional representation of a freely jointed polymer chain containing 50 identical units. The quantity of interest is the distance between the two ends of the chain (r), since this gives us some idea about the size of the molecule. Calculations based on statistical approach, averaged over many different random walks, show that the average of the square of the distance, r, is given by

$$\overline{r^2} = nl^2 \tag{21.16}$$

---

* It is important to distinguish here between the two terms that often cause confusion, configuration and conformation. The *configurations* of a molecule are stereo arrangements which are related to one another by symmetry but which cannot be converted into one another without rupturing bonds. For example, *cis-* and *trans-*dichloroethylene are the two configurations of the $C_2H_2Cl_2$ molecule. Other examples are optical isomers and derivatives of alicyclic compounds such as 1,2-*cis-* and 1,2-*trans-*dibromocyclopropane. On the other hand, *conformations* are arrangements of atoms in three-dimensional space. In ethane, the molecule can exist in one of the infinite number of conformations, the most stable of which is the staggered conformation discussed in Section 20.3.

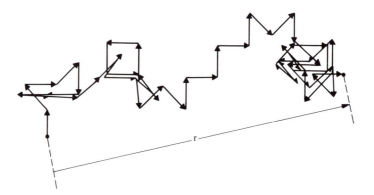

Figure 21.6 Two-dimensional 50 steps random-walk model. (Reprinted from Paul J. Flory; *Principles of Polymer Chemistry*. Copyright 1953 by Cornell University. Used by permission of Cornell University Press.)

where n is the number of bonds and l the length of each bond. The root-mean-square distance is

$$r_{rms} = \sqrt{\overline{r^2}} = l\sqrt{n} \qquad (21.17)$$

For example, a chain containing 1000 C—C bonds each having a length of 1.54 Å would give

$$r_{rms} = 1.54\sqrt{1000} = 48.7 \text{ Å}$$

which is considerably shorter than a completely stretched out chain of length $1000 \times 1.54 = 1540$ Å. We can also estimate the volume occupied by our polymer chain. If we take the diameter as 48.7 Å, the volume is given by $\frac{4}{3}\pi(48.7 \text{ Å}/2)^3 = 6.1 \times 10^4 \text{ Å}^3$. The actual volume occupied by the chain is only a small fraction of the measured sphere. What, then, is the significance of the volume of this sphere? The polymer chain in solution does not remain stationary but is constantly changing its shape and size because of thermal motion. The calculated volume represents the space within which the polymer is contained, on the average.

Another important quantity is the probability $P(r)$ of finding the two ends of a polymer chain separated by distance r. This is a well-known mathematical problem; the exact solution takes the form

$$P(r) = Ar^2 \exp\left(-\frac{3r^2}{2nl^2}\right) \qquad (21.18)$$

where A is a constant. A plot of $P(r)$ versus r is shown in Figure 21.7. As the curve shows, the probability of two ends meeting at $r = 0$ and $r \to \infty$ is zero. The value of r for the maximum value of $P(r)$ is called the most probable distance, r_{mp}.

The simple results obtained give a fairly good picture of what a polymer chain in solution might look like. The random-walk model, beside being a two-dimensional representation of a three-dimensional object,* is oversimplified in other ways. First, the bonds in the chain are not free to take

* Equation (21.18) does apply to a three-dimensional model.

Figure 21.7

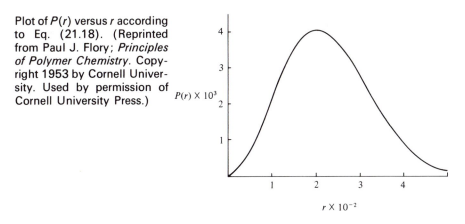

Plot of $P(r)$ versus r according to Eq. (21.18). (Reprinted from Paul J. Flory; *Principles of Polymer Chemistry*. Copyright 1953 by Cornell University. Used by permission of Cornell University Press.)

any direction with respect to one another. For example, in polyethylene the bond angles must all be close to 109°:

Second, the steric interaction between hydrogen atoms on adjacent carbons must also be considered (Figure 21.8). Both of these effects tend to reduce the compactness of the chain and increase the value of r_{rms}. Fortunately, the corrections can be easily made as follows:

$$\overline{r^2} = Cnl^2 \tag{21.19}$$

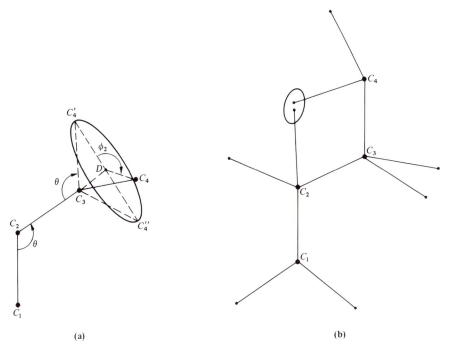

(a) (b)

(a) Conformational variations of the carbon chains in polyethylene. (Reprinted from Paul J. Flory; *Principles of Polymer Chemistry*. Copyright 1953 by Cornell University. Used by permission of Cornell University Press.) (b) Steric interaction between hydrogen atoms on neighboring carbon atoms in polyethylene.

Figure 21.8

599

where C is a constant for a particular type of polymer. Its value lies between 2 and 10 in most cases. Third, while the drunk on a random-walk model recrosses his path as many times as he desires, no two atoms in a polymer chain can occupy the same space at the same time. This *volume exclusion effect* is difficult to account for in theory. For our purposes, Eq. (21.19) is adequate for most calculations.

Obviously, the actual conformation of the polymer must depend on the nature of the solvent. In a "good solvent," that is, one that has a zero or negative heat of mixing with the polymer (an exothermic process), the chain will be in the extended form. In a "poor solvent," that is, one that has a positive heat of mixing with the polymer (an endothermic process), the chain tends to roll up like a ball of string.

PROTEINS

The simplest polymeric systems are those having identical units, for example, polyethylene. Many natural polymers possess structures far more complex than that of polyethylene. Theoretical treatment of these systems is quite formidable. An important group of complex polymers are protein molecules.

Proteins can be thought of as a polymer of amino acids, formed by the condensation reaction shown in Figure 21.9. Twenty amino acids are found in proteins (see Table 12.4). All except glycine are optically active, having the L configuration. Even for a small protein, such as insulin, which contains only 50 amino acid residues, the number of chemically different structures that could be formed is 20^{50} or 10^{65}. This is an enormously large group, since Avogadro's number is only 6×10^{23}. The actual number of proteins found to date, however, is far smaller than this figure.

In the 1930s, Pauling, Corey, and their coworkers carried out a systematic investigation of protein structure, relying mainly on the X-ray diffraction

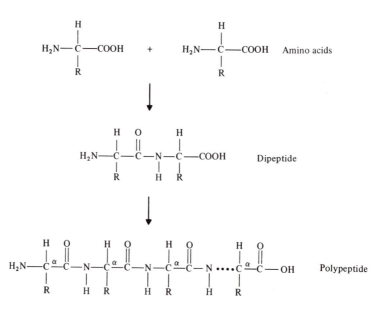

Figure 21.9 Condensation of amino acids to form dipeptides and polypeptides.

Dimensions of the planar amide group. (Reprinted from Linus Pauling; *The Nature of the Chemical Bond*, 3rd ed. Copyright 1939 and 1940, third edition © 1960, by Cornell University. Used by permission of Cornell University Press.)

Figure 21.10

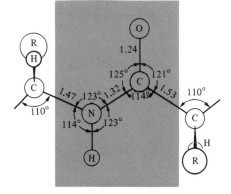

technique. As model compounds, they studied amino acids, dipeptides and tripeptides. They arrived at the following conclusions:

1. The amide group is essentially planar as a result of the resonance:

The $C-N$ bond is about 30 to 40% double-bond in character; consequently, rotation about this bond is restricted, requiring an activation energy of about 40 to 80 kJ mol^{-1}. Rotations about the $C-C_\alpha$ bonds are not restricted. Figure 21.10 shows the basic dimensions of the amide group.

2. The trans configuration is more stable than the cis configuration because of the steric interaction between the groups of the C_α atoms in the cis isomer.

3. The backbone of the polypeptide, the $-C-C-N-C-C-N-$ linkage, is more important than the side chains. Further, all residues are considered to be equivalent.

4. Hydrogen bonds play an essential role in stablizing the polypeptide chain conformation. In all cases studied, the hydrogen atom was found to lie within 30° of the N···O line. In the most stable forms, all four atoms are collinear:

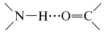

From these findings, Pauling and Corey predicted that a stable arrangement for the polypeptide chain is the α-helix structure, since it allows for a maximum number of hydrogen bonds and introduces the least amount of distortion of bond length and bond angle (Figure 21.11). The important features of α helix are (1) the number of residues per complete turn is 3.6 rather than an integer, and the pitch, the distance between successive turns, is 5.4 Å; (2) the $>N-H$ group hydrogen bonds with the $>C=O$ group on the third amino acid residue along the polypeptide chain; and (3) right-handed helixes are inherently more stable than left-handed helixes.

Because the $C-N$ bond is essentially planar, the polypeptide chain has only two degrees of rotational freedom, the rotation about the $C_\alpha-N$ bond, characterized by angle ϕ and the rotation about the $C_\alpha-C$ bond, characterized by angle ψ. This situation is shown in Figure 21.12. Both ϕ and ψ are

601

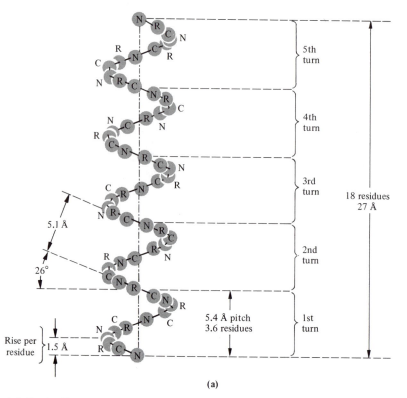

Figure 21.11 (a) Basic dimensions of an α helix. The spheres labeled R represent the carbon atoms containing the amino acid side chains, that is, the C_α atoms. (From *The Chemical Foundations of Molecular Biology* by R. Steiner. © 1965 by Litton Educational Publishing, Inc. Reprinted by permission of Van Nostrand Reinhold Company.)

zero when the two adjacent peptide bonds are coplanar and the amide hydrogen and carbonyl oxygen are away from the side chain of the C_α atom. In fact, the three-dimensional structure of the polypeptide chain can be completely defined in terms of these two angles. For an α helix, $\phi = 132°$ and $\psi = 123°$. Not all possible combinations of ϕ and ψ yield stable conformations. For example, severe steric hindrance between amide hydrogen atoms would result if $\phi = 0$ and $\psi = 180°$ or $\phi = 180°$ and $\psi = 0$.

Another important structure predicted from X-ray diffraction studies is the β-pleated structure, or simply the β structure. This structure results from intermolecular hydrogen bonding between extended polypeptide chains. In contrast to the α helix, the hydrogen bonds in the β structure are oriented roughly perpendicularly to the long axis of the polypeptide chain. Two different arrangements, called *parallel* and *antiparallel*, are known (Figure 21.13).

One way to study the conformational stability of the polypeptide chain is to use a *Ramachandran plot*, shown in Figure 21.14. This plot lays out the values of ϕ and ψ between 0 and 360°. The enclosed areas represent possible combinations of ϕ and ψ that yield least steric hindrance. We see that the dots cluster about the three regions of stability, namely, the left- and right-handed α helix and the β structure although a number of them are also scattered around, indicating a lack of regular structure.

Proteins are arranged in both the α helix and the β structure to varying degrees, confirming Pauling and Corey's theory. Generally, proteins are

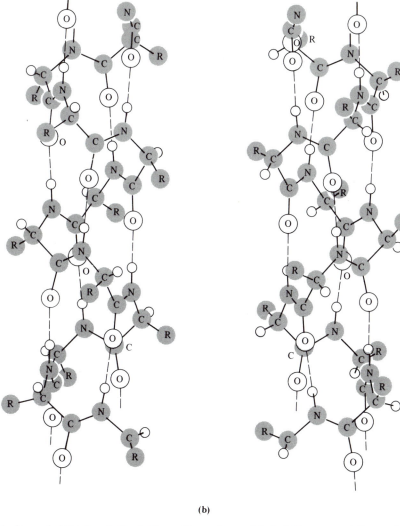

(b)

(b) Left- and right-handed α helixes. The spheres labeled R represent the amino acid side chains. (Reprinted from Linus Pauling; *The Nature of the Chemical Bond*, 3rd ed. Copyright 1939 and 1940, third edition © 1960, by Cornell University. Used by permission of Cornell University Press.)

Figure 21.11

divided into two categories, *globular* and *fibrous*. *Globular proteins* are characterized by their compactness. In these molecules, the polypeptide chain is folded up to fill most of the space within its domain, leaving relatively little empty volume. Figure 21.15 shows a schematic diagram of myoglobin, a globular protein. The presence of α helix is quite evident. In fact, both myoglobin and hemoglobin contain over 80% α-helical structure. This is not true of all globular proteins, however, for lysozyme has only about 40% α-helical content (see Figure 21.14), papain, 20%, and there is practically no α helix in cytochrome *c*. All this shows that factors other than Pauling and Corey's criteria contribute to protein conformation. We shall return to this point in the next section.

Fibrous proteins are present in wool, hair, and silk. The fibrous proteins in wool and hair are called *keratins*. The α helix in keratins accounts for the

603

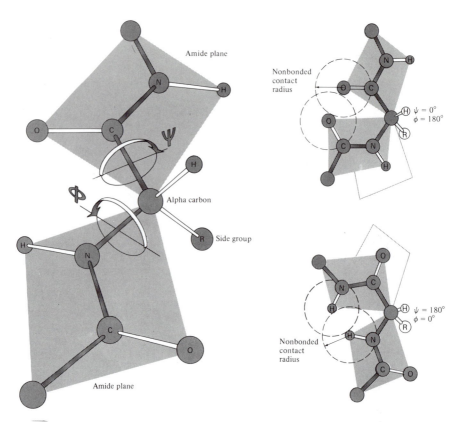

Figure 21.12 Rotational freedoms of two amide planes. The chain is shown in a fully extended conformation and the residue is of the L-configuration. Note that the convention here differs from the recommendation of the IUPAC-IUB Commission on Bio-chemical Nomenclature [see *Biochemistry* **9**, 3471 (1970).] (From R. E. Dickerson and I. Geis, *The Structure and Action of Proteins*, W. A. Benjamin, Inc., Menlo Park, Calif. Copyright 1969 by Dickerson and Geis.)

flexible and elastic properties of wool. Because interactions do not occur among different chains, wool fibers are not very strong. Silk possesses the β structure. Since the polypeptide chains are already in extended forms, silk lacks elasticity and extensibility, but it is quite strong, as a result of the intermolecular hydrogen bonds. Collagen is another example of fibrous protein. It accounts for about one-third of all proteins in the human body. Collagen is the most important component of connective tissues, such as cartilage, ligament, and tendon. The fundamental structural unit of collagen is tropocollagen, an elongated triple helix. Three polypeptide chains are in-tertwined to form a superhelix with a diameter of about 15 Å and length 3000 Å. The most notable properties of collagen are its rigidity and resistance to deformation, both of which are vital to transmitting mechanical force generated by muscles.

It is customary to divide the structure of proteins into four levels of organi-zations. The *primary structure* refers to the unique amino acid sequence of the polypeptide chain. The *secondary structure* refers to those parts of the polypeptide chain stabilized by hydrogen bonds, for example the α helix. The term *tertiary structure* is given to the three-dimensional structure stab-ilized by the disulfide bonds between cysteine residues, as well as to such

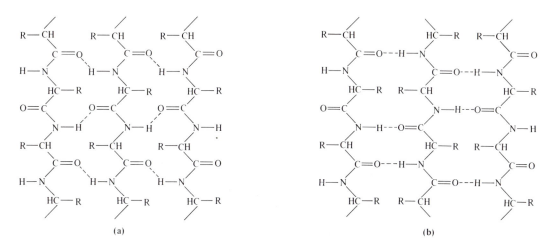

(a) Parallel β-pleated structure: the polypeptide chains all run in the same direction; and (b) antiparallel β-pleated structure: adjacent polypeptide chains run in opposite directions. (From A. G. Walton and J. Blackwell, *Biopolymers,* Academic Press, Inc., New York, 1973.)

Figure 21.13

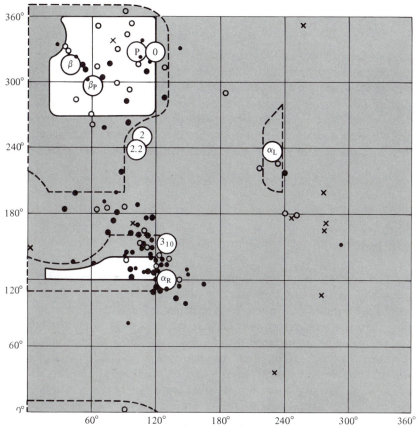

Ramachandran plot of polypeptide chain conformation in lysozyme. Hydrophobic residues are small black dots: hydrophilic, large black dots; ambivalent, open dots. Glycine is denoted by crosses. The abscissa is ϕ; the ordinate is ψ. (From R. E. Dickerson and I. Geis, *The Structure and Action of Proteins,* W. A. Benjamin, Inc., Menlo Park, Calif. Copyright 1969 by Dickerson and Geis.)

Figure 21.14

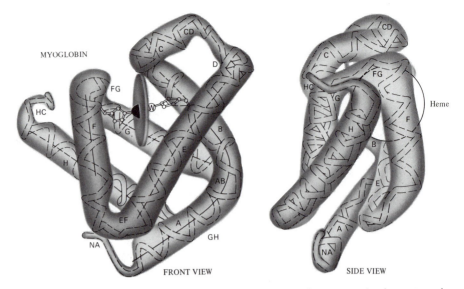

FRONT VIEW   SIDE VIEW

Figure 21.15   Front and side views of myoglobin. In the absence of oxygen, the iron atom in the heme group binds to a water molecule and the histidine residue. Other letters denote the various helixes. (From R. E. Dickerson and I. Geis, *The Structure and Action of Proteins*, W. A. Benjamin, Inc., Menlo Park, Calif. Copyright 1969 by Dickerson and Geis.)

noncovalent forces as van der Waals forces, hydrogen bonding, and electrostatic forces. The hydrogen bonds that exist in the tertiary structure are between residues which are far apart in the polypeptide chain but are brought into proximity as a result of the folding of the chain. A *quarternary structure* refers to interaction between polypeptide chains, for example, the four polypeptide chains or four subunits in hemoglobin. These chains are held together by noncovalent forces such as dispersion forces and electrostatic forces.

21.3 Stability of Proteins in Solution

Pauling and Corey's work was a great triumph in protein chemistry. The story was still incomplete, however, for although their theory successfully predicted the structure of a number of proteins, many others, such as cytochrome c, were found to possess very little or no α helix or β structure. Chemists now believe that the importance of hydrogen bonding and backbone polypeptide chain has been overemphasized, and that a complete description of protein structure must consider other types of interaction, such as electrostatic forces and van der Waals forces. The nature of side-chain residues also appears to play an important role in the overall stability of proteins. Figure 21.16 shows many of the noncovalent interactions present in a protein molecule.

THE HYDROPHOBIC BOND

So far we have discussed several types of intermolecular forces that play a role in determining the three-dimensional structures of proteins. These are elec-

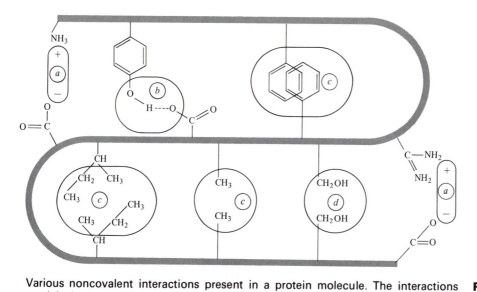

Various noncovalent interactions present in a protein molecule. The interactions are (a) electrostatic forces; (b) hydrogen bonding; (c) hydrophobic interactions; and (d) van der Waals interactions and hydrogen bonding. (From C. B. Anfinsen, *The Molecular Basis of Evolution*, John Wiley & Sons, Inc., New York, 1959.)

Figure 21.16

trostatic forces, hydrogen bonding, and van der Waals forces (dipole–dipole, ion–induced dipole, dipole–induced dipole, and dispersion forces). In addition to these forces, there is another factor that influences protein conformation. This interaction differs from the other types of intermolecular forces because it does not arise from the interaction between a pair of atoms or two groups of a molecule; instead, it involves a considerable number of water molecules; it was called the *hydrophobic bond* by Kauzmann in 1959.

When we study protein structure, one of the first things we notice is that the nonpolar groups present in valine, leucine, isoleucine, tryptophan, and phenylalanine are usually located in the interior of the protein, having little or no contact with water. This is the arrangement found in *native* proteins or proteins in their normally functioning state. What would happen if the polypeptide chain unfolds so that the nonpolar groups are now in contact with water molecules? As an estimate, let us consider the following thermodynamic changes for the transfer of 1 mol of a hydrocarbon from a nonpolar solvent such as benzene to water at 298 K. For methane, we have $\Delta S° = -75.3$ J K^{-1}, $\Delta H° = -11.7$ kJ, and $\Delta G° = 10.7$ kJ; for ethane, $\Delta S° = -81.6$ J K^{-1}, $\Delta H° = -9.1$ kJ, and $\Delta G° = 15.2$ kJ. In each case $\Delta G°$ is positive so that the process is nonspontaneous under standard state conditions. Interestingly, however, in each case the enthalpy change is negative (heat is given off when the hydrocarbon is dissolved in water), which contributes to a negative free energy of transfer. What makes the process nonspontaneous, then, is the large decrease in entropy. In Section 17.3 we discussed the formation of clathrate compounds when nonpolar hydrocarbons are dissolved in water. It is the ordering of the water molecules around the hydrocarbons that result in the large negative ΔS values. Therefore, the formation of the hydrophobic bond is largely an entropy-driven process. The reverse process, that is, the interaction between two nonpolar groups R_1 and R_2 in aqueous solution,

$$R_1(H_2O)_n + R_2(H_2O)_n \longrightarrow R_1 \bullet R_2 + 2nH_2O$$

607

liberates the ordered water molecules originally surrounding R_1 and R_2; consequently, the entropy of the system increases. Since the binding of water by individual R groups gives off some heat, the process described above is usually endothermic; that is, $\Delta H°$ is positive. At room temperature, the $T \Delta S°$ term predominates so that $\Delta G° (= \Delta H° - T \Delta S°)$ is negative and the reaction is spontaneous from left to right. At lower temperatures, the $T \Delta S°$ term decreases and the sign of $\Delta G°$ may now be determined by $\Delta H°$. If this were the case, $\Delta G°$ could become positive at a certain temperature, breaking up the hydrophobic interaction between R_1 and R_2, with resulting unfolding of the protein. Certain cold-sensitive enzymes—enzymes that lose their activity at low temperatures, for example, pyruvate carboxylase—seem to fit this description.

PROTEIN DENATURATION

A useful key to understanding the stability of proteins in solution is the manner in which these molecules denature. Protein denaturation describes any process that results in a change in the three-dimensional structure from the native conformation to some other conformation. By *native conformation*, we mean the conformation of the protein in its normal, physiological state. Strictly speaking, the native state of a protein is found *in vivo*, for example the conformation it possesses in a cell. Since it is impossible to isolate a protein without changing its environment somewhat, it is probably true that any protein *in vitro* has undergone some degree of denaturation.

In principle, a protein has many possible conformations. In one extreme, the native state, the protein possesses the maximum number of noncovalent interactions and disulfide linkages. In the other extreme, the completely denatured state, most of the noncovalent interactions and the disulfide bonds are broken and the molecule has the shape of a random coil. Depending on the protein and other conditions, there may also be other intermediate conformations.

Protein denaturation occurs in a number of ways. Changing the temperature, pH, and ionic strength, as well as adding organic solvents and reagents such as 8 M urea, 6 M guanidine hydrochloride, and detergents to a protein solution, are common methods to denature a protein molecule. Despite widespread use, the chemical basis of many "denaturants" is not always well understood. Probably, the action of urea disrupts both hydrogen bonds and hydrophobic interactions, while organic solvents such as ethanol and ethylene glycol form hydrogen bonds with polar residues on the surface of the protein molecule. The compound β-mercaptoethanol specifically cleaves disulfide bonds as follows:

$$R_1-S-S-R_2 + 2HOCH_2CH_2SH \longrightarrow$$

$$R_1-SH + R_2-SH + \underset{\underset{S-CH_2CH_2OH}{|}}{S-CH_2CH_2OH}$$

Studies of protein denaturation have yielded much significant information on protein stability. The work of Anfinsen and his coworkers on ribonuclease is particularly interesting. Ribonuclease isolated from bovine pancrease is an enzyme having a molecular mass of 13,700. It contains 124 amino acid residues and four disulfide linkages (Figure 21.17). Its three-dimensional structure has recently been solved by X-ray diffraction. The specific action of

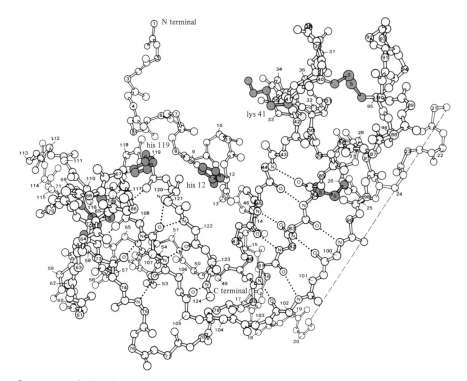

Structure of ribonuclease. Note the presence of the four disulfide linkages. The molecule also possesses a fair amount of β-pleated structure. The hydrogen atoms taking part in hydrogen bonding are not shown. (Illustration copyright by Irving Geis.)

Figure 21.17

ribonuclease is to catalyze the hydrolysis of the phosphodiester bonds in ribonucleic acids.

Ribonuclease in solution is denatured by adding β-mercaptoethanol and 8 M urea. Under these conditions all the disulfide bonds are reduced to the sulfhydryl group, that is,

$$-S-S- \xrightarrow{\;2H\;} 2-SH$$

and most or all of the secondary and tertiary structure is destroyed (Figure 21.18). The enzyme becomes completely inactive. When the denatured enzyme is oxidized with oxygen in the presence of the denaturants, a mixture of products is formed. If we assume that the probability of forming a disulfide bond between any two cysteine residues is the same, then, statistically, the total number of structurally different isomers formed is given by $7 \times 5 \times 3 = 105$. This is so because the first cysteine residue has seven choices in forming an S—S bond, the next cysteine residue has only five choices, and so on. This can be generalized to give $(N - 1)(N - 3)(N - 5) \cdots$, where N is the total number of cysteine residues present. The observed activity of the mixture or the "scrambled protein" is less than 1% of that of the native enzyme. This is consistent with the fact that only one out of every 105 possible structures corresponds to the original state. If, on the other hand, the oxidation is carried out in the absence of urea, but a trace amount of β-mercaptoethanol is retained to promote disulfide interchanges, a homogeneous product is ob-

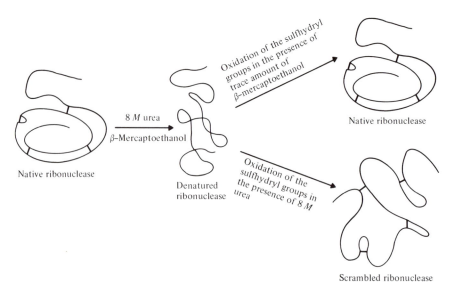

Native ribonuclease

8 M urea

β-Mercaptoethanol

Native ribonuclease

Oxidation of the sulfhydryl groups in the presence of trace amount of β-mercaptoethanol

Denatured ribonuclease

Oxidation of the sulfhydryl groups in the presence of 8 M urea

Scrambled ribonuclease

Figure 21.18   Denaturation and renaturation of ribonuclease. Depending on the conditions, we obtain either native ribonuclease or scrambled ribonuclease. The disulfide linkages are shown by short dashes.

tained that is *indistinguishable* from the native ribonuclease in activity and every other respect.

An important conclusion can be drawn from Anfinsen's work. This is the *thermodynamic hypothesis*, which states that the native state of a protein has a minimum free energy. The conformation of the native protein is determined by its overall intramolecular interactions, which in turn depend on the primary structure, that is, its amino acid sequence. Thus, while 105 different conformations are all kinetically accessible, native ribonuclease folds in such a way to give only one specific form that has the greatest thermodynamic stability. Similar observations have been obtained on other systems, lysozyme and proinsulin, for example. It is not clear at present whether the thermodynamic hypothesis really applies to all types of proteins or whether there are other factors to consider. Conceivably, some native conformation may not possess the lowest free energy. For example, a relatively high potential-energy barrier may prevent a thermodynamically less favorable state from reaching the most stable conformation. The folding of the polypeptide chain in such a case would then be kinetically controlled. Recently, it has been found that the folding of some enzymes requires the presence of certain metal ions as well as other small molecules. A fascinating example is the renaturation of pyruvate kinase, which takes place only in the presence of magnesium or manganese ions and free L-valine in solution.* Clearly, the thermodynamic hypothesis alone is inadequate to account for this behavior.

Another interesting result has emerged from the ribonuclease study. Contrary to earlier beliefs, disulfide bonds do not play an essential role in the folding process. Instead, their function is to stabilize the three-dimensional structure once it is formed. In the experiment described above, folding of the protein is entirely governed by noncovalent forces; correct cysteine pairs are brought

* See L. Bornmann, B. Hess, and H. Zimmermann-Telschow, *Proc. Nat. Acad. Sci. U.S.* **71**, 1525 (1974).

into proximity for disulfide bond formation *after* a three-dimensional network is established.

One of the major goals of a protein chemist is to predict the tertiary structure from a knowledge of the primary structure. Although the problem is undoubtedly a difficult one, rapid progress made in understanding protein structure in recent years suggests that the day will come when this goal is achieved.

The denaturation process is generally cooperative, that is, the transition from the native state, say helix, to random-coil conformation occurs over a narrow range of temperature, pH, or denaturant concentration. Experimentally, this transition can be most conveniently followed by viscosity, optical rotation, or optical absorption measurements. The simplest example of denaturation is the *two-state model*. A well-known two-state model in chemical systems is the phase transition ice $\rightleftharpoons$ water. The equilibrium constant for the transition in protein,

$$\text{N} \underset{k_r}{\overset{k_f}{\rightleftharpoons}} \text{RC}$$

is given by

$$K = \frac{k_f}{k_r} = \frac{[\text{RC}]}{[\text{N}]} = e^{-\Delta G^\circ/RT} = e^{-(\Delta H^\circ - T\Delta S^\circ)/RT}$$

where N and RC represent the native and random coil conformation. The forward and reverse rate constants k_f and k_r can be measured by suitable means, for example, the relaxation techniques discussed in Chapter 13. From the equilibrium constant K, either measured directly or from the ratio k_f/k_r, we can calculate the free energy change for the transition.

A striking thermodynamic parameter for the $\text{N} \rightarrow \text{R}$ transition is the large increase in heat capacity. From Eq. (6.15),

$$\Delta C_P = \left(\frac{\partial \Delta H}{\partial T}\right)_P$$

Hence

$$\Delta H = \Delta C_P(T_2 - T_1)$$

The large value of ΔC_P means that ΔH must itself increase greatly with temperature. Heat-capacity change at constant pressure can be obtained by calorimetric measurements, and from these data we can calculate ΔH and also ΔS if ΔG is known.

In many cases, protein denaturation cannot be described by a simple two-state model, since the transition may involve several stages and a number of intermediates could be present. A many-state model is much more difficult to handle theoretically and experimentally.

It is interesting to examine the changes in thermodynamic quantities for protein denaturation. When ribonuclease is denatured at 20°C in a solution whose pH is 2.5, we have $\Delta G^\circ = 3.97$ kJ mol^{-1}, $\Delta H^\circ = 238.5$ kJ mol^{-1}, and $\Delta S^\circ = 774$ J K^{-1} mol^{-1}. Similar values have been obtained for the denaturation of other proteins, such as chymotrypsinogen and myoglobin. The large positive value of ΔS may seem strange at first, since it contradicts our earlier discussion of the importance of the hydrophobic bond in holding proteins in the native state. We must realize that there are actually two factors that contribute to ΔS. The breaking of the hydrophobic bond does lead to a negative ΔS, as mentioned earlier. However, while each protein has only one conformation in its native state, a denatured protein may possess any one of the many possible conformations. As we saw in Chapter 7, any process that goes from a less probable state to a more probable state will result in an increase in entropy. Thus the sign of ΔS depends on the magnitude of these two opposing effects. For ribonuclease, the entropy increase for the order → disorder transition is greater than the entropy decrease when the nonpolar residues are exposed to water; therefore, ΔS is positive.

21.4 Techniques for the Study of Protein Conformation in Solution

As mentioned earlier, a large number of techniques are available for probing protein structure in solution; the information obtained from each type of measurement is then pieced together to give an overall description of the molecule. A discussion of all the major techniques is not appropriate here. Instead, we shall limit our consideration to four methods, indicating the kind of approaches employed.

LABEL TECHNIQUES

Spin Label.　Except for the occasional presence of transition-metal ions, most proteins are diamagnetic. However, it is often possible to bind a paramagnetic molecule, called a *spin label*, to a particular functional group within a protein. The nature of the environment at and near the binding site can be deduced from the esr spectra of the spin label. This technique is based on the principle that the appearance of the esr spectrum of any paramagnetic molecule depends on its rate of rotation or tumbling in solution—the slower the rate, the more asymmetric its esr spectrum (Figure 21.19).

Most common spin labels are nitroxide radicals (Chapter 18), which are stable in aqueous solutions over a large range of pH, ionic strength, and temperature. Further, their esr spectra are quite simple to interpret, consisting of just three equally intense lines that reflect interaction between the unpaired electron and the nitrogen 14 nucleus ($I = 1$ and $2I + 1 = 3$).

Consider, for example, the maleimide spin label:

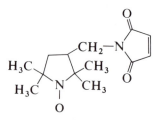

The maleimide group can bind irreversibly to the sulfhydryl group in the globular protein bovine serum albumin (BSA). When this occurs, the nitroxide radical no longer shows the three sharp lines; instead, a broad and asymmetric spectrum is obtained (see Figure 21.19c). Lowering the pH from 6 to 2, where the protein is known to undergo a conformational change, causes a change in the spin label's esr spectrum to that shown in Figure 21.19b. From this new spectrum we can infer that the sulfhydryl group is probably located in the interior region of the protein, perhaps in a crevice. In native BSA at

Esr spectra of a nitroxide radical under the following conditions: (a) free rotation; (b) rotation is partially hindered; and (c) rotation is severely hindered.

Figure 21.19

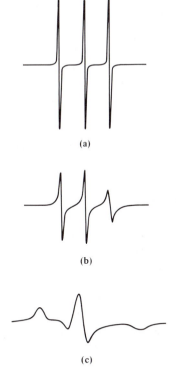

(a)

(b)

(c)

pH 6, rotation of the spin label is severely restricted by the steric interaction between the spin label and the side-chain residues. At pH 2, the protein unfolds to a certain extent and the sulfhydryl group may then be exposed on the surface of BSA, where relatively free rotation becomes possible. The esr spectrum is still noticeably asymmetric, however, since the rotation of the nitroxide is still somewhat hindered by the amino acid side chains.

Fluorescence Label. The fluorescence probe method is very similar in principle to the spin-label technique. A class of compounds possesses the interesting properties of being nonfluorescent in aqueous solution, while becoming strongly fluorescent when bound to the hydrophobic regions in proteins. Two examples are

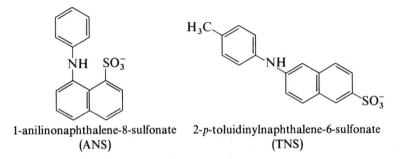

1-anilinonaphthalene-8-sulfonate (ANS)

2-p-toluidinylnaphthalene-6-sulfonate (TNS)

Again, changes in the fluorescent properties of these molecules under different conditions can be used to deduce the conformation of the protein at or near the binding site. For example, the fluorescence quantum yield of ANS in water is 0.04, whereas in the apomyoglobin–ANS complex the quantum yield is very close to unity.* From the evidence that (1) myoglobin itself does not bind ANS, (2) the stoichiometry of apomyoglobin–ANS complex is 1:1, and (3) the ANS in the complex can be expelled by reacting the complex with the heme group, it is concluded that both ANS and the heme group must bind to the same site, one which is largely hydrophobic. This is in good agreement with the X-ray data, which show that the heme group is located in the interior region of myoglobin, surrounded by nonpolar residues.

SOLVENT PERTURBATION TECHNIQUE

The absorption of uv radiation between 200 and 300 nm by proteins is due to the chromophores phenylalanine, tryptophan, and tyrosine. Both the wavelength at which the maximum absorption occurs and the molar absorptivity are sensitive to the environment immediately surrounding the chromophores. In a solvent perturbation experiment, two portions of a protein solution are placed, say, in a 1-cm-path-length cell in the sample and reference compartment of a double-beam uv spectrometer. The concentrations are equal in both cells so that the net absorption is zero. The composition of the sample solution is then modified by the addition of a small amount of perturbant such as ethanol, ethylene glycol, or glycerol, while maintaining equal concentrations of the protein in both cells. Mainly through hydrogen bonding, the perturbant can affect the tyrosine and tryptophan residues. The uv *difference spectrum* will now show either a net enhanced absorption or a dip, indicating a diminished absorption in the sample cell, depending on whether the perturbant increases or decreases the molar absorptivity of the chromophores (Figure 21.20). This technique is particularly useful in distinguishing

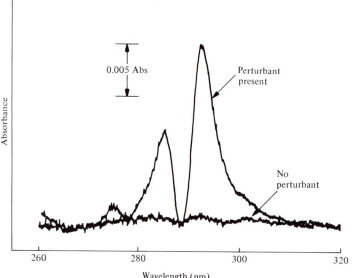

Figure 21.20 Solvent perturbation measurements of proteins, such as lysozyme, demonstrate the presence of chromophores, such as tryptophan and tyrosine, near the surface of the protein. By this method, lysozyme has been shown to possess all six tryptophan residues near the surface. The perturbant is propylene glycol. (By permission of Varian Associates, Palo Alto, Calif.)

* Apomyoglobin is myoglobin without the heme group.

the residues that are exposed at or near the surface of the protein molecule from those that are buried inside the protein, inaccessible to the perturbant. This method shows that ribonuclease in its native state has three "exposed" tyrosine residues. Denaturing the molecule determines a total of six tyrosine residues, proving that three tyrosine groups are in the interior region of native ribonuclease.

ORD

The basic principles of ORD and CD were discussed in Chapter 19. Here we shall see how optical rotation measurements can be used to estimate the α-helical contents in proteins.

Moffitt and Yang modified the Drude equation [Eq. (19.5)] to obtain the *mean residue rotation* $[m']$ as follows:

$$[m'] = \frac{a_0 \lambda_0^2}{\lambda^2 - \lambda_0^2} + \frac{b_0 \lambda_0^4}{(\lambda^2 - \lambda_0^2)^2} \tag{21.20}$$

where a_0, b_0, and λ_0 are all constants for a given system. The quantity a_0 depends on temperature and the nature of residues; b_0 depends on the level of organization; for example, b_0 is about -630 for a perfect α helix and zero for a completely random coil. The constant λ_0 is approximately 212 nm for both the α helix and the random coil. The mean residue rotation is related to the specific rotation $[\alpha]$ by the equation

$$[m'] = \frac{3[\alpha]}{n^2 + 2} \frac{\text{MRW}}{100} \tag{21.21}$$

where n is the refractive index of the medium and MRW is the mean residue mass, that is, the molecular mass of the protein divided by the total number of residues present. It is usually taken to be 115 for proteins of unknown amino acid composition. Equation (21.20) can be rearranged as follows:

$$[m'] \frac{\lambda^2 - \lambda_0^2}{\lambda_0^2} = a_0 + \frac{b_0 \lambda_0^2}{\lambda^2 - \lambda_0^2} \tag{21.22}$$

From a plot of $[m'](\lambda^2 - \lambda_0^2)/\lambda_0^2$ versus $\lambda_0^2/(\lambda^2 - \lambda_0^2)$, both a_0 and b_0 can be determined from the intercept and the slope of the line. Figure 21.21 shows such a plot for a synthetic copolymer of L-tyrosine and L-glutamic acid. At pH 4, the polypeptide chain is an α helix and b_0 or the slope is about -630; at pH 7, the slope of the line is close to zero, indicating that the chain is now a random coil. The value of b_0 serves as a convenient way to measure the percentage of α-helical content in proteins.

INFRARED DICHROISM

The techniques described so far are all suitable for studying protein conformation in solution. The ir technique cannot be readily applied to protein solutions, for the following reasons.

1. The polypeptide chain gives a complex spectrum because it contains a very large number of atoms. However, the amino acid residues share a common feature in that they all contain the amide bond $-CO-NH-$. The ir absorptions of this bond occur at about 1535 cm^{-1}, 1650 cm^{-1}, and 3290 cm^{-1}, as a result of the $>N-H$ bending $>C=O$ stretching,

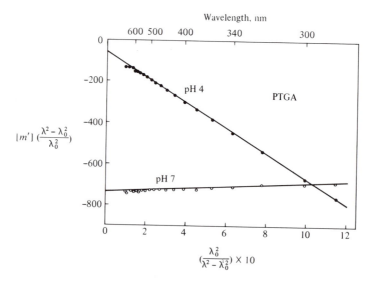

Figure 21.21 Plot of Eq. (21.22) for rotatory dispersion of a synthetic polypeptide, a copolymer of 5% L-tyrosine with L-glutamic acid (PTGA) using $\lambda_0 = 212$ nm. Helical form at pH 4; random coil at pH 7. [From P. Urnes and P. Doty, "Optical Rotation and the Conformation of Polypeptides and Proteins," *Advan. Protein Chem.* **16**, 433 (1961), Academic Press, Inc., New York, 1961.]

and $>$N—H stretching, respectively. Generally, these are the only regions where straightforward analyses can be made of proteins.

2. Water molecules, being polar, also absorb strongly in the regions of interest, namely, around 1600 cm^{-1} and 3400 cm^{-1}, obscuring much useful information. This difficulty can be partly overcome by studying protein solutions in D_2O, since its absorption bands are shifted to lower frequencies with the presence of the heavier deuterium isotope [see Eq. (18.27)]. However, since amide protons can readily exchange with deuterium, corresponding shifts of the amide bands to lower frequencies will also be observed.

Fortunately, some specific and important information about protein conformation can still be observed by the *infrared dichroism technique*. This approach involves the use of polarized infrared radiation on a thin film of oriented protein molecules. Under these conditions, absorption of the incident radiation depends on whether the electric vector component of the electromagnetic wave is parallel or perpendicular to the group responsible for absorption. Maximum absorption occurs when the electric vector is parallel to the electric dipole moment vector of the chromophore. When the electric vector is perpendicular to the same electric dipole moment vector, little or no absorption will occur.* In practice, we measure the absorption intensity with the electric field vector polarized parallel and perpendicular to the long axis of the polypeptide chain and determine the *dichroic ratio*, that is, the ratio of absorbance for perpendicular orientation to that for parallel orientation. If

---

* The reader should keep in mind that the electric vector is perpendicular to the direction of ir propagation (Figure 15.4).

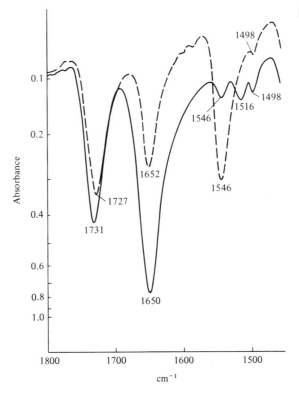

Polarized infrared spectra of poly-γ-benzyl-L-glutamate (film unidirectionally oriented from choroform solution). The solid curve represents the case in which the electric vector vibration is parallel to the orientation direction; the broken curve represents the case in which the electric vector vibration is perpendicular to the orientation direction. The peak at 1650 cm^{-1} is due to carbonyl stretching and the peak at 1546 cm^{-1} is due to $>$N—H bending. [Reprinted with permission from T. Miyazawa and E. R. Bloult, *J. Amer. Chem. Soc.* **83**, 712 (1961). Copyright by the American Chemical Society.]

Figure 21.22

the ratio is greater than unity, the absorption is said to show perpendicular dichroism; if smaller than unity, parallel dichroism.

Infrared dichroism is valuable in estimating the contents of the α helix and β structure in proteins. As Figure 21.11 shows, for the α helix, the $>$N—H$\cdots$O$=$C$<$ group is parallel to the long axis of the molecule. Consequently, both the $>$C$=$O and $>$N—H stretching modes should show parallel dichroism, while the $>$N—H bending mode should show perpendicular dichroism. Exactly the opposite case holds for the β structure, since the $>$N—H$\cdots$O$=$C$<$ group there is perpendicular to the long axis of the polypeptide chain (see Figure 21.13). Figure 21.22 shows the polarized infrared spectra of poly-γ-benzyl-L-glutamate, a synthetic polypeptide chain that is known to be α-helical. The molecules are oriented in a two-dimensional array with their long axes parallel. As we can see, this material shows parallel dichroism for the $>$C$=$O stretching mode and perpendicular dichroism for the $>$N—H bending mode. An extra peak, at about 1730 cm^{-1}, is due to the stretching of $>$C$=$O in the benzyl ester. Since this mode exhibits little dichroism, we may conclude that this particular group is randomly oriented.

INTRODUCTORY–INTERMEDIATE

Suggestions for Further Reading

BILLMEYER, F. W., JR. *Synthetic Polymers.* Doubleday & Company, Inc., Garden City, N.Y., 1972.

A readable and useful descriptive text.

BLOOMFIELD, V. A., and R. E. HARRINGTON, eds. *Biophysical Chemistry*. W. H. Freeman and Company, San Francisco, 1975.
   A collection of interesting articles from *Scientific American*.

DICKERSON, R. E., and I. GEIS. *The Structure and Action of Proteins*. Harper & Row, Publishers, New York, 1969.
   Beautifully illustrated and written in an interesting style. One of the best introductory texts on proteins.

LIGHT, A. *Proteins: Structure and Function*. Prentice Hall, Inc., Englewood Cliffs, N.J., 1974.
   A very useful and readable introductory text.

MANDELKERN, L. *An Introduction to Macromolecules*. Springer-Verlag, New York, 1972.
   A clearly written text that covers both the natural and synthetic polymers.

WOLD, F. *Macromolecules: Structure and Function*. Prentice-Hall, Inc., Englewood Cliffs, N.J., 1971.
   Discusses the structure and function of a number of macromolecules in considerable detail.

INTERMEDIATE–ADVANCED

FLORY, P. J. *Principles of Polymer Chemistry*. Cornell University Press, Ithaca, N.Y., 1953.
   Although somewhat outdated, it remains a standard text in the field. It does not cover proteins and nucleic acids.

FREIFELDER, D. *Physical Biochemistry*. W. H. Freeman and Company, San Francisco, 1976.
   A very readable, up-to-date text on the various physical techniques applied to biological systems.

FREIFELDER, D. *The DNA Molecule: Structure and Properties*. W. H. Freeman and Company, San Francisco, 1978.
   A collection of classic papers plus comments and problems.

HASCHEMEYER, R. H., and A. V. HASCHEMEYER. *Proteins*. John Wiley & Sons, Inc., New York, 1973.
   Covers most important aspects of protein chemistry. It has many recent references.

LAPANJE, S. *Physicochemical Aspects of Protein Denaturation*. John Wiley & Sons, Inc., New York, 1978.
   A very readable text.

MORAWETZ, H. *Macromolecules in Solution*, 2nd ed. John Wiley & Sons, Inc., New York, 1975.
   A useful advanced text.

SCHULZ, G. E., and R. H. SCHIRMER. *Principles of Protein Structure*. Springer-Verlag, New York, 1979.
   A very readable and informative text.

TANFORD, C. *Physical Chemistry of Macromolecules*. John Wiley & Sons, Inc., New York, 1961.
   A standard text in the field. Strongly recommended for advanced students.

VAN HOLDE, K. E. *Physical Biochemistry*. Prentice-Hall, Inc., Englewood Cliffs, N. J., 1971.
   A rather brief, but clear, treatment of many of the topics discussed in this chapter.

WALTON, A. G., and J. BLACKWELL. *Biopolymers*. Academic Press, Inc., New York, 1973.

A comprehensive and up-to-date treatment of many topics on biomacromolecules.

SECTION A   Reading Assignments

"Big Molecules," M. Morton, *Chemistry* **37**(1), 13 (1964).

"Design and Formation of Long Chain Polymers," M. Morton, *Chemistry* **37**(3), 6 (1964).

"Understanding Unruly Molecules," P. J. Flory, *Chemistry* **37**(5), 6 (1964).

"Measuring the Weight of Giant Molecules," F. W. Billmeyer, *Chemistry* **39**(3), 8 (1966).

"How Giant Molecules Are Measured," P. J. W. Debye, *Sci. Am.*, Sept. 1957.

"Average Quantities in Colloid Science," J. T. Bailey, W. H. Beattie, and C. Booth, *J. Chem. Educ.* **39**, 196 (1962).

"Molecular Weight Distributions of Polymers," A. Rudin, *J. Chem. Educ.* **46**, 595 (1969).

"Conformation of Peptides," S. Lande, *J. Chem. Educ.* **45**, 587 (1968).

"Electrophoresis," G. W. Gray, *Sci. Am.*, Dec. 1951.

"Electrophoresis in Protein Analysis," D. H. Leaback, *Chem. Brit.* **10**, 376 (1974).

"Errors in Representing Structures of Proteins and Nucleic Acids," R. A. Day and E. J. Ritter, *J. Chem. Educ.* **44**, 761 (1967).

"Hydrophobic Interaction," G. Némethy, *Angew. Chem. Int. Ed.* **6**(3), 195 (1967).

"Principles That Govern the Folding of Protein Chains," C. B. Anfinsen, *Science* **181**, 223 (1973).

"Proteins," P. Doty, *Sci. Am.*, Sept. 1957.

"Collagen," J. Gross, *Sci. Am.*, May 1961.

"Three-Dimensional Structure of a Protein," J. C. Kendrew, *Sci. Am.*, Dec. 1961. (Myoglobin)

"The Hemoglobin Molecule," M. F. Perutz, *Sci. Am.*, Nov. 1964.

"The Anatomy of Hemoglobin," M. F. Perutz, *Chem. Brit.* **1**, 9 (1965).

"The Three-Dimensional Structure of an Enzyme Molecule," D. C. Phillips, *Sci. Am.*, Nov. 1966. (Lysozyme)

"Keratins," R. D. B. Fraser, *Sci. Am.*, Aug. 1969.

"The Structure and History of an Ancient Protein," R. E. Dickerson, *Sci. Am.*, Apr. 1972. (Cytochrome *c*)

Chemistry, June 1978 (Vol. 51, No. 5). (The entire issue is on polymers.)

"The Macromolecular Concept and the Origins of Molecular Biology," R. Olby, *J. Chem. Educ.* **47**, 168 (1970).

"Amino Acid Sequence Diversity in Proteins," D. Blackman, *J. Chem. Educ.* **54**, 170 (1977).

"Structural Rules for Globular Proteins," G. E. Schulz, *Angew. Chem. Int. Ed.* **16**, 23 (1977).

619

"Electrostatic Effects in Proteins," M. F. Perutz, *Science* **201**, 1187 (1978).

"Blood Clots, Enzymes, and Antifreeze," R. Kilker, Jr., *Chemistry* **51**(6) 6 (1978).

"DNA Helix to Coil Transition: A Simplified Model," G. L. Baker, *Am. J. Phys.* **44**, 599 (1976).

"The Buoyant Density of DNA and the G + C Content," B. Panijpan, *J. Chem. Educ.* **54**, 172 (1977).

"SDS–Polyacrylamide Gel Electrophoresis," J. Svasti and B. Panijpan, *J. Chem. Educ.* **54**, 560 (1977).

"Some Physical Probes of Enzyme Structure in Solution," S. N. Timasheff, in P. D. Boyer, ed., *The Enzymes*, Vol. 2, Academic Press, Inc., New York, 1970, Chapter 8.

SECTION B

"Spatial Configuration of Macromolecular Chains," P. J. Flory, *Science* **188**, 1268 (1975).

"Conformation of Macromolecules," D. H. Napper, *J. Chem. Educ.* **46**, 305 (1969).

"Some Stereochemical Principles from Polymers," C. C. Price, *J. Chem. Educ.* **50**, 744 (1973).

"Cyanate and Sickle-Cell Disease," A. Cerami and C. M. Peterson, *Sci. Am.*, Apr. 1975.

"Principles of Electrophoresis and Ultracentrifugation," J. R. Cann, in *Interacting Macromolecules: The Theory and Practice of Their Electrophoresis, Ultracentrifugation and Chromatography*, Academic Press, Inc., New York, 1970, pp. 19–44.

"The Viscosity of Macromolecules in Relation to Molecular Conformation," J. T. Yang, *Advan. Protein Chem.* **16**, 323 (1961).

"Conformation of Polypeptides and Proteins," G. N. Ramachandran and V. Sasisckhamen, *Advan. Protein Chem.* **23**, 283 (1968).

S. N. Timasheff and G. D. Fasman, eds., *Biological Macromolecules*: Vol. 2, *Structure and Stability of Biological Macromolecules*, Marcel Dekker, Inc., New York, 1968.

"Protein Denaturation," C. Tanford, *Advan. Protein Chem.* **23**, 121 (1968); **24**, 1 (1970).

"Molecular Model-Building by Computer," C. Levinthal, *Sci. Am.*, June 1966.

"Computer Analysis of Protein Evolution," M. O. Dayhoff, *Sci. Am.*, July 1969.

"The Automatic Synthesis of Proteins," R. B. Merrifield, *Sci. Am.*, Mar. 1968.

"Polymer Models," C. E. Carraher, Jr., *J. Chem. Educ.* **47**, 581 (1970).

"Precisely Constructed Polymers," G. Natta, *Sci. Am.*, Aug. 1961.

"Protein Interactions with Small Molecules," I. M. Klotz, *Acc. Chem. Res.* **7**, 162 (1974).

"Protein–Solvent Interactions and Protein Conformation," S. N. Timansheff, *Acc. Chem. Res.* **3**, 62 (1970).

"Ion Effects on the Solution Structure of Biological Macromolecules," P. H. von Hippel and T. Schleich, *Acc. Chem. Res.* **2**, 257 (1969).

"An Introduction to Polyelectrolytes via the Physical Chemistry Laboratory," P. Ander, *J. Chem. Educ.* **56**, 481 (1979).

"Cytochrome *c* and the Evolution of Energy Metabolism," R. E. Dickerson, *Sci. Am.*, March 1980.

21.1 A polydisperse solution has the following distributions:

Number of Molecules	Molar Mass (g mol^{-1})
10	25,000
7	17,000
24	31,000
16	49,000

Calculate both $\bar{M}_n$ and $\bar{M}_w$ and the polydispersity of solution. (Polydispersity is defined by $\bar{M}_w/\bar{M}_n$.)

21.2 Elementary analysis of hemoglobin shows that it contains 0.34% Fe. What is the minimum molar mass of hemoglobin?

21.3 In an electrophoretic study of an aqueous solution of proteins it was found that there were two species with molecular masses 60,000 and 30,000. The solution contains 1.85% of protein by weight. If the fraction of the larger protein is found to be 70%, calculate $\bar{M}_w$ and $\bar{M}_n$.

21.4 Depending on experimental conditions, the measurement of the molecular mass of hemoglobin in an aqueous solution may show that the solution is monodisperse or polydisperse. Explain.

21.5 Ceruloplasmin is a protein present in the blood plasma. It contains 0.33% copper by weight. (a) Calculate its minimum molar mass. (b) The actual molar mass of ceruloplasmin is 150,000 g mol^{-1}. How many copper atoms are present per protein molecule?

21.6 At pH 6.5, the electrophoretic mobility of normal carboxyhemoglobin is 2.23×10^{-5} cm s^{-1}/V cm^{-1} and that of sickle-cell carboxyhemoglobin is 2.63×10^{-5} cm s^{-1}/V cm^{-1}. Calculate how long it will take to separate these two proteins by 1 cm if the potential gradient is 5.0 V cm^{-1}.

★21.7 The extent to which a protein molecule behaves like a spherical molecule can be tested by the frictional ratio f/f_0, where f_0 is the frictional coefficient in Stokes law [Eq. (5.18)] and f is the frictional coefficient obtained from the diffusion coefficient [Eq. (5.17)]. For spherical molecules, $f/f_0 = 1$; deviations from unity can be used as a measure of the nonspherical shape of the molecule. Consider hemoglobin and human fibrinogen (molecular mass 339,700, $s = 7.63 \times 10^{-13}$ s, $D = 1.98 \times 10^{-7}$ cm^2 s^{-1}, and $\bar{v} = 0.725$ ml g^{-1}). What conclusions can you draw about the shape of the molecules? (The radius of an assumed spherical molecule r can be obtained from the equation $M = 4\pi N_0 r^3/3\bar{v}$, where M is the molar mass.) Assume that $T = 298$ K.

21.8 In a sedimentation equilibrium experiment of a certain protein molecule carried out at 293 K, the following data are obtained: $\omega = 19,000$ rpm, $s = 2.15 \times 10^{-13}$ s, $\bar{v} = 0.71$ ml g^{-1}, and $\rho = 1.1$ g ml^{-1}. The relative concentrations at distances r_1 and r_2 from the center of rotation are $c_1 = 4.72$ ($r_1 = 5.95$ cm) and $c_2 = 12.98$ ($r_2 = 6.23$ cm). What is the molar mass of the protein?

21.9 A protein with $\bar{v} = 0.74$ ml g^{-1} is sedimented in water at 20°C. If $s_{20,w} = 3.0 \times 10^{-13}$ s and $D = 1.5 \times 10^{-6}$ cm^2 s^{-1}, what is the molar mass of the protein? The density of solution is 0.998 g ml^{-1}.

21.10 What are the units for the various viscosities defined in Eqs. (21.10) to (21.13)?

21.11 Will dissolving 1×10^{-3} g of glucose in water or 10% glycerol result in a greater relative viscosity?

21.12 The intrinsic viscosity of ribonuclease is 3.4 at 20°C and 6 at 50°C. What can you say about the change in structure?

21.13 The relative electrophoretic mobilities of a number of protein–SDS complexes in a polyacrylamide gel are as follows:

Protein	Molar Mass (g mol^{-1})	Relative Mobility
Myoglobin	17,200	0.95
Trysin	23,300	0.82
Aldolase	40,000	0.59
Fumarase	49,000	0.50
Carbonic anhydrase	29,000	0.73

Plot log (molar mass) versus relative mobility. The relative mobility of creatine kinase is 0.60. What is its molar mass? Compare your result with the molar mass of 80,000 obtained by ultracentrifugation. What conclusions can you draw?

21.14 Proteins generally have widely different structures, whereas nucleic acids have quite similar structures. Explain.

21.15 Referring to Figure 21.14, answer the following questions: (a) How can you account for the fact that the glycine residues have bonds whose ϕ and ψ angles are in the energetically unfavorable regions? (b) What can you conclude about the predominant secondary structure in lysozyme? (c) Does the molecule possess any β-pleated sheet structure? (d) Are the helical regions exposed to water in solution? (e) How about the β-pleated sheet regions?

21.16 Referring to Figure 21.17, state whether the β-pleated sheet structure in ribonuclease is parallel or antiparallel.

21.17 The enthalpy change in the denaturation of a certain protein is 125 kJ mol^{-1}. If the entropy change is 397 J K^{-1} mol^{-1}, calculate the minimum temperature at which the protein would denature spontaneously.

21.18 Consider the formation of a dimeric protein

$$2P \longrightarrow P_2$$

At 25°C we have $\Delta H° = 17$ kJ and $\Delta S° = 65$ J K^{-1}. Is the dimerization favored at this temperature? Comment on the effect of lowering the temperature. What general conclusion can you draw about the so-called "cold labile" enzymes?

21.19 How does a hydrophobic bond differ from both covalent and noncovalent bonds? What role does it play in protein structure and stability?

21.20 A large value for ΔC_P is frequently observed for protein denaturations. Suggest a reason for this observation.

21.21 What is the implicit assumption made in the label experiments and solvent perturbation techniques discussed in the chapter?

21.22 As Figure 21.11 shows, the average turn length of an α helix is 5.4 Å. Assuming this to be the case in human hair and that the hair's growth rate is 0.6 inch/month, how many turns of the α helix are generated each second?

Appendix 1

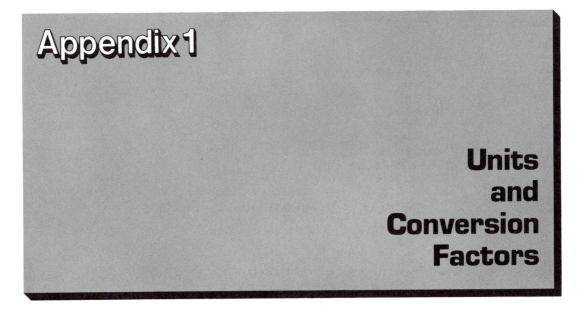

Units and Conversion Factors

Basic Units

Table A1.1

Physical Quantity	CGS Unit	SI Unit
Length	Centimeter (cm)	Meter (m)
Mass	Gram (g)	Kilogram (kg)
Time	Second (s)	Second (s)
Current	Ampere (A)	Ampere (A)
Temperature	Kelvin ($^{\circ}$K)	Kelvin (K)
Luminous intensity	—	Candela (cd)

Derived Units in the SI System

Table A1.2

Physical Quantity	Name	Symbol	Units
Energy	Joule	J	$kg\ m^2\ s^{-2}$
Force	Newton	N	$kg\ m\ s^{-2}$
Power	Watt	W	$kg\ m^2\ s^{-3}$
Electric charge	Coulomb	C	$A\ s$
Electrical resistance	Ohm	Ω	$kg\ m^2\ s^{-3}\ A^{-2}$
Electrical potential difference	Volt	V	$kg\ m^2\ s^{-3}\ A^{-1}$
Electrical capacity	Farad	F	$A^2\ s^4\ kg^{-1}\ m^{-2}$
Frequency	Hertz	Hz	s^{-1}

Fundamental Constants

Constant	CGS Units	SI Units
Avogadro's number (N_o)	6.02217×10^{23}	6.0225×10^{23}
Boltzmann's constant (k)	1.3805×10^{-16} erg K^{-1}	1.3805×10^{-23} J K^{-1}
Electron charge (e)	4.803×10^{-10} esu	1.602×10^{-19} C
Electron mass (m_e)	9.1096×10^{-28} g	9.1096×10^{-31} kg
Faraday constant (F)	$96{,}486.7$ coulombs g equiv^{-1}	$96{,}486.7$ C mol^{-1}
Gas constant (R)	8.314×10^7 ergs K^{-1} mol^{-1}	8.314 J K^{-1} mol^{-1}
Planck constant (h)	6.6256×10^{-27} erg s	6.6256×10^{-34} J s
Proton mass (M_P)	1.6725×10^{-24} g	1.6725×10^{-27} kg
Rydberg constant (R_H)	1.09737×10^5 cm^{-1}	1.09737×10^7 m^{-1}
Speed of light in vacuum (c)	2.99795×10^{10} cm s^{-1}	2.99795×10^8 m s^{-1}

Useful Conversion Factors

1 angstrom (Å) $= 10^{-8}$ cm $= 0.1$ nm $= 10^{-10}$ m
1 atm $= 760$ mm Hg $= 760$ torr $= 1.01325 \times 10^6$ dyn cm^{-2}
$\qquad\qquad\qquad\qquad = 101{,}325$ N m^{-2}
1 bar $= 10^6$ dyn cm$^{-2} = 0.987$ atm $= 100{,}007.8$ N m^{-2}
1 cal $= 4.184$ J $= 4.184 \times 10^7$ ergs
1 coulomb $= 2.9979 \times 10^9$ esu
1 dyn $= 10^{-5}$ N
1 erg $= 2.39 \times 10^{-8}$ cal $= 10^{-7}$ J
1 eV $= 23.06$ kcal mol$^{-1} = 1.602 \times 10^{-12}$ erg
$\qquad = 1.602 \times 10^{-19}$ J $= 8066$ cm^{-1}
1 F $= 96{,}487$ coulombs equiv^{-1}
$\qquad = 23{,}062$ cal V^{-1} equiv^{-1}
1 R $= 8.314$ J K^{-1} mol$^{-1} = 1.987$ cal K^{-1} mol^{-1}
$\qquad = 82.06$ cm^3 atm K^{-1} mol^{-1}
$\qquad = 0.08206$ liter atm K^{-1} mol^{-1}
1 liter atm $= 24.22$ cal
$\qquad\qquad = 101.34$ J

Small and Large Factors

Pico (p)	10^{-12}
Nano (n)	10^{-9}
Micro (μ)	10^{-6}
Milli (m)	10^{-3}
Centi (c)	10^{-2}
Deci (d)	10^{-1}
Deka (da)	10
Hecto (h)	10^2
Kilo (k)	10^3
Mega (M)	10^6
Giga (G)	10^9
Tera (T)	10^{12}

"SI Units," G. Socrates, *J. Chem. Educ.* **46**, 710 (1969).

"SI Units for Chemists," N. H. Davies, *Chem. Brit.* **6**, 344 (1970).

"Policy for NBS Usage of SI Units," *J. Chem. Educ.* **48**, 569 (1971).

"SI Units in Physiochemical Calculations," A. C. Norris, *J. Chem. Educ.* **48**, 797 (1971).

"SI Electric and Magnetic Units for Chemists," N. H. Davies, J. W. Moore, and R. W. Collins, *J. Chem. Educ.* **53**, 681 (1976).

"SI Electric and Magnetic Units for Chemists," N. H. Davies, *Chem. Brit.* **7**, 331 (1971).

Physiochemical Quantities and Units, M. L. McGlashan, The Royal Institute of Chemistry, London, 1971.

"Magnetochemistry in SI Units," T. I. Quickenden and R. C. Marshall, *J. Chem. Educ.* **49**, 114 (1972).

"International System of Units (SI)," M. A. Paul, *Chemistry* **45**(9), 14 (1972).

"Electrolyte Theory and SI Units," R. I. Holliday, *J. Chem. Educ.* **53**, 21 (1976).

"SI Stands for Student Improvement," W. G. Davies, J. W. Moore, and R. W. Collins, *J. Chem. Educ.* **53**, 681 (1976).

"Equations of Electromagnetism from CGS to SI," T. Critas and N. Kally, *J. Chem. Educ.* **54**, 530 (1977).

"SI Units: A Camel Is a Camel," A. W. Adamson, *J. Chem. Educ.* **55**, 634 (1978).

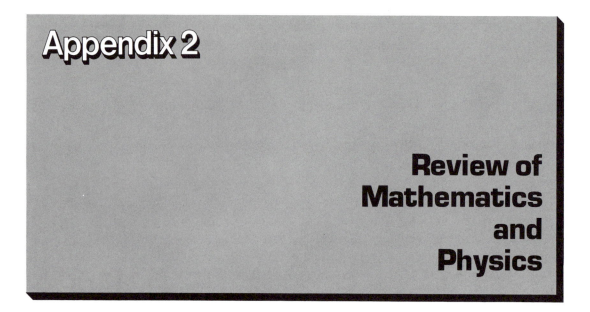

Appendix 2

Review of Mathematics and Physics

This appendix will briefly review some of the basic equations and formulas that are useful in physical chemistry.

A2.1 Mathematics

EXPONENTS AND POWERS

Many numbers are more conveniently expressed as powers of 10. For example,

$$1 = 10^0$$
$$0.1 = 10^{-1}$$
$$0.00023 = 2.3 \times 10^{-4}$$
$$100 = 10^2$$
$$100,000 = 10^5$$
$$3.1623 = 10^{0.5}$$

In general, we write a^n, where a is called the *base* and n the *exponent*. This expression is read as "a to the power of n." The following relations are useful:

Operation	Example
$a^m \times a^n = a^{m+n}$	$10^{0.2} \times 10^3 = 10^{3.2}$
$(a^m)^n = a^{m \times n}$	$(10^4)^2 = 10^8$
$\dfrac{a^m}{a^n} = a^{m-n}$	$\dfrac{10^3}{10^7} = 10^{-4}$

Note that a^0 (a to the power of zero) is equal to unity for all values of a except for $a = 0$; that is, $0^n = 0$ (for all values of n). Further, we have $1^n = 1$ for all values of n.

626

The concept of logarithm is a natural extension of exponents. The logarithm to the base a of a number x is equal to the exponent y to which the base number a must be raised so that $x = a^y$. Thus if

$$x = a^y$$

then

$$y = \log_a x$$

For example, since $3^4 = 81$, we have

$$4 = \log_3 81$$

Similarly, for logarithm to the base 10 we write

Logarithm	Exponent
$\log_{10} 1 = 0$	$10^0 = 1$
$\log_{10} 2 = 0.3010$	$10^{0.301} = 2$
$\log_{10} 10 = 1$	$10^1 = 10$
$\log_{10} 100 = 2$	$10^2 = 100$
$\log_{10} 0.1 = -1$	$10^{-1} = 0.1$

The logarithm to the base 10 is called the *common logarithm*. By convention, we use the notation $\log a$ instead of $\log_{10} a$ to denote the common logarithm of a.

Since the logarithms of numbers are exponents, they have the same properties as exponents. For simplicity we express the following relations in terms of common logarithms:

Logarithm	Exponent
$\log AB = \log A + \log B$	$10^A \times 10^B = 10^{A+B}$
$\log \dfrac{A}{B} = \log A - \log B$	$\dfrac{10^A}{10^B} = 10^{A-B}$
$\log A^n = n \log A$	

Logarithms taken to the base e are known as *natural logarithms*. The quantity e is a number given by

$$e = 1 + \frac{1}{1!} + \frac{1}{2!} + \frac{1}{3!} + \cdots$$

$$= 2.71828182845 \cdots$$

$$\simeq 2.7183$$

In physical chemistry, the exponential function $y = e^x$ is of great importance. Taking the natural logarithm on both sides, we get

$$\ln y = x \ln e = x$$

where $\ln$ represents $\log_e$. The relation between natural logarithm and common logarithm is as follows. We start with the equation

$$y = e^x$$

Taking the common logarithm on both sides, we obtain

$$\log y = x \log e$$
$$= \ln y \log e$$

since $x = \ln y$. Now $\log e = \log 2.7183 = 0.4343$; thus

$$\log y = 0.4343 \ln y$$

or

$$2.303 \log y = \ln y$$

SIMPLE EQUATIONS

Linear Equation.   A linear equation is represented by

$$y = mx + b$$

A plot of y versus x gives a straight line with slope m and an intercept (on the y axis, that is, at $x = 0$) b.

Quadratic Equation.   A quadratic equation takes the form

$$y = ax^2 + bx + c$$

where a, b, and c are constants and $a \neq 0$. A plot of y versus x gives a parabola. Let us consider a particular quadratic equation

$$y = 3x^2 - 5x + 2$$

A plot of y versus x is shown in Figure A2.1. The curve intercepts the x axis ($y = 0$) twice at $x = 1$ and $x = 0.67$. Alternatively, we can solve the equation as follows. By setting the equation to be zero (that is, $y = 0$), we get

$$3x^2 - 5x + 2 = 0$$

$$x = \frac{-b \pm \sqrt{b^2 - 4ac}}{2a}$$

$$= \frac{5 \pm \sqrt{25 - 4 \times 3 \times 2}}{2 \times 3} = 1.00 \text{ or } 0.67$$

Figure A2.1

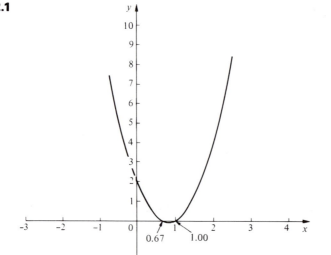

If we repeat a measurement of an experiment, we often obtain a value that is different from the previous reading, and it is appropriate to represent the result as a mean of these two numbers. The most common mean value is the *arithmetic mean*. For two readings a and b the arithmetic mean is given by $(a + b)/2$. There are occasions when the readings do not vary randomly. In such cases we may try the *geometric mean*. The geometric mean of two numbers a and b is given by $\sqrt{ab}$.

SERIES AND EXPANSIONS

Arithmetic Series

$$1, 2, 3, 4, \ldots$$

or

$$a, 2a, 3a, 4a, \ldots$$

Geometric Series

$$1, 2, 4, 8, \ldots$$

or

$$a, 2a, 4a, 8a, \ldots$$

Binomial Expansion

$$(1 + x)^n = 1 + nx + \frac{n(n - 1)}{2!} x^2 + \frac{n(n - 1)(n - 2)}{3!} x^3 + \cdots$$

Exponential Expansion

$$e^{\pm x} = 1 \pm \frac{x}{1!} \pm \frac{x^2}{2!} \pm \frac{x^3}{3!} \pm \cdots$$

$$e^{\pm ax} = 1 \pm \frac{ax}{1!} \pm \frac{(ax)^2}{2!} \pm \frac{(ax)^3}{3!} \pm \cdots$$

Trigonometric Expansions

$$\sin x = x - \frac{x^3}{3!} + \frac{x^5}{5!} - \frac{x^7}{7!} + \cdots$$

$$\cos x = 1 - \frac{x^2}{2!} + \frac{x^4}{4!} - \frac{x^6}{6!} + \cdots$$

Logarithmic Expansion

$$\ln (1 + x) = x - \frac{x^2}{2} + \frac{x^3}{3} - \frac{x^4}{4} + \cdots$$

630
ANGLES AND RADIANS

App. 2

Review of
Mathematics
and Physics

The common unit of angular measure is the *degree*, which is defined as $\frac{1}{360}$ of a complete circle. Often in physical chemistry we find it more convenient to use another unit, called the *radian* (rad). The relation between angle and radian can be understood as follows. Consider a certain portion of the circumference of a circle of radius r. The length of the arc (s) is proportional to the angle θ and the radius r so that

$$s = r\theta$$

where θ is measured in radians. Thus 1 radian is defined to be the angle subtended when the arc length s is exactly equal to the radius.

If we consider the entire circle as the arc, then

$$s = 2\pi r = r\theta$$

or

$$2\pi = \theta$$

This means that $\theta = 2\pi$ radians corresponds to $\theta = 360°$. Thus

$$1 \text{ rad} = \frac{360°}{2\pi} \simeq \frac{360°}{2 \times 3.1416} = 57.3°$$

On the other hand,

$$1° = \frac{2\pi}{360°} \simeq \frac{2 \times 3.1416}{360°} = 0.0175 \text{ rad}$$

It is important to keep in mind that although the radian is a unit of angular measure, it does not have physical dimensions. For example, the circumference of a circle of radius 5 cm is given by 2π (rad) $\times$ 5 cm = 31.42 cm.

AREAS AND VOLUMES

Triangle. Consider a triangle with sides a, b, and c and height h (with side a as base). The semiperimeter s is given by

$$s = \frac{a + b + c}{2}$$

The area (A) of the triangle is

$$A = \tfrac{1}{2}ah = \sqrt{s(s - a)(s - b)(s - c)}$$
$$= \tfrac{1}{2}ab \sin C$$

where angle C is opposite side c. If a, b, and c are the sides of a right-angled triangle, c being the hypotenuse, then

$$c^2 = a^2 + b^2$$

which is the Pythagorean theorem.

Rectangle. The area of a rectangle of sides a and b is ab.

Parallelogram. The area of a parallelogram of sides a and b is ah, where h is the perpendicular distance between the two sides whose lengths are a.

Circle. The circumference of a circle is $2\pi r$ and the area of the circle is πr^2, where r is the radius.

Sphere. The area of the curved surface of a sphere of radius r is $4\pi r^2$ and the volume of the sphere is $\frac{4}{3}\pi r^3$.

Cylinder. The area of the curved surface of a cylinder of radius r and length h is $2\pi rh$ and the volume of the cylinder is $\pi r^2 h$.

Cone. The area of the curved surface of a cone is πrl, where r is the radius of the base and l is the slant height. The volume of the cone is $\frac{1}{3}r^2 h$, where h is the vertical height (from the apex to the base).

OPERATORS

In Section 15.6 we mentioned the use of operators. An operator is a mathematical symbol that tells us specifically what to do to a number or a function. Some examples of operators are as follows:

Operator	Function or Number	Final Form
log	24.1	$\log 24.1 = 1.382$
$\sqrt{}$	974.2	$\sqrt{974.2} = 31.21$
sin	61.9°	$\sin 61.9° = 0.882$
cos	x	$\cos x$
$\dfrac{d}{dx}$	e^{kx}	$\dfrac{de^{kx}}{dx} = ke^{kx}$

DIFFERENTIAL CALCULUS

Functions of Single Variables. The following are derivatives of some common functions.

$y = f(x)$	dy/dx
x^n	nx^{n-1}
e^x	e^x
e^{kx}	ke^{kx}
$\sin x$	$\cos x$
$\sin(ax + b)$	$a\cos(ax + b)$
$\cos x$	$-\sin x$
$\cos(ax + b)$	$-a\sin(ax + b)$
$\ln x$	$1/x$
$\ln(ax + b)$	$\dfrac{a}{ax + b}$

Partial Derivatives. If a function has more than one variable, then we must use partial derivatives to see how this function varies with a particular variable. For example, the van der Waals equation [Eq. (2.12)] can be rearranged as (assume $n = 1$)

$$P = \frac{RT}{V - b} - \frac{a}{V^2}$$

Thus P is a function of both T and V. To see how the pressure varies with temperature, we write

$$\left(\frac{\partial P}{\partial T}\right)_V = \frac{R}{V - b}$$

where ∂ reminds us that it is a partial derivative; that is, the differentiation is carried out at constant volume. Generally, if we have

$$y = f(x_1, x_2, x_3, \ldots)$$

then the partial derivative of y with respect to x_1 must be written as

$$\left(\frac{\partial y}{\partial x_1}\right)_{x_2, x_3, \ldots}$$

Total Derivatives. Consider the case in which the independent variables of a function must be varied simultaneously. Let us again use the van der Waals equation. We have

$$P = f(V, T)$$

A theorem, which we shall not prove, relates the *total* differential dP to the differentials dV and dT as follows:

$$dP = \left(\frac{\partial P}{\partial V}\right)_T dV + \left(\frac{\partial P}{\partial T}\right)_V dT$$

Since

$$\left(\frac{\partial P}{\partial T}\right)_V = \frac{R}{V - b}$$

and

$$\left(\frac{\partial P}{\partial V}\right)_T = -\frac{RT}{(V - b)^2} + \frac{2a}{V^3}$$

we obtain

$$dP = \left[-\frac{RT}{(V - b)^2} + \frac{2a}{V^3}\right] dV + \frac{R}{V - b} dT$$

Both partial and total derivatives find extensive application in physical chemistry, particularly in thermodynamics.

Exact and Inexact Differentials. The expression

$$M(x, y) \, dx + N(x, y) \, dy$$

is said to be an *exact* differential if the following condition is satisfied:

$$\left(\frac{\partial M}{\partial y}\right)_x = \left(\frac{\partial N}{\partial x}\right)_y$$

This test is known as *Euler's theorem*.

It is easily shown that the expression dP or

$$\left[-\frac{RT}{(V-b)^2} + \frac{2a}{V^3}\right] dV + \frac{R}{V-b} dT$$

is an exact differential, since

$$\left(\frac{\partial\left[-\dfrac{RT}{(V-b)^2} + \dfrac{2a}{V^3}\right]}{\partial T}\right)_V = -\frac{R}{(V-b)^2}$$

and

$$\left(\frac{\partial\left(\dfrac{R}{V-b}\right)}{\partial V}\right)_T = -\frac{R}{(V-b)^2}$$

In general, if we are given a function $f(x, y)$, it is easy to show whether df is an exact or *inexact* differential. Suppose that we have

$$df = (y^2 + 3x)\, dx + e^x\, dy$$

Now

$$\left(\frac{\partial(y^2 + 3x)}{\partial y}\right)_x = 2y$$

$$\left(\frac{\partial e^x}{\partial x}\right)_y = e^x$$

Hence df is an inexact differential.

The significance of exact and inexact differentials is that if df is an exact differential, then the value of the following integral depends only on the limits of integration; that is,

$$\int_{f_1}^{f_2} df = f_2 - f_1$$

However, if $đf$ is an inexact differential, then

$$\int_{f_1}^{f_2} đf \neq f_2 - f_1$$

Unless the functional relationship between the variables x and y in $đf$ is known, the integral of $đf$ cannot be carried out. We saw in Chapter 6 that dU and dH are exact differentials while $đW$ and $đQ$ are inexact differentials. This means that the amount of work done or heat change in a process depends on the path or manner in which the process is carried out and not just on the initial and final states of the system. The symbol $đ$ reminds us that we are dealing with an inexact differential.

$$\int x^n \, dx = \frac{1}{n+1} x^{n+1} + C$$

$$\int \frac{dx}{x} = \ln x + C$$

$$\int \frac{dx}{ax+b} = \frac{1}{a} \ln (ax+b) + C$$

$$\int \sin x \, dx = -\cos x + C$$

$$\int \cos x \, dx = \sin x + C$$

$$\int \ln x \, dx = x \ln x - x + C$$

$$\int e^x \, dx = e^x + C$$

$$\int e^{kx} \, dx = \frac{e^{kx}}{k} + C$$

Since all these integrals are indefinite integrals, a constant term C must be added to the results.

A2.2 Physics

MECHANICS

We now summarize the important physical quantities in mechanics.

Velocity. Velocity (v) is defined as the rate of change of position with time; that is,

$$v = \frac{\Delta x}{\Delta t}$$

It has the units cm s^{-1} (CGS) or m s^{-1} (SI). The terms "velocity" and "speed" are often used interchangeably, although they are different quantities. Velocity is a *vector* quantity; it has both magnitude and direction. Speed is a *scalar* quantity; it has only magnitude but no direction. The distinction between these two quantities is also discussed in Chapter 3.

Acceleration. Acceleration (a) is the rate of change of velocity with time; that is,

$$a = \frac{\Delta v}{\Delta t}$$

In CGS units, a is in cm s^{-2}; in SI units, a is in m s^{-2}.

Linear Momentum. Linear momentum (p) or simply momentum of an object is the product of its mass and its velocity:

$$p = mv$$

In CGS units, p is in g cm s^{-1}; in SI units, p is in kg m s^{-1}.

Angular Velocity and Angular Momentum. Consider the motion of a particle of mass m about a circle of radius r shown in Figure A2.2. If the particle describes an angle θ in time t, then the angular velocity ω is given by

$$\omega = \frac{\theta}{t}$$

where θ is in radians and ω is in rad s^{-1}.

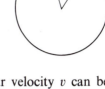

Figure A2.2

A relation between the angular velocity and the linear velocity v can be derived as follows. The linear velocity is the instantaneous velocity of the particle; its direction is always tangential to the circle at every instant. From Section A2.1,

$$s = r\theta$$

Since distance = velocity × time, we have

$$s = vt$$

Thus

$$r\theta = vt$$

or

$$\frac{r\theta}{t} = r\omega = v$$

The angular momentum of the particle is given by $m\omega r^2$ or mvr. Sometimes it is expressed as $I\omega$, where I is the moment of inertia of the particle about the center of the circle (equal to mr^2).

In classical mechanics, the angular momentum of a system can vary continuously. However, quantum mechanics imposes the restriction that the angular momentum of an atomic or molecular system is quantized; that is, it can only have certain allowed values. This is one of the fundamental postulates of Bohr's theory of the hydrogen atom, discussed in Chapter 15.

Force. The familiar definition of force (F) is Newton's second law of motion,

$$F = ma$$

This equation says that the force acting on an object is equal to the product of the mass of the object and its acceleration.

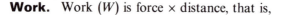

CGS units: $1 \text{ g cm s}^{-2} = 1 \text{ dyn}$
SI units: $1 \text{ kg m s}^{-2} = 1 \text{ newton (N)}$
Conversion factor: $1 \text{ N} = 10^5 \text{ dyn}$

Alternatively, force can be defined as the rate of change of momentum. Thus

$$F = \frac{\Delta p}{\Delta t}$$

Work. Work (W) is force $\times$ distance, that is,

$$W = Fs$$

CGS units: $1 \text{ g cm s}^{-2} \text{ cm} = 1 \text{ g cm}^2 \text{ s}^{-2}$
$$= 1 \text{ dyn cm} = 1 \text{ erg}$$
SI units: $1 \text{ kg m s}^{-2} \text{ m} = 1 \text{ kg m}^2 \text{ s}^{-2}$
$$= 1 \text{ N m} = 1 \text{ J}$$
Conversion factor: $1 \text{ J} = 10^7 \text{ ergs}$

Newton's Law of Gravitational Attraction. According to Newton's law of gravitational attraction, the force between two masses m_1 and m_2 separated by distance r is given by

$$F \propto -\frac{m_1 m_2}{r^2}$$

$$= -G\frac{m_1 m_2}{r^2}$$

where G is the universal gravitational constant. The negative sign indicates that the force is *always* attractive. We have

CGS units: $G = 6.673 \times 10^{-8} \text{ dyn cm}^2 \text{ g}^{-2}$
SI units: $G = 6.673 \times 10^{-11} \text{ N m}^2 \text{ kg}^{-2}$

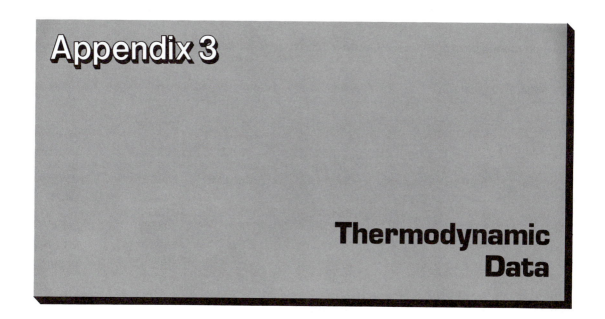

Appendix 3

Thermodynamic Data

Table A3.1

Thermodynamic Data of Selected Elements and Inorganic Compounds at 298.15 K and 1 atm[a]

Substance	State	$\Delta \overline{H}_f^\circ$ (kJ mol^{-1})	$\Delta \overline{G}_f^\circ$ (kJ mol^{-1})	$\overline{S}^\circ$ (J K^{-1} mol^{-1})	$\overline{C}_p^\circ$ (J K^{-1} mol^{-1})
Ag	s	0	0	42.71	25.49
Ag$^+$	aq	105.90	77.11	73.93	37.66
AgCl	s	−127.04	−109.72	96.11	50.79
AgBr	s	−99.50	−95.94	107.11	52.38
AgI	s	−62.38	−66.32	114.22	54.43
AgNO$_3$	s	−123.14	−32.18	140.92	93.05
Al	s	0	0	28.32	24.34
Al^{3+}	aq	−524.67	−481.16	−313.38	
Al$_2$O$_3$	s	−1669.79	−1576.41	50.99	78.99
Ar	g	0	0	154.74	20.79
As	s	0	0	35.15	24.98
AsO$_4$$^{3-}$	aq	−870.27	−635.97	−144.77	
AsH$_3$	g	171.54			
H$_3$AsO$_4$	s	−900.40			
Au	s	0	0	47.70	25.23
Au$_2$O$_3$	s	80.75	163.18	125.52	
AuCl	s	−35.15			
AuCl$_3$	s	−118.41			
B	s	0	0	6.53	11.97
B$_2$O$_3$	s	1263.57	1184.07	54.02	62.26
H$_3$BO$_3$	s	−1087.84	−963.16	89.58	82.05
	aq	−1067.76	−963.32	159.83	
Ba	s	0	0	66.94	26.36
Ba^{2+}	aq	−538.36	−560.66	12.55	
BaO	s	−558.15	−528.44	70.29	47.45
BaCl$_2$	s	−860.06	−810.86	125.52	75.31
BaSO$_4$	s	−1464.4	−1353.11	132.21	101.75
BaCO$_3$	s	−1218.80	−1138.88	112.13	85.35
Be	s	0	0	9.54	17.82
BeO	s	−610.86	−581.58	14.10	25.40
Br$_2$	l	0	0	152.30	
Br$^-$	aq	−120.92	102.82	80.71	−128.45
HBr	g	−36.23	−53.22	198.48	29.12
C	s (graphite)	0	0	5.69	8.64
	s (diamond)	1.90	2.87	2.44	6.06
CO	g	−110.52	−137.27	197.91	29.14
CO$_2$	g	−393.51	−394.38	213.64	37.13
	aq	−412.92	−386.23	121.34	
CO$_3$$^{2-}$	aq	−676.26	−528.10	−53.14	
HCO$_3$$^-$	aq	−691.11	−587.06	94.98	
H$_2$CO$_3$	aq	−699.65	−623.17	187.44	
CS$_2$	g	115.27	65.06	237.82	45.65
	l	87.86	63.60	151.04	75.73
HCN	aq	105.44	112.13	128.87	

[a] The data are from *Selected Values of Chemical Thermodynamics Properties*, by F. D. Rossini et al., published by the National Bureau of Standards (NBS Circular 500), 1952, and NBS Technical Notes 270-3 (1968) and 270-4 (1969), D. D. Wagman et al. (eds.). All data refer to 1 mole of material in the indicated state at 1 atm pressure and 298.15 K. The values for ions in aqueous solution (1 m), such as Li$^+$(aq), are based on the convention that all the properties listed for H$^+$(aq) are equal to zero. Thus, by adding values for neutral combinations of oppositely changed ions, one obtains correct values for salts in aqueous solution.

Substance	State	$\Delta \overline{H}_f^{\circ}$ (kJ mol^{-1})	$\Delta \overline{G}_f^{\circ}$ (kJ mol^{-1})	$\overline{S}^{\circ}$ (J K^{-1} mol^{-1})	$\overline{C}_p^{\circ}$ (J K^{-1} mol^{-1})
CN$^-$	aq	151.0	165.69	117.99	
CNO$^-$	aq	−140.16	−98.74	130.12	
NH$_4$HCO$_3$	s	−852.28			
CO(NH$_2$)$_2$	s	−333.19	−197.15	104.6	93.14
	aq	−319.24	−203.84	173.85	
Ca	s	0	0	41.63	26.28
Ca^{2+}	aq	−542.96	−553.04	−55.23	
CaO	s	−635.55	−604.17	39.75	42.80
Ca(OH)$_2$	s	−986.59	−896.76	76.15	84.52
CaF$_2$	s	−1214.62	−1161.90	68.87	67.03
CaCl$_2$	s	−794.96	−750.19	113.81	72.63
CaSO$_4$	s	−1432.69	−1320.30	106.69	99.58
CaCO$_3$	s	−1206.88	−1128.76	92.89	81.88
Cd	s	0	0	51.46	25.90
Cd^{2+}	aq	−72.38	−77.74	−61.09	
CdO	s	−254.64	−225.06	54.81	43.43
CdCl$_2$	s	−389.11	−342.59	118.41	
CdSO$_4$	s	−926.17	−820.2	137.24	
Cl$_2$	g	0	0	222.95	33.93
HCl	g	−92.31	−95.27	185.9	29.12
Co	s	0	0	28.45	25.56
Co^{2+}	aq	−67.36	−51.46	155.23	
CoO	s	−239.32	−213.38	43.93	
Cr	s	0	0	23.77	23.35
Cr^{2+}	aq	−138.91			
Cr$_2$O$_3$	s	−1128.43	−1046.84	81.17	118.74
CrO$_4^{2-}$	aq	−863.16	−706.26	38.49	
Cr$_2$O$_7^{2-}$	aq	−1460.63	−1257.29	213.80	
Cs	s	0	0	82.84	31.05
Cs$^+$	aq	−247.69	−282.04	133.05	
Cu	s	0	0	33.31	24.47
Cu$^+$	aq	51.88	50.21	−26.36	
Cu^{2+}	aq	64.39	64.98	98.74	
CuO	s	−155.23	−127.19	43.51	44.35
Cu$_2$O	s	−166.69	−146.36	100.83	69.87
CuCl	s	−134.73	−118.83	91.63	
CuCl$_2$	s	−205.85			
CuS	s	−48.53	−48.95	66.53	47.82
CuSO$_4$	s	−769.86	−661.91	113.39	100.83
F$_2$	g	0	0	203.34	31.46
F$^-$	aq	−329.11	−276.48	−9.62	−123.43
HF	g	−268.61	−270.71	173.51	29.08
Fe	s	0	0	27.15	25.23
Fe^{2+}	aq	−87.86	−84.94	−113.39	
Fe^{3+}	aq	−47.70	−10.54	−293.30	
Fe$_2$O$_3$	s	−822.16	−740.99	89.96	104.6
Fe(OH)$_2$	s	−568.19	−483.55	79.50	
Fe(OH)$_3$	s	−824.25			
H	g	−218.0	203.24	114.61	20.79
H$_2$	g	0	0	130.59	28.84
H$^+$	aq	0	0	0	0

Table A3.1 (*cont.*)

Substance	State	$\Delta \overline{H}_f^{\circ}$ (kJ mol^{-1})	$\Delta \overline{G}_f^{\circ}$ (kJ mol^{-1})	$\overline{S}^{\circ}$ (J K^{-1} mol^{-1})	$\overline{C}_p^{\circ}$ (J K^{-1} mol^{-1})
OH$^-$	aq	−229.94	−157.30	−10.54	−133.89
H$_2$O	g	−241.83	−228.60	188.72	33.58
H$_2$O	l	−285.84	−237.19	69.94	75.30
H$_2$O$_2$	l	−187.61	−118.11		
H$_2$O$_2$	aq	−191.13			
He	g	0	0	126.05	20.79
Hg	l	0	0	77.40	27.82
Hg^{2+}	aq		−164.38		
HgO	s (red)	−90.71	−58.53	71.97	45.73
HgCl$_2$	s	−230.12			
Hg$_2$Cl$_2$	s	−264.93	−210.66	196.22	101.67
HgS	s (red)	−58.16	−48.83	77.82	
HgSO$_4$	s	−704.17			
Hg$_2$SO$_4$	s	−741.99	−623.92	200.75	132.01
I$_2$	s	0	0	116.73	54.98
I$^-$	aq	55.94	51.67	109.37	−129.70
HI	g	25.94	1.30	206.33	29.16
K	s	0	0	63.60	29.16
K$^+$	aq	−251.21	−282.28	102.5	
K$_2$O	s	−361.50			
KOH	s	−425.85			
KCl	s	−435.87	−408.33	82.68	51.51
KClO$_3$	s	−391.20	−289.91	142.97	100.25
KClO$_4$	s	−433.46	−304.18	151.04	110.17
KBr	s	−392.17	−379.20	96.44	53.64
KI	s	−327.65	−322.29	104.35	55.06
KNO$_3$	s	−492.71	−393.13	132.93	96.27
K$_2$CO$_3$	s	−1146.12			
Kr	g	0	0	163.97	20.79
Li	s	0	0	28.03	23.64
Li$^+$	aq	−278.46	−293.80	14.23	
Li$_2$O	s	−595.80			
LiOH	s	−487.23	−443.92	50.21	
Mg	s	0	0	32.51	23.89
Mg^{2+}	aq	−461.96	−456.01	−117.99	
MgO	s	−601.83	−569.57	26.78	37.41
Mg(OH)$_2$	s	−924.66	−833.75	63.14	77.03
MgCl$_2$	s	−641.83	−592.33	89.54	71.30
MgSO$_4$	s	−1278.21	−1173.61	91.63	96.27
MgCO$_3$	s	−1112.94	−1029.26	65.69	75.52
Mn	s	0	0	31.76	26.32
Mn^{2+}	aq	−218.82	−223.43	−83.68	
Mn^{3+}	aq	−100.42			
MnO$_2$	s	−520.91	−466.10	53.14	54.02
N$_2$	g	0	0	191.49	29.12
N$_3^-$	aq	245.18			
NH$_3$	g	−46.19	−16.64	192.51	35.66
NH$_4^+$	aq	−132.80	−79.50	112.84	
NH$_4$Cl	s	−315.39	−203.89	94.56	84.10
NH$_4$OH	aq	−366.1	−263.76	181.17	−68.62
N$_2$H$_4$	l	50.42			
NO	g	90.37	86.69	210.62	29.86

Substance	State	$\Delta \overline{H_f^\circ}$ (kJ mol^{-1})	$\Delta \overline{G_f^\circ}$ (kJ mol^{-1})	$\overline{S}^\circ$ (J K^{-1} mol^{-1})	$\overline{C_p^\circ}$ (J K^{-1} mol^{-1})
NO$_2$	g	33.85	51.84	240.46	37.91
N$_2$O$_4$	g	9.66	98.29	304.30	79.08
N$_2$O	g	81.56	103.60	219.99	38.70
HNO$_2$	aq	−118.83	−53.64		
HNO$_3$	l	−173.22	−79.9	155.60	109.87
NO$_3^-$	aq	−206.57	−110.50	146.44	
Na	s	0	0	51.05	28.41
Na$^+$	aq	−239.66	−261.87	60.25	
Na$_2$O	s	−415.89	−376.56	72.80	68.20
NaCl	s	−411.00	−384.03	72.38	49.71
NaBr	s	−359.95			52.3
NaI	s	−288.03			54.39
Na$_2$SO$_4$	s	−1384.49	−1266.83	149.49	127.61
NaNO$_3$	s	−466.68	−365.89	116.32	93.05
Na$_2$CO$_3$	s	−1130.94	−1047.67	135.98	110.50
NaHCO$_3$	s	−947.68	−851.86	102.09	87.61
Ne	g	0	0	146.22	20.79
Ni	s	0	0	30.13	25.98
Ni^{2+}	aq	−64.02	−46.44	159.41	
NiO	s	−244.35	−216.31	38.58	44.35
Ni(OH)$_2$	s	−538.06	−453.13	79.50	
O	g	249.4	230.10	160.95	21.91
O$_2$	g	0	0	205.03	29.36
O$_3$	aq	−12.09	16.32	110.88	167.4
	g	142.26	163.43	237.65	38.16
P	s (white)	0	0	44.00	23.22
	s (red)	−18.40	13.82	29.31	
PO$_4^{3-}$	aq	−1284.07	−1025.59	−217.57	
P$_2$O$_7^{4-}$	aq	−2275.68			
P$_4$O$_{10}$	s	−3012.48			
PH$_3$	g	9.25	18.24	210.04	
HPO$_4^{2-}$	aq	−1298.71	−1094.12	−35.98	
H$_2$PO$_4^-$	aq	−1302.48	−1135.12	89.12	
H$_3$PO$_4$	s	−1281.14			
H$_4$P$_2$O$_7$	s	−2250.99			
Pb	s	0	0	64.89	26.82
Pb^{2+}	aq	1.63	24.31	21.34	
PbO	s (yellow)	−217.86	−188.49	69.45	48.53
PbO$_2$	s	−276.65	−218.99	76.57	64.43
PbCl$_2$	s	−359.20	−313.97	136.40	76.99
PbS	s	−94.31	−92.68	91.21	49.50
PbSO$_4$	s	−918.40	−811.24	147.28	104.18
Pt	s	0	0	41.84	26.57
PtCl$_4^{2-}$	aq	−516.31	−384.51	175.73	
Rb	s	0	0	69.45	30.42
Rb$^+$	aq	−246.44	−282.21	124.27	
S	s (rhombic)	0	0	31.88	22.59

Table A3.1 (*cont.*)

Substance	State	$\Delta \overline{H}_f^\circ$ (kJ mol^{-1})	$\Delta \overline{G}_f^\circ$ (kJ mol^{-1})	$\overline{S}^\circ$ (J K^{-1} mol^{-1})	$\overline{C}_p^\circ$ (J K^{-1} mol^{-1})
	s (monoclinic)	0.30	0.10	32.55	23.64
SO_2	g	-296.06	-300.37	248.52	39.79
SO_3	g	-395.18	-370.37	256.22	50.63
SO_3^{2-}	aq	-624.25	-497.06	43.51	
SO_4^{2-}	aq	-907.51	-741.99	17.15	
H_2S	g	-20.15	-33.02	205.64	33.97
HSO_3^-	aq	-627.98	-527.31	132.38	
HSO_4^-	aq	-885.75	-752.87	126.86	
H_2SO_4	l	-811.32			
SF_6	g	-1096.21			
Se	s	0	0	42.44	25.36
SeO_2	s	-225.35			
H_2Se	g	29.71	15.90	218.91	34.73
H_2SeO_4	s	-530.11			
Si	s	0	0	18.70	19.87
SiO_2	s (quartz)	-859.30	-805.00	41.84	44.43
Sr	s	0	0	54.39	25.10
Sr^{2+}	aq	-545.51	-557.31	39.33	
$SrCl_2$	s	-828.43	-781.15	117.15	79.08
$SrSO_4$	s	-1444.74	-1334.28	121.75	
$SrCO_3$	s	-1218.38	-1137.63	97.07	81.42
W	s	0	0	33.47	24.98
WO_3	s	-840.31	-763.45	83.26	81.50
WO_4^-	aq	-1115.45			
Xe	g	0	0	169.58	20.79
Zn	s	0	0	41.63	25.06
Zn^{2+}	aq	-152.42	-147.21	106.48	
ZnO	s	-347.98	-318.19	43.93	40.25
$ZnCl_2$	s	-415.89	-369.26	108.37	76.57
ZnS	s	-202.92	-198.32	57.74	45.19
$ZnSO_4$	s	-978.55	-871.57	124.68	117.15

Thermodynamic Data of Selected Organic and Biological Molecules at 298.15 K and 1 atm[a]

Compound	Formula	State	$\Delta \overline{H_f^\circ}$ (kJ mol^{-1})	$\Delta \overline{G_f^\circ}$ (kJ mol^{-1})	$\overline{S^\circ}$ (J K^{-1} mol^{-1})	$\overline{C_p^\circ}$ (J K^{-1} mol^{-1})
Acetic acid	CH_3COOH	l	−484.21	−389.45	159.83	
		aq	−485.26	−404.09	205.43	
		eq buf	−485.60	−417.52	249.37	
Acetate	CH_3COO^-	aq	−485.60	−376.89	112.97	
Acetaldehyde	CH_3CHO	g	−166.35	−139.08	264.22	56.07
Acetone	CH_3COCH_3	l	−246.81	−153.55	198.74	126.78
Acetylene	C_2H_2	g	226.73	209.2	200.83	43.93
Adenine	$C_5H_5N_5$	s	97.07	−300.41	151.04	143.09
DL-Alanine	$C_3H_7O_2N$	s	−563.59	−371.96	132.21	121.75
L-Alanine	$C_3H_7O_2N$	s	−562.75	−370.20	129.20	122.26
		eq buf	−554.80	−371.16	158.99	
L-Alanine ion	$C_3H_8O_2N^+$	aq	−557.94	−384.55	192.05	288.70
L-Alanine dipolar ion	$C_3H_7O_2^{+-}$	aq	−554.80	−371.16	158.99	141.00
L-Alaninate ion	$C_3H_6O_2N^-$	aq	−509.61	−314.85	121.75	71.55
L-Arginine	$C_6H_{14}O_2N_4$	s	−621.74	−656.89	250.62	233.47
L-Arginine dipolar ion	$C_6H_{14}O_2N_4^{+-}$	aq	−615.47			
DL-Aspartic acid	$C_4H_7O_4N$	s	−976.96	−729.27	154.39	
L-Aspartic acid	$C_4H_7O_4N$	s	−972.53	−729.36	170.12	155.27
		eq buf	−943.41	−736.51	291.63	
L-Aspartic acid ion	$C_4H_8O_4N^+$	aq	−955.17	−733.87	229.28	
L-Aspartic acid dipolar ion	$C_4H_7O_4N^{+-}$	aq	−947.43	−718.06	216.31	
L-Aspartic acid dipolar ion	$C_4H_6O_4N^{+2-}$	aq	−943.41	−695.88	155.65	
L-Aspartate ion	$C_4H_5O_4N^{2-}$	aq	−905.84	−638.69	89.96	
L-Asparagine	$C_4H_8O_3N_2$	s	−790.36	−530.95	174.47	160.67
L-Asparagine dipolar ion	$C_4H_8O_3N_2^{+-}$	aq	−766.09	−525.93	238.91	
L-Asparagine monohydrate	$C_4H_{10}O_4N_2$	s	−1085.75	−767.35	209.62	207.95
Benzene	C_6H_6	l	49.04	124.52	172.80	
Butyric acid	$C_4H_3O_2$	l	−535.13	−379.07	226.35	179.49
		aq	−532.62	−399.57	303.76	
		eq buf	−535.55	−412.54	320.91	
Butyrate ion	$C_4H_2O_2^-$	aq	−535.55	−372.04	201.67	
Citric acid	$C_6H_8O_7$	s	−1543.90			
		eq buf	−1515.78	−1288.63	512.54	
Dihydrogen citrate ion	$C_6H_7O_7^-$	aq	−1520.88	−1226.33	286.19	187.86
Hydrogen citrate ion	$C_6H_6O_7^{2-}$	aq	−1518.46	−1199.18	203.34	0.84
Citrate ion	$C_6H_5O_7^{3-}$	aq	−1515.11	−1162.69	92.05	−254.81
Citric acid monohydrate	$C_6H_{10}O_8$	s	−1838.45	−1473.56	283.47	268.15
Creatinine	$C_4H_7ON_3$	s	−237.65	−28.45	167.36	138.91
Creatinine ion	$C_4H_8ON_3^+$	aq		4.14		
Creatine	$C_4H_9O_2N_3$	s	−536.47	−264.01	189.54	171.96
		aq		−259.20		
		eq buf		−259.20		

[a] The data are from the appendix to R. C. Wilhoit, "Thermodynamic Properties of Biochemical Substances," Chapter 2, in *Biochemical Microcalorimetry*, H. D. Brown, ed., Academic Press, Inc., New York, 1969. The symbol "eq buf" is for an equilibrium mixture of species in an aqueous solution buffered to pH 7. The concentrations of aqueous solutions are all 1 *m*.

Compound	Formula	State	$\Delta \overline{H}_f^\circ$ (kJ mol^{-1})	$\Delta \overline{G}_f^\circ$ (kJ mol^{-1})	$\overline{S}^\circ$ (J K^{-1} mol^{-1})	$\overline{C}_p^\circ$ (J K^{-1} mol^{-1})
Creatine ion	$C_4H_{10}O_2N_3^+$	aq		-274.39		
Creatine ion	$C_4H_8O_2N_3^-$	aq		-177.82		
Creatine monohydrate	$C_4H_{11}O_3N_3$	s	-833.03	-504.59	234.30	213.38
L-Cysteine	$C_3H_7O_2NS$	s	-532.62	-342.67	169.87	173.22
		eq buf		338.95		
L-Cysteine ion	$C_3H_8O_2NS^+$	aq	-349.36			
L-Cysteine dipolar ion	$C_3H_7O_2NS^{+-}$	aq		-338.82		
L-Hydrogen cysteinate ion	$C_3H_6O_2NS^-$	aq		-290.99		
L-Cysteinate ion	$C_3H_5O_2NS^{2-}$	aq		-229.58		
L-Cystine	$C_6H_{12}O_4N_2S_2$	s	-1044.33	-685.76	280.58	261.92
		eq buf		-668.19		
L-Cystine ion	$C_6H_{14}O_4N_2S_2^{2+}$	aq		-684.50		
L-Cystine dipolar ion	$C_6H_{13}O_4N_2S_2^{2+-}$	aq		-678.23		
L-Cystine dipolar ion	$C_6H_{12}O_4N_2S^{2+2-}$	aq		-666.51		
L-Cystinate ion	$C_6H_{10}O_4N_2S^{2-}$	aq		-562.33		
Ethanol	C_2H_5OH	l	-276.98	-174.18	161.04	111.96
Ethyl acetate	$C_4H_8O_2$	l	-481.99	-337.65	262.76	105.44
Ethane	C_2H_6	g	-84.68	-32.89	229.49	52.66
Ethylene	C_2H_4	g	52.3	68.12	219.45	43.56
Formic acid	HCOOH	l	-409.20	-346.02	128.95	99.04
Formate ion	HCOO$^-$	aq	-410.03	-334.72	91.63	
Fumaric acid	$C_4H_4O_4$	s	-810.65	-653.25	166.11	142.26
		aq	-774.88	-645.80	261.08	
		eq buf	-777.39	-684.38	381.99	
Hydrogen fumarate ion	$C_4H_3O_4^-$	aq	-774.46	-628.14	203.34	
Fumarate ion	$C_4H_2O_4^{2-}$	aq	-777.39	-601.87	105.44	
α-D-Glucose	$C_6H_{12}O_6$	s	-1274.45	-910.56	212.13	218.87
		aq	-1263.06	-914.54	264.01	
β-D-Glucose	$C_6H_{12}O_6$	s	-1268.05	-908.89	228.03	
		aq	-1264.24	-915.79	264.01	
α,β-D-Glucose	$C_6H_{12}O_6$	aq	-1263.78	-916.97	269.45	305.43
α-D-Glucose monohydrate	$C_6H_{14}O_7$	s	-1571.09	-1149.55	252.30	
Glucose 1-phosphoric acid	$C_6H_{13}O_9P$	aq		-1789.50		
		eq buf		-1828.83		
Glucose 1-hydrogenphosphate ion	$C_3H_{12}O_9P^-$	aq		-1783.22		
Glucose 1-phosphate ion	$C_3H_{12}O_9P^{2-}$	aq		-1746.11		
Glucose 6-phosphoric acid	$C_6H_{13}O_9P$	aq		-1797.45		
Glucose 6-hydrogenphosphate ion	$C_6H_{12}O_9P^-$	aq		-1789.08		
Glucose 6-phosphate ion	$C_6H_{11}O_9P^{2-}$	aq		-1753.51		
Glycerol	$C_3H_8O_3$	l	-670.70	-479.49	204.60	216.73
L-Glutamic acid	$C_5H_9O_4N$	s	-1009.18	-730.95	188.20	175.23
		eq buf	-981.98	-721.87	248.95	

Compound	Formula	State	$\Delta \overline{H_f^\circ}$ (kJ mol^{-1})	$\Delta \overline{G_f^\circ}$ (kJ mol^{-1})	$\overline{S^\circ}$ (J K^{-1} mol^{-1})	$\overline{C_p^\circ}$ (J K^{-1} mol^{-1})
L-Glutamic acid ion	$C_5H_{10}O_4N^+$	aq	−981.57	−734.63	293.72	
L-Glutamic acid dipolar ion	$C_5H_9O_4N^{+-}$	aq	−981.99	−721.87	248.95	
L-Glutamic acid dipolar ion	$C_5H_8O_4N^{+2-}$	aq	−979.89	−697.47	174.05	
L-Glutamate	$C_5H_7O_4N^{2-}$	aq	−939.73	−643.50	127.61	
L-Glutamine	$C_5H_{10}O_3N_2$	s	−825.92	−532.21	195.10	183.80
		eq buf	−805.00	−528.02	251.04	
L-Glutamine dipolar ion	$C_5H_{10}O_3N_2^{+-}$	aq	−805.00	−528.02	251.04	
Glycine	$C_2H_5O_2N$	s	−537.23	−377.69	103.51	99.20
		eq buf	−523.00	−379.91	158.57	
Glycine ion	$C_2H_6O_2N^+$	aq	−527.18	−393.30	189.54	171.54
Glycine dipolar ion	$C_2H_5O_2N^{+-}$	aq	−523.00	−379.91	158.57	36.82
Glycinate ion	$C_2H_4O_2N^-$	aq	−478.65	−324.09	120.50	54.81
Glycylglycine	$C_4H_8O_3N_2$	s	−746.01	−491.50	189.95	163.59
		eq buf	−734.25	−493.08		232.21
Glycylglycine ion	$C_4H_9O_3N_2^+$	aq	−735.72	−510.87	286.60	288.70
Glycylglycine dipolar ion	$C_4H_8O_3N_2^{+-}$	aq	−734.25	−492.08	231.38	158.99
Glycylglycinate ion	$C_4H_7O_3N_2^-$	aq	−689.90	−445.76	222.17	
α-Ketoglutaric acid	$C_5H_6O_5$	s	−1026.34			
α-Ketoglutarate ion	$C_5H_4O^{2-}$	aq		−793.41		
L(+)-Lactic acid	$C_3H_6O_3$	s	−694.04	−523.25	143.51	127.61
		aq	−686.22	−538.77	221.75	
		eq buf	−686.64	−557.35		
L(+)-Lactate ion	$C_3H_5O_3^-$	aq	−686.64	−516.72	146.44	
DL-Lactic acid	$C_3H_6O_3$	l	−673.62	−518.82	192.05	211.29
α-Lactose	$C_{12}H_{22}O_{11}$	s	−2221.70			
		aq	−2232.37	−1564.90	394.13	
β-Lactose	$C_{12}H_{22}O_{11}$	s	−2236.77	−1566.91	386.18	410.45
α,β-Lactose	$C_{12}H_{22}O_{11}$	aq	−2233.09	−1567.33	399.57	
DL-Leucine	$C_6H_{13}O_2N$	s	−649.78	−358.57	207.11	195.39
L-Leucine	$C_6H_{13}O_2N$	s	−646.85	−356.48	209.62	208.36
		eq buf	−643.37	−352.25	206.69	
L-Leucine ion	$C_6H_{14}O_2N^+$	aq	−645.01	−365.56	246.44	−660.24
L-Leucine dipolar ion	$C_6H_{13}O_2N^{+-}$	aq	−643.37	−352.25	207.53	506.26
L-Leucinate ion	$C_6H_{12}O_2N^-$	aq	−600.61	−296.60	164.43	447.69
L-Malic acid	$C_4H_6O_5$	s	−1103.32			
		aq		−891.61		
		eq buf		−925.08		
L-Hydrogen malate ion	$C_4H_5O_5^-$	aq		−871.95		
L-Malate ion	$C_4H_4O_5^{2-}$	aq	−842.66			
DL-Malic acid	$C_4H_6O_5$	s	−1105.41			
α-Maltose	$C_{12}H_{22}O_{11}$	aq	−2238.27	−1573.60	403.34	
β-Maltose	$C_{12}H_{22}O_{11}$	aq	−2237.73	−1572.18	400.41	

Table A3.2 (*cont.*)

Compound	Formula	State	$\Delta \overline{H_f^\circ}$ (kJ mol^{-1})	$\Delta \overline{G_f^\circ}$ (kJ mol^{-1})	$\overline{S^\circ}$ (J K^{-1} mol^{-1})	$\overline{C_p^\circ}$ (J K^{-1} mol^{-1})
α,β-Maltose	$C_{12}H_{22}O_{11}$	aq	−2238.06	−1574.69	407.94	
β-Maltose monohydrate	$C_{12}H_{24}O_{12}$	s	−2539.27	−1809.58	417.56	453.96
Methane	CH_4	g	−74.85	−50.79	186.19	35.73
Methanol	CH_3OH	l	−238.66	−166.31	126.78	81.59
		aq	−245.89	−175.23	132.34	
L-Methionine	$C_5H_{11}O_2NS$	s	−761.07	−508.36	231.46	290.20
L-Methionine ion	$C_5H_{12}O_2NS^+$	aq	−744.75			
L-Methionine dipolar ion	$C_5H_{11}O_2NS^{+-}$	aq	−744.33			
Oxaloacetic acid	$C_4H_4O_5$	s	−984.50			
		aq	−832.62			
		eq buf	−875.71			
Hydrogen oxalo-acetate ion	$C_4H_3O_5^-$	aq	−818.39			
Oxaloacetate ion	$C_4H_2O_5^{2-}$	aq	−793.29			
Palmitic acid	$C_{16}H_{32}O_2$	s	−890.77	−315.06	455.22	460.66
		aq		−287.86		
		eq buf		−299.16		
Palmitate ion	$C_{16}H_{31}O_2^-$	aq		−259.41		
Pyruvic acid	$C_3H_4O_3$	l	−585.76			
		aq	−607.52	−486.60	179.91	
		eq buf	−596.22	−513.38		
Pyruvate ion	$C_3H_3O_3^-$	aq	−596.22	−472.37	171.54	
2-Propanol	C_3H_7OH	l	−317.86	−180.29	180.58	154.22
		aq	−330.83	−185.23	153.55	
Succinic acid	$C_4H_6O_4$	s	−940.81	−747.35	175.73	153.97
		aq	−912.20	−746.64	269.45	230.12
		eq buf	−908.72	−772.99	369.45	
Hydrogen succinate ion	$C_6H_5O_4^-$	aq	−908.89	−722.62	199.99	96.23
Succinate ion	$C_6H_4O_4^{2-}$	aq	−908.68	−690.44	92.88	122.59
Sucrose	$C_{12}H_{22}O_{11}$	s	−2221.70	−1544.31	360.24	425.51
		aq	−2215.85	−1551.43	403.76	633.04
DL-Valine	$C_5H_{11}O_2N$	s	−617.98	−359.82	181.17	
L-Valine	$C_5H_{11}O_2N$	s	−617.98	−358.99	178.74	168.82
		eq buf	−617.98	−358.65	176.98	
L-Valine ion	$C_5H_{12}O_2N^+$	aq	−612.24	−371.71	240.16	547.27
L-Valine dipolar ion	$C_5H_{11}O_2N^{+-}$	aq	−611.99	−358.65	176.98	389.11
L-Valinate ion	$C_5H_{10}O_2N^-$	aq	−567.43	−307.40	174.89	333.88

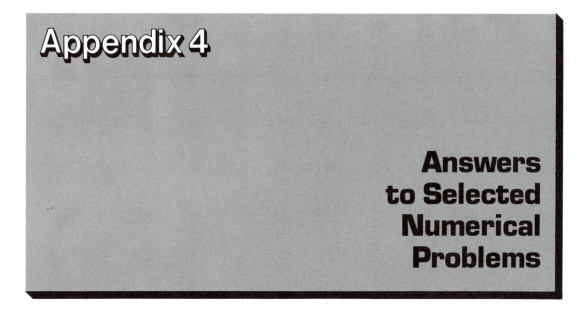

Appendix 4

Answers to Selected Numerical Problems

2.3 3.3×10^2 liters. **2.7** H_2: 11.1%; N_2: 88.9%.
2.8 (a) 0.46 liter; (b) 1200 atm. **2.11** 2.2 cm.
2.15 P_T: 1.02 atm; P_{He}: 0.72 atm; P_{Ar}: 0.30 atm. **2.18** 4.76 liters.
2.21 1130 mm Hg. **2.25** P_c: 50.4 atm; $\overline{V}_c$: 0.345 liter mol^{-1}; T_c: 565 K.
2.28 19.8 g.

3.3 c_{rms}: 4.31×10^4 cm s^{-1}; c_{mp}: 3.52×10^4 cm s^{-1}; $\bar{c}$: 3.97×10^4 cm s^{-1}.
3.7 λ: 1.46×10^{-2} cm. **3.10** 12.1 K. **3.15** 9.6 cm. **3.19** 16 g mol^{-1}.
3.28 2.3×10^5 K.

4.1 2.87 Å, 2.70 g cm^{-3}. **4.10** 1.02.

5.3 0.00788 P. **5.8** (a) 2.9 cm; (b) -1.4 cm. **5.10** 0.020 cm.
5.14 (a) 8.7×10^{-4} s; (b) 8.7×10^6 s. **5.16** 4.7 Å.
5.17 r_{Mb}: 19.1 Å; r_{Hb}: 31.3 Å.

6.7 -2.27 kJ. **6.10** 2.91×10^3 J. **6.14** 0.499 kJ mol^{-1}.
6.19 (a) 6.75 kJ, -6.75 kJ, 0, 0; (b) 2.33 kJ, -2.33 kJ, 0, 0;
(c) 0, -2.47 kJ, -2.47 kJ, -4.12 kJ; (d) 0, -1.40 kJ, -1.40 kJ, -2.33 kJ.
6.27 -1.97 kJ. **6.28** -84.6 kJ.
6.29 (a) 52.3 kJ; (b) 54.8 kJ.
6.34 (a) -64.91 kJ; (b) -9.08 kJ. **6.37** -23.2 kJ.
6.41 (a) -6.01 kJ, -6.01 kJ; (b) 37.7 kJ, 40.79 kJ.
6.42 465.2 kJ mol^{-1}.

Chapter 7 **7.3** (a) 75.1°C; (b) A: 22.7 J K^{-1}; B: -20.6 J K^{-1}; entire system: 2.1 J K^{-1}.
7.5 19.4 J K^{-1}.   **7.14** (a) -1530 J, 1530 J, 0, 0, 5.30 J K^{-1}, -1530 J;
(b) -628 J, 628 J, 0, 0, 5.30 J K^{-1}, -1530 J.
7.16 (a) 1.74 J K^{-1}, -1.74 J K^{-1}, 0; (b) 1.74 J K^{-1}, -1.27 J K^{-1}, 0.47 J K^{-1}.
7.26 -75.9 kJ.   **7.28** -5.8 K.   **7.34** 0.198 J K^{-1}.

Chapter 8 **8.3** 1.65 m.   **8.4** 500 m, 18.3 M.   **8.9** 2.7×10^{-3}.
8.12 (a) 11.5 J K^{-1}; (b) 50.5 J K^{-1}.
8.26 A: 200 mm Hg; B: 400 mm Hg.   **8.27** 0.915, 0.995.

Chapter 9 **9.1** 1.25×10^{-2} Ω^{-1} m^2 mol^{-1}.   **9.3** 0.11 M.
9.8 (a) 0.10, 0.69; (b) 0.030, 0.67; (c) 1.0, 9.2×10^{-3}.
9.12 0.295 M, -0.549°C.   **9.19** (a) 0.151 atm; (b) 0.077 atm.
9.20 K$^+$: 7.7 kJ; Na$^+$: 5.6 kJ.   **9.22** 17.6 Å.

Chapter 10 **10.3** 36.2 kJ.   **10.10** 0.115, 5.36 kJ.   **10.13** P_{trans}/P_{cis}: 3.43.
10.18 -33.8 kJ, 4.95×10^5.   **10.20** -7.3 kJ.   **10.24** 1.1×10^{-5}.

Chapter 11 **11.3** 1.69 V.   **11.4** 1.12 V, 1.11 V.   **11.10** 0.50 atm.   **11.15** 0.521 V.
11.17 (b) 4.3.   **11.18** 1.19 V.   **11.23** 118 mol.   **11.27** -43.0 kJ.
11.28 (a) -0.197 V; (b) -0.241 V.

Chapter 12 **12.3** 8.4×10^{-4}.   **12.5** 5.1.   **12.7** (a) 0.317; (b) 0.336.
12.14 (b) 6.84.   **12.15** (b) 7.9. **12.16** 6.6×10^{-5}.   **12.21** 8.95.
12.29 0, 21.4 kJ.

Chapter 13 **13.2** 0.988.   **13.5** 0.086 min^{-1}.   **13.7** Second order, 0.42 M^{-1} s^{-1}.
13.13 55 kJ mol^{-1}.   **13.18** 8.7.

Chapter 14 **14.2** 4.0×10^{-5} s.   **14.6** 6.2×10^{-6} M min^{-1}, 2.8×10^{-5} M.
14.11 (a) 1.7×10^{-5} M min^{-1}; (b) 1.7×10^{-6} M min^{-1};
(c) 1.8×10^{-6} M min^{-1}.   **14.16** 2.6×10^{-3} M.

Chapter 15 **15.5** 6.17×10^{14} Hz, 485 nm.
15.10 1.65×10^{-12} cm, 2.4×10^9 cm s^{-1}, 4.8×10^{-6} erg.   **15.11** 0.82.

Chapter 16 **16.7** 7.8×10^{-10} erg.   **16.12** 0.66.   **16.25** 240.8 kJ mol^{-1}.

Chapter 17 **17.8** -3.15×10^{-20} J/pair of H$_2$O molecules.   **17.9** 84 K.
17.16 0.26 cm, 7.7×10^6 base pairs.

Chapter 18 **18.6** 8.0×10^6 Hz, 9.8×10^{-6} nm.
18.19 (a) 1.64×10^{-39} g cm^2; (b) 1.01×10^{11} s^{-1}.
18.22 $e^{-205}/3$. **18.26** 5.50×10^{-6} M, 6.80×10^{-6} M.

Chapter 19 **19.7** 0.319.

20.2 279 nm.   **20.12** 0.022.   **20.13** 4.6.

Chapter 20

21.1 33,280, 36,770, 1.12.   **21.5** (a) 19,254 g mol^{-1}; (b) 8.

21.8 16,700 g mol^{-1}.   **21.17** 315 K.   **21.22** 10.9 turns s^{-1}.

Index